INTEGRATED MOLECULAR EVOLUTION

SECOND EDITION

INTEGRATED MOLECULAR EVOLUTION

SECOND EDITION

SCOTT ORLAND ROGERS

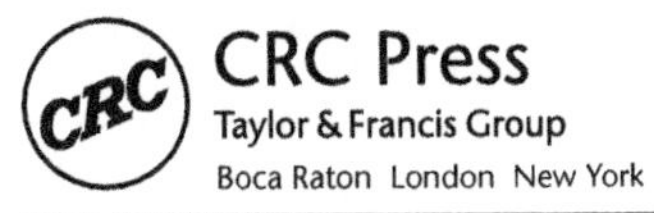

CRC Press is an imprint of the
Taylor & Francis Group, an **informa** business

CRC Press
Taylor & Francis Group
6000 Broken Sound Parkway NW, Suite 300
Boca Raton, FL 33487-2742

First issued in paperback 2019

ISBN-13: 978-1-4822-3089-5 (hbk)
ISBN-13: 978-0-367-86952-6 (pbk)

Library of Congress Cataloging-in-Publication Data

Names: Rogers, Scott O., 1953- , author.
Title: Integrated molecular evolution / Scott Orland Rogers.
Description: Second edition. | Boca Raton : Taylor & Francis, 2016. |
Includes bibliographical references and index.
Identifiers: LCCN 2015045976 | ISBN 9781482230895 (alk. paper)
Subjects: | MESH: Evolution, Molecular | Molecular Biology | Genetic Phenomena
Classification: LCC QH325 | NLM QU 475 | DDC 572.8/38--dc23
LC record available at http://lccn.loc.gov/2015045976

Visit the Taylor & Francis Web site at
http://www.taylorandfrancis.com

and the CRC Press Web site at
http://www.crcpress.com

Contents

Preface xv
Acknowledgments xvii
Author xix

SECTION I Life and Evolution

Chapter 1 Definitions of Life 3

Introduction 3
RNA and Life 6
Defining Life 9
Imagining Cellular and Molecular Dimensions 13
Key Points 16
Additional Readings 16

Chapter 2 Earth and Evolution 19

Introduction 19
What Is Evolution? 19
Earth History 22
A Short History of the Study of Evolution 30
Earth History as One Year 35
Key Points 37
Additional Readings 38

SECTION II Biomolecules

Chapter 3 DNA, RNA, and Proteins 41

Introduction 41
Nucleic Acids 42
Translation 51
Amino Acids and Polypeptides 52
Lipids 54
Carbohydrates 55
Key Points 56
Additional Readings 56

Chapter 4 The Central Dogma and Beyond 57

Introduction 57
Ribosomal RNA 57
Transfer RNA (tRNA) 64
Messenger RNA 65

Other Small Noncoding RNA 68
Beyond the Central Dogma 69
Key Points 69
Additional Readings 69

Chapter 5 Ribosomes and Ribosomal RNA 71

Introduction 71
Ribosomes as Ribozymes 71
Origin of the Ribosome 72
Translation 74
How Many rDNA Copies Are Needed? 76
Mechanisms for Increasing rRNA Gene Copy Number 77
Complexity of Ribosomes 80
Key Points 81
Additional Readings 81

Chapter 6 Structure of the Genetic Code 83

Introduction 83
Evolution of the Genetic Code 83
Why a Triplet Codon? 87
The First Genetic Code 90
Life before Translation 92
Key Points 95
Additional Readings 96

Chapter 7 DNA Replication 97

Introduction 97
Fidelity of Replication 97
Variations of Replication 101
Topology during Replication 106
Replication of Chromosomes 107
Key Points 108
Additional Readings 109

Chapter 8 DNA Segregation 111

Introduction 111
Variations on DNA Segregation in Bacteria and Archaea 111
Mitosis 113
Variations in Mitosis and the Cell Cycle 116
Variations in Chromosome Number 118
Changes in DNA Amount through the Cell Cycle 120
Meiosis 123
Sexual Reproduction 125
Key Points 127
Additional Readings 127

SECTION III Genetics

Chapter 9 Mendelian and Non-Mendelian Characters 131

Introduction 131
Alleles 132
The Basics of Mendelian Inheritance 132
Codominance, Incomplete Dominance, Overdominance, and Underdominance 137
Epistasis 138
Quantitative Trait Loci 140
Recombination and Linkage 141
Non-Mendelian Traits 141
Key Points 146
Additional Readings 146

Chapter 10 Population Genetics 147

Introduction 147
Hardy–Weinberg Equilibrium 148
Population Size 150
Life Histories 151
Modes of Reproduction 154
Key Points 158
Additional Readings 158

Chapter 11 Alleles through Time 159

Introduction 159
Natural Selection 160
Levels of Selection 162
Random Genetic Drift 165
Mating and Dispersal 166
Gene Flow 167
Other Factors Affecting Allelic Proportions 170
Key Points 170
Additional Readings 171

Chapter 12 Changes to DNA 173

Introduction 173
Classes of Mutations 173
Causes of Mutations 174
Mutation during Replication 179
DNA Repair 180
Genetic Recombination 182
Key Points 185
Additional Readings 185

Chapter 13 Infectious Changes to DNA: Viruses, Plasmids, Transposons, and Introns 187

Introduction 187
Integration into Chromosomes 191
Viruses 192
Introns 196
Transposable Elements 204
Plasmids 209
Key Points 210
Additional Readings 211

SECTION IV Multicellularity

Chapter 14 Multigene Families 215

Introduction 215
Ribosomal RNA Gene Family 217
Globin Gene Family 218
Bacterial Flagella Gene Family 222
Laccase Gene Family 223
Histone Gene Family 223
Orthologs and Paralogs 224
Polyploidization and Multigene Family Evolution 224
Key Points 227
Additional Readings 227

Chapter 15 Horizontal Gene Transfer 229

Introduction 229
Plasmids 231
Viruses 234
Symbionts and Organelles 235
Parasites and Pathogens 239
Origin of Gram-Negative Bacteria 242
Signs of HGT 243
Introns 244
Key Points 246
Additional Readings 246

Chapter 16 Development: Part I—Cooperation among Cells 247

Introduction 247
Quorum Sensing 249
Development in Animals 251
Nematode Development 252
Homeotic Genes and Proteins 253
Arthropod Development 261
Development in Vertebrates 264
Hierarchy and Evolution of Homeotic Genes 266
Key Points 268
Additional Readings 269

Chapter 17 Development: Part II—Plants 271

Introduction 271
Plant Morphology 273
Development in Plants 274
Gene Expression during Development 277
Formation of Leaves and Floral Organs 279
Plants versus Animals 287
Key Points 291
Additional Readings 292

Chapter 18 Cancer 293

Introduction 293
Progression of Cancer 295
Genes Involved in Cancer 298
Types of Cancer 301
Causes of Mutations in Carcinogenesis 301
Point Mutations 302
Recombination 302
Amplification 305
Viruses 306
DNA Viruses 309
Hormones 311
Key Points 312
Additional Readings 312

SECTION V Molecular Biology and Bioinformatic Methods

Chapter 19 Extraction and Quantification of Biological Molecules 315

Introduction 315
Extraction of Nucleic Acids Using CTAB 319
Purification of Organellar DNA 321
Extraction of RNA 322
Quantification of Nucleic Acids 324
Agarose Gel Electrophoresis 324
Extraction of Proteins 329
Quantification of Proteins 330
Polyacrylamide Gel Electrophoresis 331
Key Points 332
Additional Readings 333

Chapter 20 Recombinant DNA and Characterization of Biological Molecules 335

Introduction 335
Polymerase Chain Reaction 335
Recombinant DNA Methods 338
Southern Hybridization 343
Determination of Gene Copy Number 349

Microscopy 351
Protein Analysis 353
Key Points 355
Additional Readings 355

Chapter 21 Sequencing and Alignment Methods 357

Introduction 357
Development of DNA Sequencing Methods 357
High-Throughput Technologies 360
Next-Generation Sequencing 361
Protein Sequencing 365
Sequence Homology Searches 367
Aligning Sequences 368
Key Points 370
Additional Readings 371

Chapter 22 Omics: Part I 373

Introduction 373
Genomics 373
Transcriptomics 376
Metagenomics/Metatranscriptomics 378
Microbiomics 378
Key Points 381
Additional Readings 382

Chapter 23 Omics: Part II 383

Introduction 383
Proteomics 383
Structural Genomics 386
RNAomics 387
Epigenomics 388
Metabolomics 389
Functional Genomics 389
Key Points 391
Additional Readings 391

Chapter 24 Species Concepts and Phylogenetics 393

Introduction 393
What Is a Species? 393
Classification of Life 396
Reconstruction of Evolutionary History 399
Phylogenetics 400
Tree Terminology 401
Choosing a Genomic Region for Phylogentics 402
Other Considerations When Performing Phylogenetic Analyses 408
Models of Mutation 409
Analyzing Aligned Sequences 410

Unweighted Pair Group Method with Arithmetic Mean 410
Neighbor Joining 410
Maximum Parsimony 411
Maximum Likelihood 414
Bayesian Phylogenetic Analysis 415
Bootstrapping 415
Vertical versus Horizontal Evolutionary Events 416
Key Points 417
Additional Readings 417

Chapter 25 Phylogenetic Networks and Reticulate Evolution 419

Introduction 419
Phylogenetic Analyses of Reticulate Events 420
Advantages of Phylogenetic Networks 420
Horizontal Gene Transfers 422
Species Hybridization 423
Recombination 424
Transposition 425
Reassortment 425
Examples of Reticulate Evolution Events 426
Key Points 427
Additional Readings 428

Chapter 26 Phylogenomics and Comparative Genomics 429

Introduction 429
Improvements in Sequencing and Phylogenomics 429
What to Compare 432
Single-Nucleotide Polymorphisms 433
Microsatellites and Minisatellites 433
How to Compare 434
Testing for Selection 434
Incongruent Trees 434
Comparative Genomics 435
Synteny 440
Key Points 441
Additional Readings 442

SECTION VI Genomes

Chapter 27 RNA Viruses 445

Introduction 445
C-Value Paradox 448
Genomes and Genomics 448
RNA Virus Genomes 449
Human Immunodeficiency Virus 449
Influenza A Virus 453
Ebola Virus 457

Key Points ... 458
Additional Readings ... 459

Chapter 28 DNA Viruses ... 461

Introduction ... 461
Bacteriophage ϕX174 ... 461
Bacteriophage Lambda (λ) ... 463
Bacteriophage T4 ... 468
Mimivirus ... 471
Key Points ... 474
Additional Readings ... 474

Chapter 29 Bacteria and Archaea ... 475

Introduction ... 475
Escherichia coli ... 475
Photosynthetic Bacteria ... 477
Aquifex ... 479
Euryarchaeota ... 480
Crenarchaeota ... 482
Key Points ... 483
Additional Readings ... 483

Chapter 30 Mutualists and Pathogens ... 485

Introduction ... 485
Termite Gut Microbes ... 487
Smallest Bacterial Genome ... 487
Coresident Symbionts ... 489
Animal Parasite ... 490
Genome Mixing and Sorting ... 491
Key Points ... 492
Additional Readings ... 492

Chapter 31 Endosymbionts and Organelles ... 495

Introduction ... 495
Intracellular Endosymbionts ... 495
Mitochondria ... 496
How Many Genes Make a Functional Mitochondrion? ... 503
Chloroplasts ... 506
How Many Genes Make a Functional Chloroplast? ... 508
Differential Development and Function ... 510
Chimeric Pathways ... 510
Endosymbioses Leading to Other Organelles ... 512
Key Points ... 514
Additional Readings ... 514

Chapter 32 Protein Trafficking 515

Introduction 515
Signal Peptides in Bacteria 515
Signal Peptide Systems in Eukarya 518
Protein Trafficking in Mitochondria 520
Protein Trafficking in Chloroplasts 520
Evolution of Protein Trafficking Systems 523
Key Points 524
Additional Readings 525

Chapter 33 Eukaryotic Genomes 527

Introduction 527
Origin of the Nucleus and Mitochondrion 528
Multicellularity 531
Chromalveolata 531
Opisthokonta 532
Saccharomyces cerevisiae 532
Caenorhabditis elegans 534
Drosophila melanogaster 537
Archaeplastida 538
Arabidopsis thaliana 538
Oryza sativa 540
Key Points 543
Additional Readings 544

Chapter 34 Human Genome 545

Introduction 545
The Human Genome 545
Medical Genetics 548
Single-Nucleotide Polymorphisms 550
Forensics 551
Human Migration 551
Key Points 554
Additional Readings 554

Index 557

Preface

As with the first edition of this book, it was written with students in mind. It is meant to introduce the major topics of molecular evolution in a way that will encourage students to delve deeper into each of the topics. The book started as a series of notes, overheads, and digital slides that comprised a course in molecular evolution. The course was designed as an integrated approach to this field, for which there was no single textbook available. It draws from concepts in evolution, geology, chemistry, biochemistry, molecular biology, genetics, taxonomy, bioinformatics, various OMICS fields, and, of course, molecular evolution. Because it discusses aspects of each of these disciplines, students with broad backgrounds (as well as those with very focused backgrounds) should be able to grasp the concepts, principles, and details of this book. It presents some of the usual information regarding various aspects of cell function, but also details the variety of mechanisms that have evolved. This has been done to present a broader view of evolution that is meant to show that some processes have been approached in very different ways by the diversity of species during their evolution on Earth.

Although the first edition was organized into 18 chapters, including 197 figures, the second edition has been substantially expanded into 34 chapters, with 413 figures, essentially doubling the size of the first edition. It has been divided into six sections. Section I, *Life and Evolution*, covers the topics of evolution on Earth, prebiotic production of organic molecules, and definitions of life. Section II, *Biomolecules*, details the structures and functions of biological molecules, as well as the basic mechanisms that produce the molecules. Section III, *Genetics*, presents the basic genetic mechanisms that lead to the evolution of genomes and organisms. Section IV, *Multicellularity*, outlines the basic mechanisms of cell-to-cell communications and other processes that have led to the evolution of developmental processes in multicellular organisms. Section V, *Molecular Biology and Bioinformatic Methods*, consists of overviews of some of the methods used in molecular biological and bioinformatics research. Section VI, *Genomes*, is a survey of a set of genomes that represents a compendium of some of the important aspects of the evolution of genomes, in general.

Chapters 1 through 5 (included in Sections I and II) are nearly identical to the first five chapters in the first edition. Chapter 1 is an overview of life on Earth, as well as the possible origins of life. The definitions of life are discussed in detail. At the end of the chapter, there is an exercise designed to help the reader imagine and visualize the components of a cell in their true dimensions. Chapter 2 details the evolution of organisms on Earth. It also presents a history of the study of evolution. Chapter 3 covers the basic structures of DNA, RNA, proteins, and other biological molecules, and the syntheses of each. Chapter 4 begins with the central dogma of molecular biology, but then goes into detail about the complexities of all of the central processes of this basic principle. Chapter 5 discusses the largest ribozyme in the cell, the ribosome. This includes details about the structure and function of ribosomes, as well as a discussion about its evolution. Also discussed are the mechanisms for assuring that enough rRNA is produced to supply each cell with all of the ribosomes they need. Ribosomes have been one of the key structures in cells that have led to the success of life on Earth. The remainder of Section II includes Chapters 6 through 8. Chapter 6 is new, although it was partly covered in chapter 5 of the first edition. The new chapter details some of the recent research into the origin of ribosomes, translation, and the genetic code. Conceptually, this might be one of the more difficult to understand chapters, because an alternative organization of the universal genetic code table is presented, which is organized according to the possible evolutionary events that led to translation and the current genetic code. Although the genetic code table that has been used for several decades is informative and useful, it may not reflect the evolution of the genetic code itself. Alternative tables may more accurately reflect the evolution of the genetic code. Chapter 7 discusses the various forms of DNA replication, and how they have been important in evolution. Chapter 8 presents the common, as well as many of the uncommon, modes of separating chromosomes, from separation of chromosomes in bacteria to mitosis and meiosis in a variety of eukaryotes.

Section III includes three new chapters on genetic mechanisms. Chapter 9 is focused on Mendelian genetic mechanisms, as well as non-Mendelian mechanisms of inheritance. Chapter 10 discusses the basic concepts of population genetics, including Hardy–Weineberg equilibria, population size, life histories, and modes of reproduction, all of which affect allele proportions in populations. Chapter 11 further details some of the phenomena that affect allelic proportions in populations through time, including natural selection, random genetic drift, mating and dispersal, and gene flow. Chapters 12 and 13 present the major causes and mechanisms of mutation, including repair of mutations. These were chapters 8 and 9 in the first edition.

Section IV includes one of the first edition chapters (Chapter 11 on multigene families), which is now Chapter 14 in this second edition. The other four chapters in this section are new. Chapter 15 details horizontal gene transfers (HGTs) that occur very often, and they have been occurring for billions of years. Chapter 16 begins a discussion of development, which depends on coordinated cell-to-cell communication and precise programming of gene expression in each cell. This chapter includes details regarding quorum sensing in bacteria and development in animals. Chapter 17 continues with the details of development in higher plants. Chapter 18 is focused on the genetic changes and mechanisms in carcinogenesis. Some of the same mechanisms that cause evolutionary changes are the same mechanisms that cause cancer.

Section V outlines some of the basic methods used to study molecular evolutionary processes. Chapter 19 explains some of the methods for purifying and quantifying nucleic acids and proteins. Chapter 20 presents some of the basic recombinant and characterization methods used in molecular evolutionary studies. Chapter 21 details the various methods of sequencing of DNA (and cDNA from RNA), as well as proteins. Chapters 22 and 23 are surveys of various OMIC methods to analyze DNA, RNA, protein, and other molecular data. Chapter 24, *Species Concepts and Phylogenetics*, is an amalgamation of two of the first edition chapters (Chapters 10 and 12). This seemed to be a logical combination. Chapter 25, *Phylogenetic Networks and Reticulate Evolution*, is new in this second edition. It outlines evolutionary processes that are beyond simple bifurcating trees and events (e.g., HGTs—described in Chapter 15), and explains how they are determined. Chapter 26, *Phylogenomics and Comparative Genomics*, explains some of the processes and challenges in using genomic data in genomic studies of evolutionary processes.

The final section, Section VI, describes specific genomes. Each has been chosen either as a representative of a specific taxonomic group, or to illustrate one or more principles of the processes that occur during the evolution of the species and their genomes. Chapters 27 (RNA Viruses), 28 (DNA Viruses), 29 (Bacteria and Archaea), 30 (Mutualists and Pathogens), and 31 (Endosymbionts and Organelles) parallel Chapters 13 through 17 of the first edition. Chapter 32 is new. Its focus is on protein trafficking in cells. It begins with trafficking in bacteria, and proceeds into more complex trafficking in Eukarya. Chapter 33, *Eukaryotic Genomes*, has been edited, including the deletion of the human genome. Discussion of the human genome has been expanded in Chapter 34. Part of the additions to this chapter includes details about how the human genome has led to some practical applications of the information.

Scott Orland Rogers
Bowling Green State University, Ohio

Acknowledgments

I thank my wife Mary, daughter Liz, and son Ben for providing moral support. I could not have written this without their support. I also thank my mother, father, sister, and brother for their support throughout the years. I thank all of my students who always brought up new ideas, questions, papers, and insights, all of which made me think and rethink the ideas presented in this book. Also, thanks to Professor Arnie Bendich (my PhD Major Professor) who provided ideas, papers, enthusiasm, and critical questions about parts of this book. He taught me to think and read critically, deeply, and broadly. Special thanks to Zeynep Koçer, Lorena Harris, Amal Abu Almakarem, and Maitreyee Mukerjee for providing useful feedback on several parts of this book. Also, I want to thank the many people who provided feedback on the first edition of this book. The comments were very helpful to formulate and write the second edition. Special thanks to Chuck Crumly at Taylor & Francis Group for the time and energy that he spent assuring that this second edition would be published, and for providing comments, suggestions, and critiques. The book would not have been possible without him. Thanks also to Barbara Norwitz (also at Taylor & Francis Group) for helping to make the first edition of this book a success. Finally, I thank colleagues and students at BGSU for having patience with my frequent and lengthy absences from my office and lab while I was diligently working on this book at home.

Acknowledgements

Author

Scott Orland Rogers is a professor of molecular biology and evolution at Bowling Green State University, Bowling Green, Ohio. He received his BS (1976) and MS (1980) degrees in biology from the University of Oregon, Eugene; and PhD (1987) in plant molecular biology from the University of Washington, Seattle. He was an assistant professor and associate professor at the State University of New York College of Environmental Science and Forestry, Syracuse, NY, from 1989 through 2001, before moving to BGSU. He has taught courses in biology, botany, cell physiology, molecular biology, molecular genetics, bioinformatics, and molecular evolution. Research in his lab includes studies of microbes and nucleic acids preserved in ice, life in extreme environments, group I introns, molecular microbial phylogenetics, microbial metagenomics/metatranscriptomics, ancient DNA, and plant development.

Section I

Life and Evolution

1 Definitions of Life

INTRODUCTION

Before one can understand evolution, one must first decide on the boundaries of life, and what is at and beyond those boundaries. Complex cells, such as bacteria, probably did not spontaneously assemble from a set of chemical compounds. Fossil evidence of organisms resembling bacteria first appears in 3.5-billion-year-old rocks. Therefore, simpler organisms must have existed prior to this time. These organisms probably assembled and evolved long before one of their members evolved into the bacteria that were present 3.5 billion years ago. But, what were these first organisms? The initial answer is that no one knows, but we can make some educated guesses about the characteristics of the first organisms, as well as the processes that led to them. The first life on the Earth may have had its beginning from sets of chemicals and reactions that may have been derived in different ways, which eventually mixed in chemical pools or near undersea volcanic vents on the Earth. Stanley Miller published the first studies that produced amino acids and some other compounds from the simple molecules thought to be present on early Earth (Figure 1.1). He mixed water, ammonia, methane, and hydrogen into a sealed container, added heat and electrical discharges (to simulate lightning), and withdrew the products from time to time. He found that at least four of the amino acids found in modern cells were formed, and several precursors of nucleic acids were also produced. Since then, other experiments have been performed, including those that used different sets of gasses to more accurately reflect the early Earth, some performed under high pressure, and some done under cold conditions. In total, at least 16 of the biologically relevant amino acids, as well as fatty acids, and rudimentary nucleic acids were formed. More recently, *in vitro* experiments have been performed that have produced nucleic acids under conditions thought to have existed on the Earth early in its history. Also, amino acids and other biological compounds have been found in meteorites and comets, and peptides (linked chains of amino acids) can be formed under warm to hot conditions under high pressure, similar to the conditions in a meteorite when it passes through the atmosphere, or near deep-sea volcanic vents.

In order to begin on the pathway to an organism, amino acids, nucleic acids, fatty acids, ions, water, energy (e.g., heat or lightning), and time all are needed (Figure 1.2). The Earth is very large, and even from the beginning, it was not uniform. Some parts were hot, while other areas were cooler. The mix of chemicals and water varied. The atmospheric gasses consisted primarily of H_2, H_2O (vapor), CO_2, N_2, CH_4, and NH_3 as well as smaller amounts of CO, S_2, SO_2, and Cl_2. Pressures in the oceans varied with depth. Thus, given the nearly infinite set of conditions on early Earth, the presence of some simple chemicals (at the surface, as well as those emitted by volcanic activity), water, and energy, as well as the arrival of some chemicals on comets and asteroids, it is likely that all of the compounds necessary to produce the first truly biological compartments (which we now call cells) were present very early on the Earth. Water became more and more abundant, originating from comets and volcanoes. By 4.3 billion years ago, significant amounts of water were present in the form of worldwide oceans (which probably were not nearly as salty as they are today). The atmosphere contained huge amounts of water vapor because of the high surface temperatures. This caused convection currents that led to lightning. Vapor blown out of volcanoes, and the convection currents that were formed, also resulted in lightning. The atmosphere contained a mix of other gasses, which created reducing conditions. Sometime during the first billion years, all of the components combined to create the first biological reactions and organisms. Having the chemicals around to interact probably was not alone sufficient to produce a biological organism. Organisms carry out

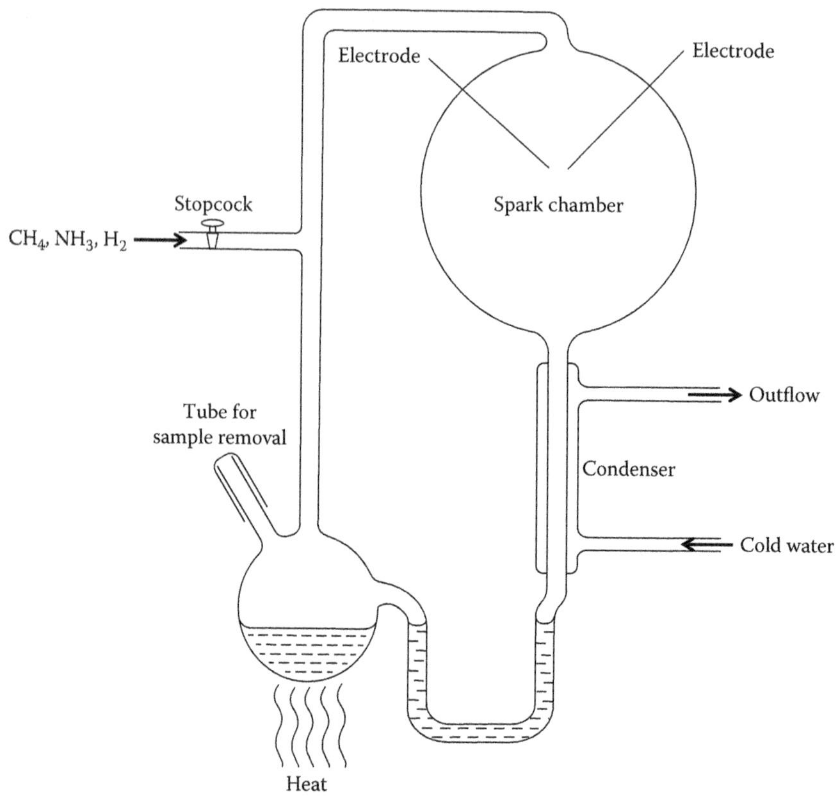

FIGURE 1.1 Diagram of the Miller–Urey apparatus used in 1953. The entire apparatus was made from glass, which is inert, except for the tungsten electrodes. Initially, the reservoir at the lower left was half-filled with distilled water, and the remainder of the apparatus was filled with hydrogen (H_2), methane (CH_4), and ammonia (NH_3) gasses. Heat was applied to the water-containing vessel, and simulated lightning was produced by sending an electrical current through the electrodes. Gasses were condensed into liquid using a cold-water condenser. As the U part of the tube was filled, its contents spilled into the reservoir. Samples were then removed from the reservoir and were tested for the presence of various chemicals, including amino acids. Many different amino acids were found, including several present in biological systems.

many chemical reactions, but they also reproduce with high fidelity. It is thought that RNA probably was one of the first molecules central to the origin of the first organisms. This is because RNA can be replicated to produce copies of itself, many RNAs are catalytic (called *ribozymes*) and can perform many different chemical reactions, and RNAs are central to all past and present organisms. They perform a myriad of functions, including encoding proteins, translation of proteins, control of gene expression, assembly of ribosomes, control of RNA concentrations, specificity of translation, initiation of DNA synthesis, addition of chromosome telomeres, and many others. It is very possible that the first self-replicating biological molecules were combinations of RNAs and peptides. These may have been concentrated and preserved by being enveloped by lipid membranes. Thus, there would be a mechanical separation between the inside and the outside, so that desirable components could be concentrated and protected inside, and waste or damaging chemicals could be kept outside of the rudimentary cell.

Because of the varied conditions on the Earth, the time involved (0.5 to 1.0 billion years), and the presence of chemicals that could become catalytic and self-replicating (e.g., some RNAs), life was initially assembled on the Earth spontaneously. While this probably was a rare event, given at least 500 million years, the 510 million square kilometers of surface area of the Earth,

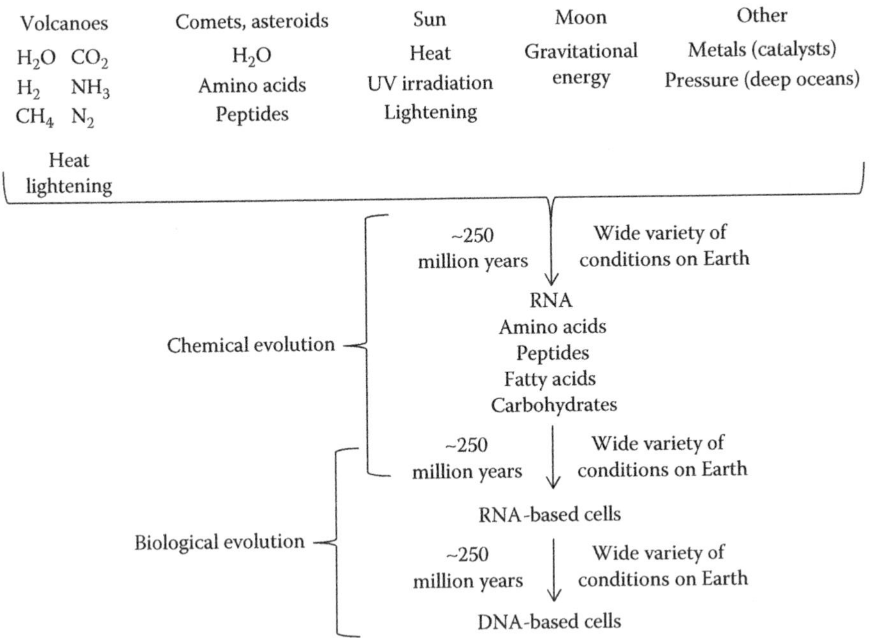

FIGURE 1.2 Summary of the components that led to chemical evolution and eventually biological evolution on the Earth. The Earth provided many of the basic components for these processes, primarily in the form of volcanism. Volcanoes released hydrogen (H_2), water (H_2O), ammonia (NH_3), methane (CH_4), and carbon dioxide (CO_2) in large quantities, and much of the ammonia was converted into nitrogen (N_2) gas. They also released carbon monoxide (CO), sulfur (S_2), sulfur dioxide (SO_2), and chlorine (Cl_2) gasses. Lightning was produced by the convection currents produced during volcanic eruptions into the atmosphere, as well as heating from the Sun. Comets and asteroids brought water and amino acids. As they entered the atmosphere, they heated up due to friction with the air, and they were exposed to high pressure. This can produce short polypeptides. Additionally, volcanoes in deep oceans under high pressures can produce of peptides. The UV irradiation arriving from the Sun may have also caused some additional reactions to occur. There was no ozone layer at that time, and UV irradiation reached the surface of the Earth at high intensity. Not only could the UV stimulate some reactions, but it could destroy some biological molecules, and therefore may have been more detrimental than beneficial. This probably means that early reactions and organisms on the Earth probably occurred mainly in deeper water, which would block the UV irradiation. Metals that existed on clay particles also could catalyze certain reactions, and many of them still are necessary for biological systems today. For example, iron is used extensively in enzymes, electron transport systems, and photosynthesis. Magnesium is required for many proteins that interact with nucleic acids and is required in some ribozymes as well. Finally, gravitational energy from the Moon was vital to the Earth. It produced tides in the oceans, but also distorted the Earth as both rotated. This helped to keep the iron core of the Earth liquid, which led to the magnetic field around the Earth. This channeled damaging high energy particles toward the poles and deflected the solar wind, which preserved the atmosphere, and protected life then, and still protects life on the Earth now.

the multitude of conditions, the huge mixture of chemicals, and energy (e.g., heat, lightning, and solar), it was inevitable that some sort of self-replicating, catalytic, self-contained life form would arise. Other such rudimentary life forms probably got started during this time, but many probably went extinct. Because extinction is almost as frequent as speciation, probably many of the original genetic lineages of these protocell species disappeared often and completely. But, at least one lineage did survive and led to the diversity of life that exists today. It is fortunate for you and I that it led to us, but if it had not, it would have eventually led to other life forms that would have been very different from the ones alive today, as well as from the ones that existed in the past.

RNA AND LIFE

All of the cellular organisms that exist today have ribosomes. These are complex ribozymes (enzymatic RNAs) that rely on sets of proteins to hold them in their catalytic conformations. Since they are found in all cellular life on the Earth and they all are genetically related to each other, they must have existed (albeit in a simpler form) in the progenote cells that were the ancestors of all subsequent cells on the Earth. The first indication of cellular life on the Earth is from 3.5-billion-year-old stromatolites (mushroom-shaped structures formed by sets of bacteria that live in shallow areas of oceans and seas). Therefore, by this time, there existed bacteria-like cells that contained ribosomes (or, at least, protoribosomes), which produced proteins using amino acids attached to tRNAs, reading a code on mRNAs. This is a complex process involving three different classes of RNA that must have taken tens or hundreds of millions of years to evolve. The cells by that time must have had a cell membrane composed of a lipid bilayer and was capable of concentrating needed chemicals and processes inside the cell and moving waste and toxic chemicals to the outside. At that time, the hereditary material may have been RNA. RNA mutates faster, so evolution could have been much more rapid in these organisms. However, because of the higher mutation rates, more lethal mutations are produced with each replication cycle, thus limiting the number of viable progeny and the sizes of the genomes.

Eukaryotes are not hugely more complex than bacteria and archaea. In fact, eukaryotes essentially are combinations of bacterial and archaeal cells. Genomic studies (studies based on the determination of the entire nucleotide sequence for the organisms) have shown that the bread and beer yeast, *Saccharomyces cerevisiae* (a single-celled eukaryote), has about 30% more genes than the bacterium *Escherichia coli.* The fruit fly (*Drosophila melanogaster*) genome has about 3.5 times more genes than *E. coli*, and the 2-mm-long worm *Caenorhabditis elegans* has about 4.4 times as many genes as the bacterium. Even humans have only about 5–6 times the number of genes as *E. coli.* Additionally, the majority of genes in the larger genomes are traceable back to bacterial and archaeal genomes.

The central pathway of information transfer in all membrane-bound organisms is from DNA to RNA to proteins (Figure 1.3). This is the so-called central dogma of molecular biology.

This characteristic ties all membrane-bound organisms on the Earth together and indicates a common origin for all. Through the process of transcription, an RNA copy of one of the two DNA strands is made. The messenger RNA (mRNA) is then translated on ribosomes into polymers of amino acids, known as *polypeptides* or *proteins.* This is the usual flow of information transfer in the cell. However, while these mRNAs (from 450 to over 50,000 different mRNAs in membrane-bound organisms) make up the complexity of a cell, they only make up roughly 5% (by mass) of the total RNA per cell. The remainder of the RNA in a cell is either transfer RNAs (tRNAs) and other small RNAs (e.g., small interfering RNAs and micro RNAs), comprising about 15% of the total, or ribosomal RNAs (rRNAs), making up about 80% of the total, respectively. These RNAs are never translated into proteins, but instead are molecules that aid in synthesizing, processing, or controlling DNA, RNA, and/or proteins. Transfer RNAs are short molecules (50–70 nucleotides) that are

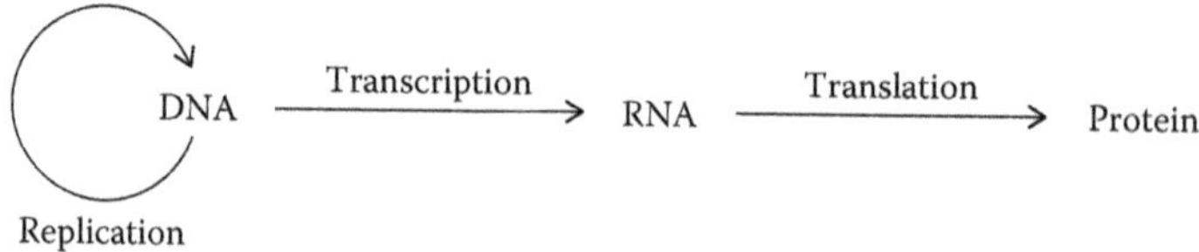

FIGURE 1.3 The central dogma of molecular biology. After the structure of DNA, transcription of mRNA, and translation of proteins using a triplet code had been discovered, Francis Crick proposed the central dogma of molecular biology. It relates the information flow from DNA to RNA via transcription and the translation of mRNA by ribosomes to produce proteins. Additionally, DNA can be replicated to make additional copies. There is much more complexity to these processes, and these processes will be discussed in later Chapters 4 through 6.

involved in translation, in that they bind specifically to the codon (made up of 3 nucleotides in a row) on the mRNA. Each carries a specific amino acid that will be added to the growing polypeptide chain.

Ribosomal RNAs, the central molecules of ribosomes, catalyze all of the reactions of translation, while the ribosomal proteins are all structural, in that they hold the rRNAs in their enzymatic conformations. This is true for all ribosomes, be they bacterial, archaeal, eukaryotic, or organellar. Ribosomes are complex, containing at least 3 rRNAs and 50 proteins in bacteria, and at least 4 rRNAs and 80 proteins in eukaryotes. All bacteria, archaea, mitochondria, plastids, and eukaryotes have ribosomes. Even viruses and viroids require ribosomes to make their proteins, although they take over host cells to accomplish this. Thus, all free-living organisms, as well as obligate parasites, mutualists, endosymbionts, and some organelles, have, or otherwise use, ribosomes. Since they are so complex, it is thought that they evolved only once on the Earth. Additionally, since they are universal to all cellular organisms on the Earth, the progenote cells that are the ancestors of these organisms also must have possessed and used ribosomes (or protoribosomes) for protein (and peptide) synthesis. Because rRNAs are so ancient, they have been used extensively in molecular evolutionary studies to answer questions of genealogy for the major groups of organisms on the Earth.

What are the signs of life? Of course, for large animals, you can look for signs of movement, breathing, growth, and a pulse. But, the absence of these does not necessarily mean that the organisms are dead or that they were never alive. There are many methods for identifying life employing culturing, microscopy, molecular biology, specific chemicals, ultraviolet light, X-rays, computers, and others. With culturing and microscopy, the results can be fairly clear. In the case of culturing, an organism either grows or fails to grow. If it grows, one can examine it in light and electron microscopes, study its growth requirements, and extract some DNA for molecular studies. Samples can also be examined directly, prior to culturing, by microscopy to look for cells, parts of cells, or groups of cells. The results are not always so clear, because some groups of cells and cell pieces look like nondescript debris, and it is often difficult to determine whether or not the cells are alive. Also, some mineral formations resemble cells.

The most common molecular method used currently is DNA sequence analysis. The sample is subjected to polymerase chain reaction (PCR) amplification to produce millions of copies of a DNA or RNA template molecule (Figure 1.4), and then, those molecules are analyzed by sequencing. Sequences can also be determined by other methods that amplify small pieces of DNA immobilized on small plates and computer chips and analyzed by detectors and computers. The DNAs can be from entire genomes, or from various samples, including those from environmental sources. The sequences can then be compared to others to determine whether the sequences match any others that have been analyzed by other researchers around the world. Sequencing of entire bacterial genomes (whole chromosome sequences) can be accomplished in a few days. For environmental samples, total RNA or DNA can be extracted from the sample and sequenced to produce millions of base pairs of information. These metagenomic studies provide an overall view of the diversity of organisms and metabolic pathways that exist in the particular environment (Figure 1.5). Most often, rRNA genes are used for broad comparisons, since these are conserved regions of DNA that are in common among bacteria, archaea, and eukaryotes, and sequences from a very large number of organisms have been determined. However, other regions of DNA can be examined. Mitochondrial genomic sequences have been determined for a large number of eukaryotes, and a large number of plastid genome sequences have been determined. Entire genomes have been determined for thousands of viruses, thousands of bacteria, hundreds of archaea, and hundreds of eukaryotes. Sequences from these can be chosen dependent on the study to be performed.

Some chemical reactions are indicative of life, although some of these have to be carefully evaluated. For example, on one of the unmanned Mariner probes that landed on Mars a few decades ago, a chemical reaction was measured by dropping some Martian soil into a specific solution. The gas that evolved appeared to indicate the presence of life. However, as was later surmised, it could also be indicative of the presence of a particular mineral that is common on Mars. Metabolic processes

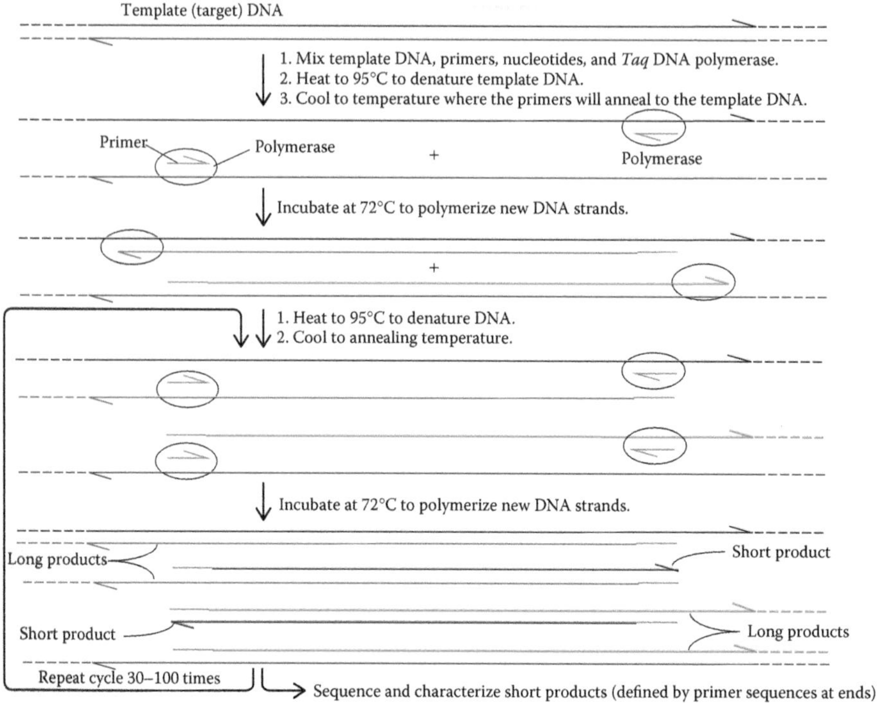

FIGURE 1.4 Diagram of the polymerase chain reaction (PCR). This procedure was first developed by Kary Mullis in 1985, but is used extensively worldwide to rapidly produce large quantities of DNA fragments of specific types. The reaction is started by mixing the sample DNA (that will include the template sequences to be copied/amplified), with short DNA primers (chosen or synthesized for each region to be copied/amplified), deoxynucleoside triphosphates (dNTPs), and a thermostable DNA polymerase (e.g., *Taq* DNA polymerase, isolated from the thermophilic bacterium *Thermus aquaticus*). Next, the mixture is heated to 95°C to denature the sample DNA. It is then cooled to a temperature that will allow the primers to anneal to the template by complementary hydrogen bonding in places where the sequences are complementary to the template strand. The temperature is increased to 72°C, which is the optimal temperature for the polymerization activity of *Taq* DNA polymerase. The temperature change steps are then repeated from 30 to over 100 times to amplify the desired fragments of DNA. During the first few such cycles, two different classes of amplification products are produced, long products and short products. The long products are DNAs of varying lengths that are being synthesized from the original template (sample) strands. The short products are being synthesized from the already copied pieces, which have ends defined by the primers that were added to the reaction. The number of long products increases linearly with each cycle, while the number of short products increases exponentially. Therefore, after 30 cycles, the reaction mixture contains primarily short products, which can be visualized by agarose gel electrophoresis (not shown).

can be measured by the addition of a radioactive substrate into an environmental sample. The products of the reaction are collected and the radioactive compounds are examined and measured. Other than the inherent problems in dealing with radioactive compounds, this method appears to be generally reliable for indicating life processes.

It is likely that organisms have been blasted off of the Earth by meteor impacts and volcanism. Therefore, there may be organisms that originated on the Earth that have fallen on other bodies in the solar system. In an analogous way, if organisms evolved first elsewhere in the solar system or beyond, this could have been the inoculum for the origination of life on the Earth. When extraterrestrial life is investigated, the first tests will likely be for ribosomal RNA or rRNA genes, since all free-living organisms are based on ribosomes. Bacteria, archaea, and eukaryotes all replicate their DNA in roughly the same way, transcribe RNA from similar DNA templates, and translate those mRNA messages into polypeptides using macromolecular assemblages of structural proteins,

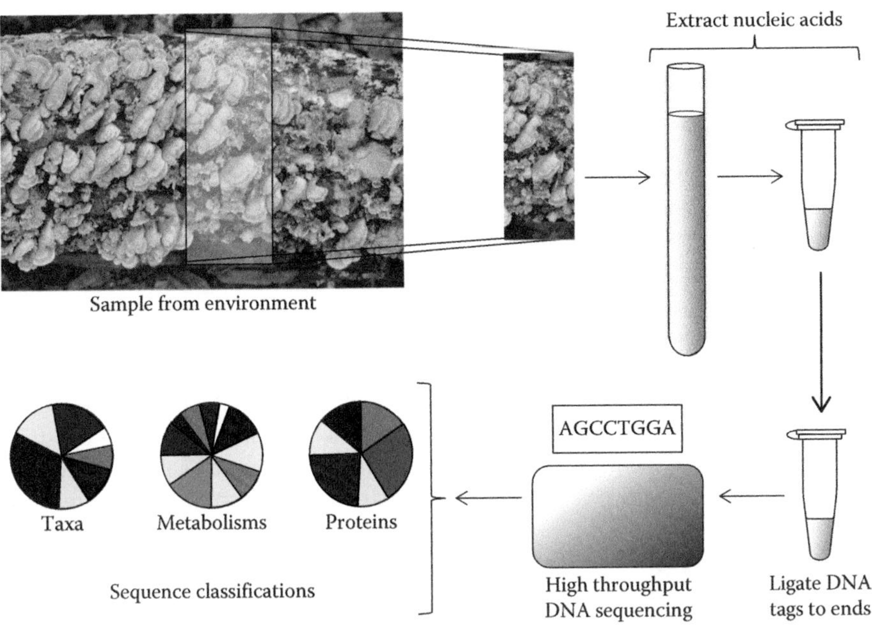

FIGURE 1.5 A brief explanation of metagenomic analysis. This process involves obtaining a sample, most often from the environment. However, there also are metagenomics projects to identify all of the microbes on and in the human body, as well as similar projects for other organisms. The process begins with a complex sample (e.g., soil, ice, water, and a log). Nucleic acids are extracted from the sample. This can be total DNA, total RNA, mRNA, or various combinations, depending on the study. If RNA is used, first a cDNA (a double-stranded copy of the RNA) must be synthesized. Then, the DNA and/or cDNA is amplified using random hexamer primers (a mixture of 6 bp primers that represent many combinations of nucleotides). Next, specific DNA pieces are ligated to the ends that are necessary for sequencing using so-called next-generation sequencing methods. These are high-throughput machines that can determine the sequences of hundreds of thousands of fragments simultaneously, producing datasets of tens of millions of nucleotides in a few hours. A computer links the fragments together, based on homologous sequences to produce longer pieces. These pieces can then be analyzed using bioinformatics methods to classify and categorize the sequences from the sample. From this, profiles of the taxa (phyla, families, genera, and species), metabolic pathways, protein groups, and other collections can be classified and characterized.

enzymatic rRNAs, and tRNAs carrying amino acids. Additionally, there are other proteins involved in turning transcription and translation on and off and positioning the DNA polymerases, RNA polymerases, mRNAs, tRNAs, and small and large ribosomal subunits. In fact, if any one of the components is altered, sometimes even slightly, it can cause fatal inhibition of the entire process. Thus, if life is found elsewhere in the solar system or in the universe that has a similar pathway of replication, transcription, and translation, it is highly likely that it has an ancestor in common with life on the Earth, including humans. Other organisms that might exist on other planets and planetesimals are viruses, specifically hardy viruses, including those that are found in the Earth ice. Although viruses are almost infinitely numerous on the Earth, they are so varied and small that they are difficult to find and identify.

DEFINING LIFE

Life is diverse, complex, and intricate. Additionally, the vast majority of life is unknown to science. It has been estimated that about 10% of the species on the Earth have been described, leaving 90% completely unknown. We know even less about all of the organisms that have gone extinct. When all of the sequences in a soil or water sample are analyzed (using metagenomic methods) and compared to the number of culturable microbes in the same sample, less than 1% of the organisms grow, making them nearly impossible to study in detail. Most are microscopic and many are virtually inaccessible for study. Thus, in many ways, it is difficult to study microorganisms and the diversity of life. Given the fact that over 99% of all species that existed on the Earth are now extinct, one would think that life is tenuous and ephemeral. However, some viruses and various bacteria that live under extreme conditions (of heat, cold, dryness, high pressure, radioactivity, etc.) indicate that some life can also be extremely hardy. Some organisms can exist in dry soils or ice in states of suspended animation for years, decades, centuries, millennia, and longer. Thus, our definition of life, and the environments that support life, must be extended. It appears that we have just scratched the surface of what comprises life and where life might be found. Just a few decades ago, it was thought that life could exist only within various narrow Earth environments and that the only possible other location for life was in the warmer locations on Mars. Today, the zone has extended to almost all environments on the Earth from the bottoms of oceans to the tops of mountains, in hot pools, encased in ice, and in clouds. Additionally, life exists miles into the Earth's crust. Thus, life may exist on or in many of the planets, planetesimals, and moons in the solar system and beyond. The challenge now is finding the best ways of detecting life in these extreme environments and then studying them.

Life is not as simple to define as one might think. Of course, we can all agree that large plants and animals are living things and that microscopic organisms moving and reproducing are living. All cellular organisms are considered living things, but what about noncellular organisms, such as viruses some of which are surrounded by a cell membrane? Life is more complex than you might initially think, and defining life is difficult (Figure 1.6). There are many organisms living on and in your body. In fact, the number of bacterial cells on and in your body is greater than the total number of your own cells. You are a living organism that carries trillions of other organisms around with you 24 h a day. Most of these are bacteria, and they are necessary for your health. In fact, you are dependent on them for your survival. This is the case for most organisms. They are interdependent on scores of other organisms. Bacteria are living things, but some are free living, and others require a host, like ourselves, for their survival. Some are pathogenic, in that they make us sick when they grow on and in us. The ones that grow in and on us still are living things, but they are dependent on other types of cells for their survival. Some bacteria have become endosymbionts that live between or inside of other cells and cannot live if they are separated from their host cells. Some of these endosymbionts have become organelles, such as mitochondria and chloroplasts. All still are living things, or at least have evolved from free-living things, and thus can be considered living organisms.

Viruses are obligate parasites, having an absolute requirement for a host organism. They cannot reproduce without host cells, but when they are in the host cells, they can reproduce rapidly. Viruses may have originated as pieces of DNA and RNA from living organisms, but eventually became the obligate parasites that they are today. Another hypothesis is that they once were free-living cellular organisms, but as they became obligate parasites, they lost most of their genomes and cellular functions. All genomes appear to include a large number of viral sequences integrated into them, including the human genome. Some bacteria, archaea, and eukaryotes also are obligate parasites, and we consider these to be living organisms. Therefore, depending on the absolute definition of life, one can draw the line between living and nonliving in different places. Transposons (mobile genetic elements), some of which cause the color variegation patterns in plants, may be defective viruses, in that they are incapable of exiting the host cells. However, they can move around within the host cell genome, and the host cell makes additional copies of these sequences. Are they living? They contain genes and gene fragments that come from truly living things. Prions, one of which is the causative agent of *mad cow* disease (bovine spongiform encephalopathy, BSE), are proteins that convert normal proteins into defective versions. Are these living? One can ask whether some molecules also satisfy parts of the definitions of life. Some RNA molecules are ribozymes, and some of these are capable of self-replication. Are these living? Some simple chemical reactions seem to indicate the presence of living things. All these factors present challenges to those studying life on

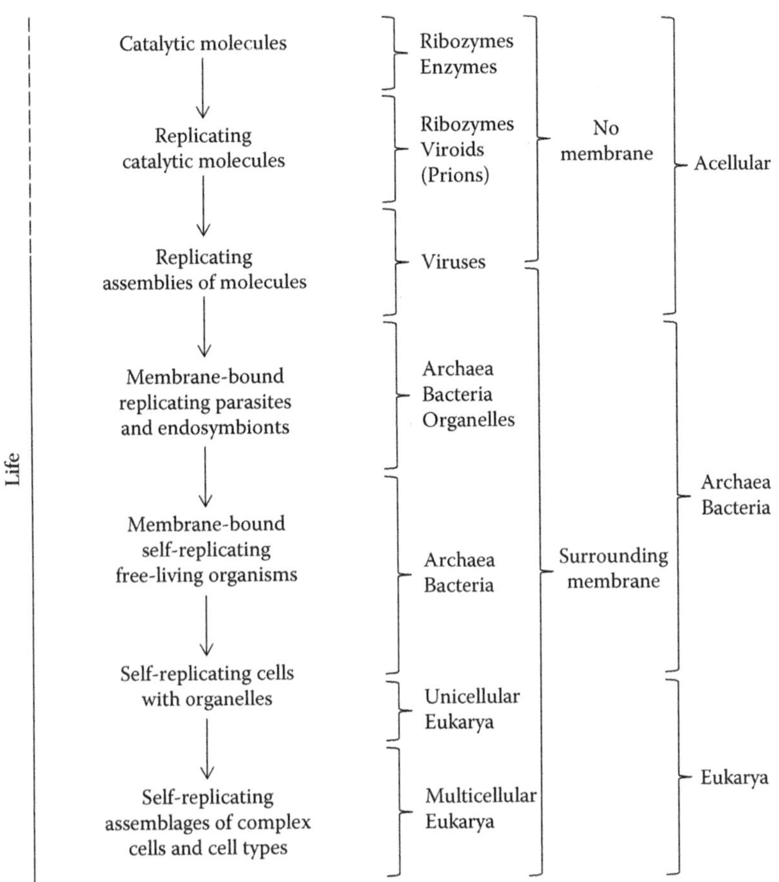

FIGURE 1.6 Continuum of life. It is clear that the cellular and multicellular organisms are living, but they can be viewed as a long continuum of entities from catalytic molecules through obligate parasites, such as viruses, some of which are surrounded by membranes (i.e., enveloped viruses). Various schemes have been constructed to identify a borderline between living and nonliving entities. However, whatever scheme is used, exceptions can usually be found. Therefore, the transition between living organisms and biological components that are nonliving is difficult to determine.

the Earth. While these molecules and assemblages of nucleic acids and/or proteins are beyond the definitions of living things, they all are indicative of the chemical processes that led to unmistakable living organisms. And many of them probably existed in the first processes of life on the Earth, as well as within the first true cells on the Earth.

The definition of life and what to look for are crucial factors in evolutionary biology. There have been many definitions of life. They change from time to time as new discoveries are made. For example, prior to the sixteenth century, microorganisms had not been observed directly and their presence was unknown. Diseases were thought to be caused by bad air (e.g., the Italian word for bad air is *malaria*), bad water, or other nonorganic contaminants of food or the body. Large organisms (we are including small insects and plants in this) were categorized as living, since they grew, moved, and/or reproduced. However, not until the end of the nineteenth century and the beginning of the twentieth century was it generally accepted that microorganisms existed and were the causes of diseases, food spoilage, wood rotting, and so on. By this time, microscopes were in widespread use, so that novel microorganisms were being identified constantly. At this time, the definition of life would have only included examples of animals, plants, fungi, and bacteria. However, at about the same time, it was realized that some diseases could be caused by extracts from diseased animals and plants. Moreover, filters could be used that would exclude all cells. With the advent of the electron microscope and continued work with filtered extracts from diseased organisms, viruses were discovered. Viruses are protein-coated genetic elements, consisting of RNA or DNA, depending on the virus species. As recently as the middle part of the twentieth century, viruses were not considered as living things and, at the same time, were not inert particles. Some scientists now view them as living things, although they all require the molecular machinery of the host cells to replicate themselves. But, bacteria that are obligate parasites also rely on the host cell for growth and reproduction. Some scientists have defined viruses as living things when they are actively infecting cells, but as inert when they are protein-coated nucleic acids outside of cells. The smallest viruses have about 10 genes (e.g., the RNA virus human immunodeficiency virus [HIV], causative agent of acquired immunodeficiency syndrome [AIDS], has about 10 genes that produce more than a dozen protein products), but are capable of killing large plant and animal hosts, such as trees and humans. On the other hand, Acanthamoeba Polyphaga Mimivirus (APMV) has a genome, the size of small bacterial genomes, and has many genes that are necessary only for cellular functions, and their presence in a virus genome is puzzling. Therefore, no clear line exists to split viruses and bacteria.

In the 1970s, naked infectious RNAs, called *viroids*, were discovered. Again, some consider these to be living things, because they replicate, although they require host cells to accomplish the replication. More recently, infectious naked proteins, devoid of any RNA or DNA, have been isolated and studied. They are known as *prions*, and they cause diseases such as scrapie, Creutzfeldt–Jakob, and *mad cow*. While they are not considered to be living, they are produced by cells, consist of amino acid sequences, and can infect other cells causing the death of those cells by converting similar proteins into prions.

During the past several decades, our ideas regarding the distinction between chemical reactions and life have blurred (Figure 1.6). If the definition of life is that of self-replication, then some proteins and RNAs can replicate themselves without the aid of being in cells. If we add on self-assembly to self-replication, some RNAs are capable of this, as well as some viruses. Catalytic, self-replicating RNAs may have been the first form of biological molecule to evolve on the Earth. All viruses, viroids, and biological molecules require cells for replication, production, and/or assembly, so at some time they must exist within host cells. The fact that all these entities require cells for their replication or production indicates that they should be considered as biological, but that they exist on the border between simple chemical assemblies and living organisms.

One of the universal characteristics of living things on the Earth is the presence of a membrane (i.e., they are membrane bound). This creates an interior and an exterior, which leads to a great deal of control over metabolic and replication processes. All bacteria, archaea, and eukarya are membrane bound. Some viruses (enveloped viruses) also have membranes surrounding the virions, but they are of host origin. Cell membranes are constructed such that they are studded with proteins that selectively include and exclude constituents inside and outside of the cell. Thus, the cell can concentrate things inside and exclude or excrete things outside. Membrane permeability is highly selective. Bacteria are membrane bound and most are capable of autonomous self-replication. However, some are parasites and must live within host organisms. In other words, they are membrane bound, but not free living. The definitions and characteristics of life are somewhat different for each type of life that is defined. Eukaryotes are the most complex organisms on the Earth, but are actually composed of blends of the progeny of a number of ancient bacteria and archaea. They all have DNA as their genetic material, and all have internal membrane-bound organelles. They have a double membrane surrounding the nucleus. Some bacteria (e.g., members of the planctomyces) have organelles that look very much like nuclei, and thus, the nucleus may have originated as a compartmentalization of the chromosomes and transcription machinery in this group of bacteria. The nucleus became specialized as a DNA storage organelle, as well as the location for DNA replication, RNA transcription, ribosome assembly, splicing by introns, and other functions.

Most eukaryotes have mitochondria that have a double membrane surrounding them and chromosomes that have genes in common with α-proteobacteria (specifically, members of the

Rickettsiales, an order containing many symbiotic species), indicating the endosymbiotic origin of these organelles. Plants and algae also have plastids (e.g., chloroplasts), again with a surrounding double membrane, and chromosomes with genes that are similar to those found in cyanobacteria, indicating an endosymbiotic origin for these organelles. Some organelles in dinoflagellates and other eukaryotes are eukaryotic cells that have been enveloped by another eukaryotic cell in secondary endosymbiotic events, and some appear to be the result of tertiary endosymbiotic events (a eukaryotic cell within a different eukaryotic cell within another eukaryotic cell, the last one being the ultimate host cell), based on the complexities of their organelles, numbers of membranes (e.g., some organelles have three and four membranes surrounding them), and genomic sequences. On the other hand, some organelles have only one membrane, possibly indicating an origin within the host cell or a loss of one or more of the membranes during evolution. Therefore, organelles are living things, or at least their ancestors were free-living organisms, but are absolutely dependent on the membrane-bound cells that surround them. They could not survive outside the host cell, but they do have membranes and they do replicate and self-assemble (with the help of the nucleus and host cell). This is very similar to parasites and endosymbionts that require several products and functions supplied by the host cells. On the other hand, the endosymbionts also provide necessary compounds to the host cell, which ultimately is dependent on the endosymbiont.

Thus, life and its definitions are varied. Life on the Earth is varied and complex. Even within an organism, there is variety from cell to cell and from nucleus to nucleus. No two nuclei in an individual are absolutely identical, primarily because with each cell division, errors in DNA replication occur. The definition of life at the level of a plant or an animal is undeniable. However, how far to go with the definition is debatable. In this book, an absolute cut-point for living versus nonliving will not be used, nor an absolute separation of species, because there is a continuum from biological molecules up to complex multicellular organisms. Instead, each organism and biological molecule will be considered as a part of a large network of molecules that are parts of life processes. This includes multicellular organisms, membrane-bound cells and organelles, viruses, and self-replicating molecules. This is because evolution of the membrane-bound forms of life likely originated from self-assembling and self-replicating molecules (likely catalytic and self-replicating RNA molecules) that got their start on the Earth very soon after it began to cool, probably about 4.3 to 3.8 billion years ago. At 3.9 billion years ago, life on the Earth was challenged by a series of meteor impacts that nearly destroyed the planet, termed *the late bombardment.* Nonetheless, life survived, spread, and diversified. Life has been challenged many times in the past 4 billion years, enduring at least six major extinction events, including the one currently occurring. It will be challenged again in the future.

IMAGINING CELLULAR AND MOLECULAR DIMENSIONS

In this book, the systems are cellular or molecular. Thus, for the most part, you are asked to use your imagination. Of course, there are some photographs and drawings of organisms and molecules, but in a way they are abstractions of reality. Before going further, you should begin to imagine at the cellular and molecular levels. If you can begin to relate to cell and molecular dimensions, you will better understand and have a deeper appreciation of the contents of this book.

In cellular and molecular biology, dimensions are used, which are beyond the resolution of the human eye. Try to visualize in your mind a micrometer (μm, 10^{-6} m, equal to approximately one-millionth of a yard), a nanometer (nm, 10^{-9} m), or an Angstrom (Å, 10^{-10} m). This is very difficult at best. After all, a micron is one-thousandth of a millimeter, well below the resolving power of your eyes. A nanometer is 1000 times smaller than that. You would barely be able to see about 25 μg (or about one-millionth of an ounce) of DNA, being about the size of the period at the end of this sentence. Try to imagine what a single microgram (μg, 10^{-6} g), nanogram (ng, 10^{-9} g), picogram (pg, 10^{-12} g), or femtogram (fg, 10^{-15} g) of DNA looks like. Again, this is very difficult because

1 μg of DNA is smaller than your eyes can resolve. Now, think of a cell. You have undoubtedly seen drawings and photographs of cells. How large is a cell? They are from 1 to 100 microns (i.e., micrometers, μm). How large is a chloroplast? On average, they are about 1 μm thick and 5 μm in length. How large is a ribosome? Ribosomes are about 30 nm in diameter. Even as you read these dimensions, it is doubtful that you are imagining their actual sizes. The numbers go out of your head about as fast as they enter.

To help you to get a feel for cellular and subcellular dimensions, imagine that we can expand a cell to become the size of a large room or a small auditorium, about one million (10^6) times larger than an average eukaryotic cell (Figure 1.7). The cell wall of a plant cell would then be about as thick as the walls of the room. Covering the inside surface of the walls, ceiling and floor would be the cell membrane. It would be 8 mm (3/8 in.) thick. This is the thickness of a thin foam pad, often used for wilderness camping. Embedded in this membrane would be pea-sized proteins that control what goes in and out of the cell. Around 25% to 75% of the membrane is made of these proteins.

The cell nucleus would be about 5 m (16 ft) in diameter (2–3 times the height of an average person). The nucleolus would have a diameter about the height of a person (6 ft or 2 m). The nucleolus is the place within the nucleus where ribosomes are produced. A typical metaphase chromosome (a condensed chromosome ready for cell division) would be the approximate size and shape of a person with arms stretched overhead and legs somewhat spread (although most peoples' central regions are larger than the centromere region of the chromosome). When the chromosome is unwound, the DNA would be 2 mm thick (the thickness of twine) and 20 km (about 12 miles) long. Humans have 46 chromosomes (23 pairs) in each of their somatic cells. Thus, the entire length of these expanded chromosomes would be 920 km (about 550 miles, or a distance between New York City and Columbus, Ohio; or San Francisco to Las Vegas). Imagine having to roll up that much twine and at the same time keep it organized so that it could be evenly distributed for cell division. The nuclear membrane enclosing the nucleus would have a thickness of 2 cm (about 1 in.). It would be covered with 6 mm (1/4 in.) wide nuclear pores, spaced 15 mm (3/4 in.) apart. These control the export of RNA from the nucleus and the import of proteins and other constituents from the cytosol. Out in the cytoplasm, the chloroplasts would be about 5 m in length and 2–3 m in width. Mitochondria would be about 2 × 1 m (the size of a tall thick person). Bacteria would be the size of an average-sized person. Typical viruses infecting the cell would be 5 cm in diameter, the size of a racquetball or baseball.

At the molecular level, a single gene on a piece of DNA would be around 0.5 m (1.5 ft) in length (Figure 1.7). If you add the controlling regions of the gene, it would range from 1 to 2 m (3–6 ft) long. The RNA encoded by the gene would be only half as thick, about the thickness of heavy string. Because RNA typically has a great deal of secondary structure, it would be twisted and folded back on itself. Ribosomes, which are about the size of golf balls, would be attached to mRNAs. A large protein or protein complex (e.g., RNA polymerase or DNA polymerase) would be about the size of a lima or fava bean. An average protein would have a diameter of about 5 mm (the size of a pea, mung bean, or soybean). An average atom would have a diameter of 0.2 mm, which might look like specks of dust floating around the inside of the room (cell).

Now imagine all of these organelles, molecules, and atoms floating around the room. Imagine the nucleus, mitochondria, chloroplasts, ribosomes, and membranes in the room. Imagine the cell being infected with a few bacteria and viruses. Now, in your mind, shrink the cell down slowly, still keeping in mind all of the enclosed constituents. Shrink it down until it is barely visible in front of you. You will need to shrink it only a bit further to achieve its actual size. In fact, if you put about 20–30 cells together, you could see them as a very small speck. Keep this exercise in mind when you are going through this book and whenever you see light or electron micrographs. Repeat as needed. Tables 1.1 and 1.2 show the actual lengths of DNA molecules, as well as the lengths if they are multiplied by one million times.

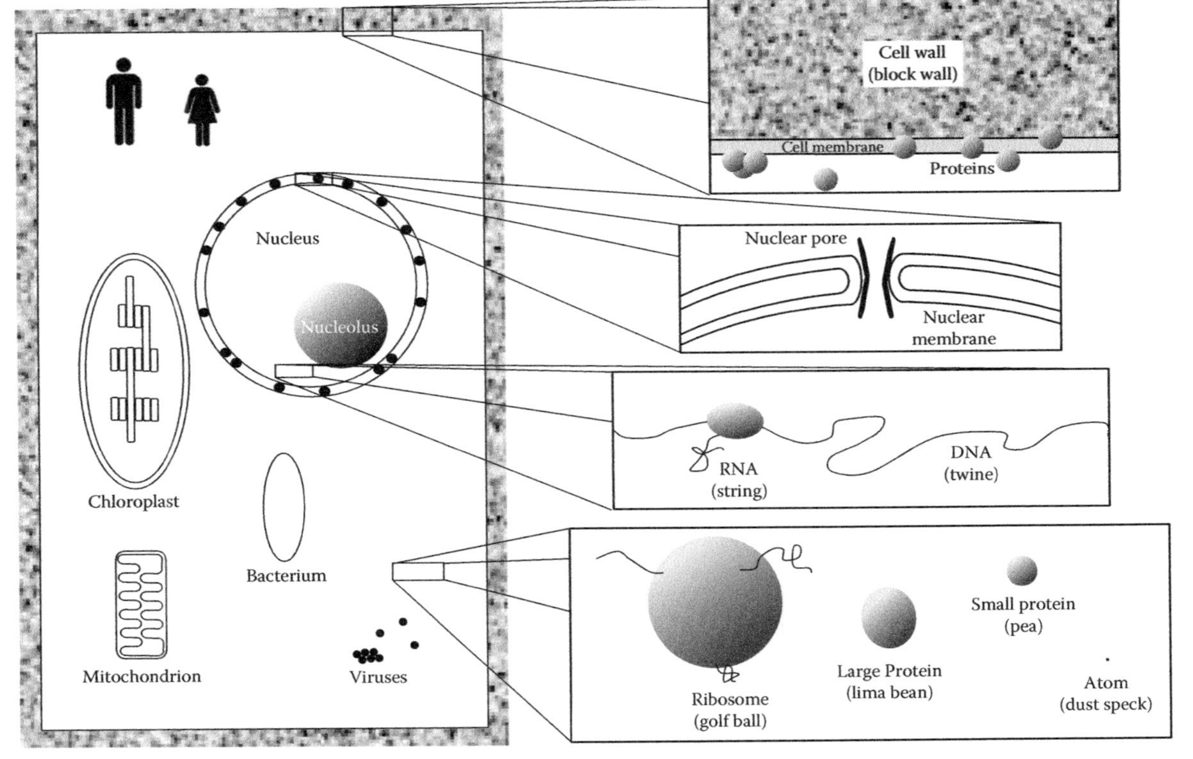

FIGURE 1.7 Cell and cellular components compared to a large room. If a cell is expanded by approximately 10^6 times, it is approximately the size of a large room. Icons for a man and a woman in the same scale are included for comparison. At this scale, a cell wall (in plants) would approximate the thickness of a cement block wall. The cell membrane would be similar in dimensions to a 1-cm-thick foam pad. Proteins, which are about the size of peas, would be embedded throughout the membrane. The nucleus would be about the size of a weather balloon inside the room and would be covered by nuclear pores. The nucleolus would be a balloon with a 2 m diameter. However, it is not surrounded by a membrane. DNA inside the nucleus would be about the thickness of the twin. The RNA being transcribed from the DNA would be the thickness of string. A chloroplast would be about the 5 m in length and about 2 m in width, while a mitochondrion would be about the size of a bacterium (1×2 m). If the cell were infected with viruses, it would be about the size of a baseball. There would be about 2,000,000 golf ball-sized ribosomes in this eukaryotic cell. Large proteins and protein complexes would be about the size of a fava or lima bean, while small proteins would be closer to the sizes of peas and soybeans. An atom would be visible, but only as a small spot, and water and other small molecules would be about the same size.

TABLE 1.1
Genome Actual Dimensions and Expanded by 10^6

Taxonomic Group	Actual Lengths	Expanded by 10^6
Mammals	0.34 to 1.36 m	340 to 1,360 km
Plants	3.4 mm to 68 m	34 to 68,000 km
Fungi	2.7 mm to 3.4 cm	27 to 340 km
Bacteria	34 μm to 3.4 mm	34 m to 34 km

TABLE 1.2
Mass of DNA Converted to Lengths and Expanded by 10^6

DNA Mass	Actual Lengths	Expanded by 10^6
1 microgram (μg)	300 km	3×10^9 km
1 nanogram (ng)	300 m	3×10^6 km
1 picogram (pg)	0.3 m	300 km
1 femtogram (fg)	0.3 mm	300 m

KEY POINTS

1. Early in the Earth's history, simple molecules were converted into some of the initial biological building block molecules, such as amino acids, nucleic acids, fatty acids, and carbohydrates.
2. The first biological systems probably contained RNA and simple polypeptides.
3. The first protocell probably contained RNA, ribosomes, proteins, lipid-based membranes, and carbohydrates. These probably were present on the Earth by about 3.9 to 4.0 billion years ago.
4. By 3.5 billion years ago, cells had begun to use DNA as their genetic material.
5. Detection of life can be difficult, and usually more than one test is necessary to confirm the presence of life.
6. Defining life is problematical. Living processes can be identified in multicellular organisms, as well as in RNA molecules. These processes form a continuum that is difficult to subdivide precisely.
7. Cellular and molecular dimensions are difficult to imagine. However, if they all are multiplied by 10^6, they can be visualized or imagined as filling a large room.

ADDITIONAL READINGS

Futuyma, D. J. 1998. *Evolutionary Biology*, 3rd ed. Sunderland, MA: Sinauer Associates.
Hall, B. K. and B. Hallgrimsson. 2008. *Strickberger's Evolution*, 4th ed. Sudbury, MA: Jones & Bartlett.
Johnson, A. P., H. J. Cleaves, J. P. Dworkin, D. P. Glavin, A. Lazcano, L. Jeffrey, and J. L. Bada. 2008. The Miller volcanic spark discharge experiment. *Science* 322:404.
Joyce, G. F. 2002. The antiquity of RNA-based evolution. *Nature* 418:214–221.
Miller, S. L. 1953. Production of amino acids under possible primitive Earth conditions. *Science* 117:528–529.
Multiple authors. 2006. *Evolution: A Scientific Reader.* Chicago, IL: The University of Chicago Press.
Pan Terra, Inc. 2003. *A Correlated History of the Earth.* Hill City, SD: Pan Terra.
Powner, M. W., B. Gerland, and J. D. Sutherland. 2009. Synthesis of activated pyrimidine ribonucleotides in prebiotically plausible conditions. *Nature* 459:239–242.

Saiki, R., S. Scharf, F. Faloona, K. Mullis, G. Horn, H. Erlich, and N. Arnheim. 1985. Enzymatic amplification of beta-globin genomic sequences and restriction site analysis for diagnosis of sickle cell anemia. *Science* 230:1350–1354.

Smith, J. M. and E. Szathmary. 1995. *The Major Transitions in Evolution.* New York: Oxford University Press.

Stearnsand, S. C. and R. F. Hoekstra. 2005. *Evolution: An Introduction*, 2nd ed. New York: Oxford University Press.

2 Earth and Evolution

INTRODUCTION

Ever since the first biological reactions and organisms became established on the Earth, evolution and the processes that drive evolution have been continuous (Figures 2.1 and 2.2). Evolution is a never-ending process, due to the mutability of DNA and the selectivity inside and outside of the organisms, whether they are unicellular or multicellular. Even when it appears to slow down, it continues, and it only stops in a particular lineage when that lineage goes extinct. The processes of mutation and selection are everywhere, including in the cells of your body and in the billions of organisms that live in and on you. Mutations may cause disease, or may be virtually undetectable, but they occur unabated throughout every cell cycle. The organisms that live in your body mutate and change continuously, and natural selection kills those that can no longer survive, while others thrive, including those that help digest the food that you eat. When you are sick, your immune system attacks the pathogenic organisms, but leaves other cells alone, which is necessary, because if it attacks all foreign cells, you would die. When you take an antibiotic, this kills many other types of cells, including the pathogens as well as the beneficial microbes. This is a type of selection, and eventually, many of the microbes may become resistant to the antibiotics due to evolutionary events, including mutation, and acquisition of resistance by the horizontal transfer of genes from other microbes. Evolution occurs around the world constantly. It was once thought that evolution was a slow process. But, we now know that evolution, and even speciation, can occur rapidly. Some of these processes can be observed in short-term, as well as long-term experiments and observations. Fossils are only one tiny example of the myriad of organisms that have existed on the Earth. Even so-called living fossils, such as crocodiles, sharks, and cycads, are changing and are not identical to the same organisms that were living millions of years ago. Evolution is a dynamic and constant process.

WHAT IS EVOLUTION?

At first glance, biological evolution seems to be a simple concept, but very few people in the world actually understand it, and no one understands all of its details and nuances. At its simplest, evolution includes mutation of genetic material (DNA for the vast majority of organisms and RNA for some viruses), and natural selection processes and genetic drift (Figure 2.3). Of course, time is involved in all of these, because there occurs a period of time where the fittest organisms survive and the unfit organisms die. Mutations often are called *random events*, but they are not random. Some of the nucleotides in DNA mutate at higher rates than others and some parts of genes mutate at different rates compared to other parts. Natural selection events act to essentially *choose* beneficial or neutral mutations and *weed out* detrimental ones. Normally, one thinks of natural selection as the process whereby an organism is able to survive in a particular environment based on nutrient utilization, color, body shape, or other obvious traits. However, selection occurs at every level, from genome replication, transcription, translation, transport, and on up in complexity within each cell. Also, selection is operational at the multicellular level in large organisms, within tissues and organs, as well as how they communicate with other tissues and organs within the organism. Therefore, evolutionary processes occur within a complex milieu of biological molecules and collections of cells (of the same or different species). Selection is constantly occurring, is continuously changing, and is a driving force of evolution, while mutations are the raw materials for these selection processes.

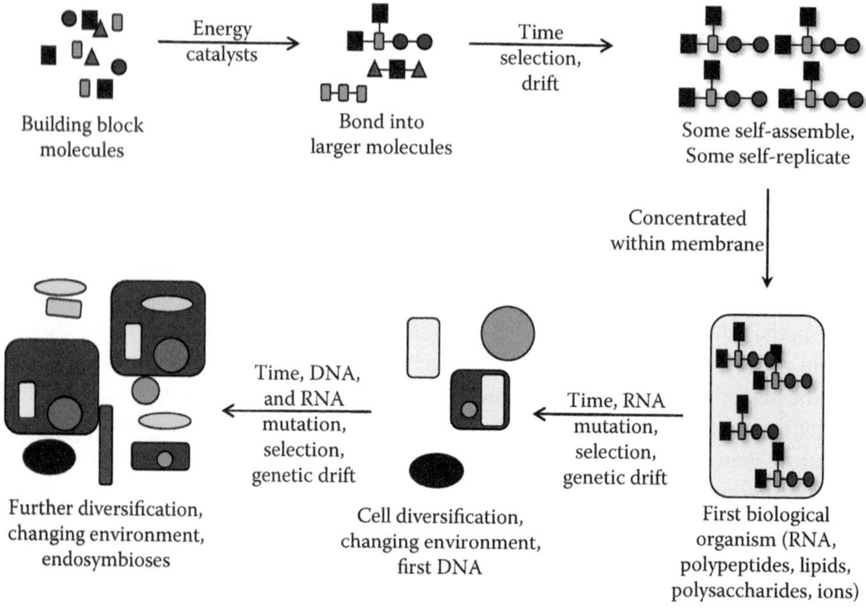

FIGURE 2.1 Evolutionary steps to life on the Earth. Early Earth was an amalgamation of compounds from the formation of the Sun and its planetary disc. As the Earth cooled, 4.0–4.2 billion years ago, chemical reactions occurred to form stable compounds, using heat and chemical energy to form chemical bonds, sometimes using natural catalysts. Other compounds formed on bodies in space and fell to the Earth. As water accumulated from volcanic activity and impacts of comets, a huge number of different conditions were present. In some of these environments, some compounds formed that could make copies of themselves given the appropriate chemical building blocks (4.0 billion years ago). Other chemicals could be produced by these self-replicating molecules. Some have hypothesized that these events occurred in pools of water. However, there was no ozone layer to block ultraviolet radiation, which could damage the sensitive molecules. Therefore, others have hypothesized that these early events occurred in the depths of the oceans where volcanic activity creates gradients of heat, chemicals, and pressures. Many have concluded that the self-replicating catalysts were RNA molecules (3.8–4.0 billion years ago). Eventually, these were surrounded by membranes (lipids and phospholipids), such that there could be an inside to the protocell and an outside. Useful compounds (carbohydrates, RNA, polypeptides) were concentrated inside and harmful chemicals and wastes were on the outside of the protocell. This initial process appears to have taken at least 500 million years. These biological cells were much simpler than the bacteria that exist now. However, if their genetic material were RNA, their mutation rates would have been very high compared to those with DNA. Therefore, mutation and selection would lead to rapid diversification of these organisms. Eventually, one organism began utilizing another nucleic acid, DNA, for its genetic material. DNA has the ability to form a stable duplex molecule. For a number of reasons, this allows for a more stable genome, as well as larger genomes. This meant that organisms could become more complex, because they could possess more genes and more gene products (e.g., proteins). For the next billion years (from 2.5 to 3.5 billion years), single-celled bacteria and archaea flourished. Some of them became dependent on one another, especially when oxygen began to build up in the atmosphere. Organisms had to adapt to the poisonous oxygen or escape from it. Those that could do neither went extinct. Some adapted by sequestering the oxygen, others detoxified it, and still others used other organisms to do this and thus formed symbiotic relationships with them. This led to specialization, diversification, and often to the loss of genes. Some of the relationships became endosymbiotic relationships that eventually led to the formation of mitochondria, chloroplasts, and other organelles. Mutation, selection, and diversification continue today, all based on events that began over 4.0 billion years ago.

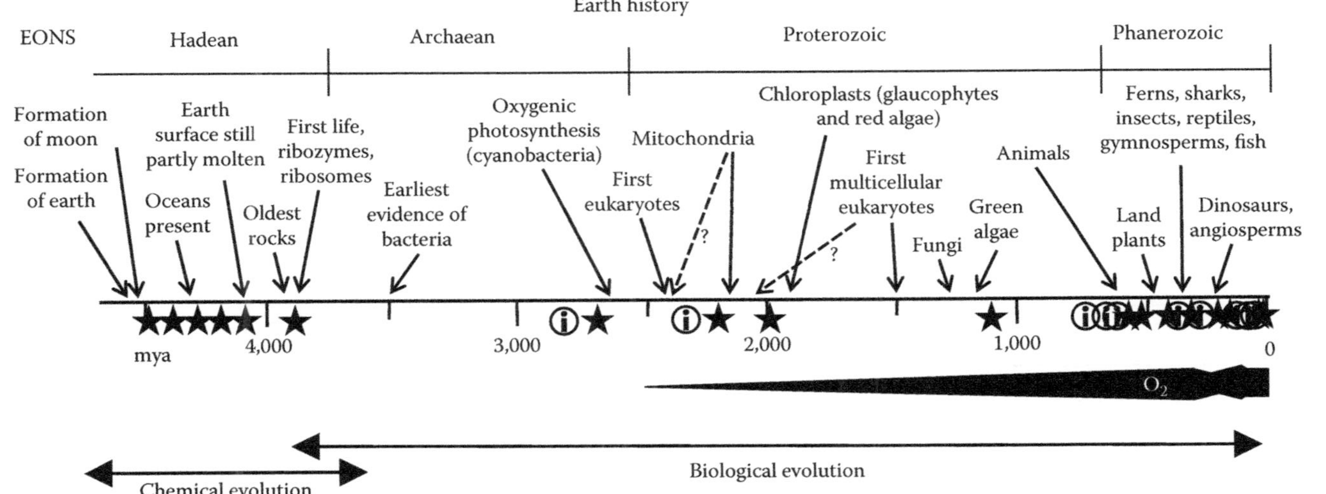

FIGURE 2.2 Natural history of the Earth. The Earth began to coalesce about 4.55 billion years ago, but about 500 million years later, it was struck by a Mars-sized object that liquefied the Earth and formed the Moon. By 4.3 billion years ago, oceans were present. RNA-based organisms may have originated between 3.8 and 4.0 billion years ago, and fossil evidence of the first bacteria had been found in rocks that are 3.5 billion years old. In this anoxygenic period, bacteria predominated (although some archaea were also present), some of them were photosynthetic. About 2.7 million years ago, a different type of photosynthesis began in cyanobacteria. This was a more efficient form of photosynthesis, but it produced O_2, which was toxic to most organisms at that time. As the levels of oxygen gas in the atmosphere increased, microbes, including a new group that would become the eukaryotes, needed to detoxify the oxygen or escape from it by going to deeper water. Ultimately, some bacteria evolved pathways to detoxify the oxygen and eventually use it for energy production. These bacteria also were enveloped by the progenote eukaryote, eventually became endosymbionts, and finally mitochondria. A few hundred million years after these events, another endosymbiont became incorporated into one group of eukaryotes and became the plastids in the first eukaryotic alga, the red algae, which diverged into glaucophytes and a later divergence became the green algae. The next 1.5 billion years witnessed diversification of single-celled organisms, the origin of multicellular organisms, and the origin and expansion of many types of organisms, including plants, fungi, and animals. Major ice ages and bolide impacts on the Earth have occurred. These greatly affected life and evolutionary processes, causing both speciation and mass extinction events. Early Earth experienced nearly continuous bombardment from asteroids and comets, some of which may have delivered significant amounts of minerals, biological molecules, and water. While the major bombardment events ended by about 3.8 billion years ago, occasional large impacts occurred (indicated by stars), some of which caused major extinction events. Worldwide ice ages (indicated by an *i* surrounded by a circle) also have occurred, many of which saw the Earth completely covered in ice, including the first known ice event about 2.8 billion years ago. Also, just prior to the Phanerozoic Eon, which started with the Cambrian period, three long and extensive ice ages occurred (between 550 and 800 million years ago), such that most of the Earth's landmasses, and large portions of the oceans were covered with ice. These have been termed the *Snowball Earth* periods.

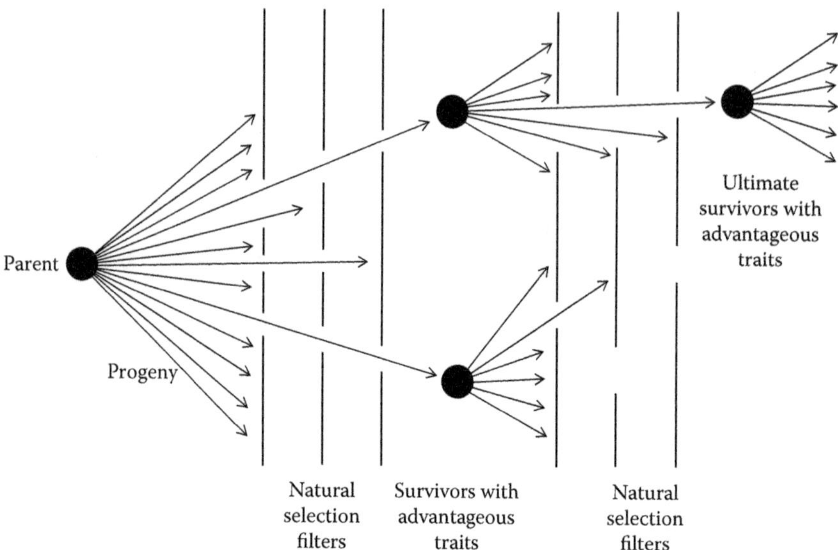

FIGURE 2.3 Model of natural selection. Organisms almost always produce many more progeny than can possibly survive. Progeny survive or die based on the genes that they inherited from their parent(s) and the environments that they encounter during their lives. A number of these progeny may survive and produce progeny of their own, which then pass through other natural environmental selective filters. Since the environments are in constant flux, natural selection also will vary with time and location.

EARTH HISTORY

The solar system was formed approximately 4.6 billion years ago. The Earth formed and began cooling about 4.55 billion years ago (Figure 2.2). While it was still forming and mainly molten, about 4.53 billion years ago, it was struck by a Mars-sized body that completely melted any rock that had begun to solidify and huge masses of material were hurled into the space around the Earth. The collision changed the composition of the Earth by stripping away some of the mantle materials. Much of this debris coalesced into two large moons that eventually collided to form the one large moon. By 4.4 billion years ago, the surface of the Earth had solidified, because mineral crystals dating to that age have been found. The Moon was so close to the Earth at that time that it nearly filled the sky. Because of this proximity, it exerted huge gravitational forces on the Earth. This would have caused crustal deformation, triggering volcanic eruptions that were common at that time.

For a time, the Earth and Moon may have appeared to be very similar in appearance, other than the difference in size. Both were pock marked by numerous impacts of meteors. The Earth and Moon experienced large showers of asteroids and comets. However, by 4.3 billion years ago the Earth contained larger amounts of water in liquid and vapor forms, and oceans were already present. The water may have come from the volcanoes and/or from comet impacts. Parts of the surface of the Earth were probably molten until about 4.1 to 4.2 billion years ago. The bombardment by large asteroids and comets, as well as internal radioactivity, kept the interior of the Earth hot enough to melt minerals, including iron, which formed the core of the Earth, and created convection currents of iron, which helped to lead to the movement of the rocky plates upon which the lighter land masses float, and allowed a magnetic field to be form inside the Earth that helped to protected the surface from damaging radiation. In addition to deflecting radiation, the magnetic field also deflected the solar wind, which allowed an atmosphere to form around the Earth. Without the magnetic field, the Earth would have a miniscule atmosphere similar to that on the Moon, and life would not have been possible. The movements of the plates provided many distinct geological and ecological zones, including those on land and those in the seas. Volcanoes released huge amounts of steam and ash, as well as molten rocks and minerals, which were lighter and thus displaced the deeper and denser minerals. These eventually became the continents and islands. Much of the steam condensed and fell as deluges of rain. Streams, rivers, lakes, seas, and oceans formed.

The chemicals necessary for life probably were present already in the form of simple lipids, precursors of nucleic acids, and some amino acids. These may have formed on clays and rock surfaces that were in water, and through the action of lightning, which was very common at that time. However, some probably arrived from space. Water had been present since about 4.4 billion years ago. Oxygen was rare and there existed a reducing atmosphere. While life may have had some beginnings at this time, the frequent bombardment by meteors may have kept the Earth sterile for much of this time. About 3.9 billion years ago, there was a heavy bombardment of asteroids, including some very large ones, that may have melted many of the rocks that had formed by that time. This may have killed all life that had begun by that time, although some researchers believe that microbes living in subsurface environments could have survived and may have been the initial seeds of all life on the Earth. Also, it is possible that if life originated in the deep oceans near volcanic vents, many types of organisms may have survived these bombardments. By 3.8 billion years, organisms that were based on nucleic acids, proteins, sugars, and lipids appear to have become established on the Earth. The basic metabolic processes that are present in cells today were established by that time. The oldest rocks that have been discovered on the Earth date to this time. By 3.5 billion years, clear signs of complex life are evident, as indicated in fossils. Stromalites have been found that date to between 3.4 and 3.5 billion years ago. The structures are very similar to stromalites that are found in modern shallow seas. They are formed by collections of cyanobacteria, as well as green non-sulfur, and other, bacteria. Modern free-living bacteria are relatively complex, having from 1 to 15 Mb (megabases or millions of bases) of DNA in their genomes containing thousands of genes. So, by 3.5 billion years ago, living organisms resembling bacterial cells were present and already might have been quite complex. Evidence of the ancestors of these organisms has not been found, but, as discussed in Chapters 1,3,5, and 6, organisms based on simpler systems and molecules must have existed prior to the appearance of bacteria. Therefore, from 3.9 to 3.5 billion years ago, life originated and became sufficiently complex that they resembled bacteria. It is believed that these early cells were photosynthetic, but used a form of photosynthesis that did not produce oxygen (i.e., anoxygenic photosynthesis). If we assume that modern bacteria have about the same constituents as the original bacteria, then by 3.4 to 3.5 billion years ago, organisms had RNA, DNA, proteins, polysaccharides, and lipids (probably including phospholipids as membrane constituents). Membranes surrounded a cytosol that included metabolic enzymes and their products. Transport proteins shuttled needed chemicals into the cell and waste products and toxins out of the cells. Ribosomes had evolved, which are complex catalytic ribozymes that translate the mRNA sequences into amino acid sequences to build proteins. This implies that rRNAs, tRNAs, and mRNAs all had become integral to the cells. Many other catalytic RNAs were likely present at this time. Many of these RNAs performed important functions in the cells, including the control of gene expression. DNA had replaced less stable RNA as the genetic material. All of the molecules for replicating DNA were present, as were the molecules for transcribing RNA from DNA genes. Although this is a large and complex set of molecules to organize into these rudimentary cells, given the 400 to 500 million years of time during which it probably occurred, it was clearly possible.

How did all of these organic chemicals form on the Earth, or did they? Various studies, such as the early experiments of Stanley Miller and Harold Urey, demonstrated that many of the organic chemicals could have been produced by nonbiological means in early Earth history. In addition, amino acids have been found in meteorites. The Earth was warm and by about 4.3 billion years ago already had great oceans of water. Much of this water may have come from comets and asteroids that collided with the Earth. Therefore, steam and clouds were plentiful. This created convection storms that produced lightning, just as occurs today. However, the atmosphere lacked oxygen and thus was a reducing atmosphere. Also, the atmosphere allowed ultraviolet irradiation to reach the surface of the Earth, because without oxygen, the protective ozone was absent. These were conditions that Miller and Urey attempted to reproduce in the lab. From this experiment, many compounds were found, including amino acids, nucleic acid precursors, and compounds resembling fatty acids. More recent attempts have been made, based on additional information about the conditions early in the Earth's history. These have produced additional compounds that are now found in organisms, including the production of at least 16 of the 20 biological amino acids. Under freezing conditions, yet other biological compounds are formed. Under heat and pressure (such as those found when an asteroid enters the atmosphere or in a deep sea volcano), small peptides (rudimentary proteins) can be formed from amino acid precursors. Additional experiments have demonstrated the formation of nucleic acids *in vitro*. Also, some ribozymes *in vitro* can recruit amino acids and small peptides to aid in their functions, and this might have led to the present nucleic acid/protein-based biological system that is present in all organisms on the Earth. Therefore, by 4.3 to 4.0 billion years ago, all of the precursors for life were present on the Earth or were delivered to the Earth by meteor passes and impacts. Some of these, as well as others, still may be arriving on the Earth, because approximately 40 tons of cosmic dust is deposited on the Earth daily.

The initial biological processes probably were based on amino acids, simple peptides, lipids, carbohydrates, and RNA (the RNA world hypothesis). There may have been many different forms of early life, but the ones that outcompeted the others were based on RNA (and eventually DNA) as the hereditary material. But RNAs also performed (and still do today) some enzymatic functions, as well as controlling gene expression. These are some of the functions that we observe in organisms living today. By 3.5 billion years ago, ribosomes and other ribozymes had probably evolved, because fossils indicating the presence of bacteria-like cells have been found. However, RNA is less stable than DNA, and there are no mechanisms to correct altered bases once they occur, and therefore, RNA mutates at rates that are much higher than DNA. Therefore, evolution during this period of time (3.5 to 3.9 billion years ago) was probably very rapid compared to evolution of DNA-based organisms. There is a drawback to these high mutation rates. This rapid mutation rate also meant that the genomes of these organisms had to be small, because fast mutation rates would not be possible for organisms with large genomes. It was about this time (between 3.8 and 3.6 billion years ago) that the first DNA-based genomes may have appeared. Because all of the organisms living today (except viruses) have ribosomes, DNA genomes, lipid bilayer, cell membranes, and similar DNA replication, transcription, and translation processes, it is concluded that the ancestor from which all of them diverged also possessed all of these. This implies that the initial diversification and divergence likely occurred prior to 3.5 billion years ago. Of course, there might have been forms of life that differed from these, in terms of the utilization of ribosomes, nucleic acids, and other constituents in the organisms we see today, but they may have gone extinct without leaving any detectable remains.

For the next billion years, single-celled bacteria diversified (Figure 2.2), including into several types that were capable of more complex photosynthesis that split water and produce oxygen (e.g., cyanobacteria), as well as the appearance of the archaea. Complex chemistry evolved to extract energy from chemical bonds, sunlight, heat, or other sources, in order to synthesize other compounds where that energy was stored and then used in subsequent metabolic processes. Many of these energy capture processes use electrons and/or protons to transfer the energy from one compound to another. However, each set of organisms produced waste products that affected them, as well as other organisms. Often, these were toxic to other organisms. This was true for oxygenic photosynthesis that is used by most cyanobacterial species. Oxygen is a highly reactive compound, and over a period of several hundred million years, the atmosphere was changed from one that contained almost no oxygen gas into one that contained more than 10% oxygen gas. It has fluctuated since then. It was approximately 13% 550 million years ago, about 35% 300 million years ago, 15% 250 million years ago, and today is approximately 21%. Another result of the increase in oxygen is that ozone (O_3) can be formed. Ozone absorbs ultraviolet irradiation that can damage the ring structures in nucleic acids and proteins. Once the ozone had increased in the upper atmosphere, much less of the damaging radiation reached the surface of the Earth, and additional niches were available for organisms to exploit, including new habitats on land. Oxygen was toxic to most organisms when it first started to build up in the atmosphere, such that some organisms were then forced to retreat into deeper aquatic systems that were anoxic. Light was limited in deeper water, making photosynthesis difficult or impossible. Eventually, some organisms evolved compounds and proteins that were able to hold onto oxygen or detoxify it by combining it with other compounds. It was during this period of time (just over 2 billion years ago) when eukaryotes first appeared. Oxygen may have been important to this evolutionary result.

By definition, the first eukaryotes had nuclei, but they may have had mitochondria as well (Figures 2.2, 2.4, and 2.5). However, because of the antiquity of these two events (appearance of nuclei and mitochondria), and their apparent close temporal association, it is difficult to determine whether the appearance of nuclei preceded mitochondria or if they appeared concurrently. It is known that they both appeared at a time when O_2 levels were climbing, which was stressful to many organisms. Some eukaryotes have no mitochondria, but they appear to have been lost, while some of these have the remnants of mitochondria. These unusual eukaryotes have some mitochondrial genes in their nuclei and thus must have had mitochondria that were lost at some point in their evolution. Some of their descendants survive today. They are the diplomonads, trichomonads, and microsporidia. One notable member is *Giardia lamblia*, an intestinal parasite of mammals (including humans) that causes a diarrheal disease known as *giardiasis*. Nuclei either originated as specialized compartments of the cell, are engulfed members of the Archaea, or were structures already present in an endosymbiont. Several species of bacteria (including *Gemmata obscurilobus* and *Pirellula marina*, within the group called *planctomycetes*) have double-membrane structures that surround their DNA that look very much like nuclei (complete with structures resembling nuclear pores), so it is possible that the ancestor of an Archaea enclosed its DNA, transcription, and ribosome assembly systems inside a double membrane that came from a planctomycete endosymbiont. Because members of the planctomycetes and the closely related spirochetes have microtubules, it is possible

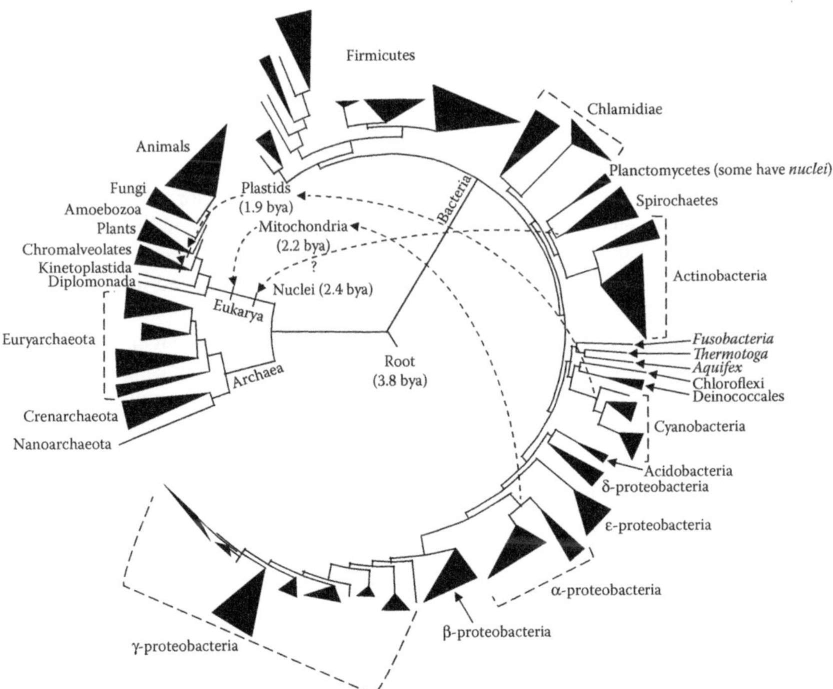

FIGURE 2.4 Circular phylogenetic tree of the major groups of organisms. Bacteria span from about the 11:00 to the 8:00 positions, with the major phyla designated. Archaea are centered at the 9:00 o'clock position (major groups are indicated), and Eukarya are centered at the 10:00 position (major groups are indicated). The mitochondria and plastids originated from α-proteobacteria and a cyanobacteria, respectively. The origin of the nucleus still is a mystery, but it either was inherited through an endosymbiotic event between a planctomycete bacterium and an archaea, or simply was the formation of a double membrane system that also formed the endoplasmic reticulum.

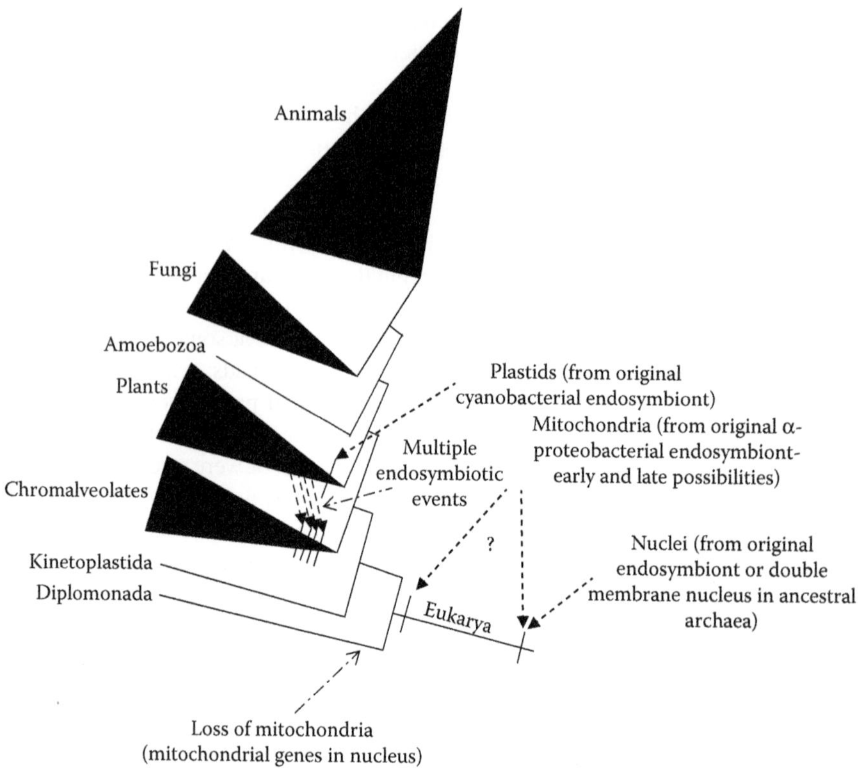

FIGURE 2.5 Section of the tree in Figure 2.4, detailing the Eukarya. Evolution of eukaryotes has occurred in a series of endosymbiotic events, beginning with the nucleus or a bacterium containing a nucleoid, and subsequently incorporation of a mitochondrion. In one lineage (diplomonads and related organisms), the mitochondria have been much reduced or lost completely, although many mitochondrial genes can be found in the nuclear genomes of these organisms, indicating that they once possessed mitochondria, but all that remain are many mitochondrial genes that migrated into the nuclei. Additional primary, secondary, and tertiary endosymbiotic events have occurred. Another primary endosymbiotic event occurred in *Paulinella*, which is a recent acquisition of a cyanobacterial endosymbiont into a unicellular heterotrophic eukaryote (not shown). Several secondary and tertiary endosymbiotic events have occurred between algal endosymbionts and chromalveolate host cells. These have produced the dinoflagellates and related organisms. Secondary events are those where a plastid-containing eukaryotic cell becomes the endosymbiont of another cell. Tertiary events are essentially an additional endosymbiotic association after a secondary endosymbiotic event.

that these were also transferred into eukaryotes via an endosymbiont from that group of bacteria. Alternatively, the eukaryotic ancestor may have independently developed a nuclear structure, but it involves a more complex pathway to do this.

The increase in oxygen levels led to a new type of eukaryote. At some point, approximately 2.2 to 2.4 billion years ago, a mutualistic association developed between a eukaryotic species and an α-proteobacterium (related to modern members of the Rickettsiales) that was capable of detoxifying oxygen, and in the process used protons that could be used to form the high energy compound adenosine triphosphate (ATP), which can be used to drive many metabolic processes. These small powerhouses not only provided protection from the effects of the oxygen, but also provided large amounts of energy (in the form of ATP and other high energy compounds) that could be used for metabolic processes. Eventually, these became mitochondria that are integral to most eukaryotic cells. This is also a vital step in the evolutionary processes that led to multicellular organisms. At about 1.5 billion years ago, the first multicellular organisms evolved, and the first organisms that clearly appeared were animals at about 600 million years ago (Figure 2.6).

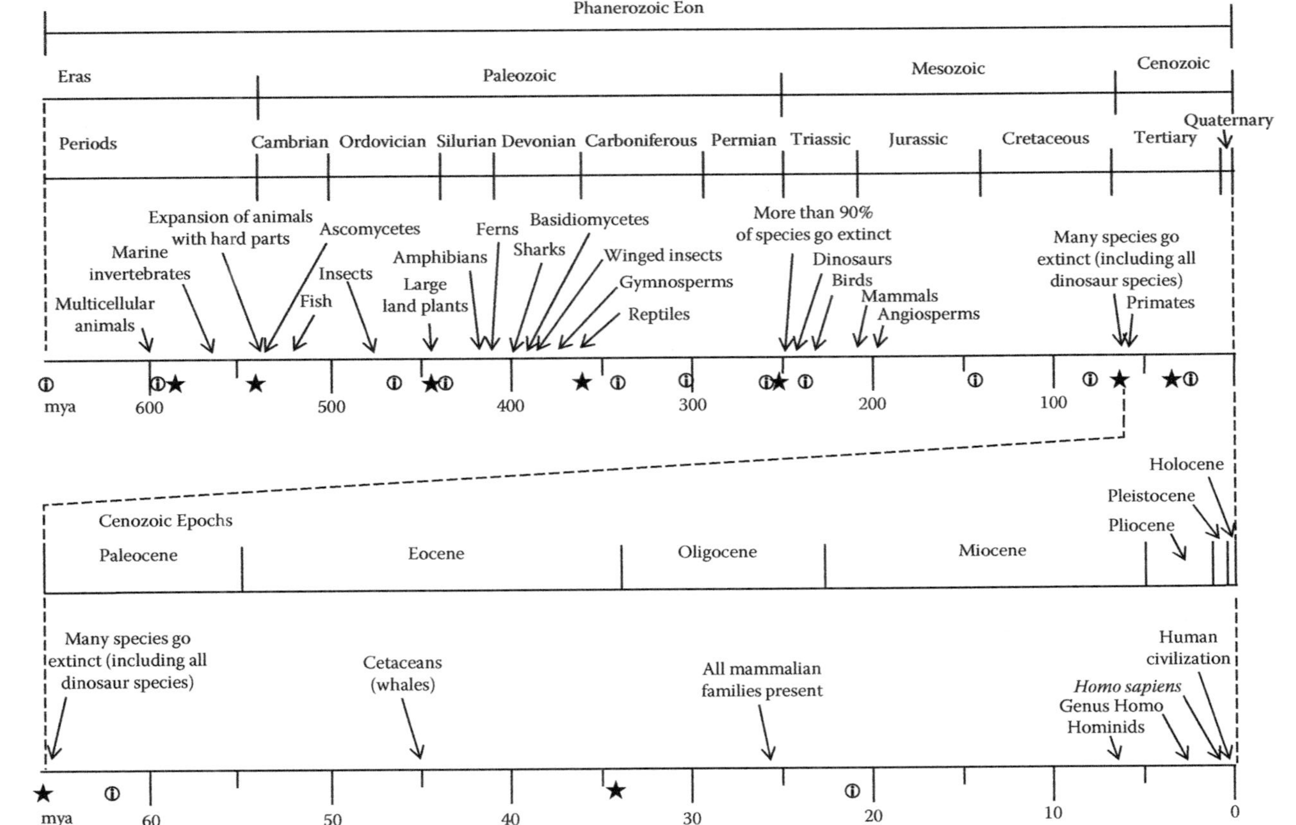

FIGURE 2.6 Details of the Phanerozoic Eon (above) and the Cenozoic Epochs (below). The first appearance of major groups of plants, animals, and fungi are indicated by arrows. Also noted are major extinctions, which are concurrent with major bolide impacts (stars), including one 245 million years ago that killed more than 90% of all species (including all species of trilobites) and the one 65 million years ago that wiped out about 70% of all species (including all of the dinosaurs). Each major ice age is indicated by an *i* within a circle. Hominids and humans are very late arrivals.

However, they left almost no fossil record because they had soft body parts. At about 530 million years ago, there is an abundance of animal fossils of great diversity. This is the so-called Cambrian explosion, which probably is only an apparent rapid increase of animal diversity. What likely occurred were two things: (1) an increase in body size (so that the fossils were clearly visible to humans examining the rocks) and (2) an increase in the number of organisms having hard body parts (probably due to increased competition and predation), which could be preserved as fossils.

Another major step in biological evolution on the Earth was the acquisition of chloroplasts (Figures 2.2, 2.4, and 2.5). These photosynthetic organelles originated from cyanobacterial endosymbionts, and this step has occurred at least two times. But, the story is complex and the history of organisms with plastids is one that proceeds through several sequential endosymbiotic events. One endosymbiotic event resulted in a cell that became the ancestor of red algae and glaucophytes (which first appeared about 1.9 billion years ago) and later green algae (which first appeared about 1.2 billion years ago). Since then, the red and green algae have participated in secondary and tertiary endosymbiotic events, the products of which have led to other major groups of organisms, such as ciliates and dinoflagellates. The green algae diversified and some of their descendants became the bryophytes and vascular plants. At least two other events have occurred that confuse this picture somewhat. A green unicellular eukaryote called *Paulinella* (in a group called the Rhizaria, phylum Cercozoa) has been described that appears to have acquired a cyanobacterial-derived organelle (termed a *chromatophore*) in an event, which occurred more than a billion years ago, and which was independent of the one that led to all other plastid-containing organisms. Additionally, it appears that the organism *Lepidodinium* was formed from a tertiary endosymbiosis between a dinoflagellate (whose plastids were inherited from red algae) and a green alga. Thus, this organism has photosynthesis genes from both red and green algal ancestors. In all of these cases, the genes of each of the participating cells have gone through mutation and selection events that have streamlined the genomes, such that neither the host organism nor the endosymbiont/organelle has a full complement of genes, and thus neither is able to survive without the other. This process of genome reduction happens very early in the association. Most of the associations may begin in a host–parasite or mutualistic relationship, and it is known that most parasites lack some genes, and therefore gene products, that they need to survive. Thus, they rely on the host to provide these. However, while the hosts incorporate many or most of the genes from the endosymbiont, they lose others as they rely on the endosymbiont to provide the functions of the gene products. Eventually, the host uses any advantageous products from the parasite, and the mutualistic/symbiotic association begins. Bacteria may also have experienced analogous events. One of the major differences between Gram-positive and Gram-negative bacteria is that the latter have two cell membranes. Because the membranes are somewhat different in composition, they may have originated from two separate species of bacteria, where one engulfed the other in an endosymbiotic event. One model that has genomic support is that a member of Actinobacteria and one from *Clostridia* produced the first Gram-negative cell, which then diversified into the large group of bacteria that exist today.

Over the past 1.5 billion years, there has been a diversification of multicellular plants, and over the past 600 million years, there has been a diversification of multicellular animals and fungi (Figure 2.6). Marine invertebrates began to appear prior to the Cambrian period (541 to 485 million years ago). Bony fish and ascomycetous fungi appeared over 500 million years ago, while the first cartilaginous fish (shark ancestors), the first plants, insects, amphibians, and marine invertebrates appeared from 400 to 500 million years ago. Seed ferns, gymnosperms, and reptiles all appeared from 300 to 350 million years ago. Dinosaurs first appeared about 245 million years ago. Birds first appeared during the Jurassic period (150 to 200 million years ago). Flowering plants and mammals first appeared during the ensuing 45 million years. Primates first appeared about 100 million years ago. Hominids and ultimately modern humans were very recent arrivals, at 3 million and 140,000 years ago, respectively.

The evolution of organisms on the Earth is defined as much by extinction as by speciation. More than 99.9% of the species that have existed on the Earth have gone extinct, and major and minor extinction events have occurred. One event about 245 million years ago wiped out approximately 90% of the species that existed at that time. This might have been a major bolide impact, but there were also huge volcanic events in Siberia, which could have released large amounts of toxic gasses into the atmosphere, and may have led to rapid changes in the amount of light reaching the Earth, as well as changes in the world temperature. Some have speculated that the mass extinction that occurred 440 million years ago might have been caused by a gamma ray burst from an exploding star far off in the galaxy. The extinction event that wiped out the dinosaurs 65 million years ago was not as bad, but still 70% of the species were expunged at that time, which was caused by a 6-mile-wide bolide impact in the Gulf of Mexico, followed by extreme volcanic activity in India, and possibly large releases of carbon dioxide from carbonates at the impact site.

Biological processes started as purely chemical processes and still are based solely on those processes. They have been elaborated over the nearly 4-billion-year history of evolution on the Earth, and they remain the foundation for all biological systems. However, they have been honed such that the successful reactions are retained if they provide a benefit to the organisms that contain them. Therefore, given the basic building blocks of RNA, amino acids, proteins, fatty acids, and sugars, the systems that contain these processes and which allow their reproduction have been selected by the Earth environments that have developed and changed over the past 4.6 billion years of its existence. These building blocks have accumulated in the cells that are alive today. Given appropriate conditions and time, evolution will continue long into the future, but extinctions will continue to occur. The average time for the establishment and extinction of a species can be from one to few million years, although the total length of time that a group of species can exist can be billions of years in duration (Figure 2.7). As was stated earlier, *Homo sapiens sapiens* has existed as a species

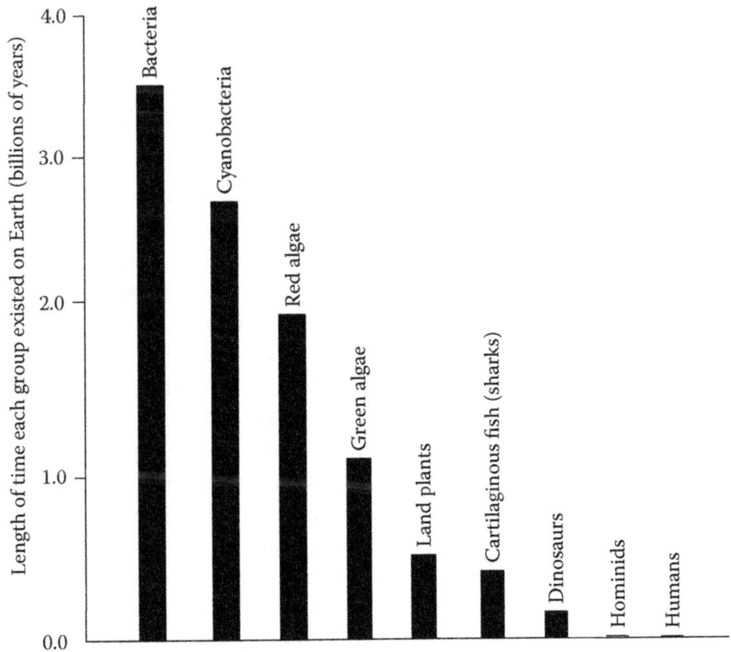

FIGURE 2.7 Length of time on the Earth for selected groups of organisms. Bacteria first appeared more than 3.5 billion years ago and have diverged greatly since that time. Oxygenic cyanobacteria have existed as a group for at least 2.7 billion years. Red algae, green algae, and land plants have existed for 1900, 1200, and 750 million years, respectively. Cartilaginous fish, including sharks, have existed on the Earth for the past 400 million years. Dinosaurs were on the Earth for about 180 million years, until an asteroid killed all of them. Hominids and humans are very recent arrivals. Hominids have existed for about 7 million years, while modern humans have existed as a species for about 140,000 years (or less).

for a mere 140,000 years. As *H. sapiens sapiens* was being established as a species 140,000 years ago, several other species in the genus *Homo* had gone extinct, but *H. sapiens sapiens* coexisted with *H. sapiens neanderthalensis*, and they interbred, such that 20% of the Neanderthal genome exists today within modern humans, and individuals alive today have 2% to 4% Neanderthal genes within their genomes. It is likely that both modern humans and Neanderthals hunted mammoths (*Mammuthus primigenius*) and other animals that are also extinct today. In fact, over the same 140,000 years, thousands of species have gone extinct, while countless speciation events have occurred. And the process continues today.

A SHORT HISTORY OF THE STUDY OF EVOLUTION

While the history of life on the Earth is nearly 4 billion years old, the study of evolution on the Earth by humans is only a few thousand years old (Figure 2.8). The initial parts of this history began when humans domesticated various plants and animals starting about 11,000 years ago. Wheat, rice, and barley were the first plants to be domesticated, which means that humans understood that they could select desirable seeds and plant them each year to guarantee a stable food source. They were practicing the process of selection. They had also domesticated microbes (yeasts) that were important in the production of beer, bread, and wine, although they did not understand that living things were performing the changes in these foods. By 10,000 years ago, goats, dogs, sheep, and pigs had been domesticated, and about 8,000 years ago, cattle and legumes (one species of which would eventually wind up in the hands of Gregor Mendel millennia later) had been domesticated. Over the next 5000 years, potatoes, maize, oats, and horses were domesticated. Towns and cities grew larger, as nutrition improved and food sources increased and became stable and predictable. Humans had realized that they could select specific individual plants and animals for breeding to improve the desired characters, which was an early form of genetics.

The Greek philosopher, Xenophanes of Colophon (570–480 BC), was the first to recognize that the Earth had undergone changes during its existence. He examined marine fossils extracted from hills near his home and concluded that the area had once been under the sea and that the fossils were long-dead organisms (although he had no idea of how long they had been dead). Aristotle (384–322 BC) and Hippocrates (460–370 BC) speculated on the mechanism of inheritance. These were the first intellectual underpinnings of what would eventually lead to studies in genetics and evolutionary biology. Aristotle used terms to indicate genus and species and divided the 500 known species into 8 classes. His student Theophrastus described 550 plant species in great detail and could probably be considered the first systematic botanist. With the fall of the Roman Empire in 476 AD, the Middle Ages began. The closing of the great academies in Athens (started by Plato and Aristotle) in 529 and the Museum of Alexandria also in 529 (and its partial destruction in 641) signaled a great slowing of scientific inquiry in Europe, a slowdown that spanned nearly a millennium. There is little to suggest that genetics and evolution were the topics of scientific discourse during this time. But, the invention of the printing press by Gutenberg in 1453 changed this in a very short time, because publications could be distributed widely, and therefore, many people could learn about new inventions and information simultaneously. Suppression of information that was considered dangerous to governing bodies (common in the Middle Ages) became more difficult. Also, multiple waves of the bubonic plague beginning in the fourteenth century decimated populations throughout Europe and Asia. This greatly weakened governmental and religious structures and great changes in social structure occurred. Over the next few centuries, major discoveries and descriptions were reported. Descriptions were published on all sorts of animals, plants, and fungi. Optics improved to the point where people like Galileo could observe the planets and moons and could describe their orbits. Hooke, Malpighi, Swammerdam, Grew, and van Leeuwenhoek could describe the microscopic parts of larger organisms, the developmental stages of some organisms, and the presence of microscopic organisms. People were discovering a previously invisible part of the world.

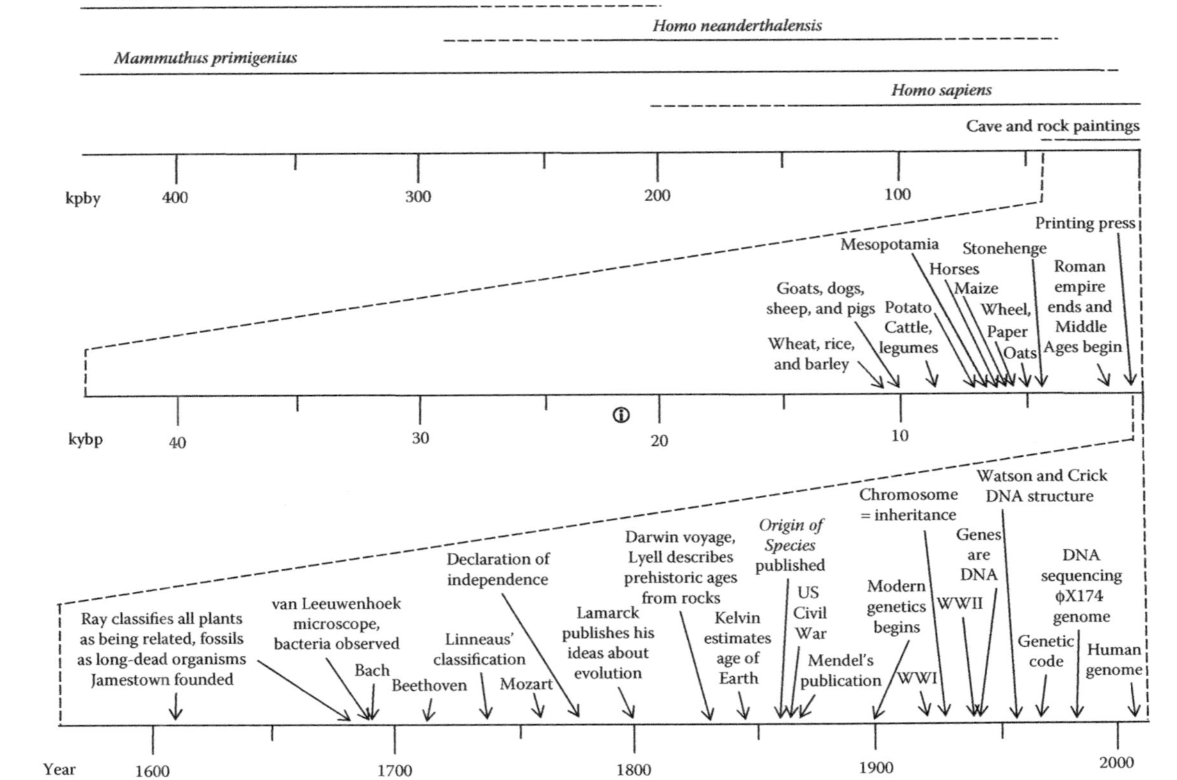

FIGURE 2.8 Details of human civilization and recent historical events related to evolutionary biology. Several species in the genus *Homo* and related genera have existed over the past 3 million years (top). By about 140,000 years ago, there only were two species, *H. sapiens neanderthalensis* (or sometimes classified as *H. neanderthalensis*) and *H. sapiens sapiens* (or sometimes classified as *H. sapiens*). By about 40,000 years ago, only *H. sapiens* remained. At about 11,000 years ago, there is clear evidence that humans started agricultural practices and began to build permanent settlements (middle). Over the next 10,000 years, many plants, animals, and microbes were domesticated. They were propagated for food, protection, commercial, and medicinal purposes. Over the past 400 years (bottom), there has been great progress in the fields of medicine, genetics, evolutionary biology, and finally molecular biology and genomics.

Pierre Belon was the first to note in 1517 that the bones from fish to mammals were analogous. That is, each bone in a fish related to a similar bone in other vertebrates all the way to mammals. This echoed what Aristotle and Theophrastus had thought about plants 1800 years earlier, when they classified plants into categories according to their similarities in structure. The most detailed comparisons were between birds and humans (published in 1555). They had concluded that these organisms fundamentally related to each other, and in particular, humans were related to animals. Botany and zoology became areas of study for scientists, and it was realized that small entities that passed from one person to another could cause diseases and thus began a flicker that would become the study of infectious diseases. The study of embryo and organ development in animals and other organisms was initiated during this period. Developmental studies, especially comparative studies among species, had major impacts on later evolutionary biology investigations. By the mid-seventeenth century, it was discovered that mammals also have eggs, which unified them with other animals, including insects. It was discovered that there exists a wide variety of microorganisms, including a group known as protists, and it was recognized that higher plants have male and female sexual organs. In 1686, John Ray published his work on the classification of plants based on common descent. In 1691, Ray also described fossils as the remains of animals from the distant past. In 1730, Otto Müller became the second person (the first was van Leeuwenhoek a 100 years earlier) to view and describe bacteria and classified them into types. In 1735, Carolus Linnaeus introduced his natural system of classification (including genus and species designations) that still is in use today.

In 1751, Pierre-Louis Moreau published a book in which he speculated on the factors involved in heredity and that species originate by chance. In 1760, Joseph Gottlieb Kölreuter described his experiments on heredity in plants. While it was known that traits are passed from parents to progeny, the mechanisms for this inheritance were completely unknown. The next year, Jean-Baptiste Robinet speculated that species form a linear progression without gaps. In 1767 and 1768, he described his experiments proving that spontaneous generation does not occur, the corollary of which is that every organism must originate from another organism of the same kind. This began to solidify the notion that organisms produce only organisms of the same kind, but that all organisms somehow were related to each other. Thus, the first formulations of evolutionary science were developing. About 1778, Jean-Baptiste Lamarck first stated his ideas that the environment causes changes in organisms that lead to the evolution of different varieties and species. Although this was incorrect, it pointed to a connection between the organism and its environment and that these related to changes in species. In 1794, Erasmus Darwin (Charles Darwin's grandfather—although Charles had not yet been born) agreed with the Lamarckian idea of the environment causing evolutionary changes in species. In that year, he stated that all warm-blooded animals have arisen from a single ancestor. In 1798, Georges Cuvier published his work on elephant and mammoth anatomy. The mammoth was found preserved in Siberian ice. He concluded that species go extinct. This was approximately 110 years after the dodo had actually gone extinct (last one observed in 1680)! So, although the process of how new species arose was not understood, their complete disappearance was.

In 1801, Lamarck began to formalize his ideas on evolution in publications, which culminated in his *Philosophie Zoologique*, published in 1809. In the same year, Charles Darwin was born. During the next two decades, Erasmus Darwin changed his views on evolution and thought that the environment somehow was the selective agent for change rather than the environment being responsible directly for change. He discussed these ideas extensively with his grandson, Charles. In 1827, Charles was admitted to the University of Cambridge to study medicine, but he fainted at the sight of blood, so he missed many of his medical classes. He became much more interested in natural history, geology, and fossils and admired the work of the geologist Charles Lyell. In 1829, Lyell published his *Principles of Geology*, which was of great interest to Darwin. It explained that layers of rocks could be followed back in time for millions of years. In 1831, Charles Darwin graduated from Cambridge with a degree in theology, not medicine. Shortly thereafter, he signed on to be the naturalist and dining companion for Captain Fitzroy on the HMS *Beagle*. This 2-year voyage ultimately became a 5-year voyage. On his voyage, he often was seasick, but loved his excursions on land. He still remained very interested in fossils and collected them everywhere he could find them, and shipped them back to the British Museum in London. Of course, he also shipped plant and animal specimens to the BM. But, he was more of a geologist than a biologist at that time. In fact, he misidentified many biological specimens, including his famous finches (found in 1835 on the Galapagos Islands). He did not label them as finches, but mistakenly identified them as many different types of birds, primarily because of the variety in their beaks. Actually, it was an ornithologist, John Gould, at the BM that accurately identified the birds after their arrival back at the BM. He determined that they all were finches, but that their beaks were distinctive based on their locations and their feeding behaviors. These facts were extremely influential in the formulation of Darwin's ideas about evolution over the ensuing decades.

A discovery in 1831 would also influence evolutionary biology. That year, Robert Brown first described the cell nucleus, thus leading the way to elucidation of the distinction between cells with and without nuclei. In 1833, Charles Lyell identified the Recent, Pliocene, Miocene, and Eocene periods of the Earth's history from rock strata. Over the next decade, more periods of the Earth's history were uncovered, and Agassiz first described Ice Ages that have covered large portions of the Earth in the past. In 1836, Darwin returned to England, and in 1839 and 1840, he published his accounts of the *Beagle* voyage collections. In the same year, Theodor Schwann published his work on cell biology, pointing out the similarities between animal and plant cells. During the next few years, Darwin published an abstract of his ideas about evolution and published books on his geological research from the *Beagle* voyages as well. The stage had been set for the presentation of his ideas, but he waited for years before publishing, knowing that it would be a controversial report and that it might upset his family life as well.

Meanwhile, in 1846, Lord Kelvin estimated that the Earth was 100 million years old, based on the current internal temperature and mass of the Earth. Although this value falls well short of what we now know, it clearly demonstrated the antiquity of the Earth. His miscalculation primarily was due to the fact that he was unaware of the natural radioactivity within the Earth, which caused additional internal heating. In that same year, Matthew Fontaine Maury charted the Atlantic Ocean and found that it was shallower in the middle, the first indication of the mid-Atlantic ridge formed between two tectonic plates. In 1856, Louis Pasteur demonstrated that fermentation is caused by microorganisms, not by chemicals. In 1857, Gregor Mendel started his experiments on garden peas. The next year, 1858, a manuscript from Charles Darwin, as well as letters between Darwin, Alfred Wallace, and Asa Gray, all describing evolution by natural selection, was published. Wallace had collected specimens around the world and came to the same conclusions as Darwin. However, Wallace was slowed considerably in his investigations because one of the ships that carried his collected specimens burned on its way back to England. In November 1859, the first edition of Charles Darwin's *Origin of Species* was published. In the following year, Gregor Mendel discovered the laws of heredity, but friends and colleagues expressed disinterest, so he dropped the project for some time. He resumed them out of an intellectual interest, and in 1865, Mendel finally published (in German) his work on heredity in peas. He presented the results at a few meetings, but it was received with indifference. Over the next 35 years, the work almost disappeared and was cited only three times in other publications. Unfortunately, Mendel died in 1884, long before his work was rediscovered in 1900 (independently by Hugo de Vries, Carl Erich Correns, and Erich Tschermak von Seysenegg) and gave rise to modern genetics.

In 1869, Johann Miescher isolated a chemical rich in phosphorus and nitrogen that he called *nuclein* (that later was determined to be DNA), because it was isolated from cell nuclei. In 1871, Darwin published *Descent of Man* indicating that humans also are a part of the story of evolution. The book was partly inspired by Charles Lyell's 1863 publication that indicated the antiquity of humans. In 1892, August Weismann proposed that eggs and sperm carry the material inherited by progeny. In that same year, tobacco mosaic disease was found to be caused by an entity too small to be seen in a microscope. Decades later, it was found to be caused by tobacco mosaic virus. In 1900, Erich von Tschermak, Carl Correns, and Hugo de Vries independently rediscovered Mendel's report on inheritance and almost immediately realized its implications. Over the next decade, many principles of genetics were determined, including that the chromosomes were the carriers of heredity, each type of gene often had many forms (called *alleles*), and that often traits could be dominant or recessive. Genetics flourished further in the next two decades, and by 1930, discoveries showed that genes were positioned on chromosomes, human ancestors (including those in different genera) had existed and gone extinct, X-rays could be used to cause mutations, and natural selection could be responsible for the distribution of gene alleles in a population. During the 1930s, Harriet Creighton, Barbara McClintock, and Curt Stern demonstrated that chromosomes could exchange adjacent pieces, creating crossovers. These were coincident with changes in traits inherited by the progeny. Furthermore, McClintock discovered that some small pieces of the chromosome appeared to jump from one location to another. A few decades later it was discovered that these were genetic elements called transposable elements that could change locations on the chromosomes. These two findings showed that inheritance was even more complex than the simple traits that Mendel had reported more than half a century before. Large and small pieces of chromosomes could be exchanged or could move from one chromosomal location to another.

Advancements in studies of genes continued, and in 1941, George Beadle and Edward Lawrie Tatum proposed a theory that for each gene, one protein was produced. We now know that this is correct for many genes, but that some genes can produce more than one protein through a process (alternative splicing) that uses different parts of a single gene to produce different proteins. Speaking of proteins, by 1940, although it had been established that the chromosomes held the hereditary material, no one knew whether genes consisted of DNA or proteins. Many thought it had to be the proteins, because there were 20 amino acids and many different ways to connect them into complex linear sequences, while DNA had only 4 bases, and hence, the number of combinations was vastly reduced. But, in 1944, Avery, MacLeod, and McCarty published the results of their experiments that indicated that the hereditary material was DNA, and not protein. Additional proof of this was provided by the elegant experiments of Hershey and Chase in 1952. Then, in 1953, Watson and Crick, with the aid of an X-ray diffraction image produced by Rosalind Franklin, published their one-page article in *Nature* (adjacent to articles by Wilkins, Stokes, Wilson, Franklin and Gosling) indicating that DNA was a double helix. They also suggested that the double-helical nature of DNA provided a means by which DNA could be readily replicated, producing two identical copies. Over the ensuing years, the mechanism of DNA replication (known as *semi-conservative replication*, which uses both DNA strands to synthesize two new double-stranded copies, and suggested by Watson and Crick) was demonstrated by Mathew Messelson and Frank Stahl.

By the mid-1960s, Crick, Sydney Brenner, Alan Garen, and others had elucidated the genetic code, and methods for cloning small pieces of DNA had been developed. The age of molecular biology had begun. Also during the 1960s, fossils of ancient species in the genus *Homo* (and related genera) were unearthed in Africa (and elsewhere), extending our knowledge of the number and variety of hominids that have existed. By the 1970s, additional hypotheses regarding evolution were espoused, including the theory that chloroplasts were the descendants of once free-living photosynthetic cyanobacteria and that the ancestors of mitochondria were free-living α-proteobacteria (proposed by Lynn Margulis). Both had become organelles, having originated as endosymbionts. In 1977, Carl Woese and George Fox published a paper indicating that there was yet another domain of life besides bacteria and eukaryotes. They called them *archaebacteria*, which are now termed *the Archaea*. While they are similar in form to bacteria (and were originally called *bacteria*), they are very different from bacteria and have some similarities to eukaryotes, indicating that they are somehow between bacteria and eukaryotes in an evolutionary sense. In the same year, two groups (Maxam and Gilbert; Sanger et al.) published methods for sequencing DNA, which was the initial step toward determining the genome structure of each species. The first complete genome of a virus (ϕX174) was completed in 1977, followed by determination of the first free-living bacterial genome (*Haemophilus influenzae*) in 1995, the first eukaryotic sequence (*Saccharomyces cerevisiae*) in 1996, the first multicellular eukaryote (*Caenorhabditis elegans*) in 1998, and, of course, the first release of the human genome in 2003.

Molecular biology methods allowed researchers to address many questions in genetics, cell biology, and evolutionary biology. Detailed studies of gene structure, organization, and expression were performed. Gene hierarchies were found, where a few genes controlled large sets of other genes. These were especially abundant in multicellular organisms, where body plans were involved. Comparisons of genes among different organisms were undertaken in many labs. This began a new field in molecular phylogenetic analyses. Evolutionary trees could then be constructed based on gene sequences. This allowed direct comparison of the hereditary material and thus was a direct view of the molecular changes of the evolutionary processes. These allowed elucidation of the relationships among similar, as well as disparate organisms, but it also allowed comparisons among similar and disparate genes. Ultimately, these methods led to the finding that mitochondria truly did originate from α-proteobacteria (specifically from one group the Rickettsiales), and plastids originated as cyanobacterial endosymbionts. It also brought us to where we are today. Molecular biology has shown that the principles set forth by Mendel, Darwin, and others are correct. At the same time, we now know that the picture is much more complex (and much more interesting) that anyone could have imagined in Darwin's time. This is because organisms and pieces of DNA are being transferred constantly by organisms large and small, and some of the organisms and pieces of DNA are being incorporated permanently into the genomes of the receiving cells. This means that inheritance occurs both in a vertical manner (i.e., from parent to progeny) and in a horizontal manner (i.e., from one species into another). It remains to be seen what additional surprises await scientists involved in genetic and evolutionary biology projects. And, remember that all of this started when humans began to domesticate plants and animals, developed into scientific disciplines in Greek and Roman times, and then expanded greatly only in the last few hundred years. Currently, we are in a time of great discoveries in the study of evolutionary processes. And, still evolution continues.

EARTH HISTORY AS ONE YEAR

The magnitude of the time scale for the Earth's history is difficult to fully comprehend. However, we can use a human time scale to aid with this. If we squeeze the Earth's history into a one-year period (Figure 2.9), we can begin to understand the comparative lengths of time for each period of time. So, starting at 12:01 a.m. on January 1, the Earth has coalesced into a sphere that is recognizable as a planet. Approximately 1.6 days later (at 4.53 billion years ago), another body about the size of Mars collides with the Earth, releasing an enormous amount of energy and throwing huge masses of materials out into space. Much of the debris first forms two moons, and then, the two moons collide to form a single moon. After nearly continuous bombardment by asteroids and comets, after about 44 days (approximately mid-February) things settle down somewhat and some parts of the Earth surface begin to solidify (4.0 billion years ago). However, on about February 21, there is a huge bombardment from space and much of the Earth's surface again becomes molten (approximately 3.9 billion years ago). Afterward, there is relative calm, and by early March, the surface of the Earth begins to cool and solidify into rock. From the end of February through the beginning of March, life begins on the Earth. By the end of March, bacteria are present (3.5 billion years ago). At this time, the Earth still would have been an inhospitable place for humans. The atmosphere consisted of ammonia, carbon dioxide, carbon monoxide, methane, hydrogen, and water vapor, and almost no oxygen was present. But, during the first few days of May (2.6 billion years ago), the oxygen levels begin to rise. These are coincident with an increase in the prevalence of oxygenic cyanobacteria species, which released oxygen as a waste product of their photosynthetic processes. Eukaryotes appear on June 21 (2.4 billion years ago), and the endosymbionts that will become mitochondria appear sometime between June 21 and July 6. These organelles (originating from an endosymbiotic α-proteobacteria) were needed to detoxify oxygen in these cells. By August 1, another important endosymbiotic relationship was formed that resulted in a photosynthetic eukaryote. Approximately 1.9 billion years ago, red algae resulted from a combination of a eukaryotic cell and a cyanobacterium. The cyanobacterium eventually became

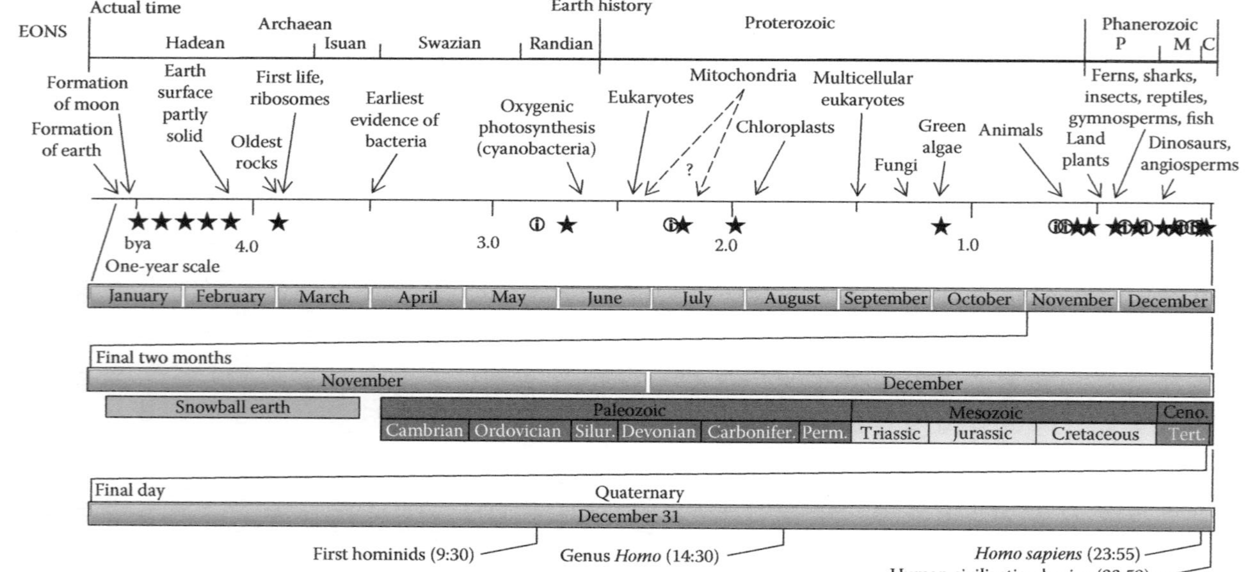

FIGURE 2.9 Earth's evolution as one year. The top timeline is a linear representation of the major events that occurred during the entire history of the Earth. The other timelines represent the history of the Earth on a one-year time scale. The lowest two bars illustrate the last two months and the last day of the year, respectively. A detailed description of the figure is within the text. P, Paleozoic Era; M, Mesozoic Era; C, Cenozoic Era; Ceno., Cenozoic Era; Silur., Silurian Period; Carbonif., Carboniferous Period; Perm., Permian Period; Tert., Tertiary Period.

highly dependent on its host cell and ultimately became the chloroplasts. The organisms with these chloroplasts diversified, as did the organelles. Over the next 50 days, three groups emerged: red algae, glaucophytes, and green algae. Red algae would appear early in August, while the green algae would not appear until late September. Green algae eventually led to plants (but that would not happen until well into November). But, by the last of October (or about 750 million years ago), the continent of Rodinia had formed near the South Pole. This caused a series of events that led to cooling of the Earth, such that ice over a kilometer thick covered the entire planet. This is known as the Snowball Earth. Several of these ultra-cold periods occurred from early to mid-November (until about 600 million years ago). It appears that this extended cold period might have ended with a huge bolide strike (or a set of collisions). And 540 million years ago (November 17), the Cambrian period began. A large diversity of animals suddenly appeared in the fossil record at this point, which has been termed the Cambrian explosion, although the diversity probably existed prior to this time, but few fossils were left behind. Because of the huge amounts of carbon dioxide and other greenhouse gasses, the planet warmed significantly, and photosynthetic organisms grew rapidly. The types of plants and animals expanded. Ozone now blocked much of the UV irradiation from the Sun and meant that organisms could now survive on land surfaces. However, at 250 million years ago (December 11), a global cataclysm occurred. It is known that this time coincides with huge volcanic eruptions in Siberia, but it may have been initiated by a large bolide strike elsewhere on the Earth. This caused enormous releases of ash and greenhouse gases and may have led to anoxic conditions in the oceans. All of this resulted in mass extinction. More than 90% of all species suddenly went extinct. But, some of the reptiles that survived gave rise to dinosaurs, which appeared shortly after the mass extinction. Temperatures were high and carbon dioxide levels were relatively high. Land plants grew fast and large. Animals grew ever larger, so that by 100 million years ago (December 23), some dinosaurs were enormous. At about 95 million years ago (December 25), the first small primates appeared. Things were going well for dinosaurs and small mammals. Then, a bolide about 6 miles across collided with the Earth near Chixilub, Mexico (65 million years ago, or December 26 on our 1-year scale). Organisms within several hundred kilometers were killed instantly. More than 70% of all species went extinct. The impact may have stimulated volcanism in India, since there were huge volcanic eruptions at exactly the same time. Also, the meteor had struck part of the continental crust that contained carbonates. This may have caused a huge release of carbon dioxide. The ash from the fires that the explosion had caused, plus the ash from the volcanoes, blocked the Sun. However, the CO_2 caused trapping of heat on the Earth. These conditions may have caused cooling followed by heating of the Earth, leading to additional extinctions. This event wiped out the dinosaurs, although birds are very closely related to dinosaurs, so not all of the dinosaur relatives went extinct at that time. Afterward, there was a rise in the number and diversity of mammals, including primates. Seven million years ago (about 9:30 a.m. on December 31), the first hominids appeared. About 4.5 million years later (about 2:30 p.m. on December 31), the genus *Homo* is first seen, and *H. sapiens sapiens* appears about 5 min before midnight on December 31. The entire human time of what we call *civilization* occurs in the final 1 min of the last day of the year, and your entire life occurs in the last 0.1–0.3 s before midnight.

KEY POINTS

1. Evolution consists primarily of mutation followed by natural selection and genetic drift.
2. The Earth formed 4.55 billion years ago, after which a Mars-sized body struck the Earth forming the Earth–Moon system. This event helped to form the molten iron core of the Earth that protects organisms from some of the harmful radiation emitted by the Sun.
3. The first chemicals necessary for the initiation and evolution of life formed very soon after the Earth cooled.
4. Early life probably got its start between 4.0 and 3.5 billion years ago.
5. After life had been initiated on the Earth, single-celled bacteria and archaea predominated.

6. The first eukaryotes appeared 2.4 billion years ago.
7. Mitochondria first appeared between 2.2 and 2.4 billion years ago. They started as endosymbionts of early eukaryotes as way to adapt to increasing levels of oxygen in the atmosphere and oceans.
8. Chloroplasts (and plastids, in general) first appeared 1.9 billion years ago. They began as endosymbionts of heterotrophic eukaryotes. The eukaryotic cells then diversified into glaucophytes and red algae, and later green algae (1.2 billion years ago).
9. Multicellular eukaryotes first appeared 650 million years ago.
10. The study of evolution by humans began with the domestication of plants and animals 10,000 to 11,000 years ago, although academic studies of evolution had their beginning in Greece about 2500 years ago, and more recently intense study beginning a few hundred years ago.

ADDITIONAL READINGS

Ciccarelli, F. D., T. Doerks, C. von Mering, C. J. Creevey, B. Snel, and P. Bork. 2006. Toward automatic reconstruction of a highly resolved tree of life. *Science* 311:1283–1287.

Darwin, C. 1859. *The Origin of Species* (Mentor Book version, 1958). Chicago, IL: The New American Library of World Literature.

Futuyma, D. J. 1998. *Evolutionary Biology*, 3rd ed. Sunderland, MA: Sinauer Associates.

Gardner, E. J. 1972. *History of Biology*, 3rd ed. Minneapolis, MN: Burgess Publishing.

Hall, B. K. and B. Hallgrimsson. 2008. *Strickberger's Evolution*, 4th ed. Sudbury, MA: Jones & Bartlett.

Hellemans, A. and B. Bunch. 1988. *The Timetables of Science*. New York: Simon & Schuster.

Johnson, A. P., H. J. Cleaves, J. P. Dworkin, D. P. Glavin, A. Lazcano, L. Jeffrey, and J. L. Bada. 2008. The Miller volcanic spark discharge experiment. *Science* 322:404.

Joyce, G. F. 2002. The antiquity of RNA-based evolution. *Nature* 418:214–221.

Miller, S. L. 1953. Production of amino acids under possible primitive Earth conditions. *Science* 117:528–529.

Pan Terra, Inc. 2003. *A Correlated History of the Earth*. Hill City, SD: Pan Terra.

Powner, M. W., B. Gerland, and J. D. Sutherland. 2009. Synthesis of activated pyrimidine ribonucleotides in prebiotically plausible conditions. *Nature* 459:239–242.

Stearnsand, S. C. and R. F. Hoekstra. 2005. *Evolution: An Introduction*, 2nd ed. New York: Oxford University Press.

Wallace, R. A., J. L. King, and G. P. Sanders. 1981. *Biology: The Science of Life*. Santa Monica, CA: Goodyear Publishing.

Wilson, E. O. 1992. *The Diversity of Life*. New York: W.W. Norton & Co.

Section II

Biomolecules

3 DNA, RNA, and Proteins

INTRODUCTION

Life on the Earth began from very simple molecules. The main characteristic of these molecules was that they normally contained a few different types of atoms that formed compounds that were somewhat stable, but were capable of carrying out simple reactions with other molecules. Primarily, they contained carbon and hydrogen, secondarily nitrogen and oxygen, and to a lesser extent phosphorus and sulfur. On early Earth, the combination of water, time, heat, solar energy, pressure, and some compounds from extraterrestrial sources produced the first biological reactions and eventually cells that were capable of absorbing nutrients for their metabolic processes and reproduction, and excreting waste products. They used amino acids to build proteins, but also relied heavily on RNAs for structural elements, enzymatic functions, and genetic capabilities. Selection favored those that reproduced reliably over long periods of time. During this early evolution, a lipid bilayer cell membrane, use of carbohydrates, reliance on proteins and RNA, and translation of proteins by ribosomes became established. However, because of the high mutation rates for RNA, genomes would have been relatively small, and even then many of the progeny would have died, because they would have inherited too many mutations to survive. But, one particular mutation, which gave rise to the utilization of a 2′-deoxyribose for at least one gene, led to an advantage for the organisms that rapidly outcompeted the RNA-based cells. The switch to DNA as the hereditary material from the original RNA had begun.

By the time bacteria had appeared on the Earth about 3.5 billion years ago, DNA probably was being used as the primary genetic material for these cellular organisms. The enzymes necessary for DNA metabolism are similar to those for RNA metabolism, and therefore, the transition to DNA synthesis required only small mutations in a small set of genes. The various RNAs retained their abilities to carry out many biological functions, including holding the genetic instructions for the daughter cells. However, RNA is much more prone to errors than is DNA, so the genomes and the cells had to be simpler at that time. Nevertheless, given rapid and prolific reproduction of new cells, survival of a few accurately copied cells would have been assured. When DNA was used as the genetic material, the organisms could retain genes for many generations without suffering as many mutations, and they did not have to produce as many progeny since more of the daughter cells would survive. This would reduce their energy requirements while bolstering their abilities to survive and reproduce. In essence, they became much more efficient at producing viable daughter cells.

The earliest forms of what could be called *biological organisms* probably were an amalgamation of RNAs, amino acids, polypeptides, lipids, and carbohydrates. All of these have been produced in laboratory experiments of one sort or another. Each might have been produced under different conditions on the Earth, as well as arriving on asteroids and comets that impacted the Earth. Some may have come from the so-called primordial soups that were on the Earth very soon after the surface began to cool and solidify by 4.3 billion years ago. But, some may have originated deep in the oceans near hydrothermal vents, where there was water and a variety of simple molecules, as well as gradients of heat and pressure. However, for the past 3.5 million years, all cellular organisms have been based on just a few major systems to produce DNA, RNA, amino acids, proteins, ribosomes, lipids, and carbohydrates, as well as variants of each. By the time organisms had begun to use DNA as the genetic material, rudimentary ribosomes were being used as the primary means to produce peptides and proteins. There still are simple mechanisms in cells today that produce some peptides without relying on ribosomes. These mechanisms may have been among the first biological processes that linked amino acids together into functional peptides.

The central dogma of molecular biology first set forward by Francis Crick states that the flow of information in cells is from DNA to RNA to protein. But, each of these processes had a beginning. Since proteins make cells what they are biologically, it may have been that the first rudimentary biological processes consisted of simple chains of amino acids. As was already discussed, amino acids, polypeptides, and RNA can be formed from simple components that existed in early Earth environments. Simple peptides and proteins can facilitate reactions, and some could hold onto other types of molecules so that they could perform specific functions or help them to be more efficient in these processes. However, proteins and sets of proteins cannot reproduce themselves in the same way that RNA and DNA can. Therefore, if proteins formed the initial biological systems, then one of the nucleic acids, probably RNA, was vital to maintaining a set of genetic instructions to make new such systems that could continue to propagate themselves. Very early in the evolution of life, these biological processes became enveloped within a lipid-based membrane. This protected the biological molecules by separating them from other molecules that could degrade them or interfere with their functions. It also concentrated the molecules leading to more efficient and reliable reactions. As the polypeptides became more complex through mutation and selection, they improved the efficiency of the reactions and aided in a wider variety of reactions. Some types of proteins are very good at stabilizing RNA. The first protein/RNA associations could have led to several important innovations, including assuring accurate replication of the genetic material (in this case, it would have been the RNAs) and assuring efficient synthesis of proteins through formation of tRNAs, protoribosomes, and eventually ribosomes. It also assured the production of other important molecules, such as adenosine triphosphate (ATP), lipids, and carbohydrates. All of these predated the fossil bacteria of 3.5 million years ago. The final major step, that of use of a variant of RNA, called *DNA* (as in 2′-deoxy-RNA), allowed greater fidelity of replication and stability of the genetic material. DNA molecules readily form double strands, which are much more stable than single-stranded RNA molecules, so that they are less prone to damage. Plus, because there are two strands, one being the reverse complement of the other, the additional strand can be checked for errors during the replication process. Therefore, in a phylogenetic sense, RNA and proteins (and amino acids) are probably the most ancient of biological molecules, while DNA is likely the most recent, having become established as the genetic material by about 3.5 billion years ago.

NUCLEIC ACIDS

While RNA and DNA contain very similar components, they perform different functions in cells. DNA is the 2′-deoxyribose relative of RNA. Both consist of nucleotides that have a base (either one of two purines or one of two pyrimidines), a sugar (ribose or 2′-deoxyribose), and a phosphate group attached to the 5′ carbon on the sugar (Figures 3.1 and 3.2). In both polymers, the sugars are joined together in the long backbone of the nucleic acid by phosphate ester bonds through the fifth and third carbons on the sugars. The atoms of the rings on the bases are designated with numbers (1–6 for pyrimidines and 1–9 for purines). The atoms on the sugars are designated by numbers with prime (′) after them in order to distinguish them from the atoms in the bases. In deoxyribose and ribose sugars, there are five carbon atoms that are numbered 1′ through 5′ (Figures 3.2 and 3.3). In RNA, the 2′ and the 3′ carbons have hydroxyl (OH) groups attached to them, while in DNA, only the 3′ carbon has a hydroxyl group. The 2′ carbon has only a hydrogen (H) group. Because of this, in DNA, the specific sugar is a 2′-deoxyrobose, and thus, the DNA in all organisms is based on 2′-deoxyribonucleic acid. While this seems like a small change in the molecules, it makes a big difference in how the molecules can pack together. This is not only due to the smaller size of the hydrogen versus the hydroxyl, but also because the C–H moiety is

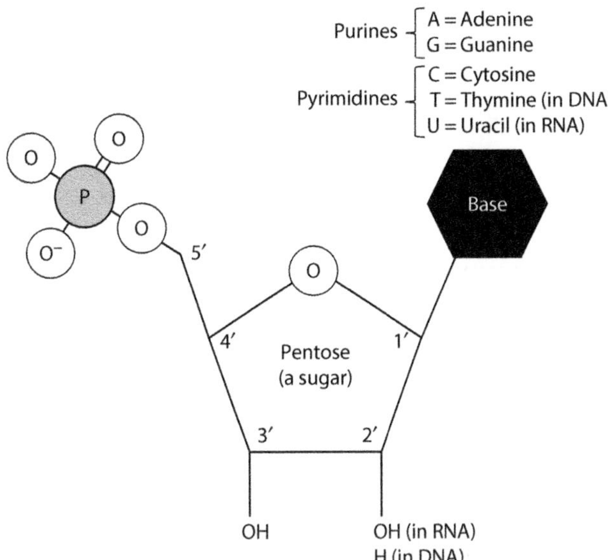

FIGURE 3.1 General form of nucleotides (in DNA and RNA). Each is built around a pentose ring: ribose in RNA and 2′-deoxyribose in DNA. A phosphate is attached to the fifth carbon (C-5′) by an ester bond. A purine or pyrimidine is attached to the first carbon (C-1′) on the pentose by a carbon–nitrogen bond. The two purines are adenine and guanine, while the pyrimidines are cytosine, thymine (in DNA), and uracil (in RNA). Additionally, methylated versions of A, G, and C are often present in some DNAs and RNAs. Several modified bases are often found in RNA, including inosine, wyosines, and pseudouridine.

nonpolar, while the C–OH group is polar and reactive. The nucleotides in DNA can pack together to form more stable double-helical structures. RNA can also form double-helical structures, but the hydroxyl on the 2′ carbon means that it cannot form exactly the same type of helix, and therefore, the helix is less stable. Because of this, the double helices in RNA generally are shorter than those formed in DNA. They also have more bulged regions (caused by unpaired nucleotides) in these helices, as well as many nucleotides that interact with other distant nucleotides on the molecule as well as adjacent nucleotides.

In DNA and RNA, the nucleotides are joined through phosphate ester bonds at two places, through the 3′ carbon and through the 5′ carbon (Figure 3.4). Thus, they are connected by phosphodiester bonds through these carbons. This is why, when discussing DNA (as well as RNA), there is a stated directionality because of the 5′ and 3′ parts of the nucleotides. The convention is that the arrowhead points toward the 3′ end, since this is the direction of DNA replication by DNA polymerase. Typically, the ends of the DNA will have a phosphate at the 5′ end (the phosphate is attached to the 5′ carbon) and a hydroxyl at the 3′ end (the hydroxyl is attached to the 3′ carbon). However, there are a few exceptions to this. Occasionally, the 5′ phosphate is lost from the end, but most often, this is replaced rapidly. The 5′ ends of most mRNAs have a special cap. This is a guanosine that is added posttranscriptionally. Instead of the standard 5′ to 3′ phosphodiester linkage joining nucleotides, this one is joined via a 5′ to 5′ triphosphate linkage (Figure 3.5). Additionally, the guanosine that is added has a methyl group attached to C-7 (carbon 7). Most mRNAs also have polyadenosines added to their 3′ ends. Both the cap and the polyA tails are added posttranscriptionally, and they are important for efficient translation and stability of the mRNAs.

DNA normally is in the form of a stable double helix that stores genetic information for accurate replication and transcription. It is very stable, and the structure is predictable and highly ordered because of the sugar–phosphate backbones, as well as the hydrogen bonds between the

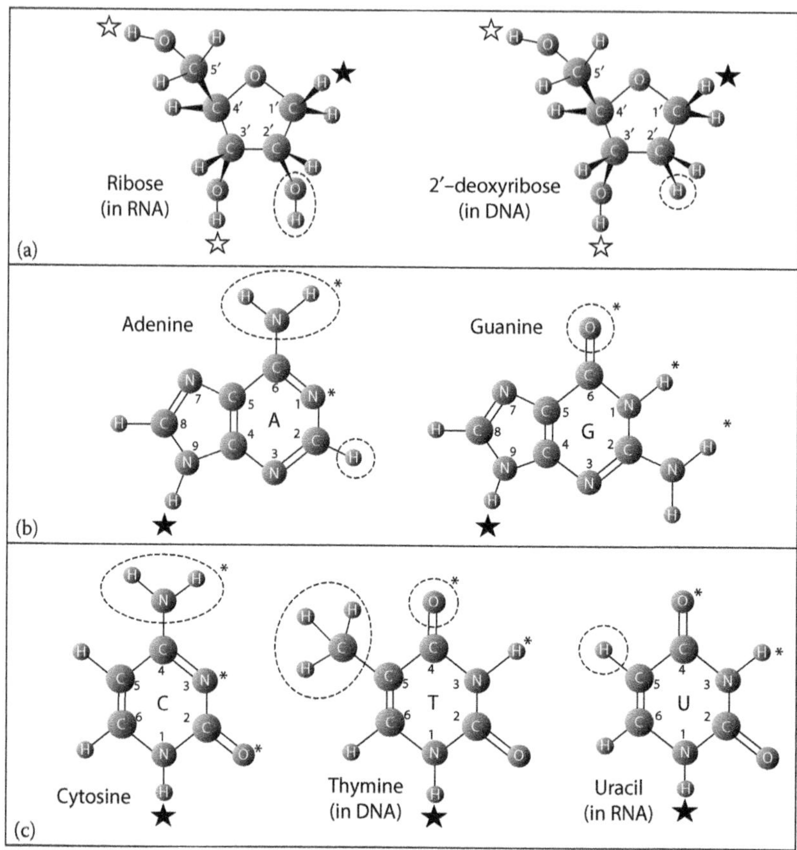

FIGURE 3.2 Components of nucleotides in RNA and DNA. (a) The two pentoses differ only at the 2′ carbon, where there is a hydroxyl in RNA and a hydrogen in DNA. Open stars indicate the attachment sites for the ester bonds joining the phosphates and the adjoining pentoses of adjacent nucleotides. The filled stars indicate the attachment sites for the bases. (b) The two primary purines in RNA and DNA are adenine and guanine. They differ at ring atoms 1, 2, and 6. At position 1 in adenine, the nitrogen is connected to C-6 by a double bond, which causes the nitrogen to be an electron acceptor in hydrogen bond interactions. In guanine, the nitrogen at the same position is joined to C-6 with a single bond and thus has a hydrogen attached to the carbon. This makes the hydrogen an electron donor in hydrogen bond interactions. At C-2, adenine has a hydrogen, while guanine has an amino group attached to the carbon. At C-6, adenine has an amino group, while guanine has a carbonyl oxygen. Solid stars indicate the location of the covalent bond to the ribose (in RNA) or 2′-deoxyribose (in DNA). Asterisks indicate the atoms involved in hydrogen bonds in Watson–Crick base pairs. (c) The three primary pyrimidines are cytosine (in RNA and DNA), thymine (in DNA), and uracil (in RNA). Cytosine differs from uracil and thymine at atoms 3, 4, and 5 on the ring. At position 3, cytosine has a nitrogen bound to C-4 with a double bond, causing the nitrogen to be an electron acceptor when forming a hydrogen bond. Both thymine and uracil have an N–H at the same site that causes the hydrogen to be an electron donor in a hydrogen bond. At C-4, cytosine has an attached amino group, while thymine and uracil each has an oxygen at that site. At C-5, cytosine and uracil have hydrogens, while thymine has a methyl group.

bases (A–T and G–C pairs) in what are termed Watson–Crick base pairs. Under physiological conditions, DNA is normally in its B-form (Figure 3.6). This is a right-handed helix, with antiparallel strands and Watson–Crick complementary bases (A–T and G–C pairings). There are approximately 10.5 bases per complete turn of the helix, and the bases are tilted slightly with respect to the major axis of the molecule. The sugars are on the same side of the helix with respect to the bases (in cross section) and thus are in a *cis* conformation. This orientation causes the

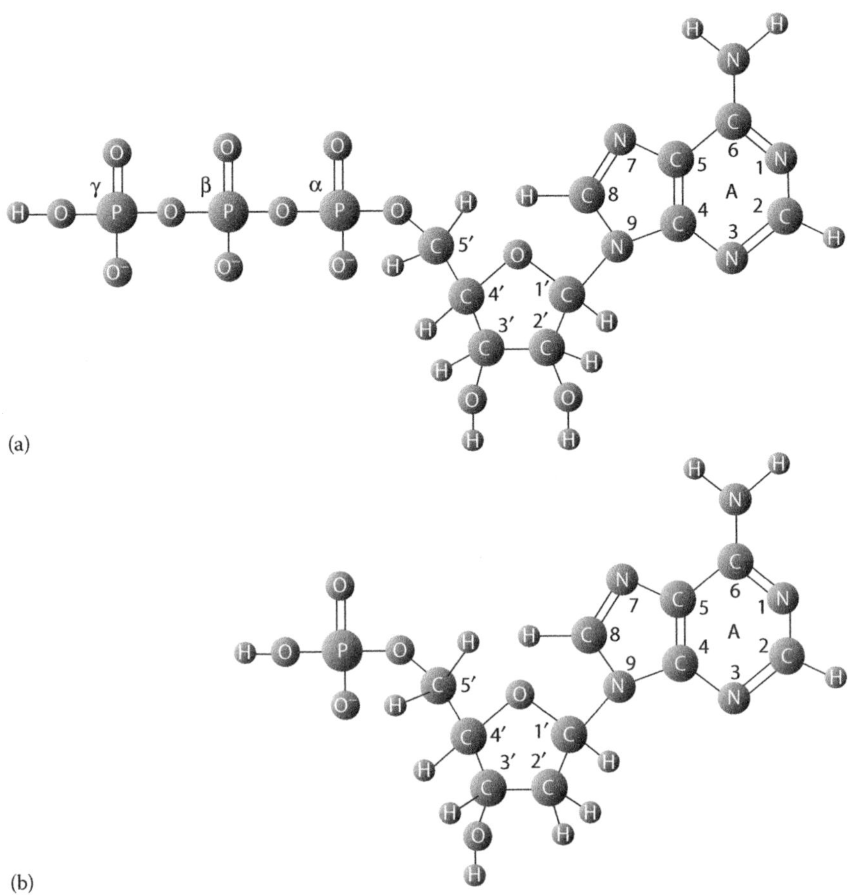

FIGURE 3.3 Two examples of nucleotides. (a) Adenosine triphosphate (ATP) is one of the primary carriers of bond energy in cells and is also used to transfer phosphates to other molecules (e.g., certain amino acids in proteins). The three phosphate bonds are high energy bonds, with the third bond (between the β and γ phosphates) being the most energetic. There are several other nucleotide-containing molecules that act in energy transfer, as well as nucleotides that are secondary messengers, such as cyclic-AMP (not shown). (b) 2′-deoxyribose adenosine monophosphate, which is one of the unit nucleotides in DNA.

bases to be more exposed on one side of the DNA than on the other, forming what are called the *major and minor grooves*. This conformation is important for many processes, because most DNA-binding proteins attach to the major groove side of the DNA. As salt and other conditions change, the DNA conformation changes. If the salt concentration is increased, the DNA will flex into its A-form (Figure 3.6). While this is also a right-handed helix, there are approximately 11 base pairs per turn of the helix, the major and minor grooves are nearly the same size, the bases are tilted to a greater angle, and the bases are less accessible to proteins. However, a hole forms in the central axis of the molecules. A-form is rare under physiological conditions. A third form, Z-form DNA, is also formed under high salt conditions in molecules with high G–C contents. In contrast to other forms of DNA, it is a left-handed helix. This causes kinks in the sugar–phosphate backbone, which resembles the letter Z. Z-form DNA has been found in living cells, but only under specific conditions. The DNA must have a high G–C content, and many of the C residues must be methylated, in the form of m^5C (cytosine with a methyl group on C-5; Figure 3.7). The methylase that adds methyl groups to the cytosine residues does so primarily when the cytosines immediately precede guanosines, called *CpG sites*

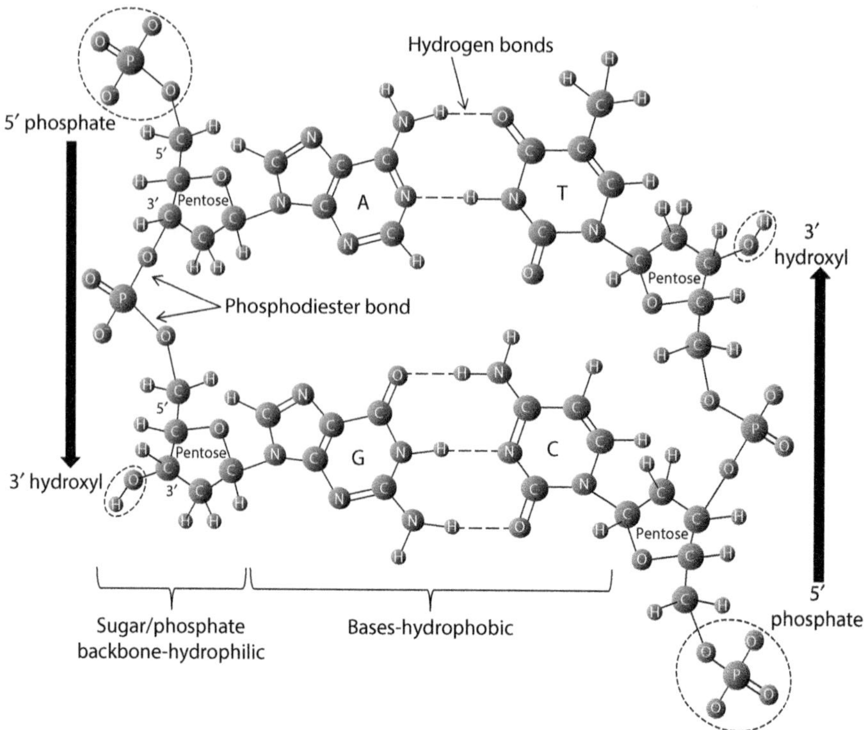

FIGURE 3.4 A section of DNA showing hydrogen bonded nucleotide pairs (standard Watson–Crick base pairs) and the sugar phosphate backbones. G–C pairs are held together by three hydrogen bonds, while A–T pairs are held together with only two hydrogen bonds. The two pentose–phosphate backbones are oriented in an antiparallel fashion. Each has a 5′ phosphate at one end and a 3′ hydroxyl at the other end. But they are oriented in opposite directions. The bases are hydrophobic and are in the center of the molecules, making that region hydrophobic, while the pentose–phosphate backbone is hydrophilic and negatively charged. The hydrophilic nature of this region comes from the hydrophilic pentose, as well as the charged phosphate groups. These areas are on the outside of the molecules in contact with the aqueous environment.

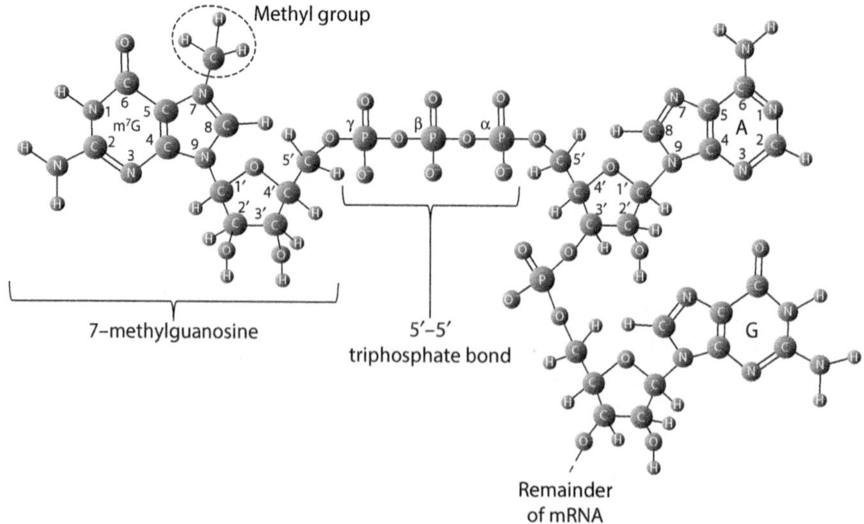

FIGURE 3.5 Illustration of a guanosine cap on the end of most eukaryotic mRNAs. The guanosine is methylated at N-7 and is added posttranslationally with a 5′–5′ triphosphate linkage.

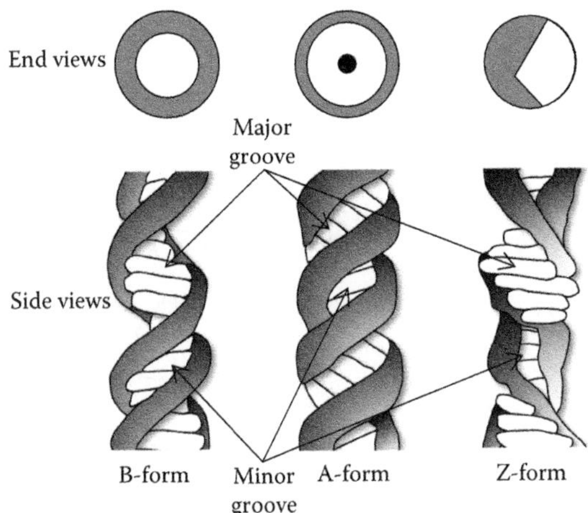

FIGURE 3.6 End and side views of the three major forms of DNA, illustrating the double-helical structure. The pentose–phosphate backbone is indicated in gray, while the bases are white. Under physiological conditions, most DNA is in the B-form. Under high salt conditions, the DNA can change into its A-form and Z-form. Both A-form and B-form DNA are right-handed helices with the bases in the middle of the molecules. However, in B-form DNA, the major groove is more pronounced and the bases are nearly perpendicular to the long axis of the molecule. The bases fill the entire center of the molecules. In A-form DNA, the major groove is less pronounced, and the bases are tilted much more than in the B-form. A pronounced hole forms in the central axis of A-form DNA (black dot in center of end view). Z-form DNA is a left-handed helix where the backbone forms a zig-zag pattern. In this case, the bases are primarily exposed on one side of the helix, and the pentose–phosphate backbone is exposed on the other side of the helix. While it forms in solution under high salt conditions, it can also form under physiological conditions when the cytosines are heavily methylated (i.e., when there is a high concentration of m^5C in the DNA). When it forms, it normally turns off gene expression in that area, since the DNA-binding proteins fail to attach to the region. However, in a few cases, Z-form DNA acts to turn on expression of some genes.

(cytosine–phosphate–guanosine). In most cases, when a gene is methylated and Z-form DNA is found in the region of the gene, the gene is down regulated, or turned off, partly because transcription factors and RNA polymerase cannot bind to the region.

Cytosines, adenines, and guanosines all can be methylated (Figure 3.7). Cytosines can be methylated on the fifth carbon of the pyrimidine ring, and guanines on the seventh carbon (see Figure 3.5). Adenines can be methylated on the amino group attached to C-6. Methylation has several important functions. First, as stated above, methylated cytosines are important in gene regulation. Second, bacteria use methylation systems to avoid DNA degradation by foreign invaders and their nucleases. Many nucleases will recognize and cut unmethylated DNA, but fail to recognize and cut methylated sequences. Many bacteria have restriction endonucleases that will cut foreign DNA that is necessary for killing other bacteria and utilizing their nucleic acids. However, in order to protect their own DNA, they have enzymes that methylate their own restriction endonuclease recognition sites, so that their restriction endonucleases will not degrade their own DNA. In bacteria, adenines primarily are methylated, but many bacteria have cytosine methylases as well. In eukaryotes, methylation primarily is on cytosines. Third, methylation has been found to aid in DNA repair during replication. DNA is normally methylated shortly after the new strands are synthesized. Since DNA replication is semi-conservative (i.e., one strand always is the *old* strand that acts as the template and the other strand is the newly synthesized strand), for a short time after a region is replicated, one strand is methylated, while the other is not. This is called *a hemi-methylated state*. During this time, repair enzymes scan the DNA for mutations. If they recognize a mutation, they will preferentially use the methylated (i.e., the old) strand as the template

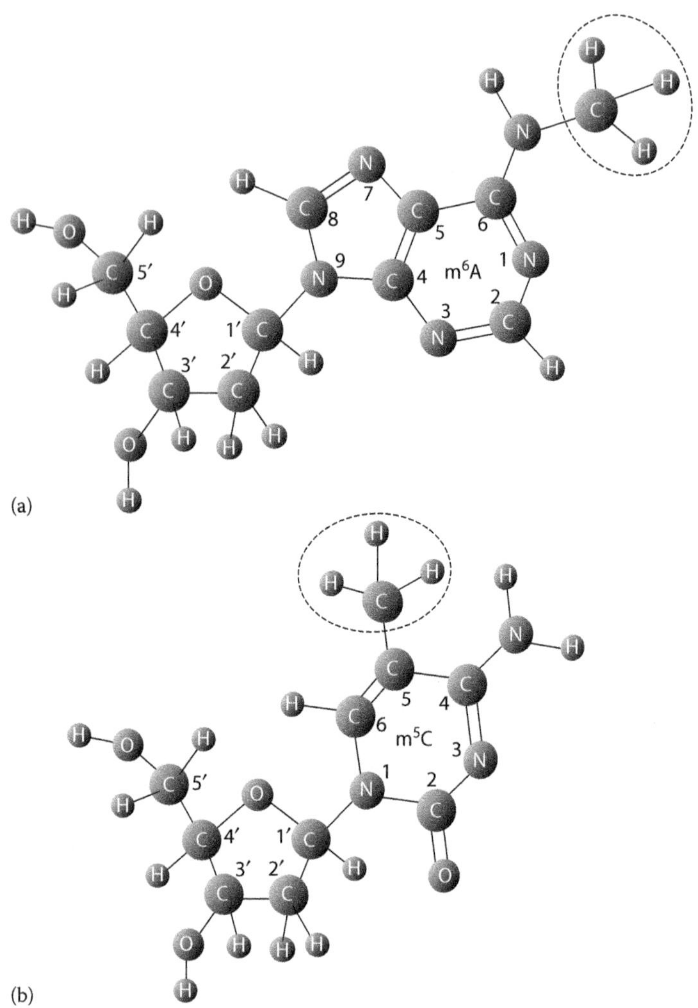

FIGURE 3.7 Two common methylated bases: (a) 6-methyladenine (m^6A) and (b) 5-methylcytosine (m^5C). Both help to protect the DNA from degradation. Additionally, m^5C is used to identify the template DNA strand from the newly synthesized strand during replication, so that the original template strand can be used to repair any mutations that occurred during replication. The m^5C bases are also used to control gene expression by causing the formation of Z-form DNA.

to repair the lesions. This increases the likelihood of eliminating any mutations that resulted from the replication process.

DNA has been retained as the primary carrier of genetic information due to its higher stability as compared to RNA, thus allowing genomes to become larger during evolution. However, the 2′ hydroxyl also leads to other interactions in RNA that do not normally take place in DNA, allowing the bases in RNA to interact in many more ways than the bases in DNA. Because of this, the variety of secondary and tertiary structures that can form in RNA is far greater. RNA can fold into structures that are chemically reactive, such that some RNAs are ribozymes. No DNAs have been found that perform analogous functions. In fact, RNAs can have many functions, including being ribozymes, structural elements, carriers of genetic information, signaling molecules, controllers of gene expression, and others. In addition, several other bases can be found in RNAs (especially tRNAs), including inosine, pseudouridine, and several wyosines (among many others) that are never found in DNA. RNAs perform a myriad of other functions in cells, including translation of proteins. In fact, they perform so many vital functions in cells that they rival proteins in the variety of their interactions.

In DNA, Watson–Crick interactions (A–T and G–C base pairs) are the primary base conformations. However, in RNA, hydrogen bonding can be found along the Hoogsteen edge and the sugar edge, as well as along the Watson–Crick edge of each base (Figure 3.8). In DNA, the Watson–Crick edge of one base forms hydrogen bonds with the Watson–Crick edge of the opposing base. However, in RNA each edge of each base has a possibility of hydrogen bonding with any other edge of another base, for example, a Hoogsteen edge may hydrogen bond with a Watson–Crick edge of another base. Therefore, there are 384 combinations (12 families of interactions between the different edges of the molecules, times 16 possible pairings of each of the 4 nucleotides, times 2 since the sugars can be in a *cis* or *trans* conformation in relation to one another). Of the 384 possibilities, more than 300 have been observed based on RNA crystallographic studies. Additionally, base triples are possible (Figure 3.9), and many of these have been found in RNA molecules as well. While some hydrogen bonds form between nucleotides that are in close proximity to one another, many of these are long-range interactions, such as those that aid in forming intricate structures in RNA (e.g., those that form in rRNAs and tRNAs). Often, specific guanosines and adenines are reactive or aid in making uracils reactive. For example, specific adenines in group II introns are highly reactive and initiate the first reaction in the process of RNA splicing. Also, GNRA (guanosine–nucleotide–purine–adenine) loops (often GAAA) are found in many RNAs that are enzymatic, including rRNAs and introns.

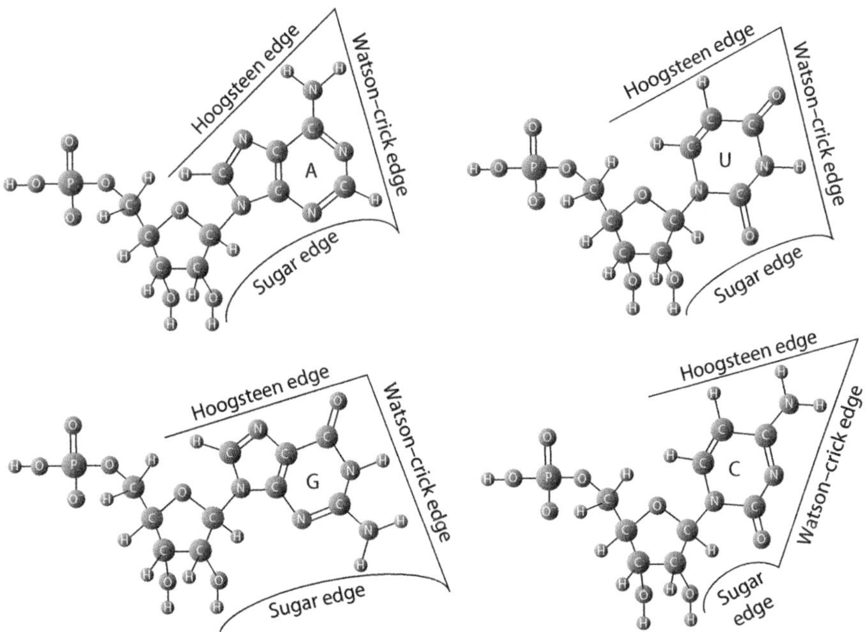

FIGURE 3.8 Edges on RNA nucleotides that can interact by hydrogen bonding. In RNA, hydrogen bonds can form on at least three edges of each molecule, known as the *Watson–Crick*, *Hoogsteen*, and *sugar edges*. In contrast, DNA usually forms only hydrogen bonds along the Watson–Crick edges. Many interactions have been detected in RNA molecules of different types. For example, hydrogen bonds can form between a Hoogsteen edge of one nucleotide and the Watson–Crick edge of another. Furthermore, two such interactions are possible, one that places the sugars in a *cis* conformation and the other that places the sugars in a *trans* conformation. In DNA, the sugars are always in the *cis* conformation, and the base interactions are always of the Watson–Crick/Watson–Crick type.

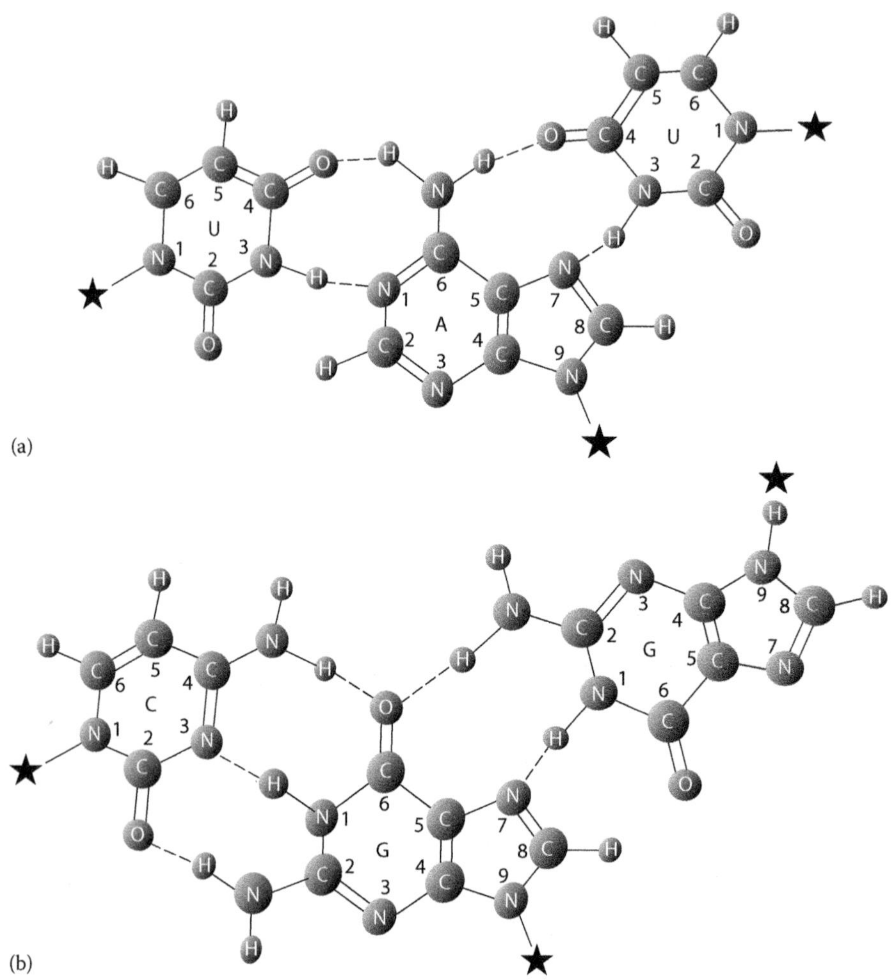

FIGURE 3.9 Two examples of base triples: (a) U–A–U and (b) C–G–G. In each case, there are hydrogen bonds along the Watson–Crick and Hoogsteen edges. Many types of base triples have been found in various RNA molecules. For the U–A–U triple, the U–A pair (U on the left) is a standard *cis* Watson–Crick/Watson–Crick interaction, while the A–U pair (U on the right) is a *cis* Hoogsteen/Watson–Crick interaction. For the C–G–G triple, the C–G pair is a standard *cis* Watson–Crick/Watson–Crick interaction, while the G–G pair is a *trans* Hoogsteen/Watson–Crick interaction.

Ribosomal RNAs are among the most complex RNAs. Ribosomes appear to have evolved in several steps, and it is likely that very early in evolution, fidelity and efficiency of translation were much lower than for modern ribosomes. The peptidyltransferase with an accompanying tRNA site appears to have been present in the earliest protoribosomes. Sometime later, the region duplicated, forming two sites where tRNAs could bind. This became the central core of the modern ribosome. It is unknown whether the triplet code or a more limited coding system existed from the beginning. However, the first and second positions of the modern triplet genetic code are much more conserved than is the third position, and therefore, it is likely that the initial coding system consisted of a single nucleotide encoding a single (or a few) amino acids, or a duplex code, where almost all of the amino acids could be represented by two nucleotides (which is close to what exists today). However, once the evolution of the triplet coding system became established, it was so successful evolutionarily that the organisms that possessed this system were the only ones that survived and diversified because it is the system that is used in all organisms and organelles today, without exception. Although there is a nearly universal code, codon usage can differ greatly from one organism to another and can differ from one cell type to another in multicellular organisms.

TRANSLATION

Translation is a complex process (Figure 3.10) that requires at least three different classes of RNA (rRNA, mRNA, and tRNA), as well as many proteins and factors (e.g., ribosomal proteins, initiation factors, and elongation factors). Translation can proceed only when all of the component RNAs and proteins are present. The ribosomes must associate with all of the ribosomal proteins and then with an mRNA to start the process. The tRNAs have a longer route. First, a nucleotidyltransferase enzyme must add CCA (cytosine–cytosine–adenine) to the 3′ end of each tRNA. Then, each tRNA is acted upon by specific aminoacyl tRNA synthetases each of which attaches a specific amino acid to each specific tRNA, based on their structure and anticodon. This is actually where the first half of the specificity of translation is set. Then, the tRNA with bound amino acid must find the correct codon on the mRNA already in a ribosome that is complementary to its anticodon. Next, its amino acid is removed from the tRNA and is attached to the growing polypeptide chain by the peptidyltransferase of the rRNA. The growing polypeptide chain is threaded through a tunnel in the large subunit and eventually it emerges once completed (i.e., when a stop codon is reached, and it dissociates from the ribosome). The mRNA and two ribosome subunits then dissociate and the process is repeated with another mRNA. This completes the pathway of information that flows from the DNA to the RNA, and then into a polypeptide. While selection for advantageous genes occurs at

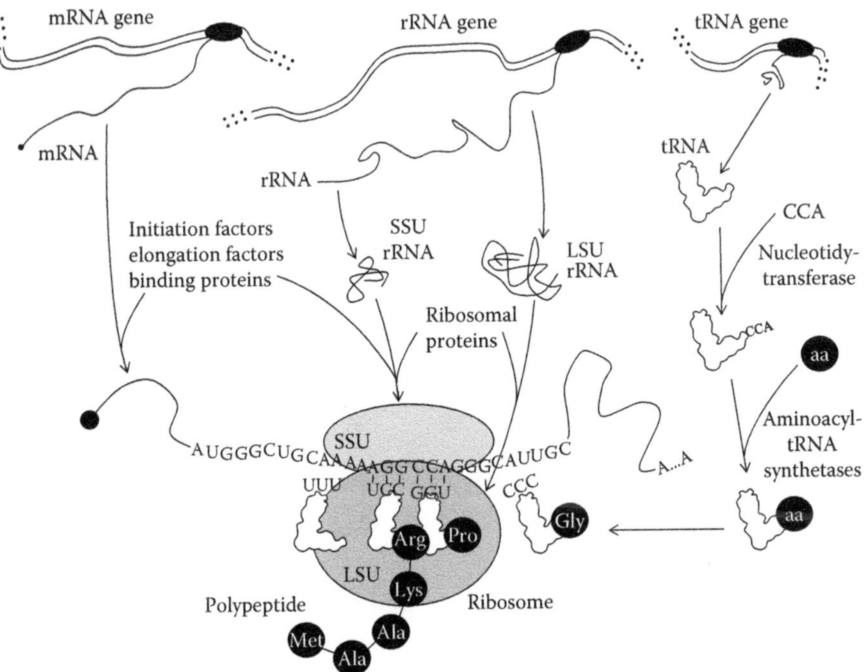

FIGURE 3.10 Overview of translation. Messenger RNAs (mRNAs) encode the messages that are decoded by the rRNAs and tRNAs. The rRNAs provide the sites into which the mRNAs and tRNAs are aligned, such that the amino acids on the tRNAs can be joined to form the encoded polypeptide. The selectivity of the genetic code originates with the aminoacyl tRNA synthetases that attach specific amino acids to specific tRNAs. Then, their anticodons will hydrogen bond to the mRNA codons, thus delivering the amino acid that corresponds to the triplet on that portion of the mRNA.

the protein level and beyond, selection also operates at each of the preceding steps. This is discussed in more detail in Chapters 5 and 6.

AMINO ACIDS AND POLYPEPTIDES

In all amino acids, the atoms around the α-carbon can be arranged in two ways (Figure 3.11). They will rotate polarized light to the left (levorotatory or L-amino acids) or to the right (dextrorotatory or D-amino acids). All amino acids that are incorporated into biological proteins by translation using ribosomes are L-amino acids. Only a few D-amino acids are present in cells, and these are incorporated into peptides in processes that do not use ribosomes. During translation on ribosomes, a peptide bond is formed as each amino acid is added to the growing polypeptide chain. The enzymatic peptidyltransferase activity exists on the large subunit of the rRNA. The polypeptides and proteins determine the characteristics of the cells to which they belong. In most cells, there are thousands to tens-of-thousands of different proteins. They provide structure, signaling, gene expression control, metabolic and catabolic activities, and other functions in the cells.

While there are many amino acids, the majority of biological organisms use only 20 of them to build their proteins (Figure 3.12) according to a genetic code that is the same for most organisms

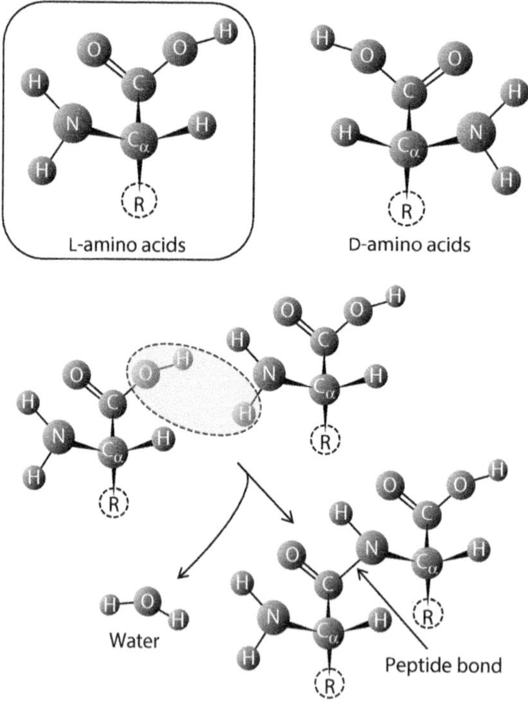

FIGURE 3.11 General structure of amino acids and peptide bonds. Amino acids all have the same central structure consisting of a central carbon atom (designated the α carbon) attached to a hydrogen, an amino group, a carboxyl group, and one other group (designated R) that is specific to each amino acid. Figure 3.12 has a diagram of each of the R groups. There are two arrangements possible around the α carbon. Arranged in one way, the molecules rotate polarized light to the left (levorotatory), while the mirror image arrangement yields molecules that rotate polarized light to the right (dextrorotatory). For amino acids, these are designated L-amino acids and D-amino acids. All proteins in all organisms that are synthesized by ribosomes contain only L-amino acids. The amino acids are joined together by the peptidyltransferase on the large subunit rRNA that catalyzes the formation of peptide bonds through a reaction between the carboxyl on one amino acid and the amino group on the incoming amino acid. A water molecule is generated in the reaction.

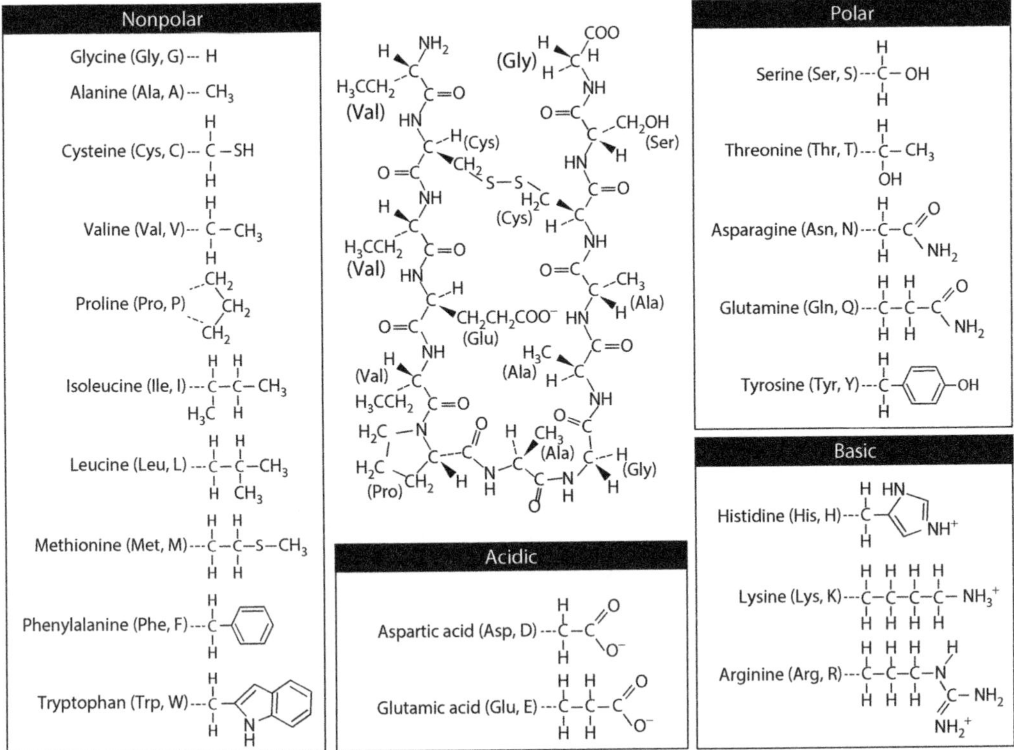

FIGURE 3.12 Structures of the side chains (R groups) of the 20 amino acids found in organisms. Three-letter code and single letter designation are indicated for each. They are categorized according to whether they are nonpolar, polar, basic (positively charged), or acidic (negatively charged). A short polypeptide is shown (middle) to indicate how the amino acids are arranged along the polypeptide chain. Nonpolar amino acids are usually found within the interior of globular proteins or within the transmembrane sections of proteins that are either embedded within a membrane or traverse the entire membrane. Polar groups are usually found within the aqueous regions of cells (e.g., the cytosol). Charged groups often are found within catalytic sites. Each of the amino acids has the ability to rotate around each of the N–C bonds (except proline), although each has a range of rotation that is dependent on the size of the side chain (or R group). There are three amino acids that have special functions within proteins. The R group in glycine is a hydrogen atom. Because it is so small, there is free rotation of the polypeptide chain around glycine. Conversely, when a proline is in the chain, a rigid bend is produced in the chain and no rotation is possible at that location. (*Note*: Based on its structure, proline is an imino acid and is the only imino acid found in proteins.) Cysteine can exist as a single amino acid, with an SH group, or disulfide bridges can form between two separated cysteines, which limits the flexibility of the protein.

and organelles. Proline actually is an imino acid, which forms rigid kinks in the polypeptide chain. Of the 20 amino acids, 8 are hydrophobic, 7 are hydrophilic, 3 are basic, and 2 are acidic. In general, the hydrophobic amino acids are found within the interior of the proteins, spanning membranes, or are involved in protein–protein interactions. The hydrophilic regions cover the surfaces of cytoplasmic proteins or are in areas that are exposed to aqueous environments. The acidic and basic amino acids often are found in the enzymatic centers of the proteins or in regions that shuttle electrons, since they can transfer electrons and protons to and from the substrates and the enzyme, as well as to and from other proteins and cofactors. Cysteines often form disulfide bridges between various parts of the protein to hold it into a specific conformation. A variant of cysteine, called *selenocysteine*, is the product of posttranslational modification of cysteine residues and is vital in some proteins. Often, it is referred to as the 21st biological amino acid.

LIPIDS

Lipid layers (primarily a phospholipid bilayer) surround all cells, cellular organelles, and many types of viruses (Figure 3.13). These layers are necessary for cellular life in order to concentrate the molecules needed for life processes and to exclude molecules that are not needed or are toxic to the vital molecules. The original cells may have had simpler membranes. Lipids can form spontaneously from simple compounds that would have been present on early Earth, including near-thermal vents in the oceans. When lipids are placed in water, the hydrophobic tails associate together pointing away from the water, while the hydrophilic heads associate with the water. They can form small beads, called *micelles*, or if they form in two layers, they can form a liposome, which will have an inside and an outside separated by the lipid bilayer. Lipids can also react with other compounds, including phosphorus-containing molecules, to form phospholipids, which have a phosphate group attached at one end to the lipid tail. These molecules are part hydrophobic lipid and part hydrophilic phosphate. Because of this, they form very stable bilayers in water, such that the hydrophobic parts are together and out of direct contact with the water, and the phosphate parts are in direct contact with the water. Although the first cells relied on the surrounding environment for their lipids, eventually, they would have had to produce their own lipids to ensure a supply in order to grow and divide. Modern cells produce enzymes that synthesize their membrane lipids, as well as other lipids needed by those cells. Lipids also serve as carbon reserves. Storage lipids are primarily long chains (sometimes branched chains) of carbons and hydrogens, often with a hydrophilic end to aid in solubility. Membrane lipids often have

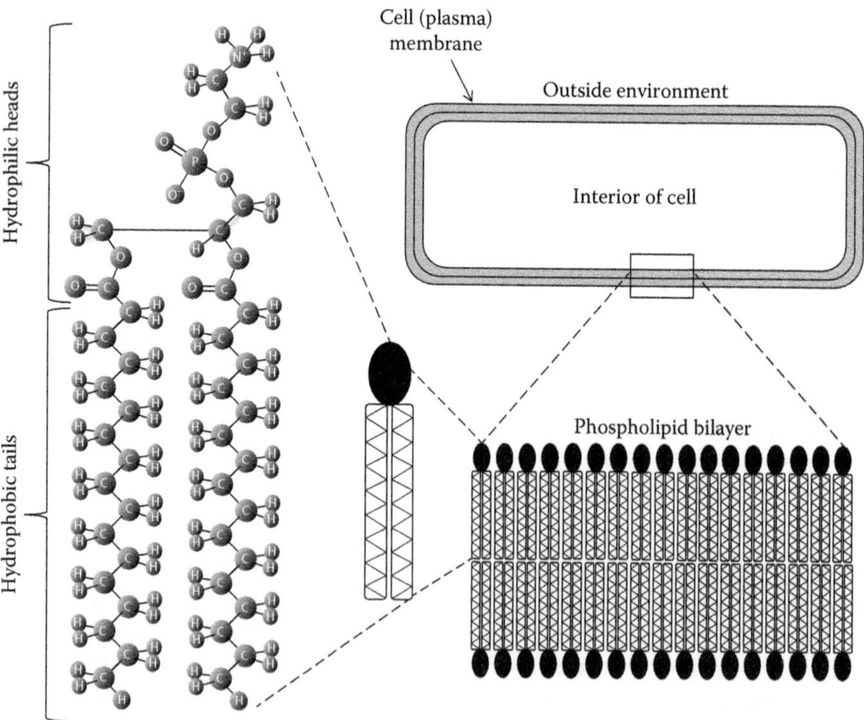

FIGURE 3.13 General structure of phospholipid bilayers. Phospholipid bilayers consist of molecules composed of two hydrophobic fatty acid chains, connected by a glycerol molecule to a hydrophobic polar head containing a phosphate group, and other polar groups as well. Linkage of the lipids to the glycerol is via glycerol–ester bonds, which are present in bacteria and eukaryotes. In Archaea, the linkages are via glycerol–ether bonds, and the lipid chains contain isoprene side chains.

similarities to the storage lipids, but they often have two long hydrophobic fatty acid (made of carbons and hydrogens) chains and a large hydrophilic head, attached through a glycerol group.

CARBOHYDRATES

Carbohydrates are another major component in cells. They consist mainly of carbon, oxygen, and hydrogen, although some are attached to other atoms and molecules as well (Figure 3.14). As with lipids, one of their major functions is to store carbon and to deliver it where needed in the cell. But, another major function is that they serve as structural elements for many biological molecules, including some of the important polymers (e.g., nucleic acids, cell walls, and starch). Carbohydrates may be in the form of monosaccharides, disaccharides, or polysaccharides. Mono- and disaccharides are almost immediately usable as carbon sources, while polysaccharides may be used to store the carbon for long periods of time. Alternatively, both types can serve as important structural components. Some polysaccharides are used as cell wall material in organisms as diverse as bacteria, archaea, plants, and fungi. Examples of common carbohydrates are glucose, sucrose, and lactose, as well as ribose, deoxyribose, and ribulose. In addition, high-energy bonds (e.g., connecting to phosphates) can be added to sugars to store or deliver chemical bond energy, such as in ATP. They can also be components of important messenger molecules, such as cyclic-AMP.

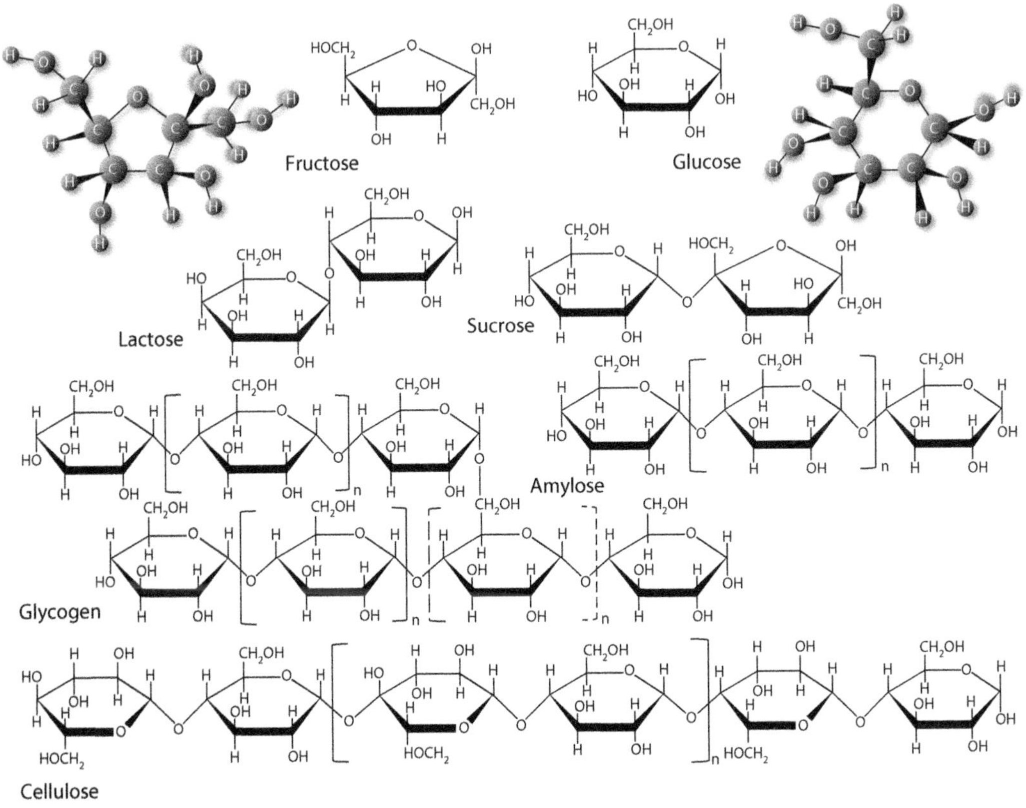

FIGURE 3.14 General structure of carbohydrates. Carbohydrates are molecules consisting primarily of carbon, hydrogen, and oxygen. Monosaccharides (e.g., fructose and glucose) have a single sugar molecule. Disaccharides (e.g., lactose and sucrose) have two saccharides joined with an ester bond. Polysaccharides (e.g., amylose, glycogen, and cellulose) have multiple saccharides joined together with ester bonds in long chains, some of which (e.g., glycogen) have branched chains.

KEY POINTS

1. All life on the Earth is characterized by having nucleic acids, proteins, lipids, and carbohydrates.
2. RNA was probably the first nucleic acid used by organisms on the Earth. It performs many functions in modern cells and can form many different conformations. The bases can interact by forming base pairs, base triples, and other long-range interactions, some of which function in ribozymes (catalytic RNAs, such as ribosomes).
3. DNA became established as the molecule that carries the hereditary information by about 3.5 billion years ago.
4. DNA forms a double helix where the two strands are in an antiparallel orientation, hydrogen bonded in the center by interactions between a purine and a pyrimidine (A pairs with T forming two hydrogen bonds, and G pairs with C forming three hydrogen bonds).
5. DNA can be found in at least three conformations: A-form, B-form, and Z-form. Under physiological conditions, most DNA is in the B-form. DNA is a right-handed helix with each turn of the helix including 10.5 bases and being 3.4 nm in length. The bases are more exposed on one side than the other, thus forming a major and a minor groove. In A-form DNA, which is formed under nonphysiological conditions, the major and minor groves are less pronounced, and there is a hole in the middle of the molecule. Z-form DNA is a left-handed helix. It is formed under physiological conditions when the DNA is highly methylated.
6. Amino acids are translated into polypeptides (proteins) by ribosomes using the triplet code contained on mRNAs and specified by tRNAs carrying specific amino acids. The arrangement of the genetic code suggests a possible evolutionary sequence of events that resulted in the present genetic code.
7. Phospholipid bilayers form membranes around all cells. Fatty acids appeared early on the Earth and are the major building blocks for phospholipids. Lipids can also be used as a form of stored carbon in organisms.
8. Carbohydrates formed early in the Earth's history. They are used in a variety of processes in organisms. There are forms that are immediately useable in metabolic processes and forms that store the saccharides for later use.

ADDITIONAL READINGS

Alberts, B., D. Bray, K. Hopkin, A. Johnson, J. Lewis, M. Raff, K. Roberts, and P. Walter. 2013. *Essential Cell Biology*, 4th ed. New York: Garland Publishing.

Blackburn, G. M. and M. J. Gait (eds.). 1996. *Nucleic Acids in Chemistry and Biology*. New York: Oxford University Press.

Bokov, K. and S. V. Steinberg. 2009. A hierarchical model for the evolution of 23S ribosomal RNA. *Nature* 457:977–980.

Darnell, J., H. Lodish, and D. Baltimore. 2007. *Molecular Cell Biology*, 6th ed. New York: W.H. Freeman and Company.

Gautheret, D., S. H. Damberger, and R. R. Gutell. 1995. Identification in base-triples in RNA using comparative sequence analysis. *J. Mol. Biol.* 248:27–43.

Karp, G. 2013. *Cell and Molecular Biology*, 7th ed. New York: John Wiley & Sons.

Krebs, J. E., E. S. Goldstein, and S. T. Kirkpatrick. 2012. *Lewin's Genes XI*, 11th ed. Sudbury, MA: Jones & Bartlett.

Stombaugh, J., C. L. Zirbel, E. Westof, and N. B. Leontis. 2009. Frequency and isostericity of RNA base pairs. *Nucleic Acids Res.* 37:2294–2312.

Zirbel, C. L., J. E. Sponer, J. Sponer, J. Stombaugh, and N. B. Leontis. 2009. Classification and energetics of the base-phosphate interactions in RNA. *Nucleic Acids Res.* 37:4898–4918.

4 The Central Dogma and Beyond

INTRODUCTION

All students of biology and many others have heard of the *central dogma* of molecular biology (see Figure 1.3). DNA is transcribed into RNA, which then is translated into proteins. This *central dogma* is memorized by countless people every year and becomes ingrained in their thoughts about all molecular biology processes. While it is a useful hook for people to remember, it is an oversimplification of what actually goes on in cells (which is what Francis Crick indicated not long after he coined the term in 1956) and leads to misconceptions. First of all, as most are aware, there are several different categories of RNA. There are messenger RNAs (mRNAs), ribosomal RNAs (rRNAs), transfer RNAs (tRNAs), and a diverse set of other small noncoding RNAs (ncRNAs) that function in the control of expression of the other RNAs. Where do these fit into this scheme? Figure 4.1 shows an overview of this more complex process.

Also, the cellular concentrations of these RNAs vary. How does this affect these processes? These questions are answered in this chapter.

RIBOSOMAL RNA

The most abundant RNAs in cells are the rRNAs (Figure 4.2). The rRNAs all are incorporated into ribosomes where they function in the process of translation. In Bacteria and Archaea, there are three types of RNA: (1) small subunit (SSU), (2) large subunit (LSU), and (3) 5S rRNAs. The SSU and LSU often are referred to as the 16S and 23S rRNAs, respectively, based on their sedimentation rates in gradient centrifugation procedures. In eukaryotes, there are four RNAs (SSU and LSU, as well as a 5S and a 5.8S rRNA). The LSU and SSU rRNAs often are termed the 18S and 25S rRNAs, respectively, again based on their sedimentation rates. The rRNAs comprise about 80% of the total cellular RNA (by mass), which converts to about 10 million molecules of each type per average eukaryotic cell (about 20,000 of each in a typical bacterial cell). This equates with the number of ribosomes in a cell. In eukaryotes, RNA polymerase I transcribes all of the rRNA, with the exception of the 5S rRNA (which is transcribed by RNA polymerase III). While these are the major RNAs in every cell, rRNAs are never translated. Their sequences encode no proteins. Ribosomes are large and complex ribozymes, having evolved over a period of approximately 3.8 billion years. They consist of an SSU (composed of the small rRNA subunit and a few dozen proteins) and an LSU (composed of the other three rRNAs and several dozen proteins). The ribosome specifically positions each mRNA onto its surface, actively moves the mRNA through the gap between the LSU and SSU, positions two tRNAs at a time for proper addition of each amino acid onto the growing polypeptide, covalently attaches each amino acid onto the polypeptide chain, and pushes the growing polypeptide through a tunnel in the LSU (see Figure 3.10). Experiments have shown that the rRNAs (and not the proteins) are responsible for all of these functions. The proteins act to hold the rRNAs into their functional tertiary structures. Therefore, these rRNAs are enzymes and thus are in a special class of molecules known as *ribozymes*. To reiterate, the rRNAs are transcribed, but are never translated, and they are the enzymes that are central to the process of translation.

Because of the huge number of rRNAs needed by each cell, transcription of rRNA has to be a very active and efficient process. A large number of RNA polymerase I complexes and attendant proteins are needed in this process. (*Note*: If rRNA were transcribed at the rate of an mRNA gene, it would take more than a year to synthesize enough rRNA for a single bacterial cell. In Chapter 5, the mechanisms that have evolved in various organisms to ensure a sufficient supply

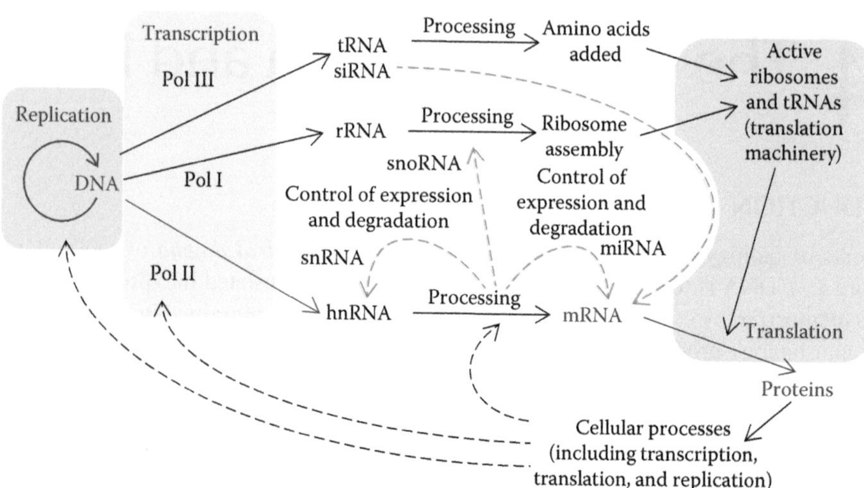

FIGURE 4.1 Expanded version of the *central dogma*. The simple version of the *central dogma* is indicated by blue arrows and lettering. Each of the synthesis processes (replication, transcription, and translation) is indicated by red lettering inside colored boxes. DNA is transcribed into rRNA, tRNA, and mRNA. In eukaryotes, there are three RNA polymerases (pol I, pol II, and pol III), while bacteria have one polymerase that transcribes all RNA types. Most of the RNAs are processed after transcription. The tRNAs, transcribed by RNA polymerase III, start as longer precursors and are folded and then cut into their final lengths (of approximately 70–90 nucleotides). Then, they have the trinucleotide, CCA, added to their 3′ ends prior to the addition of the amino acids to the final 3′ adenine. Small interfering RNAs (siRNAs), which control expression of some mRNAs, are also transcribed by RNA polymerase III. The rRNAs, transcribed by RNA polymerase I, begin as long precursor molecules and are sequentially cut and modified to form the final functional rRNA species, primarily by the action of snoRNAs (small nucleolar organizer RNAs that exist within mRNAs, primarily within introns), as well as some proteins. Eukaryotic mRNAs, transcribed by RNA polymerase II, begin as a mixture of RNAs of varying size, called *heteronuclear RNA* (hnRNA), and in general for all organisms, it is often referred to as *pre-mRNA*. They have a 7-methylguanosine cap added their 5′ ends during transcription. Those RNAs that contain introns must be spliced prior to export from the nucleus. Some of the introns and intergenic regions also contain ncRNAs, such as snRNAs (small nuclear RNAs that are incorporated into snRNPs, small ribonuclear particles, that are necessary for the splicing of spliceosomal introns in hnRNAs), snoRNAs (small nucleolar RNAs that aid in the processing of rRNA), and miRNAs (microRNAs that control the expression and degradation of mRNAs). Also, mRNAs have poly-A tails added to their 3′ ends prior to export from the nucleus. Only the mature processed mRNA is eventually translated on ribosomes into proteins. The rRNA is the catalytic portion of the ribosomes and in conjunction with tRNAs (with attached amino acids) synthesizes polypeptides based on the sequence of nucleotides in the mRNA.

of rRNA are discussed.) Parts of the genes are well conserved among all cellular organisms. This implies that rRNA in all organisms has originated from one or a small group of ancestors. These organisms have become the most successful forms of life on the Earth. Another reason that the genes are well conserved is that their functions are absolutely crucial for all organisms. The rRNAs are ribozymes that perform all of the functions of ribosomes, with the final products being nearly all of the proteins and peptides in all cells. Once ribosomes had evolved, it appears that they already were complex, and any major changes in function of one part or another probably were lethal.

In addition to the rRNA genes being conserved, the gene arrangement for each set of genes is also similar among Eukarya, Archaea, and Bacteria (Figure 4.2). Each consists of a promoter, an SSU gene, one (in bacteria and archaea) or two (in eukarya) internal transcribed spacer region, and an LSU gene. In Bacteria and Archaea, the internal transcribed spacer sometimes includes one or more tRNA genes. These may be late arrivals in these regions or may indicate some possible modes

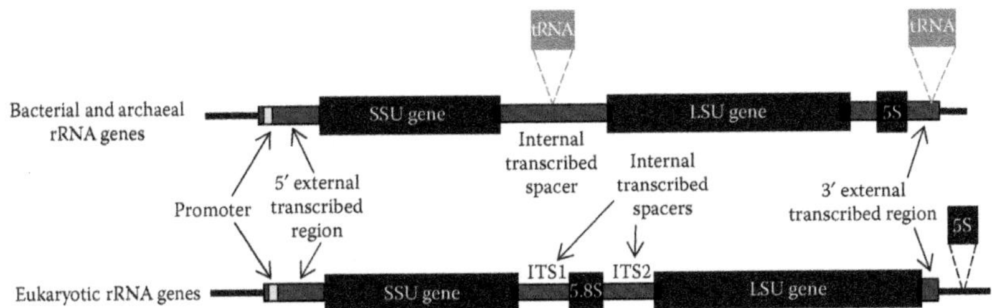

FIGURE 4.2 Arrangement of the rRNA genes in Bacteria, Archaea, and Eukarya. All three have promoters within external transcribed regions on the 5′ end. They also have 3′ external transcribed regions as well. All three have SSU genes. The SSU genes in bacteria and archaea usually are approximately 1500–1600 nucleotides in length, while the SSU genes in Eukarya are from 1700 to about 1900 nucleotides in length. All have LSU genes, which range in length from 2200 to 2500 nucleotides in Bacteria and Archaea, to 2900 nucleotides in Eukarya. Bacterial and archaeal internal transcribed spacers sometimes include tRNA genes (sometimes more than one), as do the 3′ ends of the transcript. In Bacteria and Archaea, the 3′ region also contains a 5S gene. In Eukarya, a 5S gene is sometimes found in this location, although it is transcribed by RNA polymerase III, while the other genes all are transcribed as a single unit by RNA polymerase I. The eukaryotic locus also contains a 5.8S gene surrounded by two internal transcribed spacers (ITS1 and ITS2). ITS1 corresponds to the internal transcribed spacer in Bacteria and Archaea, while ITS2 originated as an intervening sequence (possibly an intron) that interrupted the LSU gene. The 5.8S gene is homologous to the 5′ section of the LSU in Bacteria and Archaea.

of evolution of the rRNA locus. In Eukarya, the small and large rRNA subunit genes are longer than those in bacteria and archaea, but the organization is the same in both groups of organisms. The spacer is interrupted by a small gene, the 5.8S gene, which creates two internal transcribed spacers (ITS1 and ITS2). The 5.8S gene originated as a part of the LSU rRNA gene (i.e., in Bacteria and Archaea, these sequences appear in the 5′ region of the LSU gene) that has become separated from the remainder of the LSU rRNA gene. The 5.8S rRNA hybridizes to the 5′ end of the LSU rRNA in the ribosome. In Bacteria and Archaea, the analogous region curls over to hybridize with part of the LSU rRNA further along the molecule.

In Bacteria and Archaea, the 5S rRNA genes are included within the operon with the other rRNA genes and all are transcribed with the same RNA polymerase, and at the same time. In eukaryotes, the 5S genes are transcribed by RNA polymerase III and therefore can be separated from the other rRNA genes. The 5S genes occasionally are located adjacent to the main rRNA locus, but in most cases, they are in different regions of the genome. When they are located adjacent to the main rRNA locus, the 5S genes may be oriented in either direction with respect to the main rRNA genes because the polymerase and transcription systems are different. In most eukaryotic species, the 5S genes exist as tandem arrays on parts of the chromosomes that are distant from the main rRNA genes. In all organisms, rRNA is transcribed at high rates, which gives these regions a characteristic feathery appearance (Figure 4.3).

In Bacteria and Archaea, the assembly of the small and large ribosomal subunits is initiated while the rRNA still is being transcribed (Figure 4.4). Folding and attachment of the ribosomal proteins begins soon after transcription has begun. By the time the completed transcript is detached from the RNA polymerase, many of the ribosomal proteins have associated with the rRNA, and processing of the rRNA into the mature SSU and LSU rRNA molecules has begun. Finally, additional proteins attach to the SSU and LSU rRNAs, mRNA associates with the small ribosomal subunit, the large ribosomal subunit binds to the small ribosomal subunit (with associated mRNA), and translation begins. In eukaryotes, the process is more complex, since transcription and translation occur in separated cell compartments. All of the transcription, as well as most of the ribosome assembly, takes place in a region of the nucleus called *the nucleolus* (Figures 4.4 and 4.5), while translation occurs in the cytosol or on the endoplasmic reticulum.

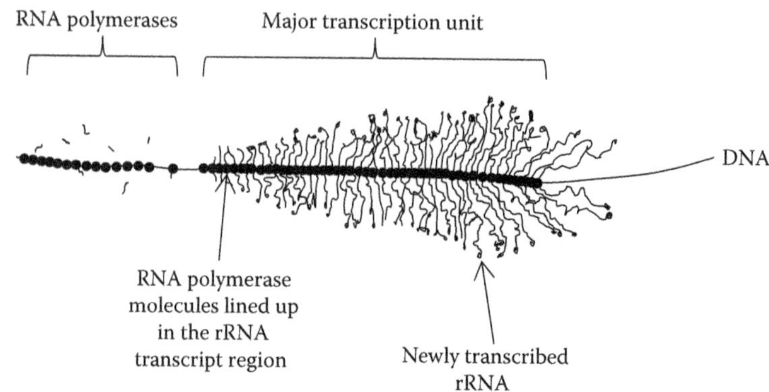

FIGURE 4.3 Diagram of transcription of a ribosomal gene. A single rRNA gene operon is depicted showing the feather-like appearance of the process of transcription in an electron microscope view. DNA is the horizontal line, with RNA transcripts (perpendicular to the DNA) emerging from RNA polymerase molecules (dots) along the DNA. In eukaryotes, RNA polymerase I molecules line up on the 5′ side of the promoter to increase the transcription rates of the operon. The polymerase I molecules are about as close together as physically possible. This ensures a steady and large supply of rRNA for the ribosomes.

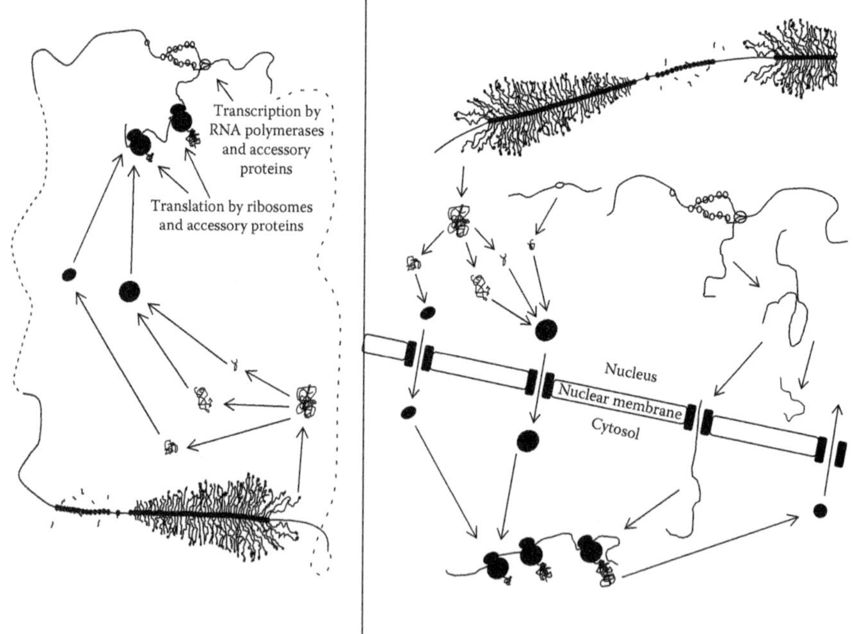

FIGURE 4.4 Cyclic view of transcription of rRNA and assembly of ribosomes in bacteria (left) and eukaryotes (right). In Bacteria and Archaea, transcription, ribosome assembly, and translation occur in the same cell compartment. However, in Eukarya, transcription and ribosome assembly occur in the nucleolus and translation occurs in the cytosol. (However, in one group of bacteria, the Planctomycetes and related bacteria, translation appears to be separated from transcription and ribosome assembly by membranes that resemble a nuclear membrane.) In each case, some ribosomal proteins begin to associate with the rRNA while transcription is underway. Following the completion of transcription, the rRNAs are processed by snoRNAs to produce the mature rRNA molecules. Ribosomal proteins associate with the rRNAs to assemble the small and large ribosomal subunits. This occurs in the cytoplasm in bacteria (except in Planctomycetes and related bacteria) and in the nucleolus of eukaryotes. The completed small and large ribosomal subunits are exported to the cytosol and associated with mRNAs. Some of the mRNAs encode ribosomal proteins, which are translated. The ribosomal proteins are then imported into the nucleus and moved to the nucleolus for assembly into ribosomes.

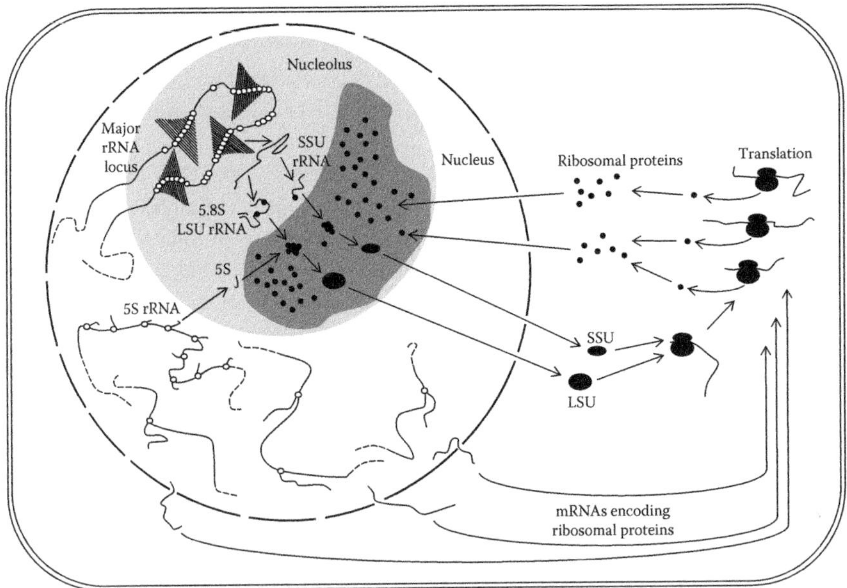

FIGURE 4.5 Broad view of ribosomal assembly in Eukarya. The major rRNA gene repeats are transcribed within the nucleolus in the nucleus. While the nucleus has a double membrane studded with nuclear pores, the nucleolus has no membrane, but does have distinct areas, including a fibrillar region (light gray) and a granular region (dark gray). As the rRNA is being transcribed, ribosomal proteins are being attached. The LSU, 5S and 5.8S, rRNAs are incorporated into the large ribosomal subunit, while the SSU is incorporated into the small ribosomal subunit. The 5S rRNAs are transcribed in the nucleus and moved to the nucleolus for incorporation into large ribosomal subunits. After the LSU and SSU are complete, they are transported out of the nucleus through nuclear pores. Then, they associate with an mRNA, join together, and begin translation of proteins. Ribosomal proteins are encoded on mRNAs that are transcribed in the nucleus and transported out into the cytosol. They are then translated on ribosomes and the resultant ribosomal proteins are transported into the nucleus and moved to the nucleolus where they attach to the rRNAs and become part of the ribosomal subunits.

The nucleolus has no membrane, but is organized into two distinct regions. The fibrillar region contains primarily rRNAs, including full-length transcripts, and processed molecules. The granular portion contains primarily ribosomal proteins. Transcription of the major rRNA gene loci, as well as separate assembly of the small and large ribosomal subunits, occurs in the nucleolus. The 5S rRNA is transcribed in the nucleoplasm and then is moved into the nucleolus for ribosome assembly. Next, the separate subunits are transported through the nuclear pores out of the nucleus into the cytosol. The ribosomes then start to associate with mRNAs and begin translation. If the emerging protein is to end up in a particular compartment of the cell (e.g., membrane and plastid), often it will have a signal sequence on its amino end. If so, this end will associate with specific proteins that will cause attachment to particular membranes in the cell. The ribosome often will be seen attached to membranes. In the case of the rough endoplasmic reticulum, the rough parts in the electron micrographs are ribosomes attached to the endoplasmic reticulum membrane. The proteins are then immediately threaded through pores comprised of other proteins in the membrane as they are translated by the ribosomes.

While the rRNA genes themselves are well conserved, the number of gene copies, the arrangements of multiple copies along chromosomes, and the ways in which the multiple copies are arranged on the chromosomes differ considerably from bacteria to eukaryotes. Among Bacteria and Archaea, there are from 1 to 15 copies of each rRNA gene set per genome. When more than one copy is present, they are separated on the chromosome, being surrounded by completely different mRNA and tRNA genes. For example, *Escherichia coli* (K-12) has seven rRNA copies, and each appears along a different portion of the *E. coli* chromosome; and, they vary somewhat. Some have one or more tRNA genes within the internal transcribed spacer, while other copies contain no tRNA genes. While the rRNA genes are separated, their expression is constitutive and at very high rates. This is needed because 20,000 rRNAs must be made in each cell every 20 min when the cells are rapidly growing and dividing.

The vast majority of eukaryotes possess multiple rRNA copies, and they are arranged in one or more long tandem arrays (Figure 4.6). Each unit in the array consists of an SSU gene, an LSU gene, a 5.8S gene, a promoter, 5′ and 3′ externally transcribed regions, two internal transcribed spacers (ITS1 and ITS2), and an intergenic spacer (IGS) that sometimes is called a *nontranscribed spacer* (NTS) (but small transcripts are produced in this region, and therefore, the term *nontranscribed* is inaccurate). In higher eukaryotes, the number of genes in these tandem arrays generally is 50 or more (e.g., in yeast, there are about 140 copies, and in human, there are 200–300 copies), up to tens of thousands per haploid set of chromosomes in some higher plants. Some species have more than one set of tandem arrays. For example, at least two separate clusters of these genes (that vary somewhat in their IGS regions) are found in pea (*Pisum sativum*). In hexaploid wheat (*Triticum aestivum*), there are three such arrays, each having originated from the three parent genomes that were joined to generate modern agricultural wheat. When genomes are mixed in this way, often one set of the rRNA genes is transcribed, while the others are not. This process, known as *nucleolar dominance*, is due to protein factors that attach polymerase I to the rRNA locus. These factors must bind to the promoter prior to binding of RNA polymerase I.

Eukaryotes must fill each cell with rRNA in order to make the millions of ribosomes that are needed. Within the IGS region, there are small tandem repeats (50–325 bp each) that include copies

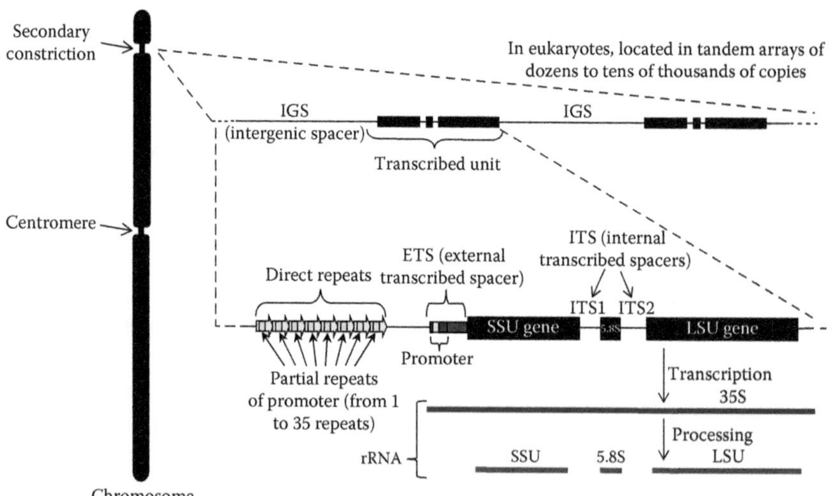

FIGURE 4.6 Organization of rRNA genes in eukaryotes (based on *Vicia faba* rRNA gene locus). Usually, all of the rRNA genes are at one chromosomal locus within a region called *a secondary constriction*. When viewed in an electron microscope, these regions actually are not constricted or condensed, but the DNA is looped out and dispersed, indicating that even when the chromosome is condensed, rRNA genes are not condensed, since they need to continue transcription even through mitosis and cell division. Within the rRNA locus, there exist from a few dozen to tens of thousands of tandem repeats. In some species, there is more than one locus. Each tandem repeat unit consists of an SSU gene, a 5.8S gene, an LSU gene, internal transcribed spacers (ITS1 and ITS2), external transcribed regions (on the 5′ end and the 3′ end), and an intergenic spacer (IGS). The IGS contains short repetitive elements, parts of which match sequences found in the promoter. These act to concentrate RNA polymerase I molecules in this region so that they can be rapidly shuttled to the promoter. The genes are synthesized as one long transcript, which is processed into the three mature rRNAs that are immediately incorporated into the large and small ribosome subunits.

of part of the promoter sequence (Figures 4.4 and 4.5). Transcription factors attach specifically to these, and in turn, these factors attach RNA polymerase I complexes. In essence, this concentrates RNA polymerase I, such that many such complexes are ready to move to the promoter and start transcription. In fact, in order to make enough rRNA, hundreds of RNA polymerase I complexes are concurrently transcribing rRNA from the same rRNA gene. Before transcription of each of the individual rRNAs is completed, processing of the rRNA already has begun. Initially, the rRNA is a large (approx. 35S to 45S) linear molecule (Figure 4.7). Within a few minutes, it has been processed by snoRNAs (small nucleolar organizer RNAs) and other factors, cutting it in a number of specific reactions to yield the SSU rRNA that is incorporated into the small ribosomal subunit, and the LSU and 5.8S rRNAs, which are incorporated into the large ribosomal subunit. The 5S rRNA is also incorporated into the large ribosomal subunit, but in eukaryotes, it is usually transcribed from a 5S rRNA gene located elsewhere in the genome. Processing is similar in Bacteria and Archaea, except that ITS2 and the 5.8S rRNA gene are absent, although the region of the LSU rRNA that is analogous to the 5.8S rRNA remains present within the 5′ end of the LSU rRNA.

Group I introns (described in more detail in Chapter 13) have been found in the rRNA loci from a broad range of eukaryotes. In fact, among fungi, often many group I introns are found in several places in the rRNA genes. In sharks, group I introns are located within ITS1 and ITS2, and in some cases, there are introns within introns. The introns must be removed during processing prior to assembly of the ribosomes and export to the cytosol. Thus, when there are introns within introns,

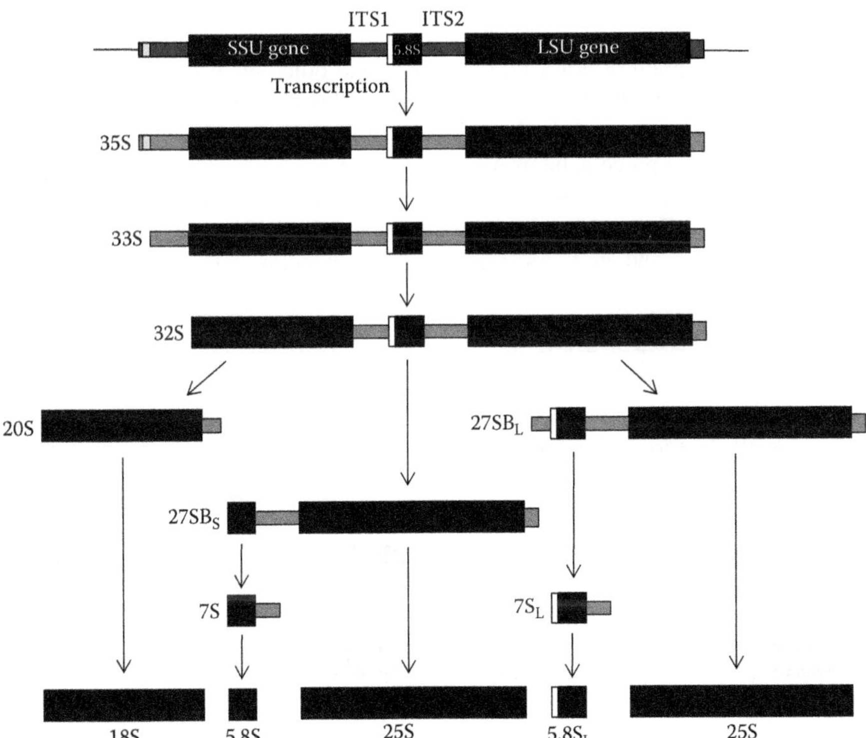

FIGURE 4.7 Processing of rRNA in eukaryotes. The initial transcript is between 35S and 45S in size and includes all of the genes as well as internal and external transcribed regions. Initially, a short piece of the 5′ region is cut, producing a 33S rRNA, followed by another cut that produces a 32S rRNA, where the 5′ end is the final end of the SSU rRNA. Next, a cut occurs within ITS1 to form a 20S and two slightly different 27S rRNAs (27SB$_L$—long, and 27SB$_S$—short). The 20S rRNA is further processed to produce the mature SSU rRNA (18S), and the 27SB rRNAs are each processed into mature 25S rRNAs and two versions of 7S rRNAs, which are further processed into two versions of 5.8S rRNA.

first the internal intron must be removed, followed by removal of the second. Group I introns are self-splicing ribozymes that rely on proteins only to help maintain the active tertiary structure of the ribozyme. Because of the widespread distribution of introns among eukaryotes, it is thought that the insertion of the introns into the rRNA locus occurred very early in the evolution of eukaryotes.

Ribosomal RNAs (as well as group I introns) are ribozymes, and as such, the tertiary structures of various sections of the molecules is absolutely vital to their activities. It has been hypothesized that ITS2 originated as an intron inserted near the 5′ end of the LSU rRNA gene, thus splitting off the original 5′ end of the LSU rRNA gene (which still is part of the LSU gene in Bacteria and Archaea). The interactions include standard Watson–Crick base pairs, as well as many more nonstandard base pairs, including the very important G•U wobble base pair. Additionally, there are many interactions between distant parts of the RNA, triple bases, A-platforms, and others. These enzymes appear to be very ancient, and there exists a wide variety of interactions between the bases that lead to structural and enzymatic functions for those regions. Analogous to the importance of RNA folding in the introns, accurate folding of the rRNA also is crucial to all of the functions of the ribosome.

TRANSFER RNA (tRNA)

The next most abundant of the RNAs are the tRNAs. They comprise about 12%–15% (by mass) of the total RNA of each cell. In an average eukaryotic cell, there are approximately 65 million tRNA molecules (about 160,000 in a bacterial cell), although there usually are only 20–30 different versions in each cell. In eukaryotes, they are transcribed by RNA polymerase III, which is also used to transcribe other small RNAs (Figure 4.8). Although these genes are transcribed in a manner that is similar to transcription of mRNAs, the tRNAs and other small RNAs are never translated into proteins. As with rRNAs, the tRNAs are involved exclusively in translation, although they are never translated. Therefore, approximately 90%–95% of the RNA (rRNA plus tRNA) in all cells is dedicated to the translation process, but is never translated.

The characteristics of tRNAs are unusual compared to rRNA and mRNA genes, in that the promoters (A box and B box, Figure 4.8) exist within the genes (5S rRNA genes also have an embedded promoter, called *the C box*, and are transcribed by RNA polymerase III). The transcription units are

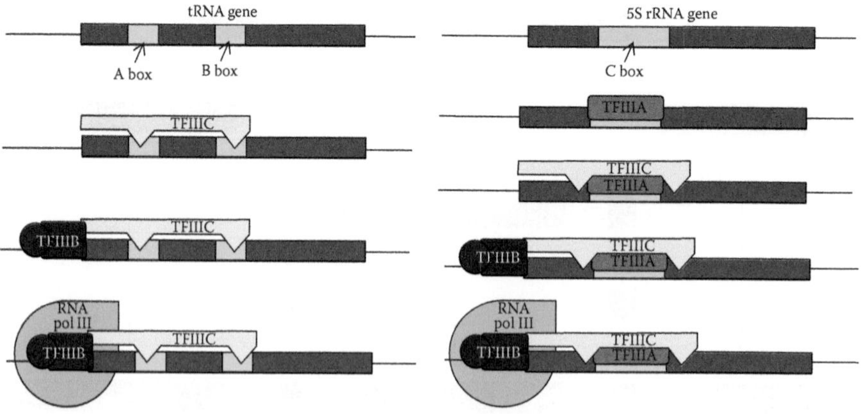

FIGURE 4.8 Details of transcription of the tRNA and 5S rRNA genes by RNA polymerase III. The tRNA genes (left) all have promoter sequences within the gene itself (A box and B box) that are recognized by TFIIIC (transcription factor IIIC, which is composed of at least six different proteins). Once TFIIIC binds to the gene, it allows binding of TFIIIB (which is composed of at least three different proteins). Finally, RNA polymerase III can bind to the site to start transcription. RNA polymerase itself consists of at least 15 different proteins. Therefore, at least two dozen proteins are required to recognize and then transcribe a single tRNA gene. For 5S rRNA genes, an additional protein complex, TFIIIA attaches to the sequence known as *the C box*. This allows the sequential binding of TFIIIC, TFIIIB, and RNA pol III, which initiates transcription of the gene.

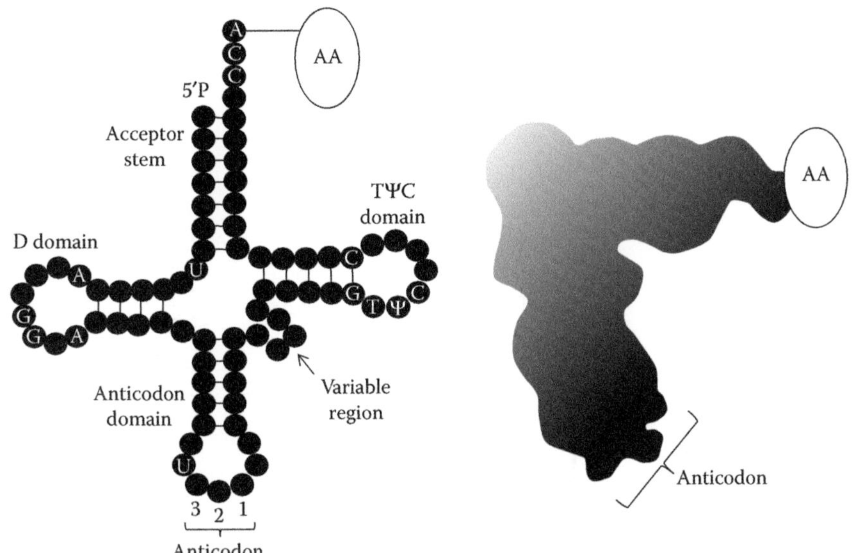

FIGURE 4.9 General structure of tRNA. Nucleotides and paired regions are shown on left. Standard nucleotides are indicated as A, G, C, and U. Nonstandard nucleotides thymine (T) and pseudouridine (Ψ) are also indicated. The D, anticodon, TΨC, and acceptor domains are found in all tRNAs. A short variable loop is located between the anticodon and TΨC domains. The three nucleotides in the anticodon loop are numbered to indicate the nucleotides that pair with the first, second, and third nucleotides of the codon on the mRNAs. The CCA is added posttranscriptionally to the 3′ end. An amino acid (specific for the tRNA and its anticodon) is added to the A of the CCA motif. The diagram on the right is the outline of the three-dimensional structure of the tRNA, with attached amino acid.

generally somewhat longer than the mature tRNA, the final pieces being trimmed by specific RNases. Some tRNAs have unique introns (some have group I introns, while others have archaeal introns, described in Chapter 13), which must be spliced out prior to folding into the final active tertiary structure. They can exist as operons in bacteria and eukaryotes, and thus, each transcription process can lead to long transcripts that contain more than one tRNA. These are then processed to final length, and each folds into the characteristic conformation. They are clipped to final lengths of 75–95 nucleotides. Then, each has a CCA sequence added to the 3′ end (Figure 4.9). This is the site for addition of the amino acid and is absolutely necessary for this process. After addition of the CCA, each tRNA can be recognized by specific enzymes (called *aminoacyl tRNA synthetases*). There are two classes of aminoacyl tRNA transferases that recognize the tRNAs either on the major groove side or on the minor groove side, and these seem to have evolved independently (the details of this are discussed in Chapter 6). Each of these enzymes has a site for a specific amino acid. Once the specific tRNA and the specific amino acid are in place, the enzyme covalently attaches the amino acid to the tRNA by binding to the A of the 3′ CCA sequence. In this way, the amino-acyl tRNA synthetases set the genetic code for the organism. They determine how each of the triplet codons is *decoded* during translation.

MESSENGER RNA

The mRNAs are among the least abundant of the RNAs, although most research involves these RNAs because they specify the proteins that exist in each cell, and therefore, they encode the characteristics for each cell. While they comprise only about 5%–10% (by mass) of the total RNA in a cell, they are the most diverse class of RNAs. In bacteria, only one type of RNA polymerase is used to synthesize all RNA species in the cells, and often, several genes are transcribed as an operon that encodes several proteins (Figure 4.10). However, in eukaryotes, the mRNAs all are transcribed

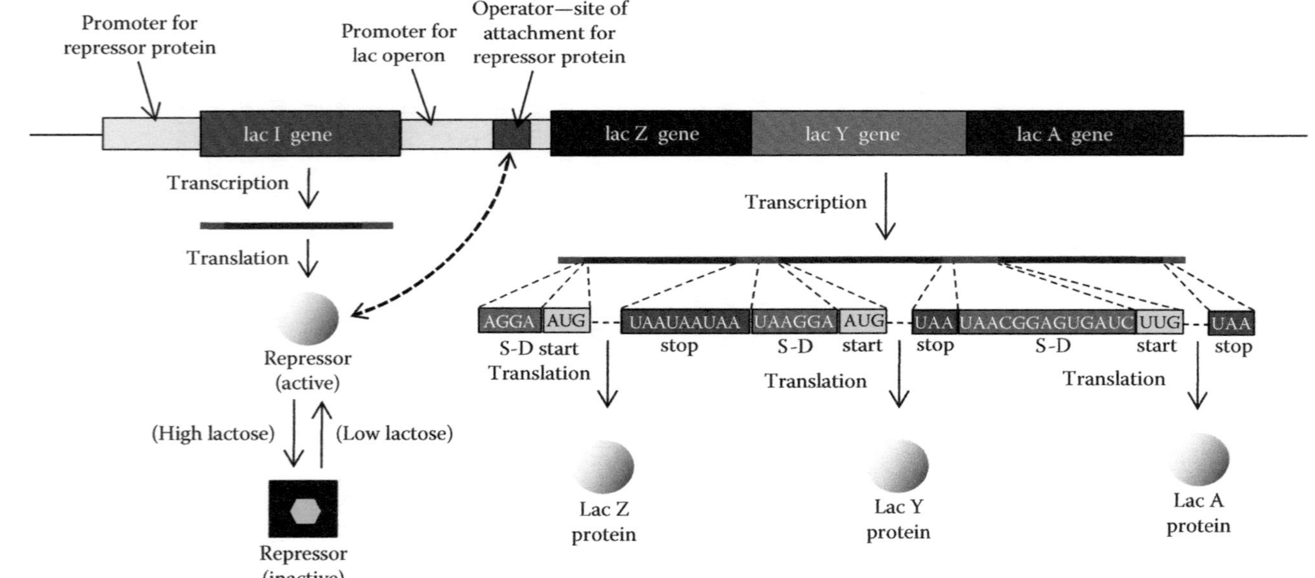

FIGURE 4.10 Organization of the lactose operon from *E. coli*. This region has one gene that is encoded on a single transcript (the *lac I* gene), as well as a transcript that encodes three different proteins (lac Z, lac Y, and lac A). The lac I repressor is able to attach specifically to the operator region of the lac operon promoter, which blocks transcription. When lactose is present, it attaches to the repressor, which changes its conformation such that it will not attach to the operator. The promoter can then initiate transcription. The resulting mRNA is loaded into a ribosome, where the first Shine–Dalgarno sequence attaches and aligns the mRNA in the ribosome to set up translation. Translation begins at the first AUG codon. This continues through the first stop sequence, which causes release of the completed polypeptide for lac Z. The next Shine–Dalgarno sequence signals a new message, and translation is initiated at the AUG codon. This continues, as above, to produce the lac Y protein, and the cycle is repeated to produce the lac A protein. Finally, the mRNA is released by the ribosome.

by RNA polymerase II. Following transcription, they must be processed, which includes the addition of a 5′ m^7G cap (see Figure 3.5) and a poly A tail and removal of introns by splicing (including alternative splicing in many cases). Only then they are transported out of the nucleus and into the cytosol, where they are translated on ribosomes to produce proteins. Therefore, these RNAs most closely conform to the *central dogma*, even though they comprise only a small fraction of the total RNA in a cell. However, while they are among the least abundant RNAs in the cell (some exist as only one or a few copies per cell), in bacteria there are from many hundred to several thousands of different kinds that encode the hundreds to thousands of proteins in each cell. In eukaryotes, there are thousands to tens of thousands of different mRNAs, and there are approximately 350,000 of them in an average cell (2,500 in a bacterial cell). Therefore, although their numbers are relatively small, their diversity is great, and they determine the overall characteristics of each cell because of the protein products that they encode.

For most mRNA genes, there are sequence elements upstream from the transcription start site (e.g., TATA elements). These are recognized by factors (e.g., sigma factors in Bacteria) that aid in attachment of RNA polymerase. The transcription start sites are also upstream from the translation start sites, and therefore, there are 5′ untranslated regions (UTRs) in the transcribed mRNA (as well as 3′UTRs). Other elements (e.g., the UP element) are sometimes found further upstream of the transcription start site. In Bacteria and Archaea, transcription and translation are not separated, and therefore, translation may begin before the completion of transcription. In Bacteria, the start of translation begins at the first AUG codon past the Shine–Dalgarno sequence (a site recognized by part of the 3′ end of the small rRNA subunit). This sequence hybridizes to an anti-Shine–Dalgarno sequence on the SSU rRNA on the ribosome. Each series of three bases is associated with the anticodons on the tRNAs, and the ribosomes add the amino acids to the growing polypeptide chain as long as there are interpretable codons. Once a stop codon is encountered, this signals the ribosome to stop the addition of amino acids and to release the completed polypeptide. If the mRNA contains only one gene, then it is released. If it is a polycystronic mRNA, an additional Shine–Dalgarno sequence is encountered, and translation of the next protein is initiated at the AUG codon. This process proceeds as before, until translation of all of the proteins encoded by the mRNA is completed.

The process is more complex for eukaryotes, primarily because there are many more proteins involved in initiating and maintaining transcription of each mRNA (Figure 4.11). In addition to RNA polymerase II, which consists of about 15 separate protein subunits, a TATA-binding protein is involved, which binds to the upstream region of the gene and then helps to position the polymerase to the promoter region. Additionally, many genes use enhancers (regions that may be several kilobases away from the gene) and activator proteins, which act together to increase transcription of the gene. Sometimes there are silencer regions and repressors that cause decreases in transcription. Because of these complexities, gene regulation has many more controls in eukaryotes, while offering a much broader degree of control, often in different cell types. Also, transcription occurs only in the nucleus and translation occurs only in the cytosol or on the endoplasmic reticulum. Therefore, in eukaryotes the two processes are separated.

Introns are common in the genes of higher eukaryotes. The introns within mRNA genes primarily are *Spliceosomal introns* (described in more detail in Chapter 13). This type of intron relies on small nuclear ribonuclearproteins (snRNPs) to accomplish splicing of the exons and removal of the introns. The presence of introns provides yet another control of expression of genes, in that the introns must be removed before the mature mRNAs are transported out of the nucleus and onto the cytosolic ribosomes. Spliceosomal introns are similar to group II introns that are found in organellar (primarily mitochondrial) genomes. It is possible that spliceosomal introns evolved from group II introns that likely originated in the proteobacterial ancestor of mitochondria. Group II introns are ribozymes that do not require spliceosomes. Otherwise, the splicing mechanism is very similar to spliceosomal intron splicing. (*Note*: Introns are discussed in detail in Chapter 13.)

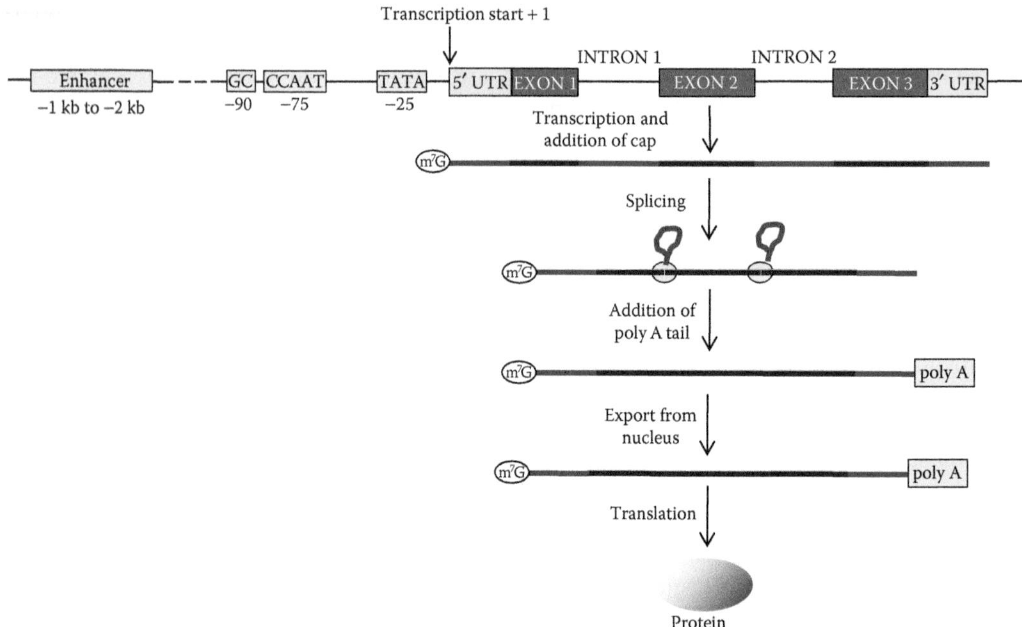

FIGURE 4.11 Organization of a typical eukaryotic mRNA gene. In general, the basic aspects of transcription are similar to those in Bacteria and more similar to those in Archaea. However, control of transcription is more complex in Eukarya. In addition to a TATA sequence at about −25, there also is a CAT box at about −75, and a GC-rich box at about −90. Each of these interacts with transcription factors that bind to these regions (including the TATA-binding protein). Enhancers also may be present. These are regions that are a kilobase or two upstream (and some have been found downstream of the genes), which interact with activator proteins that increase transcription rates. There are at least two to three dozen proteins (including subunits of RNA polymerase II) involved in this process. As transcription proceeds, an m^7G cap is added to the 5′ end of the mRNA transcript. This is recognized by a cap-binding protein that is a part of the translation initiation process. Not all eukaryotic genes contain introns, but many do. The number and lengths of the introns vary greatly. For those that do not have introns, one part of the process is eliminated, but all others occur in eukaryotic cells. A poly A tail is also added to the end of most mRNAs. This consists of sometimes hundreds of A nucleotides, the addition of which is signaled by a sequence near the end of the mRNA transcript. Once all of these steps are accomplished, the mature mRNA is transported out of the nucleus. They then associate with ribosomes, with the aid of cap-binding protein and initiation factors. Control of expression occurs at each point along the way, such that all processes must operate properly in order to result in a functional protein. Thus, selection may be active at every step of the process.

OTHER SMALL NONCODING RNA

Small RNAs also exist in eukaryotic cells. These act as control switches in some of these processes. There are several general types: (1) small interfering RNAs (siRNA), which attach to mRNAs to prevent their expression or mark the mRNAs for degradation (transcribed by RNA polymerase III, although an RNA-dependent RNA polymerase is indicated in some cases); (2) microRNAs, which attach to mRNAs to prevent their expression or mark them for degradation (transcribed by RNA polymerase II); (3) small nucleolar RNAs (snoRNA), which orchestrate the cleavage and modification of rRNA precursors into the final rRNA forms (transcribed by RNA polymerase II); and (4) small nuclear RNAs (snRNA), which are involved in splicing of spliceosomal introns in mRNA (transcribed by RNA polymerase II). Many originate as noncoding regions of mRNA and thus are synthesized by RNA polymerase II as part of an mRNA molecule. However, as noted above, some may be synthesized by RNA polymerase III or by an RNA-dependent RNA polymerase.

BEYOND THE CENTRAL DOGMA

Now, we can construct a model for the central dogma that is more accurate, informative, and complex (Figure 4.1). It includes explicit information about each type of RNA and how they interact to control and accomplish translation to synthesize proteins. It makes clear that most RNAs are not translated into proteins, but that most of the RNAs participate in translation in one way or another. In Bacteria and Archaea, the system is less elaborate, but in Eukarya all portions of the central dogma have been altered in one way or another. To accomplish translation, mRNAs, rRNAs, and tRNAs act in concert to find the start of the encoded mRNA message, decode the gene by creating a polypeptide, find the end, and release the completed protein. Again, the mRNA contains the message for the amino acid sequence that will be formed. The rRNA provides the mechanics for positioning the mRNA and the tRNAs, as well providing the enzymatic function to move the mRNA message and add amino acids to the growing polypeptide chain. The tRNAs provide the amino acids according to the genetic code.

KEY POINTS

1. While the central dogma is a useful tool for understanding the flow of information in a cell, it is a simplified version of what actually occurs in cells.
2. The majority (95%, by mass) of the RNA in a cell provides the machinery for the process of translation.
3. The majority of RNA in a cell is never translated.
4. Only the mRNA is translated into protein. Although it is only 5% of the total RNA in a cell, it is the most complex portion in terms of sequence diversity. From a few thousand to tens of thousands of different versions exist in each cell.

ADDITIONAL READINGS

Alberts, B., D. Bray, K. Hopkin, A. Johnson, J. Lewis, M. Raff, K. Roberts, and P. Walter. 2013. *Essential Cell Biology*, 4th ed. New York: Garland Publishing.

Alberts, B., D. Bray, A. Johnson, J. Lewis, M. Raff, K. Roberts, and J. D. Watson. 1994. *Molecular Biology of the Cell*. New York: Garland Publishing.

Darnell, J., H. Lodish, and D. Baltimore. 2007. *Molecular Cell Biology*, 6th ed. New York: W.H. Freeman and Company.

Gargas, A., P. T. DePriest, and J. W. Taylor. 1995. Positions of multiple insertions in SSU rDNA of lichen-forming fungi. *Mol. Biol. Evol.* 12:208–218.

Karp, G. 2013. *Cell and Molecular Biology*, 7th ed. New York: John Wiley & Sons.

Rogers, S. O. and A. J. Bendich. 1987a. Heritability and variability in ribosomal RNA genes of *Vicia faba*. *Genetics* 117:285–295.

Rogers, S. O. and A. J. Bendich. 1987b. Ribosomal RNA genes in plants: Variability in copy number and in the intergenic spacer. *Plant Mol. Biol.* 9:509–520.

Smith, J. M. and E. Szathmary. 1995. *The Major Transitions in Evolution*. New York: Oxford University Press.

Vanrobays, E., P.-E. Gleizes, C. Bousquet-Anonelli, J. Noaillac-Depeyre, M. Caizergues-Ferrer, and J.-P. Gelugne. 2001. Processing of 20S pre-rRNA to 18S ribosomal RNA in east requires Rrp10p, an essential non-ribosomal cytoplasmic protein. *EMBO J.* 20:4204–4213.

5 Ribosomes and Ribosomal RNA

INTRODUCTION

Ribosomes were first described in the middle of the twentieth century using electron microscopy and biochemical methods. Not long thereafter, it was discovered that they sometimes were linked together like a beaded necklace in what were termed *polysomes*. Eventually, it was found that ribosomes were attached to mRNA, and when several ribosomes were attached to the same mRNA, these were termed *the polysomes*. Over the next few decades, both bacterial and eukaryotic ribosomes were studied. All of them consisted of a small subunit (SSU) and a large subunit (LSU) and all performed translation of the mRNA into polypeptides by attaching amino acids to the growing polypeptide chain one at a time according to a linear triplet nucleotide code that specifies the next amino acid to be added to that chain.

They all consisted of three to four RNA molecules and dozens of proteins (Figure 5.1). In Bacteria, Archaea, and organelles (mitochondria and plastids), the SSU is composed of a SSU ribosomal RNA (SSU rRNA), which ranges from about 1200 to 1600 nucleotides (nt) in length and two dozen proteins, while the LSU contains an LSU rRNA from 2300 to 3000 nt in length, a 5S rRNA (approximately 120 nt), and three dozen proteins. Eukaryotes have even larger ribosomes, with SSU rRNAs (between 1800 and 2000 nt) and LSU rRNAs (between 3500 and 5000 nt), and have two small rRNAs, a 5.8S (about 150–160 nt) and a 5S (approximately 120 nt). The SSU has approximately 33 ribosomal proteins, and the LSU has about 50. None of the ribosomal proteins is known to perform any enzymatic function. Many, called *chaperones*, associate with the rRNA molecules to hold them in stable conformations that maintain the efficiency of the translation process. Others associate with non-ribosomal proteins to perform other functions, such as attachment to membranes and folding of the emerging polypeptides. The SSU rRNA molecules are highly conserved, such that bacterial ribosomal rRNA can be compared with archaeal SSU rRNA and eukaryotic SSU rRNA. This characteristic led Carl Woese and George Fox to characterize the rRNA from a broad range of organisms to form the first phylogenetic tree based on molecular characters and led to finding an entirely new taxon of life, the Archaea (Figure 5.2).

RIBOSOMES AS RIBOZYMES

The rRNAs are responsible for all of the enzymatic functions of the ribosome, including covalently binding the amino acids to the growing polypeptide chain (peptidyl transferase activity), detaching the amino acid from the tRNA, and moving the mRNA through the ribosome. Additionally, the rRNA forms the sites where the tRNAs associate with the ribosome, thus positioning them to deliver their amino acid cargo and then forcing them off of the ribosome. Initially, it was thought that the important enzymatic functions of the ribosomes were due to the ribosomal proteins. Almost all of the ribosomal proteins in the *Escherichia coli* ribosomes were tested to determine whether they were essential for growth of the mutant bacteria. In experiments to determine where the catalytic components were on ribosomes, each protein was analyzed genetically and often biochemically. Mutant strains of *E. coli* were produced, each of which lacked one of the ribosomal proteins. Also, some mutants were produced that lacked more than one of the ribosomal proteins. It was found that none of the proteins were essential for ribosome function, although some mutants reduced the efficiency of translation. In subsequent studies, it was shown that most of the ribosomal proteins hold

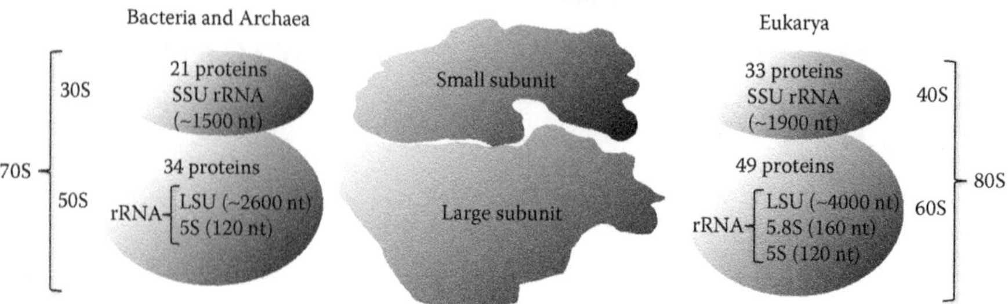

FIGURE 5.1 General structure of bacterial/archaeal and eukaryotic ribosomes. Bacterial and archaeal ribosomes are approximately 70S (S = Svedberg units, which are a measure of how the molecules sediment in centrifuged density gradients). The SSUs sediment at 30S, while the LSUs sediment at 50S. The SSU consists of a SSU rRNA and approximately 21 ribosomal proteins, while the LSU consists of a LSU rRNA, a 5S rRNA, and about 34 ribosomal proteins. Eukaryotic ribosomes sediment at 80S, which include a 40S SSU and a 60S LSU. The SSU contains an SSU rRNA and approximately 33 ribosomal proteins, while the LSU has an LSU, 5S, and 5.8S rRNAs, and approximately 49 proteins. The diagram in the center is the outline of the LSU and SSU based on tertiary structure from X-ray crystallography studies.

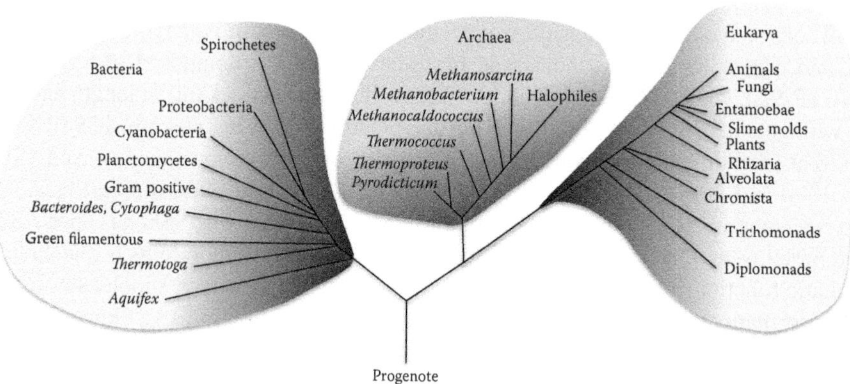

FIGURE 5.2 Basic phylogenetic tree of life based on SSU rRNA sequences. The tree, first determined in the studies of Carl Woese, indicated that there were three major branches of life on the Earth. What had been classified as bacteria could be split into two major groups, bacteria and archaea, based on their SSU rRNA sequences. Furthermore, the eukarya were phylogenetically closer to the archaea than to the bacteria. The root to the ultimate progenote is located between the bacteria and archaea.

the rRNAs in specific conformations that allow them to carry out their functions more efficiently. Finally, in the 1980s, it was discovered that some RNA molecules had enzymatic activities. The facts all came together at that time and it was concluded that ribosomes are ribozymes. In fact, they appear to be the largest and most complex of all ribozymes.

ORIGIN OF THE RIBOSOME

Ribosomes are large RNA and protein complexes, and translation is a finely regulated process. The evolution of this RNA–protein complex began as much smaller and simpler molecules. While ribosomes are the largest ribozymes in cells, they also are the largest RNA–protein complexes in cells. How did they grow so large and complex? Evidence indicates that they may be much simpler than their present sizes otherwise suggests. Experiments have shown that many parts of the rRNA and the ribosomal proteins can be removed without seriously affecting the translation functions of the ribosomes. Some have compared it to an onion, consisting of about 10 layers, each of which consists of 1–20 RNA loops and/or stems. Nearly 20 of the outermost loops of RNA (with attached proteins) can be removed without abolishing ribosome function. Moving into the next layer of the ribosome, nearly a dozen of those regions can be removed, again with no serious effects. The next layers of loops also are expendable. At its core, there exists a length of rRNA about 110 nt in length that appears to be the section that might represent the original center of the ribosome (Figure 5.3). During its early evolution, it was duplicated, which started a process that led to the much larger molecule that is seen today in ribosomes. The duplicated regions form the A-site and the P-site of the ribosome that bind the tRNAs that hold the incoming amino acid and the growing polypeptide chain, respectively. The initial 110 nt fragment might not have joined amino acids to form proteins, but probably did associate with some of the early proteins at that time. After the duplication, it may have joined together various building blocks, eventually specializing in peptide bonds between amino acids, containing the essential rRNA peptidyl transferase activity. The peptidyl transferase in association with the P-site comprises the essential primordial functional portion of the ribosome. Additional loops provided more specific and accurate reactions, and the ribosomal proteins provided stability to the active form of the rRNA. Additionally, some of the proteins on the outer surface of the ribosome associate with other proteins that aid in specific functions of the ribosome, such as binding to membranes, subunit interactions, and folding the emerging polypeptide chain.

Once ribosomes had begun their evolution, the organisms that possessed them had distinct advantages over their competitors. Protein synthesis became more accurate and efficient, thus increasing the chances for survival of those organisms that possessed ribosomes and translation of mRNA. This is evident in the fact that cellular organisms on the Earth all have ribosomes. This means that they probably evolved only once and that once they evolved, they conferred such an advantage to their cells that they completely outcompeted anything else that was around at that time. Since all organisms appear to have evolved from a common organism that existed more than 3.5 billion years ago, the ribosome with A, P, and E sites, peptidyl transferase activity, and the ability to decode a trinucleotide codon must have predated that time. Current estimates indicate that protoribosomes may have originated more than 3.6 billion years ago, and the complete bacterial ribosome was present from approximately 3.3 billion years ago. The late heavy bolide bombardment may have sterilized the surface of the Earth at 3.9 billion years ago. However, if some life forms were living deep in the oceans, they may have been protected from the effects of the bombardment. In any case, sometime prior to 3.6 billion years ago, RNAs became self-replicating, self-perpetuating molecules, some of which were ribozymes, from which the ribosome arose. Sequence conservation across all forms of life on the Earth indicates that major changes are not well tolerated in rRNA, especially in the SSU

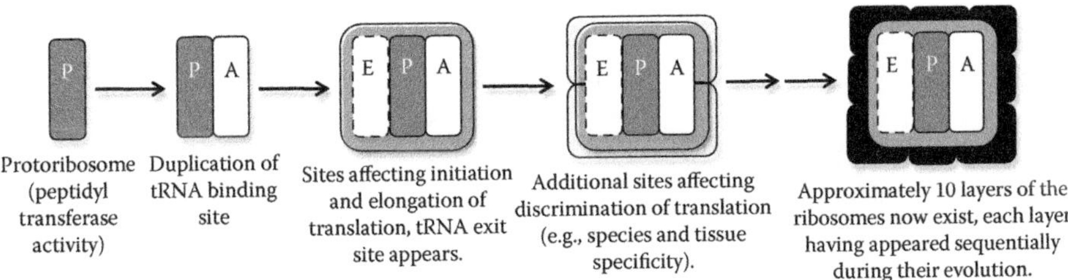

FIGURE 5.3 Possible evolution of the ribosome. Most of the LSU rRNA can be deleted, and translation will continue, although sometimes at a reduced rate. However, a core of about 110 nt is necessary to hold the tRNA with attached amino acid in order to form a peptide bond with another amino acid (the peptidyl transferase activity). Additionally, the 110 nt is repeated in the ribosome. These two regions form the A-site and the P-site and probably were formed by a duplication of the initially formed protoribosome. Further out on the ribosome is the E-site and accompanying regions. Out from this are other regions of the ribosome, which form successive layers of the ribosome. At least 10 such layers have been found. Each layer may have been added during evolution to provide more stability to ribosomes and to allow interactions with additional proteins, such as those that aid in translational control and fidelity.

rRNA molecule and surrounding the central portions of the LSU rRNA. Parts of these molecules are so highly conserved that changes to those parts are probably lethal to the ribosome and to the cell. Therefore, heavy selection continuously maintains the present sequence of much of the rRNA.

TRANSLATION

All of the peptidyl transferase activity (the enzymatic function that adds additional amino acids to the growing polypeptide chain) and the tRNA-binding sites are in the LSU rRNA (Figure 5.4).

Therefore, it is likely that the large ribosomal subunit represents the basic unit of the original ribozyme. The small ribosomal subunit appears to have had an independent origin, possibly representing a molecule that efficiently shuttled RNA molecules, such as the mRNAs. The SSU initially attaches to the mRNA and then associates with the large ribosomal subunit by interactions between the SSU and LSU rRNAs, as well as between some of the proteins. Once both subunits began to associate with one another, the efficiency and fidelity of translation would have improved almost immediately, since there would be a specific way to deliver and position the mRNAs to the translation machinery of the LSU rRNA.

During translation, rRNAs perform several functions. First, the rRNA recognizes (by binding or hybridization) either sequences (e.g., the Shine–Dalgarno sequence on the 5′ ends of mRNAs in bacteria), other structural regions, or guanosine caps with attendant cap-binding proteins on the 5′ ends of eukaryotic mRNAs. The mRNA and rRNA become associated with one another in a very specific orientation. Next, the ribosome small and large subunits join together in order to initiate translation. They do this because there are nucleotides on the SSU rRNA and the LSU rRNA that form numerous hydrogen bonds to hold the two subunits together. They also must move the mRNA along in order to find and decipher the message encoded in the mRNA sequence. Once an AUG start codon is found, the first tRNA with an amino acid (methionine) is in position to begin translation of the mRNA (Figure 5.5). This tRNA is shifted from the A (amino acid) to the P (polypeptide) site, and the tRNA with the complementary anticodon enters the A site. Next, there are specific interactions between the tRNAs (that hold the individual amino acids) to be added and the rRNA. The peptidyl transferase catalyzes the reaction to covalently bind the two amino acids together with a peptide bond. This process repeats to form the growing polypeptide chain until a stop codon is reached, in which case the polypeptide is severed from the last tRNA to release it from the ribosome through a tunnel in the LSU. The small and large subunits separate, the mRNA is released, and the process begins again once the SSU has attached to another mRNA.

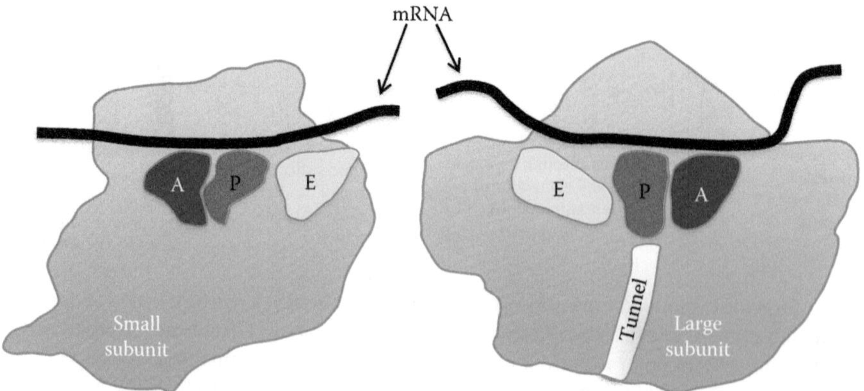

FIGURE 5.4 View of the interfaces between the small and large ribosomal subunits. The outlines are based on the crystallographic structures of each subunit. The three tRNAs (one each at the A, P, and E sites) are shown. An mRNA is shown threaded through the ribosome. The synthesized polypeptide is threaded through the tunnel (or channel) during translation.

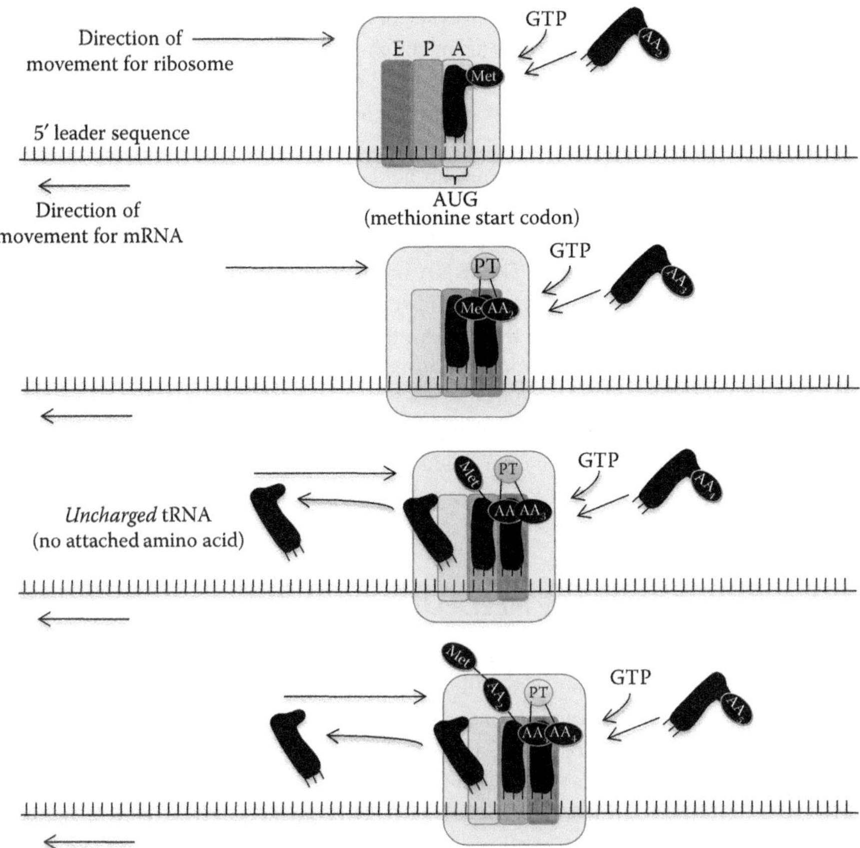

FIGURE 5.5 Schematic of the process of translation. After the 5′ region of the mRNA associates with the ribosome, it is moved along the ribosome until a start codon (AUG) is located. Then, a tRNA with a CAU anticodon (complementary to the AUG), with attached methionine amino acid, hybridizes with the codon in the A-site and is moved to the P-site, as the mRNA is also moved in the same direction. The next codon is now at the end of the A-site, and the tRNA with the complementary anticodon moves into the A-site. Then, the peptidyl transferase of the LSU rRNA covalently binds the amino acid held in the A-site to the amino acid held in the P-site, and the amino acid is released from the tRNA in the P-site. Immediately thereafter, there is a shift of the ribosome (with respect to the mRNA), moving the uncharged tRNA to the E-site, where it dissociates from the ribosome. At the same time, the tRNA that now holds the growing polypeptide chain is moved to the P-site, and the cycle repeats until a stop codon is encountered. At this point, the stop is recognized and the polypeptide is cut from the final tRNA, thus releasing the completed polypeptide.

Translation consists of an intricate interaction between the mRNA, the rRNA, tRNAs, GTP, the growing polypeptide chain, and the incoming amino acid. There are three positions for tRNAs on the ribosomes: the A (amino acid), P (polypeptide), and E (exit) sites. The A-site holds the incoming tRNA with a single amino acid covalently bound to a CCA motif on the 3′ end of the molecule to the amino end of the amino acid. The P-site holds a tRNA attached to the polypeptide chain being synthesized. The E-site briefly binds the tRNA that has lost its attachment to the polypeptide chain that was connected while it was in the P-site. Initiation factors are needed to begin the translation process at the correct site. Also, the start codon, which usually is an AUG is recognized by the anticodon CAU on the methionyl-tRNA that is held initially in the A-site. GTP is needed to lock the tRNA into the A-site. The ribosome then shifts relative to the mRNA, positioning the methionyl-tRNA in the P-site. This also slides the next codon into one end of the A-site, and hydrogen bonding to the corresponding tRNA increases the stability of the correct tRNA moving into the A-site.

Again, this is locked into the A-site by hydrolysis of GTP. Attack of the carboxyl group on the methionine (in the P-site) by the amino group of the A-site amino acid occurs, stabilized in a tetrahedral intermediate by a specific adenine on the LSU rRNA. The two amino acids are covalently bound as the P-site tRNA is separated from its amino acid. A shift of both tRNAs (as well as the mRNA) occurs, moving the free tRNA into the E-site (exit site), and the other tRNA (now attached to an amino acids that is linked to another amino acid by a peptide bond) to the P-site. The next tRNA that can hydrogen bond to the mRNA codon then moves into the A-site, again utilizing GTP. The amino group on the tRNA in the A-site attacks the carboxyl group of the growing polypeptide chain attached to the tRNA in the P-site. The tRNA in the A-site is covalently bound to the growing polypeptide chain, while the tRNA in the P-site is separated from the growing polypeptide chain, which then moves into the E-site, while the tRNA attached to the polypeptide chain moves to the P-site. Then, the process repeats until a stop codon (usually UAA, UAG, or UGA) is encountered on the mRNA. There are no tRNAs that bind to this codon, and so the A-site is left empty. This allows binding of release factors access to this site, which detach the polypeptide chain from the P-site tRNA, thus releasing the tRNA and the completed polypeptide chain. Other factors then cause the release of the ribosome from the mRNA and dissociation of the small and large ribosomal subunits. Then, they are ready to start the process on other mRNA molecules ready for translation.

HOW MANY rDNA COPIES ARE NEEDED?

In order to produce a sufficient number of proteins to maintain viability in each cell (approximately 10^6 protein molecules per bacterial cell, and 10^9–10^{10} proteins per cell in eukaryotes), cells must contain a very large number of ribosomes. A typical bacterium (e.g., *E. coli*) contains about 20,000 ribosomes (with a range of 6,800–72,000 per cell). A typical eukaryotic cell contains about 10,000,000 or more ribosomes (yeast range from 150,000 to 350,000 per cell, while human cells usually have tens of millions of ribosomes in each cell). This presents cells with a challenge—that of being able to produce enough rRNA for all of those ribosomes. RNA polymerase synthesizes RNA at a rate between 50 and 150 nt/s. If there is a single rRNA gene, and each gene is transcribed by one RNA polymerase at a time, how long would it take to make 20,000 ribosomes for a bacterium or 10,000,000 for a eukaryotic cell? Ribosomal rRNA genes are transcribed in long pieces, which consist of both the SSU and the LSU molecules, as well as sections that are discarded prior to complete assembly of the ribosomes, and the 5.8S molecules in eukaryotes. In bacteria, the total length for each rRNA transcript is 5–6 kb. Eukaryote transcripts range from 6 to 14 kb. Therefore, for bacteria, RNA polymerase needs to incorporate 100–120 million nucleotides into rRNA molecules in each cell. If we use the value of 100 nt/s for the RNA synthesizing rate of RNA polymerase, the bacterium had only a single rRNA gene, and only a single polymerase per gene was producing RNA, approximately 100,000 s would be required to synthesize all of the rRNA needed for a single bacterial cell. Since there are 86,400 s in a day, it would require approximately 1 day, 3 h, and 47 min to synthesize enough rRNA for each bacterial cell. Since some bacterial cells can divide in less than 20 min, clearly, this rate of synthesis is insufficient. However, when rRNA genes are viewed during transcription in electron microscopes, a feather-like structure is seen (see Figure 4.3), which indicates concurrent transcription is occurring by approximately 100 RNA polymerase molecules that are packed onto the gene. The RNA molecules range from short (near the 5′ end of the SSU rRNA gene) to full length (just past the end of the LSU rRNA gene). By packing approximately 100 RNA polymerase molecules onto the gene, the 20,000 rRNAs that are required for each cell can be synthesized in about 15–30 min. Genomic and genetic studies have shown that bacteria have from 1 to 15 rRNA gene repeats within their genomes. Therefore, for some, they may be capable of reducing the time more than this, although each of the rDNA loci may not be active at the same time or under the same conditions.

For eukaryotes, producing a sufficient number of rRNAs for their ribosomes is even more problematic. A typical eukaryotic cell has 10,000,000 ribosomes (including SSU, LSU, and small rRNAs).

TABLE 5.1
Number of rRNA Genes per Haploid Genome in Various Organisms

Organism	rRNA Genes per Haploid Nucleus
Tetrahymena (a ciliate)	1
Mitochondria	1 or 2
Chloroplasts	1–3
E. coli	7
Saccharomyces cerevisiae	140 (mean, varies)
Homo sapiens	200 (mean, varies)
Salamanders	5,000–10,000
Plants (e.g., *Vicia faba*)	200–22,000

If we go through the same calculations, 10,000,000 ribosomal RNAs, with a transcript length of 10 kb would require the polymerization of 100 billion nucleotides to produce the requisite number of rRNA molecules for a single cell. At a polymerization rate of 100 nt/s, it would take one billion seconds (11,570 days or more than 30 years) to produce this amount of rRNA, where a single gene and a single polymerase perform the synthesis. Again, if approximately 100 RNA polymerases were packed onto each gene, synthesizing concurrently, this would reduce the number to about 116 days. Most eukaryotic cells are able to reproduce at rates much higher than one division per 116 days, so clearly a single rRNA gene is insufficient for eukaryotes. However, there are some species (e.g., some ciliated protozoans) that apparently have only a single rRNA gene in their primary genomes, while other eukaryotes have solved the problem by retaining hundreds to thousands of repeats of the entire rRNA gene region (Table 5.1). Often, these occur in long tandem repetitive regions in one or more chromosomal loci. Many fungi have about 150 rRNA gene copies per haploid set of chromosomes, while the human genome contains approximately 200 copies (400 per diploid cell). The numbers are approximate, because the number of repeats is variable in most species. For example, in broad bean (*Vicia faba*), the number of repeats per haploid set of chromosomes ranges from 230 to 22,500 in a single population of plants. Clearly, most of the repeats may not be functional at any one time. Some eukaryotes have approximately 50 repeats, but this appears to be a minimal number in most cases. In insect development experiments, numbers as low as 20–25 copies can be tolerated, but if there are only 15 or fewer copies, development is arrested. At 20 copies, if everything is operating about as fast as possible, approximately 2–3 days (per diploid cell) would be required to produce enough rRNA to allow each cell division to occur. Therefore, the rate of cell division is at least partly dependent on the number of rRNA genes available for transcription.

MECHANISMS FOR INCREASING rRNA GENE COPY NUMBER

Most eukaryotes appear to retain high copy numbers of rRNA genes within their genomes (usually hundreds), thereby assuring that a sufficient number of genes will be present when additional ribosomes are needed. They do this by maintaining their genes in long tandemly repeated arrays (see Figure 4.6). This allows pairing of the chromosomes during meiosis and mitosis. Unequal crossovers then lead to one chromosome gaining a number of gene repeats and the other losing the same number of repeats. These changes have been demonstrated in somatic cell divisions in plants, animals, and fungi, which can lead to large differences in rRNA gene repeats from cell to cell within individuals, and from individual to individual within a single population. Therefore, in eukaryotes recombination maintains rRNA genes at levels that are sufficient to assure an adequate number of rRNAs and ribosomes in each cell at all times.

However, as was pointed out above, there are some organisms with only one copy of the rRNA genes and others that have evolved other mechanisms to increase the number of rRNA gene copies when needed. One of these occurs in ciliated protozoans, such as *Tetrahymena*. These organisms have two nuclei. One is small and diploid. It is called *the micronucleus*. It carries only one rRNA gene on each of the two homologous chromosomes. But, as already mentioned, this would be insufficient to provide all of the rRNA needed for each of the cells. The primary function of this nucleus is to store the genetic material for each of the cells from one generation to the next. The large nucleus, called *the macronucleus*, is much more active. The macronucleus undergoes a DNA synthesis phase whereby it becomes polyploid, and some genes are replicated more than others. Replication of the rRNA genes occurs extrachromosomally. The single chromosomal copy of rDNA is surrounded by several special sequences, called *A′ repeats* and *M repeats* (Figure 5.6). Upstream from the gene, there are two direct A′ repeats and two inverted M repeats. Downstream from the gene there is one M repeat and one A′ repeat. As the macronucleus begins to enlarge, the rRNA gene plus surrounding regions are cut out, with the cut occurring between one of the M and A′ repeats on each side. Next, telomeres are added to three of the four cut ends, thus protecting those ends from damage. Then, DNA replication begins on the upstream end of the fragment containing the rRNA gene. The ends fuse, and replication proceeds along both strands to the ends, which produces a duplexed symmetrical molecule that has two rRNA genes and four of the M repeats. These can then undergo additional replication cycles, thus producing a much larger number of rRNA genes that are then transcribed into the required rRNAs for construction of ribosomes.

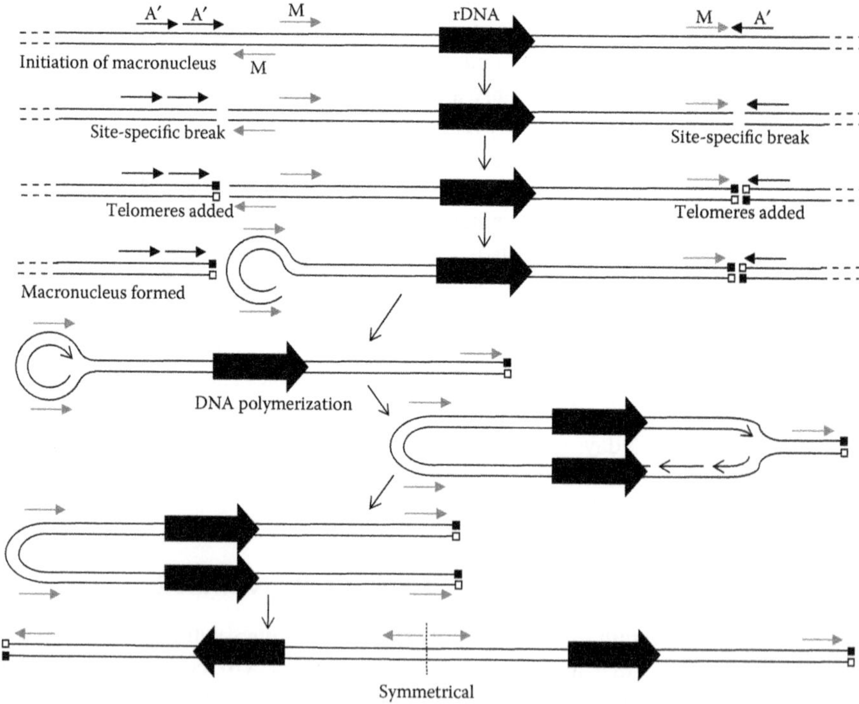

FIGURE 5.6 Extrachromosomal amplification of rRNA genes in *Tetrahymena* (a ciliate). Ciliates have a micronucleus (genomic nucleus) and a macronucleus (includes multiple copies of a subset of the genomic sequences, including rRNA genes). As the macronucleus is initiated, the rRNA gene segment is cut by an endonuclease. Telomeres are added to three of the DNA ends, but not to the 5′ end containing the rRNA genes locus. This end curls around and invades the same DNA further downstream, which is followed by DNA replication to form a long linear symmetrical molecule with two copies of rRNA. This is followed by additional rounds of replication to form hundreds of rRNA gene copies.

Mitochondria and chloroplasts (or, more generally, plastids) also transcribe their own rRNA using genes on their chromosomes. Mitochondria generally have one copy of rDNA, while chloroplasts have either one or two, depending on the taxonomic group. In higher plants, usually two copies are present on a long inverted repeat, while in lower organisms, usually only one copy is present. Extrachromosomal copies of rDNA have never been observed in any of these organelles, but polyploidization of organellar genomes is common. Multiple copies of the genome, sometimes in a way that produces more of some regions than others, lead to an increase in the number of rRNA genes to a level that is sufficient to produce enough rRNA to produce the ribosomes needed in these organelles.

In amphibian oocytes, there is yet another pathway that has evolved to provide enough ribosomes for particular cells. Amphibian oocytes are very large cells. Often, they are several millimeters in diameter and therefore can be seen with the unaided eye. They have approximately 10^{12} ribosomes and therefore 10^{12} sets of rRNA molecules. One of the reasons, the cells are so large and contain so many ribosomes, is that the egg cell undergoes rapid cleavage and cell divisions following fertilization. There simply is not sufficient time to produce such large numbers of ribosomes and cells following fertilization to satisfy the requirements for the complex developmental changes that occur at that time. As one might imagine, this presents these large oocytes with additional survival obstacles, since they must produce such a large number of ribosomes. There are two mechanisms involved. The first is one that many animals, as well as plants, utilize. Egg cells in these organisms are the products of asymmetrical cell divisions during meiosis, such that only a single mature egg cell is produced from each meiotic division (Figure 5.7). The other cells are very small and degenerate into nonfunctional cells (e.g., polar bodies). In producing one large egg cell in this process, almost all of the rRNAs and ribosomes end up in one cell. Therefore, although there is only

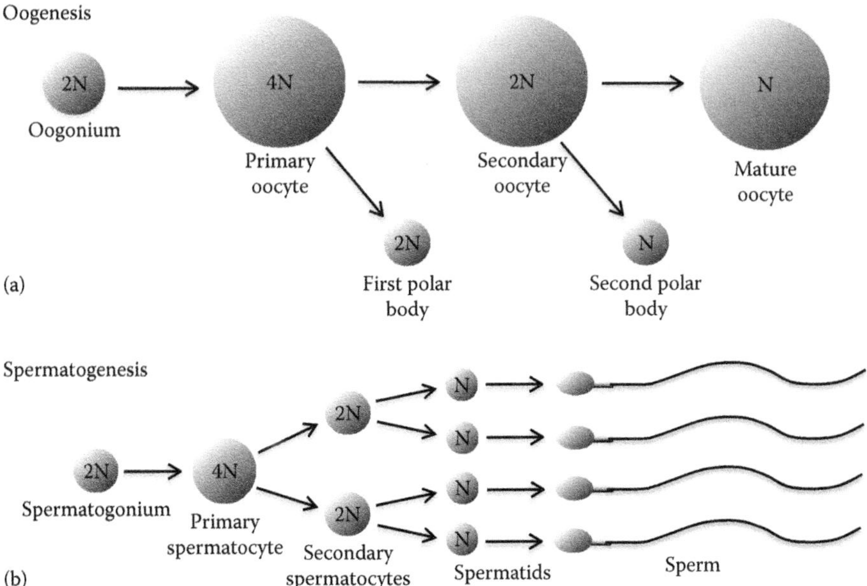

FIGURE 5.7 Formation of oocytes and sperm in animals. Some cells, such as oocytes undergo asymmetric cell divisions. Often, the smaller cells die, while the larger cells survive. The primary reason for this is to increase the amount of specific components in those cells. For example, some amphibian oocytes contain 10^{12} ribosomes (and thus 10^{12} copies of each of the rRNAs). (a) In oocytes (including human oocytes) meiosis begins with a large cell with a tetraploid amount of DNA (4N), which undergoes two successive divisions that results in one large haploid oocyte and two or three small cells, called polar bodies that usually die. (b) In contrast, meiosis to produce sperm cells yields four equivalent haploid sperm cells.

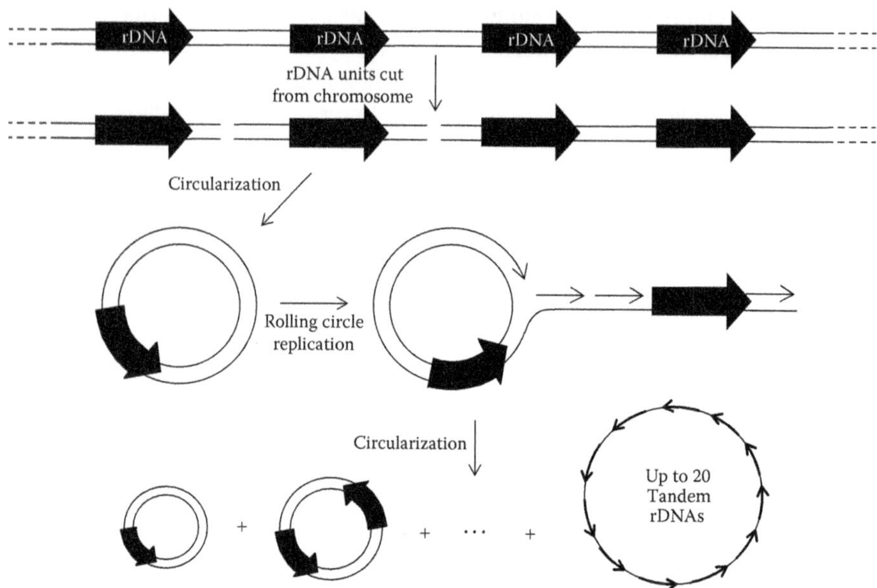

FIGURE 5.8 Extrachromosomal amplification of rRNA genes in amphibian oocytes. Because of the need for an extremely large number of ribosomes in amphibian (e.g., *Xenopus laevis*) oocytes, a mechanism to amplify the number of genes has developed. First, some of the rRNA gene repeats are cut away from the chromosome. The repeats circularize and rolling circle replication is initiated. Pieces of the replicated DNA are cut and circularized to form a diversity of molecules having from one to up to twenty rRNA gene repeats.

a haploid number of rRNA genes in the nucleus, roughly four times the number of rRNAs and ribosomes are present. Although amphibians have several hundred tandemly arranged rRNA gene repeats in their genomes, this is insufficient to produce the 10^{12} rRNAs needed in the mature oocyte. A second mechanism has developed in oocytes to boost the number of rRNA genes in these cells. The most detailed studies of this are from the African clawed frog, *Xenopus laevis*. In this case, one or more of the rRNA gene repeats are cut from the chromosome (Figure 5.8). These fragments circularize and then are replicated via a rolling circle mode of DNA replication. Ultimately, the replicated fragments circularize, such that a variety of circular DNAs are formed that have from 1 to 20 rRNA genes on each. Thousands of these molecules are formed and are used to transcribe all of the rRNA needed to produce the 10^{12} ribosomes needed in the mature oocyte.

COMPLEXITY OF RIBOSOMES

Ribosomes probably began to evolve during the time when RNA and proteins performed most of the cellular functions (i.e., the RNA world that spanned from about 3.8 to 3.6 billion years ago), and DNA was not yet present in cells. Ribosomes may have gone through several stages of initial evolution during that time (the details of this are presented in Chapter 6), but by about 3.5 billion years ago, they were very similar to current ribosomes. They had LSU and SSU, which had SSU rRNA, LSU rRNA, 5S rRNA, peptidyl transferase activity, A, P, and E sites for attachments of the tRNAs. They used a triplet code for translating proteins. These are assumed, because all modern cellular organisms have all of these parts and functions in common, and because of their complexity, it is likely that they evolved only once on the Earth. Since then, there have been changes, such as the appearance of the 5.8S gene in eukaryotes, which was derived from a part of the LSU of the progenote LSU rRNA. Additional RNA sequences and proteins have been added since then, as evidenced by the differences seen between bacteria and eukaryotes, and differences among various groups of organisms. Also, as cells became larger, they have evolved mechanisms to increase the number of rRNAs (and therefore, ribosomes) per cell. These RNA–protein complexes have been absolutely vital to evolution on the Earth.

KEY POINTS

1. Ribosomes are present in all cells and some organelles. They consist of LSU and SSU and perform the majority of protein synthesis in cells.
2. Ribosomes consist of enzymatic RNAs (ribozymes) and stabilizing proteins.
3. During translation, a tRNA with attached amino acid associates with the ribosome in the A-site by hydrogen bonding with the codon of the mRNA, followed by peptide bond formation with the growing polypeptide chain held by the tRNA in the P-site of the ribosome. Next, the tRNA in the A-site, which now holds the growing polypeptide chain, moves to the P-site, and the uncharged tRNA exits by way of the E-site on the ribosome. This process repeats until a stop codon is reached. Then, the polypeptide is released.
4. Ribosomes may have evolved from a core of about 110 nt, but have grown and become more efficient over 3.5–4.0 billion years.
5. Ribosomes are needed in high concentrations in cells. Single rRNA gene copies are insufficient to provide all of the rRNA transcripts needed to construct all of the ribosomes. Organisms have evolved several mechanisms that result in high copy numbers of rRNA genes, including maintaining high copy numbers within the genome, extrachromosomal amplification of rRNA genes, asymmetric cell division, and producing multiple copies of all or parts of the genome (becoming polyploid).

ADDITIONAL READINGS

Alberts, B., D. Bray, K. Hopkin, A. Johnson, J. Lewis, M. Raff, K. Roberts, and P. Walter. 2013. *Essential Cell Biology*, 4th ed. New York: Garland Publishing.

Alberts, B., D. Bray, A. Johnson, J. Lewis, M. Raff, K. Roberts, and J. D. Watson. 1994. *Molecular Biology of the Cell*. New York: Garland Publishing.

Bleichert, F., S. Granneman, Y. N. Osheim, A. L. Beyer, and S. J. Baserga. 2006. The PINc domain protein Utp24, a putative nuclease, is required for the early cleavage steps in 18S rRNA maturation. *Proc. Natl. Acad. Sci. USA* 103:9464–9469.

Bokov, K. and S. V. Steinberg. 2009. A hierarchical model for the evolution of 23S ribosomal RNA. *Nature* 457:977–980.

Darnell, J., H. Lodish, and D. Baltimore. 2007. *Molecular Cell Biology*, 6th ed. New York: W.H. Freeman and Company.

Karp, G. 2013. *Cell and Molecular Biology*, 7th ed. New York: John Wiley & Sons.

Rogers, S. O. and A. J. Bendich. 1987a. Heritability and variability in ribosomal RNA genes of *Vicia faba*. *Genetics* 117:285–295.

Rogers, S. O. and A. J. Bendich. 1987b. Ribosomal RNA genes in plants: Variability in copy number and in the intergenic spacer. *Plant Mol. Biol.* 9:509–520.

Rogers, S. O., S. Honda, and A. J. Bendich. 1986. Variation in ribosomal RNA genes among individuals of *Vicia faba*. *Plant Mol. Biol.* 6:339–345.

Smith, J. M. and E. Szathmary. 1995. *The Major Transitions in Evolution*. New York: Oxford University Press.

Vanrobays, E., P.-E. Gleizes, C. Bousquet-Anonelli, J. Noaillac-Depeyre, M. Caizergues-Ferrer, and J.-P. Gélugne. 2001. Processing of 20S pre-rRNA to 18S ribosomal RNA in yeast requires Rrp10p, an essential non-ribosomal cytoplasmic protein. *EMBO J.* 20:4204–4213.

6 Structure of the Genetic Code

INTRODUCTION

The genetic code was first decoded during the middle part of the twentieth century. For these early experiments, the length of the codon was unknown. No one knew whether a single nucleotide, dinucleotide, trinucleotide, or longer codon encoded each of the amino acids. Initially, poly-U RNAs were synthesized and used in cell-free *in vitro* translation systems. Only poly-phenylalanine peptides were produced. Similar experiments using poly-C and poly-A RNAs produced poly-proline and poly-lysine polypeptides, respectively. Analogous experiments using RNAs with alternating bases (e.g., UCUCUCUC) produced polypeptides with alternating amino acids (e.g., threonine–histidine). These all indicated that the codon consisted of a triplet of nucleotides. For example, in the previous example of the UCUCUCUCUC synthesized RNA, the triplets would be UCU (which encodes threonine) and CUC (which encodes histidine). While some of the codons were more challenging to deduce, eventually all of the codons were determined to elucidate the genetic code, including the fact that most amino acids were encoded by more than one triplet codon. Also, it became clear that the first and second nucleotides were the primary determiners of the amino acids, while the third nucleotide could vary without altering the amino acid sequence. During the first few decades after the genetic code was solved, it was thought to be a *universal genetic code*, and for the majority of organisms, it is universal (Figure 6.1). However, while most bacterial, archaeal, and nuclear genes use the universal genetic code, there are many organisms whose nuclear and/or organellar genes utilize slightly different codons for some amino acids, as well as different start and stop signals (Figure 6.2). For example, some of the stop codons in the universal code encode for amino acids in some genomes, and the standard methionine start codon is instead isoleucine for some genomes. Additionally, a number of different start and stop codons have been identified in a variety of organisms. However, these variations in the genetic code likely occurred after the evolution of the *universal genetic code*, because the variations usually appear in the organellar or nuclear genomes of a limited number of organisms.

Translation is a complex process that currently uses a triplet code. However, the first forms of translation were likely simpler, and this is indicated in both the modern triplet code and in the structure of ribosomes. The fact that the middle nucleotide of the codon is the most conserved for a particular amino acid, and the third nucleotide is the most variable hints to a simpler genetic code consisting of fewer than three nucleotides per codon. Also, while the core structure of ribosomes is highly conserved, there is great variation in size and structure of the outer portions of the large and small subunits in bacteria versus eukaryotes, indicating evolutionary changes in size and structure. Additionally, the pools of tRNAs in cells differ, even within the cells of multicellular organisms, such that some codons are used to greater extents than alternative codons for the same amino acid, thus allowing for a greater flexibility in the expression and reducing the redundancy of the genetic code, and again hinting at a simpler process and structure in the past.

EVOLUTION OF THE GENETIC CODE

The universal genetic code (Figure 6.1) is a part of translation that all students learn, but the evolutionary processes that led to what we see today are deeply hidden because many of the changes occurred over 3.5 billion years. However, a few things can be gleaned from what is known. The code is easily understood once the starting site is found on the gene. Most often, this is an AUG codon, which starts the protein off with a methionine. Then, the next three nucleotides are recognized by the next tRNA through complimentary hydrogen bonding to the codon, and the next amino acid is

First position	Second position U	Second position C	Second position A	Second position G	Third position
U	Phe (F)	Ser (S)	Tyr (Y)	Cys (C)	U
U	Phe (F)	Ser (S)	Tyr (Y)	Cys (C)	C
U	Leu (L)	Ser (S)	STOP	STOP	A
U	Leu (L)	Ser (S)	STOP	Trp (W)	G
C	Leu (L)	Pro (P)	His (H)	Arg (R)	U
C	Leu (L)	Pro (P)	His (H)	Arg (R)	C
C	Leu (L)	Pro (P)	Gln (Q)	Arg (R)	A
C	Leu (L)	Pro (P)	Gln (Q)	Arg (R)	G
A	Ile (I)	Thr (T)	Asn (N)	Ser (S)	U
A	Ile (I)	Thr (T)	Asn (N)	Ser (S)	C
A	Ile (I)	Thr (T)	Lys (K)	Arg (R)	A
A	Met (M)	Thr (T)	Lys (K)	Arg (R)	G
G	Val (V)	Ala (A)	Asp (D)	Gly (G)	U
G	Val (V)	Ala (A)	Asp (D)	Gly (G)	C
G	Val (V)	Ala (A)	Glu (E)	Gly (G)	A
G	Val (V)	Ala (A)	Glu (E)	Gly (G)	G

FIGURE 6.1 The primary genetic code used for most organisms and organelles. The first, second, and third positions of each triplet are indicated. The three stop codons also are indicated (in white boxes). Molecular structures for each R group are indicated within the boxes corresponding to the amino acid. Colors indicate the class of amino acid: orange, hydrophobic; blue, polar; aqua, basic; and green, acidic. Three letter abbreviations for the amino acids (left in each box) and single letter codes (right in each box) are provided.

added according to that codon. Although it often is referred to as the *universal genetic code*, there are other versions of the code. For example, as already mentioned, some organelles use slightly different codes (Figure 6.2). So, even though this is a complex system, there have been evolutionary changes in some genetic lines that have led to successful changes in the system.

Although the precise sequence of events that led to translation using a genetic code is unknown, some conclusions can be made based on what is known about the characteristics and organization of the genetic code. First of all, the nucleotide in the second position of the triplet codon is the most conserved determiner of the selected amino acid (Table 6.1). The first nucleotide is also important, but it is not as strong a determiner as the second position nucleotide. The third position nucleotide is the least important, and in fact, it can be changed in several ways without changing the amino acid that is added for that codon. Amino acids are categorized by being either hydrophobic, polar, basic, or acidic. If these characteristics are displayed on the genetic code, then clear patterns are observed (Figure 6.1). For example, if a U is in the second position, the amino acid always is hydrophobic, and U is the only base that signifies a start codon, whether in the universal genetic code or in variations of the code (Figure 6.2). Half of the codons with a C in the second position encode hydrophobic amino acids, while the other half are polar. Codons with an A in the second position always are either polar or charged, and thus, they are hydrophilic.

Codons that have a G in position one and an A in position two always encode acidic amino acids. Codons with either C or A in position one, followed by an A or G in position two, encode basic or polar amino acids. Thus, there is a structure in the genetic code that definitely points to the strong affect in the second position, and secondarily in the first position, and the relatively weak importance of position three, which leads to the addition of amino acids of particular categories. While the third position is relatively unimportant for amino acid determination, it is important with respect to translation start and stop signals, as well as proportions of various tRNAs present in specific cell types.

First position	Second position: U	C	A	G	Third position
U	F	S	Y	C	U
	F	S	Y	C	C
	L (S)	S (STOP)	STOP *(Y,Q)	STOP *(W,C)	A
	L (GO)	S	STOP *(Q,L)	W	G
C	L	P	H	R	U
	L	P	H	R	C
	L	P	Q	R	A
	L (GO)	P	Q	R	G
A	I (GO)	T	N	S	U
	I (GO)	T	N	S	C
	I (M, GO)	T	K (N)	R (S,G, STOP)	A
	M GO*	T	K	R (S,G,K, STOP)	G
G	V	A	D	G	U
	V	A	D	G	C
	V	A	E	G	A
	V (GO)	A	E	G	G

FIGURE 6.2 Genetic code with canonical and alternative codes. The most common start codon is AUG (indicated by a GO with an asterisk), which encodes a methionine. However, other start codons (indicated as GO) have been observed in some of the genes for a variety of species. This includes UUG, CUG, and GUG, which are start codons that encode leucine or valine, as well as AUU, AUC, and AUA, all of which encode isoleucine. In a few cases, AUA is a start codon that encodes methionine. Similarly, the most common stop codons are UAA, UAG, and UGA (each indicated by a STOP with an asterisk). However, in some of the genes of a variety of organisms and mitochondria, the same codons sometimes encode tyrosine, tryptophan, or cytosine. For some genes, UCA, AGA, and AGG act as stop codons (indicated as STOP). Additionally, in a few cases, codons have been found that differ from the standard code. Specifically, in some genes, UUA encodes serine, rather than leucine; AGA encodes serine or glycine rather than arginine; and AGG encodes serine, glycine, or lysine rather than arginine.

TABLE 6.1
Amino Acid Characteristics Compared with Nucleotide Usage for Each Codon

Category of Amino Acid	Codon Position[a]: First	Second	Third
Hydrophobic	N	U, c, g	N
Polar	U, a, c	C, A, g	N
Acidic	G	A	N
Basic	C, A	A, G	N

[a] Upper case letters indicate invariant or predominant nucleotides. Lower case letters indicate nucleotides used at lower frequencies.

An additional and important characteristic of the genetic code is that the amino acid RNA synthetase (aaRS; also called *aminoacyl tRNA synthase*) is regularly arranged within the code. The aaRSs represent a second genetic code, in that they covalently bind the specific amino acids to the corresponding tRNAs. There is one aaRS for every type of tRNA in each cell. Moreover, there are two classes (I and II) of aaRSs, based on their evolution and the side of the tRNAs to which they

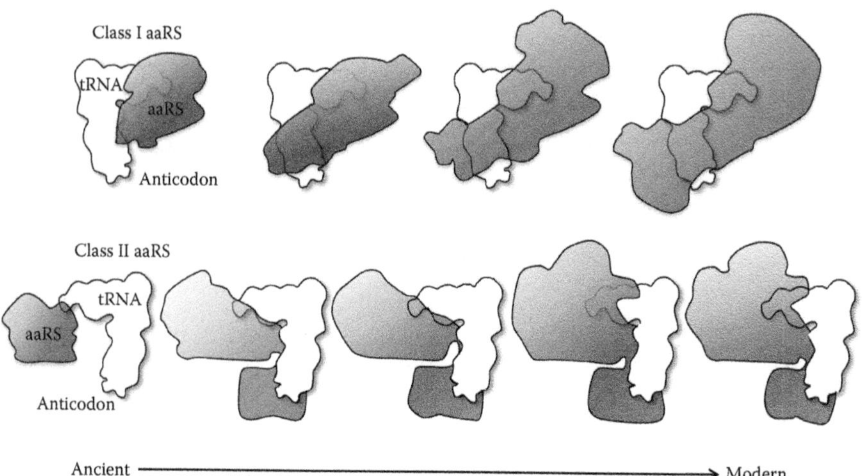

FIGURE 6.3 Model of the evolution of class I (upper series) and class II (lower series) of aaRSs (based on known aaRSs and the frequencies of the corresponding amino acids in ancient proteins). Most class I aaRSs approach the tRNAs from the minor groove side, while most class II aaRS approach from the major groove side (although the portion that contacts the anticodon loop portion twists around to the other side of the tRNA). The most ancient aaRSs (which are class II) only make contact with the acceptor arm portion of the tRNAs. More evolved forms of both class I and class II aaRSs make contact with both the acceptor arm and the anticodon loop.

bind (Figure 6.3). Almost all of the class I aaRSs bind to the minor groove side of the tRNA. The only exceptions are the aaRSs for tyrosine and those for some of the arginine tRNAs (with AGY codons), which are class I aaRS that bind to the major groove sides of the tRNAs. Most of the class II aaRSs bind to the major groove side of the tRNA. The only exceptions are the aaRSs for phenylalanine and tryptophan, which bind to the minor groove sides of the tRNAs. One other curious difference is that the aaRSs for lysine are class II in most organisms, but class I aaRSs for lysine are present in some archaea. Overall, half of the aaRSs in cells are class I, and half are class II. This indicates a nonrandom organization of the genetic code. Furthermore, the two classes evolved independently. There are no DNA or amino acid homologies, and their modes of aminoacylation differ. Class I aaRSs aminoacylate at the 2′ OH of the terminal adenosine on the tRNA, and they are either monomeric or dimeric. Class II aaRSs aminoacylate at the 3′ OH of the terminal adenosine on the tRNA, and they are either dimeric or tetrameric.

While the standard way of presenting the genetic code indicates structure related to the two classes of aaRSs, it fails to confer the possible reasons for the structure. However, it has been noted that when the genetic code table is reorganized to group the codons according to their reverse complementarity to other codons, the groupings of the aaRSs form a distinct pattern (Figure 6.4). The proposed reason for this pattern is that during the initiation of translation and the genetic code more than 3.5 billion years ago, the first genes (likely RNA genes) were double stranded and self-complementary, such that each gene encoded two polypeptides, one reading from one of the strands and the other reading in the reverse direction on the other strand. Thus, each triplet codon would have a self-complementary version on the opposite strand that would code for a different amino acid. One immediate implication of this is that the middle codon would become the most conserved, while the first and third codons would experience more variation, which is exactly the situation with the modern genetic code. Additionally, this would allow one of the two to vary greatly, while retaining more sequence conservation for the other. This is true for the modern genetic code, which contains a conserved nucleotide in position one of the codon and a variable codon in the third position of the codon.

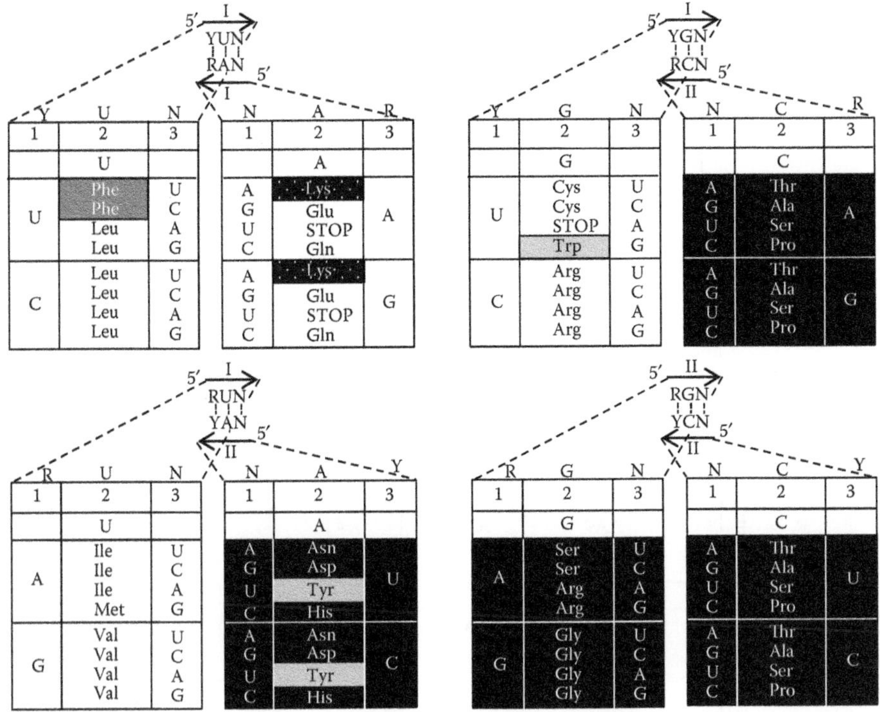

FIGURE 6.4 Genetic code rearranged according to aaRS class, approach of the aaRS, and complementarity of codons. Amino acids in black font with white background are added to tRNAs by class I aaRSs (indicated as I), most of which approach the tRNA from the minor groove side. Those in white font with black background are added to tRNAs by class II aaRSs (indicated as II), most of which approach the tRNA from the major groove side. The exceptions are tryptophan and tyrosine, which are added to tRNAs by class I aaRSs, but the enzymes approach the tRNA from the minor groove side; phenylalanine, which is added to the tRNA with class II aaRSs, but approach the tRNA from the minor groove side; and lysine, which is added by class II aaRSs in most organisms, but is added by class I aaRSs in some archaea. Codons for each amino acid are indicated. Arrows indicate the direction of the codons in the mRNAs. Complementary triplets are grouped into the four quadrants.

WHY A TRIPLET CODON?

If the genetic code were dependent on a one-nucleotide code, then only four amino acids could be possible (Figure 6.5). If the code were based on a dinucleotide system, then 16 (4^2) amino acids would be possible. If unified start and stop codons were used, then this would drop to 14 or lower, although some codons could code for more than a single amino acid. The triplet code allows for 64 (4^3) possible amino acids. This allows for encoding of all 20 amino acids, including some redundancy, as well as use of a few of the codons for translation start and stop signals (Figure 6.6). The redundancy has been used by various species and cell types that utilize one of the codons over the others for specific amino acids. Early in the Earth's history, other genetic coding systems may have evolved, but the one that is universal today succeeded over all others.

The current genetic code consists of triplets of nucleotides, each of which hydrogen bonds to three nucleotides on tRNA anticodon loops. Each triplet is specific for a certain tRNA that carries a specific amino acid. A triplet code is complex, and the first genetic code probably consisted of single nucleotides that stood for single amino acids (Figures 6.5 and 6.7). However, with only four different nucleotides, only four amino acids would be encoded, although each nucleotide could have designated more than one amino acid. This limited and error-prone system then could progress into a duplex code. In this case, 16 different amino acids could be encoded (15, if you set one aside as a stop

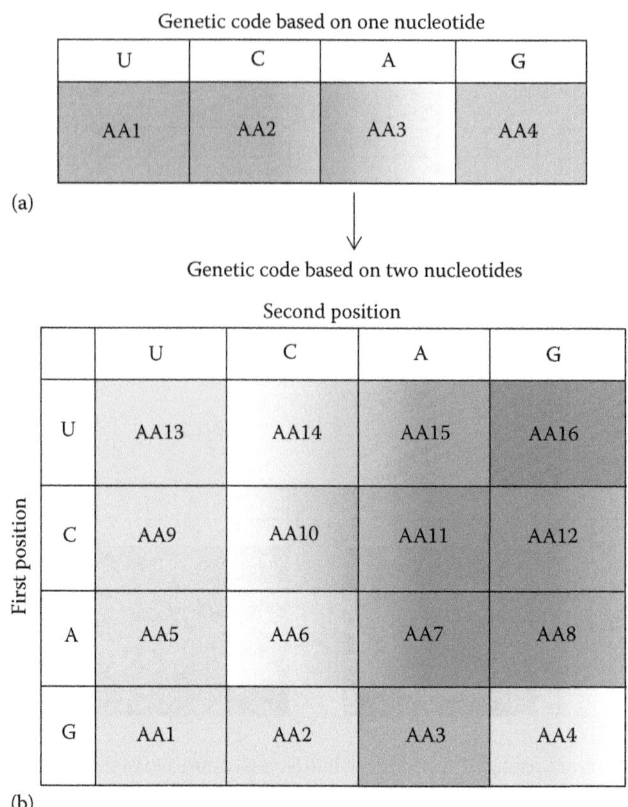

FIGURE 6.5 Evolution of the universal code indicating amino acid categories. The first step in this process (a) probably was simply a 1-to-1 relationship between each of the nucleotides in RNA and each amino acid at that time. Each nucleotide could represent a category of amino acid available at that time. Orange (U, C, or G) would have coded for hydrophobic amino acids. Green (A) would have coded for hydrophilic amino acids. The next step might have been a dinucleotide code (b). This would have allowed more diversity in the number and types of amino acids that could be incorporated into the proteins, and all current types (hydrophobic, orange; polar, purple; basic, aqua; and acidic, green) would have been possible.

codon), which is very close to the 20 present today. An indication that this might have occurred can be seen by viewing the genetic code (Figures 6.1 and 6.2) based on category of amino acid. However, evolution from a single nucleotide to a triplet code is also possible, thereby skipping a duplex code.

The code is organized according to the type of amino acid. Most of the amino acids on the left half of the chart are hydrophobic, while the hydrophilic amino acids are mainly on the right half. Also, only a minority of codons encode acidic or basic amino acids. The current genetic code could be used to encode for as many as 63 amino acids (plus one stop codon). Also, the genetic code probably still is evolving, which is indicated by redundancy in the code. However, each codon and each type of tRNA is not used to the same degree in each species or in each tissue within one individual. Currently, there are 20 amino acids encoded by 61 of the codons. The remaining three of the codons are stop codons that end translation. However, as mentioned earlier, this *universal* genetic code is not quite universal (Figure 6.2). Some mitochondria use a slightly different code. Specifically, UGA, which is a stop codon in most cells, signifies the addition of tryptophan in mitochondria. Also, AGA and AGG are stop codons in mitochondria, while they encode arginine in the universal code, and AUU and AUA are methionine codons, while in the universal code these signify isoleucine. Another variation on the universal genetic code has been demonstrated in the ciliated protozoan *Tetrahymena thermophila*, where there is only one stop codon, UGA. Also, the UGA codon is

<table>
<tr><th rowspan="2">First position</th><th colspan="4">Second position</th><th rowspan="2">Third position</th></tr>
<tr><th>U</th><th>C</th><th>A</th><th>G</th></tr>
<tr><td rowspan="4">U</td><td rowspan="2">Phe</td><td rowspan="4">Ser</td><td rowspan="2">Tyr</td><td rowspan="2">Cys</td><td>U</td></tr>
<tr><td>C</td></tr>
<tr><td rowspan="2">Leu</td><td rowspan="2">STOP</td><td>STOP</td><td>A</td></tr>
<tr><td>Trp</td><td>G</td></tr>
<tr><td rowspan="4">C</td><td rowspan="4">Leu</td><td rowspan="4">Pro</td><td rowspan="2">His</td><td rowspan="4">Arg</td><td>U</td></tr>
<tr><td>C</td></tr>
<tr><td rowspan="2">Gln</td><td>A</td></tr>
<tr><td>G</td></tr>
<tr><td rowspan="4">A</td><td rowspan="3">Ile</td><td rowspan="4">Thr</td><td rowspan="2">Asn</td><td rowspan="2">Ser</td><td>U</td></tr>
<tr><td>C</td></tr>
<tr><td rowspan="2">Lys</td><td rowspan="2">Arg</td><td>A</td></tr>
<tr><td>Met (START)</td><td>G</td></tr>
<tr><td rowspan="4">G</td><td rowspan="4">Val</td><td rowspan="4">Ala</td><td rowspan="2">Asp</td><td rowspan="4">Gly</td><td>U</td></tr>
<tr><td>C</td></tr>
<tr><td rowspan="2">Glu</td><td>A</td></tr>
<tr><td>G</td></tr>
</table>

FIGURE 6.6 Current genetic code showing amino acid categories. The structure of the universal genetic code is clear. All codons beginning with U and half of those beginning with C code for hydrophobic amino acids. GAN codons all code for acidic amino acids. The upper right two-thirds of the codons code primarily hydrophilic amino acids (polar and basic amino acids). Colors are as in Figure 6.5.

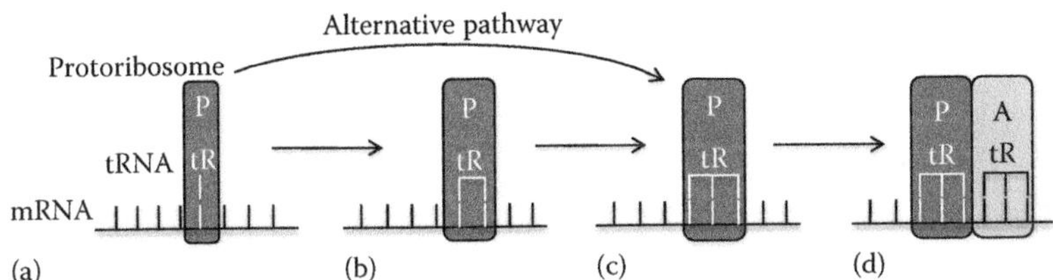

FIGURE 6.7 Possible evolution of the triplet code. (a) Protoribosome may have consisted of a peptidyl transferase (P) and a single base that would specify each of four amino acids (possibly more if there was some nonspecificity in pairing). Although U can pair with A, and G with C, U also can pair with G, and C with A. This would allow for the incorporation of four to eight different amino acids. This probably occurred when biological systems primarily included RNA as the main genetic and enzymatic molecules. This system assured that there was a molecular memory (i.e., inheritance) of reproduction of beneficial polypeptides. (b) A slight shift, and expansion of the pairing region on the protoribosome could change the codon to a dinucleotide system. This would increase the number of possible amino acids to 16 (more if there is some nonspecificity in pairing), which is close to the 20 that are seen today in biological polypeptides. This probably occurred at a time when proteins were becoming more important in biological systems and possibly when additional amino acids were becoming available in the environment or from rudimentary internal biochemical reactions. (c) A final shift would expand the pairing region to a trinucleotide codon, which is the present situation. This would allow translation of codons for all 20 amino acids, plus redundancy for most. Also, it would increase the number of different polypeptides that could be made and would allow for changes in translation under varying conditions, leading to altered cell morphology and physiology, and eventually multicellularity. (*Note*: It is possible that the two-nucleotide codon step was skipped [the alternative pathway], because the triplet code appears to have originated due to complementarity of the codons, with the third base of each pair being flexible to allow for more specificity of the first base in the codon.) (d) Duplication of the tRNA-binding site created a P-site and an A-site. This increased the efficiency, specificity, and velocity of translation by assuring accurate delivery and stability of the next tRNA set to have its attached amino acid added to the growing polypeptide chain. This protoribosome formed the core that would become the ribosomes in all of the cellular organisms present today.

occasionally used to signify the addition of selenocysteine (a variant of cysteine that contains selenium instead of sulfur) rather than a stop. This means that in this case all 64 codons may be used to translate mRNAs into proteins.

THE FIRST GENETIC CODE

The genetic code exhibits an organization that has evolved over the billions of years of its existence. The organization as well as the frequencies of certain amino acids in proteins points to its origin and evolution. Ancient proteins have been identified in contemporary organisms as those proteins that are common to all organisms, especially those that relate to central metabolic processes. Tabulation of the amino acid frequencies in these proteins indicates that the four most frequent amino acids are alanine, glycine, valine, and glutamic acid (Figure 6.8). Interestingly, these exclusively occur on the bottom row of the standard genetic code table (Figure 6.1), each being encoded by triplets where the middle nucleotide differs for each one (A for alanine, G for glycine, U for valine, and C for aspartic acid, or glutamic acid), and the first nucleotide always is a G. Therefore, the first genetic code might have been one in which each nucleotide designated a single amino acid. Experiments using prebiotic conditions on the Earth have produced many biologically relevant amino acids, including glycine, alanine, valine, aspartic acid, and glutamic acid. Therefore, it is likely that these amino acids were available during the formative years of RNA-based life and thus provided the raw materials for the evolution of the production of peptides and proteins. Eventually, a single-nucleotide code would have become more elaborate to include three nucleotides per codon. Again, from the frequencies of amino acids, this expanded code appears to have begun with G in the first position and another nucleotide in the second position (Figure 6.8). The third position was more variable and could change without altering the amino acid specification for many of the codons. This allowed more flexibility in the code, but also may have led to greater fidelity in the process because of the increase in hydrogen bonding between the codon and anticodon. The expansion of the genetic code appears next to have included A and C in the first position, thus adding the

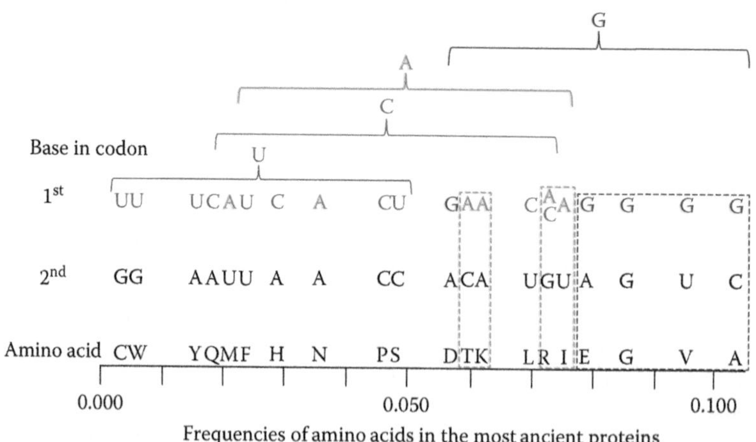

FIGURE 6.8 Amino acids arranged horizontally according to frequencies within ancient proteins. The four most common amino acids within ancient proteins are alanine (A), valine (V), glycine (G), and glutamic acid (E). The four least common are cysteine (C), tryptophan (W), tyrosine (Y), and glutamine (Q). Based on the first two letters of the triplet codon for each (and from other data), the initial single base codons probably were those for the four most frequent amino acids: C for alanine, G for glycine, U for valine, and A for glutamic acid. During the evolution of the triplet codon, the first codons appear to have had G as the first base. Later, A and C appeared in the first position, and U was the last to be added to the first position of the triplet code.

potential for at least 10 more amino acids, bringing the total to 14. The last nucleotide to appear in the first position was U, which ushered in the evolution of the modern genetic code that is still present today. Based on codon/anticodon sequences, amino acid frequencies in ancient proteins, and aaRS class, an evolutionary reconstruction is possible. As stated earlier, the first four amino acids in the code were probably alanine, glycine, valine, and aspartic acid (and/or glutamic acid). The next to be added were likely isoleucine, threonine, serine, proline, leucine, methionine, arginine, glutamine, and histidine. Later arrivals were cysteine, tyrosine, lysine, asparagine, phenylalanine, and tryptophan (Figure 6.9).

The genetic code appears to have had two parts to its evolution, the first being the evolution of the aaRSs, and the second being codon evolution based on triplet codes. The symmetry of the genetic code by aaRS class (Figure 6.6) and approach of the aaRSs to the tRNAs, as well as the complementarity of the codons and anticodons that mirrors this symmetry, indicate the coevolution of the two processes. The symmetry also hints to a time when genes consisted of double-stranded RNA, with one of the strands encoding a peptide in one direction and the other strand encoding a different protein on the other strand. Buttressing this is the step-by-step changes in the codons and anticodons (Figure 6.9) that coincide with the frequencies and appearances of each of the amino acids that are found in ancient proteins common to all organisms (Figure 6.8).

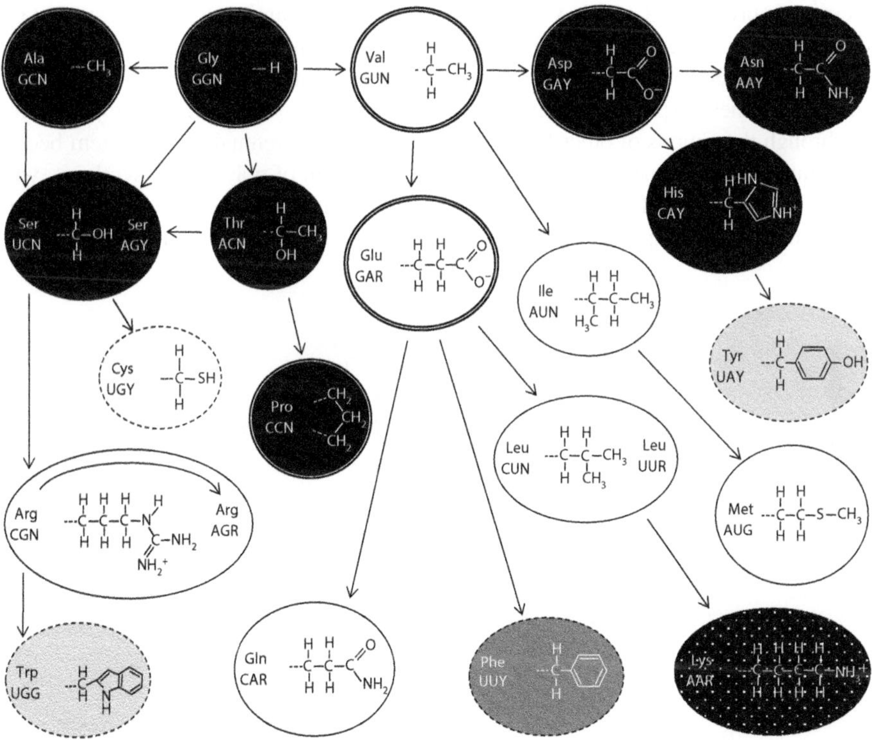

FIGURE 6.9 Model of the evolution of the genetic code. The model is based on single base changes in the triplet codon, aaRS class, approach of the aaRS, frequencies of amino acids in ancient proteins, and complementarity groups of the codons/anticodons (as shown in Figure 6.4). Amino acids that can be produced from prebiotic conditions on the Earth are surrounded by double lines. Four of these are found in the highest frequencies in ancient proteins (i.e., aspartic acid, alanine, glycine, and valine; in addition to glutamic acid and proline) and were likely among the first amino acids utilized in biological systems starting approximately 3.5–3.8 billion years ago. Amino acids with single solid lines appeared somewhat later during the evolution of the genetic code, while those with dashed lines were the last to appear in the genetic code. Shading is the same as in Figure 6.4. Differences in shading are explained in Figure 6.4 legend.

LIFE BEFORE TRANSLATION

Although ribosomes became the major process for synthesizing peptides and proteins in cells, the complex process of translation of mRNA using ribosomes and tRNAs might have had its origins in other processes that predated these molecules. Models of these primordial processes were deduced through studies of nonribosomal peptide synthesis, which have been described in many types of unicellular and multicellular organisms. Synthesis of these peptides requires specific synthetases that catalyze reactions that produce chains of amino acids to create each specific peptide. The synthetases appear to have originated at about 3.8 billion years ago, when other peptide systems also may have been evolving. They are built from modules consisting of enzymes that add one amino acid to the growing chain to produce a mature peptide that functions in the cell, or outside of the cell, which often are toxins. Therefore, in this case, peptides are formed by linking together modules that add amino acids (many of which differ from standard amino acids used in ribosomal translation of mRNA) in specific sequences. No mRNA is involved in the process.

Another indication of the primordial processes that led to translation using ribosomes came from analyses of dipeptide frequencies in ancient proteins. Dipeptides containing primarily alanine, valine, isoleucine, and glycine (and to a lesser extent lysine, arginine, glutamic acid, and proline) have been found in higher numbers than expected in ancient proteins. This hinted to a process of first producing dipeptides using a synthetase, followed by joining of the dipeptides in chains using a ligase (Figure 6.10). This process probably utilized an RNA molecule that carried the amino acid to the synthetase, which may have been the predecessor of the tRNAs. Initially, the synthetases and ligases may have been separated, but ultimately they may have formed a complex that had both the synthetase and ligase functions, allowing them to produce peptides (Figure 6.11). These peptides could be joined into longer molecules through the actions of other ligases to form proteins. Eventually, this system became more elaborate, forming both a nonribosomal and a ribosomal system of polypeptide synthesis, which both first appeared between 3.5 and 3.8 billion years ago. For nonribosomal peptide synthesis, several synthetases evolved, with each capable of adding a specific amino acid to the growing chain. For ribosomal

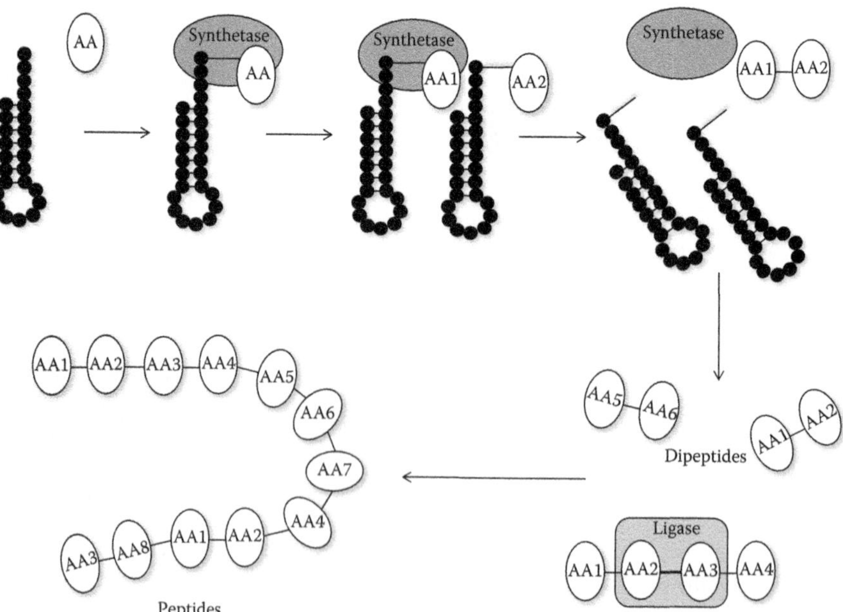

FIGURE 6.10 Model for the evolution of peptide-producing reactions in early cells. Initially, a synthase that formed peptide bonds produced dipeptides. The amino acids were delivered to the synthetases by small RNAs that predated tRNAs. The dipeptides were then ligated together to form small peptides (i.e., polypeptides).

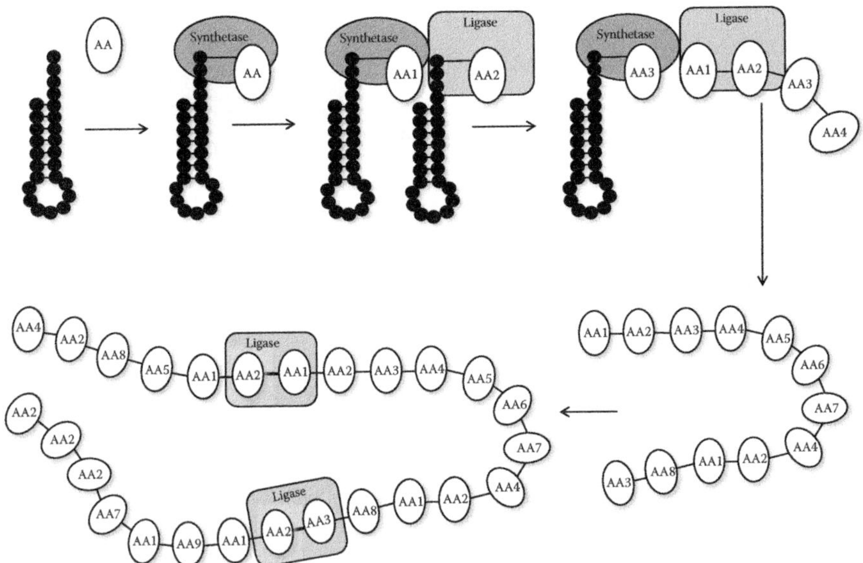

FIGURE 6.11 Model for the evolution of protein-producing reactions in early cells. As a refinement of the process shown in Figure 6.10, the synthase and ligase formed a complex, such that longer peptides could be produced. Additional ligases could join the peptides together to form longer more complex peptides (i.e., proteins). This complex diverged into the two main processes now found in cells. One became translation (which relies on ribosomes, tRNAs, and mRNAs), and the other became nonribosomal peptide synthesis that produces a wide variety of polypeptides in cells. Peptide synthesizing systems analogous to this process have been found in some modern cells.

protein synthesis, the aaRSs provided one part to the process, while rRNAs provided the peptidyl transferase function, and eventually mRNAs evolved to carry the code for the sequence of amino acids to be added to polypeptide chain. The entire process from simple systems that produced dipeptides and short polypeptides to a fully functioning ribosome appears to have taken from 300 to 600 million years.

Part of the complexity of the genetic code is that it is a combination of two separate codes that became one. The first code is determined by aaRSs that attach specific amino acids to the tRNAs. The second code is on the other end of the molecule, the anticodon that interacts with the codon on the mRNA. The two appeared at different times. The acceptor arm (where amino acids attach) of tRNA is the most ancient (Figure 6.12). It expanded to include the TΨC domain, and eventually, the entire piece duplicated as a reverse complement to form the D domain and anticodon loop, thus completing the general form of all tRNAs (Figure 6.13). Part of the D domain and TΨC domain remain complementary and pair in the tRNAs today. The entire system of tRNAs, mRNAs, and rRNAs coevolved into the complex translation machinery that exists in all cells today (Figure 6.14). Evidence from the analysis of the frequencies of amino acids in ancient proteins indicates that the initial code was determined by the acceptor arm of the tRNA attached to the amino acid. Peptidyl transferase on the protoribosome bound the amino acid on the tRNA in the original A-site to the growing polypeptide chain on the tRNA at the P-site. Other than the peptide and amino acid, all other molecules were RNAs. It is thought that the aaRSs also were RNAs initially, but were replaced by more efficient and stable proteins very early in the evolution of translation. The second code originated from a duplication of the acceptor domain of the tRNA. An indication of this is that the anticodon of each tRNA is complementary to three nucleotides near the end of the acceptor arm. Ultimately, the two codes coevolved to produce the complex and intricate genetic code that has led to the evolution of all of the organisms extant and extinct on the Earth, including single-celled organisms and complex multicellular organisms in which codon usage often varies from one cell type to another.

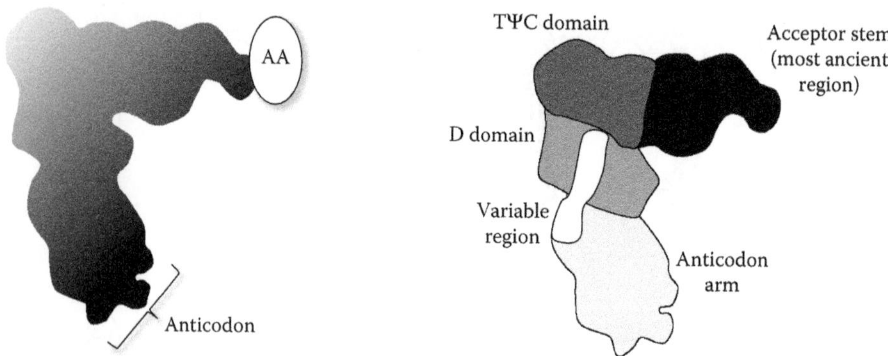

FIGURE 6.12 Model of parts of a tRNA (based on tRNA sequenced and amino acid antiquity). The acceptor arm was the first part of tRNA to appear approximately 3.6–3.8 billion years ago. This was followed by the addition of the TΨC domain. The next oldest portion is the D domain, followed by the anticodon domain, and finally the variable domain. Shading indicates age, with darker being older.

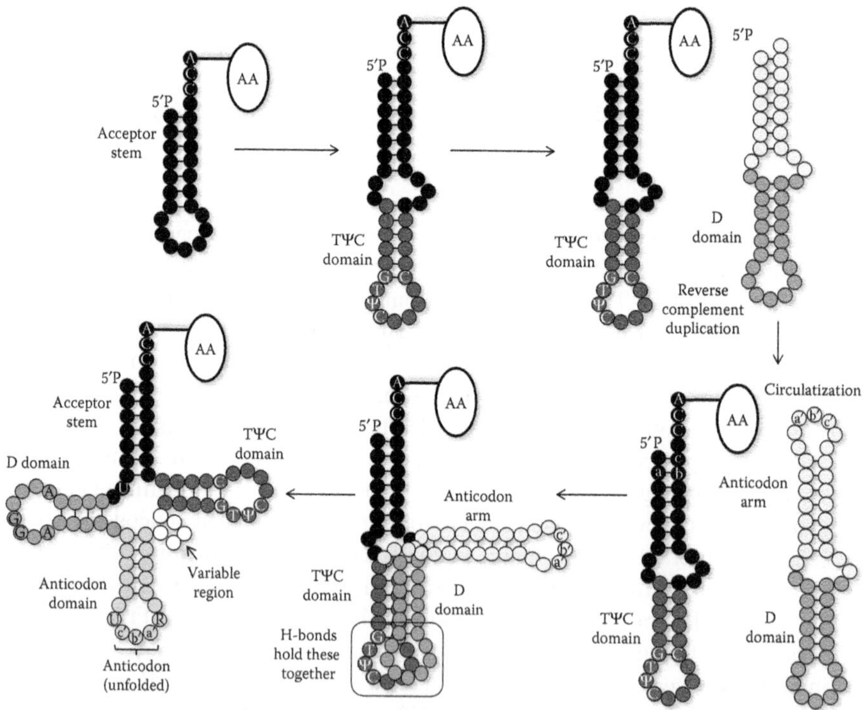

FIGURE 6.13 Details of the evolution of tRNA. The acceptor arm that holds the amino acid (AA) is the oldest portion of the tRNA. The extension that became the TΨC domain was next to appear as an extension of the acceptor arm. The entire piece (acceptor arm plus the TΨC domain) was duplicated as a reverse complement of itself that became fused into one longer RNA molecule. Hydrogen bonds formed between the TΨC and D domains, which held the acceptor arm and anticodon domains (respectively) in a conformation that was necessary for translation. The last region to evolve was the variable region that is specific for each tRNA species. Shading indicates age, with darker being older.

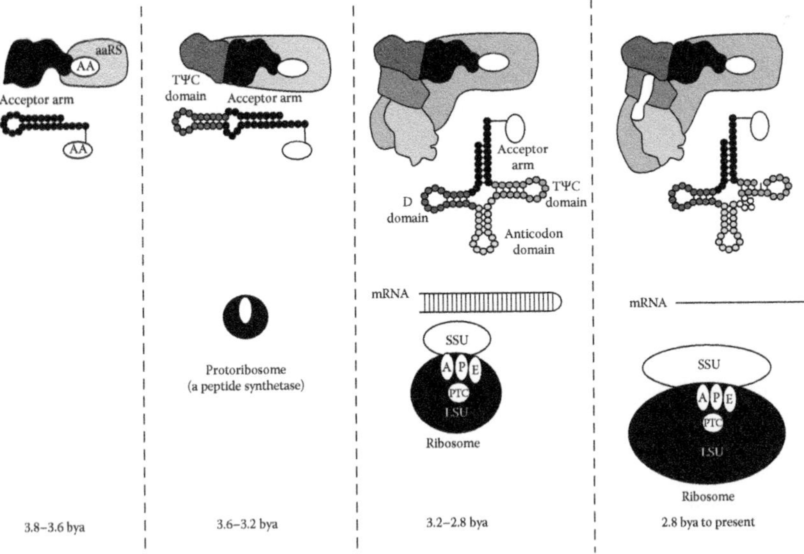

FIGURE 6.14 Model of the evolution of the integral parts of translation. Approximately 3.8 billion years ago the precursor to acceptor arm of the tRNAs and aaRS proteins formed the basis of peptide synthesis by cooperating to attach amino acids to the acceptor arms. Space filling (top) and ball (bottom) models are presented for each tRNA and tRNA precursor. By 3.6 billion years ago, the acceptor arm had grown in length to include the TΨC domain, and the aaRSs had also increased in size. Synthetases served to join the amino acids together. They contained at least one site that held the tRNA so that the attached amino acid could be correctly positioned for peptide bond formation with another amino acid. One of these synthetases formed the core of the modern ribosome. Sometime between 3.6 and 3.2 billion years ago, modern tRNAs evolved, likely due to a duplication of the acceptor arm and the TΨC domain, to form the anticodon loop and the D domain. At the same time, one of the synthetases (the peptidyl transferase) had evolved into a rudimentary ribosome. The large subunit (LSU) contained the A, P, and E sites for tRNAs. Also, by this time, short double stranded RNAs served as messages (the initial mRNA), which increased the efficiency of peptide and protein synthesis. These genes were double stranded because they appear to have encoded polypeptide messages on both strands. The complementarity is evidenced by the complementarity observed in the modern genetic code (see Figure 6.4). The small subunit (SSU) assured proper delivery of the mRNAs to the LSU. Since that time, ribosomes have become larger, which has increased the complexity and fidelity of protein synthesis, as well as assuring accurate delivery of the proteins to the various parts of the cells.

KEY POINTS

1. The genetic code originated more than 3.5 billion years ago from a simpler system.
2. While AUG is the most common start codon, and UAA, UAG, and UGA are the most common stop codons, other start and stop codons are used in a variety of organisms and organelles.
3. Aminoacyl-tRNA synthases (aaRSs) set the genetic code by attaching specific amino acids to specific tRNAs.
4. Rearrangement of the genetic code according to class I and class II aaRSs indicates a possible origin of the genetic code.
5. The triplet genetic code probably evolved from a simpler system, which may have been based on one or two nucleotide codes.
6. The current genetic code is structured in a way that gives clues as to its evolutionary history.

7. The first genetic code might have begun with single or doublet nucleotide codons that encoded each of the amino acids.
8. Studies of the most ancient proteins have indicated that alanine, glycine, valine, aspartic acid, and glutamic acid were among the earliest amino acids used in proteins.
9. Both ribosomal protein synthesis and nonribosomal peptide synthesis may have evolved from a simpler synthetic pathway for joining amino acids.
10. The most ancient parts of tRNAs are the acceptor arms, where amino acids are attached. A later sequence duplication added the anticodon arm, which initiated the evolution of translation.

ADDITIONAL READINGS

Alberts, B., D. Bray, K. Hopkin, A. Johnson, J. Lewis, M. Raff, K. Roberts, and P. Walter. 2013. *Essential Cell Biology*, 4th ed. New York: Garland Publishing.

Alberts, B., D. Bray, A. Johnson, J. Lewis, M. Raff, K. Roberts, and J. D. Watson. 1994. *Molecular Biology of the Cell*. New York: Garland Publishing.

Bada, J. L. 2013. New insights into prebiotic chemistry from Stanley Miller' discharge experiments. *Chem. Soc. Rev.* 42:2186–2196.

Bokov, K. and S. V. Steinberg. 2009. A hierarchical model for the evolution of 23S ribosomal RNA. *Nature* 457:977–980.

Caetano-Anollés, G., M. Wang, and D. Caetano-Anollés. 2013. Structural phylogenomics retrodicts the origin of the genetic code and uncovers the evolutionary impact of protein flexibility. *PLoS ONE* 8(8): e72225.

Darnell, J., H. Lodish, and D. Baltimore. 2007. *Molecular Cell Biology*, 6th ed. New York: W.H. Freeman and Company.

Karp, G. 2013. *Cell and Molecular Biology*, 7th ed. New York: John Wiley & Sons.

Miller, S. L. 1953. Production of amino acids under possible primitive earth conditions. *Science* 117:528–529.

Rodin, S. N. and A. S. Rodin. 2008. On the origin of the genetic code: Signatures of its primordial complementarity in tRNA and aminoacyl-tRNA synthetases. *Heredity* 100:341–355.

Smith, J. M. and E. Szathmary. 1995. *The Major Transitions in Evolution*. New York: Oxford University Press.

7 DNA Replication

INTRODUCTION

In this chapter and in Chapter 8, DNA replication, segregation of DNA, and variations of chromosome partitioning are addressed. The cell cycle is a good place to start the discussion. All cellular organisms have cell cycles. In Bacteria and Archaea (Figures 7.1 and 7.2), the cycle includes a G_1 (gap) phase where each cell has one genome of DNA, followed by an S (synthesis) phase where the chromosomal DNA is replicated to produce duplicate copies of the genome, proceeding into a G_2 phase, and finally through binary fission to produce two cells, again each having one full set of genes. Other than dying, a cell can exit from the cell cycle in two ways. The first is to form spores, which occurs in many species. This normally occurs with cells in G_1. The cell encases some cytoplasm and its genomic DNA inside a hard shell that will protect the contents for extended periods, sometimes for many years. Upon germination of the spore, the cell again enters a normal cell cycle. The second is through differentiation. For example, members of the genus *Caulobacter* form two types of cells: stalk cells and swarmer cells. Stalk cells are attached to surfaces by a holdfast, and therefore, these cells cannot move. Swarmer cells are motile, and eventually settle onto a surface, grow a holdfast and stay there, thus differentiating into stalk cells. But, they divide to form two cells. One remains as the stalk cell, and at the other end is a swarmer cell. The stalk cell eventually dies, and the swarmer cell swims away to find a place to attach. Therefore, there is a cell cycle, but one of the cells always differentiates, divides, and eventually dies.

Among Eukarya, the cell cycle is more elaborate, because it includes mitosis (Figures 7.3 and 7.4). Again, there is a G_1 phase where cells are usually diploid, although there are many types of cells (e.g., in fungi) that are normally haploid in G_1 (i.e., most fungi undergo haploid cell cycles for most of their life cycles). The cells proceed through the S phase, where the DNA is completely replicated, and on to G_2, where cells are tetraploid if they started as diploid, or diploid if they started as haploid. Finally, they proceed through mitosis (Figures 7.4 and 7.5), which includes a prophase, where the chromosomes condense; prometaphase, where the microtubules polymerize and the nuclear membrane disintegrates; metaphase, when the chromosomes are lined up in the middle of the cells; anaphase, where the chromatids separate to become chromosomes; telophase, where the chromosomes decondense and the nucleus is reconstructed; and finally cytokinesis, where the cytosol divides to produce two daughter cells.

FIDELITY OF REPLICATION

Before each cell divides, it must replicate its DNA. Fidelity of replication is of paramount importance, especially for organisms with larger genomes. If replication is an error-prone process, then the genomes must be of limited size. For example, the haploid human genome consists of more than 3 billion base pairs, so that each diploid cell in your body has about 6 billion base pairs of DNA in its nucleus. Since DNA is double stranded, this equates to 12 billion nucleotides per diploid nucleus. If DNA replication makes one mistake every 100,000 nucleotides, then there could be as many as 12,000 mutations generated for each replication and cell division cycle. Even if we limit this to the regions of the genome that encode genes (about 1.5% of the genome), there still would be about 180 mutations created among the 23,000 genes in the human genome. This would mean that on average, almost 1% of the genes would have new mutations every time the cell divided. Since you have billions of cells in your body, and they all originated from cell divisions, you would have an extremely high number of mutations in your body, and many would be fatal mutations, at least for the cells that contained the mutations. Clearly, this error rate is too high for organisms with large

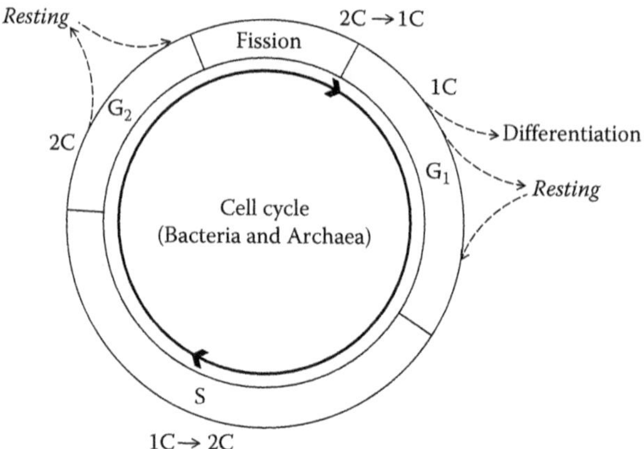

FIGURE 7.1 General cell cycle for Bacteria and Archaea. Beginning in G_1 (gap 1), the cells contain a single copy of their genome (1C). Then, they proceed through the S phase where the chromosome is replicated by DNA polymerase and accessory proteins. The cells then proceed through G_2 where they have a 2C level of DNA. In both G_1 and G_2, some types of cells can proceed into *resting* stages, normally by forming spores or differentiating into other forms of cells. During G_2, in preparation for fission, the chromosomes are pulled to the poles of the cells by proteins, some of which are embedded in the membrane. Following fission, the cells are in G_1 phase.

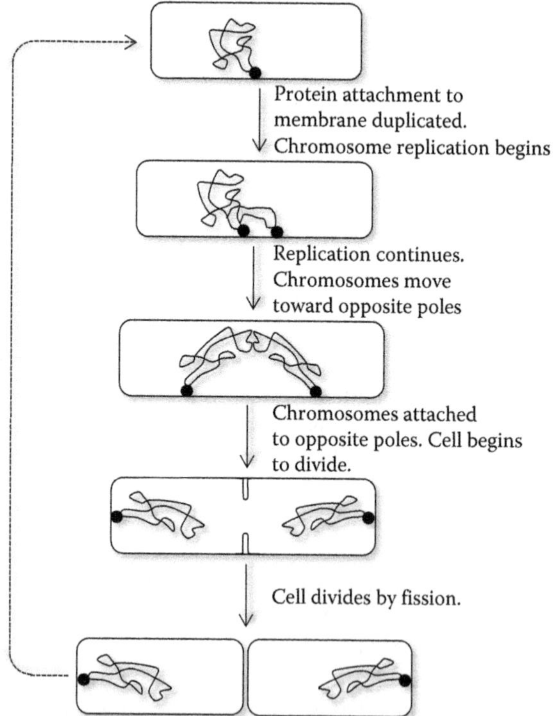

FIGURE 7.2 Diagram of the cell cycle for single-celled bacteria and archaea. The chromosomes remain attached to the membrane by proteins embedded in the cell membrane. When replication begins, more of the attachment proteins are synthesized and the replicating chromosome becomes attached in two places, one for each of what will become two complete copies of the genome. If more than one chromosome is present, each of the replicating chromosomes is attached similarly to the protein and membrane. The proteins migrate to the poles pulled by other proteins that exist at the poles of the cell. They pull the attached chromosome with them, thus separating the chromosomes assuring equal distribution to each of the two future daughter cells. Finally, the cell divides by invagination of the membrane and cell wall in the center of the cell to form the two separate daughter cells.

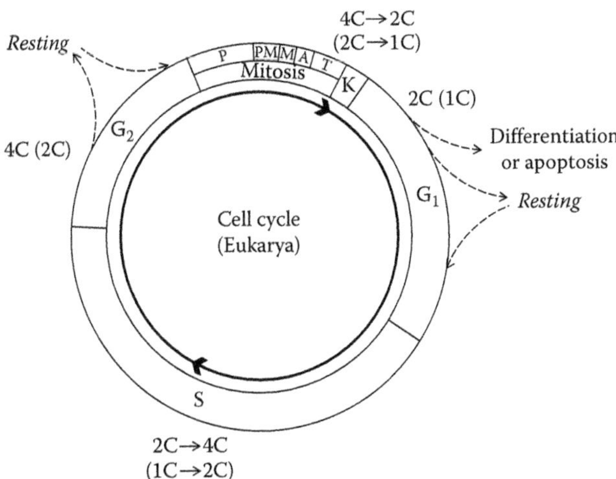

FIGURE 7.3 General cell cycle for Eukarya. The cell cycle is more complex than for archaea and bacteria, since the chromosomes are larger (in most cases). Although G_1, S, and G_2 are similar, this cell cycle includes mitosis, which is a more elaborate process whereby chromosomes are segregated to each of the resulting daughter cells. Additionally, in many taxa diploid cells (2C) predominate and therefore increase to a 4C level during S and G_2. In other taxa (C levels in parentheses), haploid cells (1C) predominate and thus increase to a 2C level during S and G_2. Also, cells that differentiate seldom can reenter the cell cycle, including cells that normally die (undergo apoptosis) as a part of their normal developmental pathway in multicellular organisms. Mitosis consists of the following phases: prophase (P), prometaphase (PM), metaphase (M), anaphase (A), and telophase (T), followed by cytokinesis (K).

genomes, although this error rate (10^{-5}, or one change every 10^5 nucleotides) is a typical error rate for DNA polymerases, whose error rates range from a bit lower than 10^{-5} to approximately 10^{-6}. The origin of DNA polymerase was probably from an RNA-dependent RNA polymerase. One evolutionary remnant of this origin is that DNA polymerase uses short RNAs as the molecules that prime the process. That is, short RNA molecules are used to start DNA polymerization, and then, the RNA is later removed from newly synthesized DNA. Both DNA polymerases and RNA polymerases use polymers of nucleic acids as templates, require nucleoside triphosphates as substrates, and add nucleotides in the 5′–3′ direction. Both have similar error rates as well. Although the error rates seem high for organisms with large genomes, the mutations that have been generated over the past 3.5 billion years have led to the biological evolution that has occurred during that time. The larger genomes that exist today are only possible because DNA repair enzymes and systems correct many of the mutations that are generated by DNA polymerase.

DNA replication on a short linear fragment of DNA is straightforward, although somewhat complex, because it involves more than just DNA polymerase. First, remember that DNA is wound in a double helix that is inaccessible to DNA polymerase, a large enzyme having multiple parts. Therefore, the two strands of DNA need to be separated to allow access to the strands by one or more DNA polymerase molecules and attendant proteins (Figure 7.6). Some of the attendant proteins include topoisomerases and helicases, proteins that unwind the DNA, as well as single-strand binding proteins that stabilize single-stranded portions of the DNA as it is unwound. Next, the short RNA primers must find complementary regions to form DNA–RNA hybrid portions on the DNA strands. They accomplish this with the help of proteins that form primeosomes. Finally, the DNA polymerase can associate with the primeosome and the RNA primer. The polymerase starts moving down the DNA–RNA hybrid region in the 5′–3′ direction until it encounters the end of the RNA. It then begins by allowing the bases on the nucleoside triphosphates to hydrogen bond to the base on the DNA, which will be used as the template strand. When hydrogen bonding in a standard Watson–Crick orientation occurs, the DNA polymerase catalyzes the reaction between the 5′ alpha

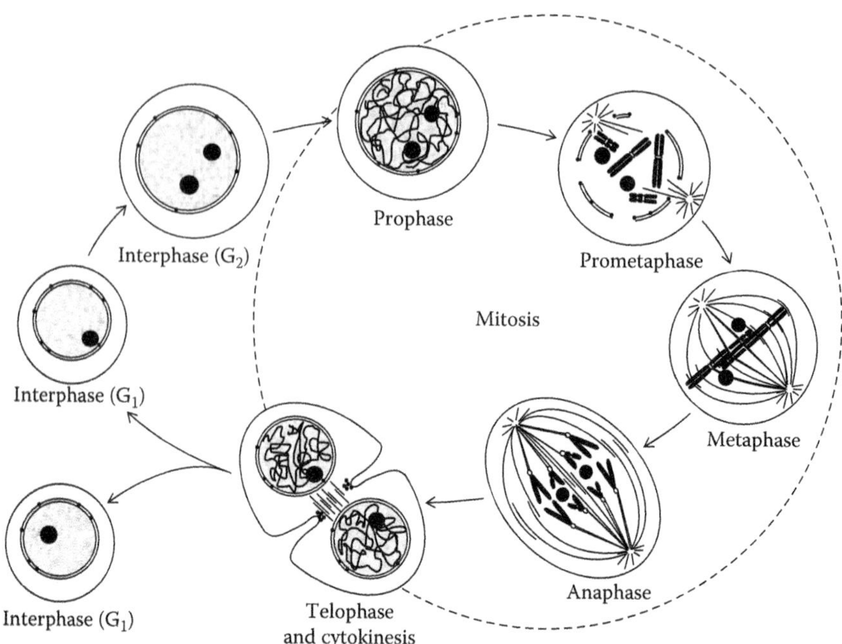

FIGURE 7.4 Cell cycle for a typical animal cell, with emphasis on mitosis. Cells in G_2 usually are larger than cells in G_1 and have larger nuclei, due to the increase in DNA content. During prophase, the chromosomes begin to condense within the nucleus. Next, microtubules begin polymerizing on opposite sides of the nucleus, and the nuclear membrane begins to disintegrate during prometaphase. Metaphase is reached when the microtubules attaching each chromosome to both poles push the chromosomes to the center of the cell, called the *metaphase plate*. Anaphase begins with separation of the chromatids and movement of the chromosomes to opposite poles, caused by the actions of microtubule depolymerization and movement of kinesin on the microtubules. Additionally, the polar microtubules are moved by dynein, which forces the poles of the cell apart, thus stretching the cell. Finally, during telophase the nuclear membrane is reformed around the chromosomes, as the chromosomes decondense. The cell is pinched in the center by a ring of actin, thus forming the two daughter cells.

phosphate on the incoming nucleoside triphosphate with the 3′ hydroxyl on the last nucleotide of the newly synthesized strand (or initially with the first nucleotide of the new strand). One of the polymerases thus replicates one of the two strands, while another proceeds in the opposite direction along the opposite strand of DNA. The polymerases continue to add nucleotides one at a time until they reach the end of the DNA segment.

While the two polymerases attach to both DNA single strands and move in opposite directions (i.e., both are moving in the 5′–3′ directions on their respective strands), there is a problem with this arrangement. How is the DNA upstream (in the 5′ direction) replicated? The polymerases moving toward the forked region where the double-stranded region is being separated into two single-stranded regions can remain on the template strand continuously, as long as they continue to synthesize in the 5′–3′ direction. These polymerases are moving along what is termed the *leading strand* in DNA replication. The other strand is known as the *lagging strand*. This is because it must be polymerized in short segments due to the fact that polymerase can only polymerize in the 5′–3′ direction (and therefore is moving along the template strand in the 3′–5′ direction), so there is a delay because it cannot begin replication until a sufficient length of DNA is available in that region. Thus, short segments are formed in this region, known as *Okazaki fragments* (named after their discoverer). Later, they are joined together by another enzyme, called *DNA ligase*.

The second problem with linear DNAs is that polymerase cannot physically complete polymerization of the lagging strand at the ends. There needs to be adequate room for the primer,

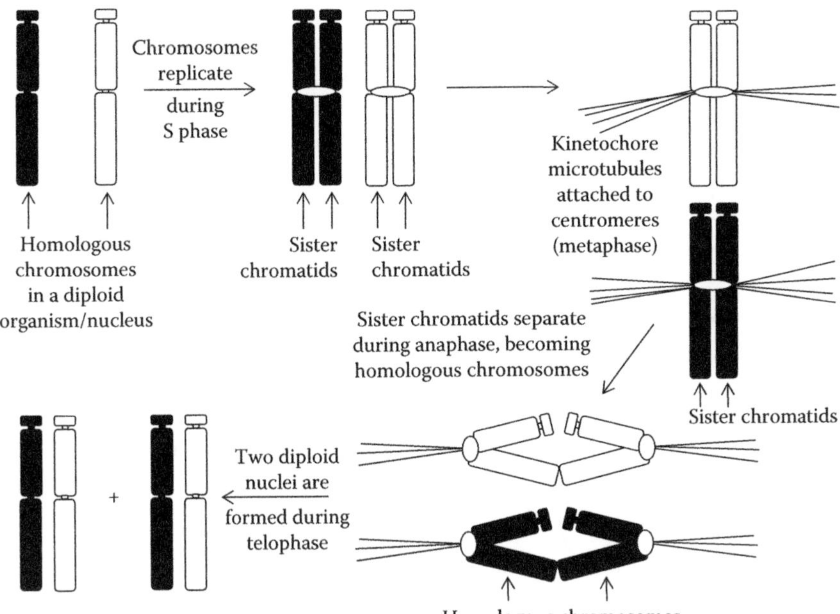

FIGURE 7.5 Chromosome movements during mitosis. Homologous chromosomes in diploid cells replicate during the S phase but normally do not separate, and therefore, each consists of two sister chromatids. During prometaphase and metaphase, the centromeres of each chromosome attach to microtubules from each pole. During anaphase, the sister chromatids from each homologue are pulled apart, one to each pole. Once they are separated, they become homologous chromosomes. Each diploid cell receives a pair of homologous chromosomes.

primeosome, and polymerase, but these regions are too short. However, if these regions are not replicated, then the piece of DNA (e.g., the ends of chromosomes) gets shorter with each replication cycle. In order to overcome this problem, organisms have developed at least two different strategies. First, in most bacteria and in some organelles, the chromosomes are circular. Therefore, there are no chromosome ends, and the polymerases can circulate completely around the chromosome during replication (Figure 7.7). The remaining nicks are then sealed by DNA ligase, as well as by topoisomerases. Most other organisms have linear chromosomes. The ends of their chromosomes have telomeres that are added after replication has been completed. They are added by an enzyme called *telomerase*, which is related to reverse transcriptases (and other RNA-dependent RNA polymerases). However, telomerases are DNA polymerases that rely on RNA primers (Figure 7.8). The RNA primers are from 5 to 23 nucleotides in length and vary in sequence from one species to another. Telomerases add multiples of these sequences to the ends of the chromosomes, thus protecting them from the shortening that ultimately would occur from the replication process. DNA completes the process by replicating the opposite strand.

VARIATIONS OF REPLICATION

Replication can be accomplished in several ways. One is by rolling circle replication (Figure 7.9). This is common for some plasmids and bacteriophage (e.g., lambda). The replication cycle begins with a double-stranded circular DNA molecule. One of the strands is cut, while the other strand is continuous. The replication initiation complex (including DNA polymerase) attaches to the nicked site and begins synthesizing a new strand in the standard way. New double-stranded DNA is continuously formed as the polymerase on the leading strand proceeds around the circle. The 3′ end of the single-stranded portion is displaced, and other polymerase molecules attach to form new DNA

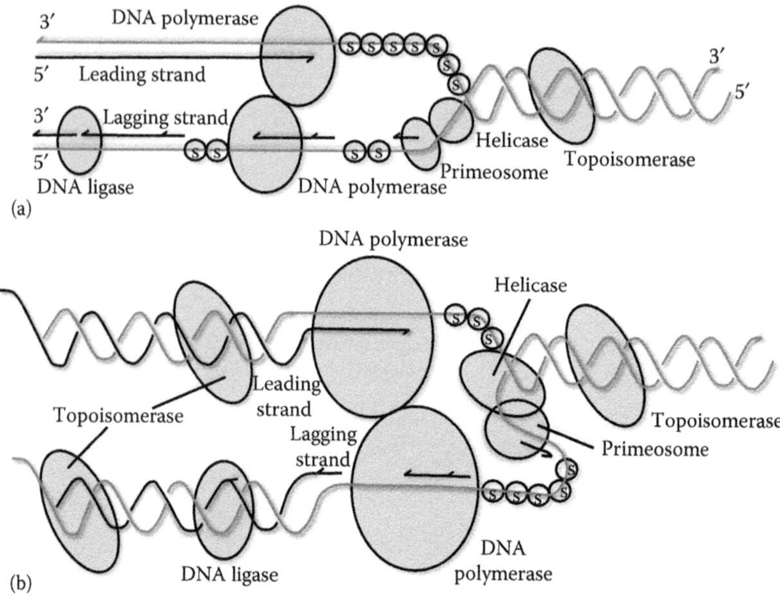

FIGURE 7.6 Summary of DNA replication. (a) The first step in replication is the formation of a single-stranded region (gray lines indicate template strands), which usually requires attachment of single-strand-binding proteins (S) to stabilize the single-stranded DNA. Topoisomerases and helicase with associated primeosomes then act in concert to move along the DNA relaxing and denaturing the DNA as they move along the double-stranded molecule. The primeosome binds RNA primers (thin black lines) that are used to initiate DNA replication by DNA polymerase. Since DNA polymerase only adds nucleotides to the 3′ end of DNA, one strand (the leading strand) is synthesized in long continuous strands (thick black lines). In order to complete the other strand (the lagging strand), it must be repeatedly primed by primeosomes and RNA primers. The DNA polymerase synthesizes short segments of the lagging strand (short thick black lines), again adding to the 3′ end of the new strand. It displaces the RNA primers of the previously synthesized fragment, and the final nucleotide is covalently joined to the next nucleotide by DNA ligase. (b) During DNA replication, the DNA polymerases on the leading and lagging strands are physically associated with one another. This forces the lagging strand to loop somewhat until the primeosome finds a priming site, which then activates the lagging strand DNA polymerase to begin synthesizing new DNA on that strand.

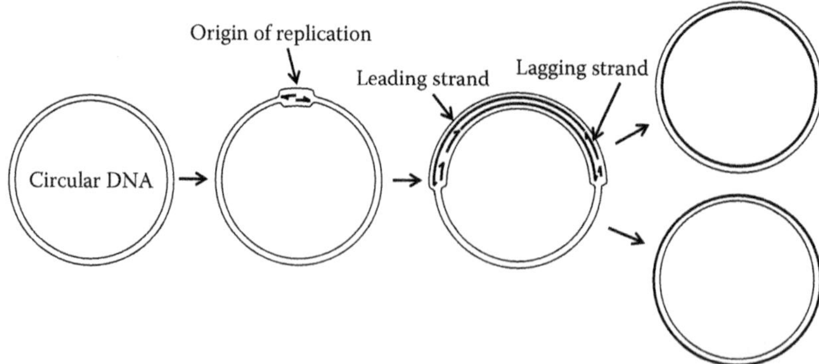

FIGURE 7.7 Replication of a circular DNA molecule. The first step in the process is the denaturation of the DNA at an origin of replication. Because of its appearance in electron microscope images, it is called a *D-loop*. Replication proceeds in both directions on the leading strand, and then, the lagging strand is synthesized as the primeosomes, RNA primers and DNA polymerase have enough space on the DNA to begin polymerization. Polymerization proceeds around the circle until the termination region is reached. Separation of the two newly synthesized molecules is then completed by topoisomerases, ligase, and other enzymes. Thin black lines indicate original strands. Thick black lines indicate newly synthesized DNA.

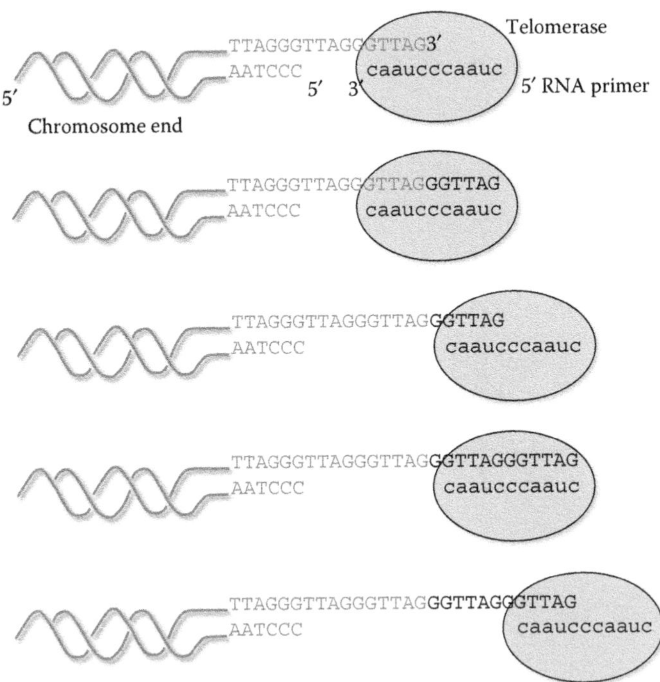

FIGURE 7.8 Addition of telomeres by telomerase. Because DNA polymerase cannot complete the ends of linear chromosomes, telomerase is needed to add DNA to the ends of these chromosomes. Telomerases use RNA primers for this process. Primers from a wide range of organisms are very similar in sequence, indicating a common origin for all of the telomerases. The first step in the reaction is that the telomerase with the primer attaches to the end of the chromosome where it recognizes the telomere sequence through complementary hydrogen bonding. Telomerase then adds bases to the end of the strand (in the 5′–3′ direction) using the RNA as the template. Once that has been completed, the telomerase (with primer) moves along the DNA until the bases all match again, and then, the cycle is repeated. After long stretches of the repetitive telomere sequence are added, DNA polymerase can fill in the other strand (and ligase seals the gap), except for the distal end of the chromosome. Here, the telomerase folds the single strand over to form a loop on the end, which is then covalently sealed. Gray lettering and lines indicate the end of the original chromosome after replication by DNA polymerase. Uppercase black letters indicate nucleotides added by telomerase. Lowercase black letters indicate the RNA telomerase primer.

on this template. This forms long linear concatemers of the genome. Some of the bacteriophages that have rolling circle replication can begin new replication on the newly replicated molecules, such that the replicating molecules become branched structures containing many copies of the genome.

Another variation of replication occurs in several viruses (including Gemini and T4 viruses) and appears to be used to replicate some chromosomes, as well, including those in plant mitochondria and chloroplasts. This is known as *recombination-dependent replication* (RDR, Figures 7.10 and 7.11). It begins with pairing of regions on two parts of the chromosome(s) and double-stranded breaks, followed by recombination. Essentially, this begins as a recombination event with strand invasion, often beginning with a double-strand break. However, it involves extensive DNA replication. The end results are highly branched DNAs of varying lengths, as well as linear molecules, and a small number of subgenomic circular molecules. The branched molecules contain many copies of the genome, while the linear molecules are of variable length, and include those that have only small portions of the genome up to pieces that are composed of more than one genome.

As stated above, RDR appears to be prevalent in the organelles of plants. While the genome in plant mitochondria and chloroplasts appears to be circular based on mapping with restriction

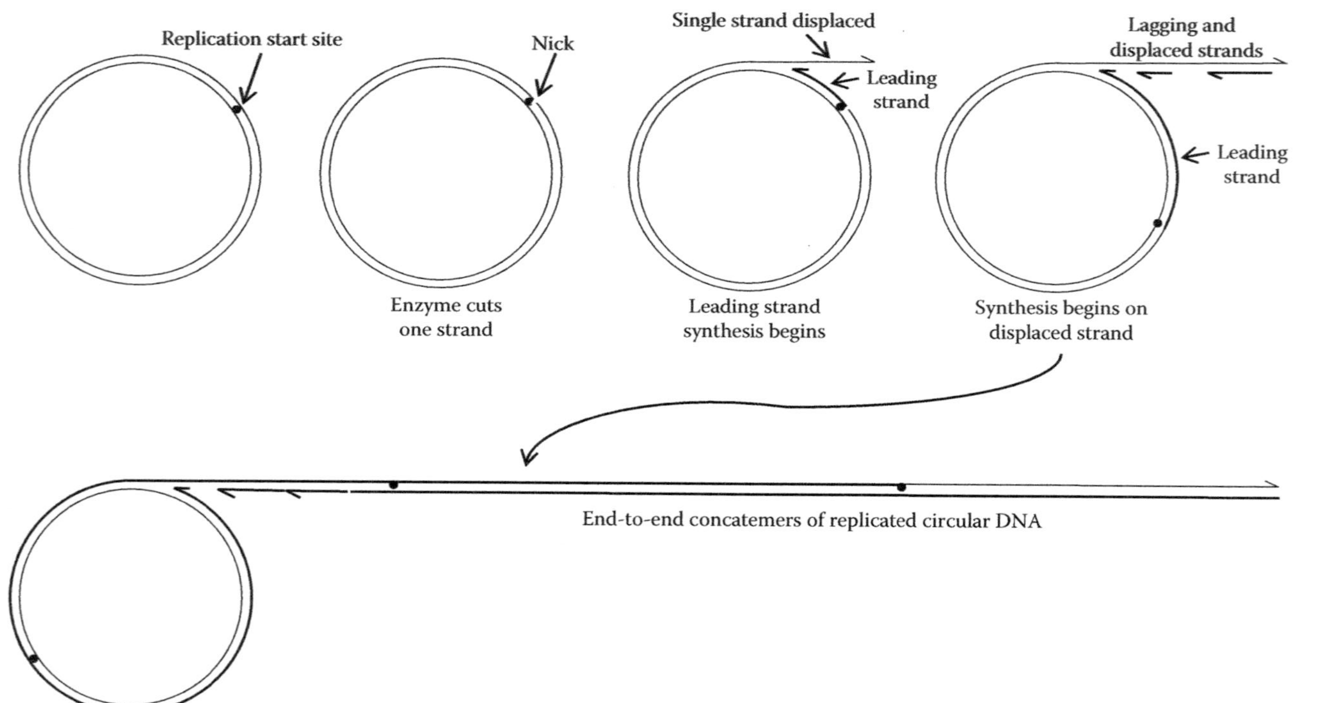

FIGURE 7.9 Rolling circle replication. Some viral DNA, as well as other DNA, molecules are copied using rolling circle replication. First, one strand is cut, forming a nick in the DNA. The nicked strand then is able to dissociate from the other strand through stabilization by single-strand-binding proteins. Topoisomerases, helicases, primeosomes, and DNA polymerase then act in concert to move along the circular strand to synthesize a new DNA strand (leading strand). The displaced strand is replicated by DNA polymerases and accessory proteins to become the lagging strand. Eventually, the circular DNA is copied many times to form long linear concatamers of the genome. For many such DNA molecules, the long concatamer is cut at specific sites to form genome-sized pieces. These can later circularize to begin another replication cycle. Thin black lines indicate the original DNA strands. Thick black lines indicate newly synthesized DNA.

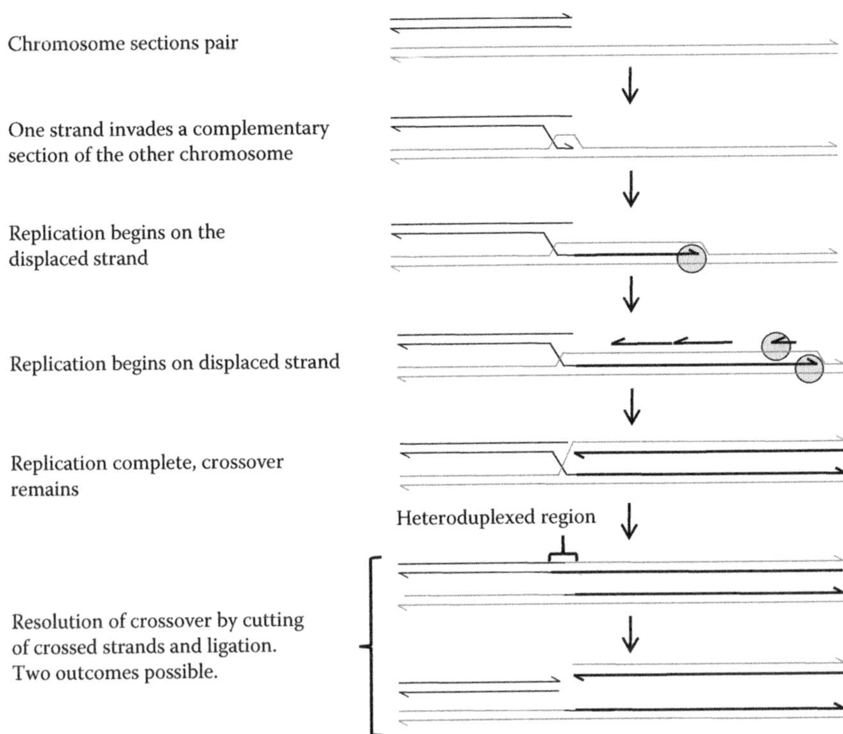
Chromosome sections pair
One strand invades a complementary section of the other chromosome
Replication begins on the displaced strand
Replication begins on displaced strand
Replication complete, crossover remains
Heteroduplexed region
Resolution of crossover by cutting of crossed strands and ligation. Two outcomes possible.

FIGURE 7.10 Recombination-dependent replication (RDR). Some viruses, mitochondria, and chloroplasts (and possibly bacteria) rely on this mode of replication. It begins with single strand invasion, exactly as in a recombination event. However, in this case, replication ensues from the 3′ end of the invading strand. Replication on the displaced strand is accomplished by the lagging strand replication process. When replication is completed, a crossover remains in the region of the original strand invasion. This can be resolved in two ways. First, the crossed strands can be cut and attached to the replicated strands to form two longer molecules. Second, the crossed strands can be cut, but only one is rejoined with the original molecules. This leaves three pieces of DNA, including the original long molecule (part of which underwent replication) and two smaller molecules. A heteroduplexed region is found in the region of the original strand invasion. Thin black lines represent the strands of the invading DNA molecule. Thin gray lines indicate the strands of the invaded DNA molecule. Thick black lines indicate newly synthesized DNA. Gray circles are the DNA polymerase complexes.

enzymes and sequencing, electron microscopy and observations of mitochondria and chloroplasts in an electric field indicate that the genome consists of linear fragments of different parts of the genome, as well as large multimers of the genome (Figure 7.11). In fact, in each organelle there can be many different fragments carrying copies of the same genes, making them polyploid. Replication in these organelles appears to be based on RDR (Figure 7.11). The genome is being replicated as large networks of DNA, many times larger than the genome. Also, these large molecules are fragmenting (possibly because of additional recombination events), to produce fragments of many sizes, containing partial genes, single genes, and multiple genes. On paper, when all of the fragments are aligned, they form a map of the genome that appears to be circular (Figure 7.11, lower right). However, no genome-size circular molecules have ever been documented in plant mitochondria or plastids. The arrangement of the genomic DNA appears to consist of networks of linear branched concatemers, formed by RDR, and smaller fragments of sections of the genome. It has been suggested that similar mechanisms may be present in some bacteria, as well.

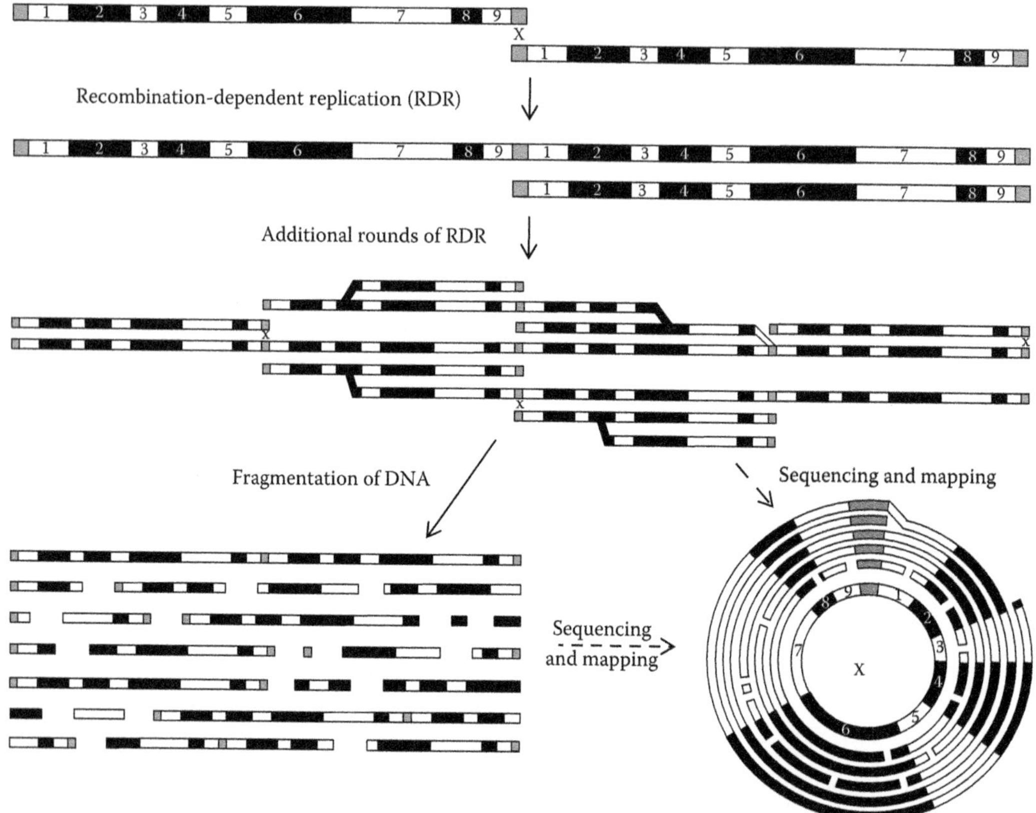

FIGURE 7.11 The overall process of RDR (also refer to Figure 7.10). The process begins with strand invasion (a crossover), followed by replication. Arbitrary numbers and shading have been added for reference. RDR can begin on molecules that already have started the RDR process, and therefore, large networks of DNA can be formed (middle). Molecules such as these have been observed by fluorescence and electron microscopy in mitochondria and chloroplasts. Fragments of DNA also have been observed. Therefore, fragmentation of the DNA also appears to occur (lower left). Most molecules are linear, but a few are subgenome-sized circles. When the genomes of these mitochondria and chloroplasts have been mapped with restriction enzymes or sequenced (lower right), the reconstructions of the genomes appear to be circular, yet no large circles have been demonstrated. The conclusion is that because of repetitive sequences within the genome, the map appears to be circular, but the physical form of the genome is a combination of linear, netted, and fragments of DNA.

TOPOLOGY DURING REPLICATION

Topological issues arise when replicating long stretches of DNA. First of all, most bacterial chromosomes are circular. The first problem is to find a starting location for replication. Origins of replication have been identified on circular plasmids and bacterial chromosomes. Although these have been located on many circular DNA molecules, it appears that there are no universal consensus sequences, and under varying conditions, different start points for replication may be used. Apparently, the starting points for replication are those parts of the DNA that can readily open up to accommodate the initiators of replication, namely, the topoisomerases, helicases, single-strand-binding proteins, primeosomes, and DNA polymerase. But, a problem still exists. DNA molecules are not only twisted double helices, but the double helices are twisted in a conformation called *supercoiling*. Most DNA is negatively supercoiled. Therefore, as the polymerases proceed in opposite directions on the opposite strands around the circle, the DNA will become increasingly negatively supercoiled to a point where the DNA will break, or the replication process will stop. This is one of the main purposes for the topoisomerases in the replication process. At least two types of topoisomerase exist. Type I topoisomerases attach to the DNA and cut and reseal only one strand at a time. In this way, torsional strain can be released or added by rotation of the one strand around the other. Type II topoisomerases cut both strands simultaneously and then spin one part of the DNA relative to the other part. Therefore, during replication the topoisomerases are responsible for maintaining a constant torsional strain in the DNA molecule, such that replication can be completed without breakage of the chromosomes. Without topoisomerases, replication of DNA, especially circular molecules, would be difficult, if not impossible.

REPLICATION OF CHROMOSOMES

Chromosomes in most eukaryotes are linear, and the DNA is normally negatively supercoiled B-form DNA. Large chromosomes are organized into domains, each of which is coated with proteins to protect the structure and maintain the DNA in a conformation that assures accurate gene expression. Additionally, replication timing becomes an issue. DNA polymerases synthesize DNA at rates averaging 50 nucleotides per second. To replicate the human genome of 3 billion base pairs (6 billion per diploid cell) in a single cell starting from one end and proceeding to the other end would take approximately 1.7 million hours (about 200 years). Even if each of the 46 chromosomes were replicated simultaneously starting from one end and proceeding to the other end, it still would require more than 4 years to replicate the DNA in one cell. Therefore, instead of having a single starting point for replication, there are multiple replication start sites on each chromosome (Figure 7.12). For large chromosomes, thousands of such sites may exist. This makes it possible for the chromosomes to be replicated in a shorter amount of time, usually between 30 min and 24 h (depending on species, genome size, and developmental stage).

Large chromosomes present additional challenges. In addition to supercoiling, they are coiled around proteins that are further attached to other proteins that maintain order in the chromosomes (Figure 7.13). In Eukarya and Archaea, the DNA is coiled around histones, while Bacteria have analogous basic proteins that attach to the DNA to keep it compact and ordered. These must be removed during replication (as well as transcription), since DNA polymerase cannot pass through DNA that is wrapped around the histone proteins. All of the other proteins involved in replication (i.e., topoisomerases and helicases) must also pass through the DNA that is devoid of histones. The histones are removed and reattached through methylation/demethylation and acetylation/deacetylation reactions. Finally, once the DNA is replicated, the two replicated double-stranded molecules (e.g., chromosomes) must be separated. This occurs through the actions of the histones,

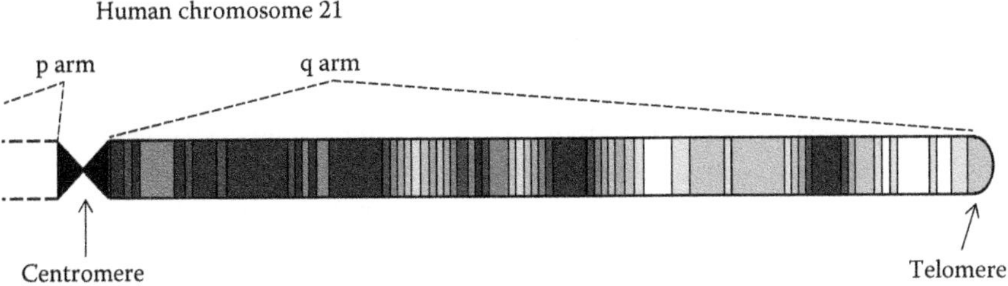

FIGURE 7.12 Replication timing on human chromosome arm 21q. This is an example of the way in which larger chromosomes are replicated. Replication begins at many locations simultaneously, and some portions of the chromosome are replicated earlier in the S phase than others. The white regions are replicated first, and the black areas are the last ones to be replicated. Areas in gray are intermediate, with the lighter gray areas being replicated earlier, and the darker gray areas being replicated later.

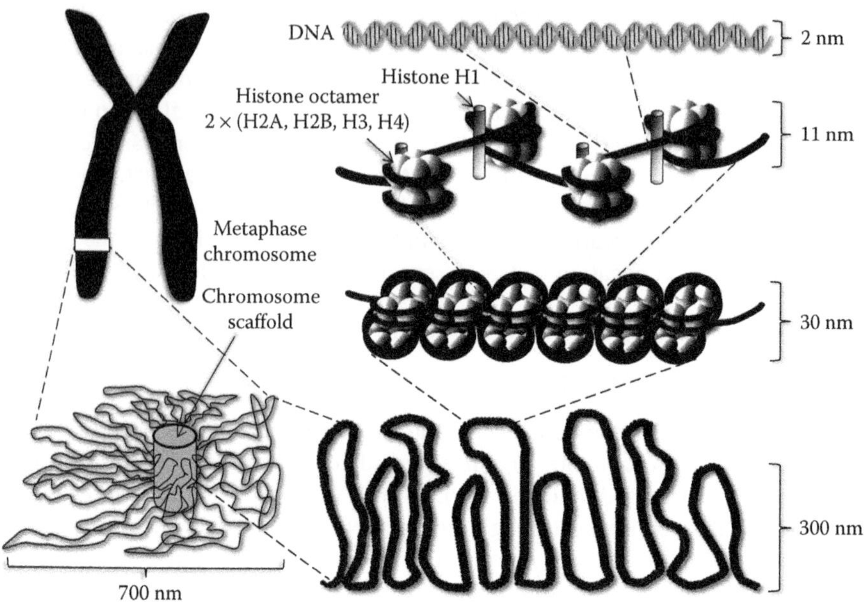

FIGURE 7.13 Chromosome packaging for eukaryotes. Successive degrees of chromosome condensation are illustrated. Starting with the chromosome (upper left), a segment is expanded showing that the chromatin (DNA plus proteins) is attached to a central matrix, called *the scaffold* (lower left). A metaphase chromosome arm is approximately 700 nm in diameter. Each one of the loops of chromatin spreads out an average of 300 nm from the scaffold (lower right). Each loop is composed of 30 nm fibers that consist of DNA wrapped around histones. Specifically, the DNA is wrapped around octamers consisting of histones H2A, H2B, H3, and H4 (two of each type of histone). These octamers are brought together by a central section consisting of histone H1. When the H1 histones are pulled apart, the histone octamer with DNA is 11 nm in diameter. Finally, a DNA molecule is shown (upper right) without attached histones has a diameter of 2 nm.

DNA-binding proteins, structural elements (e.g., microtubules), topoisomerases, nucleases, and other proteins, to ensure that the two DNA molecules are separated and untangled. For a small plasmid, this seems relatively simple. However, for a chromosome that contains many millions of base pairs, this can seem impossible. Nonetheless, it occurs during each cell cycle. For example, in human somatic cells, each of the 46 chromosomes contains millions of base pairs of DNA, and each is replicated and separated accurately during each cell division. When all goes well, each cell receives one copy of each of the replicated chromosomes. When the chromosomes fail to separate and/or segregate correctly, the results can be from relatively minor to fatal.

KEY POINTS

1. DNA polymerase synthesizes new strands of DNA using a template strand and adds only to the 3′ end of the growing molecule (i.e., synthesizes in the 5′–3′ direction).
2. DNA polymerase in living systems uses RNA primers to begin replication. Because of this, other similarities and phylogenetic analyses, it has been concluded that DNA polymerases were derived from RNA-dependent RNA polymerases.
3. Accessory proteins (primeosomes, helicase, topoisomerases, and ligase) are necessary for replication.
4. Telomerase (an RNA-dependent DNA polymerase) is necessary to add telomeres to the ends of linear chromosomes. If they are not added, the chromosome will shorten with each replication cycle.

5. There are several variations on the replication process, including rolling circle replication and RDR.
6. For small molecules, replication may start and end at single sites. For large chromosomes, in order to complete replication in a reasonable period of time, replication begins at many sites in a coordinated fashion.

ADDITIONAL READINGS

Alberts, B., D. Bray, K. Hopkin, A. Johnson, J. Lewis, M. Raff, K. Roberts, and P. Walter. 2013. *Essential Cell Biology*, 4th ed. New York: Garland Publishing.

Alberts, B., D. Bray, A. Johnson, J. Lewis, M. Raff, K. Roberts, and J. D. Watson. 1994. *Molecular Biology of the Cell.* New York: Garland Publishing.

Beard, W. A. and S. H. Wilson. 2003. Structural insights into the origins of DNA polymerase fidelity. *Structure* 11:489–496.

Bendich, A. J. 2004. Circular chloroplast chromosomes: The grand illusion. *Plant Cell* 16:1661–1666.

Constantini, M. and G. Bernardi. 2008. Replication timing, chromosomal bands and isochores. *Proc. Natl. Acad. Sci. USA* 105:3433–3437.

Darnell, J., H. Lodish, and D. Baltimore. 2007. *Molecular Cell Biology*, 6th ed. New York: W.H. Freeman and Company.

Karp, G. 2013. *Cell and Molecular Biology*, 7th ed. New York: John Wiley & Sons.

Kreuzer, K. N., M. Saunders, L. J. Weislo, and H. W. E. Kreuzer. 1995. Recombination-dependent DNA replication stimulated by double-strand breaks in T4. *J. Bacteriol.* 177:6844–6853.

8 DNA Segregation

INTRODUCTION

Most students learn that bacteria separate their chromosomes as part of the process of binary fission and eukaryotes separate their chromosomes by mitosis or meiosis followed by cytokinesis (cell division). However, they rarely learn that there are many other common variations on these three themes. First, the basics will be described, and then, variations will be explained. Together, the diversity in the replication and partitioning of chromosomes illustrates some of the evolutionary steps in these processes. In bacteria, the chromosomes are attached to the cell membrane by specific proteins. As replication and cell division are initiated, these proteins double, thus attaching the replicating chromosome to the membrane in two places (see Figure 7.2). These proteins then begin to migrate toward opposite poles of the growing bacterial cells. By the time the chromosome has been completely replicated, the proteins are at opposite poles, and as the cell grows, the chromosomes are pulled apart completely, which leaves one complete chromosome on each side of the cell as the two cells are pinched apart to form the two daughter cells.

VARIATIONS ON DNA SEGREGATION IN BACTERIA AND ARCHAEA

Bacteria and archaea are diverse in their characteristics, including the arrangements of their chromosomes and other DNA elements (Figures 8.1 and 8.2). In most bacterial and archaeal cells, there is a single chromosome that is replicated and segregated to the two daughter cells (Figure 8.1a). However, some species have more than one chromosome. In these cases, it is assumed that each of the chromosomes proceeds through a similar process of replication for each chromosomes, attachment to the membrane by specific proteins, and movement of the chromosomes to the poles caused by movement of the attached proteins. However, the additional chromosomes mean that the process must be orderly and the movements of the chromosomes must be coordinated so that each of the resulting daughter cells possesses one copy of the complete genome. Another variation is polyploidy. In some bacteria and organelles, the genome is replicated multiple times without cell (or organelle) division in the process of endoreduplication of the chromosome (Figures 8.1b and 8.2). While the purpose for this is unknown, it has been hypothesized that some of the genes (ribosomal RNA genes, photosynthesis genes, or others) are needed in multiple copies, and in order to make these copies, the entire genome is replicated. While this might initially seem to be a waste of energy, the alternative might be fatal to the cell or organelle.

A third variation of replication and division of DNA does not involve the entire chromosome (Figures 8.1c and 8.2). It was discovered somewhat by accident, and at first glance, it appeared to represent Lamarckian evolution. Some bacterial culture plates had been left in the incubator for extended periods of time, such that the plates were overgrown with bacteria (a specific strain of *Escherichia coli* in this case). When these were plated on selective medium that should have killed them, some of them grew. In fact, more of them grew than a fresh culture that had not been grown under the stress of crowding and nutrient limitation caused by overgrowth of the culture. Experiments were carried out to test this further. It was found that when cells are grown under stress (primarily when grown past the log-growth phase and the cells are crowded) and then placed on a medium lacking specific nutrients required by that strain of bacteria, more cells mutated to survive on the medium than unstressed cells. Furthermore, it appeared that they could mutate specifically to grow in the specific medium. That is, the bacteria seemed to react specifically to the environmental conditions to survive in those conditions. And, the trait was heritable. That is, the organism reacted to the environment by rapid mutation of specific genes. Upon further study, it was found that

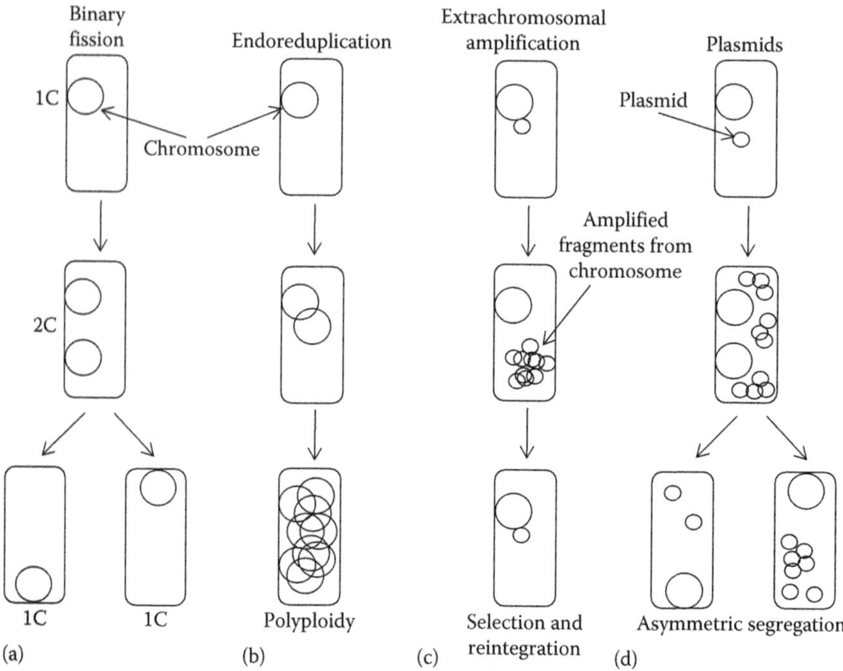

FIGURE 8.1 Replication pathways for bacteria and archaea. (a) For general cell duplication, bacteria and archaea replicate their chromosomes and divide by fission. (b) Some bacterial cells undergo multiple rounds of replication without cell division, becoming polyploid, in a process called *endoreduplication.* (c) Experiments indicate that bacteria can produce multiple copies of portions of their genomes in times of stress to increase their rates of mutation to attempt to overcome the stressful environment. The additional copies are replicated (amplified) extrachromosmally, and with each replicated molecule, the probability of mutation (including beneficial mutations) increases. Eventually, within the population, some individuals survive, ultimately incorporating the vital gene into their genome and continuing to grow and divide. (d) Plasmids are similar to viral nucleic acids in that they replicate autonomously separate from the chromosome, and no known proteins attach them to the cell membrane to assure equal segregation. Instead, they often exist in multiple copies (sometimes hundreds per cell), such that after cell division, each cell is likely to have some copies of the plasmid, although often the distribution is asymmetric.

Lamarckian evolution was not the cause. When bacteria are grown under stressful conditions, they loop out short portions of the genome that then can replicate autonomously and rapidly with high mutation rates. This has the effect of producing many mutant proteins per cell, and in a large population of bacteria, huge numbers of mutant genes and proteins are produced. When placed in the specific media, the number of mutant genes is many times that in a standard culture, and therefore, the probability of having mutants that can survive the medium is increased. Furthermore, in the bacteria that grew, the mutant genes were found both on small pieces of extrachromosomal DNA, but also integrated into the chromosome. The pieces of DNA had excised from the chromosome and replicated, but they apparently reintegrated into the chromosome. Other studies indicate that activation of an error-prone correction and replication system also is involved in this process, which caused increases in mutation rates.

A fourth mechanism for replication and separation of DNAs is seen with plasmids (Figure 8.1d). These are extrachromosomal autonomously replicating DNAs. Some are only a few kilobases in size, while others are the size of small bacterial chromosomes. Therefore, there actually are some overlaps between chromosomes and plasmids, at least based on size. Plasmids can contain a wide variety of genes, including those that confer resistance to antibiotics and heavy metals. They do not appear to attach to the cell membrane, but instead they are *free floating* in the cells. They often exist

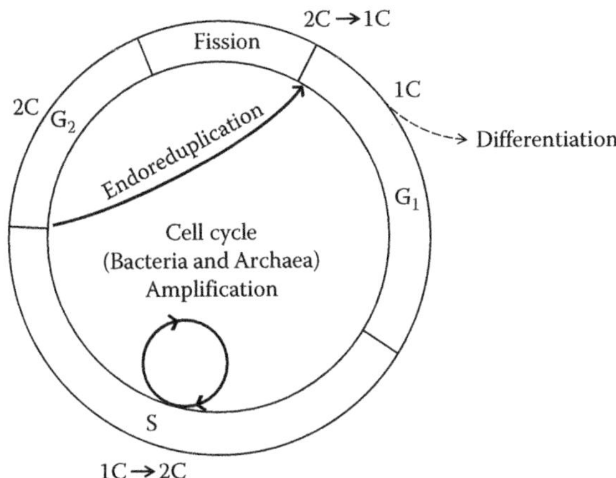

FIGURE 8.2 Cell cycle for bacteria and archaea indicating alternate pathways. Endoreduplication involves bypassing fission. The cells may proceed through part of G_2, but fission is bypassed. Amplification involves replication of parts of the genome to a greater extent than other parts, usually during the S phase. The figure indicates that most of the cell cycle remains intact, such that the usual expression of proteins may occur. However, gene expression patterns would be expected to change specifically in certain parts of the cell cycle in order to produce altered patterns for specific sections.

in many copies (sometimes hundreds of copies), such that as the cell divides, each of the daughter cells receives one or more copies of the plasmid. However, this is not always the case, and often, daughter cells contain no copies of the plasmid.

MITOSIS

Mitosis in eukaryotes is similar to chromosome separation in Bacteria and Archaea, in that the chromosomes are attached to proteins that are attached to parts of a membrane, although it is the nuclear membrane in eukaryotes, whereas it is part of the cell membrane for bacteria and archaea. Also, there are some bacteria, specifically Planctomycetes, that have structures that look very similar to nuclei, and the DNA in these bacteria may be attached to the internal membrane corresponding to a nuclear membrane in eukaryotes. Some hypotheses indicate that the nucleus originated in this group. Also, Spirochaetes are in the same phylogenetic clade. This group has microtubules (as do Planctomycetes), protein structures that move chromosomes in eukaryotes. Therefore, an endosymbiotic event between a member of the Archaea and a bacterium in the Spirochaete/Planctomycete phylogenetic clade may have been the origin of the first eukaryote that contained both a nucleus and microtubules.

Although it is simple to memorize the phases of mitosis, it is a complex process. On the other hand, the signals for the various phases in mitosis are fairly simple. Mitosis can be split into prophase, prometaphase, metaphase, anaphase, and telophase (see Figures 7.4 and 7.5). At all other times, the nucleus is in interphase. During interphase, the nucleus has a characteristic organization. Portions of the chromosomes are covered with proteins, and hence in histological sections, these regions, termed *heterochromatin*, stain heavily with cytological stains. These regions are concentrated around the nuclear membrane and are regions of the chromosome that usually are not involved actively in transcription. Other parts of the nucleus stain lightly, which are known as *euchromatic regions*, and are composed of DNA that has fewer attached proteins. In these areas, the DNA usually is being transcribed, sometimes at high rates. Nuclei also contain nucleoli (as described in Chapter 5). These are integral parts of the nucleus. Ribosomal RNA is transcribed and ribosomes are partly assembled in the nucleolus. Part appears to have a fibrous portion, called *the fibrillar portion*. It is where long strands of rRNA are being synthesized and processed. The granular portion is where ribosomes are being assembled. The ribosomal proteins are imported from the cytosol, assembled into the large and small subunits (with incorporation of the rRNAs), which are exported out of the nucleus to the cytosol.

Prior to mitosis, the DNA in the nucleus must be replicated, after which time the nucleus has a 4*C* (tetraploid) level of DNA (for originally diploid cells, Figure 8.2). By the time each chromosome has been replicated (during the S phase), each of the duplicated chromosomes (called *sister chromatids* at this point) are mainly separated from one another, but still connected at their centromeres. The G_2 phase follows in the cell cycle where the cell often remains for some time at the 4*C* level. For example, many of your skin cells are 4*C*. When you cut yourself, these cells are ready to proceed through mitosis to rapidly repair the wound. They have already completed the S phase. Once cells enter mitosis, they proceed through without stopping. Primarily, this is because during prophase, the chromosomes condense. That is, additional proteins attach to the chromosomes that concentrate the DNA into smaller structures until they form the familiar metaphase chromosomes, the most condensed form of the chromosomes. In fact, the reason that chromosomes got their name is that they stained darkly in cells prepared for microscopy (i.e., chromo—for colored or stained; and soma—for body). Again, it is primarily the additional proteins that attached to the stain molecules, thus causing the coloration of the chromosomes. When the chromosomes are condensed, most of the genes are not being transcribed because they are so tightly condensed that they are unavailable to the transcription machinery. This is why the cell must proceed through mitosis as rapidly as possible because most genes are not being transcribed. One set of genes that continue to be transcribed are the ribosomal RNA genes. They are one of the only regions that are not condensed during mitosis, and often, the nucleolus remains intact throughout mitosis. These normally are in regions called *secondary constrictions* (the centromere being the primary constriction). In reality, rather than being constrictions, these regions are uncondensed regions on the chromosome that are active in transcription. Because they are uncondensed, they contain fewer of the proteins that adhere to the cytological stains, and therefore in the microscope, they appear clear, while the remainder of the chromosome is densely stained.

A lengthy mitosis could mean death for the cell, because transcription of much of the genome is halted at that time. Therefore, the cells must proceed rapidly through mitosis so that the chromosomes can decondense and begin transcription again. Mitosis is usually the shortest part of the cell cycle in most organisms. As the chromosomes condense, the nuclear membrane also is changing. Proteins called *lamins* that are attached to the membranes are being phosphorylated. When phosphorylated, they change shape that causes them to pull the membrane into pieces that still are attached to parts of the chromosomes. This causes the membrane to become fragmented, and by prometaphase, the nucleus has lost its integrity. At the same time, microtubules are beginning to gather at what will become the two poles of the cell (Figure 8.3). In the cells of many animals, these form around a centrosome in a star pattern in what is termed *astral mitosis* (Figure 8.3a). At their centers are centrioles that consist of two microtubule bodies in a specific arrangement (with nine triplets of microtubules surrounding a central element) that are at right angles to one another. In other organisms (e.g., plants and diatoms), no asters form, but a microtubule organizing center (MTOC) forms at each pole. This is known as *anastral mitosis* (Figure 8.3b). In both astral and anastral mitoses, these microtubules are known as *polar microtubules*. The kinetochore microtubules continue to lengthen during prometaphase until some of them begin to interact with the kinetochore of the centromere on each chromosome. Kinetochores consist of protein complexes that form on the centromere (primary constriction) of each chromosome. They attach the microtubules to the chromosomes. These microtubules are known as *kinetochore microtubules*. Once the microtubules make contact with the kinetochore, the chromosome is attached to one of the two cell poles. At this point, movement of the chromosome toward the middle of the cell (as well as some occasional movement toward the pole) can

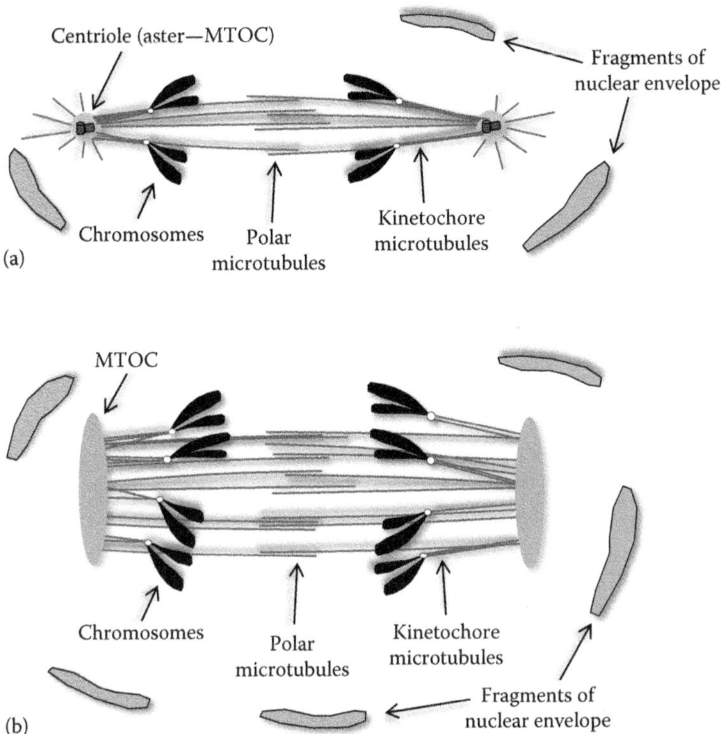

FIGURE 8.3 Anaphase for (a) astral and (b) anastral mitosis. In both types of mitosis, kinetochore microtubules join the chromosomes to the poles, polar microtubules join both poles, and MTOCs are present. In astral mitosis, centrioles are at the center of the MTOCs. For many species, the nuclear membrane dissociates, but fragments of the membrane remain and reassemble during telophase.

be observed. In addition to microtubule polymerization, it is moved along the microtubules by *kinesins*, protein motors that use adenosine triphosphate (ATP) to move (*walk*) along the microtubules. Shortly thereafter, microtubules that are lengthening from the other pole contact the other kinetochore on the chromosome (attached to the homologous chromatid), and the chromosome is stabilized in the center of the cell. At the same time, other microtubules from both poles have grown past one another and overlap in the middle of the cell. Microtubule-associated proteins (primarily dynein) connect the opposing microtubules throughout the middle portions of the cell. The cell has now reached metaphase. This phase is recognized by the alignment of the chromosomes in the middle of the cell along what is termed the *metaphase plate*. At this point, the chromosomes exhibit small movements, presumably due to the opposing forces of the microtubules attaching them to both poles, and controlled by the actions of dyneins and kinesins that move in opposite directions along the microtubules.

The onset of anaphase is rapid and appears to be modulated by calcium levels. Upon a sudden increase in calcium and protein kinase activity, there is an immediate deopolymerization of the kinteochore microtubules on the kinetochore end that causes them to shorten rapidly (Figure 8.4). This pulls each of the kinetochores on each chromosome and the attached chromatids to opposing poles. In addition, *kinesins*, proteins that *walk* along the microtubules using ATP as an energy source, actively move the kinetochore along the microtubule toward the pole. At the same time, dyneins on the polar microtubules begin to move these microtubules with respect to one another, which begin to push the two poles apart. The chromatids then

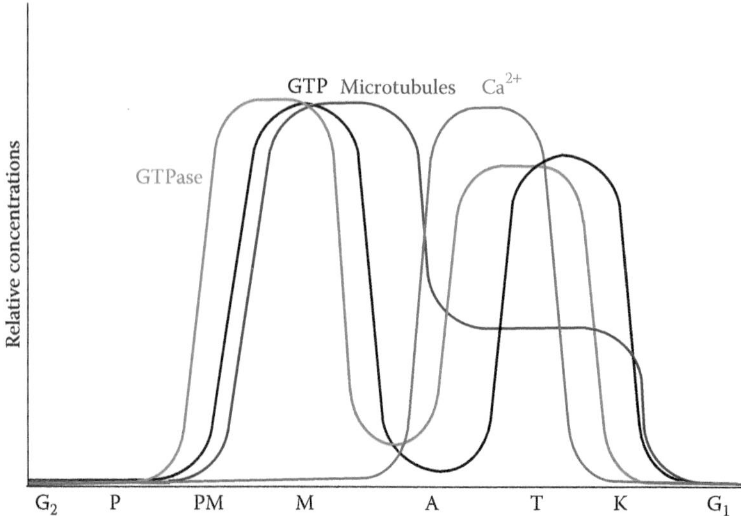

FIGURE 8.4 Time course of various cell components during mitosis. GTPase activity begins to rise during prophase, which signals an increase of GTP in the cell. GTP is required for microtubule polymerization, which occurs during prometaphase. All three continue to increase and reach peaks during metaphase. During anaphase, there is a sudden release of calcium ions, which causes rapid depolymerization of the microtubules, which causes the chromatids to pull apart and the resultant chromosomes to be pulled to the poles by the shortening kinetochore microtubules. A second smaller increase in GTPase and GTP causes the polar microtubules to continue to polymerize, which pushes the cell poles apart. Eventually, these all decrease, and the cells divide.

separate, becoming individual chromosomes. (Note the change in terminology. Although the chromosome/chromatid has not physically changed, each double-stranded DNA is termed *a chromatid* when two or more are physically attached to one another, but when they separate they are called *chromosomes*.) As the chromosomes reach the poles, the lamins are dephosphorylated and the nuclear membranes begin to coalesce into integrated nuclei, as the chromosomes begin to decondense. This is called *telophase*. The polar microtubules continue to push against one another to push the cell poles further apart. At this point, mitosis ends and cytokinesis (cell division) begins. In cells without cell walls, an actin ring forms around the center of the cell just under the plasma membrane. This shrinks rapidly to pinch the cell into two daughter cells. In cells with cell walls, microtubules, as well as actin, are involved in laying down a cell plate or septum in the middle of the cell. This eventually forms into new cell wall material with plasma membranes on both sides. Thus, two daughter cells are formed, although in many of these kinds of cells cytoplasmic passages (e.g., plasmodesmata in plants, and pores in fungal septa) remain between the daughter cells.

VARIATIONS IN MITOSIS AND THE CELL CYCLE

For many eukaryotic cells, the nuclear membrane dissociates during mitosis (Figure 8.3), and the chromosomes are attached at one point (the centromere) to the kinetochore microtubules that pull the chromosome to each of the poles. As stated earlier, there are variations on this main theme of mitosis, as well as the replication and chromosome separation process. One variant of mitosis occurs in some animals and plants. Instead of being attached at a centromere and kinetochore to the microtubules, they are attached along the entire length of the chromosomes (Figure 8.5).

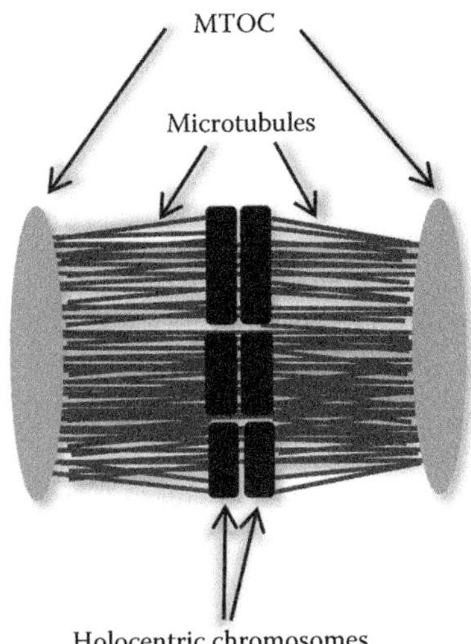

FIGURE 8.5 Holocentric (holokinetic) chromosomes. In some organisms, no distinctive centromere or kinetochore forms. The chromosomes are attached along their entire lengths. During anaphase, they are pulled toward the MTOCs by large sets of microtubules.

These are called *holocentric* (*holokinetic*) chromosomes. They line up parallel to the metaphase plate during metaphase and are pulled to the poles along their length during anaphase. Another variant occurs in many different types of organisms, including diatoms, dinoflagellates, and fungi. In these species, the nuclear membrane never fragments (Figure 8.6). Intact nuclei always are present. In most dinoflagellates, the nucleus remains intact, but the chromosomes within condense and are attached to the nuclear membrane by proteins (Figure 8.6a). This is reminiscent of the arrangement in bacteria (Figure 8.6b), although in eukaryotes the chromosomes are attached to the nuclear membrane, while in bacteria they are attached to the cell membrane. Another difference is that microtubules are involved. However, only polar microtubules are present. No microtubules are attached to the kinetochores. The microtubules form tunnels through the nucleus without disturbing the integrity of the nuclear membrane. As mitosis continues, the chromosomes move to the poles of the nucleus and the nucleus divides in a process that resembles binary fission in bacteria. Cytokinesis completes the process.

In other dinoflagellates that undergo astral mitosis, the nucleus remains intact, but both polar and kinetochore microtubules are present (Figure 8.6c). The polar microtubules form tunnels through the nucleus (again without breaking through the nuclear membrane), but the kinetochore microtubules attach to proteins that are attached to the centromeres of the chromosomes that continue to reside within the nucleus. As the kinetochore microtubules shorten, the chromosomes are drawn to the poles while remaining within the nucleus. The nucleus and cytoplasm then divide to produce the two daughter cells. Another variation on mitosis occurs in yeasts and diatoms (Figures 8.6d and 8.7a). Again, the nuclear membrane remains intact, but the microtubules form inside the nucleus rather than in the cytosol. Both polar and kinetochore microtubules are present and mitosis is very much like that in other organisms. However, all of this occurs inside the nucleus. An additional step in this process can be seen in basidiomycetous fungi (Figures 8.6e and 8.7b). The nuclear membrane partly dissociates, but what remains partly surrounds the chromosomes as they separate. Both polar and kinetochore microtubules are present, and if the nuclear membrane were to dissociate completely, it could not be distinguished from mitosis in animal cells. Each of these may be evolutionary steps in processes that began billions of years ago. Binary fission must have occurred in the bacteria 3.5 billion years ago, and presumably, mitosis existed by the time eukaryotes first appeared about 2.3–2.4 billion years ago.

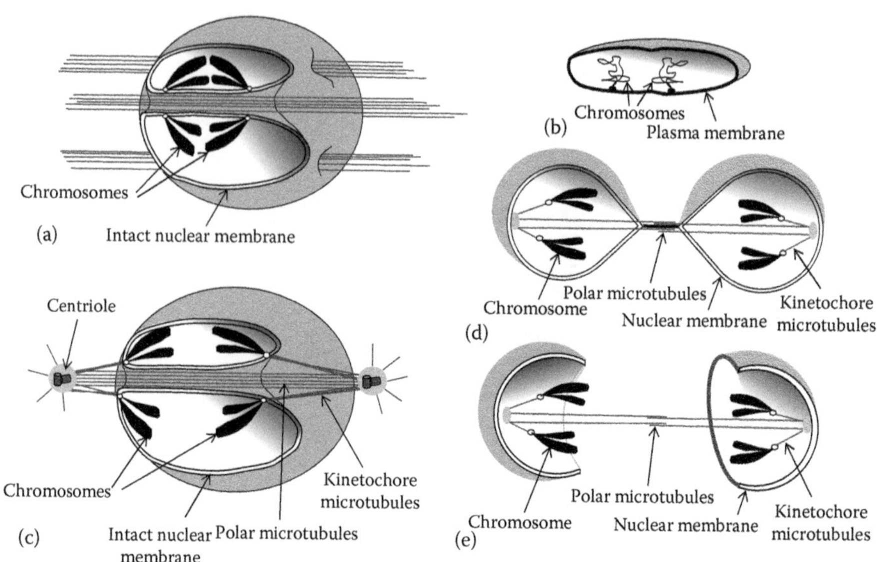

FIGURE 8.6 Variations of mitosis compared with bacterial fission. (a) In most dinoflagellates, the chromosomes are attached to the membrane (nuclear membrane, in this case) by specific proteins and are moved to opposite poles during mitosis. The nuclear membrane never disintegrates during mitosis, but simply splits in a process that resembles fission. Another difference is that microtubules are present, but they pass through channels in the nucleus. They are responsible for elongating the nucleus and the cell during mitosis and cytokinesis. (b) In bacteria, the chromosomes are attached to the cell membrane by specific proteins. Although the proteins differ, the arrangement is similar to that in the dinoflagellate nuclei. (c) In other types of dinoflagellates, a similar situation occurs. The chromosomes are attached to the nuclear membrane by specific proteins, which are attached to kinteochore microtubules on the cytosol side of the membrane. This type of mitosis is similar to mitosis in animals, except that the nuclear membrane never dissociates. (d) In ascomycetous fungi, mitotic division of the nucleus resembles fission in bacteria. The nuclear membrane never dissociates, and the microtubules form within the nucleus rather than in the cytosol. Both kinetochore and polar microtubles are present and act to move the chromosomes to the nucleus poles and stretch the nucleus (respectively). (e) In basidiomycetous fungi, the chromosomes are pulled out of the nucleus, but during anaphase, the nuclear membranes partly form around the chromosomes and at telophase completely form around each set of chromosomes. Each of the variants may indicate steps in the evolution of the process of accurate and efficient chromosome segregation.

VARIATIONS IN CHROMOSOME NUMBER

The C-value of a species is defined as the *constant* amount of DNA found in each haploid nucleus, although it is now known that these amounts can vary. Therefore, a haploid cell has a 1C level of DNA, a diploid cell has a 2C level, a triploid cell has a 3C level, and so on. Sometimes this equates with the usual chromosome number for a species, although not always. For humans, $C = 3.1$ billion base pairs for a female gamete and about 3.0 billion base pairs for a male gamete. The characteristic chromosome number for a species is designated N (or n). Therefore, N equals the usual number of chromosomes for a haploid cell, 2N for a diploid cell, 3N for a triploid cell, and so on. In humans $N = 23$, and therefore, in most of your diploid somatic cells $2N = 46$, with either 6.2 billion base pairs of DNA for females or 6.1 billion base pairs for males. So, usually in the cell cycle for a diploid cell, the cell starts at 2C and 2N in G_1, replicates its DNA in the S phase to produce a cell that has a 4C amount of DNA, but the number of chromosomes is still 2N (since the chromatids have not separated). The cell becomes 4C and 4N during anaphase, and finally, two cells result, each being 2C and 2N after cytokinesis.

Fungi exhibit variation in the quantity of DNA that they have in each cell. Many grow as yeasts, single celled round or oblong cells, or as multicellular filamentous organisms that can be haploid

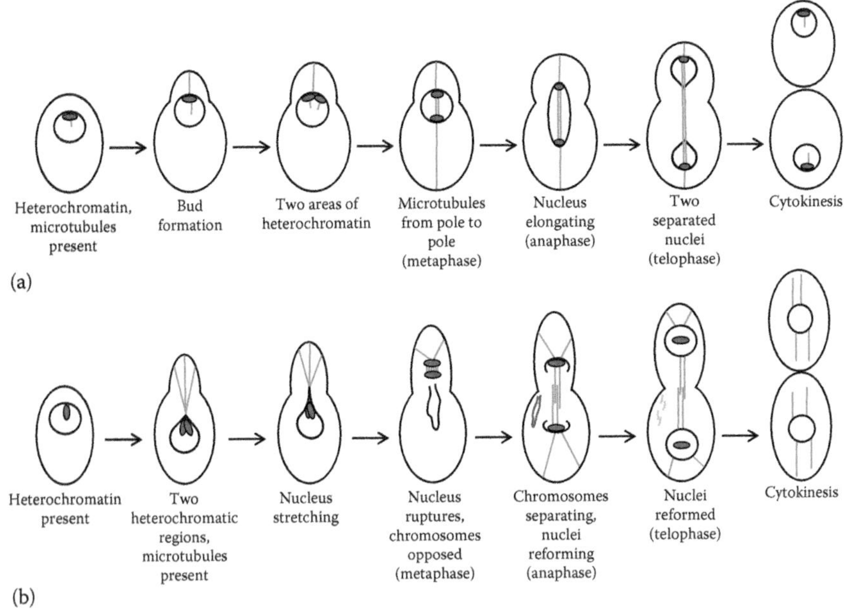

FIGURE 8.7 Mitosis in fungi. (a) In ascomycetous yeasts, the chromosomes are partially condensed during most of the cell cycle. Also, microtubules are present within the nucleus. After duplication of the chromosomes, microtubules begin to push out a bud, which is also attached to the nucleus. The chromosomes move to the poles, as cytosolic microtubules attach the nucleus to both poles. Then, the microtubules within the nucleus push the poles of the nucleus apart, which eventually divides. This is followed by cytokinesis. (b) In basidiomycetous yeasts, mitosis begins with microtubules pushing out a bud, as with ascomycetes. However, the microtubules then begin pulling the chromosomes out of the nucleus, leaving a distorted nucleus devoid of chromosomes. Microtubules then begin pushing the two sets of chromosomes apart, while new nuclear membranes begin to form around each set of chromosomes. By telophase, the nuclei have formed completely, and this is followed by cytokinesis.

or diploid, and some become multinucleate. All types of cells divide using mitotic separation of chromosomes. For a diploid cell, they go through the normal progression of $2C/2N$ in G_1, $4C/2N$ by G_2, $4C/4N$ in anaphase, and back to $2C/2N$ following cytokinesis. On the other hand, haploid cells would have the following progression: $1C/1N$ in G_1, $2C/1N$ in G_2, $2C/2N$ in anaphase, and back to $1C/1N$ following cytokinesis. However, many fungi grow as hyphae, long multicellular linear and branched chains of cells. These can be haploid, diploid, dikaryotic, or multinucleate syncitia (singular—syncitium). Dikaryotic cells contain two haploid nuclei (termed $N + N$, and/or $C + C$), each originating from a different parent during sexual reproduction. During mitosis, both nuclei proceed through prophase, prometaphase, metaphase, anaphase, and telophase simultaneously, but separately (Figure 8.8). Their progress would be: $1C/1N + 1C/1N$ in G_1, $2C/1N + 2C/1N$ in G_2, $2C/2N + 2C/2N$ in anaphase, and $1C/1N + 1C/1N$ after cytokinesis. In ascomycetes, this simply appears to be two mitotic events occurring in the same cell (Figure 8.7a). Two spindles are formed (one for each set of chromosomes), and after telophase, a new cell plate forms to divide the four nuclei. However, in basidiomycetes, the process is more elaborate (Figure 8.8b). While one spindle forms in a usual fashion, being pointed toward both poles, the other spindle pushes a protrusion (a clamp) out of one corner of the mother cell, but still is enclosed within the cell membrane and cell wall. The protrusion extends into a hook that eventually touches the future daughter cell, and the membranes and cytosols join between the two cells. As telophase concludes, one of the haploid nuclei ends up in one daughter cell and the other is pushed into the other daughter cell by its spindle through the hook, called *a clamp connection*. The cell membranes and walls then close within the

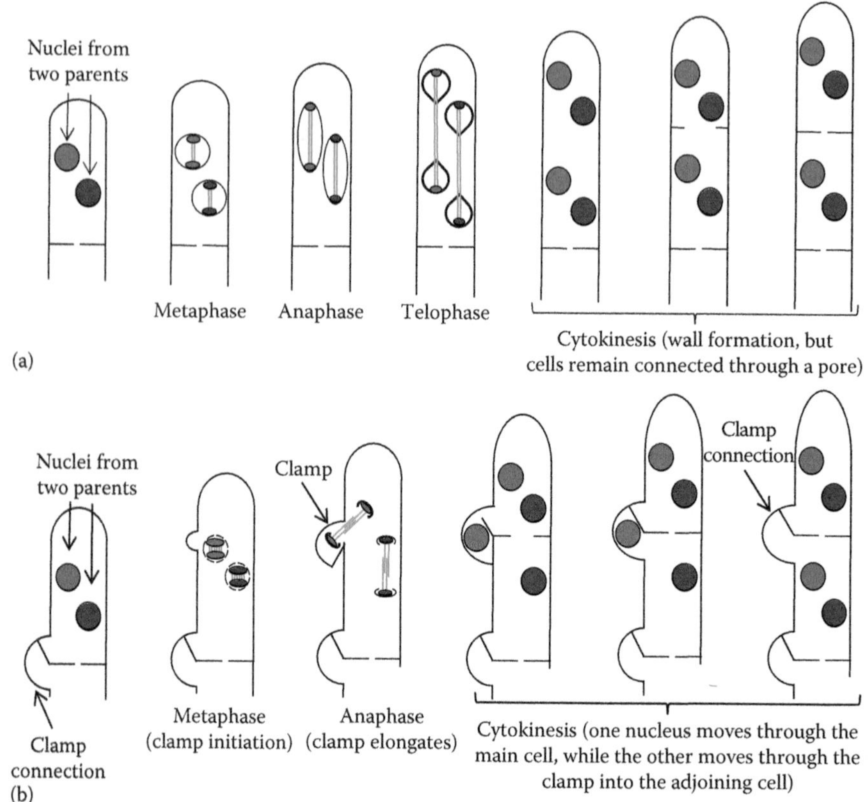

FIGURE 8.8 Mitosis in dikaryotic filamentous fungi. Filamentous fungi can be haploid, diploid, dikaryotic, or polyploid during parts of their life cycles. When two mating types join, in what is called anastamosis, two separate nuclei exist per cell, which is called a dikaryon. Mitosis in these cells takes place for each of the two nuclei in a coordinated manner. (a) In filamentous ascomycetes, mitosis in dikaryotic cells essentially entails coordinated mitosis of each haploid nucleus. After mitosis, cytokinesis divides the cells such that each cell has two haploid nuclei, one of each type. However, in most cases, the cell separations, called *septa*, often have pores that maintain connections between adjacent cells. (b) In filamentous basidiomycetes, mitosis is more complex. First, the nuclear membrane dissociates, and microtubules act to move the chromosomes and lengthen the cell. However, the orientations of the two sets of chromosomes differ. One set of chromosomes moves directly toward the cell poles. The other set is oriented at an angle, such that one haploid set is pushed out through a cell protrusion called *a clamp*. Eventually, microtubules move these chromosomes into what has become one of the two daughter cells. During telophase, the nuclear membranes form and the cell walls are completed. In this case, a bulge is evident at the border between the cells, known as *the clamp connection*. The main septa have pores that allow a permanent connection between the cytosols of the cells.

clamp connection and between the main parts of the two daughter cells, resulting in two daughter cells with a closed clamp connection and a wall separating them.

CHANGES IN DNA AMOUNT THROUGH THE CELL CYCLE

There are several variations in the cell cycle that can lead to nuclei that contain amounts of DNA that differ from those of the starting cell (Figures 8.9 and 8.10). Of course, one of these is meiosis, and this will be discussed in detail below. Essentially, at the start of meiosis the cell is at a $2C/2N$ level (Figure 8.9b). It then proceeds through the S phase to become a cell at a $4C/2N$ level in G_2. However, in division I of meiosis, two cells are produced that are at $2C/1N$ levels (rather than $2C/2N$, as in standard mitosis). Division II produces four truly haploid cells ($1C/1N$). Another alteration on

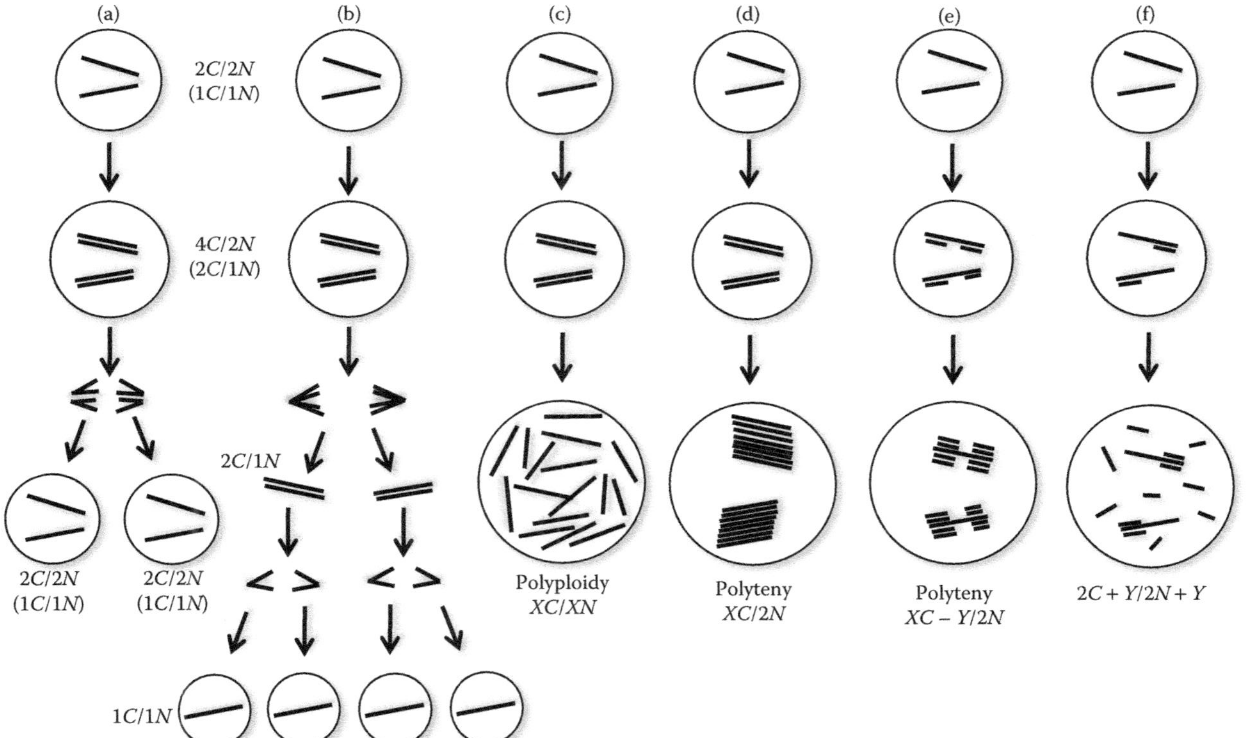

FIGURE 8.9 Variations of DNA amounts and chromosome numbers in eukaryotic nuclei. (a) Mitosis of either diploid or haploid nuclei is shown (haploid amounts are in parentheses). Nuclei begin with a diploid amount of DNA (2*C*) and a diploid number of chromosomes (2*N*). Each chromosome duplicates in the S phase resulting in a 4*C* amount of DNA, but with the chromatids paired, the number of chromosomes still is diploid (2*N*). After mitosis, the daughter cells each are at 2*C*/2*N* levels. If the nuclei begin as haploid (1*C*/1*N*), in G_2 they are 2*C* and 1*N*, and after mitosis the two daughter cells are at 1*C*/1*N*. (b) In meiosis, the nuclei begin as 2*C*/2*N*. After the S phase, the nuclei are at 4*C*/2*N*, as in mitosis. However, after the first division, the two daughter nuclei are at 2*C*/1*N* (as opposed to 2*C*/2*N* in mitosis). After the second division, each of the four nuclei is haploid (1*C*/1*N*). (c) In the process of endomitosis, the DNA is replicated repeatedly, and in each cycle, the chromosomes separate. However, no cytokinesis occurs, forming a polyploidy nucleus. (d) Polytene chromosomes are formed by repeated replication without chromosome separation in a process called *endoreduplication*. (e) Polytene chromosomes also can be produced through differential rounds of replication, where some parts of the chromosome are replicated less than other sections (called *underreplication*). (f) Replication and dissociation of multiple extrachromosomal copies of those regions, called *amplification*.

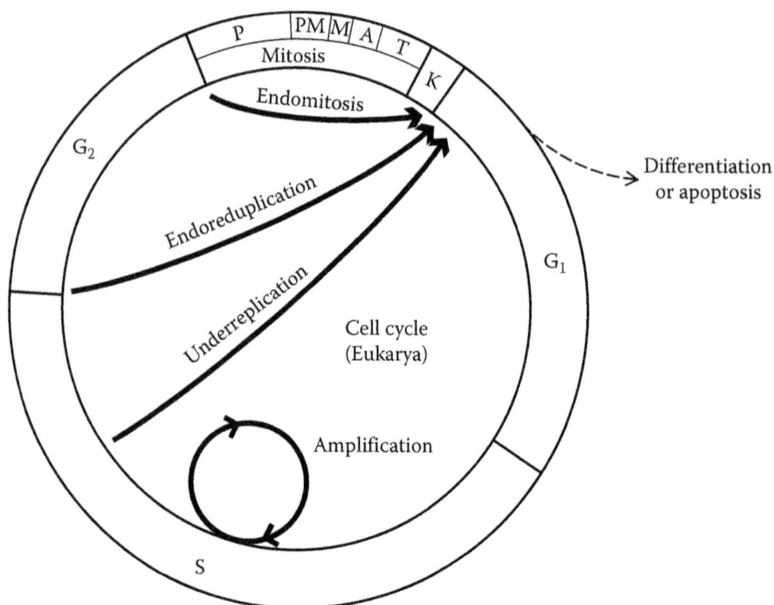

FIGURE 8.10 Cell cycle for eukaryotes indicating alternate pathways. Endomitosis is the process where mitosis and cytokinesis are bypassed, leading to polyploid nuclei. Endoreduplication produces polytene chromosomes through the elimination of G_2, mitosis, and cytokinesis. Underreplication also produces polytene chromosomes, but parts of S are bypassed, as well as G_2, mitosis, and cytokinesis. Amplification is repeated replication of one or more parts of the genome to generate extrachromosomal copies of parts of the chromosome(s).

the cell cycle is in a process called *endomitosis* (Figures 8.9a and 8.10). In essence, the cell cycle is used without mitosis. Therefore, DNA synthesis occurs many times without mitosis and cytokinesis. This forms cells that have multiple copies of the genome and multiple copies of each chromosome, designated XC, XN. These are polyploid cells and can be very large due to the high DNA content, which also leads to larger nuclei and cells in general. Many types of cells in plants that are primarily in storage tissues are polyploid, such as cotyledons and endosperm. Contrary to some descriptions of plant endosperm tissues as triploid, many of these tissues begin as triploids, but then proceed through many rounds of replication so that ultimately they are polyploid. Another variation on this pattern is the formation of polytene chromosomes by a process called *endoreduplication* (Figures 8.9d and 8.10). In this case, mitosis is skipped, but other parts of the cell cycle also are eliminated because the chromosomes never dissociate from one another, but continue to replicate still connected to one another. In the end, very large chromosomes are produced, known as *polytene chromosomes*. Therefore, the cells are at an XC, $2N$ level. This is a specific form of polyploidy. *Polyteny* also occurs when there is unequal endoreduplication, that is, when some of the sequences are replicated to a greater extent than others, termed *underreplication* (Figures 8.9e and 8.10). In this case, they would be designated $XC - Y$ (multiples of the genome, but less of some parts), but still $2N$. Polytene chromosomes can be found in certain cells in plants and animals. They have been found in salivary gland cells of some insects and in suspensor cells of some plants. Another variant is called *amplification* (Figures 8.9f and 8.10). In this case, specific sequences are replicated many times. In some cases, they are replicated on the chromosomes, but in most cases they are amplified extrachromosomally. These are common for ribosomal RNA genes that are needed in very high copy numbers. In this case, the designations would be $2C + Y$ (a diploid amount of DNA, plus additional amplified pieces), and $2N + Y$ (a diploid number of chromosomes, plus other extrachromosomal pieces).

MEIOSIS

While bacteria do participate in exchange of genetic material through conjugation and recombination, they do not have exchange through sex using meiosis. This innovation probably developed after the first eukaryotes appeared on the Earth. However, the advent of meiosis may not have required much time to appear after mitosis. Meiosis involves two cell divisions (Figures 8.11 and 8.12). The first part of meiosis is unique to this process, while the second part resembles a mitotic division. In fact, the collection of proteins involved in the second division is almost identical to that found in a standard mitotic division. Therefore, it appears that meiosis was an addition of steps prior to a mitotic division that included many recombination events. This is the unique part of meiosis that led to its success evolutionarily. Meiosis provided a great advantage to organisms, because it increased the rate of gene exchanges on the chromosomes. This had the effect of getting rid of some deleterious gene combinations while retaining others and yielded new combinations of genes, some of which would lead to increases in fitness.

Meiosis is divided into two major parts (Figures 8.11 and 8.12), division I and division II. Division I is further divided into prophase I, prometaphase I, metaphase I, anaphase I, and telophase I. Additionally, prophase I has several parts: leptotene, zygotene, pachytene, diplotene, and diakinesis. Prophase I is when the chromosomes are brought together specifically for recombination between all four chromatids of each chromosome pair. As the cell enters leptotene, the chromosomes begin to condense and a specific set of proteins that will form part of the synaptonemal complex form a scaffold for the entire length of each chromatid (Figure 8.13). In zygotene, other proteins connect to the scaffold to complete the synaptonemal complex that aligns all four matched chromatids for each chromosome pair. At this point, the synaptonemal complex looks like a zipper between the chromatids. During pachytene, a small number of protein complexes attach along each of the synaptonemal complexes. These are recombination nodules that break and join DNA segments to produce crossovers between the chromatids. In diplotene, the synaptonemal complex degrades and disappears. The chromatids begin to decondense and chiasma (crossover points) can

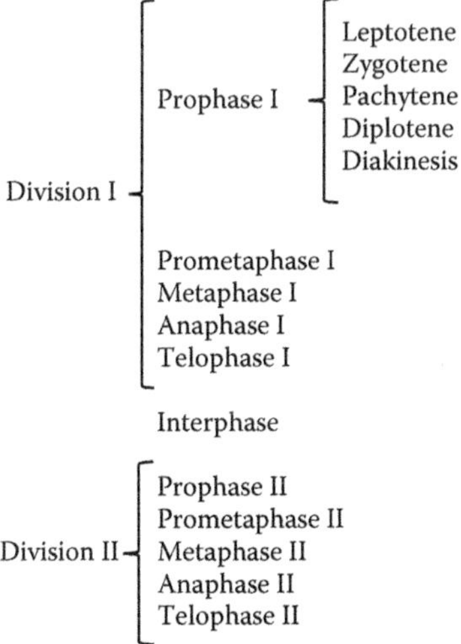

FIGURE 8.11 Stages of meiosis. Division I consists of an extended prophase that includes leptotene, zygotene, pachytene, diplotene, and diakinesis, which leads to crossovers between chromatids. This is followed by prometaphase I, metaphase I, anaphase I, and telophase I, in which homologous chromosomes are separated into the resultant daughter cells. Then, there is a brief interphase, followed by division II, which is similar to a mitotic division, and includes prophase II, prometaphase II, metaphase II, anaphase II, and telophase II, where sister chromatids are separated to form haploid nuclei.

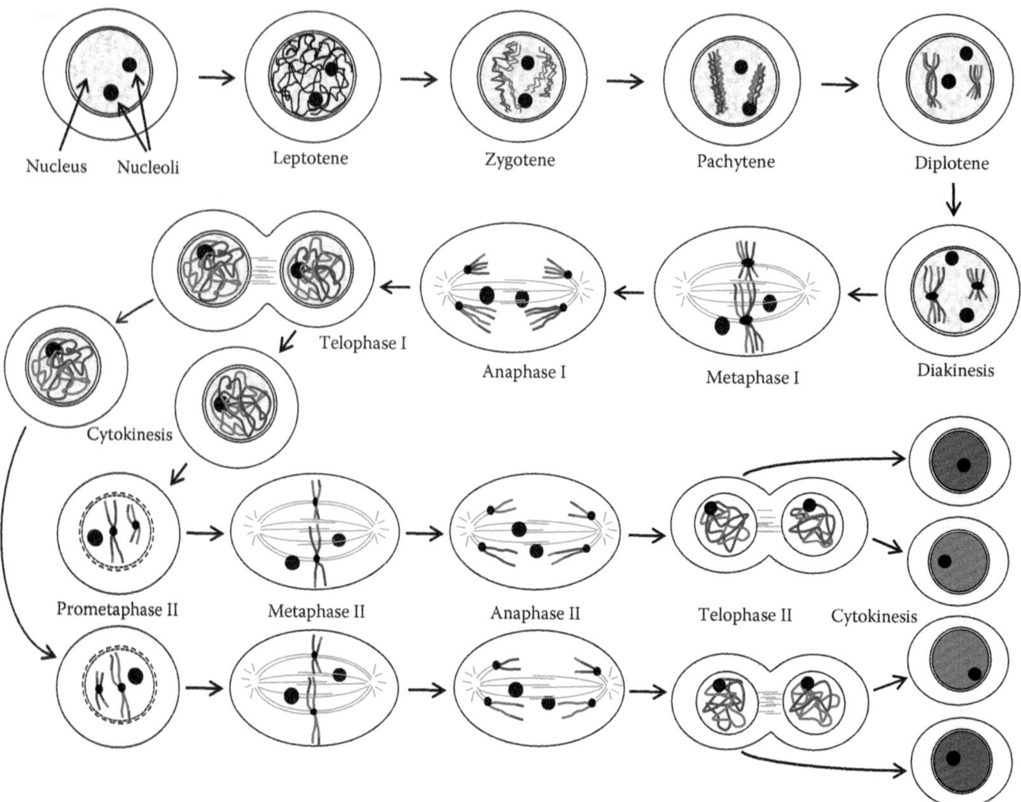

FIGURE 8.12 Diagram of meiosis. The process begins with a lengthy prophase where the four chromatids from each of the two homologous chromosomes are condensed (during leptotene) and brought into contact with one another (zygotene) by the synaptonemal complex. During pachytene strands of the four chromatids undergo recombination events through the actions of recombination modules on the synaptonemal complex. The crossovers (chiasma) are resolved during diplotene and diakinesis. The homologous chromosomes are then separated by microtubules. The stages of chromosome separation are metaphase I, anaphase I, telophase I, and cytokinesis. At this point, each of the cells has a 2*C* amount of DNA, but have a 1*N* number of chromosomes. The second cell division is similar to mitoses in haploid organisms. The nuclei proceed through prophase II, prometaphase II, metaphase II, anaphase II, and telophase II, followed by cytokinesis, resulting in four haploid cells (1*C*, 1*N*).

be seen. *Diakinesis* is characterized by the further separation of the chromatids, such that all four are visible, as are the chiasmata. Next, the cell enters prometaphase I, where polar and kinetochore microtubules begin to polymerize and span from pole to pole and to connect the poles to the kinetochores on the chromosomes. However, instead of attaching each of the two chromatids to opposite poles, as in mitosis, each of the two homologous chromosomes is attached to opposing poles (Figures 8.12 and 8.14).

During metaphase I, the chromosomes move to the metaphase plate by the actions of microtubules, dyneins, and kinesins. At anaphase I, each of the homologous chromosomes in each chromosome pair moves to opposite poles. During telophase I, the microtubules depolymerize and the nuclear membranes reform around the chromosomes, which partially decondense. As was stated previously, division II is a mitotic division, except that only one homologous chromosome is in each cell (Figures 8.12 and 8.14), and therefore, the resulting four cells are pseudohaploid after diakinesis, that is, they are 2*C*/1*N*. The final nuclear and cell division results in four 1*C*, 1*N* cells.

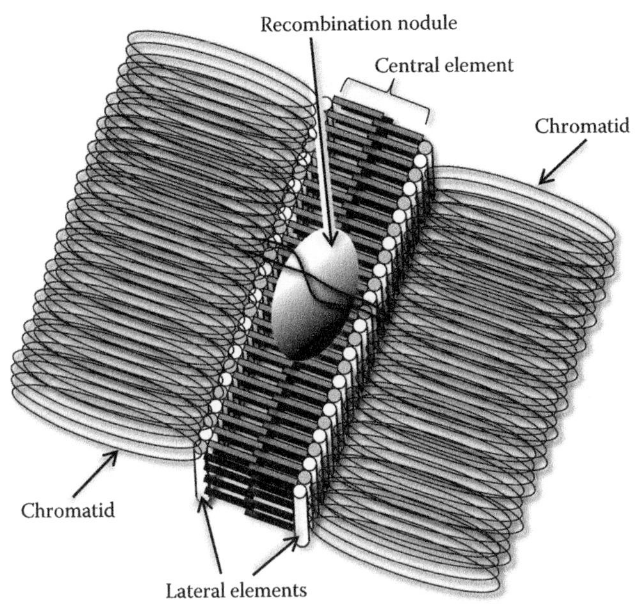

FIGURE 8.13 Synaptonemal complex that forms during prophase in meiosis. This consists of a central element made from proteins that form a structure that resembles a zipper, in that the linear proteins overlap in the middle. Lateral element proteins attach to the central element and also bind to the chromatids (chromatin), which are organized in spirals. Recombination modules are spaced along the synaptonemal complex and are responsible for generating crossovers between the chromatids. The proteins involved in this structure are unique to prophase I of meiosis.

SEXUAL REPRODUCTION

Meiosis evolved as a way to produce new combinations of alleles, but was coupled with sexual reproduction. Questions still arise about how and why sexual reproduction evolved and was so successful in eukaryotes. Of the many theories as to the successes of sexual reproduction and diploidy, one of the most compelling is that it allows new combinations of genes to be tested, and it allows beneficial alleles to survive even when a deleterious mutation has occurred in another allele. Therefore, the genome is buffered somewhat from the effects of new mutations that might be disadvantageous for the organism. Another evolutionary benefit, which is evident from the sizes of eukaryotic genomes, is that it allows genomes to expand because more than one allele for each locus is present, thus buffering against a disadvantageous mutation. In essence, this implies that mutation rates can be higher in diploid sexual organisms compared to a haploid organism with the same genome size.

As to how sexual reproduction evolved, some parts of the process are present in some bacteria. They can exchange DNA either by plasmids, viruses, or pili. Plasmids are coated or uncoated circular DNAs ranging in size from a few kilobases to several hundred kilobases. Some transfer their DNA in ways that are similar to the transfer of viruses from one cell to another. Viruses can also transfer bits of DNA, by first incorporating DNA from one cell, integrating it into its genome, and then transforming another cell through infection. Pili are connections between bacteria that bring bacteria together in order to transfer lengths of DNA, sometimes more than 100 kb long. This has been termed *bacterial sex*, and while it differs from true sexual exchange in eukaryotes, it has some analogous properties and benefits.

Many types of bacteria contain more than one copy of their genomes, becoming diploid or polyploid under some conditions, and some are often polyploid, at least during some parts of their life cycles. This has at least two advantages. First, it allows them to produce all of the ribosomal RNAs

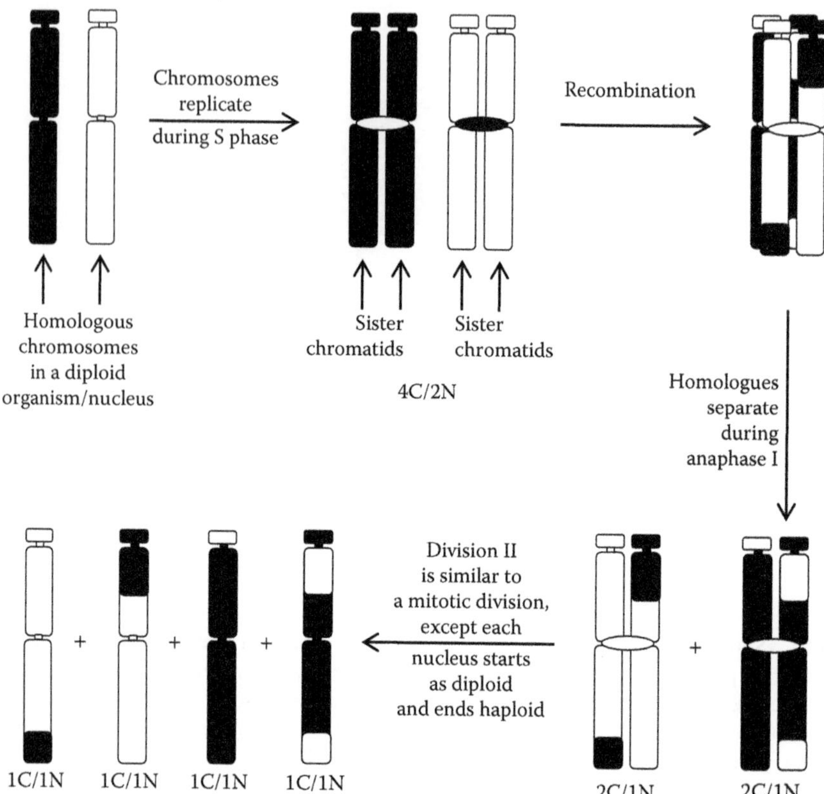

FIGURE 8.14 Movements of chromatids during meiosis. Once the chromosomes condense during prophase in division I, crossovers occur between the four chromatids. During the first division, the homologues separate (which differs from mitosis where the sister chromatids separate). The second division is similar to a mitotic division, in that the sister chromatids separate. However, when the cells begin this division, they are at $2C/1N$ levels and therefore are half of the values of nuclei during prophase in mitosis ($4C/2N$).

and ribosomes that they need to produce the proteins they need for life and reproduction. This is because of the need to increase gene copy number to produce large numbers of ribosomes. Second, it allows the organisms to survive, even if some mutations have occurred in some of the alleles of various gene loci. Therefore, producing more than one copy of the genome, and exchange of pieces of DNA, as well as recombination, all are present in Bacteria (as well as Archaea), which all are precursors to sexual reproduction.

Sex appears to have begun formally in some of the first eukaryotes, because it has been observed in most eukaryotic phyla that have been carefully studied. In multicellular organisms, there often are two (or more, in some cases) sexes, often with dimorphic characteristics. However, in many of the unicellular organisms, while there are two mating types and they reproduce by forming haploid gametes that combine to form the diploid zygotes, the cells are often morphologically indistinguishable from one another. Mitosis was present in early eukaryotes, although they could have been haploid or diploid organisms. Meiosis essentially was an addition to mitosis in diploid organisms that was a means to produce recombinant chromosomes and to reduce the number of chromosomes to the haploid level in preparation to fertilization to form a diploid zygote. Selection caused this retention of diploidy, because the formation of tetraploids is usually lethal to the cell. Thus, meiosis to form gametes, and joining of gametes to form a zygote evolved as a single system.

Not all eukaryotes have been demonstrated to have sexual stages. In fact, many fungal species have been found only in haploid forms. Whether these species have lost their ability for sexual reproduction, or it occurs so rarely that it has never been observed, is unknown. Some species exhibit variations of the sexes. For example, in humans there are normally just males and females (although hermaphrodites sometimes occur). In dioecious plants, there are male plants and female plants. However, many plants are monoecious, having both male and female flowers on the same plant or male and female organs within the same flower. Likewise, for some animals, there are hermaphrodites, having both female and male organs, and they can self-fertilize. In many cases, the males of these species are haploid organisms. In some arthropods, females can produce females and/or males parthenogenetically (i.e., without sex). In general, unicellular and multicellular eukaryotes exhibit a variety of patterns of reproduction, from haploid formation of spores, parthenogenesis, and sexual reproduction that includes sexes other than only males and females. Therefore, there are many ways to segregate, duplicate, and join DNA in various organisms. Each process has provided some advantage to the particular species during evolution, such that a variety of systems now exist.

KEY POINTS

1. Eukaryotes have true mitosis and cytokinesis; there are similarities to fission in Bacteria and Archaea. Specific proteins attach the chromosomes to membranes in all the cells. In mitotic cells, the chromosomes also are attached to microtubules, which are responsible for moving the chromosomes to the poles of the cell during anaphase.
2. The amounts of DNA in Bacteria and Archaea can vary greatly depending on the types of cells and their response to environmental conditions.
3. Many variations of mitosis exist in various groups of organisms.
4. For a given species of eukaryote, DNA amount and number of chromosomes per cell and per nucleus may vary depending on the type of cell, developmental stage, and responses to environmental conditions.
5. Meiosis consists of two divisions, including an extended prophase (where recombination events occur) leading up to the first division. During the first division, homologous chromosomes are separated (in contrast to mitosis, where sister chromatids are separated). Division II is similar to a mitotic division. In an evolutionary sense, the sequence of appearance was: fission, mitosis, and then meiosis.
6. Sexual reproduction coevolved with meiosis to provide a means to join new combinations of genes and alleles to provide evolutionary advantages to the progeny.

ADDITIONAL READINGS

Alberts, B., D. Bray, K. Hopkin, A. Johnson, J. Lewis, M. Raff, K. Roberts, and P. Walter. 2013. *Essential Cell Biology*, 4th ed. New York: Garland Publishing.

Alberts, B., D. Bray, A. Johnson, J. Lewis, M. Raff, K. Roberts, and J. D. Watson. 1994. *Molecular Biology of the Cell*. New York: Garland Publishing.

Alexopoulos, C. J. and C. W. Mims. 1979. *Introductory Mycology*, 3rd ed. New York: John Wiley & Sons.

Becker, W. M., J. B. Reece, and M. F. Poenie. 1996. *The World of the Cell*, 3rd ed. New York: The Benjamin/Cummings Publishing.

Gibson, J. L., M.-J. Lombardo, P. C. Thornton, K. H. Hu, R. S. Galhardo, B. Beadle, A. Habib et al. 2010. The σ^E stress response is required for stress-induced mutation and amplification in *Escherichia coli*. *Mol. Microbiol.* 77:415–430.

Gladfelter, A. and J. Bermen. 2009. Mitosis in filamentous fungi. *Nat. Microbiol. Rev.* 7:875–886.

Nagl, W. 1981. Polytene chromosomes in plants. *Int. Rev. Cytol.* 73:21–53.

Stahl, F. W. 1988. Bacterial genetics. A unicorn in the garden. *Nature* 335:112–113.

Straube, A., I. Weber, and G. Steinberg. 2005. A novel mechanism of nuclear envelope break-down in a fungus: Nuclear migration strips off the envelope. *EMBO J.* 24:1674–1685.

Tropp, B. E. 2008. *Molecular Biology, Genes to Proteins*, 3rd ed. Sudbury, MA: Jones & Bartlett.

Section III

Genetics

9 Mendelian and Non-Mendelian Characters

INTRODUCTION

Gregor Mendel presented and published his genetic studies of garden peas only a few years after Charles Darwin published his book on natural selection. However, while Darwin became a celebrity, Mendel's publication was met with puzzlement and skepticism and was essentially ignored until well after his death. When rediscovered early in the twentieth century, his single publication, supported by publications from three other scientists, became the foundation of classical genetics. Mendel began his experiments long before he published the work, and at one point, he was convinced to give up on the experiments. He resumed the experiments a few years later and ultimately presented his results at a scientific meeting on March 8, 1865. The work was published in 1866. Most who listened to, or read, his results failed to understand the significance, and many dismissed the findings, as they were contrary to some ideas about inheritance at the time. To be fair, his presentation and discussion of the results failed to clearly define the tenets of what is now called *Mendelian genetics* and *Mendelian inheritance*. Also, the relatively lengthy publication was in German, which limited its distribution and exposure. Over the next several decades, the publication was cited three times in the scientific literature. But, in 1900 (14 years after Mendel's death), Hugo de Vries, Carl Correns, and Erich von Tschermak-Seysenegg independently rediscovered, confirmed, and publicized Mendel's work. However, while the term *genetics* was coined in 1905 by William Bateson, the units of inheritance (i.e., the genes), their organization, and their molecular nature were not clarified until later in the twentieth century. Today, classical Mendelian genetics is focused on determining how genes are segregated and inherited each generation. While it basically follows genes as they exist in diploid organisms, its principles are applicable to bacteria, archaea, and eukarya (in general).

Mendel, as a monk, had time to work in the garden. While he observed and catalogued the physical characteristics of many plants, he chose garden peas (*Pisum sativum*) for his published experiments. Peas can self-pollinate or cross pollinate. Thus, inbred lines can be allowed to self-pollinate such that with repeated selection and self-pollination, specific traits reliably appear in every individual of the next generation (i.e., the traits breed true). While Mendel knew nothing about DNA, genes, segregation of genes, ploidy, mitosis, or meiosis, he did observe that the traits appeared in predictable patterns. Traits, such as flower color, seed coat color, seed morphology, plant height, and others, were chosen because their inheritance was stable and predictable. The various traits appeared to be independent of one another, and the proportions of the traits that appeared in each generation could be anticipated. He also determined that some traits could be hidden for a generation, only to appear again in the following generation. By allowing self-pollination for several generations, he was selecting for homozygous plants. We now know that the hidden traits were the result of recessive traits that were masked by the effects of the dominant alleles in heterozygous plants. Although it seems simple now, Mendel performed extensive experiments for many years and analyzed huge amounts of data before he presented the results. Even then, his findings were controversial, and only a few scientists understood the significance of the results. Unfortunately, he died more than a decade before his work became the spark that initiated the inception of genetics.

ALLELES

Alleles are different versions of genes that perform the same function (i.e., they are orthologs) that are generally in the same genetic locus for all of the individuals/taxa being examined. Their frequencies and distributions are dictated by several factors, including mutation rate, genetic interactions among individuals within the population (e.g., bacterial conjugation, fungal anastomosis, and mating), natural selection, gene flow (migration; introduction of alleles into or out of a population), and genetic drift (allele variation within the population without selection). Population size and mode of reproduction are important modulators of all of the factors. While the alleles vary in their DNA sequences, in many cases they also vary with respect to their expression characteristics, which sometimes cause changes in the functions and/or concentrations of the corresponding proteins. Some alleles are under selection, whether positive or negative, but most may be under no selection at all. They are said to be neutral, thus conveying neither positive nor negative fitness to the organism (in their current environment). These alleles change randomly as the individuals arise in the population, and thus, the alleles may become established within the population or the alleles may disappear rapidly if the individuals fail to reproduce. Some alleles may simply hover at a particular proportion in the population for many generations without much change (Figure 9.1). These processes represent random genetic drift. They are changes that are ongoing continuously within the population, and which do little to change the dynamics of the individuals in the population. However, occasionally environmental changes produce selective pressures on some of the previously neutral alleles that cause them to become under either positive or negative selection.

THE BASICS OF MENDELIAN INHERITANCE

Mendelian inheritance is based on the segregation and independent assortment of alleles in diploid organisms. For example, in garden peas, the allele for purple flowers is dominant (denoted with a capital letter, e.g., B), and the allele for white flowers is recessive (denoted with a lower case letter, e.g., b). A plant with white flowers must have a bb genotype (homozygous for the recessive allele), while a plant with purple flowers can be either BB (homozygous for the dominant allele) or Bb (heterozygous for the flower color alleles; Figure 9.2). During segregation of the alleles, a BB parent

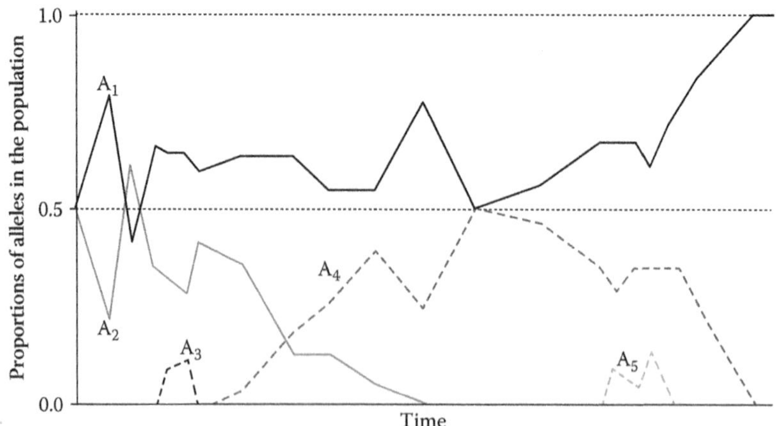

FIGURE 9.1 Time course of the proportions of various alleles within a population. Most of the alleles vary in their proportions due to genetic drift caused by random sampling from one generation to the next. The most prominent allele is A_1 through most of the time course and becomes the only allele late in the time course. Allele A_2 shows more instability in its proportion and eventually goes extinct. Alleles A_3 and A_5 appear and each go extinct after a short time. The A_4 allele appears early and appears to be headed toward becoming established in the population, but eventually goes extinct.

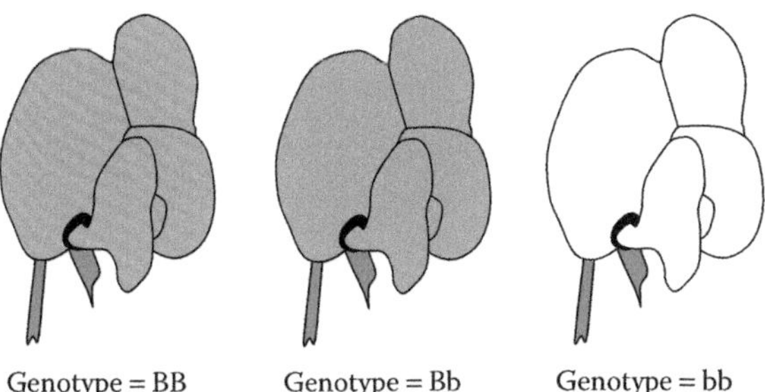

FIGURE 9.2 Phenotypes and genotypes of flower color in garden peas, the subject of part of Mendel's experiments.

can only contribute the dominant B gametes, a bb parent can contribute only the recessive b alleles, but a heterozygous parent can contribute both B and b alleles in their gametes. This can be illustrated using Punnett squares to predict the genotypes of the progeny. When a homozygous BB plant self-pollinates, it will produce only progeny that also produce purple flowers, all of which will have a BB genotype (Figure 9.3). Similarly, when a homozygous bb plant self-pollinates, it will produce only progeny that produce white flowers, all having the bb genotype (Figure 9.4). When a heterozygote Bb plant self-pollinates, approximately 75% of the progeny will produce purple flowers, while approximately 25% of the progeny will produce white flowers (Figure 9.5). The progeny that produce purple flowers will consist of two different genotypes, homozygous BB plants and heterozygous Bb plants, while the progeny with white flowers will be bb homozygotes (Figure 9.6). Furthermore, the Bb heterozygotes will comprise approximately 50% (two of the four) of the progeny that are

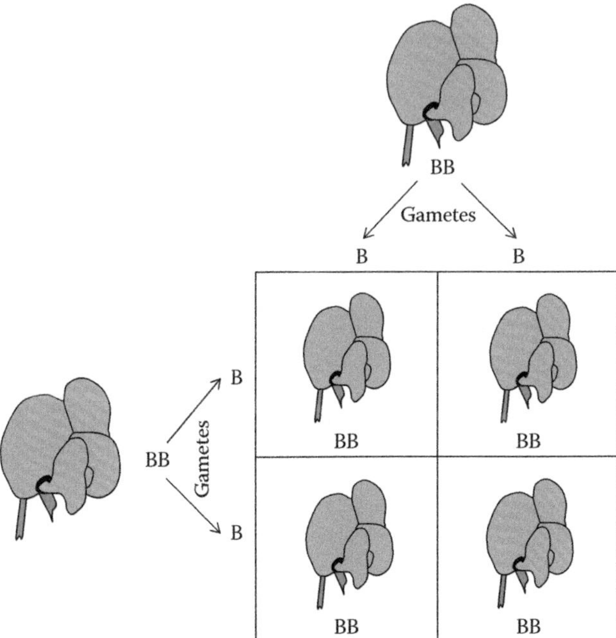

FIGURE 9.3 Punnett square showing the breeding of the dominant trait for flower color, which is purple. Mendel bred plants with purple flowers until all of the plants in each generation produced purple flowers. Although he did not know it, he was producing plants that were homozygous for the dominant allele for flower color.

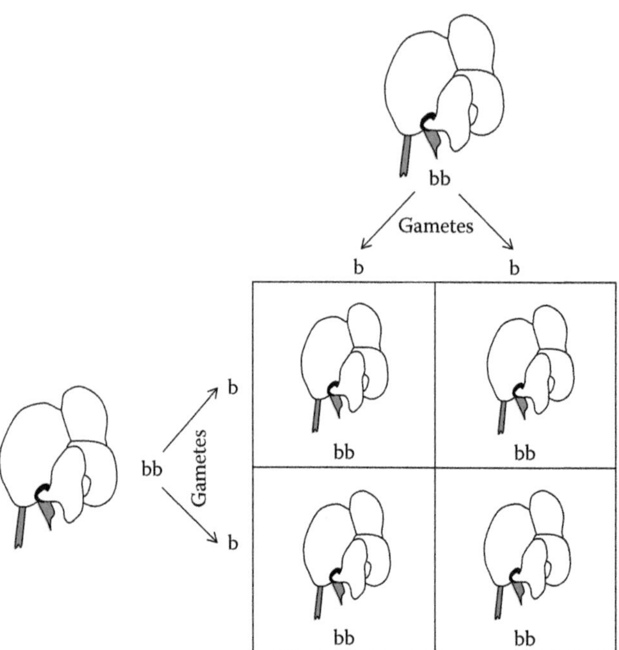

FIGURE 9.4 Punnett square showing the breeding of the recessive trait for flower color, which is white. Mendel bred plants with white flowers until all of the plants that were always produced white flowers in every generation. He was producing plants that were homozygous for the recessive allele for flower color.

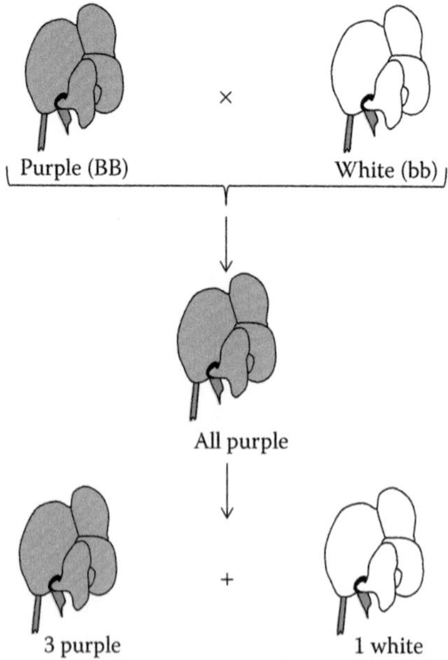

FIGURE 9.5 Phenotypic analysis of one of the experiments performed by Mendel. When true breeding plants with purple flowers (homozygous for the dominant allele) were crossed with true breeding plants with white flowers (homozygous for the recessive allele), all of the progeny of the first generation (F_1) produced purple flowers. When those plants were allowed to self-pollinate, plants with purple flowers and those with white flowers were produced in a 3:1 ratio (respectively).

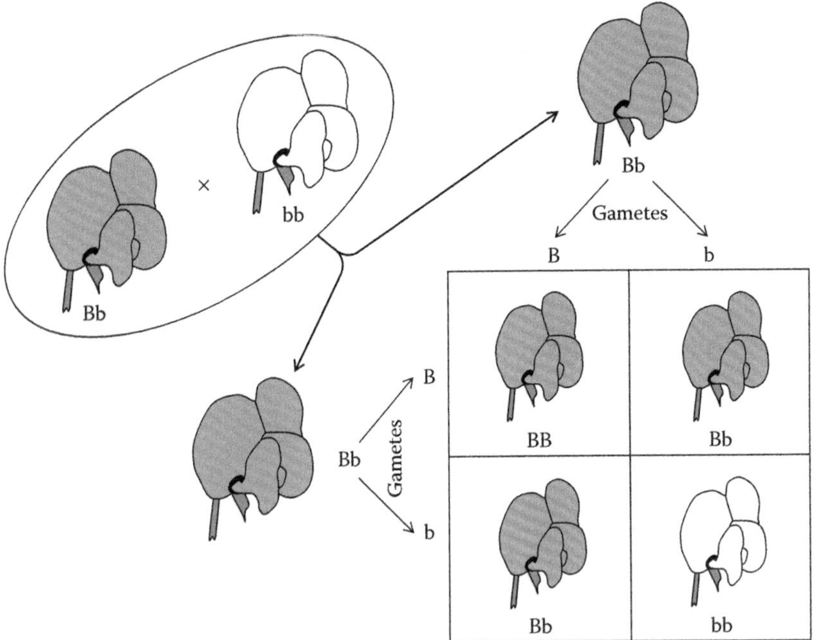

FIGURE 9.6 One of the experiments performed by Mendel on the inheritance of flower color. After he bred plants for several generations to produce purple flower and white flower lines, he crossed them, which produced all purple flower plants. They all were heterozygotes (which was unknown to Mendel). When these were allowed to self-pollinate, progeny with purple flowers and those with white flowers were produced in a 3:1 ratio. Punnett square analysis shows that the genotypes were 1BB:2Bb:1bb.

produced, while the BB homozygotes will be 25% (one of the four) of the progeny and the bb homozygotes will be 25% (one of the four) of the progeny.

Mendel knew nothing of genes and alleles. He observed traits, or characters (he used the German term *merkmal*). He allowed plants with purple flower phenotypes to self-pollinate for several generations until all of the progeny every time all had purple flowers. He also allowed plants with white flower phenotypes to self-pollinate for several generations until all of the progeny produced white flowers every time. Then, he gathered pollen from the plants with the purple flowers and pollinated white flowers, and took pollen from white flowers and pollinated the purple flowers. He allowed them to fruit, collected the seeds, grew the plants, and recorded their phenotypes for many characters, including flower (i.e., petal) color. All of the plants in this F_1 (first filial) generation produced purple flowers. When he allowed these to self-pollinate to form fruit, the plants in the F_2 generation produced flowers that were either purple or white. Furthermore, there were three plants that produced purple flowers for each plant that produced white flowers (i.e., a 3:1 ratio). Mendel found the same proportion for many characters, including plant height, pod color, position of flowers on the plant, and many others. These were traits with two alleles, one of which was dominant, and the other that was recessive (Figure 9.7).

More than one set of genes can be analyzed using Mendel's methods and Punnett squares as long as there are only two alleles for each gene locus (for diploid organisms), the alleles segregate independently, and one allele for each gene locus is dominant over the other. For example, two of the characters that Mendel followed were seed color and seed shape. He found that the yellow seed color was dominant over green and that the round seed character was dominant over a wrinkled shape. He again bred plants so that they had either all dominant (round yellow seeds) or all recessive (wrinkled green seeds) characters for several generations. Then, he cross-pollinated the plants that displayed the dominant characters with the plants that had all recessive characters. In the

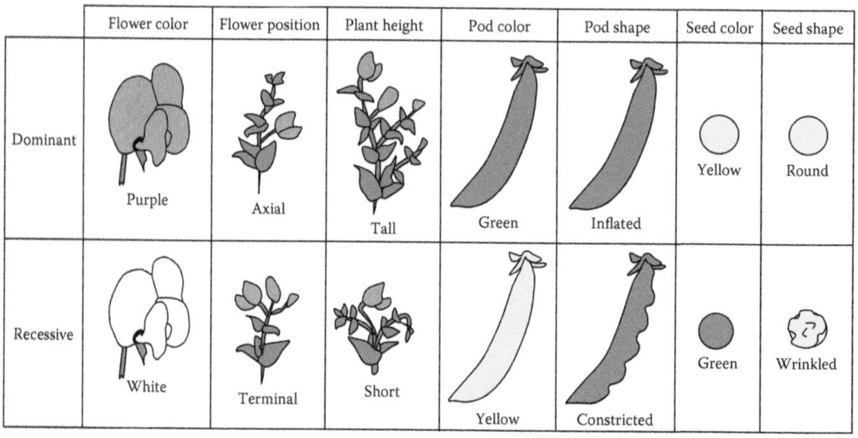

	Flower color	Flower position	Plant height	Pod color	Pod shape	Seed color	Seed shape
Dominant	Purple	Axial	Tall	Green	Inflated	Yellow	Round
Recessive	White	Terminal	Short	Yellow	Constricted	Green	Wrinkled

FIGURE 9.7 Dominant and recessive characters chosen by Mendel for his experiments with garden peas.

F_1 generation, all of the plants had all dominant characters. They all produced round yellow seeds. In the F_2 generation (following self-pollination), four phenotypes were observed in ratios of nine plants that produced yellow round seeds, three plants that produced yellow wrinkled seeds, three that produced green round seeds, and one that produced green wrinkled seeds (i.e., a ratio of 9:3:3:1; Figure 9.8). It is now known that gene alleles at each of the two genetic loci on the chromosomes

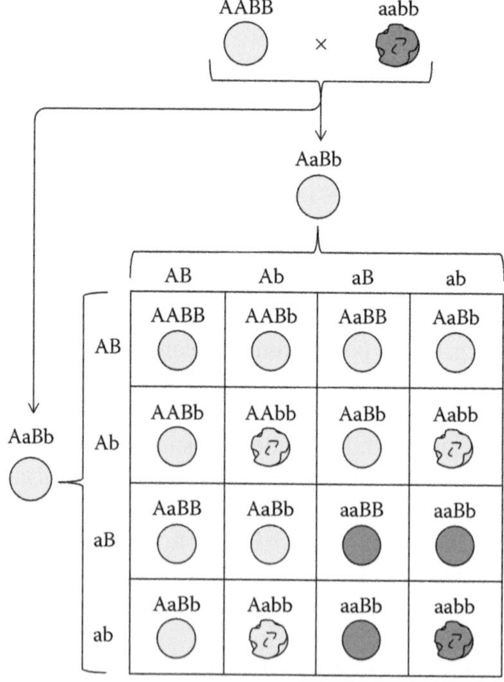

FIGURE 9.8 Inheritance analysis with two alleles. Mendel bred lines of plants so that they had either all dominant or all recessive traits and then crossed them. In this case, plants with dominant alleles for seed color and shape (yellow and round, respectively) were crossed with plants with recessive alleles (green and wrinkled, respectively). The resulting F_1 progeny all had yellow round seeds. When these were allowed to self-pollinate, four phenotypes were produced: plants that produced yellow round seeds, those that produced yellow wrinkled seeds, those that produced green round seeds, and those that produced green wrinkled seeds. They were consistently produced in a 9:3:3:1 ratio (respectively). Punnett square analysis shows the resulting genotypes that explain the phenotypes.

are responsible for the phenotypes, and that each of the alleles segregates during meiosis, and are independently inherited during the formation of the zygote.

CODOMINANCE, INCOMPLETE DOMINANCE, OVERDOMINANCE, AND UNDERDOMINANCE

While many genetic loci are occupied by alleles that exhibit dominance over other alleles, and therefore are amenable to analysis by Mendelian methods, most have more blended characters and are more difficult to discern. Some exhibit incomplete dominance, where one allele is only partially dominant over the other allele(s). For example, some of the genes that determine feather colors in birds may produce a mottled, intermediate, or blended pattern of color when two different alleles are expressed in the same individual. Similarly, some alleles are codominant, which results in the production of the characteristics of both alleles. One example is the ABO blood group genotypes (Figure 9.9). The A homozygotes produce only the A antigen, and B homozygotes produce only the B antigen, while the AB heterozygotes produce both A and B antigens, and those that are

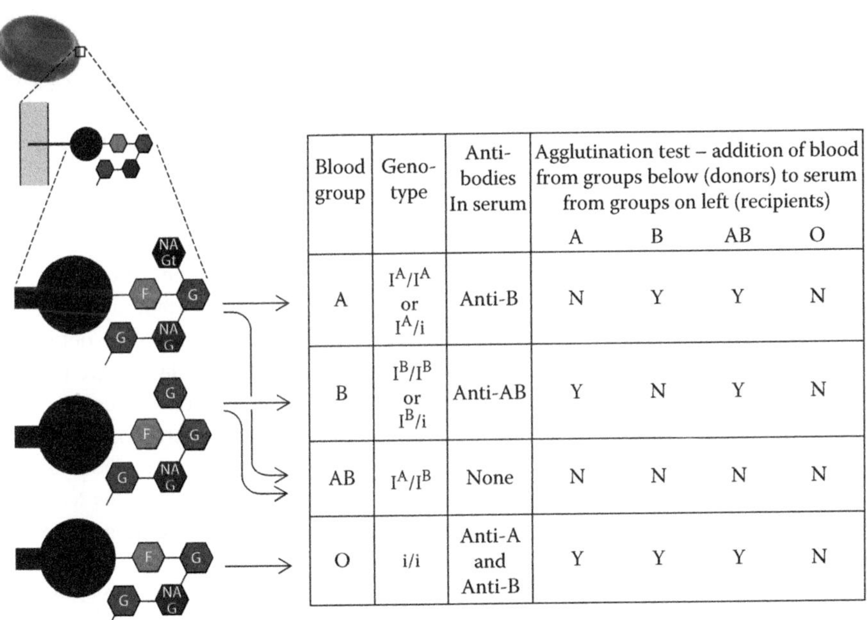

Blood group	Genotype	Antibodies In serum	Agglutination test – addition of blood from groups below (donors) to serum from groups on left (recipients)			
			A	B	AB	O
A	I^A/I^A or I^A/i	Anti-B	N	Y	Y	N
B	I^B/I^B or I^B/i	Anti-AB	Y	N	Y	N
AB	I^A/I^B	None	N	N	N	N
O	i/i	Anti-A and Anti-B	Y	Y	Y	N

FIGURE 9.9 Structure, phenotypes, and genotypes of the ABO blood groups. The basis of the ABO blood groups is a cell surface protein that is glycosylated to varying degrees (top left). A cell surface protein is first glycosylated by an enzyme produced by a gene that exists in most (but not all) human populations. The sugar group contains a fucose (F), two glucose moieties (G), and an *N*-acetyl-glucosamine (NAG). Individuals in the A blood group (genotypes I^A/I^A, or I^A/I, see table) produce an enzyme that adds an *N*-acetyl-galactosamine (NAGt) to this glycosylated protein. Individuals in the B blood group (genotypes I^B/I^B, or I^B/I, see table) produce an enzyme that adds a glucose moiety, rather than a NAGt to the sugar group. Those in the AB blood group are heterozygous (I^A/I^B), and have both enzymes, and thus have both glycosylated cell surface proteins on their red blood cells. Individuals of type O are homozygous for defective A and/or B genes. They have the basic glycosylated protein, but without the addition of the NAGt or additional glucose. After transfusion with foreign blood, agglutination can occur, due to antibodies against the cell surface glycoproteins. Individuals that are type A have anti-B antibodies in their blood serum and therefore can accept blood from type A or type O individuals. Those that are type B have anti-A antibodies and can accept blood from type B or type O individuals. Those that are type AB can accept blood from any blood group, while those that are type O can donate blood to those of type A, B, AB, and O. This system is an example of codominance.

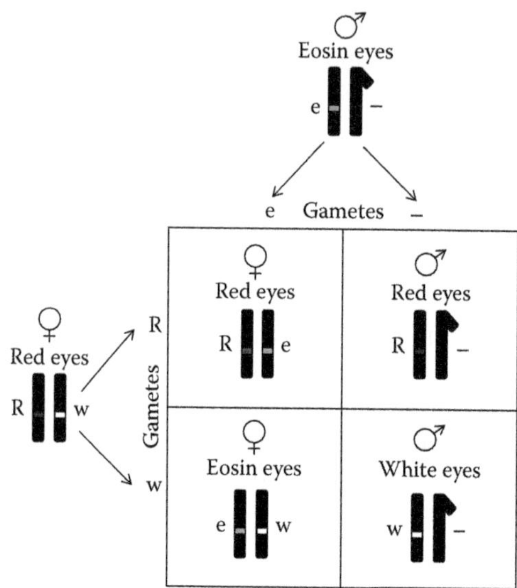

FIGURE 9.10 Punnett square showing the inheritance of eye color in *Drosophila melanogaster*. The gene locus for eye color is sex linked, being on the X chromosome. Thus, females can be homozygous or heterozygous, while males are hemizygous (having the gene on only the X chromosome of the XY pair). Red is the wild type dominant allele. When a heterozygous R/w (red/white) female is crossed with a male with eosin eyes (e allele, or any recessive allele for eye color), the resulting progeny are one female with red eyes (R/e), one female with eosin eyes (e/w), one male with red eyes (R/–), and one male with white eyes (w/–); a 2:1:1 phenotypic ratio of red:eosin:white (respectively) eye color.

homozygous for the i allele produce neither the A nor the B antigens, and thus are classified as type O. Another example of this is eye color in *Drosophila melanogaster* (fruit fly). While there are numerous alleles for eye color, for each cross only three alleles participate in the determination of eye color. This is because the gene is on the X chromosome, of which females have two and males have one, for a total of three possible alleles. When a male with the red eye color allele (dominant) is mated to a heterozygous female (e.g., the white allele on one X chromosome and the eosin allele on the other X chromosome), all of the female progeny will be heterozygous with red eyes (Figure 9.10). However, half of the males will have white eyes and the other males will have an eye color based on the other recessive allele. Therefore, instead of resulting in a Mendelian 3:1 ratio of phenotypes, the phenotypic ratios are 2:1:1 (red:eosin:white, respectively).

Overdominance and underdominance occur when the survival of the homozygotes differs significantly from the survival of the heterozygotes. This can occur from interactions of the products of each of the two alleles, or with their interactions with other molecules and biological systems within the organisms. Overdominance is observed when the heterozygotes survive to a greater extent than do the homozygotes (dominant or recessive alleles). Therefore, the fitness of both of the homozygous genotypes is lower than that of the heterozygous genotypes. Conversely, underdominance occurs when there are fewer heterozygous progeny than expected. This implies that the fitness of both types of homozygotes is higher than that expected of the heterozygotes. For overdominance and underdominance, the fitness of the heterozygote genotype differs from the average fitness of the homozygous dominant genotype and the homozygous recessive genotype.

EPISTASIS

The products of each of the alleles of each gene locus are never isolated from those produced by other genes, and there often are interactions between them, and in some cases, the products interact directly with the genes to change their expression. In many cases, this leads to situations that produce progeny whose phenotypic ratios differ from the standard Mendelian ratios. For example, coat color in some rodents is controlled by two different genes (denoted A and C), each with more than one allele, and the gene products interact to produce the coat color. When a mouse is homozygous A/A C/C, it is a banded color (called *agouti*). If it is crossed with a mouse that is homozygous a/a and c/c, which has a solid white coat, all of the F_1 progeny are banded (i.e., agouti), and all have an A/a C/c genotype. When male and female heterozygotes from the F_1 generation are bred, banded and solid white progeny are produced in a 9:4 ratio, and there are also three solid black progenies produced, for a 9:4:3 ratio of agouti:white:black phenotypes. The genotypes of the agouti progeny are A/A or A/a, and C/C or C/c. The solid white progeny all are homozygous c/c, but of the four, only one is a/a, and the others are either A/A or A/a. The three solid black mice are all homozygous a/a, but are either C/C or C/c. Therefore, instead of recovering a Mendelian ratio of 9:3:3:1, the ratio is 9:4:3, due to the interaction between the products of the alleles from two loci (Figure 9.11). In this case, the C locus is responsible for producing the dark pigment, while the A locus modulates the amount of pigment that is produced.

A similar example of epistasis occurs with flower color in some species of sweet pea. In this case, the genetic loci are designated C and P. When the loci are homozygous recessive for either locus, that is, either c/c or p/p, the flowers are white, no matter the genotype at the other locus. However, when both loci are either homozygous dominant, that is, C/C and P/P, or heterozygous (i.e., they each have at least one dominant allele at each locus), then the flowers are purple. Therefore, if a plant with white flowers and the genotype C/C p/p is crossed with a plant also with white flowers, but the genotype c/c P/P, all of the F_1 progeny will be C/c P/p, and they produce purple flowers (Figure 9.12).

AaCc (Agouti) × AaCc (Agouti)	AC	aC	Ac	ac
AC	AACC (Agouti)	AaCC (Agouti)	AACc (Agouti)	AaCc (Agouti)
aC	AaCC (Agouti)	aaCC (Black)	AaCc (Agouti)	aaCc (Black)
Ac	AACc (Agouti)	AaCc (Agouti)	AAcc (White)	Aacc (White)
ac	AaCc (Agouti)	aaCc (Black)	Aacc (White)	aacc (White)

FIGURE 9.11 Punnett square analysis of epistasis of alleles coat color in rodents. In this case, the products of the A and C genes interact to cause an inheritance pattern that differs from a standard Mendelian pattern. In this example, a male and a female, both of which are heterozygous for both loci, are mated, and their progeny are scored for coat color. The A gene product controls how much pigment is produced by the product of the C gene. When both dominant A and C alleles are present (homozygous or heterozygous for each), the coat color is tan and banded along the animal, called *agouti*. However, when the homozygous recessive a allele is present with the dominant C allele (homozygous or heterozygous), pigment is produced in higher amounts and the animal is gray to black without banding. In an individual that is homozygous for the recessive c allele, no pigment is produced, regardless of the A allele, and therefore, the individual is solid white. Therefore, the proportions of coat color in the progeny is 9:4:3, for agouti:white:black (respectively).

CcPp (Purple) × CcPp (Purple)

	CP	cP	Cp	cp
CP	CCPP (Purple)	CcPP (Purple)	CCPp (Purple)	CcPp (Purple)
cP	CcPP (Purple)	ccPP (White)	CcPp (Purple)	ccPp (White)
Cp	CCPp (Purple)	CcPp (Purple)	CCpp (White)	Ccpp (White)
cp	CcPp (Purple)	ccPp (White)	Ccpp (White)	ccpp (White)

FIGURE 9.12 Punnett square analysis of epistasis in sweet pea flower color. In this case, the product of the C gene and that of the P gene are enzymes within the same pathway that produces pigment. The recessive alleles produce nonfunctional enzymes. When at least one dominant allele is present at each of the two loci, then purple pigment is produced. However, for plants that are homozygous recessive for either locus, no pigment is produced, and the flowers are white. This produces progeny that produce purple or white flowers in a 9:7 ratio (respectively).

If these plants are allowed to self-pollinate, some of the F_2 progeny they produce will have purple flowers, and some will have white flowers, in a ratio of 9:7 (respectively). The nine with purple flowers will have genotypes of C/C or C/c, and P/P or P/P, while the seven with white flowers will each have genotypes where at least one of the loci is homozygous for the recessive allele (c/c and/or p/p). In this case, both the P and C genes normally produce enzymes within a pathway that produces purple pigment. If either enzyme is missing, the pathway is incomplete, and no purple pigment is produced (Figure 9.13). The recessive genes produce nonfunctional enzymes (or produce no product), and therefore, if either locus is homozygous recessive, the pathway is truncated.

QUANTITATIVE TRAIT LOCI

Many phenotypic traits vary in a continuous manner. That is, there are gradations of the trait, rather than an all-or-nothing phenotype. This results from the expression of several or many genes and their alleles, each of which contribute to the final phenotype. They are termed quantitative trait loci, or QTLs. The majority of genes within a genome are of this type. The genes produce enzymes within multistep biochemical pathways, or interconnecting pathways. Some produce DNA binding proteins that control the expression of other genes. Some of these trait systems consist of fewer than a dozen genes, while it has been estimated that others may be the result of the action of over 100 genes. The shift from one allele to another at just one locus sometimes produces changes that can be readily observed. However, some produce changes that are so subtle they cannot be observed. Because of this, discerning changes in these systems is complex and difficult.

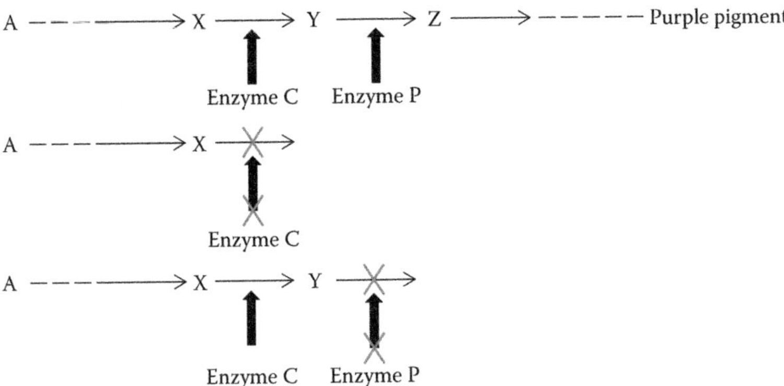

FIGURE 9.13 Pigment pathway (with functional enzymes) and those that are truncated. When the C enzyme and the P enzyme are expressed, purple pigment is produced, and therefore, purple flowers result. When a plant is homozygous c/c or p/p, the pathway is truncated, resulting in no purple pigment being produced, and therefore, white flowers result.

RECOMBINATION AND LINKAGE

As was discussed in Chapter 8, recombination occurs in almost all organisms and can create new alleles, new combinations of alleles, eliminate detrimental alleles, repair broken chromosomes, and perform other functions. Alleles are segregated during meiosis both by separation of the chromatids/chromosomes, as well as via recombination. If the genes/alleles are far from one another or they are on different chromosomes, then they segregate independently. However, if they are on the same chromosome, they are linked, at least at some level. If there is no recombination, as occurs in fruit fly males, then the genes/alleles on the same chromosome cannot segregate independently and therefore always are found together in the next generation. For most organisms, recombination does occur between homologous chromatids and chromosomes, and therefore, if the genes/alleles are far from one another on the same chromosome, they can segregate in an independent fashion. However, if they are close together on the chromosome, the probability of a crossover becomes less likely, because there are on average only one or a few crossovers per chromosome during meiosis, and even fewer during mitotic divisions. Therefore, genes/alleles that are close to one another on the chromosome are tightly linked, and therefore often are inherited together, and do not segregate independently.

NON-MENDELIAN TRAITS

In many cases, genetic loci in diploid eukaryotes have at most two different alleles, just as Mendelian genetics predicts. However, this is not true for Bacteria, Archaea, and some Eukarya, which normally have one allele per locus present in each cell, because they all have only one copy of their genomes per cell. Because of this, Mendelian genetic systems, per se, do not exist in these organisms and cell types. Instead, the cell populations must be assessed for the presence and proportions for the various alleles. Some bacteria maintain two copies or more of their chromosomes within each cell, either constantly or under specific conditions. However, lacking meiotic systems, segregation of the alleles, and formation of zygotes never occur, which are necessary in a Mendelian system.

Many genetic loci in diploid eukaryotes have more than two alleles. This can occur when there has been an unequal crossover within a locus, which often results in two alleles occurring in tandem on one chromosome. If this is combined in a zygote with a chromosome that has a single allele at the same locus, then the organism may have three functional alleles. If it is paired with a chromosome that also has two tandem copies of the allele, then four alleles could be present. This situation might seem rare, but this is often how multigene families are formed. Early in the process, the alleles may be expressed at the same time, creating a situation where there is a blended phenotype. Later in the evolution of a multigene family, the various versions of each gene are expressed in a specific pattern. They are usually found in multicellular organisms and are important components in developmental processes.

Multigene families also exhibit non-Mendelian segregation in multicellular organisms. Ribosomal RNA, histone, and other genes that are tandemly repeated on chromosomes undergo frequent unequal crossovers in somatic cells, such that each cell in the organism can be from slightly different to extremely different than other cells in the organism with respect to the number of gene copies and the number of different alleles that it contains. Although this has been documented in plants, fungi, and animals, it may occur in most eukaryotic species. For example, in broad bean (*Vicia faba*), up to a 95-fold difference in the number of rRNA gene copies per haploid set of chromosomes was found in a single population of plants (Figure 9.14). The range was from 230 copies to more than 22,000 at a single locus. Furthermore, the plants showed no obvious phenotypic differences. Within an individual plant, the rRNA gene copy number range was as high as 12-fold, indicating the presence of frequent unequal crossovers in the somatic cells (i.e., during mitotic divisions) of the plant. There were also allelic differences in the rRNA gene repeat units, and when the alleles were tracked in the F_1 and F_2 generations, the pattern of inheritance differed significantly from a simple Mendelian pattern (Figure 9.15). Again, the results indicated recombination events occurred frequently in somatic cells (Figure 9.16). Similar experiments were carried out for histone genes, which also occur in tandem repeats, with similar results. However, because the copy numbers for these genes are lower than for rRNA genes, the differences were smaller than those found for rRNA genes. These frequent changes are caused by frequent recombination events in the somatic cells of the plants (Figure 9.17), but also occur in most eukaryotes that have been studied.

In plants, rRNA genes, as well as other genes, can be induced to change. In flax (*Linum usitatissimum*), several genotypes (called *genotrophs*) can occur within a single population. Phenotypically, one is tall, one is short, and one is intermediate in height. When the total amount of DNA is measured in cells of each of the genotrophs, the amounts vary significantly (Figure 9.18). The tall genotroph has the most DNA per nucleus, the small genotroph has the least, and the intermediate genotroph has an amount between the two. The differences are in the tens of millions of base pairs, so there are very large differences in the amounts of DNA in each of the genotrophs. The intermediate genotroph is called the *plastic genotroph* because if it is treated in certain ways (e.g., growing it in certain fertilizers, changing the soil pH), the characters of the plant and the amount of DNA in the nuclei of the plant will change, such that the progeny will be all of the tall type or all of the short type. These changes occur gradually as the plant grows, and these changes can be traced up the stem of the plant while the plant is growing in the various conditions. After many generations of growing under standard conditions, both the tall genotrophs and the small genotrophs gradually

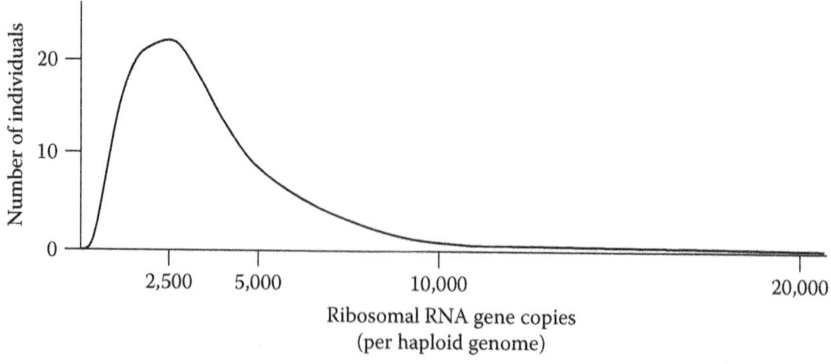

FIGURE 9.14 Range of ribosomal RNA gene copies in broad bean (*Vicia faba*). When a population of more than 400 individual plants was assayed for rRNA gene copy number, a 95-fold range was measured. While the mean rRNA gene copy number was approximately 2500, the range was from 230 to nearly 22,000 copies per haploid genome. All of the plants were similar, with no obvious phenotypic differences.

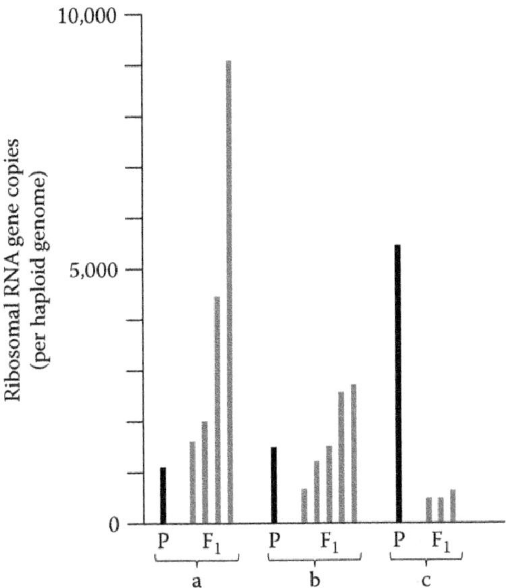

FIGURE 9.15 Examples of the range of rRNA gene copies among F_1 progeny produced through self-pollination of the parent. In the first example (a), the parent had approximately 1100 rRNA gene copies per haploid genome, while all of the progeny had greater numbers of rRNA genes, including one that had approximately 9000 copies. In the second example (b), some of the progeny had smaller numbers than the parent, while others had greater numbers of rRNA genes. In the third example (c), the parent had approximately 5500 rRNA gene copies per haploid genome, while each of the progeny had only 500–600 copies.

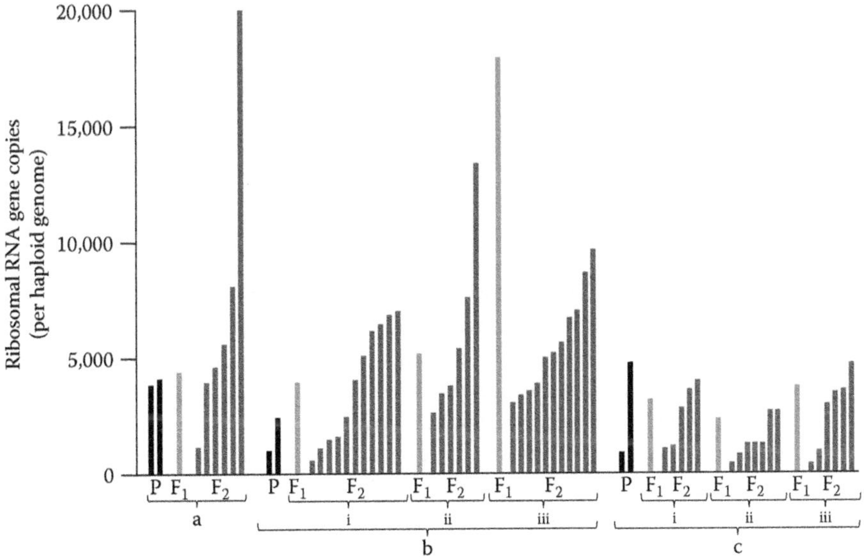

FIGURE 9.16 Examples of the range of rRNA gene copies among F_1 progeny produced through cross-pollination of parents with different rRNA gene coy numbers, as well as among the F_2 progeny produced through self-pollination of the F_1 plants. In the first example (a), while the parents and the F_1 plant all had relatively similar rRNA gene copy numbers (approximately 4,000 copies each), the range in the F_2 progeny was from 1,500 to 20,000. In the second example (b), each the three F_1 individuals had rRNA gene copy numbers that were higher than for either of the parents. In the F_2 progeny, the rRNA gene copy number ranged from approximately 500 to 13,000. In the third example (c), the ranges in rRNA gene copy number are similar for the parents, F_1 plants, and F_2 plants.

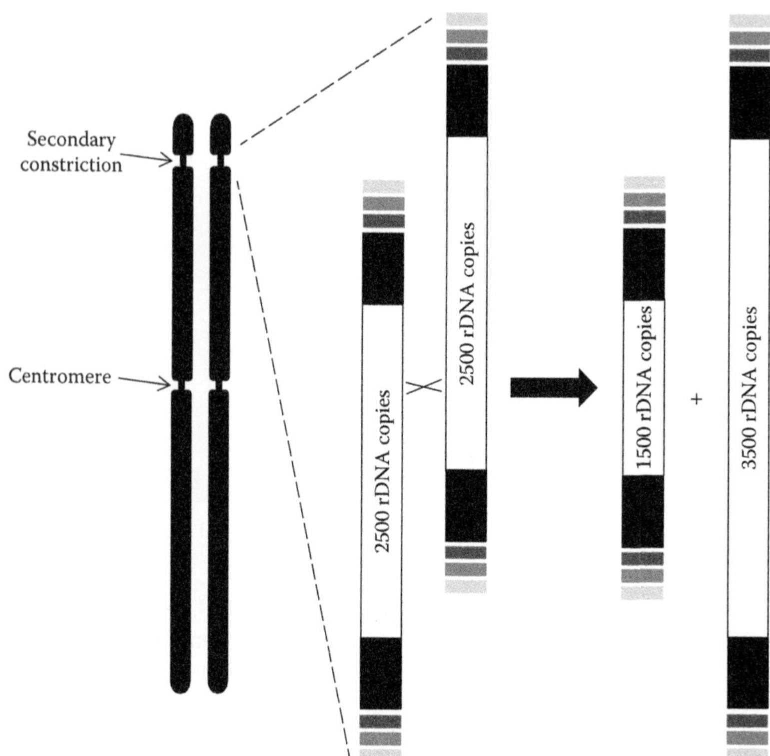

FIGURE 9.17 Model for changes in rRNA gene copy number by recombination. The genes for rRNA exist in long tandem arrays in most eukaryotic genomes. Most occur at a single locus. The number of genes is known to change rapidly in the somatic cells of many eukaryotes. The mechanism for this change is probably via unequal crossovers during mitotic divisions. This can occur because each of the tandem repeats is very similar or identical to each of the other repeats. Therefore, they can line up shifted from one another by one or more repeats during a crossover event, which produces two chromosomes that differ from the original two chromosomes. One chromosome will gain rRNA gene repeat units, while the other will lose the same number. When there are allelic differences among the repeats (usually the differences are in the spacer regions), the distributions of alleles may also change at each locus during a recombination event.

revert back to plastic genotrophs. These continue to produce progenies that are plastic genotrophs, unless they are grown under conditions that will then induce them to produce tall or short progeny.

Some species are the result of hybridization events between two different species. Often, many of the genes are similar among the two genomes, although not all of the duplicated genes are expressed, and many mutate and become nonfunctional genes (i.e., pseudogenes). Many times, entire chromosomal regions are lost during the evolutionary sorting out process in the hybrids. Many plant, amoebae, and other species are polyploid. Sometime in their evolutionary history, they each replicated their chromosomes without undergoing mitosis and cytokinesis, and in doing so, additional sets of chromosomes resulted. This created genomes that originally had multiple copies of all genes and alleles, all carried on separate copies of the chromosomes. Each of the copies of the genes and alleles may mutate isolated from all of the other copies creating new alleles of those genes. Many often mutate to the point of being nonfunctional, but many also have slightly different functions than the original gene. Some have become nonfunctional pseudogenes. In many polyploidy species, large sections of some parts of one or more of the duplicated chromosomes have undergone many mutation events, including large deletions. This was a consequence of sorting out of the most advantageous duplicated genes from those that were less advantageous, neutral, or detrimental (Figure 9.19). Some multicellular organisms also contain polyploid cells as a normal

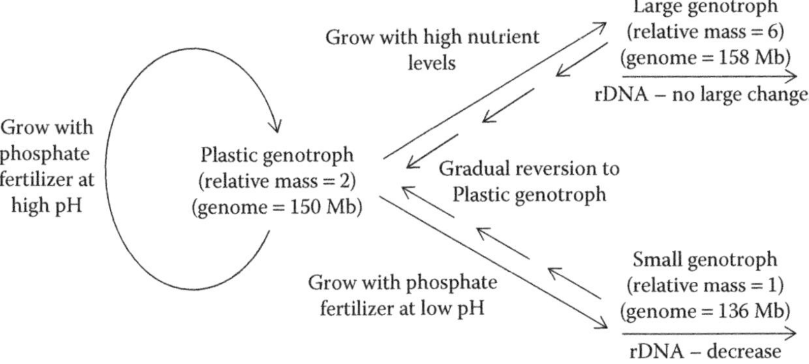

FIGURE 9.18 Genomic and phenotypic changes induced in flax (*Linum usitatissimum*) by various growth conditions. Within a population of flax plants, three different forms (called *genotrophs*) can be found. The large genotroph is more massive than the other two genotrophs and has more DNA per nucleus than in the other two genotrophs. The small genotroph is the smallest in mass and has the least DNA in its nuclei. The plastic genotroph is intermediate between the two in mass and in the amount of DNA in its nuclei. However, the plastic type exhibits rapid phenotypic and genotypic changes in response to environmental conditions. When grown with ammonium as the sole nitrogen source, a plastic genotroph begins to change, and all of its progeny are small genotrophs, with smaller mass and less DNA per nucleus. When grown under standard conditions and without ammonium as the sole nitrogen source, these small genotrophs produce progeny that are also small genotrophs. However, after many generations, the progeny revert back to the plastic type. Similarly, if the plastic genotrophs are grown with a phosphate fertilizer at low pH, the plants will change, and their progeny all will be large genotrophs, with larger mass and more DNA per nucleus. Again, these progeny will stably produce progeny of the large type for many generations. When grown in non-stressful conditions, their progeny will eventually revert back to the plastic type.

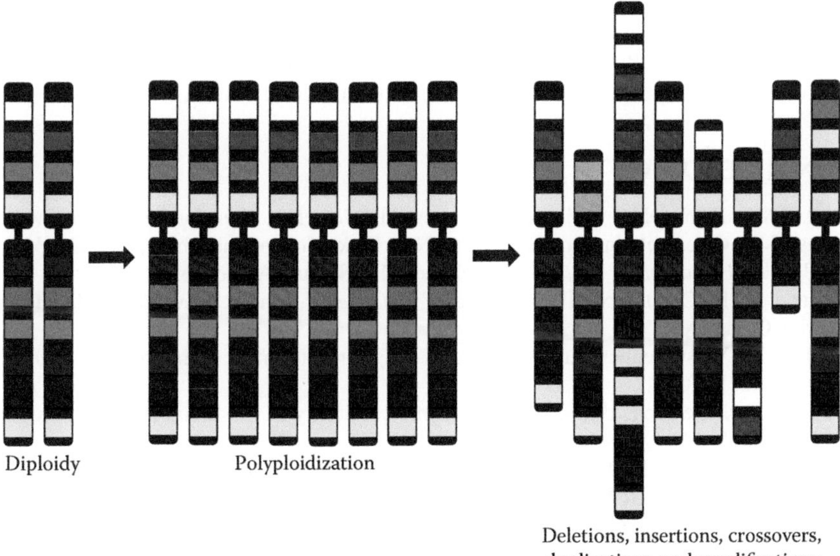

FIGURE 9.19 Model for the evolution of a polyploid genome. Polyploid organisms have evolved from diploid relatives through duplication of chromosomes without cytokinesis. Following polyploidization, some parts of the duplicated chromosomes are free to mutate and change, including experiencing small and large deletions, insertions, duplications, amplifications, and crossovers.

consequence of development of particular tissues, such as the polytene chromosomes in the salivary glands of insects, the endosperm of grass seeds, the cotyledons of many plant seeds, and others. While polyploidization of the genome in gametes does become part of the evolutionary history of an organism, the polyploidization events that occur mainly in somatic tissues do not cause general polyploidization of all of the cells of the progeny and thus do not become a part of the evolutionary history of these organisms.

KEY POINTS

1. Gregor Mendel's research on the inheritance of traits in garden peas was only appreciated 35 years after it was published, and 14 years after his death.
2. Mendel's research dealt with genes that had two alleles, one of which was dominant and one that was recessive, in plants that could be both cross-pollinated and self-pollinated.
3. Alleles are versions of genes that perform the same function (i.e., they are orthologs).
4. Mendelian genetics applies mainly to diploid organisms that may possess two identical alleles or two different alleles at a specific genetic locus.
5. Alleles segregate during meiosis and then are brought together (usually in different combinations) during fertilization when the zygote is formed.
6. In a typical cross between individuals, each of which is heterozygous for alleles (one that is dominant, and one that is recessive), a 3:1 ratio (dominant character to recessive character) of traits will usually result in the F_1 progeny. In crosses with individuals that are heterozygous for two traits, the ratio usually is 9:3:3:1 (dominant/dominant to dominant/recessive to recessive/dominant to recessive/recessive).
7. Some genetic loci, such as the ABO blood group alleles, exhibit traits that are codominant; and others exhibit incomplete dominance.
8. Epistatic traits are those that are affected by more than one gene locus, the products of which interact to result in the final phenotype.
9. Most phenotypic traits are determined by sets of traits, called *quantitative trait loci* (QTLs).
10. Recombination and genetic linkage can affect the segregation of genetic loci.
11. Non-Mendelian inheritance is characteristic of many multicopy genes, especially those that are tandemly repeated and experience frequent recombination events.

ADDITIONAL READINGS

Cullis, C. A. 1986. Plant DNA variation and stress. In *Genetics, Development, and Evolution*. eds. J. P. Gustafson, G. L. Stebbins, and F. J. Ayala. New York: Plenum Publishing, pp. 143–155.

Futuyma, D. J. 1998. *Evolutionary Biology*. Sunderland, MA: Sinauer Associates.

Klug, W. S. and M. R. Cummings. 1994. *Concepts of Genetics*, 4th ed. New York: Macmillan College Publishing Company.

Mendel, G. J. 1866. VersucheüberPlanzen-Hybriden (Experiments on plant hybrids). *Verhand. Naturfor. Vereines Brünn* IV:3–47.

Rogers, S. O. and A. J. Bendich. 1987a. Heritability and variability in ribosomal RNA genes of *Vicia faba*. *Genetics* 117:285–295.

Rogers, S. O. and A. J. Bendich. 1987b. Ribosomal RNA genes in plants: Variability in copy number and in the intergenic spacer. *Plant Mol. Biol.* 9:509–520.

Rogers, S. O. and A. J. Bendich. 1992. Variability and heritability of histone genes H3 and H4 in *Vicia faba*. *Theor. Appl. Genet.* 84:617–623.

Russell, P. J. 1998. *Genetics*, 5th ed. Menlo Park, CA: Benjamin/Cummings Publishing Company.

10 Population Genetics

INTRODUCTION

Genes and organisms do not exist in isolation. Cells and biological molecules are always surrounded by other cells and biomolecules. Organisms exist in populations, consisting of members of the same species, and communities, consisting of populations of many species. Biomolecules interact with other organic molecules, as well as inorganic molecules and ions. This has been the case since the first protocells appeared on the Earth. Because the replication process was (and still is) imperfect, frequent mutations occurred. Some of the mutations were beneficial, in that they improved replication frequency or fidelity, or a change to an RNA molecule increased or changed the function of a ribozyme (e.g., rRNA). While many mutations decrease the fidelity of the various biochemical reactions in a cell, often leading to their extinction, a large number of mutations have no effect at all. These are neutral mutations. Because they have no net effect on the resulting molecule, these mutations can accumulate, thus creating large set of gene versions, in a process called *genetic drift*. In a population of heritable units (RNAs, DNAs, or organisms), the sequences can vary, sometimes to a large extent, while the molecules or organisms appear to function normally and uniformly. Thus, a population of molecules or organisms can have varying degrees of genetic diversity for extended periods of time and exhibit few, if any, phenotypic differences.

Some of the molecules or organisms increase in the population, while others may decrease or go extinct without the influence of any selective pressures (Figure 10.1). But, when environmental conditions change, each of the different genetic types might react differently to emerging selective pressures. Statistically, a genetically diverse population has an increased probability for reproductive success than less diverse populations, especially as environmental conditions change. This is likely how life originated and survived on the Earth. As stated in earlier chapters, RNA was likely the initial biological molecule, although it probably interacted with amino acids and peptides very early in the evolution of life on the Earth. However, not all RNA molecules were a part of this path to biology. Of the enormous number of reactions between RNA molecules with other RNA molecules, fatty acids, amino acids, carbohydrates, water, and so on, only a small set were capable of both catalysis and self-replication. Out of this small group, several were probably somewhat successful at reproducing themselves, and therefore, there probably were populations of RNA protocells that varied with respect to their total genetic contents.

RNA mutates more rapidly than DNA, such that many mutations would have been generated in much less time than in the DNA-based organisms alive today. Through genetic drift, as well as natural selection, and eventual changing local, regional, and global events on the Earth, a few of the RNA-based protocells became more capable of capturing the energy and nutrients that they needed to outcompete other protocells. Eventually, somewhere between about 3.9 and 3.5 billion years ago, one (or a small group) of protocells began to use a chemically altered type of nucleic acid that was missing the hydroxyl group on the 2′ carbon of the ribose, known as *2′-deoxyribonucleic acid* or *DNA*. This was likely caused by a mutation that might have had little functional effect at first. But, this started the transition from using RNA as the genetic material to using DNA as the genetic material. DNA had two distinct advantages over RNA as the holder of genetic information. DNA was much less reactive than RNA, because the 2′-hydroxyl is a reactive group on RNA molecules, and second, because DNA was missing the bulky oxygen atom on the 2′ carbon, it could form smooth extended double helices that provided both protection of the genetic information (the bases) and two copies of each region of the chromosome. Once DNA was adopted as the genetic material, genomes were more stable and could become larger and more complex. While genetic mutation was slower in DNA, and thus genetic drift was a slower process, the stability and increased replication fidelity probably outweighed genetic diversity.

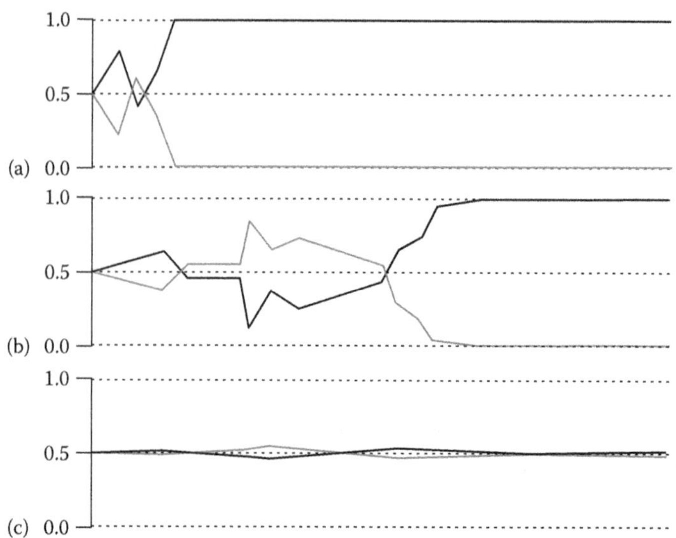

FIGURE 10.1 Time course of the proportions of two alleles in a population. In small populations (a), the allele proportions change rapidly, and often, one allele becomes fixed in the population, while the other disappears. As populations grow (b), the allele proportions are slower to change, and fixation and extinction of alleles occurs less often. In large populations (c), allele changes are less frequent and tend to hover around particular proportions.

HARDY–WEINBERG EQUILIBRIUM

In an ideal unchanging environment, the alleles will remain in a constant equilibrium with each other. For example, at time zero, if allele A_1 is present in 50% of the cells/organisms, and allele A_2 is present in 50% of the cells/organisms, then at some time later, the proportions of the alleles will also each be at 50%. However, if the proportions change, then the alleles are out of equilibrium because something has perturbed the conditions or simply because of random genetic drift. These can be calculated and considered using the Hardy–Weinberg equation. It assumes that the alleles will be at equilibrium as long as the following are true: (1) The population is infinitely large; (2) There is no mutation (including no recombination or gene duplication); (3) There is no selection; (4) There is no migration (i.e., no gene flow); (5) Mating is random; (6) Allele frequencies are equal among the sexes (for sexual organisms only); (7) All genotypes are of equal viability and fertility; and (8) The allele frequencies remain static from one generation to the next.

However, there is no ideal situation in which allele frequencies remain the same over time. But, the Hardy–Weinberg equation is useful in indicating what might be changing, and by how much, which can lead to some educated guesses about what might be occurring to change the allele frequencies. The Hardy–Weinberg equation states that the frequencies of the two alleles in a large population of diploid organisms that only reproduce sexually, and mate randomly, will remain constant as long as there are no biases with regard to allele preference. If there is a change in allele probabilities within the population, then the alleles may change in frequency; that is, they will differ from the Hardy–Weinberg equilibrium, possibly due to one of the assumptions (above) being false. Given two alleles (in a monoecious diploid, that is, one with two sexes), A and a, with frequencies of p and q, respectively, then the occurrence in the population of AA, Aa, and aa genotypes will be p^2, $2pq$, and q^2, respectively. This is analogous to a Punnett square analysis (Figure 10.2). If allele A occurs at a frequency of 0.50 and allele a occurs at a frequency of 0.50, then the proportions would be $p^2 = 0.25$, $2pq = 0.50$, and $q^2 = 0.25$. This can be graphically represented as well (Figure 10.3). Also, the sum of the frequencies equals 1 (i.e., 100%), which means that $p^2 + 2pq + q^2 = 1$. This implies that $q = 1 - p$ and $p = 1 - q$. If the alleles fail to follow this equation, then the alleles are said

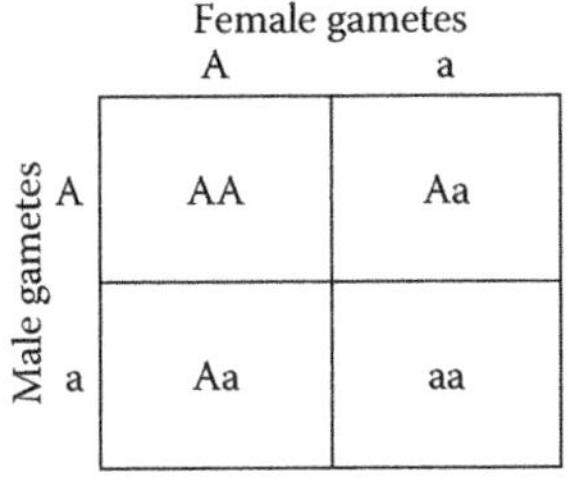

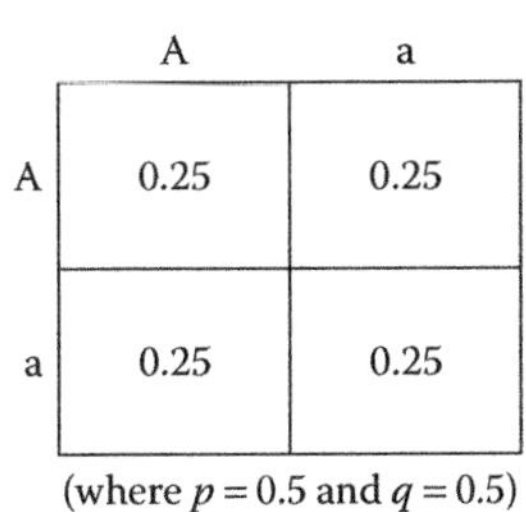

FIGURE 10.2 Comparison of Punnett square of allele combinations in progeny (left) with the proportions of the allele combinations (right). Gametes (top and left), A and a, contribute to the diploid progeny in each to produce AA, Aa, and aa individuals. The proportions of each combination (0.25, 0.50, and 0.25, respectively) of alleles are presented in the right square.

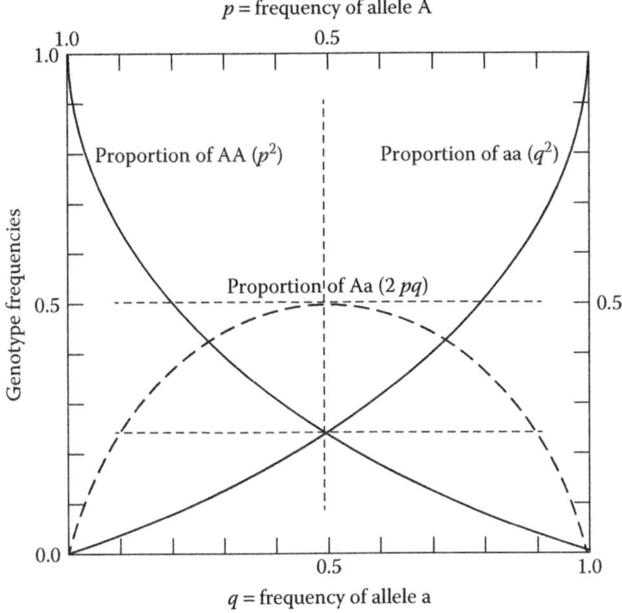

FIGURE 10.3 Graphical representation of the Hardy–Weinberg equilibrium for two alleles. Proportions for each of the homozygotes (p^2 for AA and q^2 for aa; solid line curves) and for the heterozygotes (pq for Aa; dashed line curve) are presented. When the proportion of each of the alleles is 0.50, then $p^2 = 0.25$, $q^2 = 0.25$, and $pq = 0.50$ (indicated by dashed straight lines).

to be in disequilibrium, which indicates that various factors (e.g., selection, gene flow, or multiple alleles) may be acting on one or both alleles. Although the Hardy–Weinberg equation was devised for diploid organisms that reproduce sexually, it can be extended to haploid organisms, bacteria, and archaea. However, the alleles in those organisms are each in separate cells, and therefore, the inheritance patterns may differ significantly from those for diploid organisms.

For three alleles, A_1, A_2, and A_3, the frequencies of alleles would be $p^2 + q^2 + r^2 + 2pq + 2pr + 2qr = 1$. Equations for populations with more alleles can be similarly constructed. For organisms that are polyploid, a term for the ploidy level must be added, such that for polyploid organisms with two alleles for a particular gene, the gene frequencies would be $(p + q)^a$, where a is the ploidy level. For example, for a population where the individuals are tetraploid with two alleles, the expected frequencies would be AAAA $= p^4$, AAAa $= 4p^3q$, AAaa $= 6p^2q^2$, Aaaa $= 4pq^3$, and aaaa $= q^4$. The equations continue to become more complex as the numbers of alleles and the ploidy levels increase.

The Hardy–Weinberg equilibrium equation can indicate when there are effects on the proportions of the alleles in the population. Mutation, genetic drift, gene flow, nonrandom mating, natural selection, and other phenomena can affect the proportions of the alleles in a population. Within a population, it is sometimes difficult to elucidate what phenomenon is affecting the allele frequencies. For example, mutations are not truly random. However, many models for mutation assume that they are. If there are certain types of mutations that occur more often, for example, C-to-T transitions, and an allele has an abundance of cytosine residues, then the allele frequencies could be skewed. This could first be indicated by an analysis using a Hardy–Weinberg model. Additionally, some regions of certain alleles may be more exposed to mutagens, which can affect the mutation frequencies for that allele. At the population level, random genetic drift can affect allele frequencies due to either random effects or phenomena that appear to be random.

POPULATION SIZE

All species must produce more individuals every generation than can possibly survive to reproductive age. This is necessary because few individuals survive long enough to reproduce. For some species, the number of progeny produced is only slightly more than the current population. These are the organisms whose parents usually spend a great deal of time protecting and teaching their progeny. In these cases, many (but not all) of the progeny reach reproductive maturity, although all of the mature individuals may not participate in reproduction to produce progeny (Figure 10.4). Humans, and many other mammals, have life histories of this type. On the other end of the spectrum, there are species that produce huge numbers of progeny per generation, most of which do not survive to reproductive maturity. Many types of arthropods have this type of life history. One mating may produce thousands of offspring. However, only a small proportion survives to eventually reproduce. Predation, disease, pesticides, storms, and other events reduce their numbers to levels that are sometimes barely sufficient to assure a robust population for the next generation.

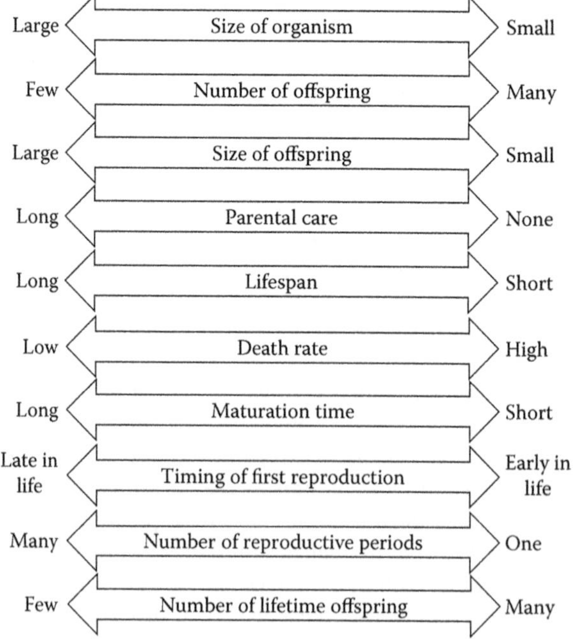

FIGURE 10.4 Range of characteristics of life histories. Some organisms (left) produce few progeny, nurture the progeny, and many survive to reproduce. Other species (right) produce large numbers of progeny, have high death rates, and low numbers of individuals that survive to reproduce.

Even with humans, the probability of being born is extremely small. A human male produces approximately 525 billion sperm during his lifetime. You might think that this is enough to produce at least millions of offspring. However, a human female releases only about 400 viable eggs during her lifetime. The chance of the one sperm and one egg that formed the zygote that produced you is approximately one in 210 trillion (525 billion × 400). And, out of all of that genetic material, couples usually only produce one to several children. You are one lucky person. Why produce so much sperm? Why produce so many eggs? One part of the answer is that not all eggs and not all sperm are viable, and not all combinations of eggs and sperm will produce a viable offspring. Not all matings produce a zygote. Not all sperm cells and egg cells ever get a chance to meet. So, although the numbers seem large, evolution has led to the production of huge numbers of sperm and egg cells because the chances of a successful fertilization event are so small. In fact, in couples where the amount of sperm produced by the male is low, but well above zero, often the couple is infertile. For cells and organisms, in general, excess is needed to increase the probability of survival. Also, there is competition among sperm, and this is especially true when more than one male mates with a female for many species. The semen of each contains chemicals and components that will inactivate other sperm cells. So, there are competitive forces that act not only at the level of the entire organism to find a mate, but there is competition between sperm cells to find and fertilize an egg cell. Evolutionary forces have led to the production of large numbers of sperm and egg cells in sexual organisms in order to overcome all of the forces restricting fertilization. Presumably, the species that had fewer of these cells disappeared because they could not maintain a population sufficient to prevent their extinction.

Most plants produce huge amounts of pollen and large numbers of oocytes (egg cells). In fact, clouds of pollen grains are released from many tree species in the spring. The number of pollen grains and egg cells far exceeds the number of individual seeds that are produced. And, the number of seeds that are produced far exceeds the number of seeds that eventually germinate and grow into mature plants. Plants are constantly exposed to stressful conditions. They are exposed to heat, cold, drought, floods, poor soils, UV irradiation, high fluxes of light, insufficient light, animals eat them, microbes infect them, other plants shade them, and people spray them with herbicides. Additionally, the seed must land in a spot with sufficient moisture, heat, nutrients, and light to germinate and grow to maturity. It is amazing that any of them survive, but a few do grow to maturity. However, it is because of all of these challenges that plants have evolved to produce so much pollen, numerous egg cells, and large numbers of seeds. This is the only way that they have been able to produce enough different genetic versions of themselves to allow a small proportion to survive into each successive generation to reproduce.

Fungi produce huge quantities of spores. As with plants, they have evolved this strategy because they are exposed to a large number of challenging conditions that lead to the failure of most spores to survive. Fungi can reproduce asexually through the production of haploid spores (called *conidia*) that are produced by haploid hyphae (a filamentous growth form), or sexually, when two haploid hyphae of different mating types meet and fuse (anastomose) to form a dikaryotic hypha. They then form a fruiting body, undergo recombination and meiosis, and form new genetically distinct haploid spores. Because mating only occurs between hyphae of different mating types, there must be sufficient numbers of spores to produce the large numbers of individual hyphal nets. Most spores produced by fungi are spread through the air, and some are spread in water, which has led to extremely broad distributions for some fungi. This has been a very successful trait for some species that have worldwide distributions.

LIFE HISTORIES

Allele proportions in a population are dependent on the type of organism, its life cycle, and the gene(s) being evaluated. Most bacteria have only one copy of their genome, and therefore, single alleles of each gene are usually present in each cell (see Figure 6.2). However, several different alleles can exist within the population. Some bacteria maintain more than one copy of their chromosome (see Figure 7.1), and therefore, two or more alleles for a gene are possible in a particular cell. Some bacterial and archaeal populations of cells can be studied under laboratory conditions, and their genomes are relatively small compared to eukaryotic genomes, such that genetic experiments of mutation rates, selection, allele frequencies, and gene flow can be readily measured and/or manipulated. However, because they have no sexual reproduction, segregation of alleles does not occur in these organisms as it does in diploid eukaryotes. Nonetheless, population studies of bacteria have provided a great deal of information regarding mutation rates, natural selection, gene flow, fitness, and genetic drift, most of which occur by the same mechanisms as those in both unicellular and multicellular eukaryotes. For example, average mutation rates for genes in bacteria are similar to those for eukaryotes, and when mutagens are used on bacteria, the effects are similar to those when eukaryotic cells are exposed to the same mutagens. Additionally, various genotypes can be grown together in various concentrations to study gene flow, selection, and random genetic drift within a population. The conditions can also be altered to study aspects of each of the processes. Because the cells grow so rapidly (doubling times are minutes or hours for most species), replicates of the experiments are readily possible. Long-term experiments lasting decades have been performed with *Escherichia coli* to study various phenomena. These experiments are impossible to perform on multicellular eukaryotes, although the results apply to these organisms. A great deal has been learned about mutation rates under standard conditions, as well as how mutation rates can greatly increase under stressful conditions, such as cell crowding, nutrient limitation, temperature changes, exposure to toxins, and other stressors. Additionally, selective pressures and interactions of various genotypes and alleles have been studied. Many of the fundamental aspects of cell-to-cell interactions, reaction to stress, causes of mutations, DNA repair mechanisms, and other cell and molecular aspects of evolution have been elucidated by experimentation using bacterial cells (primarily *E. coli*).

Unlike bacteria and archaea, many eukarya exist primarily as diploids, and therefore, it is often the case that two or more alleles for the same gene locus are present in each cell. However, some eukaryotes spend most or all of their lives as haploids, a condition that is closer to that of bacteria than to what is considered a standard eukaryote. Also, some eukaryotes are polyploid or hybrids (crosses between different varieties or species), and often, their genomes contain many alleles for the same gene locus. Additionally, some eukaryotes rarely (if ever) reproduce via sexual reproduction. The asexual forms for a large number of species are known, while the sexual forms are completely unknown (and may not exist) for many of them. Nonetheless, studies using unicellular eukaryotes, such as amoebae and fungi, have proven useful in determining some of the fundamental tenets of population genetics, many of which are applicable to multicellular eukaryotes. *Saccharomyces cerevisiae* has been especially useful in this regard, because this yeast undergoes meiotic divisions and mating and also often grows as a haploid (Figure 10.5).

Therefore, although the cells are small and have genomes that are closer in size to bacteria than to most multicellular eukaryotes, most of the genetic and molecular mechanisms that are present in multicellular eukaryotes are also present in *S. cerevisiae*. Yeast cells grow primarily as haploids, so some aspects of population genetics are more similar to those for bacteria than to organisms that are primarily diploids. Studies of these haploid yeast cells have shown that the mechanisms of evolution are nearly identical, although somewhat more complex, to those in bacteria. However, because yeasts undergo meiosis and sexual reproduction, recombination and allele segregation in progeny is much more analogous to those processes in multicellular eukaryotes.

Multicellular organisms present yet another twist in the generation, distribution, and evolution of alleles within a population of a species. Although a few species of bacteria form filaments, and therefore can be considered marginally multicellular, many different phyla of eukaryotes exhibit multicellularity as haploids, as diploids, and some as polyploids (Figure 10.6). When discussing multicellular organisms, it is important to recognize that there are different levels of complexity among eukaryotes. In some, the cells of the organism are mainly alike and simply form filaments or

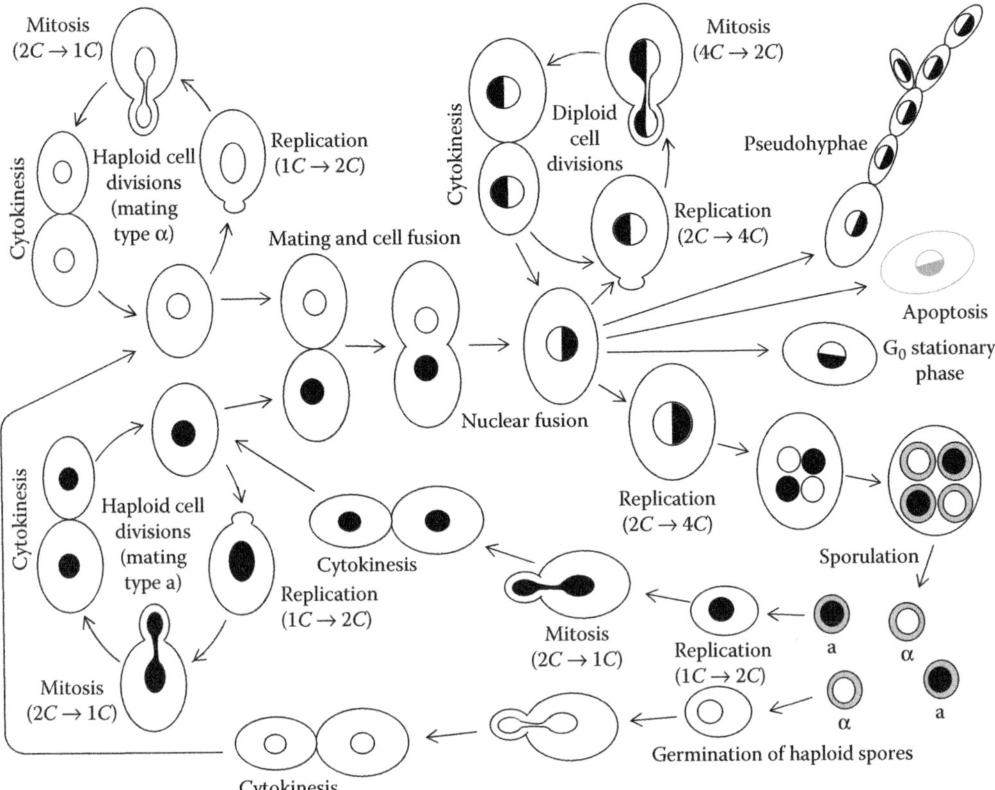

FIGURE 10.5 Life cycle of *Saccharomyces cerevisiae*. This yeast has both extensive haploid and diploid phases. The haploid cells go through many cycles of budding, mitosis, and cytokinesis to produce new haploid cells. When haploid cells of opposite mating type (one that is type a and one that is type α) meet, they mate by fusing together. The two nuclei then fuse to form a diploid nucleus. One of four possible outcomes is possible. The diploid cells can enter a diploid cell cycle that includes replication, budding, and cytokinesis. Alternatively, instead of producing yeast cells, pseudohyphae can be produced, where the cells do not separate after cytokinesis, and they have elongated cell shapes. The diploid cells also can proceed through apoptosis (programmed cell death). The final possible pathway is through sexual reproduction. The nucleus first undergoes DNA replication to form a 4C nucleus. The nucleus then undergoes meiosis, which includes recombination. Sporulation produces the four haploid spores, in which the alleles have recombined (two of each mating type). The spores later germinate to produce additional haploid cells.

sheets of nearly identical cells. As such, each cell is essentially already compatible with other cells in the filaments or sheets. However, often some of the cells become specialized for reproduction, and this occurs both with haploid and diploid organisms. Therefore, if these multicellular organisms are each considered as cell populations, their positions from a population genetics standpoint are mainly limited, because only a select few participate in reproduction of the species, and therefore, it is these cells that determine the genetic makeup of the succeeding generations. In more complex organisms, many cell types are produced, each of which must be compatible with the other cell types, even while they have different gene expression patterns. If the expressed products are incompatible with one another, toxic combinations may result that could be debilitating or lethal to the entire organism. Therefore, although only a select few cells participate in forming the next generation, their genes must interact properly in the entire organism in the next generation or they will perish without reproducing. In this way, multicellular organisms have increased selective pressure on some of their genes, specifically those that are crucial for proper tissue and organ functioning, as well as intercellular compatibility within the organism. This requires that large multicellular

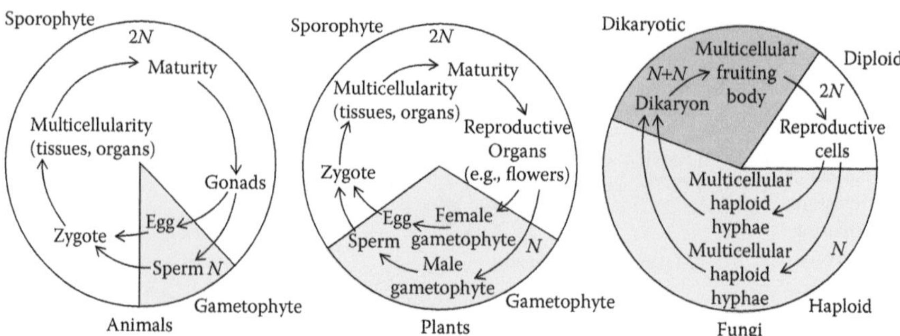

FIGURE 10.6 General life cycles for animals, plants, and fungi. For most animals, the sporophyte (diploid stage) is the predominant form, while the gametophyte consists primarily of the sperm and egg cells. In plants, the sporophyte is the dominant form for vascular plants, while the gametophyte is small, consisting of only a few cells, including an oocyte in the female gametophyte and sperm cells in the male gametophyte. For lower plants, the gametophytes are small, but are the predominant form, while the sporophytes are usually smaller and short lived. In fungi, cells grow mainly as haploids. Haploid hyphae of different genotypes grow together to form dikaryotic cells, which then form a fruiting body. Eventually, in some cells, the two nuclei fuse, the DNA is replicated, and the nucleus undergoes meiosis to form haploid spores.

organisms have genomes that are more tolerant of mutations, which is likely one of the reasons that multicellular organisms are diploid. If one of the alleles is somehow faulty in some tissues, the other allele may compensate for the deficiency. This is less likely to occur if the cells are haploid. While eukaryotes all have haploid stages, the haploid generations are often small and produce only one or a few cell types.

MODES OF REPRODUCTION

For sexual species, the amount of allelic diversity is dependent not only on the population size, but also on whether the species exhibits inbreeding, outbreeding, frequent mating, and other factors. Inbreeding limits the genetic diversity and can lead to problems with homozygosity of detrimental alleles. However, it may lead to less incompatibility among various allelic combinations. On the other hand, outbreeding usually produces more allelic diversity and less chance of homozygosity of detrimental alleles. However, it may lead to some allelic combinations that cause disease or they may be lethal.

Many animal species reproduce mainly by parthenogenesis, essentially producing clones of themselves. In some cases, this is an extreme example of inbreeding. Many arthropod species reproduce this way during much of their life cycle. For example, female aphids produce only female progeny during the summer. All are clones of the female parent, and all are born as nymphs that grow into adults (Figure 10.7). That is, the female does not lay eggs, but gives birth to live female nymph progeny (i.e., they are viviparous) that are identical to the mother, and they may number in the thousands from a single mother. Recombination and gene flow are close to zero, such that allelic proportions within a clonal population are constant during the summer. In the fall, there is a switch, and the females produce more females, and some of them lay eggs that hatch the following spring, again producing clones of the female parent. Some species of aphids produce both females and males during the fall. Furthermore, males and females often both have wings, and therefore, they can fly to other locations carrying their component of genes to mix with those of another population some distance away. At this time of year, the males deposit their sperm in the females, and the females lay the fertilized eggs, which develop and hatch the following spring. Because of mainly parthenogenesis during the summer, the allele proportions and combinations during the summer may be very different than those in the fall and early spring, when migration (i.e., gene flow) and sexual reproduction are more frequent.

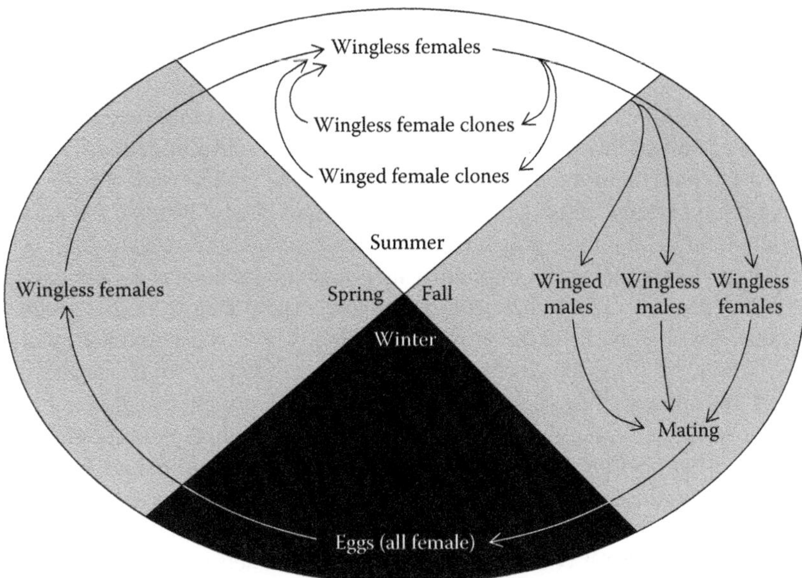

FIGURE 10.7 Life cycle for many species of aphids. When eggs hatch in the spring, all the emerging aphids are female. During the summer, these females produce clones of themselves that are born as fully formed nymphs. Thousands of these female clones may be produced by a single female. Most develop into wingless females, while a few develop wings and migrate to other plants. During the fall, females, as well as wingless and winged males, are produced. After mating between a male and a female, the female lays fertilized eggs that survive through the winter. The cycle then repeats beginning in the spring.

While most multicellular organisms are dioecious (producing both female and male sexes), some, such as the nematode *Caenorhabditis elegans*, produce no true females, but produce only hermaphrodites (that produce both eggs and sperm) and males, and the hermaphrodites outnumber the males by 1000-to-1. Most of the progeny are produced by the hermaphrodites, which produce both eggs and sperm. The eggs are fertilized with sperm while still inside the hermaphrodite, and then the fertilized eggs are laid. Therefore, the progeny are clones of the hermaphrodite parent. Occasionally, a male is produced, which lacks one of the sex chromosomes. Males migrate to find a hermaphrodite and copulate to transfer sperm that fertilizes the eggs, thus creating new individuals that generally produce more diverse allelic combinations. Another twist on this type of sexual system has been described for species from another nematode genus, *Mesorhabditis*. As with *C. elegans*, there are only hermaphrodites and males, but reproduction is only via parthenogenesis. Copulation between a hermaphrodite and a male must occur before the female begins producing egg cells and fertilized eggs. However, the sperm from the male only stimulates production of eggs in the hermaphrodite. The sperm from the male does not fertilize the egg cells. The eggs are fertilized by the sperm of the hermaphrodite, and therefore, the fertilized eggs are clones of the hermaphrodite. While this may seem to be an evolutionary dead end, species of this genus have existed for millions of years.

Many plants can reproduce vegetatively (Figure 10.8), and about 8% of plant species propagate exclusively in this manner, which is somewhat analogous to parthenogenesis in animals, because only the clones of the parent plants are produced. As with animals, this limits allelic diversity. However, plants have very active recombination systems in somatic cells, and when under stress, mutations increase as well (see Figures 9.14 through 9.18). Therefore, changes in allelic diversity may change in a plant population even in the absence of sexual reproduction. Also, they generally have more genes and larger genomes than most animal species, thus creating the possibility of greater genetic diversity. Many plants are at least partially self-pollinating, which also tends to limit allelic variation within populations. However, most plants are also cross-pollinating, and therefore, they are somewhat analogous to some animal species that can reproduce parthenogeetnically, as well as sexually. Many plant species are monoecious (Figure 10.8) and are capable of self-pollination. Most of them have evolved mechanisms to avoid self-pollination or decrease its frequency. For those species that are monoecious, primarily producing flowers that have both male and female reproductive tissues, several mechanisms have evolved to limit the amount of self-pollination. The most common mechanism is differential maturation of the sexual organs. The anthers (male floral organs) and the carpels (female floral organs) mature at different times. Individual flowers in the area also mature at slightly different times, and thus, pollen from one flower cannot pollinate the carpel on the same flower, but it can pollinate the carpel of another flower whose carpel has a receptive pistil (which is where the pollen grain sticks and begins to germinate). The pollen from that second flower has already been shed long before the carpel was fully mature, so that those grains cannot fertilize the carpel on that flower. In other species, the carpels mature before the anthers, and therefore, the carpels of one flower already have been fertilized by pollen from another flower by the time the pollen is released from the former flower. For some species, individual plants produce male flowers and female flowers, but also produce flowers that have both male and female organs. Depending on the flowers being produced, the flowers can be self-pollinated or cross-pollinated. Many plant species are dioecious, that is, there are individuals that produce only male flowers and others that produce only female flowers. Of course, these are incapable of self-pollination and therefore must rely on proximity to a plant of the opposite sex for fertilization. More than 50% of the plant species

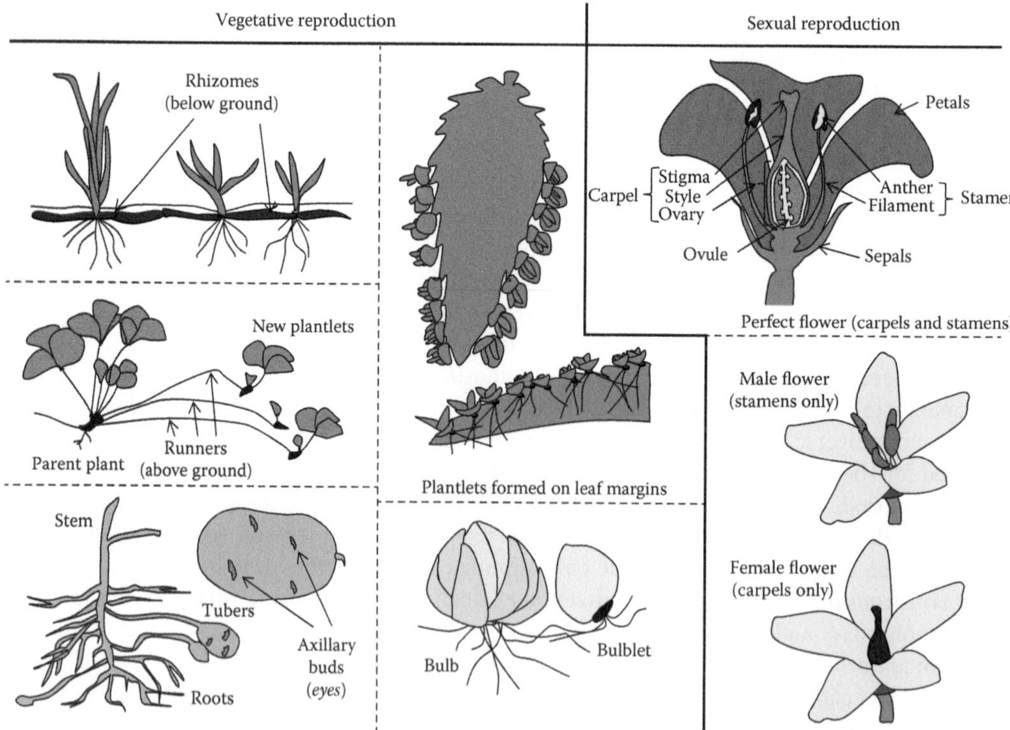

FIGURE 10.8 Examples of modes of vegetative and sexual reproduction in vascular plants. Many species of plants can reproduce vegetatively (left), including those that form rhizomes, runners, tubers, and bulbs, as well as those that form complete plantlets on the margins of their leaves. Flowering plants produce a wide variety of flowers (right), including those that have both male and female organs (i.e., perfect flowers), and those that produce either male or female flowers. Depending on the species, male flowers may be born on one plant, while female plants are produced on another plant (i.e., they are dioecious), or they can be produced on different portions of one plant (monoecious). Plants with perfect flowers are all monoecious.

that have been studied primarily produce seeds through outcrossing, mostly because their anthers and carpels mature at different times. Outcrossing potentially leads to the highest degree of allelic diversity, while vegetative propagation leads to clonal populations.

Species in the Dikarya, or true fungi, have lifestyles that are unique among eukaryotes. Most of their life cycle consists of growing as haploid hyphae and hyphal nets, some of which can span several square kilometers. Additionally, the cells of hyphae often have cytosols that are connected through controlled passages through the cell walls. Presumably, the allelic diversity is low in these large nets, although they can be several thousand years old, and therefore, many mutations would be likely to have occurred among the nuclei in these hyphal nets. Occasional interactions with adjoining hyphal nets occur, but it is unknown whether genetic materials are passed from one haploid hyphal net to another. However, if the cells are compatible and of different mating types, they may fuse, or anastomose, where the cytosols of the two cells are joined. Therefore, when the hyphae anastomose, the two organisms are fused together. Next, the cell formed from the region of anastomosis begins to grow, the DNAs are replicated, the nuclei divide, and the new cell formed becomes multinucleate (Figure 10.9). The haploid nuclei divide independently and asynchronously, and they do not fuse with one another. New cells are formed from this cell, and each of the new cells contains two nuclei, one from each of the two hyphae that anastomosed. These binucleate cells (the source of the name *Dikarya* = *binucleate*) then continue to divide to form the fruiting body. Once the fruiting body is complete, the two nuclei fuse in specific cells. After replication that brings the nucleus to a 4C level, the nucleus transitions into meiosis to form four haploid spores. In some types of fungi, the four haploid cells proceed through one or more mitotic divisions to form additional haploid spores.

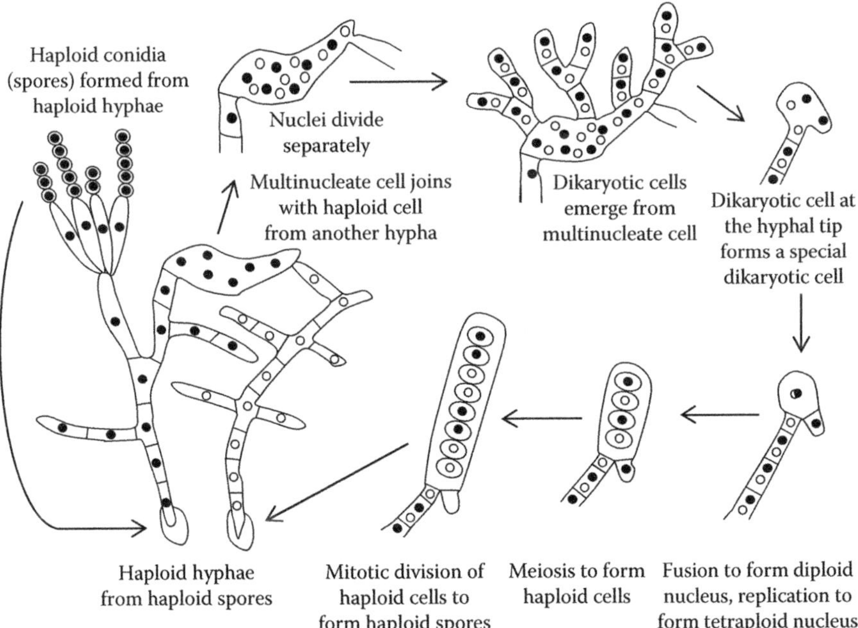

FIGURE 10.9 Generalized life cycle for Dikarya (ascomycete). In true fungi (Dikarya), most of the life cycle is spent as haploid hyphae. Hyphal cells of a different genotypes fuse to form a new cell that often is multinucleate. Dikaryotic cells are produced from this cell and develop into a fruiting body. In specific cells, the two nuclei fuse to forma diploid cell, which immediately replicates to form a tetraploid nucleus. This nucleus then undergoes meiosis to form four haploid cells. In many fungi, these haploid cells undergo either 1 or 2 mitotic divisions that develop into 8–16 haploid spores. The spores are then dispersed and germinate into haploid hyphae. For basidiomycetes, the phases are similar, although the details of each of the phases differs from those of ascomycetes.

They are released and dispersed by wind, water, animals, and so on and germinate to form new haploid hyphae. There is no evidence that any exchange of DNA occurs between nuclei in the multinucleate or dikaryotic cells. However, recombination does occur during meiosis. Therefore, allelic diversity may change during the meiotic phase of the life cycle. The other main source of allelic change occurs through spore dispersal, which can sometimes span thousands of miles.

KEY POINTS

1. Evolution is a consequence of populations of organisms, their genes, and gene products interacting and changing over time.
2. The Hardy–Weinberg equilibrium equation can be used to calculate the proportions of each of the alleles in a single generation, or over many generations, in order to examine trends in allele frequencies.
3. Population size is vital to the survival of each species, and there are many strategies used by various species that assure progression of the species into the next generation and beyond.
4. Life histories of species vary greatly from species that produce few young, where high percentages survive to reproductive age, to those that produce enormous numbers of progeny, of which few survive long enough to reproduce. Each of the strategies has been honed by evolutionary processes to yield species that can be sustained for long periods of times.
5. While many organisms have only a single copy of their genomes in each cell, many others have two or more copies per cell. Some eukaryotic species live primarily as haploids, while others exist primarily as diploid, and some are dikaryotic or polyploidy.
6. Eukaryotes utilize a variety of means to propagate, from exclusively asexual to completely sexual, and many use both.
7. In many plants and animals, both male and female sexual organs exist in a single individual (i.e., they are monoecious), while for others, there are separate male and female individuals (i.e., they are dioecious). Some have combinations of these extremes.
8. Many types of organisms (e.g., fungi and unicellular plants) have different mating types, rather than true sexes.

ADDITIONAL READINGS

Alexopoulos, C. J. and C. W. Mims. 1979. *Introductory Mycology*, 3rd ed. New York: John Wiley & Sons.

Becker, W. M., J. B. Reece, and M. F. Poenie. 1996. *The World of the Cell*, 3rd ed. New York: The Benjamin/Cummings Publishing.

Bidlack, J. and S. Jansky. 2010. *Stern's Introductory Plant Biology*, 12th ed. Columbus, OH: McGraw-Hill Education.

Futuyma, D. J. 1998. *Evolutionary Biology*. Sunderland, MA: Sinauer Associates.

Gladfelter, A. and J. Bermen. 2009. Mitosis in filamentous fungi. *Nat. Microbiol. Rev.* 7:875–886.

Graur, D. and W.-H. Li. 2000. *Fundamentals of Molecular Evolution*, 2nd ed. Sunderland, MA: Sinauer Associates.

Griffin, D. H. 1996. *Fungal Physiology*, 2nd ed. New York: Wiley-Liss & Sons.

Hartl, D. L. 2012. *Essential Genetics: A Genomics Perspective*, 6th ed. Burlington, MA: Jones & Bartlett.

Klug, W. S. and M. R. Cummings. 1994. *Concepts of Genetics*, 4th ed. New York: Macmillan College Publishing Company.

Reece, J. B., L. A. Urry, M. L. Cain, S. A. Wasserman, P. V. Minorsky, and R. B. Jackson. 2010. *Campbell Biology*, 9th ed. New York: Benjamin/Cummings Publishing Company.

Russell, P. J. 1998. *Genetics*, 5th ed. Menlo Park, CA: Benjamin/Cummings Publishing Company.

11 Alleles through Time

INTRODUCTION

How do the changes occur and become established in populations such that specific genes survive? Whether an allele becomes established at high frequency or disappears depends on many factors, one being the fitness of the allele for the organisms in a particular environment. The process differs somewhat in sexual eukaryotes compared to asexual organisms. Among the latter organisms, once a mutant gene arises, it either has a detrimental effect on the organism, an advantageous effect on the organism, or a neutral effect on the organism. If it is detrimental, it may not survive in the population unless it is linked to a gene that is beneficial (Figures 8.1 and 11.1), or it may survive at low levels if its negative characteristics are mild. If it is advantageous, then it has a higher probability of being inherited by its progeny, because the organisms that possess the allele have a higher probability of surviving to a stage where they can produce daughter cells or progeny. Eventually, the allele might become the most frequent allele in that population. If it is a neutral allele, then it can go in either direction, and its frequency may change due to random genetic drift. It may persist for some time within the population, but it is equally likely that it will disappear. All of this is dependent on the size of the initial population, random events, and the comparative rates of cell division and cell death in that environment (Figure 8.1). In sexual eukaryotes, their nuclei usually contain at least two copies of each gene, and each might be a different allele. Therefore, the various alleles have a higher probability of remaining in the population, even if an individual allele is detrimental (especially when recessive), because it might be paired with an allele that is advantageous for that organism. Additionally, crossovers can alter the linkage to the detrimental allele, which can sometimes eliminate it prior to reproduction to produce the next generation. Therefore, in these eukaryotes, there is some buffering for those genomes that contain detrimental alleles, while in haploid eukarya, bacteria, and archaea, natural selection is much more immediate, because there is little to mitigate the effects of detrimental alleles.

The numbers and proportions of alleles within a population are rarely static. Novel alleles are appearing while others are disappearing from the populations. These changes are caused by a number of factors, including natural selection, genetic drift, mating factors, population size, and migration (i.e., gene flow). Individual genes may be acted upon by one or more of the factors, and in some cases, the entire genome might be affected by one or more of the factors. Discussions of selection usually focus on environmental factors that affect a particular genotype or phenotype. For example, the passenger pigeons went extinct because many humans liked to kill them, and they were easily killed. However, passenger pigeons had thousands of genes, some of which had originated in prehistoric bacteria, and most of which had evolutionary histories millions or billions of years long. The organs, tissues, and cells within each passenger pigeon had to cooperate and collaborate successfully with each other. Each cell had to function properly. While extinguishing the passenger pigeons was a relatively easy and rapid process, the evolution of the passenger pigeon was a complex and long process. And, those processes were ongoing even in the very last surviving individual.

Natural selection is dependent on both the internal and the external environments of the cells or multicellular organism. An allele that is advantageous under one set of conditions might be neutral or detrimental under another set. Additionally, some alleles of one gene might confer an advantage to the organisms in a particular environment, while the opposite might be true for an allele of a different gene. For example, for some plants, such as maize, an increase in atmospheric CO_2 can lead to increased photosynthetic productivity. However, increases in atmospheric CO_2 lead to an increase in global temperature, because CO_2 is a greenhouse gas that prevents heat from radiating into space.

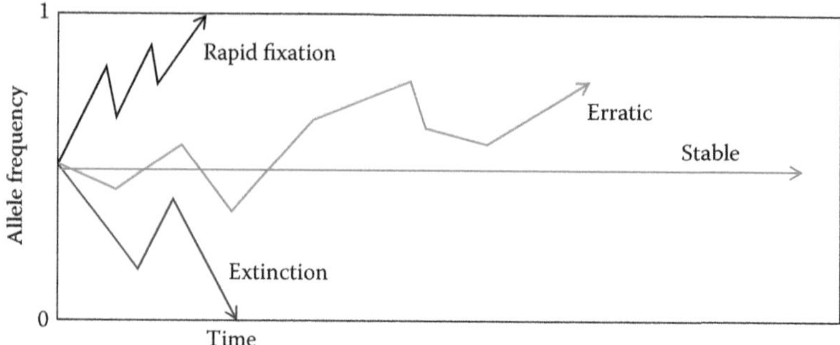

FIGURE 11.1 Examples of allele proportion changes over time. Beneficial alleles often are rapidly fixed (black), while detrimental alleles may experience rapid extinction (red). Neutral alleles may exhibit erratic changes in proportion when in small populations (green), but they may remain at stable levels when in moderate-to-high proportions in large populations (blue).

This is not a problem for photosynthetic processes per se, but it is a problem for reproduction. The formation of the gametes and fertilization are negatively affected even by an increase of a degree or two. Thus, the plants may be more vigorous in vegetative growth, but they become defective in their abilities to reproduce. This will lead to the extinction of these and other plants unless they are able to migrate to higher altitudes or latitudes, and some already have begun their migrations. For example, during the latter portions of the twentieth and early portions of the twenty-first century, many species of forest trees have been migrating due to temperature increases. They have been receding from lower elevations and parts of their historically southern ranges, partly because of the failure to propagate in the higher temperatures, and partly because of changes in precipitation. Their distributions have moved northward and to higher elevations on mountains. However, once they spread to the tops of the local mountains, or they reach an area that is incompatible with their growth and reproduction requirements, their migration possibilities will come to an end.

NATURAL SELECTION

Natural selection is more complex than it initially seems. First discussed by Charles Darwin, Alfred Russel Wallace, and others in the nineteenth century, its central tenets are that there are more organisms of a species produced than can survive, and those that are most suitable in a particular environment or circumstance are those that survive to reproduce. The most famous example of natural selection is that of the peppered moth, *Biston betularia*, in industrial areas of England during the nineteenth century. The common form prior to 1811 was a mottled white and gray form, variety *typica*. From 1811 to 1848, individuals of this variety decreased in number, while a new completely black variety, *carbonaria*, appeared and increased. The cause was deduced as being natural selection, based on predation. Prior to 1811, the tree bark onto which the moths were normally found had a mottled appearance, very similar to the patterns on the wings of the moths. Therefore, their wings provided camouflage to the moths such that predators (mainly birds) could not see them easily. However, after 1811, black soot from increasing amounts of industrial factory smoke coated the tree trunks, which meant that the whitish-gray moths could clearly be seen by predators, which found and ate them in large numbers, thus causing their numbers to rapidly decline. Mutants with completely black coloration were now camouflaged on the sooty tree bark, and therefore, their numbers increased due to decreased predation. Conversely, pollution controls that were put into place in England starting in the 1960s caused a reduction in soot in the air, which led to a reduction in the amount of dark soot on the tree bark. This caused sharp decreases in the number of black moths and a concomitant increase in the numbers of whitish-gray moths. This is a natural selection at the level of the individual phenotype. However, selection within a species, as well as selective processes that occur among the cells within a multicellular individual, can affect the overall fitness of an allele, individual, or a population. If the allelic composition of an individual negatively affects fitness at all, those alleles may be underrepresented or disappear from the population. In other words, the presence of a few alleles that cause decreases in the numbers of individuals carrying those alleles also affects all of the other genes in the organism (including those that are beneficial in other circumstances) and therefore can cause decreases in other unrelated genes.

For those genes that are under selective pressures, they carry a certain potential to be either beneficial or detrimental to the organism under a specific set of conditions. They may provide a high degree of benefit to the organism, in which case the organism may be better able to survive in those conditions, and therefore has an increased probability of passing the beneficial genes (as well as all of the others in its genome) onto its progeny. In the opposite sense, detrimental genes lessen the probability of passing on the detrimental gene (as well as all of the other genes in its genome) on to progeny, because the chances of producing any progeny are decreased (in ranging degrees).

The relative fitness of an allele and how it affects other alleles can be measured in some cases. In two-allele systems, there can be various combinations of alleles that affect the fitness of the individuals in different ways (Figure 11.2). Both alleles can have positive or negative fitness values, and these can vary in their benefits or detriments to the organism. Alternatively, one allele may be beneficial, while the other might be detrimental. In some instances, one allele has a positive or negative effect on fitness, while the other allele is neutral. Finally, in some cases, the two alleles interact to cause a higher fitness value than the sum of the fitness values for the two alleles, or conversely, they can have a lower fitness value than the sum of the fitness values for the two alleles. A variation of the Hardy–Weinberg equation can be used to determine the relative fitness values of the two alleles, both in the homozygotes and in the heterozygotes. If w_{AA} = fitness of the AA homozygote, w_{Aa} = fitness of the Aa heterozygote, w_{aa} = fitness of aa homozygote, p = frequency of the A allele, and q = frequency of the a allele, then $p^2w_{AA} + 2pqw_{Aa} + q^2w_{aa} = \mathbf{W}$(mean fitness). This can be simplified, as in the Hardy–Weinberg equation, to $p^2(w_{AA}/\mathbf{W}) + 2pq(w_{Aa}/\mathbf{W}) + q^2(w_{aa}/\mathbf{W}) = 1$. For the same equation, the selection coefficients can be substituted for the fitness values. The selection coefficient (designated s) is equal to 1 minus fitness ($s = 1 - w$; also, $s + w = 1$, and $w = 1 - s$), and therefore, selection can be calculated by substituting for the fitness value.

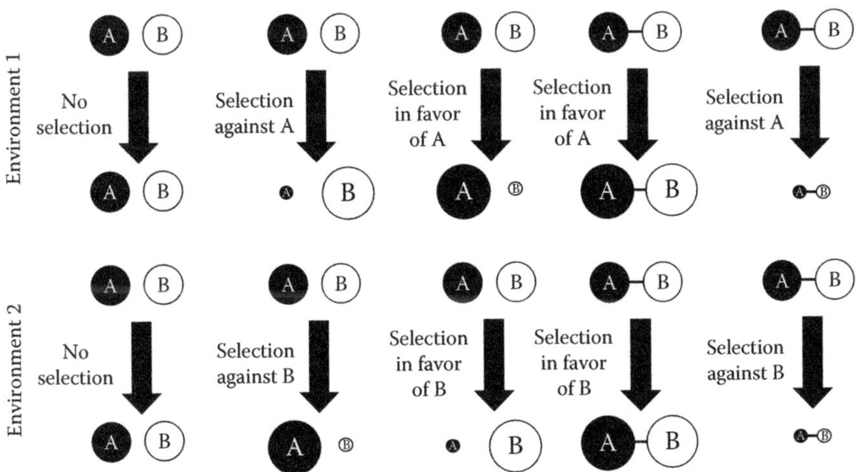

FIGURE 11.2 Trends in gene or allele proportions in various scenarios. If no selection is influencing either A or B, then the two would change only due to random drift. Under selection pressures, each of the genes/alleles would increase in proportion if they were beneficial to individuals in the population or would decrease in proportion if they were detrimental to individuals in the population. Exceptions to this are when the genes/alleles are genetically linked or when they act together (either positively or negatively). In these cases, the genes/alleles tend to increase or decrease at the same time.

LEVELS OF SELECTION

Most consider selection to be operating at the level of the individual multicellular organism, or at the level of the whole cell, in cases where the organisms are unicellular. However, selection occurs within individual cells, in order to maintain all of the vital cellular functions (Figure 11.3). In cells, especially in those that have compartments (i.e., nuclei, mitochondria, plastids), there are selective pressures within the cell, as well as selective pressures coming from the environment. Therefore, maintaining a certain amount of genetic diversity is vital to the organism as a whole, as well as the populations of cells and molecules. For organisms with organelles and compartments, each compartment will exert different types of selective pressures on the expression of the genes within. Again, maintaining an amount of genetic diversity is important to the function and survival of the cells, organelles, and molecules.

Mutations are defined as changes in the genetic material (DNA for most organisms and RNA for many types of viruses) that are inherited by the progeny. There are many correction systems for DNA, and therefore, a large percentage of the DNA changes are corrected before they have any effects on the cells or be passed to progeny. Therefore, there is an inheritance filter (i.e., selection) that removes any changes in the DNA prior to separation of the chromosomes and cell division (Figure 11.3). For RNA, the situation differs in that there are few mechanisms to correct errors. This is the primary reason that RNA viruses usually have higher rates of mutation than do DNA viruses, or for that matter all DNA-based organisms. It is also the reason that no large-genome RNA organisms exist, because mutation rates are too high to allow a large set of genes to be accurately maintained. Once the mutation is set, selective pressures begin to act. If the mutation occurs in the promoter region of a gene, or changes the ability of RNA polymerase to attach to the promoter, this can affect transcription of the gene. This can be a change that increases or decreases transcription. When either of these occurs, the difference for the cell or the organism can range from no detectable

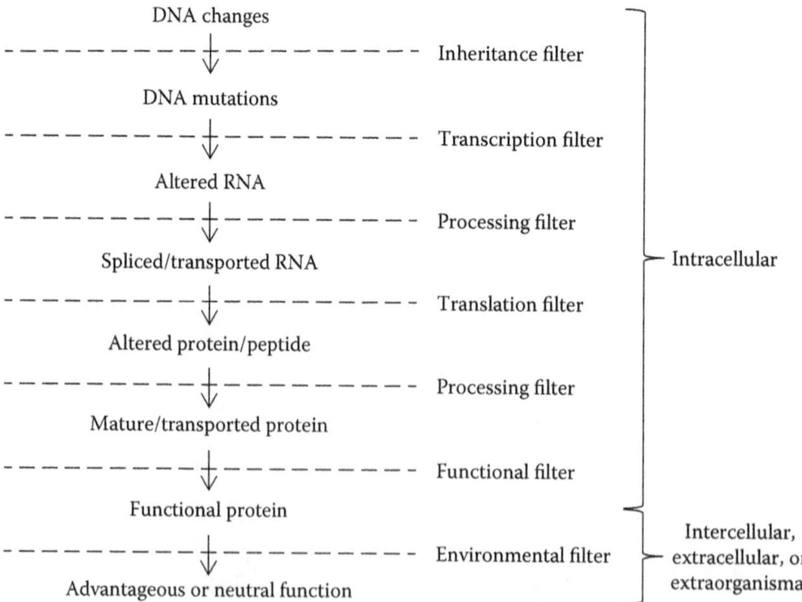

FIGURE 11.3 Levels of selection starting with changes in DNA. When there is a change in the genome, it can be corrected by repair systems. If not, it becomes a mutation. If the mutation does not affect transcription, then the gene is transcribed into RNA. Succeeding steps can affect the expression of the gene in RNA processing, translation, protein processing, transport, and functioning of the protein (or RNA, in the case of an RNA gene). All of this selection occurs at the intracellular level. If the mutation affects fitness of the cell, then there is selection at the intercellular, extracellular, or extraorganismal (i.e., environmental) levels.

change to a lethal change. Therefore, there is a selective screen that acts at this point (Figure 11.3). If there is a sufficient amount of transcription, then the RNA that is produced can proceed to the next steps in gene expression. If the amounts of RNA are insufficient to proceed through the next steps of expression, expression of that gene ends at that point. If the gene normally produces a vital gene product, then the cell dies, and therefore, the mutation disappears.

Many genes contain introns that must be removed prior to translation. In Bacteria and Archaea, introns are rare. However, in Eukarya, introns can be present in nuclear and organellar RNAs, and many result in alternative splicing to produce more than one RNA. At least four different classes of introns exist, but all must be removed prior to full maturation of each of the RNAs in which they occur. Therefore, yet another selective screen exists at this point (Figure 11.3). For ribosomal RNA (rRNA) and transfer RNA (tRNA), the precursor RNAs must be processed (i.e., cut) in specific ways to produce the final active RNAs. For messenger RNA (mRNA) in bacteria and archaea, translation can begin immediately. However, in eukaryotes, the mRNA must first be processed. Processing involves several steps, depending on the type of RNA. First, a derivative of guanosine (7-methylguanosine, m^7G), known as *a cap*, is covalently added to the 5′ end of the mRNA through a 5′–5′ triphosphate bond (see Figure 3.5). This end later binds to cap-binding proteins (CBP) that are important to proper binding of the mRNA to the small ribosomal subunit. In eukaryotes, a poly-A tail protects the 3′ end of the RNA from degradation, thus controlling the half-life of the RNA, which controls how long the RNA is available for translation. Additionally, it is important in export of the mRNA out of the nucleus. Polyadenylation begins by removal of a section of the 3′ end of the mRNA, usually near a specific sequence (AAUAAA). Then, an enzyme adds the tail of adenosines one at a time, up to about 250. A mutation in this sequence can alter the addition of poly-A and therefore can affect export from the nucleus as well as long-term stability of the mRNA. In Bacteria and Archaea, polyadenylation is used to label the mRNA for degradation. Therefore, it also controls the half-life of an mRNA, but in the opposite sense from a eukaryotic mRNA. Organelles are variable. In some, a poly-A tail increases the half-life of the mRNA, while in others, polyadenylation signals degradation of the mRNA. Therefore, processing of RNA is another selective screen in the process of gene expression (Figure 11.3).

Once the mRNA is exported out of the nucleus, it attaches to ribosomes to begin translation of the message into the protein products. At each step in this process, mutations can affect the expression of the RNA. If the attachment site (usually an adenosine or guanosine) to the m^7G is mutated, addition of the m^7G might be inhibited or inefficient. This will affect several steps in the process of transport and attachment of the mRNA. The cap is recognized by the CBP, which ushers the mRNA out of the nucleus through the nuclear pore. The cap also appears to provide some protection against RNA degradation at the 5′ end, is involved in promotion of intron excision near the 5′ end, and is important in efficient binding of the mRNA to the ribosome. Thus, a mutation in this region may greatly affect transport, processing, and translation of the mRNA. Additional sites in the 5′ untranslated region (5′ UTR) of the mRNA interact directly with the SSU rRNA (the Shine–Dalgarno sequence in bacteria) or with specific proteins (initiation factors) that aid in positioning the mRNA on the ribosome. Specific sites on the introns cause changes in splicing. If the intron fails to be removed from the mRNA, the molecule usually is not exported from the nucleus. Some mutations cause aberrant splicing, so an intron is removed, but not at the usual sites. Therefore, the coding region on the 5′ side of the intron is normal, while the portion past the 3′ side of the splice point is abnormal. If the coding region is in the same reading frame, insertion or deletion of a stretch of amino acids will be present in the protein. However, approximately two-thirds of the time, the splice will be such that the reading frame is shifted causing a nonsense mutation to form a different amino acid sequence in the protein. Therefore, the resulting mRNA successfully attaches to ribosomes, and a protein is produced, but the protein often is faulty. If the poly-A tail is not added, the half-life of the mRNA is affected, such that expression of the resulting protein is decreased from normal. Mutations that affect each of these processes may or may not allow the RNA to proceed to the next step, which is translation (Figure 11.3).

Thus far, expression of the gene has passed through several selection filters, and if the RNA has finally reached the ribosome, additional filters lie beyond this step. If the mRNA successfully attaches to the ribosome, then it has passed all of the selective screens to reach that point, and translation has been initiated. The start codon almost exclusively is AUG, which means that a methionyl-tRNA is the first amino acid on the polypeptide chain that is being synthesized. The first barrier to translation may be created by a mutation that either abolishes this codon in some way or creates a new AUG upstream from the wild-type start site. In the majority of cases this either will fail to produce a protein or will result in a mutant protein that probably will fail to function properly. In rare cases, a novel protein may be produced that is advantageous to survival of the organism. Next, mutations in the coding region have varying effects. Synonymous mutations cause a change (usually in the third position of the codon) in the DNA and RNA that does not change the amino acid during translation. However, the change still can have an effect, because all cells make varying amounts of each of the tRNAs, and therefore, if the mutation results in the use of a tRNA which is in lower (or higher) amounts, this could result in a change in the amount of the protein being synthesized. Missense mutations cause a change in the amino acid sequence (nucleotide changes in the first or second position). If the change has no effect on the charge, tertiary structure, or active site (for enzymes) of the protein, the mutation may have little effect on the cell. However, if it affects any of these, the changes could range from minor to lethal, depending on the protein being made. Nonsense mutations cause premature termination of translation by creating a stop codon (UAG, UAA, and UGA) in the middle of the coding sequence. Therefore, a truncated protein is produced. In most cases, the protein that is produced is nonfunctional. Therefore, translation presents yet another powerful selective screen in the process of evolution (Figure 11.3).

Many proteins are modified after translation. Some are cleaved, some are glycosylated, some are phosphorylated, some are cross-linked, and others go through other processes. Also, most need to associate or interact with other proteins and cellular molecules to carry out their functions. Therefore, there is yet another selective screen after translation. They must be properly processed and assembled with other proteins. If they are improperly processed or contain changes that do not allow efficient association with other molecules, or lack the amino acids that are usually phosphorylated, they may be either partially functional or nonfunctional. In some cases, they may bind to other molecules and substrates in a way that is nearly irreversible, which also affects their functionality. They must be transported to the proper cellular location as well. For example, many eukaryotic proteins have signal peptides on their amino ends. These signal peptides associate with signal recognition proteins that shuttle the proteins to the compartment in the cell into which they will be imported. This includes insertion into a membrane, and importation into the mitochondrial matrix, mitochondrial intermembrane space, chloroplast stroma, lumen, and intermembrane spaces, as well as into the endoplasmic reticulum, where some proteins are modified by the Golgi and then exported to the cell surface. A mutation causing an alteration in protein transport can cause an otherwise functional protein to end up in the wrong compartment, thus rendering it nonfunctional. Mutations allowing complete processing and transport may pass through this selective filter (Figure 11.3).

Finally, if the protein ends up in the correct location in the cell, environmental conditions and reproductive opportunities will determine whether or not the protein functions in a way that allows the cell to survive (Figure 11.3). Therefore, mutations (including the expression of their products) proceed through a myriad of selective sieves, most of which are intracellular. Some mutations are never observed, because the cell dies before it can be observed. Other mutations allow the cell to divide a few times before it and its progeny expire. Only the mutations that allow a measurable number of cells to survive are observed and can be recorded are true mutants. Whether a mutation is observed depends on the location of the mutation, the base change, the condition of the cell, and finally, the environment around the cell. In the case of Bacteria, Archaea, and single-celled Eukarya, the environment might be the physical environment or a consortium of organisms surrounding the individual cell. In multicellular organisms, the environment consists of contact with other cells of the organism, contact with other organisms living in and on the organism, and the internal and external environments of the organism. All of these factors and levels of selection determine the ultimate mutation rate for a particular genome, portion of a genome, gene, non-coding region, or nucleotide.

RANDOM GENETIC DRIFT

Most geneticists and evolutionary biologists once thought that most genes and their corresponding alleles were under selective pressure, which was thought to be the major driver of evolution. However, many genes appeared to have no measurable selection acting upon them. Additionally, their survival to the next generation appeared to be a consequence of random sampling. Various alleles would disappear from the populations merely due to the fact that not all individuals participated in producing the next generation. By the second half of the twentieth century, some proposed that most alleles were subject to this semi-random process that came to be called *random genetic drift* and was part of what was termed *neutral theory.* It was proposed that the proportions of most alleles in the population increased and decreased in the population due to genetic drift, not by selection (Figure 11.4). Since that time, it has been supported in many studies. This implies that most genes are mutating to produce new alleles and that they frequently increase and decrease in the population, without the influences of natural selection.

Genes that are not under selection change in semi-random way and thus can appear, increase, decrease, remain unchanged, or disappear without any effect on the population (Figure 11.1). Also, if the allele occurs in a small proportion of the population, it is more likely to disappear, whereas if

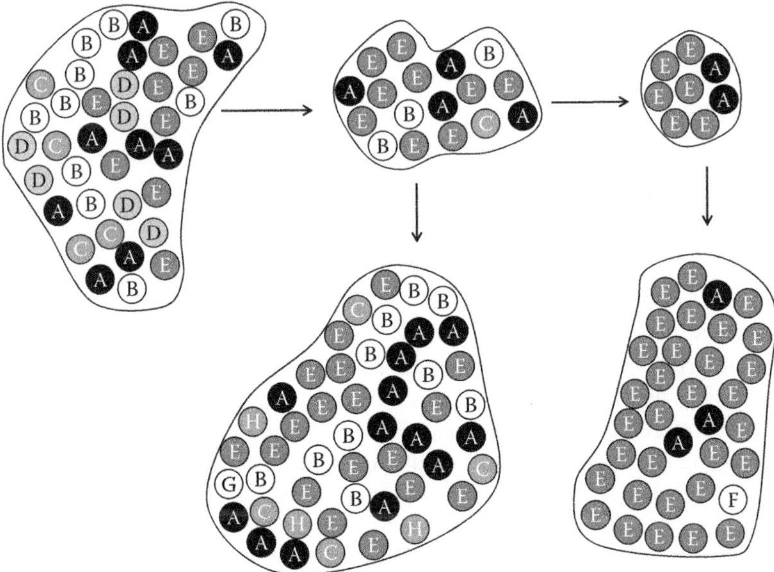

FIGURE 11.4 Changes in allele proportions due to random genetic drift. The population begins (upper left) with a mixture of five alleles (A–E), with higher proportions of alleles A, B, and E; and lower proportions of alleles C and D. Through random sampling of alleles in the next generation (upper middle), both the numbers and proportions of alleles changes, such that the E allele is in the highest proportion, and A and B are in lower proportions, C has almost disappeared, and D has gone extinct. Two populations are started from this population. In one (lower left), allele E has increased further, and A and B have increased somewhat. C remains low, and two new alleles, G and H, have emerged. The second population was established from a small portion of the original population and thus contains two alleles, E (the major allele) and A (the minor allele). This becomes the founding population of a new population (lower right), which contains mainly allele E, and two minor alleles, A and F.

it occurs in a large proportion of the population, it is more likely to become fixed in the population. The probability of an allele to increase or decrease in a population is related to the proportion of individuals that have the allele, as well as the total number of individuals participating in reproduction within the population. If the allele is present in few individuals within a small population of reproducing individuals, then it is more probable that it will be absent from the next generation. However, if it is present in most of the individuals in a small population of reproducing individuals, then it is more probable that it will increase in the next generation. For larger populations, sampling is also important, although the probability of extinction of a rare allele or the increase in a common allele will usually take more generations to occur. Of course, if few individuals are produced in the next generation, then very rapid changes in allele frequency can occur, again due to the fact that fewer individuals with a narrow range of alleles contributed their alleles to the next generation. Therefore, in general, and without natural selection, the allele proportions are dependent on population size, number of alleles, sample size (the proportion of the population that participate in reproduction), migration (i.e., gene flow, described below), and the number of generations.

Random genetic drift appears to be a major driver of evolution. Specific allele combinations may disappear due to random sampling to produce the next generation (Figure 11.4). This can have large consequences in subsequent generations. Should some alleles become subject to selection due to changes in local or internal conditions, the proportions of those alleles may change rapidly in the population. This is one reason that it is important for a species to maintain a high level of genetic diversity, so that the probability of survival of the species is increased should conditions change. *Homo sapiens* may have nearly gone extinct approximately 200,000 years ago, when their numbers and their population size and genetic diversity were reduced. This is known as *a genetic bottleneck*. Because of the small population, the survival of each of the individuals was nearly random and less dependent on the genotype of the individual. Fortunately for us, somehow some of the individuals survived, possibly by crossing with other hominids and possibly through the development of an extensive social order. The species then diversified, migrated, and eventually became successful, spreading to all continents on the Earth.

MATING AND DISPERSAL

For animals, mating choice is usually a complex matter, based on developmental state, the availability of mates, as well as olfactory, behavioral, and morphological cues. These can affect whether there is inbreeding or outbreeding, and whether one animal has one, many, or no mates. For multicellular organisms, nonrandom mating can affect allele proportions. For example, many birds exhibit sexual dimorphy, where the males develop colorful plumage, distinctive songs, and often other displays, such as nest building and dances. The female chooses a male based on multiple cues in the displays. Therefore, selection based on behavior is active in these animals. Some birds parasitize nests, by laying their eggs among those already laid by the parents that built the nest. When the parasite bird egg hatches, the chick either pushes out the other eggs or kills the chicks as they hatch. The parents seem oblivious to this and continue to feed the parasitic chicks, thereby abolishing spread of their genomes and enhancing the spread of the parasitic genomes during that breeding season. On the other hand, many fish species have a more randomized mode of reproduction. The female fish lay eggs, and males (sometimes multiple males) then swim over and fertilize the eggs with their sperm. Each of these reproductive strategies affects the allelic makeup of the population as a whole.

For higher plants, mating and spread of alleles is accomplished by both dispersion of pollen and seeds. For both gymnosperms and angiosperms, a common mode of pollen distribution is through the air. Most pollen grains are small and light and can be carried by wind currents for miles, sometimes for thousands of miles. Pollen grains have been found in alpine lake sediments and polar ice from species that do not grow in those regions. Animals, including arthropods, disperse pollen for a broad range of plant species, especially angiosperms. Ants, bees, birds, beetles, and other groups of animals are extremely important in spreading pollen from one flower to another, which is a more targeted strategy for pollination than by wind. Dispersal can be from local to many miles, and sometimes is intercontinental. Seeds also have similar systems of dispersal. Some are wind dispersed, while many are dispersed by animals. Some of these require that the seeds travel through the digestive systems of animals. Others are dispersed by water; in the case of some coconuts, they may travel over thousands of miles of ocean before coming to rest on a beach where the plant can begin to grow. Therefore, a diversity of mating and dispersal mechanisms has evolved in plants that assure their propagation. However, dependence on dispersal by specific animals can be risky, especially if the animals become extinct. Most plants can be pollinated by multiple animal species and therefore have evolved a protection against the disappearance of their means of reproduction.

Many species of fungi rely on animals for part of their reproduction and dispersal. Two examples are those that are said to be *sexually transmitted* and those that cause morphological changes in plants that are normally pollinated by bees. Members of the fungal Order Laboulbeniales (a group of ascomycetes) grow on the exoskeletons of many species of insects, but do not normally kill their hosts. Although their spores may be transferred to a new individual by casual contact, most are transferred during copulation, and thus, they are sexually transmitted. The targeted transmission between individuals of the same species, and mostly among flying insects, assures efficient dispersal of diverse genotypes among members of each species. Other types of fungi, specifically the basidiomycete, *Puccinia monoica*, grow on plants (usually a species of *Arabis*). In this case, the fungus grows into the plant tissues to utilize its nutrients, and also renders it sterile, in that it cannot form normal flowers. However, compounds emitted by the fungus cause a change in the morphology of the plant such that the leaves turn yellow and resemble flowers, although they contain no stamens or carpels. These pseudoflowers look normal enough to attract pollinators (e.g., bees), which visit the flowers long enough to brush against the fungal spores, and when they visit a real flower they deposit the diploid spores onto the plant, or when they visit another pseudoflower they deposit either a male or female fungal spores to produce the diploid form of the fungus. Thus, the fungus uses the plant and the pollinating insect to assure adequate dispersal and mixing of its alleles.

GENE FLOW

Organisms do not exist in isolation of all other organisms, and often, two or more populations of the same species meet and exchange genes (Figures 11.5 and 11.6). This is known as *gene flow* (Figure 11.5). Populations of bacterial, archaeal, or eukaryotic species may separate, mutate, and then rejoin again. Whether an allele survives or not is dependent on its initial concentration in the population, mutation rates, as well as its fitness (Figure 11.7). Fitness simply indicates whether or not it has a positive, negative, or neutral effect on the survival of the organism at a point of time in a particular environment. If a gene enters the population in high numbers and it either has a positive or neutral effect on the cell or the population, it is more likely to survive. If it enters in a low amount or has a negative effect on the cell or the population, then it is probable that it will disappear. Fitness is dependent on the environmental conditions at the time, and therefore, an allele can be neutral in some conditions, but might be advantageous or disadvantageous in other situations. Therefore, in migratory situations, the fitness of an allele may differ among the environments visited by the organisms. For example, movement from a cold climate into a warm climate may influence the organisms differently in the two climates, based on the sets of alleles that they carry. Long-term selective pressures for some of the genes that allowed an organism to survive in the ocean 2 billion years ago might be lethal in the same types of organisms living in the ocean today. This also implies that organisms are exposed to varying environmental conditions whether or not they migrate. Evolution on the molecular scale is very fluid and can change rapidly based on the immediate environment.

Since genes and genomes are constantly changing, species are constantly changing. Therefore, species exist only for short periods of time. In fact, depending on the amount of genetic change that is being considered, they may exist for a very short time. Although the concept of species differs

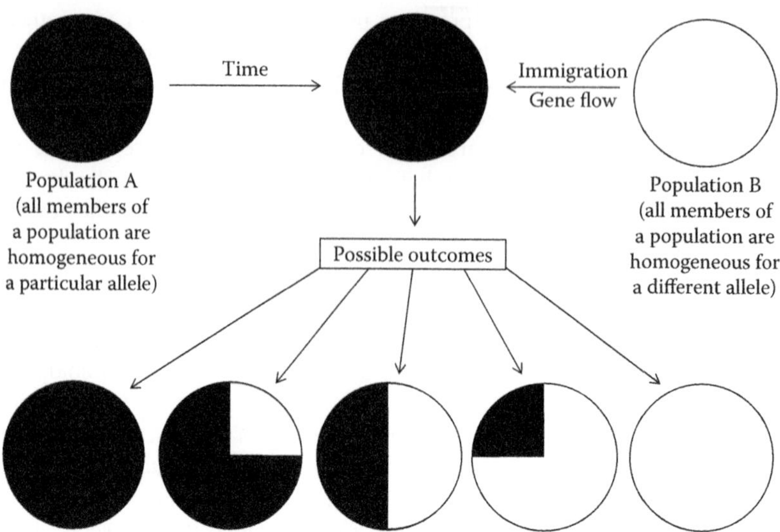

FIGURE 11.5 Gene flow between two populations, each with a single (but different) allele. If both alleles are neutral, then the outcomes depend on the number of individuals that participate to produce the resulting population. If one allele or the other is under selective pressure, it may affect the influence of gene flow.

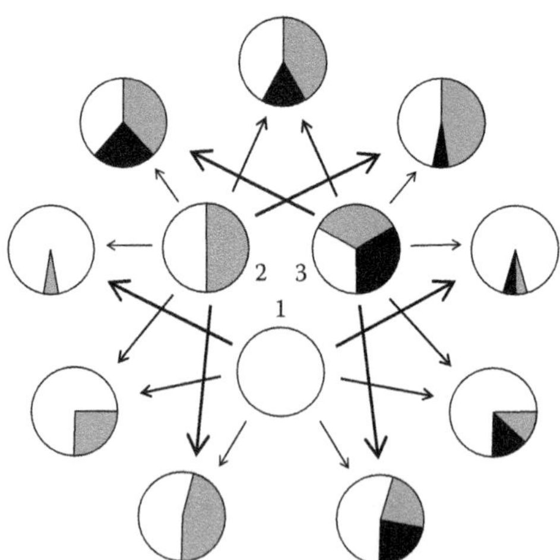

FIGURE 11.6 Gene flow predictions when there are one, two, or three alleles in the populations. In the center, the three circles (1, 2, and 3) represent three possible populations: 1, with one allele of a specific gene; 2, with two alleles of the gene; and 3, with three alleles of the gene. Arrows indicate gene flow among each pair of populations. Arrow thickness indicates the proportional contribution for each gene flow event. For example, when there is gene flow between a population with one allele of the gene (white) and one with two alleles (gray and white, including the one found in the other population), the resulting population could have a range of proportions of the gray and white alleles depending on the relative proportions of individuals in each of the original populations and the relative fitness of the alleles. The same principles apply to gene flow when there is gene flow between populations having one allele and another having three alleles (gray, white, and black) and populations of two and three alleles (top).

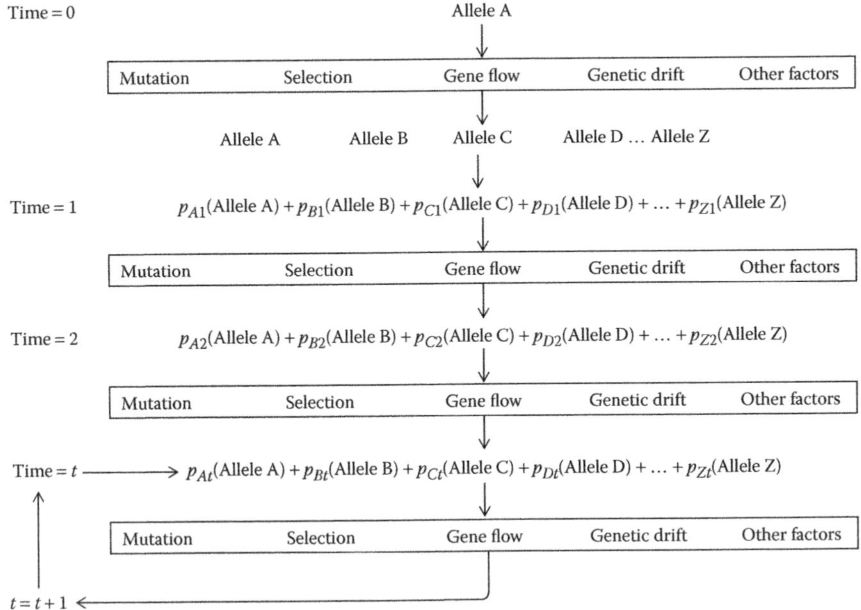

FIGURE 11.7 Model for allele proportion changes in a population over time. At $t = 0$, one allele, A, is present. At a later time, several alleles are present, generated by various processes, including one or more of the following: mutation, selection, gene flow, genetic drift, and other factors. Each is present in a particular proportion (designated p). At time $t = 2$, more changes occur to change the proportions for each of the alleles. At each step, the relative proportions of each of the influencing processes may differ that is, for a time, mutation and genetic drift may have the greatest influence on allele proportions, while at a later time, selective pressures may have greater influence.

from one discipline to another, and from one time to another, it still is a useful concept that can be used to discuss evolutionary processes in terms of species and speciation. Once one population is isolated from another population of the same species, the two populations can begin to diverge to an extent that they become subspecies or separate species. This is because genetic drift and natural selection will act in slightly different ways on the populations such that the mutant alleles that are retained will differ between the populations. Eventually, they will become so different that gene flow and reproduction between the two populations ceases. However, separation of the two species is dependent on mutation and selection rates, as well as the population sizes, geographic barriers, and biological barriers. Even after speciation has occurred, some hybridization is possible, as seen in the genus *Canis* and many other taxa (described in Chapter 14). Therefore, although we speak of species as being distinct entities, they are in constant states of flux because of ongoing evolutionary processes. Also, genes (sometimes many in one event) may be transferred from one species to another, which is yet another type of gene flow.

Gene flow can cause changes in allele frequency. Given a population with stable allele frequencies, an equilibrium situation will be maintained. However, if there are introductions of alleles from one population to another, then the alleles will be in disequilibrium for a period of time. This may persist over long periods of time or merely be episodic. For some plants and animals, hybridization between individuals of related species can mate to produce viable offspring. These can then breed with one of the two species, thus introducing sets of alleles in a process called *introgression*. One extreme form of gene flow is when an individual or a set of individuals colonizes a new location. They carry with them a set of alleles that will determine the long-term genetic components of the population into future generations. The genetic consequence of this founding population (or individual) is termed *the founder effect.* The founding population can have a large effect on the ultimate success, failure, and genetic composition of the population long into the future.

OTHER FACTORS AFFECTING ALLELIC PROPORTIONS

Genes are linked when they are on the same chromosome, but because of recombination, the degree of linkage depends on the distances between the linked genes. If they are very close, then the linkage will be tight, and most crossovers will not change that linkage. If they are far apart on the chromosome, they often are exchanged between the homologous chromosomes (and occasionally between nonhomologs). For species with sexes and sex chromosomes, certain genes are sex linked, because they are physically linked on the sex chromosomes. When the products from these genes interact with genes on autosomal chromosomes, expression of certain alleles may be accentuated or suppressed depending on the sex of the individual. Because of this, the mode of sexual reproduction can affect the allelic proportions in a population and can also alter the phenotypes of the individuals for certain traits.

In addition to interactions of gene products within an organism, other organisms can change gene expression within an organism to greatly affect them. For example, many species of arthropods reproduce sexually and normally produce males and females in roughly equivalent numbers. However, when the individuals are infected with certain bacteria (usually symbionts or endosymbionts; one group being species of *Wollbachia*), after mating, the females will produce females and males, but will also produce hermaphrodites in species that normally do not produce hermaphrodites. But, this occurs only when the female is infected with the bacteria, and while the females and hermaphrodite progeny are infected with the bacteria, the males are never infected. When females of other arthropods species are infected with these bacteria, the females only produce clones of themselves by parthenogenesis, and all of the progeny are infected with the bacterium. In another set of arthropod species, if the female is infected, but not the male, after mating with the uninfected male, only female progeny are produced, and they all are infected with the bacterium. No males are produced. In yet another group of arthropods, if the female is infected with the bacterium, but not the male, then after mating, equal numbers of males and females are produced, but they all are infected with the same bacterium. However, if the male is infected, but not the female, then after mating, no progeny is produced. Each of these may also cause changes in allele proportions in the populations, especially those that are sex linked.

KEY POINTS

1. Evolution is caused by a number of factors, including mutation, selection, random genetic drift, gene flow, and others.
2. Natural selection leads to differential survival of individuals and their genes and therefore can cause the rapid fixation of beneficial gene combinations and the rapid extinction of detrimental gene combinations.
3. Natural selection is active from the level of the gene up to the level of the individual organisms in its current environment.
4. Random genetic drift (i.e., random sampling) causes allele proportions to change within a population, even when selection is inactive. It may be the major mechanism of allele changes in a population.
5. Mating and dispersal of gametes and progeny vary greatly among species and can greatly affect allelic change within a species.
6. Gene flow occurs when a new allele or set of alleles is introduced into a population via immigration of an entire organism or cells (e.g., plant pollen or fungal spores). When sufficient gene flow is present, the allele proportions in the population can be altered.

7. Genetic linkage of genes, as well as sex linkage, can affect the distribution of alleles within a population.
8. Symbiotic or infectious organisms can affect the allele proportions, as well as the sexes of the affected species.

ADDITIONAL READINGS

Becker, W. M., J. B. Reece, and M. F. Poenie. 1996. *The World of the Cell*, 3rd ed. New York: The Benjamin/Cummings Publishing.

Futuyma, D. J. 1998. *Evolutionary Biology*. Sunderland, MA: Sinauer Associates.

Graur, D. and W.-H. Li. 2000. *Fundamentals of Molecular Evolution*, 2nd ed. Sunderland, MA: Sinauer Associates.

Hartl, D. L. 2012. *Essential Genetics: A Genomics Perspective*, 6th ed. Burlington, MA: Jones & Bartlett.

Klug, W. S. and M. R. Cummings. 1994. *Concepts of Genetics*, 4th ed. New York: Macmillan College Publishing Company.

Reece, J. B., L. A. Urry, M. L. Cain, S. A. Wasserman, P. V. Minorsky, and R. B. Jackson. 2010. *Campbell Biology*, 9th ed. New York: Benjamin/Cummings Publishing Company.

Rogers, S. O. and M. A. M. Rogers. 1999. Gene flow in fungi. In *Structure and Dynamics of Fungal Populations*. ed. J. Worrall. Dordrecht, the Netherlands: Kluwer Academic Publishers, pp. 97–121.

Russell, P. J. 1998. *Genetics*, 5th ed. Menlo Park, CA: Benjamin/Cummings Publishing Company.

12 Changes to DNA

INTRODUCTION

At its essence, the process of evolution is simple. Mutations, the raw materials of evolution, are passed to progeny cells. The cells (whether a unicellular or multicellular organism) survive to reproduce or they die without passing on those mutations. In other words, evolution boils down to mutation combined with many other factors that interact with the results of those mutations. However, the details of evolutionary processes are numerous, varied, and often subtle. When DNA mutates, it is transcribed to form altered RNAs that may be subsequently translated into aberrant proteins. Conditions inside the cell, as well as the environment around the cell, then act to determine the effects of mutations on the mutant cells. Although this seems simple, each step in the process, and each region of DNA, each RNA, and each protein has different selective pressures (or no pressures whatsoever) acting on them. Therefore, far from being simple, the process of evolution involves an intricate system of interactions that ultimately leads to the survival or death of the cell or organism. In this chapter, the processes of mutation as they affect DNA, RNA, and proteins in cells and organisms are discussed.

CLASSES OF MUTATIONS

There are several different classes of mutation (Figure 12.1). The most common are *transitions* (Figure 12.1b). These occur when one purine (A or G) is substituted for the other purine (G or A, respectively) or one pyrimidine (C or T) is substituted for the other pyrimidine (T or C, respectively). These occur more often than *transversions* (Figure 12.2), which are changes from a purine to a pyrimidine, or vice versa (Figure 12.1c). Depending on the sequence, the rate of transition is often between 1.5 and 3.0 times higher than transversion. Another general type of mutation is an *indel* (Figure 12.1d and e), where either insertions or deletions occur. Often, it cannot be determined whether an insertion or deletion occurred, because the data only show that one sequence has additional nucleotides, while the other sequence has fewer nucleotides in the same position. Therefore, it is sometimes difficult to determine whether one sequence gained a set of nucleotides or the other sequence lost a set of nucleotides, unless other genealogical or evolutionary information is known for the samples or sequences. Indels can occur as a result of replication, insertion of a virus (or other) sequence, or by recombination. When it is generated by recombination, this is known as an *unequal crossover*, and one chromosome experiences an insertion, while the other chromosome experiences a deletion of the same size as the two chromosomes exchange segments. Another type of mutation is called *an inversion* (Figure 12.1f). This is where a segment of DNA has flipped 180° with respect to the other parts of the chromosome or larger segment. This inverts the sequence from the other strand onto the opposite strand. One might think this would be a difficult feat. However, inversions do occur sometimes during replication and during integration or excision of pieces of DNA, as well as during some recombination events. Apparently, in these processes, the DNA is looped around itself, thus facilitating breakage and rejoining to form an inversion.

(a) AGTCTGCTTTGAGGACGAGCCTATG

(b) AGTCTGCTTT**A**AGGACGAGCCTATG

(c) AGTCTG**G**TTTGAGGACGAGCCTATG

TTGAGG
(d) AGTCTGGTACGAGCCTATG

TCGGCAAT
(e) AGTCTGCTTTGAG**TCGGCAAT**GACGAGCCTATG

(f) AGTCTGCTT**CGTCCTCA**AGCCTATG

FIGURE 12.1 Major types of mutation in DNA. Changes are shown relative to the top sequence (a). (b) A G-to-A transition, (c) a C-to-G transversion, (d) deletion of six nucleotides (TTGAGG), (e) insertion of eight nucleotides (TCGGCAAT), and (f) inversion of eight bp sequences (TGAGGACG). In an inversion, the double-stranded region is flipped by 180°, which inserts the reverse complement of the opposite strand into the top strand, thus becoming CGTCCTCA.

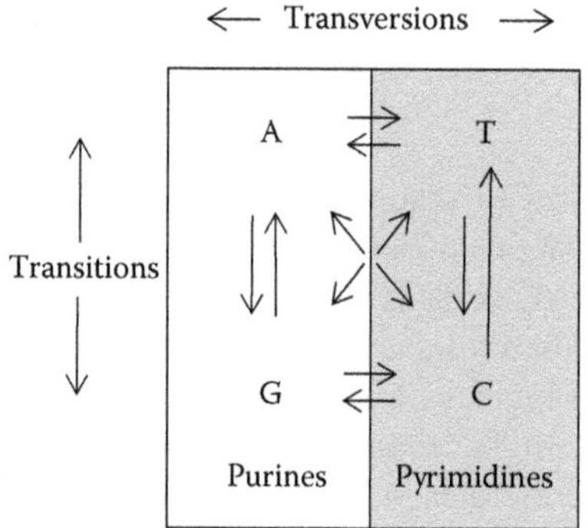

FIGURE 12.2 Comparison of the frequencies of transitions and transversions. Transitions (a purine-to-purine or pyrimidine-to-pyrimidine change) occur more often than do transversions (purine-to-pyrimidine or vice versa), as indicated by the lengths of the arrows. The ratio of transitions to transversions usually ranges from about 1.5:1.0 to more than 3.0:1.0. The most frequent changes are C to T changes, primarily because the change from m^5C to T is not corrected by repair systems.

CAUSES OF MUTATIONS

Changes to DNA and RNA can be caused by many factors (Figure 12.3 and Table 12.1). Certain chemicals, radiation, ultraviolet (UV) irradiation, heat, enzymes, slippage of DNA polymerase, nucleotide triphosphate concentrations, recombination, replication, infections, and other processes are known to cause these changes. There are also many repair processes that can correct these in DNA-based organisms. However, once the change has become established and the cell containing the original change has replicated, the mutation is fixed and can only be corrected by a subsequent mutation that restores the original genotype (called *a back mutation*). The only other way to remove the mutation from the cell population is for the cell to die. The major chemical changes to DNA are of three types: hydrolytic attack, oxidation, and *S*-adenosyl methionine transfer of a methyl (Figure 12.4 and Table 12.2). The major hydrolytic reactions cut the glycosidic bonds attaching the purines (adenine or guanine) to the deoxyribose sugars. This leads to a

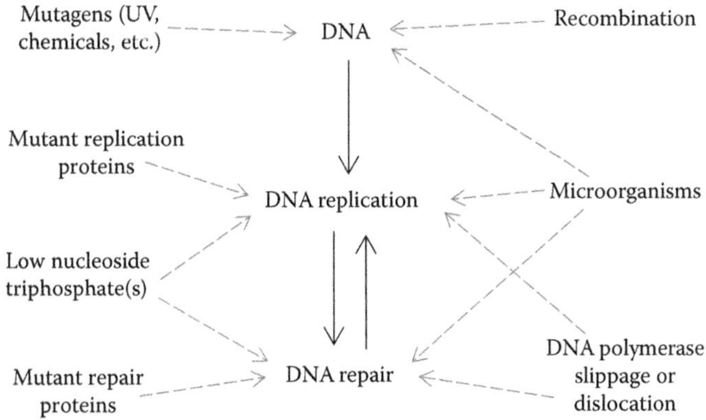

FIGURE 12.3 Factors that influence mutation. DNA can be changed directly by chemicals, UV irradiation, ionizing radiation, microorganisms (e.g., viruses), and other factors. Additionally, recombination and other processes may introduce mutations. Once DNA replication is initiated, the fidelity of replication can be affected by many factors, including the presence of mutant proteins, low concentrations of one or more of the dNTPs, microorganisms, or by slippage of DNA polymerase (normally in regions containing sequence repeats). Many of these changes can be reversed by DNA repair. However, repair can be affected by low dNTP concentrations, mutant repair proteins, microorganisms, and DNA polymerase slippage (since DNA replication is involved in DNA repair). However, even if the DNA changes are repaired, they are subject to new mutations by the replication process.

TABLE 12.1
Causes of Mutations and Their Effects

Categories and Types	Effect of Mutation
A. Mutagens	
UV irradiation	Pyrimidine dimers
Chemicals (some)	Many, from point mutations to cutting of the sugar-phosphate backbone
Ionizing radiation	Bond breakage, including cutting of the sugar-phosphate backbone
Infections (some)	Disruption of genes and control of cell cycle, chromosomal translocations
Enzymes (some)	Cutting of the sugar-phosphate backbone, disruption of replication and repair
B. Replication Errors	
Mutant DNA polymerase	Slow, inefficient, or inaccurate replication and repair
Low dNTP pools	Increase in rate of point mutations
Mutant primeosomes	Slow, inefficient, or inaccurate replication and repair
Mutant gyrase, helicases, and topoisomerases	Slow, inefficient, or inaccurate replication and repair
Mutant accessory proteins	Slow, inefficient, or inaccurate replication and repair
C. Altered Base Pairing	
G–T pairing	Point mutations
A–C pairing	Point mutations
Other nonstandard pairs	Point mutations
D. Repair Pathway Errors	
Mutant base excision repair	Varies from point mutations to gene loss
Mutant Nucleotide excision repair	Varies from point mutations to gene loss
Mutant mismatch repair	Varies from point mutations to gene loss
Mutant glycosylases	Point mutations

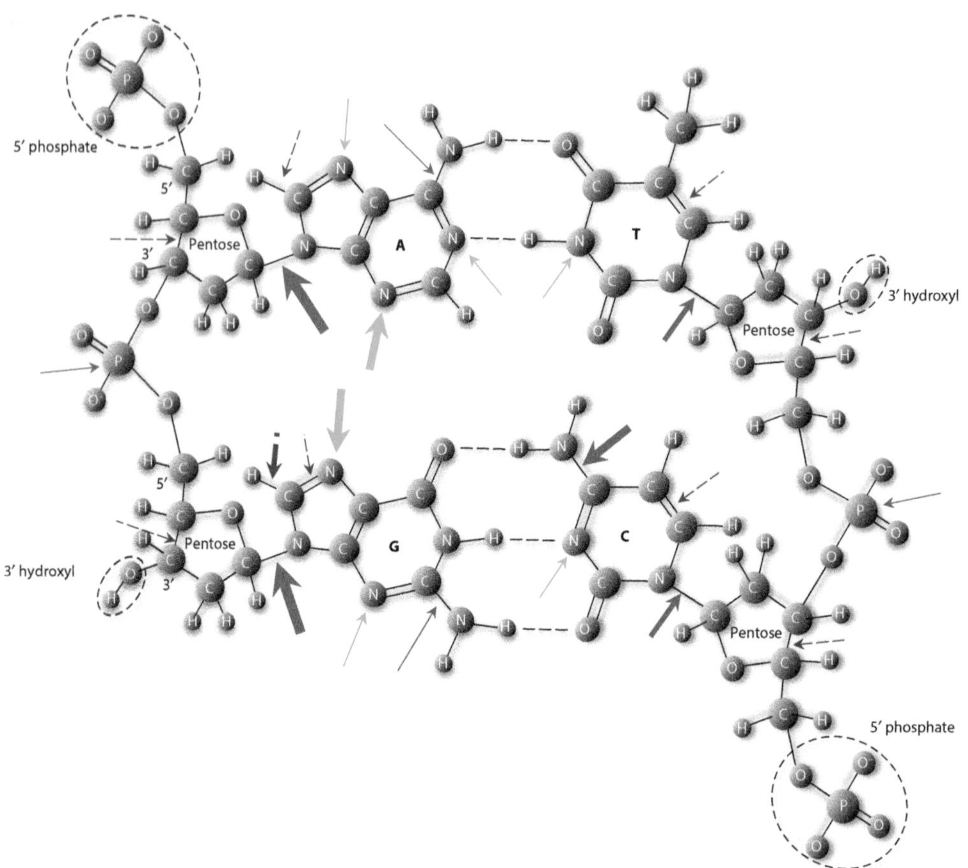

FIGURE 12.4 Locations where DNA is subject to degradative processes. The main reactions that degrade DNA are hydrolytic attack (blue arrows), oxidation (red arrows), and *S*-adenosyl methionine transfer of a methyl (green arrows). The thickness of the arrows indicates the degree of susceptibility of the chemical bonds and therefore, the frequency of occurrence of these reactions. For example, hydrolytic attack of the glycosidic bonds that attach the purines (A and G) to the deoxyribose sugars (thick blue arrows) are among the most frequently observed reactions. These occur readily when DNA is subjected to acidic conditions. Another frequent hydrolytic reaction is the deamination of C. When the C is a m^{5}C, deamination produces a T.

TABLE 12.2
Major DNA Mutations (in Order of Decreasing Frequency)

Original Nucleotide	Mutation	Resulting Molecule	Pairs With	*In Vivo* Repair
C	Deamination	U	A	Yes
m^{5}C	Deamination	T	A	No
A	Deamination	Hypoxanthine	C	Yes
G	Deamination	Xanthine	C	Yes
A	Depurination	No base	N	Yes
G	Depurination	No base	N	Yes
G	Oxidation	8-OH-G	A	No
A	Cyclic oxidation	8,5 cyclic A	Unknown	No
Y/Y	Dimerization	Pyrimidine dimers	Nothing	Yes
N	Hydroxymethylation	H-N-CH_2OH	Unknown	Unknown

depurinated site, which are very often corrected by repair enzymes. If they are not repaired, then any of the four nucleotides will be added to the new strand across from the apurinic site, since there is no base with which to pair. Another frequent hydrolytic reaction is deamination of cytosines. When a standard cytosine is deaminated, it becomes a uracil. Repair enzymes recognize the uracil and rapidly cut it out and other systems repair the strand. However, often cytosines are methylated at the C5 position. When an m^5C (5-methyl cytosine) is deaminated, it becomes a thymine (Figure 12.5 and Table 12.2), and there are no known repair systems that recognize or repair this mutation. It is for this reason that C to T changes are the most frequent changes in DNA (Figure 12.2). Occurring less often are hydrolytic attacks on the glycosidic bond of pyrimidines, which result in apyrimidinic sites. Again, these often are repaired, but if they are not, a mutation can become established during replication of the DNA. Finally, hydrolysis of purines can lead to deamination of adenine to form hypoxanthine and deamination of guanosine to form xanthine (Table 12.2). Both are recognized and converted back to the original base by repair enzymes. Hydrolysis of the phosphate backbone also can occur, especially when the DNA is exposed to acidic conditions. This breaks the DNA chains and can cause serious damage to DNA integrity. However, repair enzymes that repair single-strand breaks and others that repair double-strand breaks are present in most cells.

Oxidation of DNA can occur at several sites on the molecule, but these are less frequent than the hydrolytic reactions (Figure 12.4 and Table 12.2). These reactions occur more often in areas of high oxygen concentration, including areas of oxidative phosphorylation (e.g., in bacteria and mitochondria) and photosynthesis (e.g., in chloroplasts). One of the most frequent changes is hydroxymethylation of guanosine to form 8-hydroxy-guanosine (8-OH-G). It is unknown whether repair systems can recognize and repair these mutations. However, the 8-OH-G pairs most often with adenines, which can cause a G to T change in DNA during replication. This specific change is frequently found in mitochondrial genomes. It may be one cause for ageing in eukaryotic cells. Adenines also can be oxidized, but to a lesser extent. Also, the double bonds of purines and pyrimidines can be oxidized, fundamentally changing the base. In all of these, repair enzymes probably can recognize and convert the nucleotides back to the original condition. Finally, the C3 to C4 bond of the deoxyriboses can be oxidized, which forms a kink in the backbone at that site. This kink signals a set of repair enzymes that corrects this change. Several other physical processes can cause mutations. *S*-adenosyl methionine is known to cause transfer of a methyl group to particular compounds within the cell (Figure 12.4). One of the major effects in DNA is methylation of amino groups in purines, to form methylated and hydroxymethylated nucleotides (Table 12.2). It is unknown whether there are repair mechanisms for these mutations.

Radiation can cause major changes in DNA. Alpha particles are mostly harmless to DNA, because of their low energy. They may cause some damage if they are from radioactive chemicals that are incorporated into the DNA itself, but otherwise, normally they are not hazardous. On the other hand, beta and gamma particles can damage DNA by breaking chemical bonds in DNA. Beta particles are high-energy electrons, while gamma particles are similar to X-rays. Both can be very damaging to DNA, essentially breaking or disrupting any chemical bond with which they come into contact. UV irradiation is capable of damaging DNA and is evident in the disease known as *xeroderma pigmentosum*. The rings of both purines and pyrimidines absorb UV irradiation and thus become somewhat unstable and reactive. This is especially true for pyrimidines, such that adjacent pyrimidines react to form a butyl (four carbon) ring that joins the pyrimidines (Figure 12.6). This happens most often with adjacent thymidines, but also occurs with adjacent C–T pairs and less often with C–C pairs (the ratio of occurrence of T–T:T–C:C–T:C–C dimers is 68:16:15:1). In people with xeroderma pigmentosum, one or more of the proteins in the pathway that recognizes and repairs these lesions is faulty. When these people are exposed to even small amounts of UV irradiation, they exhibit blisters and sores (due to cell damage and cell death) on all of the areas of skin that were exposed to the UV. Additionally, they have high frequencies of skin cancers. Therefore, it is clear that each time when humans are exposed to sunlight and other sources of UV irradiation,

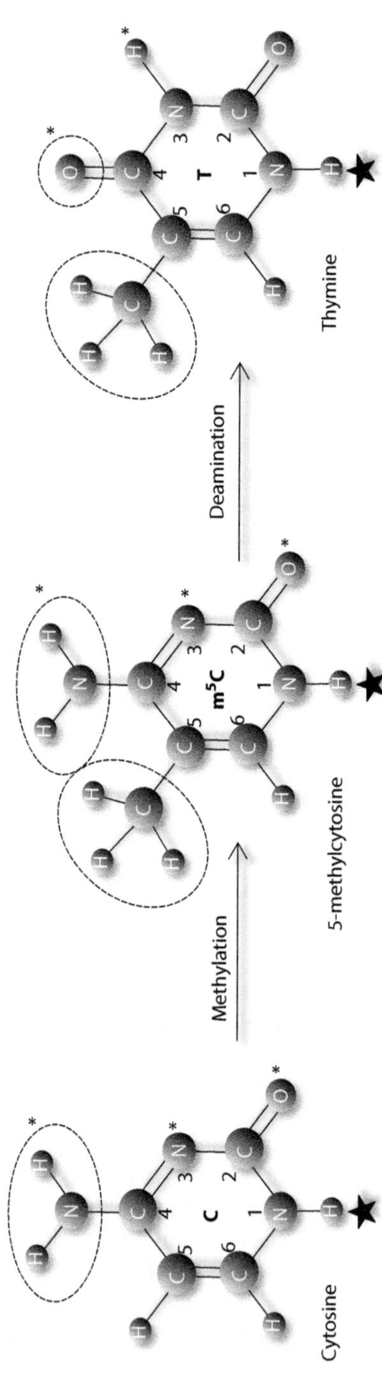

FIGURE 12.5 Production of a T from a C. In high G/C regions, many of the C's may be methylated on the fifth carbon to become m^5C. When this is deaminated, it forms a T, which is indistinguishable from any other T in the DNA and therefore is not recognized as a mutation by repair systems.

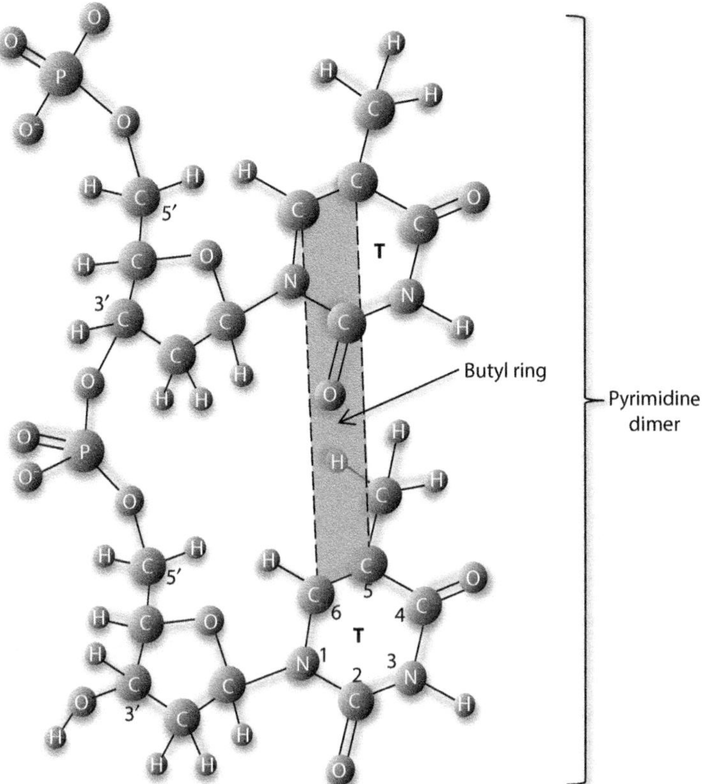

FIGURE 12.6 Thymidine (pyrimidine) dimer. When UV irradiation is absorbed by adjacent thymidines in the DNA, the double bond between C-5 and C-6 is broken and the two thymidines are joined through C-5 and C-6 by a butyl (four carbon) ring.

a large number of nucleotides are damaged in the exposed cells. In most people, the damage is rapidly recognized and corrected by repair enzymes.

MUTATION DURING REPLICATION

Many mutations occur during replication. Each step in the process can generate errors, although repair systems normally correct most of the errors (Table 12.3). In the first step of the process, base pairing alone can generate a mutation every 10–100 nucleotides. This is primarily caused by base mispairing, because an A can sometimes pair with C instead of T, and G can sometimes pair with T instead of C. Even when DNA polymerase is added to the synthesis process, many mutations can

TABLE 12.3
Mutation Rates for Several Steps in the Replication and Repair Pathways

Mechanism (Step)	Cumulative Errors
Base pairing	10^{-1} to 10^{-2}
DNA polymerase (base selection plus 3′–5′ proofreading)	10^{-5} to 10^{-6}
All of the above, plus accessory proteins	10^{-7}
All of the above, plus post-replication mismatch repair	10^{-10}

TABLE 12.4
Number of Cumulative Mutations for Replication and Repair Processes in 10^9 Nucleotides

Mechanism (Step)	Cumulative Errors
Base pairing	10^8 to 10^9
DNA polymerase (base selection plus 3′–5′ proofreading)	10^4 to 10^5
All of the above, plus accessory proteins	10^3
All of the above, plus post-replication mismatch repair	<10

be introduced, although the number of errors is reduced. DNA polymerase introduces errors in the replicated strand at a rate of 10^{-5} to 10^{-6} (one change in every 100,000–1,000,000 bases). Actually, DNA polymerase alone has a higher rate than this, but part of the enzyme contains proofreading portions that rapidly correct any detected mistake. Accessory proteins (e.g., single-stranded-binding proteins) decrease the mutation rate further (to about 10^{-7}) by stabilizing and protecting the DNA in the region that is being replicated. Finally, post-replication repair systems reduce the rate of mutation to as low as 10^{-10}. Consider how many mutations would occur per replication cycle if these correction systems were absent (Table 12.4).

The haploid human genome consists of three billion base pairs. In a diploid human genome, there would be six billion base pairs. Therefore, in a somatic cell, each strand of DNA on each chromosome would have to be replicated, producing a total of more than 12 billion base pairs of DNA (or 24 billion bases, 12 billion of which are incorporated into the newly synthesized strands of DNA). The number of cumulative errors per replication cycle would be approximately 6×10^7 to 6×10^8 mutations if replication were only based on base pairing. If replication were only based on the functions of DNA polymerase without accessory proteins and post-replication repair, there would be approximately 6×10^3 to 6×10^4 mutations per replication cycle. Because the human genome has about 20,000–25,000 genes, then a large number of genes would be expected to experience mutations during each round of replication and cell division, and in fact, some genes would have more than one mutation. Clearly, maintenance of the human genome would be impossible at these rates of mutation. However, when the accessory proteins and post-replication repair systems are added to the replication process, the number of mutations per replication and cell division cycle on average is less than 10 per replication cycle. Many organisms have much larger genomes than do humans, and therefore, it is expected that their genomes would experience tens to hundreds of mutations per replication cycle, unless those organisms have additional repair systems. Fortunately, for the larger eukaryotic genomes, less than 1% of their genomes contain genes, so that most mutations occur in nongenic regions. Nonetheless, this is an amazing rate of fidelity for the replication process. Each part of the system has evolved so that these large genomes are possible. Without each one of these systems ensuring accuracy, organisms with genomes of one million base pairs or more would be impossible.

DNA REPAIR

Damage to DNA can be corrected by a set of repair systems (Figure 12.7). Direct reversal is possible for several types of lesions. Photolyases can reverse pyrimidine dimer damage directly, although a specific enzyme-based repair system also exists that first removes the section of the DNA containing the dimerized pyrimidines and then synthesizes a new corrected DNA strand. One reason for the two systems is that the UV irradiation from the Sun generates thousands of lesions in skin cells every day. If left uncorrected, many skin cells would die, and others would form tumors, some of which would be cancerous. Methylation of some bases also can be reversed. Demethylation of some cytosines and adenines, as well as direct demethylation of guanines, can occur. Most correction is accomplished using repair

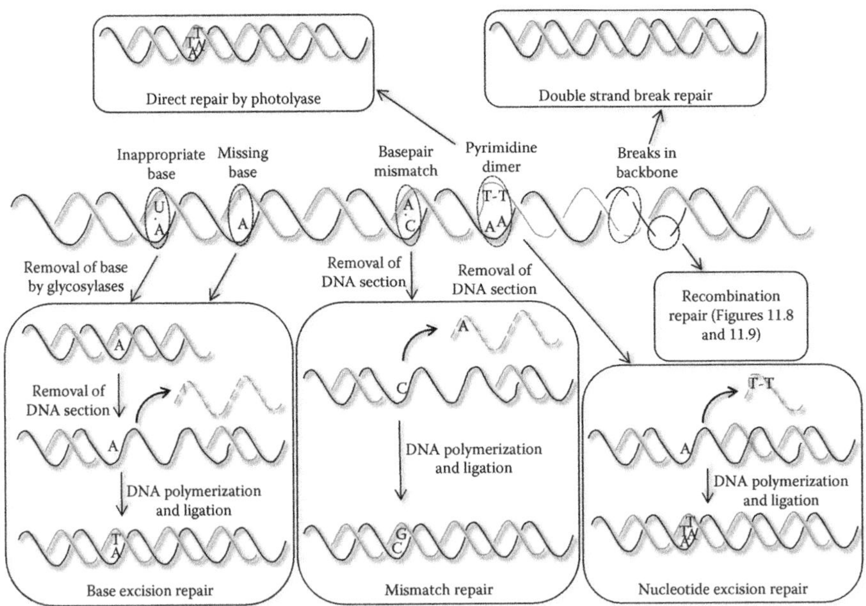

FIGURE 12.7 Primary DNA repair systems. Thymidine dimers can be directly repaired by photolyase (top left), an enzyme that is activated by light energy to remove the butyl ring from thymidine dimers. Double strand breaks can be repaired by a multimeric set of proteins that recognizes the break, brings the two ends together, and then ligates the ends (top right). Base excision repair (lower left) is initiated when an inappropriate base (e.g., a U in DNA) or an AP (apurinic or apyrimidinic) site are detected by specific proteins. If it is an inappropriate base, a glycosylase first removes the base by cutting the glycosidic bond. Next, a section of the DNA is removed and DNA polymerase then fills in the section of the strand that was removed and ligase forms a bond between the final two bases. Mismatch repair (lower middle) is initiated when an unusual base pair is detected. Specific proteins move along the DNA, and as long as it finds G–C and A–T base pairs, it does not trigger repair. However, when it detects a nonstandard base pair (usually by its three-dimensional form), it will signal excision of the lesion and surrounding region of one of the strands. Then, DNA polymerase and ligase repair the incomplete strand. Nucleotide excision repair is the main pathway that repairs pyrimidine dimers. The dimers create a bulge in the DNA that is recognized by specific proteins, which act in conjunction with many other proteins to remove the dimer, plus the regions surrounding this lesion. DNA polymerase and ligase then repair the incomplete strand. Recombination repair is initiated when one or more of the strands are broken (e.g., by ionizing radiation). The nicked regions can invade homologous chromosome regions to repair the lesion(s). Recombination is detailed in Figures 12.8 and 12.9.

pathways, most of which begin with enzymes that recognize and cut out the abnormal base (or bases) using endonucleases or glycosylases. In base excision repair, specific bases are recognized in the DNA. For example, uracil is a nonstandard base in DNA. A specific enzyme, called *uracil glycosylase*, cuts the uracil base from the attached sugar, creating an apyrimidinic site. Also, under acidic conditions, purines may be removed from the sugar. These create apurinic sites. Specific enzymes called *apurinic/apyrimidinic (AP) endonucleases* recognize sugars without bases and remove those sites as well as sections of the DNA strand upstream and downstream from the lesion. In most of these processes, more than a single base is removed, and sometimes hundreds to thousands of bases are removed. All of them use DNA polymerase to fill in the gap(s) created by the nucleases and finally DNA ligase to seal the final gap by forming a phosphodiester bond between adjacent bases. Another type of repair system is *mismatch repair* (Figure 12.7), where nonstandard base pairs are recognized. When an A–C, G–T, purine–purine, or pyrimidine–pyrimidine pair forms, a kink or a bulge is formed in the DNA. Enzymes (e.g., MutL, MutS, and MutH) traverse along the DNA, and when they come upon one of these mismatches, they are able to remove the mismatched base, as well as stretches of the DNA upstream and downstream from the lesion. DNA polymerase then fills in the long gap, and ligase binds the final two bases together along the sugar–phosphate backbone. Nucleotide excision repair is a similar process, but it uses a different set of recognition and endonuclease enzymes. Some of these are Uvr–A, Uvr–B, and Uvr–C, because they were originally observed when mutants of these repair enzymes were more susceptible to UV irradiation damage. This system primarily recognizes pyrimidine dimers caused by UV damage. As with base excision repair, the enzymes cut out a large region of the DNA strand. After excision, DNA polymerase and ligase fill in the gap to complete the repair. Finally, there are repair systems that detect and correct single- and double-stranded breaks. Of course, if there are any mutations in any of the genes of these repair systems, mutations generally will increase because the lesions will not be repaired or may be repaired at slower rates.

GENETIC RECOMBINATION

Genetic recombination can both cause and repair DNA damage. It is often involved in repair of single and double strand DNA breaks (Figures 12.7). Recombination creates a crossover of two double-stranded DNA molecules. However, in the region of the crossover, the DNA strands are not simply exchanged (Figures 12.8 and 12.9). They form a region that consists of one strand from one double helix paired with the opposite strand from the other double helix. This creates a region called *a heteroduplex* and is noticeable when the crossover occurs within a region where different gene alleles exist on each of the two chromosomes. As the crossover event continues, DNA polymerase moves in and uses one strand or the other as the template and displaces the other strand as it polymerizes a new strand so that the two strands are completely complementary. Eventually, it falls off,

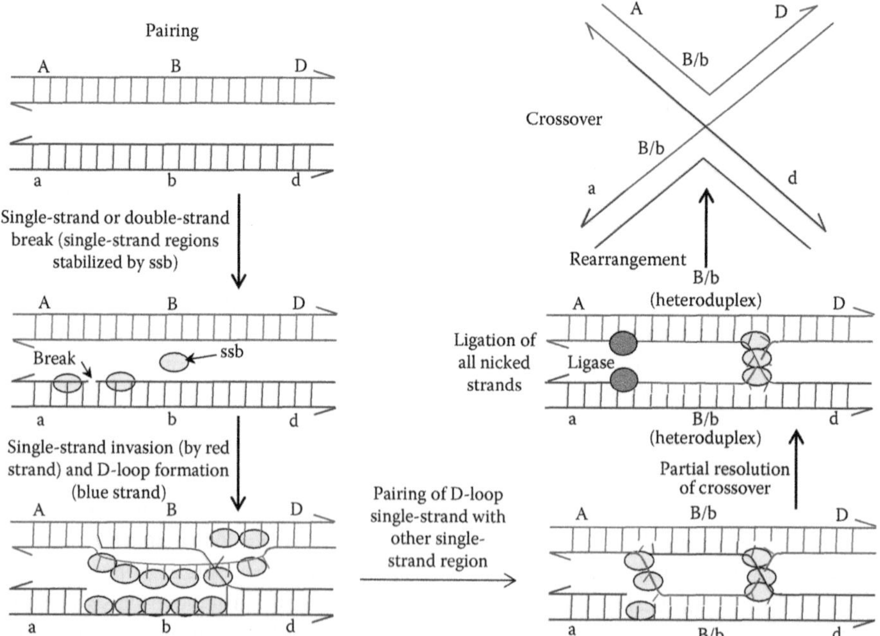

FIGURE 12.8 Formation of a crossover by genetic recombination. Sections of two chromosomes that differ in three genetic loci (A/a, B/b and D/d) are shown. A crossover is initiated when one or more of the strands are broken. The single strand (stabilized by single-strand binding proteins, ssb) can dissociate from the opposite strand and invade the other chromosome. This displaces the analogous strand on the other chromosome forming a D-loop (named because it resembles the letter D when observed in an electron microscope). The D-loop can pair with the opposite strand on the other chromosome. A break to sever the D-loop on one side is followed by ligation. At this point, the two chromosomes are joined by a single crossover where each chromosome has a heteroduplexed region.

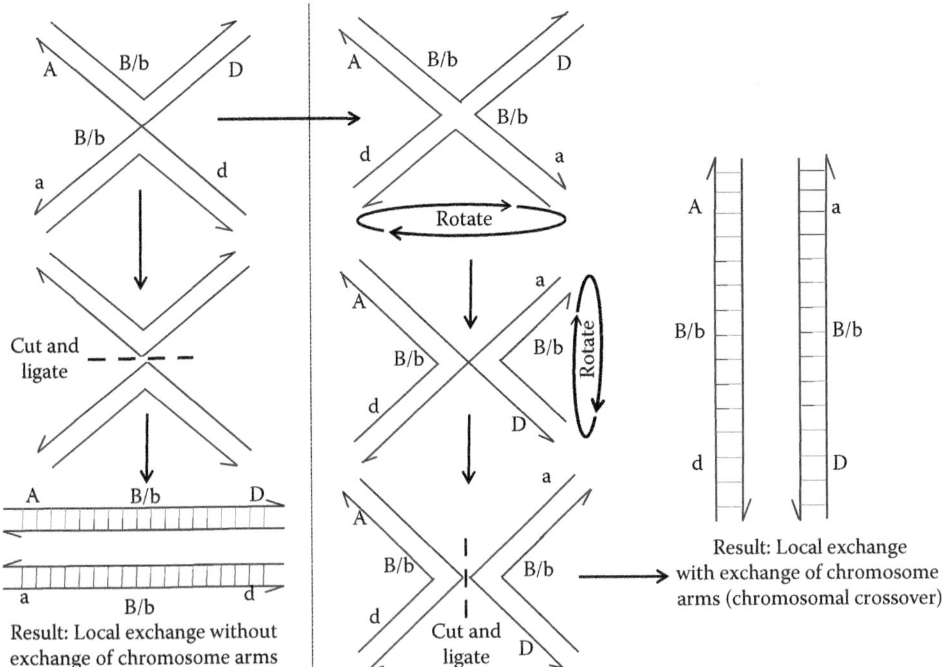

FIGURE 12.9 Two ways to resolve the crossover. If the chromosomes are separated by cutting at the crossover (starting at the top left and proceeding down to the bottom left), the original chromosomes have the same alleles surrounding the crossover (i.e., the A allele is connected with the D allele, as in the original chromosome). However, the B gene region contains one strand from the B allele and one strand from the b allele. Therefore, this is a heteroduplexed region. There is a second way to separate the chromosomes at the same crossover. If the chromosome ends are rotated (moving from top left to the right, and then down), followed by cutting, the ends of the chromosomes are exchanged. That is, the A allele is physically connected to the d allele on the same chromosome, and the other chromosome has the a and D alleles. Again, the B gene region (where the crossover occurred) is heteroduplexed.

and ligase seals the last small gap. This can often alter the allele ratios in meiosis. This effect first was noticed in fungi when segregation of certain alleles of genes did not fit the typical Mendelian segregation patterns. That is, normally during meiosis, usually there is a 1:1 segregation of alleles. In some ascomycetous fungi (e.g., *Neurospora crassa*, a rust that infects wheat), eight spores are formed in the ascus (fruiting body). After crossing individuals with two different genotypes, four of the resulting spores should be of one genotype and the other four should be of the other genotype. However, ratios of 5:3, 6:2, and even 7:1 and 8:0 resulted, and this happened with many types of genes. Eventually, it was discovered that the gene conversions were occurring when the genes were involved in a crossover. Specifically, it occurred when the heteroduplexed region was within or near the gene that was converted. In the heteroduplexed region, the polymerase sometimes used one strand (containing one of the alleles) or the other (containing the other allele) as the template. Therefore, occasionally, it would convert one allele into the other allele, and vice versa, and occasionally, conversion resulted in an allelic ratio that differed from the expected 4:4 segregation.

Crossovers can occur between homologous chromosomes and sister chromatids during meiosis, but also in somatic cells between sister chromatids, homologs, and occasionally between nonhomologs (Figure 12.10). This can result in chromosomal translocations, where parts of two nonhomologous chromosomes exchange parts of each. These occur in many types of cancer, as well as in other diseases, such as Down's syndrome and schizophrenia. Genetic exchanges can also occur when viruses integrate into host chromosomes, as well as when they excise from those chromosomes. These may

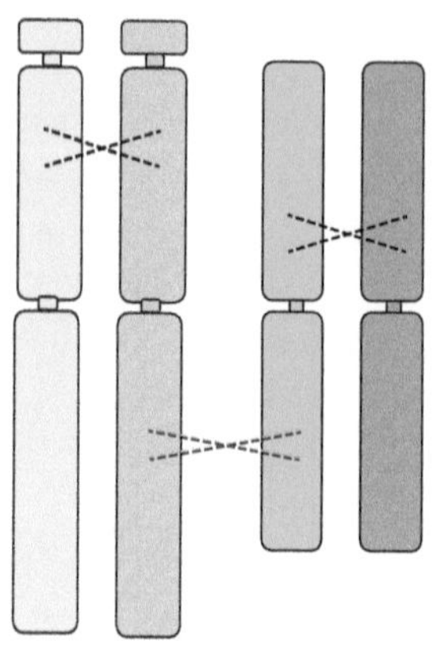

In somatic diploid cells, crossovers between homologous and non-homologous chromosomes
(a)

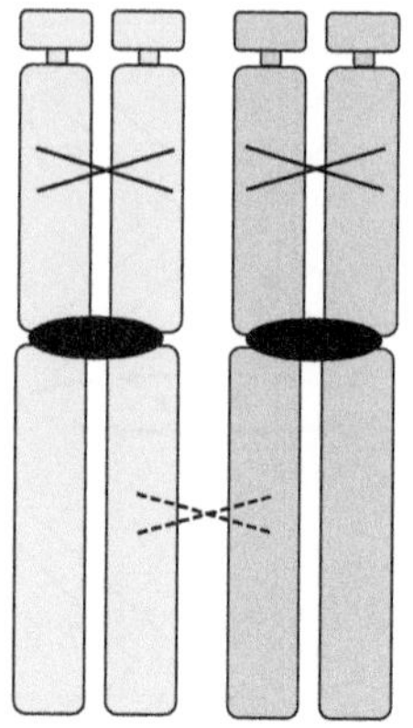

During mitosis, crossovers between sister chromatids and homologous chromosomes
(b)

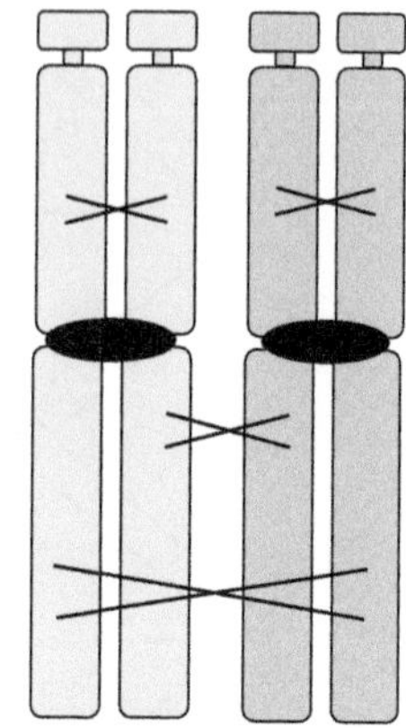

During meiosis, crossovers between sister chromatids and homologous chromosomes
(c)

FIGURE 12.10 Crossovers possible among sets of chromosomes. In diploid somatic cells (a), homologues can undergo recombination, although at low frequency (as indicated by black dashed crosses). Rarely, non-homologues can recombine (red dashed crosses), although this often causes problems in gene expression (e.g., some cancers are caused by these chromosomal translocations). During mitotic divisions (b), crossovers between sister chromatids can occur (black crosses). Exchanges between homologous chromosomes occur at lower frequency (black dashed crosses). The frequency of crossovers is highest during meiosis (c). All four chromatids are brought together by the synaptonemal complex, and crossovers are produced by recombination modules. Crossovers between all four chromatids are possible.

cause mutations in the host chromosome, or conversely, the viruses can integrate mutated versions of the host genes. Some of these are characteristic of some types of cancer. Integration of mobile elements also can cause mutations in an analogous way, although the insertion and excision events differ.

In general, genetic recombination leads to reassortment of alleles, but it can also lead to insertion of additional sequences, such as integration of plasmid, virus, and other DNAs. It can lead to deletions of sequences, sometimes many thousands of nucleotides at a time. These mutations can have large effects on cells, as in the case of integration of a DNA copy of the HIV viral RNA into human T cells. The integration ultimately leads to disfunction of the host chromosomes, converting the cells into virus factories, rather than the immune cells that they once were. Recombination also causes changes in gene copy number in multicopy regions, such as in the ribosomal RNA gene locus. Crossovers in this region are the mechanisms that maintain high gene rRNA copy numbers in eukaryotic genomes and are linked with the process of gene conversion that causes most or all of the rRNA gene copies to be identical. A recombination event in meiosis involves the synaptonemal complex that forms the synapses of the chromatids resulting in chasmata. This region also contains recombination nodules that contain the enzymes responsible for cutting and rejoining the strands of DNA, and heteroduplexed regions are found in these areas. In other types of crossovers, such as in eukaryotic somatic or asexual cells and bacteria, other proteins are present that initiate crossovers, although the overall process and result are the same.

Most recombination events begin with single-strand invasion by a region of DNA from the adjacent DNA duplex. Usually, this is initiated by a single-strand break, followed by migration and invasion by the single-stranded DNA (Figure 12.8). Alternatively, in some cases, it begins as a single-stranded region known as *a D-loop* where no break in the DNA has yet occurred, although the end of another single strand invades the region when there is sufficient homology. The single-strand regions are stabilized by single-strand-binding proteins (e.g., SSB, recA, and rec1). Stabilization is necessary because the base regions of the DNA are hydrophobic and associate with other hydrophobic regions, such as the base regions of other DNA strands. These hydrophobic interactions, plus hydrogen bonds between the bases, allow the single-stranded DNA to invade the other DNA double helix. The interactions are stronger when the two DNA molecules have identical sequences because the base pairing favors pairing of the complementary bases. However, nonhomologous regions occasionally pair to form a crossover. Once there is strand invasion, the pairing region grows as long as the bases are complementary (or nearly so). The homologous strand on the other duplex (which has been displaced by the invasion) begins to pair with the complementary strand on the other duplex, and eventually, there is a single-strand break on that strand. Therefore, at this point, there is a region where one strand of each duplex is paired with the anti-parallel strand on the other duplex. In this region, therefore, each duplex consists of a heteroduplex. There are two ways (Figure 12.9) in which the crossover can be resolved to separate the two duplexed DNAs (e.g., chromosomes). If the two strands are crossed in the middle of the crossover, and the strands are cut and then ligated to the strand on the opposite duplex, the ends of the duplexed molecules still are on the original piece of DNA to which they were attached. The only difference is that a section of heteroduplexed DNA exists on each of the resulting duplexed molecules. If there are differences in the two strands, the mismatch repair system recognizes and changes these so that there are no mismatched pairs. This is where a change in the 1:1 (or 4:4 during meiosis) ratios in gene ratios may change. The crossover can result in a change of the four ends of the duplexed DNAs as well. If the DNAs are twisted just a bit, and then the strands are cut and then ligated (Figure 12.9), the ends of the duplexed molecules are switched. In this case, each duplexed DNA molecule (e.g., chromosome) has a new combination of alleles and also has a heteroduplexed region at the site of the crossover. These heteroduplexed regions often extend for hundreds of nucleotides, and again, polymerase and ligase operate to produce a fully complementary set of continuous DNA strands.

KEY POINTS

1. Mutations can have a range of effects. Some cause no detectable changes, others are lethal, and still others have varying degrees of effects.
2. Many DNA changes can be repaired before they become mutations, but if the repair systems are faulty in some way, many more mutations can accumulate, increasing the likelihood of lethal or damaging effects.
3. Once a mutation is established, then it is subject to selection.
4. Selection occurs within the cell for various cellular processes; it can be important in cell-to-cell interactions or may change interactions of the organism with its environment.
5. The other way that mutations can be generated is through infection by viruses and other organisms (including bacteria and eukaryotic parasites).

ADDITIONAL READINGS

Alberts, B., D. Bray, K. Hopkin, A. Johnson, J. Lewis, M. Raff, K. Roberts, and P. Walter. 2013. *Essential Cell Biology*, 4th ed. New York: Garland Publishing.

Alberts, B., D. Bray, A. Johnson, J. Lewis, M. Raff, K. Roberts, and J. D. Watson. 1994. *Molecular Biology of the Cell*. New York: Garland Publishing.

Darnell, J., H. Lodish, and D. Baltimore. 2007. *Molecular Cell Biology*, 6th ed. New York: W.H. Freeman and Company.

Freiberg, E. C., G. C. Walker, W. Siede, R. D. Wood, R. A. Schultz, and T. Ellenberger. 2005. *DNA Repair and Mutagenesis*, 2nd ed. Washington, DC: ASM Press.

Graur, D. and W.-H. Li. 2000. *Fundamentals of Molecular Evolution*, 2nd ed. Sunderland, MA: Sinauer Associates.

Higgs, P. H. and T. K. Attwood. 2005. *Bioinformatics and Molecular Evolution*. Carleton, Australia: Blackwell Publishing.

Karp, G. 2013. *Cell and Molecular Biology*, 7th ed. New York: John Wiley & Sons.

Lewin, B. 2008. *Genes IX*. Sudbury, MA: Jones & Bartlett Publishers.

Li, W.-H. 1997. *Molecular Evolution*. Sunderland, MA: Sinauer Associates.

Li, W.-H. and D. Graur. 1991. *Fundamentals of Molecular Evolution*. Sunderland, MA: Sinauer Associates.

Lindahl, T. 1993. Instability and decay of the primary structure of DNA. *Nature* 362:709–715.

13 Infectious Changes to DNA

Viruses, Plasmids, Transposons, and Introns

INTRODUCTION

In this chapter, infectious changes to nucleic acids are discussed, including those caused by viruses, plasmids, transposons, and introns. They are grouped in this way because some of them appear to be related, and similar processes are involved when they infect and change genomes. Infections by some viruses are well known for their ability to mutate the host genome (e.g., human immunodeficiency virus [HIV] and human papilloma virus). However, some bacteria and some eukaryotic parasites (e.g., trypanosomes) have also been documented to alter host genomes, causing mutation by inserting sequences into the host genome. Viruses, plasmids, and parasites are the major agents of horizontal gene transfers, that is, movements of DNA from one species to another species. Likewise, transposons and mobile introns have the ability to insert themselves into specific sites of the genomes where they exist, and some introns have the ability to invade other genomes, and thus are similar to viruses. In fact, some viruses, introns, and transposons have genes that are very similar. The reverse transcriptases of disparate viruses, such as HIV and cauliflower mosaic virus, some retrotransposons, and some types of introns, all have reverse transcriptases that are similar and related phylogenetically. Additionally, these reverse transcriptases have similarities to telomerases in eukaryotes (Figure 13.1).

Often, viruses have target sequences on the host chromosomes into which they integrate (Figure 13.2). In many instances, the viruses express a gene that encodes an endonuclease that recognizes and cuts the host DNA at one or more specific locations, based on the nucleotide sequence of these locations. Endonuclease genes such as these are also found in some transposons and some introns, again indicating some similarities among these elements. An argument can be made that some of these infectious agents and elements appear to be related, and therefore form a phylogeny whose members can be viewed as parts of a large evolutionary process (Figure 13.3). Simply stated, the process is as follows: At stage 1, viruses can infect cells and many integrate into host chromosomes; at stage 2, some introns can be *infectious* in that they can convert intron-less strains of organisms into intron-containing strains when the two strains are crossed; at stage 3, transposons have the ability to invade various parts of a genome once they are introduced into those genomes; and finally at stage 4, some transposons and introns eventually lose all of their mobility, and some still can have effects on the surrounding genes.

Changes caused by these elements have had great effects on evolution. In humans and other organisms with larger genomes, the presence of introns allows for alternative splicing. In many cases, a single gene can encode more than one protein product. For example, human genomes have from 23,000 to 25,000 genes, but they can produce more than 90,000 different proteins. The presence of introns can also cause diseases when inaccurate splicing occurs as a result of a mutation. This is the case in the disease beta thalassemia, where altered splicing produces abnormal hemoglobins that cannot bind oxygen, and as a result, the affected individuals experience severe anemia. However, alternative splicing appears to aid in rapid evolution in many species, including humans. The reason for this is that new proteins can be produced from a single gene, thus eliminating the need for the evolution of a completely new gene. Therefore, although alternative splicing can and does cause some problems when mutations alter the splicing sites, it can be of great evolutionary benefit as well.

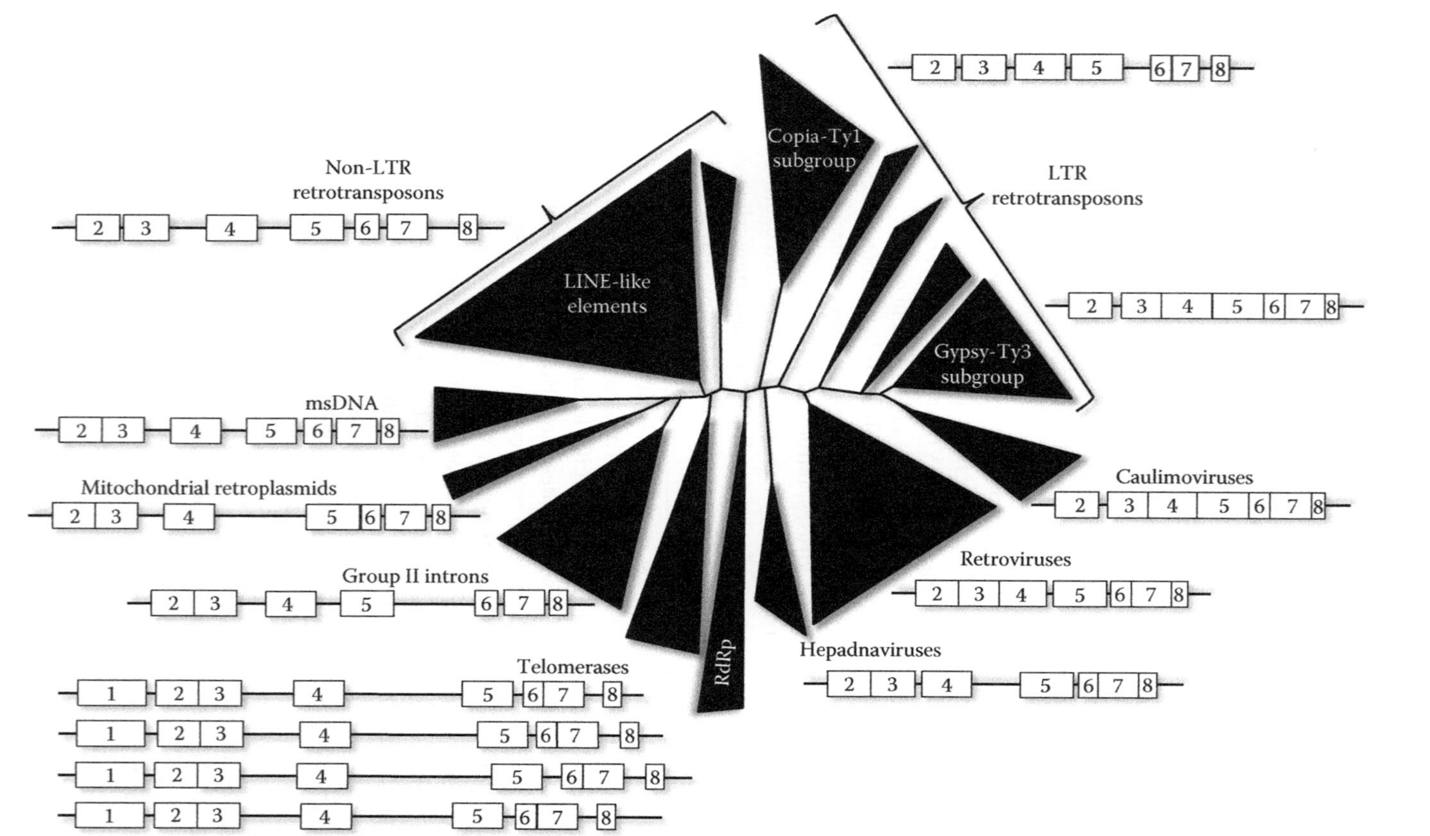

FIGURE 13.1 Phylogenetic tree of reverse transcriptases and related gene sequences among viruses, transposons, and telomerases. Diversity of each group is indicated by the size of the triangles. Conserved regions (numbered 1 through 8) are indicated beside each group of genes. Region 1 is unique to telomerases (lower left). LINE, long interspersed repetitive elements. The tree is rooted to RNA-dependent RNA polymerases (RdRp). LTR, long terminal repeat; msDNA, multicopy single-stranded DNA.

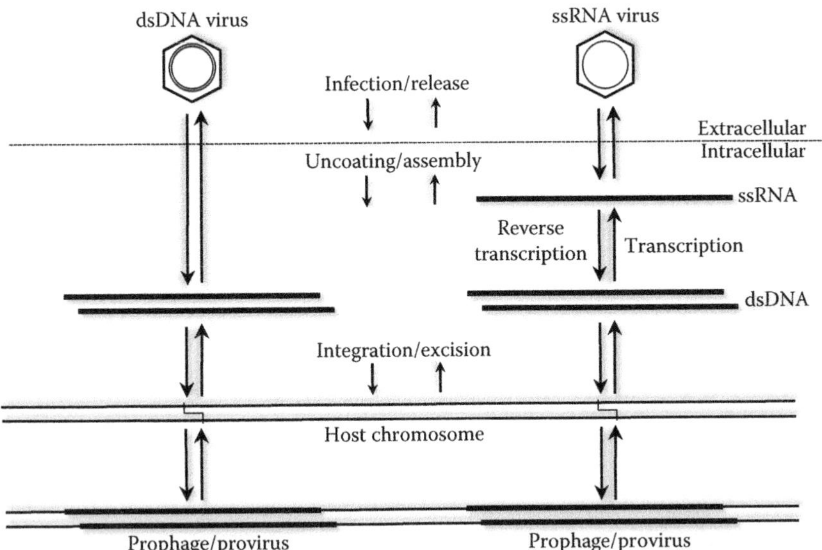

FIGURE 13.2 General pathway of virus integration into and excision from host chromosome. DNA viruses can integrate directly, whereas RNA viruses must first produce a DNA copy of their genome prior to integration to become a prophage. Usually, some viral genes are expressed although integrated as a prophage. Upon excision, different sets of genes are expressed to initiate the viral infectious cycle.

Although viruses, transposons, and introns can cause major genetic changes, not all of them do. Many viruses do not integrate into the host chromosome, and thus cause no mutations. Some transposons are incapable of moving to other locations. Some that transpose to other locations integrate into genomic regions that cause no phenotypic change in the organism. Although a few introns are capable of converting intron-less genes into those with introns, most are incapable of movement or any type of conversion. For introns, the major changes occur when the normal splicing positions mutate into sites that no longer function in splicing. In these cases, alternate cryptic splice sites may be utilized, which cause the production of an altered messenger RNA (mRNA) that encodes a mutant protein.

Integration of bacteriophage into bacterial genomes is well documented. Bacteriophage usually insert into specific sites of the bacterial genome (Figure 13.4). They accomplish this using a homologous recombination pathway. A site on the bacteriophage chromosome is identical (or nearly so) to a site on the bacterial chromosome. During the lysogenic cycle, specific integration proteins that are expressed from the phage genome act to cut the bacterial chromosome and insert the phage genome into the break site. Bacteria have evolved to accommodate this integration event in an evolutionary sense. They survive the lysogenic cycle, and this aids the virus because during each replication cycle, the phage genome is replicated by being an integral part of the bacterial genome. This is not only tolerated by the bacterium, but in some cases the presence of the integrated phage confers a selective advantage to the bacterial cell, because it confers protection against infection by other viruses. About 1%–2% of bacterial genomes are composed of these virus sequences integrated into the chromosome(s). When the level of particular proteins (repressors of the lytic cycle) decreases, often related to the bacteria being stressed, the phage DNA excises from the bacterial chromosome, again using specific proteins that cut out and circularize the phage DNA, and the lytic cycle begins.

Viruses also invade eukaryotic genomes. For example, approximately 8%–9% of the human genome is composed of virus sequences. The human genome is just over three billion base pairs, or six billion base pairs per diploid cells. Therefore, approximately 500 million base pairs in each human somatic cell nucleus are integrated virus sequences. This is equivalent to approximately 100–1000 bacterial genomes. Virus sequences probably exist in all eukaryotic genomes. Integration events probably have been occurring for billions of years. Many of the integrated virus sequences

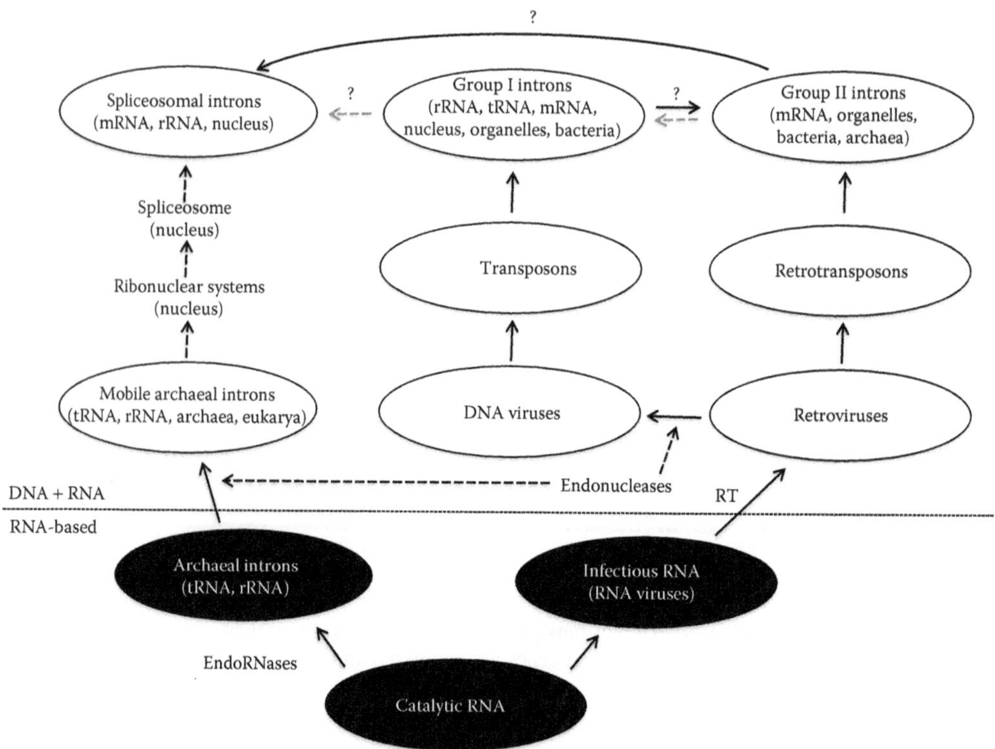

FIGURE 13.3 Proposed model for the evolution of mobile genetic elements. Catalytic RNAs appeared early on the Earth and were the likely precursors of mobile genetic elements. Some of these may have become infectious elements, others became the archaeal introns, which rely primarily on other RNAs for processing and activity. As DNA became the prevalent form of genetic material, some of the elements could evolve into those that produced DNA from RNA templates to become retroviruses, and later DNA viruses. Similarities in endonucleases and reverse transcriptases indicate that the transposons and retrotransposons may have evolved from similar viruses and retroviruses. Further evolutionary reduction may have led to the evolution of the various introns from transposons and retrotransposons. The possible phylogenetic relationships among the group I, group II, and spliceosomal introns remain unclear, although the most likely scenario (indicated by solid black lines) is that group II introns were derived from group I introns, and spliceosomal introns were derived from group II introns. Although spliceosomal introns splice in a manner that is similar to group II introns, group I and spliceosomal introns have been found overlapping with one another in the rRNA genes of some organisms, indicating a possible common origin. Spliceosomes may have originated from primordial groups of ribonuclear systems. Question marks indicate possible pathways that have not been confirmed.

are incapable of excising from the host chromosomes. Therefore, essentially they are remnants of ancient integration events. For some, none of the genes appear to be expressed. In others, some of the genes are expressed. Integration may occur sporadically through the life of an individual organism, and for some of the recent integration events, the virus sequences are expressed to some extent. For some of these (e.g., herpesviruses), skin lesions occasionally are produced. In others (e.g., simian virus SV40), integration can cause tumors in the host. Some cause death (e.g., HIV). In some individuals, a case of measles does not end with the recovery from acute disease symptoms. Years after the disease symptoms have disappeared, brain lesions begin to appear, eventually leading to neurological disorders and ultimately death. This is caused by the expression of measles virus genes in the brain cells of the affected patient, presumably from a copy of the virus genome that is integrated into the genome of the patient years previously during the initial infection phase. HIV is another virus that integrates into host cell genomes, and in this case, takes over the T lymphocytes to produce more viruses, which renders the T cells useless as immune cells.

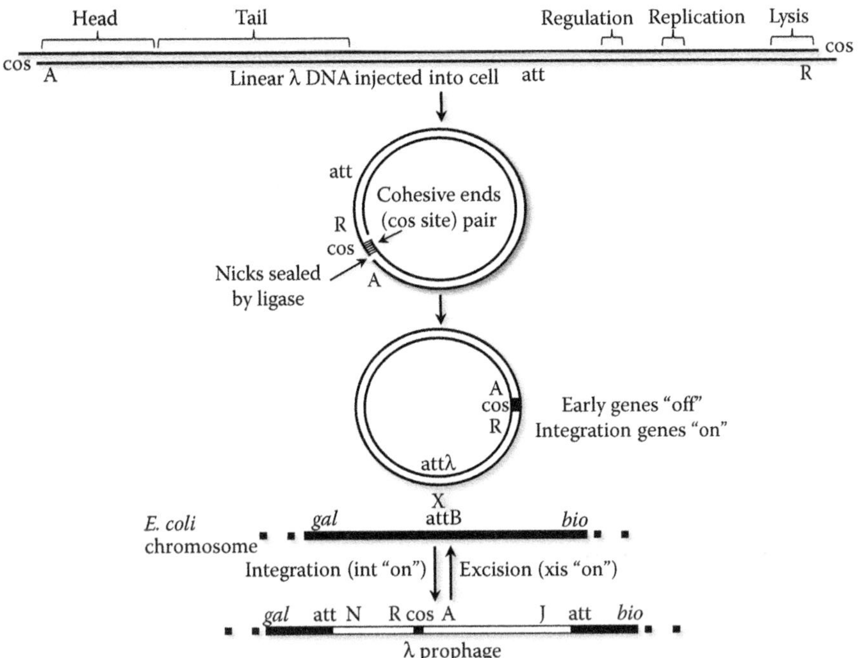

FIGURE 13.4 Integration and excision pathway for bacteriophage lambda (λ). The λ chromosome is injected into the host (*E. coli*) cell as a linear dsDNA. The ends of the chromosome have 5′ protruding ends (known as cos, or cohesive, sites) that are complementary to one another, which facilitates circularization of the chromosomes. The λ chromosome then associates with the *E. coli* chromosome near the *gal* and *bio* genetic loci. This association occurs at homologous regions known as att sites (for attachment sites). Proteins produced by λ and *E. coli* cut the chromosomes, produce a crossover, and then reseal the nicks in the DNA. The λ chromosome is then a prophage, expressing a few genes that maintain the prophage until a signal is sensed to excise the λ chromosome from the *E. coli* chromosome to begin a lytic (infectious) cycle.

INTEGRATION INTO CHROMOSOMES

Bacteriophage (bacterial viruses) of *Escherichia coli*, such as lambda (and other lambdoid phage) have been studied in great detail. Lysogeny (integration into the host chromosome) begins with injection of the phage chromosome and circularization of the chromosome through ligation of the cos sites at the ends of the chromosome (Figure 13.4). Depending on the conditions inside the bacterium, the phage proceeds through either the lytic cycle or the lysogenic cycle. If the bacterium is not under nutritional or other stress, the phage normally will begin the process of lysogeny. Its lysogenic genes are expressed and the lytic genes are repressed. The primary lysogenic genes encode proteins that are responsible for integration of the phage genome into the bacterial chromosome, as well as an antisense RNA that acts to repress transcription of the genes in the lytic pathway. All phage that have been studied integrate at specific sites in the host genomes. For lambda, there is a sequence on the phage genome (the attλ locus) that matches a site on the bacterial chromosome (attB), that is between two operons (*gal* and *bio*). A crossover occurs, which is resolved by the phage Int protein and the bacterial integration host factor protein that act to cut, align, and ligate the phage chromosome into the host chromosome. Then, each time the bacterial chromosome replicates, the integrated phage DNA is also replicated as a part of the same process, using the bacterial replication machinery. Also, a few of the phage genes are expressed, although it is a lysogen. The gene products (primarily the cro repressor) continue to repress the lytic pathway genes. During times of stress, this process breaks down and the xis protein is produced, triggering excision of the phage chromosome from the bacterial chromosome and the initiation of the lytic pathway.

Some of the general processes of integration are similar to the insertion of certain introns and transposons. Some mobile introns and transposons encode endonucleases that target specific sites on the host chromosome. These become the sites of integration. The endonuclease of the yeast omega (ω) introns that integrate into ribosomal RNA (rRNA) genes encodes an endonuclease that recognizes a site within the rRNA gene, but cuts the DNA a few dozen nucleotides beyond that site. Again, this is an indication that certain types of introns and transposons have features in common with viruses, although transposons often encode more than one gene product (e.g., endonuclease, transposase, resolvase). All can be classified as mobile genetic elements at some level, although many of these appear to have lost their ability to insert and/or excise from the chromosomes, and thus are unable to move. Nonetheless, although they are mobile, they have the ability to cause mutations. Additionally, even when they do not move, if there are certain mutations within these elements, these mutations can cause serious genetic and functional problems for the organisms, as in the case of introns that suffer a mutation that causes problems in splicing of the surrounding exons.

VIRUSES

Viruses can be conceptualized as mobile infectious genetic elements. Some viruses carry their genes as RNA versions, whereas others use DNA. Although they are incapable of growth and replication without the aid of a host cell, some are as complex as some cellular organisms. For example, some viruses can infect other viruses, as well as cells. The genomes of Mimiviruses exceed 1.2 Mb, and thus are larger than some bacterial genomes, whereas the genome of HIV is 9.7 kb, which encodes only 14 gene products, but that is sufficient to kill a human, whose genome encodes well over 23,000–25,000 genes. The origin of viruses is still unknown. Some have suggested that viruses represent some of the earliest biological entities on the Earth, whereas others consider viruses as pieces of nucleic acids that originated from cellular organisms, or originated as parasites and continue as parasites today.

Viruses come in a variety of forms. The shapes range from filamentous to round to icosahedral to conical to irregular (Figure 13.5). Most have protein coats (i.e., capsids) that surround the particle, but some additionally surround themselves with parts of the host membrane (called enveloped viruses). Some viruses kill the host cells by lysing them, whereas others simply bud off of the host cells without killing them. Chromosomes in RNA viruses primarily are linear, but there are some that are circular. They can be single stranded or double stranded (Figures 13.6 and 13.7). Some are positive sense (mRNA versions, ready for translation), whereas others are negative sense, requiring an RNA-dependent RNA polymerase (RdRp) to transcribe the positive sense strand. In this case, the viruses need to package the RdRp into the virus particle so that the first step of infection is to produce RNA molecules to begin the production of virus proteins and additional copies of the virus chromosome(s). Some viruses carry single chromosomes, whereas others have split genomes and have up to eight separate pieces of RNA that compose their genomes. Influenza A is one of these. It has eight pieces of negative-sense RNA in each particle that carry one gene each, although through differential splicing or processing, some can produce more than a single gene product. The chromosomes of DNA viruses can be linear or circular, and all contain a single chromosome. For those with linear chromosomes, often the chromosome circularizes shortly after infecting the host cell.

A unifying taxonomic scheme (the Baltimore classification system) has been adopted for all eukaryotic viruses. There are seven major types of viruses in this system, based on their modes of replication (Figures 13.6 and 13.7). However, some of the characters and genes are overlapping, and thus, it appears that many or most of the viruses may have a common origin. Type I viruses use double-stranded DNA (dsDNA) as their genetic material, and therefore use many of the same replication, transcription, and translation pathways common to all of their host organisms. Type II viruses are based on single-stranded DNA (ssDNA). Upon infection, the ssDNA is converted to dsDNA, and it proceeds through replication, transcription, translation, and assembly. Often, production of new copies of the chromosome is via rolling-circle replication. Type VII viruses also have

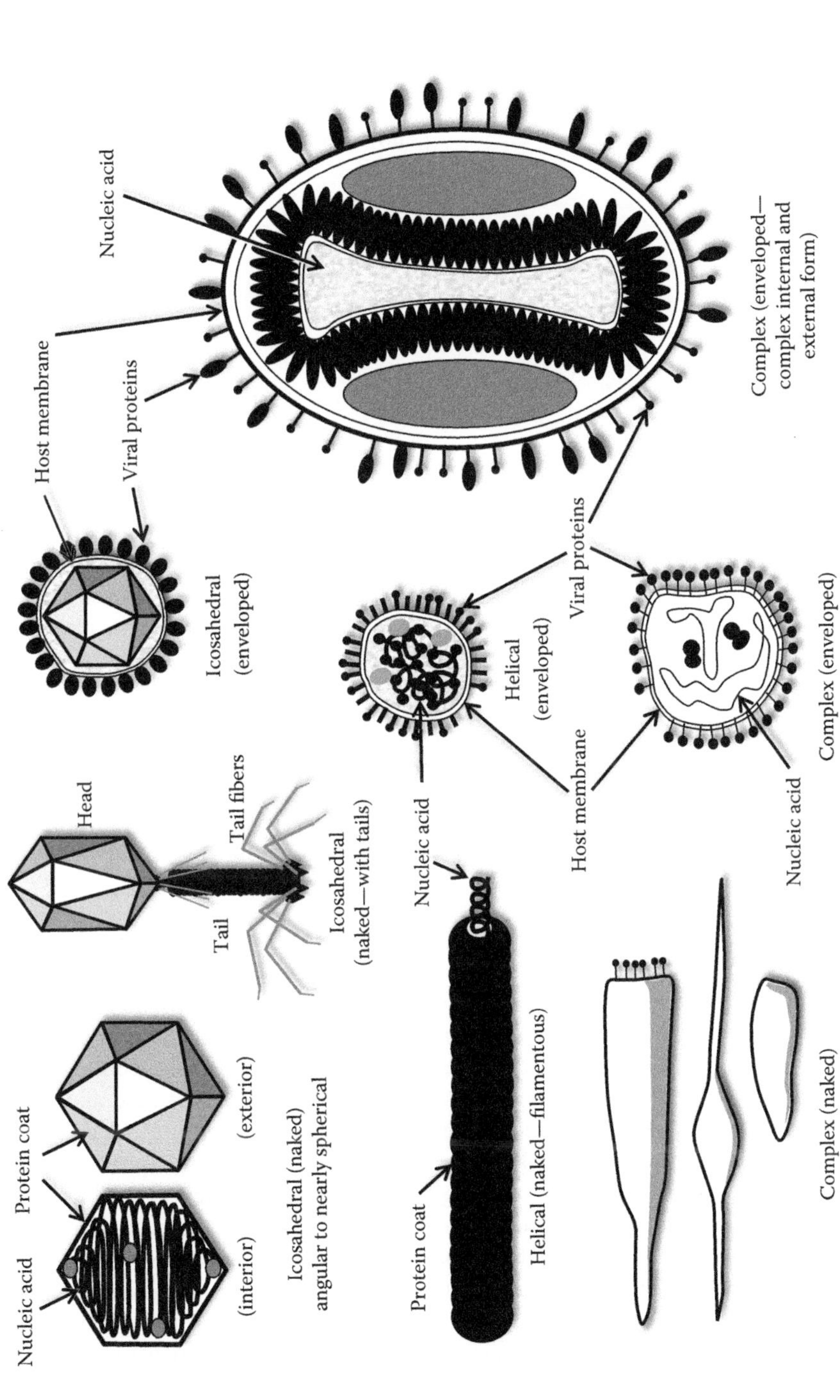

FIGURE 13.5 Illustration of the variety of virus forms. Many that infect bacteria and eukaryotes are either icosahedral or filamentous. Some have more complex shapes, including many that infect archaea and eukaryotes. Among all groups, some are enveloped inside a membrane that originates from the host cells. Finally, there are some, such as poxviruses, that have complex virus shapes, as well as complex surrounding structures and envelopes.

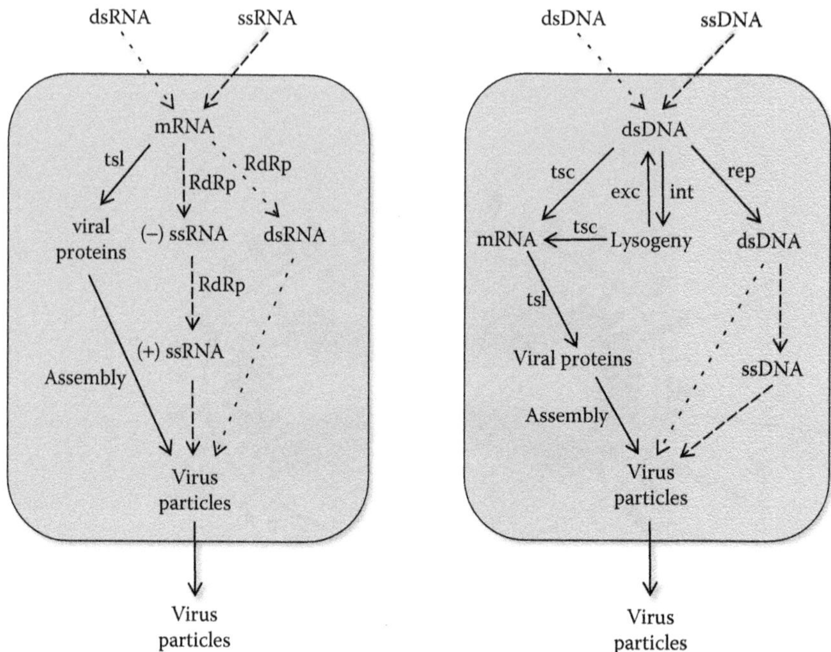

FIGURE 13.6 General pathways among viruses that infect Bacteria and Archaea. Essentially, there are single-stranded and double-stranded RNA and DNA viruses that must express their genes, replicate their chromosomes, assemble new virus particles, and exit from the host cell. RNA viruses normally use RdRps to produce more RNA (although some use reverse transcriptases to make DNA copies of their genomes during infection), including additional copies of the viral chromosomes. The mRNAs are translated on host ribosomes and the translated proteins then interact with the viral chromosomes to assemble into new virions. For DNA viruses, the viruses have a choice as to whether or not to proceed through a lysogenic (integrates into the host chromosome) or a lytic (production and release of more virus particles). In this case, host proteins can perform almost all of the functions to produce virus proteins and additional copies of the virus chromosome. Arrows are solid, dashed, or dotted to indicate specific pathways for each type of virus.

genomes based on DNA, but the virus DNA is formed from mRNA rather than from the standard replication of the DNA chromosome. Therefore, these are known as dsDNA–reverse transcriptase viruses. Many of the DNA-based viruses can also form lysogenic states in which the virus chromosome integrates into the host chromosome. There are four types of RNA viruses (types III, IV, V and VI). Type III viruses use dsRNA as their genetic material. When the virus infects the host cell, the particle is disassembled and the dsRNA is unwound, releasing the sense strand and antisense strand of RNA, as well as an RdRp. Translation begins, and new strands of viral RNA are produced using the RdRp. New viruses are then assembled. Type IV viruses carry a sense (positive) strand of RNA as the genetic material. Therefore, they can begin to translate new proteins upon infection, but also use an RdRp to produce antisense and additional sense strands of RNA, the latter of which are packaged into newly assembled virus particles. Type V viruses use the antisense (negative) RNA strand as the genetic material. Therefore, they must polymerize the sense strands using an RdRp before they can replicate their chromosomes and synthesize viral proteins. Although these viruses are not known to carry a reverse transcriptase, some have been demonstrated to integrate DNA copies of their genome into the host chromosome, as is the case with measles virus. Type VI viruses are retroviruses, such as HIV. The genetic material is ssRNA, and the viruses usually carry with them a reverse transcriptase in order to produce a DNA copy once they have invaded a cell. They can then either produce more virus particles, using the host transcription and translation machinery, or integrate into the host chromosome to initiate a lysogenic state.

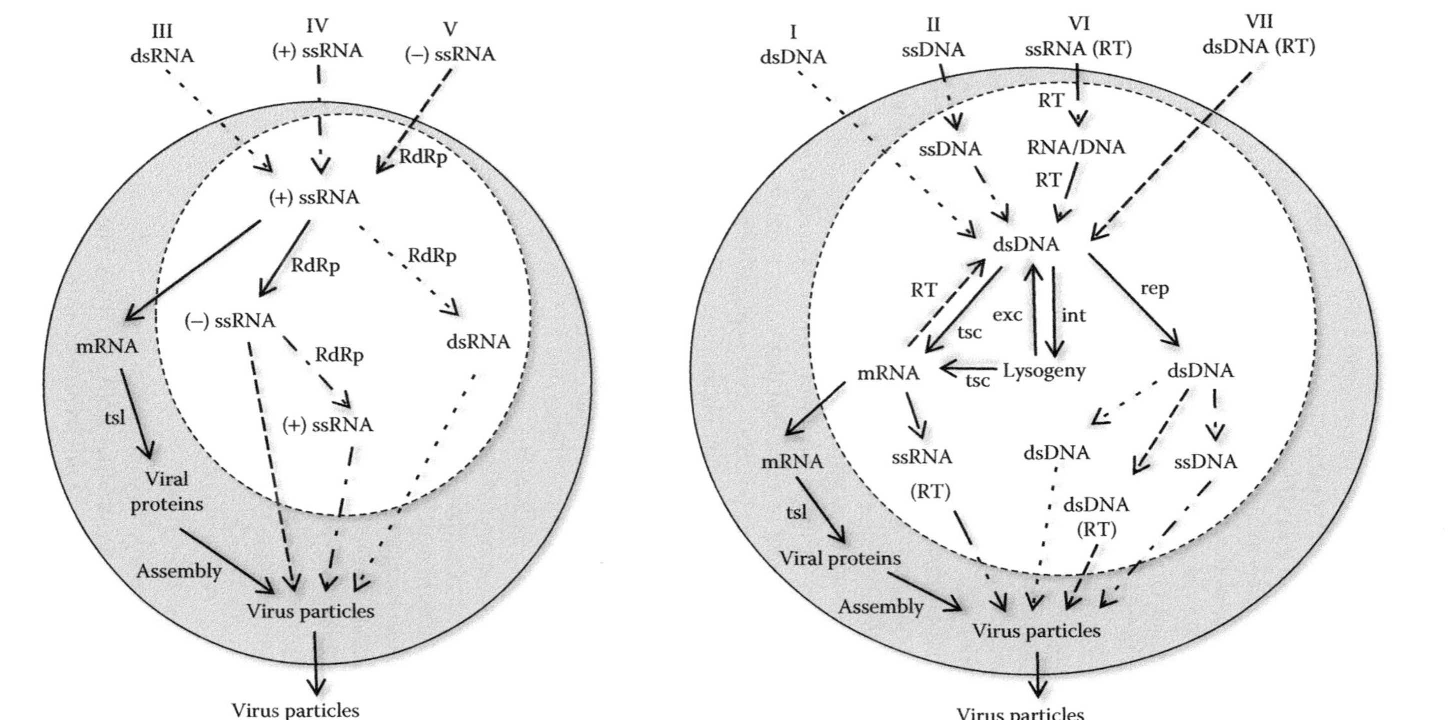

FIGURE 13.7 General cycles for viruses that infect Eukarya. There are seven basic types of viruses, as outlined in the Baltimore system of virus taxonomy. Types III, IV, and V essentially are similar to RNA viruses that infect bacteria, except that a part of the life cycle occurs in the nucleus and the other part is in the cytosol. As with bacterial RNA viruses, these viruses utilize RdRps. Most of the viruses that form lysogens are DNA viruses. However, one type of RNA viruses (type VI), those that possess a reverse transcriptase, can also form a DNA copy of their genomes and then can integrate into the host chromosome. One other group of viruses (type VII) also use a reverse transcriptase, but they produce chromosomes that are dsDNA. RT, reverse transcriptase; tsl, translation; tsc, transcription. Arrows are solid, dashed, or dotted to indicate specific pathways for each type of virus. Nucleus is dashed to indicate that some viruses do not enter the nucleus (e.g., many of the RNA viruses).

Several other infectious particles exist. Satellite viruses usually are reduced versions of a large virus. They must coinfect with the larger virus in order to reproduce, but often the resulting infection produces more satellite viruses than parent viruses. Tobacco necrosis virus (TNV), which infects a large number of plant species, is often found with its satellite virus (sTNV). When they coinfect a plant, the majority of virus particles that are produced are sTNV particles. The sTNV genome consists of only the coat protein gene, whereas the TNV genome is composed of several genes that control replication and assembly of the viruses, as well as the TNV coat protein gene. Mimiviruses often infect amebae with a satellite virus called a virophage (called Sputnik). When they coinfect a cell, the virophage takes advantage of the Mimivirus functions to produce copies of itself. In this case, all of the particles that are produced contain the virophage, and no Mimivirus particles are produced. Viroids are naked RNA molecules whose replication is similar to circular ssRNA viruses. Some of the plant viroids move from cell to cell being transported by a mechanism that moves other RNAs from cell to cell as a normal function in plants. Another group of infectious agents have no known genetic system. Prions are composed of only protein. They are identical in amino acid sequence to specific normal host cell proteins but have an altered conformation. They are capable of converting the normal proteins into the altered conformation, which makes them ineffective in their normal cellular function, which kills the affected cells. In the case of bovine spongiform encephalopathy, the infected cells die, which essentially form cavities in the affected brain tissue. Although the viruses and viroids may have common origins, prions appear to arise spontaneously in certain organisms. Therefore, although they have some evolutionary effects (because selection would act at the level of whether the protein is convertible or not), they are not known to form a phylogenetic grouping, as do viruses and other genetic elements.

There are three major hypotheses as to the origin of viruses. Viruses might be remnants of ancient cells that have reduced the number of genes that they carry, because they rely on host cells to provide the functions for the deleted gene products. This is plausible because bacterial parasites tend to lose genes as they become more dependent on the host cells. Also, the Mimiviruses can be infected by other viruses, and their genomes are larger than some bacterial genomes, including many of the parasitic bacteria. They may simply be reduced versions of once free-living cellular organisms. The extreme examples of gene loss are in mitochondria, which have fewer genes than many types of bacteria, although the ancestors of mitochondria were free-living bacteria with a full set of genes (numbering in the thousands). The second hypothesis is that viruses are mobile segments of larger genomes. Therefore, parts of the genomes of cellular organisms have become able to replicate separate from the remainder of the genome and have become infectious agents. In this case, viruses may have had multiple origins, each arising from unique host organisms, some of which adapted to other hosts. The third hypothesis is that viruses and cellular organisms coevolved very early on the Earth. This theory supposes that shortly after the first cells appeared, complexes of protein and nucleic acids became adapted to entering the cells and using their protein machinery for their replication. This is a plausible hypothesis because viruses appear to infect virtually every species on the Earth, indicating a very early evolution for these infectious genetic elements. Some cellular organisms may owe their evolution to viruses.

INTRONS

Introns are common in eukaryotic pre-mRNAs and rRNAs, but are rare in bacterial and archaeal, genes, being found primarily in rRNAs and transfer RNAs (tRNAs). The first introns were discovered when it was found that mRNAs in the cytosol were shorter than the corresponding RNAs in the nuclei (Figure 13.8). These RNAs were termed heteronuclear RNAs, because often they were of various lengths. Subsequently, transmission electron micrographs of DNAs hybridized to the corresponding mRNAs indicated that the DNAs often had regions that did not correspond to any part of the mature mRNAs, forming loops of various sizes (Figure 13.9). The loops represented portions of the gene that were also transcribed into RNA, but were subsequently removed from the RNA.

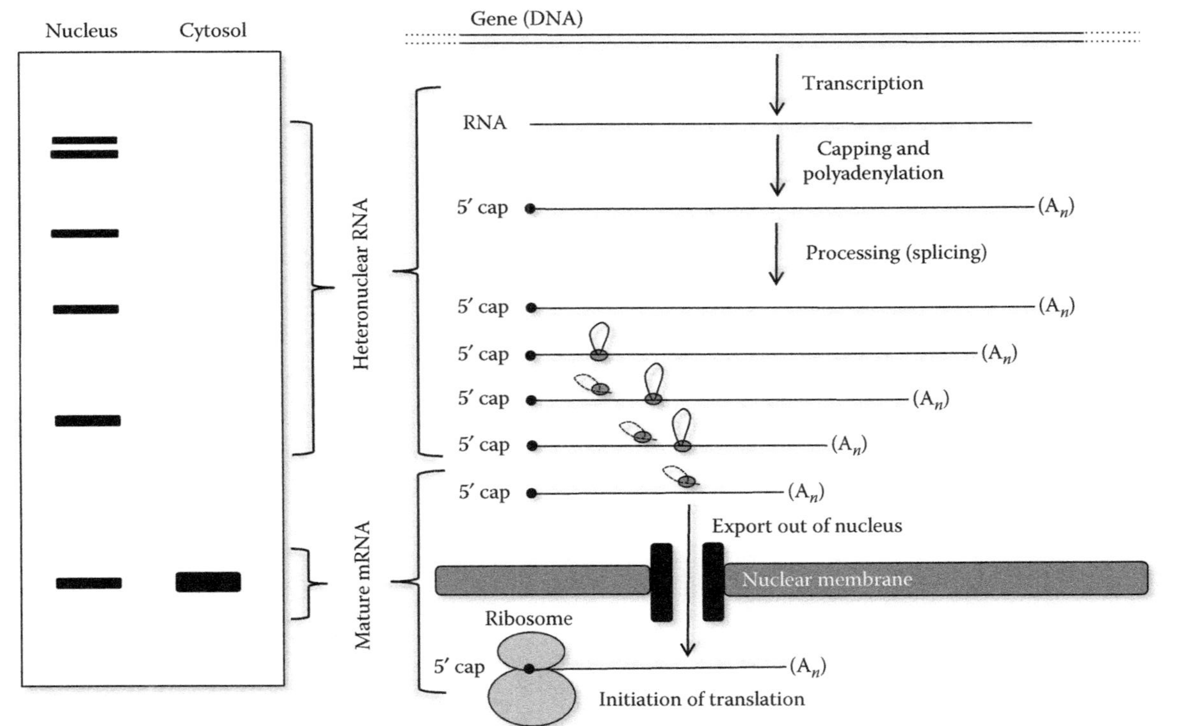

FIGURE 13.8 Introns were first discovered because for some mRNAs that they found in the cytosol had several larger versions that were located only in the nuclei (gel diagram on the left). These were termed *heteronuclear RNAs* to indicate both the variety of sizes and their location. Eventually, it was determined that pieces of the transcript were being chopped out of the RNA, and the two surrounding sections were being ligated (spliced) together. These are the introns and exons, respectively.

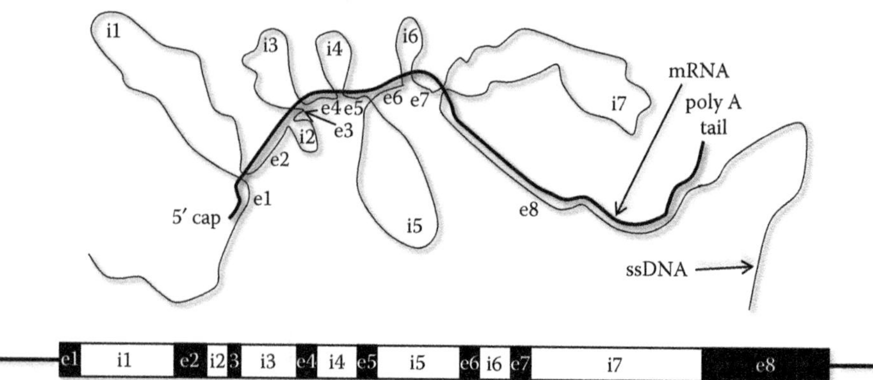

FIGURE 13.9 Introns were confirmed by electron microscopy. DNA was denatured to produce single-stranded molecules. These were mixed with the RNA transcripts from the cytosol. They then were allowed to anneal together to form heteroduplexes consisting of a genomic DNA strand and an mRNA strand. Where they hybridized, due to complementary base pairing, they formed a stable duplexed molecule. Where they differed, a single-stranded portion remained. These formed loops where portions of the DNA were not represented in the mature mRNA. The loops were the introns (i1–i7), and the hybridized regions were the exons (e1–e8).

During excision, the surrounding segments of RNA (representing the exons, or expressed translated regions) were joined in the process of splicing. Introns have since been identified in many genes and organisms. Based on their modes of splicing, there are four main groups of introns: group I, group II, spliceosomal, and archaeal (Figures 13.10 and 13.11).

The first two are also ribozymes (i.e., enzymatic RNAs), and because of this, it is thought that they are very ancient, having first appeared when biological systems depended on RNA for enzymatic and genomic functions. Two additional classes of introns are group III and twintrons. Group III introns are much like group II introns, whereas twintrons essentially are introns within introns. Examples of group I, group II, group III and twintrons have been described. These are not simply biological curiosities but indicate that introns can invade many chromosomal sites, including sites that already contain an intron. Some contain open reading frames that encode an endonuclease. Thus, some are mobile elements, in that they insert into DNA genomes, although they retain the ability to splice and remove themselves when in the form of RNAs. Most contain no open reading frames, but many contain small RNAs, including small nuclear RNAs, small nucleolar RNAs, and microRNAs that are important in control of expression of mRNAs and rRNAs.

The omega (ω) group I intron in some strains of the yeast *Saccharomyces cerevisiae* can insert into intron-less rRNA genes using an endonuclease that is encoded in a gene within the intron. This has been termed *intron homing* because the introns insert into very specific sites, based on the recognition sequence of the endonuclease. Many other homing introns have been described. It is possible that this mechanism was the cause for the original spread of introns in the genomes of organisms. Whether the original introns were infectious or were carried by viruses is unknown. However, several bacteriophage are known to carry group I introns, suggesting that viruses may have aided in their movements.

The mode of splicing is characteristic for each class of intron. Both group I and group II introns have the ability to splice *in vitro* without the aid of any proteins, and therefore are considered to be ribozymes. However, it appears that many rely on chaperone proteins that stabilize the active conformation of the RNA *in vivo*. For group I introns, the intron folds into a characteristic tertiary structure (Figure 13.12) that lines up the exons using an internal guide sequence (IGS). Several pairing regions, termed P1 through P10, act to hold the exons together while maintaining the active conformation in the catalytic site of the ribozyme. P1 and P10 line up the exons, whereas P4, P5, and P6 form a supporting structure, and P3, P7, P8, and P9 form another supporting structure that contains the enzymatic portion of the ribozyme. In addition, P7 holds a single guanosine

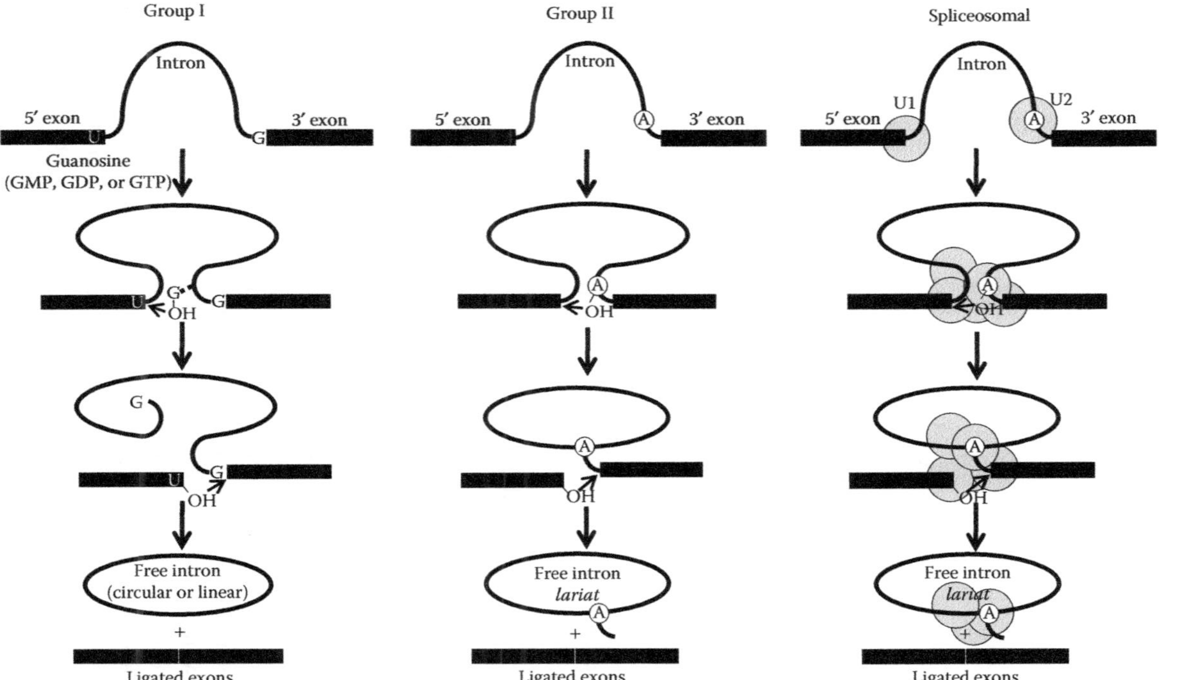

FIGURE 13.10 General splicing mechanisms for group I, group II, and spliceosomal introns. For group I introns, the 3′ hydroxyl of the free guanosine (in the form of guanosine monophosphate [GMP], guanosine diphosphate [GDP], or guanosine triphosphate [GTP]) held by P7 attacks the phosphodiester bond of the uracil in the G•U pair at the 5′ exon–intron border. The guanosine becomes covalently bound to the intron, whereas the 5′ exon is separated from the intron. The 3′ hydroxyl of the uracil then attacks the phosphodiester bond of the nucleotide (usually a G, but occasionally a U) at the 3′ end of the intron. This reaction separates the intron from the 3′ exon and ligates the two exons together. Often, the intron circularizes. For group II introns, the initial reaction occurs when the 2′ hydroxyl on an internal adenine attacks the phophodiester bond at the 5′ exon–intron border. This separates the intron from the 5′ exon, whereas the adenine is then attached at three locations to the intron, through the 2′, 3′, and 5′ carbons. This forms what has been termed a *lariat*. The 3′ hydroxyl at the end of exon then attacks the phosphodiester bond at the end of the intron to ligate the two exons and free the lariat-form intron. Both group I and group II introns are capable of proceeding through the splicing reactions without the addition of any proteins/enzymes. This is because they are ribozymes. However, *in vivo*, there are chaperone proteins that stabilize the active conformations of the ribozymes, and thus increase the efficiency of the reactions. The splicing reactions for spliceosomal introns are very similar to those of group II introns. However, these introns require spliceosomes (U1–U6), which are small particles composed of RNA and proteins that are found in eukaryotic nuclei. They are also called small nuclear ribonuclear particles.

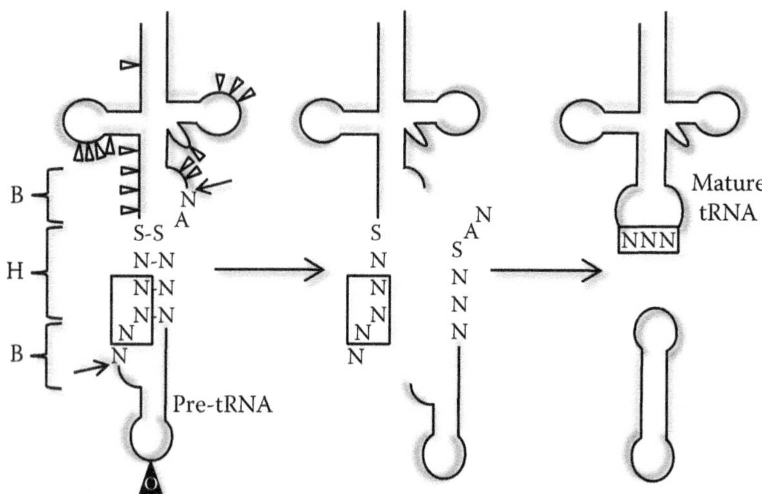

FIGURE 13.11 General splicing mechanism for archaeal introns, shown here in a tRNA. An endoribonuclease, encoded in an open reading frame (O inside the black triangle), recognizes a bulge–helix–bulge (BHB) region in a double-stranded region of the RNA. Conserved nucleotides are indicated. Some archaeal introns lack the open reading frame, but the endoribonuclease can be supplied by any such gene in the genome. Once the endoribonuclease cuts the RNA (arrows), the tRNA and the intron circularize. The insertion site shown is the most common site within the anticodon loop. Other insertion sites are indicated by open triangles. For both rRNAs and tRNAs, after the endoribonuclease has cut the strands, the ends of the molecules become covalently bonded. For rRNAs, circular molecules are formed, which include the SSU rRNA and LSU rRNA, separately. Subsequently, these are processed to produce the mature SSU and LSU rRNAs. For the tRNAs, fusing of the ends forms the mature tRNA molecule. N = A, G, C, or U; S = G or C.

(that is not covalently attached to the intron initially), which initiates the first part of the reaction. The guanosine becomes covalently bound to the 5′ end of the intron as the 5′ exon is separated from the intron (Figure 13.10). Next, the hydroxyl on the 3′ end of the 5′ exon reacts with the guanosine (or a uracil in some group I introns) at the 3′ end of the intron to separate the intron from the 3′ exon and ligates the two exons together.

Group I introns are widespread, having been found in bacteria, bacteriophage, and many groups of eukaryotes. At least five major subgroups exist (IA, IB, IC, ID, and IE), based on sequence and structure characteristics. Additionally, each subgroup is further subdivided into subgroups based on the secondary structure and genomic locations. For example, almost all eukaryotic nuclear group I introns are located in the rRNA genes (in the large and small subunit genes), and most are in the same subgroup, IC1, although a few in the large subunit (LSU) rRNA gene are in group IC3. Another set of subgroup IC1 introns has been found in the two internal transcribed spacers (ITS1 and ITS2) in almost all lineages of sharks, and possibly exists in ITS1 and ITS2 in other animal groups as well. This indicates that these introns inserted into the rRNA gene locus more than 400–500 million years ago, because this predates the appearance of all of the groups of organisms that contain subgroup IC1 introns. The IA1 introns are the most ancient and may have existed in bacteria and bacteriophage for a much longer time. These introns normally are from a few hundred to a few thousand nucleotides in length, and some contain genes that confer mobility. However, a few have been described from fungi that are from 64 to 74 nucleotides in length (Figure 13.13), which occur near the 3′ end of the small subunit (SSU) rRNA gene. They are missing most of the supporting scaffolds but have retained P1, P10, as well as part of P9, which has a site that is similar to the P7 region that holds the initiating guanosine in larger group I introns. Based on the missing portions of the intron, the introns would be expected not to splice, but they do. Part of the rRNAs (specifically ITS1) may help to support these introns so that they can splice. They appear to have been formed as a result of

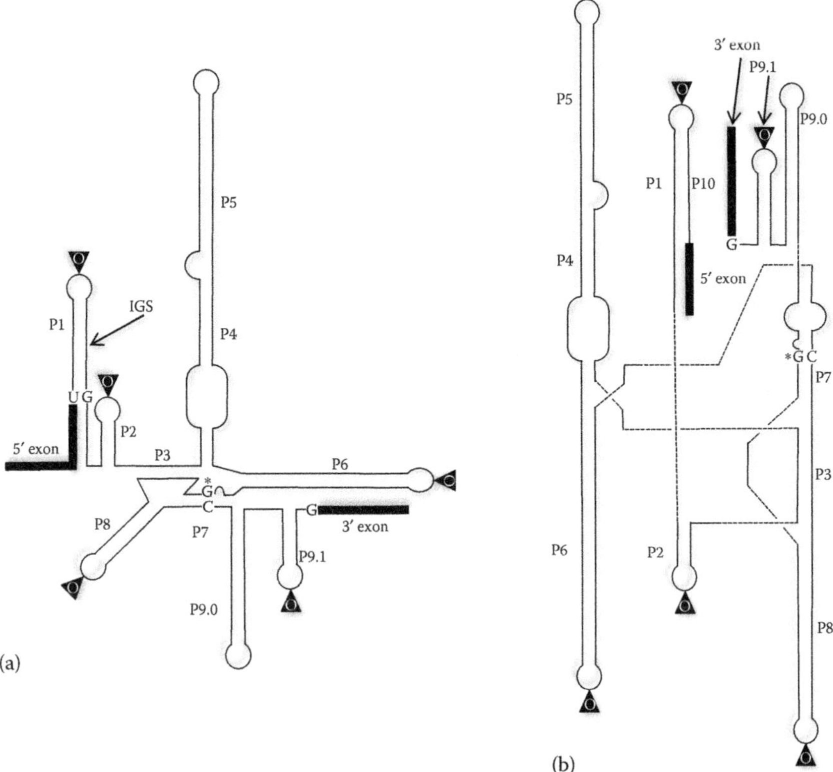

FIGURE 13.12 General secondary structure of a group I intron: (a) represents the original forms that were determined for these introns and (b) is based on the functional parts of the same intron. Pairing regions P1 through P9.1 are shown. P10 also forms between the IGS (internal guide sequence) in P1 and part of the 3′ exon. P9.1. P1, P10, and P9 line up the exons, whereas P7 holds the guanosine to begin the splicing reaction. P4, P5, and P6 form a scaffold that holds the other portions of the intron. P8 and P3 hold P7 near the central portion of the intron. P1 lines up the 5′ exon with the 3′ exon, which is held by P10 and P9.1. Black triangles with an *O* for both forms indicate optional open reading frames that are present in various group I introns. Most often, these encode homing endonucleases.

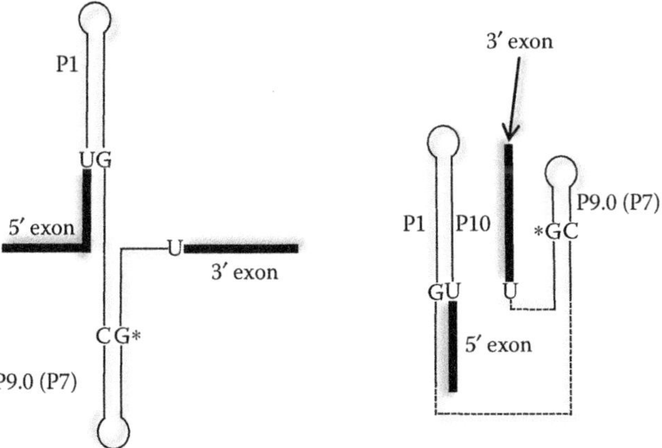

FIGURE 13.13 Small group I intron that has been found in the SSU rRNA of a few species of fungi. In this intron, almost all of the standard pairing regions have been deleted. Only P1, P9.0, and P10 remain. Furthermore, a portion of P9.0 is capable of coordinating the free guanosine, and therefore replaces the missing P7.

mispairing during recombination. Because no introns of intermediate size (between approximately 80 and 200 nt) have been found in nature, the small introns may be the only forms that still retain splicing ability, whereas introns of intermediate size cannot splice, and therefore no SSU rRNA can be formed, which would be lethal for cells containing the faulty introns. Another interesting feature of group I introns in rRNA genes is that some of the introns in these genes have similarities to mRNA spliceosomal introns. These may be evolutionary intermediates between group I introns and spliceosomal introns that may also have been produced by recombination and gene conversion events that are very active in the rRNA gene loci.

Group II introns have a very different tertiary structure (Figure 13.14). However, like group I introns, they consist of several paired regions that align the two exons together with the enzymatic portion of the molecule. In addition to being introns, some are also considered retrotransposable elements. Some encode a reverse transcriptase (located in the D4 loop) and can move from one location to another via an RNA intermediate. They have been found in the rRNA genes of eubacteria, archaea, and eukaryotic mitochondria and chloroplasts. Instead of a free guanosine, group II (and group III) introns contain a reactive adenosine that is within the intron (Figure 13.10). In the

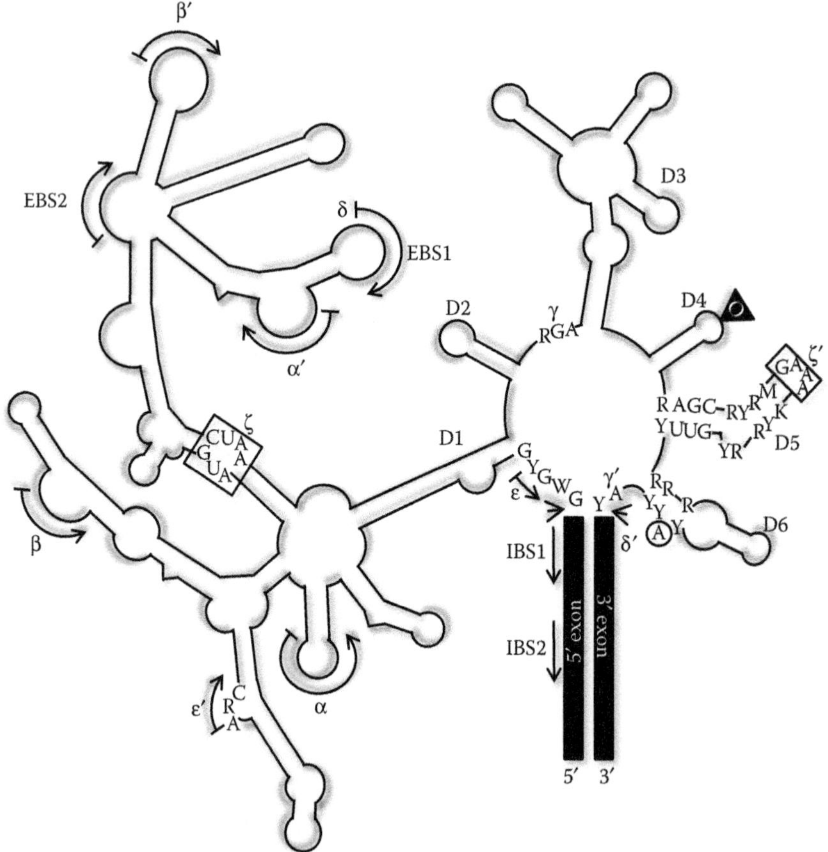

FIGURE 13.14 General form for a group II intron. The 5′ and 3′ exons are aligned with the aid of domains D1 through D6 (also sometimes indicated with Roman numerals I–VI). Several parts of the intron form associations because of base pairing of distant nucleotides. These are indicated by Greek symbols (α–α′, β–β′, δ–δ′, γ–γ′, ε–ε′, and ζ–ζ′). exon-binding sequence 1 (EBS1) and EBS2 hydrogen bond to intron-binding sequence 1 (IBS1) and IBS2. The adenine that initiates the first reaction of splicing and forms the lariat structure is circled in D6. Conserved nucleotides are indicated. Black triangle with an *O* indicates the location of an optional open reading frame that usually contains a reverse transcriptase gene. R = A or G; Y = C or U; W = A or U; M = A or C; K = G or U.

first part of the reaction, a hydroxyl at the 5′ exon–intron border reacts with the adenosine, which creates a covalent bond within the adenosine residue to form a loop, or lariat, structure around the adenosine residue. At the same time, the 5′ exon is separated from the intron. The second part of the reaction is identical to the second part of the reaction for group I introns. The hydroxyl on the 3′ end of the 5′ exon attacks the 3′ intron–exon border, resulting in the ligation of the two exons and the release of the intron, which is in the form of a lariat.

Spliceosomal introns have a structure that is similar to group II introns, and the general mechanism for splicing is also nearly identical (Figure 13.10). However, spliceosomal introns are not self-splicing *in vitro*. Instead, they require several spliceosomes that are composed of short RNA molecules and proteins. As with other introns, the reactions are carried out by the RNAs, not the proteins. The proteins act to hold the RNAs in their catalytic conformations. These introns can range in size from a few dozen nucleotides to tens of kilobases in size. As with group II introns, the end products are the spliced exons and a free intron in the form of a lariat with parts of the spliceosome still attached.

Archaeal introns have been found in a broad range of organisms, including bacteria, archaea, and eukarya. They are found primarily in tRNA genes and secondarily in rRNA genes of bacteria and archaea. Also, in the Crenarchaeota, they are found in some mRNAs as well. Some contain a coding region for an endoribonuclease that cuts the pre-RNA. The cleavage site is characterized by having a bulge–helix–bulge structure, with a central recognition site (Figure 13.11). Most have inserted into the same site in tRNA that contains the endonuclease recognition sequence. However, in the Crenarchaeota, there are many other sites of insertion. Splicing of archaeal introns is very different from that of other introns. First, two regions flanking the splice junctions hybridize to form a double-stranded region. The endoribonuclease recognizes a site in this double-stranded region and makes a staggered cut, leaving a 3′ end with six unpaired nucleotides. The action of the endoribonuclease also leaves a hydroxyl on the 5′ end of the 3′ exon and a phosphate on the 3′ end of the 5′ exon that is attached simultaneously to the 2′ and 3′ carbon atoms. It detaches from the 3′ carbon leaving a hydroxyl on the 3′ carbon. The hydroxyl on the 3′ exon is phosphorylated by a kinase. Then, the hydroxyl on the 3′ end of the 5′ exon reacts with the phosphate on the 5′ exon to ligate the two exons together. Finally, the phosphate on the 2′ carbon (now at the exon–exon border) is removed.

An interesting feature of archaeal introns is that the endoribonuclease appears to aid in the processing of rRNA, whether or not the intron is present at that locus (Figure 13.15). When the two external transcribed spacers, as well as the internal transcribed spacer (which sometimes contains a tRNA gene), pair up, a double-stranded recognition site is created. The endoribonuclease cuts the two sites, which circularize to produce two circular RNAs. One contains the SSU rRNA and the other contains the LSU rRNA. These are then processed by other mechanisms to produce the mature rRNAs. This indicates that archaeal introns have played an important role in the processing of rRNAs during evolution, hinting at a very ancient origin for these introns.

Introns have had several important effects in evolution. First, as mobile elements, they have caused deleterious mutations by insertion into genes. However, although most group II introns in Bacteria and Archaea still retain mobility as retrotransposons, most group I and spliceosomal introns have lost that ability. Once they are established in a genome, they have effects on the cell and on the evolution of the cell and organism. Because introns must be removed and the exons must be spliced prior to accurate functioning of the RNA, expression of those RNAs is affected by the rate and accuracy of the splicing process. This adds another level of control to the expression of RNAs containing introns. RNA from some introns acts to reduce expression of certain genes by hybridization to the mRNAs (microRNAs, small interfering RNAs, etc.). Additionally, if the RNA contains more than one intron, then different patterns of splicing, called alternative splicing, can result (Figure 13.16). If two introns are present, then at least two different mature RNAs can be produced. The first would be the result of splicing by the two introns separately. The second would result from the use of the first exon–intron border and the last exon–intron border. In this case, the mature RNA would be shorter, because both introns and the middle exon would

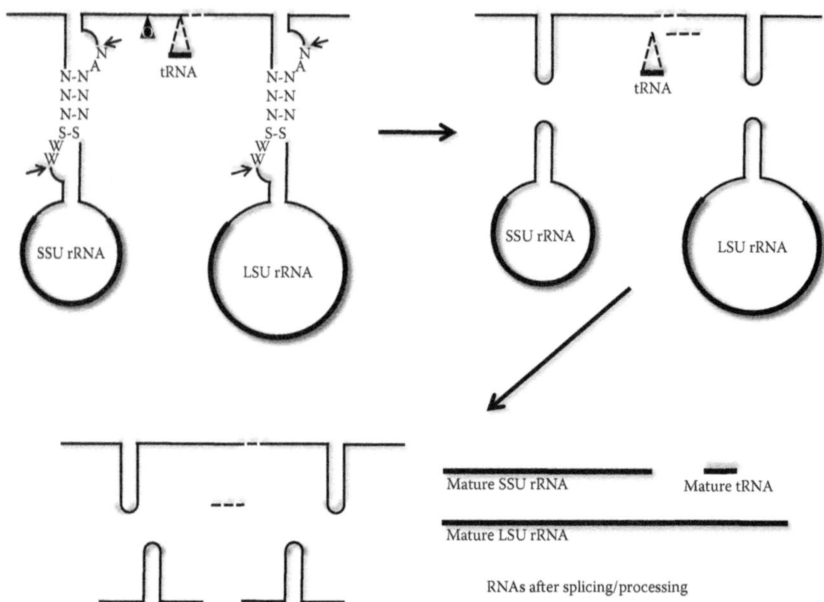

FIGURE 13.15 Processing of rRNA by an archaeal splicing mechanism in Archaea. An endoribonuclease recognizes a bulge–helix–bulge (BHB) region in a double-stranded region of the RNA and cuts at the bonds marked by arrows. After the endoribonuclease has cut the strands, the ends of the molecules become covalently bonded. For rRNAs, circular molecules are formed, which include the SSU rRNA and LSU rRNA, separately. Subsequently, these are processed to produce the mature SSU and LSU rRNAs. In this molecule, a tRNA also is produced, which may or may not have an additional archaeal intron.

be removed from the precursor RNA. If more introns are present, additional alternative splicing events could occur. The main implication for this is that more than one RNA can result from a single gene. For mRNA genes, this means that more than one polypeptide can be formed from a single gene. In animals, including humans, the diversity in auditory and olfactory neurons is partly due to alternative splicing. This has led to the development of large sets of neurons with sensitivities to a large number of sensory inputs. Another result of introns is that successful recombination events can occur between alleles, genes, and gene domains (Figure 13.17). If crossovers occurred in the coding regions, they would be expected to be deleterious most of the time. Because most introns have no reading frame (because there is no coding region), there are many positions within the intron for recombination events that would have little or no effect on splicing. This allows the possibility for new combinations of exons to be formed. Therefore, increased rates of evolution and adaptations could result.

TRANSPOSABLE ELEMENTS

Transposons are mobile genetic elements that have the ability to move from one location in a genome to another. Several types exist (Figure 13.18), including those that move as DNA elements (transposons) and those that move as RNA intermediates (retrotransposons). At one time, they were known as *jumping genes*. The first hints of these elements came in the 1930s and 1940s. Barbara McClintock was a geneticist studying recombination in *Zea mays* (maize), when she noticed some plants that exhibited unusual patterns of inheritance for many genetic loci. In one genetic background, the patterns were consistent with typical Mendelian inheritance, whereas in another genetic background, fruits on the same ear of corn exhibited broad ranges of pigment patterns, starch production, and other traits (Figure 13.19). The patterns of inheritance were non-Mendelian and

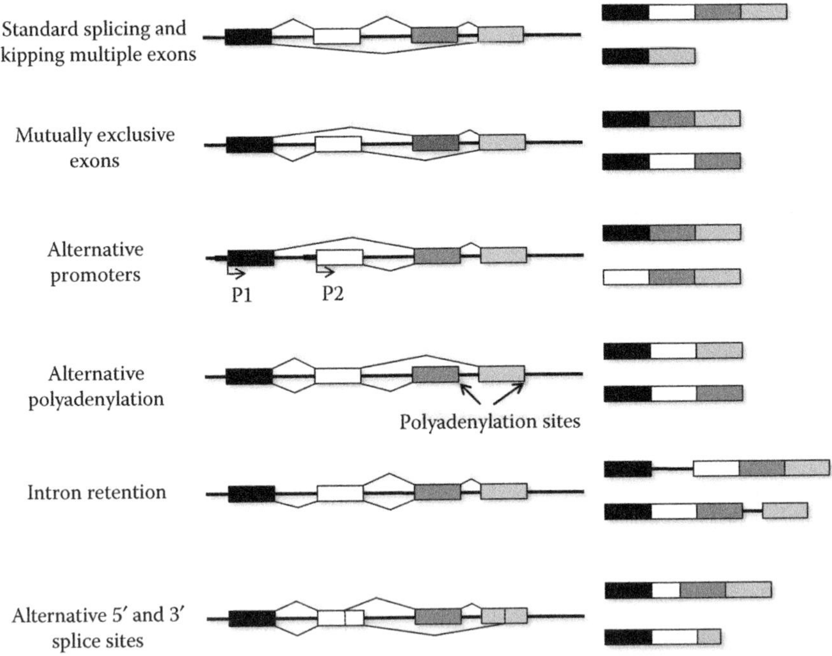

FIGURE 13.16 Various forms of alternative splicing of mRNAs. The mature mRNA may contain all of the exons, or only a few of them. In some genes, multiple promoters lead to more than a single transcription start site. At the other end, there may be multiple polyadenlyation sites in separate exons, such that the mRNA ends on different exons. In some genes, the introns are retained and become additional exons. Finally, cryptic splice sites within exons can be used to cut within sections that are exons in other mRNAs. Each of the different mRNAs will form different forms of related polypeptides. Some have very different functions. Therefore, more than one protein can be formed from a single gene or transcript.

somewhat unpredictable. Expression of traits could vary, producing fruits with spots, streaks, and patches of color. McClintock determined that there were two basic types of elements that were causing the changes. One type, which she called Ac (activator) elements, could disrupt various genes in the genome and appeared to be able to hop in and out of these loci. The other type, which she called Ds (dissociation) elements, could not move unless an Ac element was also present. They could hop into and out of various loci but also produced frequent chromosome breaks. These chromosomal dissociations could be observed cytologically, which was the reason they were termed dissociation elements. McClintock's studies were a curiosity at first. Eventually, elements with similar genetic characteristics were found in other organisms, including other plants, yeasts, bacteria, and insects. However, it was not until the advent of molecular biology methods that the elements were sequenced and understood. It turned out that Ac elements had short terminal inverted repeats on each end and at least one gene in the middle, which was termed a transposase (Figure 13.20). However, Ds elements all had the same terminal repeats, but parts of the transposase genes were missing. In one case, almost the entire transposon was missing except for the terminal repeats. Also, it was discovered that these transposons moved as DNA versions, normally during replication cycles. We now know that there are large numbers of transposons in the maize genome. In fact, they comprise much of the genome.

Other transposons have different characteristics (Figure 13.18). One type has long terminal direct repeats (LTRs) and contains a reverse transcriptase. Furthermore, these move as RNA molecules that are then reverse transcribed into DNA during insertion into another locus. A third type resembles a DNA version of an mRNA. They have no terminal repeats but encode a reverse transcriptase and have a poly A region at the end that resembles the poly A tails of mRNAs. These move as RNA

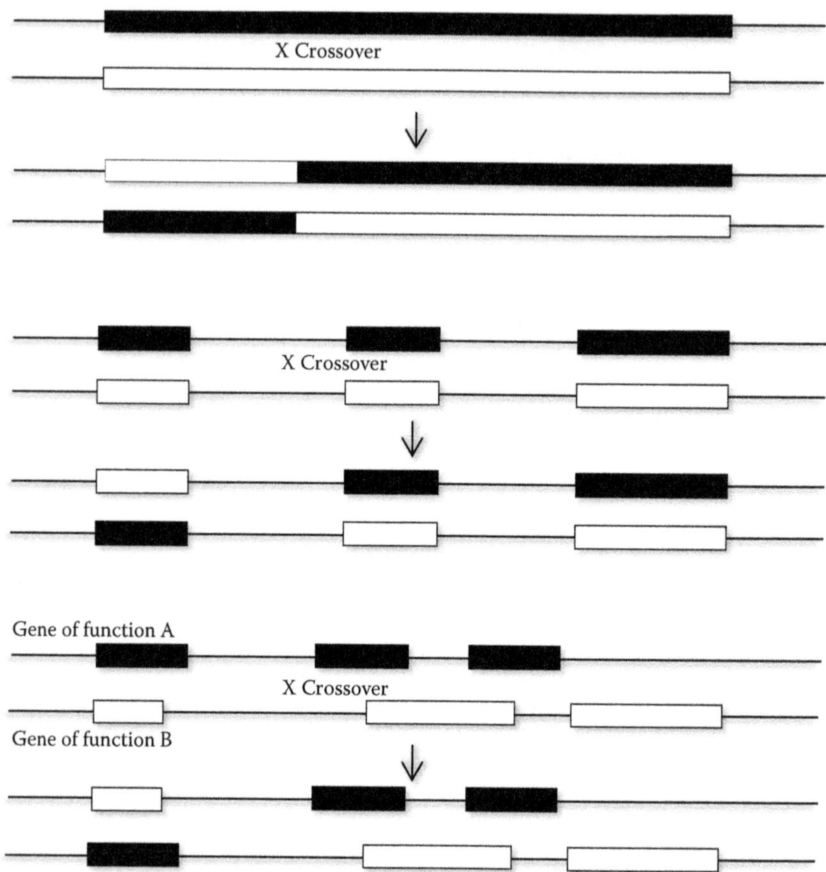

FIGURE 13.17 Introns can increase recombination success within genes. In an intron-less gene, the two alleles must align perfectly or be offset by multiples of three nucleotides in order to maintain the reading frame following a crossover. However, in genes that contain introns, a crossover can occur within an intron, and still maintenance of the reading frame is assured by accurate splicing. If a crossover occurs within the intron, the entire portions of the gene can be interchanged without affecting the reading frame. Additionally, recombination between nonhomologous genes is possible, again by crossing over within the intron regions. Therefore, mixing of gene functions and protein domains is possible. For example, a gene for a cytosolic protein may recombine with a membrane protein gene, with the possible result of forming an isoform of a cytosolic protein that becomes embedded in the membrane and a membrane protein that becomes a cytosolic version.

intermediates, so it is likely that they originated from mRNAs. As mentioned earlier, some of the reverse transcriptases from the retrotransposons are similar to those in group II introns and some viruses, including the reverse transcriptase of HIV.

All transposons have the ability to move to a large number of sites. In fact, in maize about 80% of the genome is made up of transposons. It is likely that the transposons have moved and increased in number during the evolution of maize and its wild relatives. Transposons have been found in many other plants, as well as in insects, fungi, bacteria, and other organisms, including humans. As stated before, transposons can insert at many locations, although the patterns are not random. They appear to insert either during replication or into areas where there is open chromatin (euchromatic regions) on the chromosomes whether or not the cell is in the S-phase of the cell cycle. This allows more access to enzymes that are necessary to accomplish the insertion and resolution of the transposition event. When the transposons insert into genes, they may cause a variety of effects. If they interrupt genes or their promoters, the genes may not be expressed (Figure 13.21). In the case of

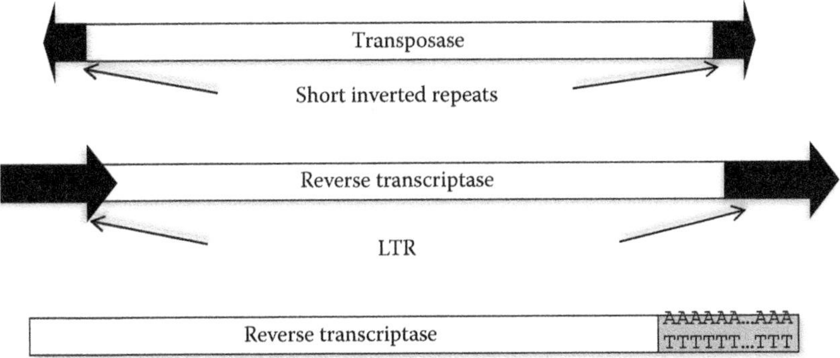

FIGURE 13.18 The three basic types of transposons. The first type (top) has two short inverted repeats on the ends and usually at least one gene, called a transposase that accomplishes insertion of additional copies of the transposon. These move as DNA, excising and inserting into additional sites or following a replicative pathway. Examples are Ac–Ds elements (maize), P elements (fruit fly), tn3 and IS1 (*E. coli*), Tam3 (*Antirrhinum*), and Spm (maize). The second type of transposon has LTRs and usually at least one gene that encodes a reverse transcriptase. These move via an RNA intermediate produced by a promoter in the LTR. Examples are copia (fruit fly), Ty (yeast), THE-1, and bs1. The third type of transposon resembles a DNA copy of a mature mRNA and usually includes at least one gene encoding a reverse transcriptase. As with the second type of transposon, they move via RNA intermediate. Examples are P elements (fruit fly), L.1 (human), and cin4 (maize).

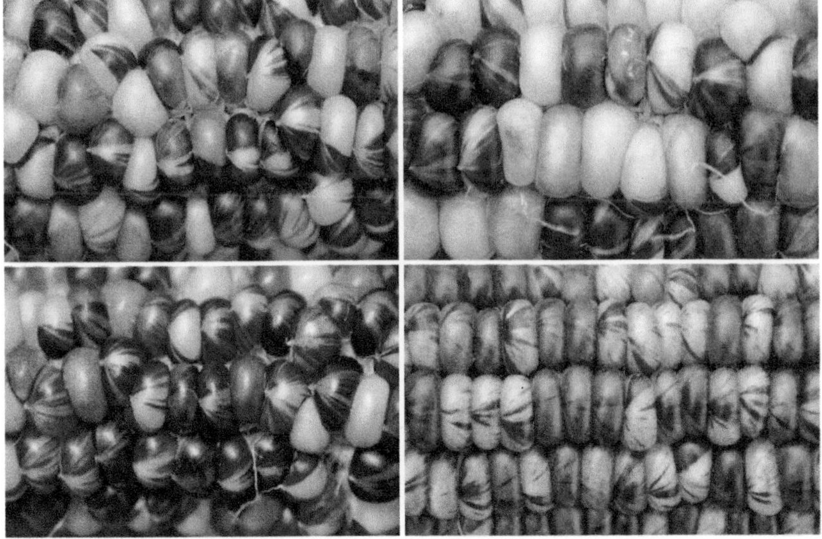

FIGURE 13.19 The effects of transposable elements can be seen in these maize fruits (kernels). The somewhat random pattern of pigmentation is caused by interruptions of pigment-producing genes that are interrupted and shut off by frequent insertions of transposons into those genes. These form uniform lighter pigmentation in some fruits, as well as colorless, striped, and spotted fruits. Although the pigmentation changes are obvious to see, a large number of other genes that have no obvious phenotypes are also affected by these insertions.

a gene that is involved in pigment formation, the affected cells would lack the pigment. Cells that have functional genes would exhibit the pigment so that the tissue would appear spotted, mottled, or striped (Figure 13.19). If the cell is hemizygous, the amount of pigment would be reduced below wild-type levels, thus producing at least three shades of pigmentation. Metabolic and other genes are also similarly affected, producing smaller tissues or misshapen tissues (which is the case when insertion is in one of the genes responsible for starch production). Of course, when a vital gene is

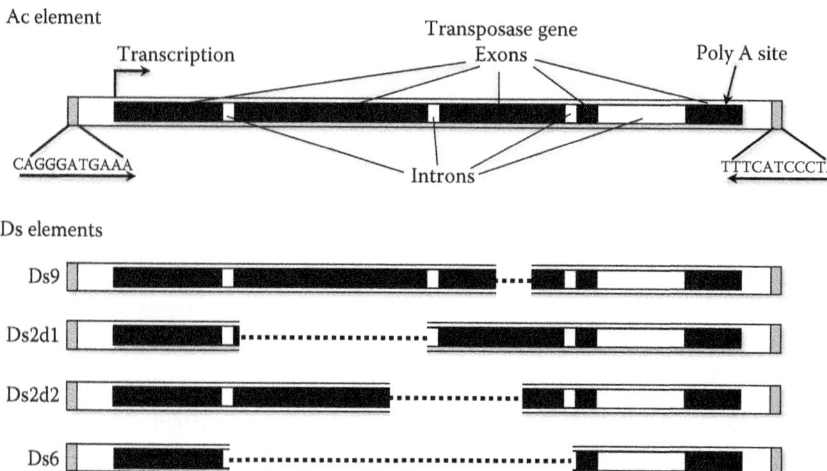

FIGURE 13.20 Ac and Ds elements are versions of the same family of transposons. Ac elements have the short inverted repeats as well as a complete transposase gene. The gene is able to act on any fragment of DNA that contains the short inverted repeats. Ds elements have the inverted repeats, but various parts of the transposase gene have been deleted, and thus produce no transposase enzyme. Therefore, these elements cannot move unless there is an intact version of Ac in the genome.

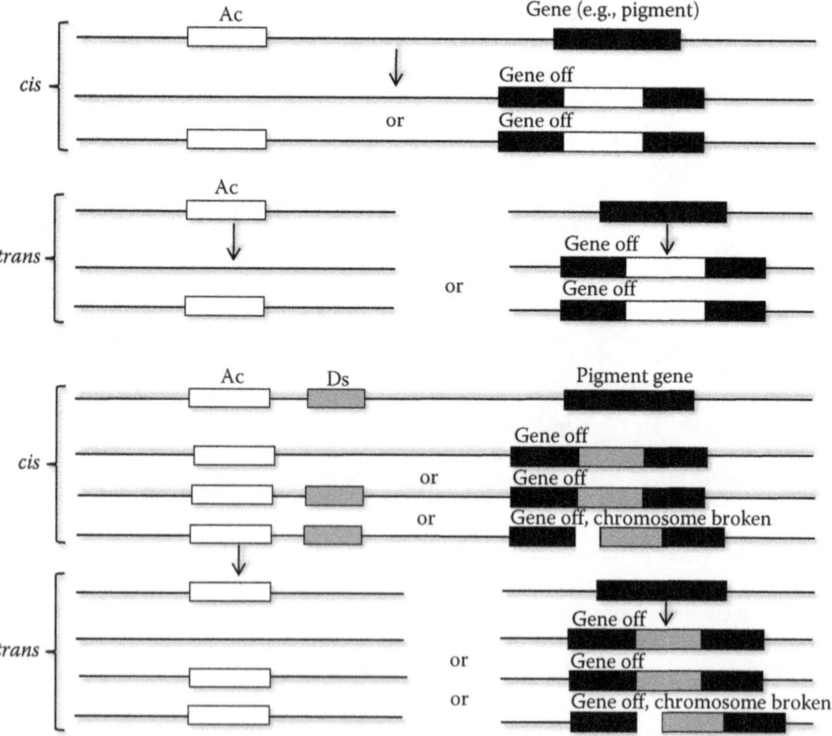

FIGURE 13.21 The effects of transposition. Ac elements can transpose from one location to another, excising from one place and inserting into another locus. They can also move a copy of themselves to other locations. Normally, both occur during DNA replication. Transposition can be accomplished whether the new location is on the same piece of DNA (*cis*) or on another piece of DNA (*trans*). The movement of Ds elements by Ac elements is shown in the lower half of the figure. Again, Ac can act in *cis* or in *trans*.

affected, the cells often die. The transposons also occasionally are removed from a particular locus. This can be caused not only by a crossover event but also by excision of the transposon from the site. If it exits perfectly from the site, the gene can be restored. However, often the removal is not perfect, and some nucleotides from the transposon remain after excision. This usually leaves the gene nonfunctional, or additional amino acids may be encoded in the gene. A third event sometimes occurs where there is invasion of a site by the transposon, but resolution of the insertion is incomplete, causing breakage of the chromosome. This was what McClintock originally observed in her experiments with Ds elements in maize in the 1930s.

PLASMIDS

Plasmids are autonomously replicating DNAs that are common in bacteria but are found in many other organisms and organelles. Many are circular, but some linear plasmids have also been described (Figure 13.22). They can be less than a kilobase to more than a megabase. When they are large, they may act as secondary chromosomes. In bacteria, some are transferred by conjugation, which is a rudimentary form of sex in bacteria. The bacteria form pili that are thin extensions of the bacterium, which connect with an adjacent bacterium. The plasmids are then transferred from one cell to another across the membranes. The F plasmid of *E. coli* has been studied extensively. A replicative cycle begins where a single strand is produced. The leading end is threaded into the recipient bacterium, and this continues until the end of the plasmid is reached. As the single-stranded molecule enters the recipient cell, DNA polymerase synthesizes the other strand until the entire plasmid is double stranded. Then, the ends are ligated to form a full-length double-stranded circular plasmid. Depending on the plasmid, new genes or alleles may be transferred in this way. Many genes for antibiotic resistance genes are carried and transferred on plasmids, which is why

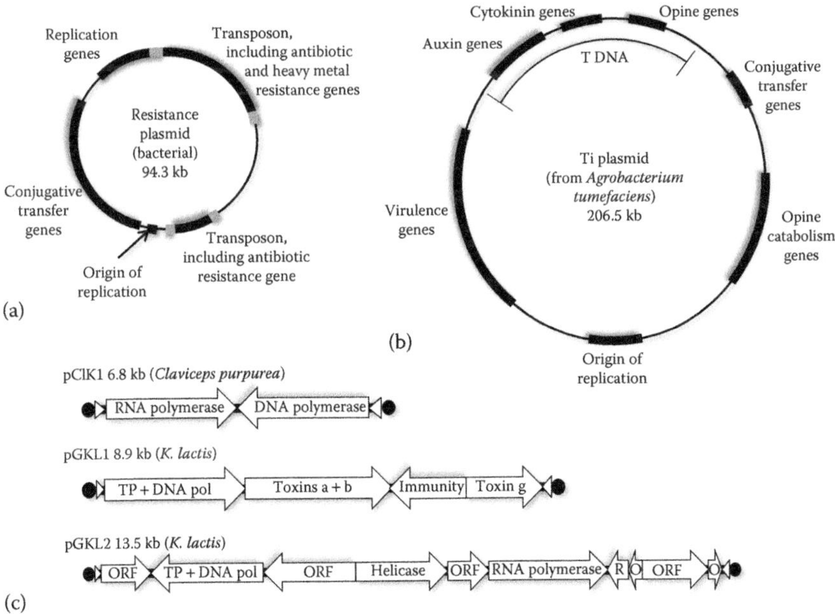

FIGURE 13.22 Examples of circular and linear plasmids. Circular plasmids range in size from about a kilobase to more than a megabase. (a) An *E. coli* plasmid containing antibiotic and heavy metal resistance genes. (b) The Ti (tumor inducing) plasmid from *A. tumefaciens*. The T DNA portion of this plasmid is transferred into plant cells during infection by this bacterium. (c) Linear plasmids have been found in a number of organisms. These all are from fungi. Some appear to have no affect on the organism, whereas others are capable of killing organisms that do not contain the plasmid, but confer immunity to the toxins in cells that harbor the plasmid. Others confer resistance to specific antibiotics. ORF, open reading frame.

bacteria rapidly become resistant to all of the antibiotics that are used to treat bacterial infections. Also, some plasmids carry genes for resistance to heavy metals, and therefore, this trait can also be transmitted through a population of bacteria or from one species to another.

As mentioned earlier, some viruses act somewhat like conjugative plasmids. Bacteriophage M13 has two forms: One is a double-stranded form that exists in the bacterium, and the other is a coated single-stranded form that can be transferred from one bacterium to another. Therefore, this form is very much like a conjugative plasmid, but without pilus formation. The plasmids of *Agrobacterium tumefaciens* and *A. rhizogenes* are also capable of altering plant genomes and cells. When the bacterium enters a wound in the plant, it transfers parts of the plasmid by coating it, producing a particle that resembles a virus. The DNA is transferred into the plant cell nucleus, where it integrates into the chromosome and begins to produce plant hormones and other products that significantly alter the plant cell and all of its daughter cells, which causes a tumor to form. However, *killer* linear plasmids in some yeasts act somewhat like viruses in that they can be transferred from one cell to the next. If the plasmid is present in a cell, then the toxin it encodes does not kill the cell. The toxin will kill all cells that lack the plasmid. Thus, the plasmid can rapidly increase its concentration in two ways: by transfer into new cells (i.e., transformation) or by killing plasmid-less cells.

One plasmid, or at least a piece of DNA called a *minicircle* in *Trypanosoma cruzi*, is capable of integrating into human chromosomes causing the chronic disease, known as Chagas disease. The organism has a genome consisting of maxicircles and minicircles that exist in separate cell compartments. When it infects the blood of its hosts, it releases some of the minicircles, which are transferred into host cells, often into muscle cells, and primarily cardiac muscle cells. These pieces of DNA integrate in specific places (primarily into long interspersed repetitive elements, which make up approximately 21% of the human genome) in the genome and often produce aberrant mRNAs and proteins. Eventually, the affected cells begin to die and the patients suffer a long decline as their heart muscles deteriorate. Not only can these sequences be transferred by the parasite, but if they become incorporated into the cells of an embryo or fetus, they can be inherited vertically for generations.

KEY POINTS

1. Many types of viruses, plasmids, transposons, and introns are related and are members of a large group of mobile (or formerly mobile) genetic elements.
2. Mobile genetic elements can cause a number of deleterious, as well as useful, changes in genomes.
3. Some viruses integrate into the host chromosome(s). Some of these are prophage and proviruses that can excise at a later time. Others are simply integrated sequences that many remain permanently integrated into the genome.
4. Viruses exist in a variety of forms and can have either RNA- or DNA-based genomes.
5. Viruses are categorized into seven types (I through VII, based on the Baltimore system), according to the nucleic acid type and the mode of infection and replication.
6. Introns are of four types: group I, group II, spliceosomal, and archaeal. Group I and II introns are self-splicing *in vitro*. Spliceosomal introns resemble group II introns but require spliceosomes for splicing. Archaeal introns rely on an endoribonuclease for splicing.
7. More than one RNA and protein can be produced from a single gene due to alternative splicing.
8. Transposons are intragenomic mobile elements. At least three types have evolved: one that moves as a DNA version, with short inverted repeats, and two that move as RNA versions (retrotransposons), which must be copied into a DNA version during integration into the chromosome.

9. Plasmids can be transferred from one cell to another, and from one species to another, often in a form that resembles a virus. Some plasmids arc large enough to be considered as small chromosomes.

ADDITIONAL READINGS

Alberts, B., D. Bray, K. Hopkin, A. Johnson, J. Lewis, M. Raff, K. Roberts, and P. Walter. 2013. *Essential Cell Biology,* 4th ed. New York: Garland Publishing.

Alberts, B., D. Bray, A. Johnson, J. Lewis, M. Raff, K. Roberts, and J. D. Watson. 1994. *Molecular Biology of the Cell.* New York: Garland Publishing.

Becker, W. M., J. B. Reece, and M. F. Poenie. 1996. *The World of the Cell,* 3rd ed. New York: Benjamin/Cummings Publishing.

Darnell, J., H. Lodish, and D. Baltimore. 2007. *Molecular Cell Biology*, 6th ed. New York: W. H. Freeman and Company.

Eichbush, T. H. 1997. Telomerase and retrotransposons: Which came first? *Science* 277:911–912.

Federoff, N. V. 1984. Transposable genetic elements in maize. *Sci. Amer.* 250:84–98.

Gregory, T. R. 2005. *The Evolution of the Genome*. San Deigo, CA: Elsevier Academic Press.

Harris, L. and S. O. Rogers. 2008. Splicing of an unusually small group I ribozyme. *Curr. Genet.* 54:213–222.

Karp, G. 2013. *Cell and Molecular Biology*, 7th ed. New York: John Wiley & Sons.

Lehmann, K. and U. Schmidt. 2003. Group II introns: Structure and catalytic versatility of large natural ribozymes. *Crit. Rev. Biochem. Molec. Biol.* 38:249–303.

Lyke-Anderson, J., C. Aagaard, M. Semionenkov, and R. A. Garrett. 1997. Archael introns: Splicing, intercellular mobility and evolution. *TIBS* 22:326–331.

Maniatis, T., E. F. Fritch, and J. Samboook. 1982. *Molecular Cloning: A Laboratory Manual.* Cold Spring Harbor, NY: Cold Spring Harbor Laboratory.

Marck, C. and H. Grosjean. 2003. Identification of BHB splicing motifs in intron-containing tRNAs from 18 archaea: Evolutionary implications. *RNA* 9:1516–1531.

Moreira, S., S. Breton, and G. Burger. 2012. Unscrambling genetic information at the RNA level. *WIREs RNA.* doi:10.1002/wrna.1106.

Nakamura, T. M., G. B. Morin, K. B. Chapman, S. L. Weinrich, W. H. Andrews, J. Lingner, C. B. Harley, and T. R. Cech. 1997. Telomerase catalytic subunit homologs from fission yeast and human. *Science* 277:955–959.

Perlman, P. S. and M. Podar. 1996. Reactions catalyzed by group II introns in vitro. *Meth. Enzymol.* 264:66–86.

Podar, M., L. Mullineaux, H. R. Huang, P. S. Perlman, and M. L. Sogin. 2002. Bacterial group II introns in a deep-sea hydrothermal vent environment. *Appl. Environ. Microbiol.* 68:6392–6398.

Rogers, S. O., Z. H. Yan, K. F. LoBuglio, M. Shinohara, and C. J. K. Wang. 1993. Messenger RNA intron in the nuclear 18S ribosomal DNA gene of deuteromycetes. *Curr. Genet.* 23:338–342.

Rogozin, I. B., L. Carmel, M. Csuros, and E. V. Koonin. 2012. Origin and evolution of spliceosomal introns. *Biol. Direct* 7:11.

Shinohara, M. L., K. F. LoBuglio, and S. O. Rogers. 1996. Group-I intron family in the nuclear ribosomal RNA small subunit genes of *Cenococcum geophilum. Curr. Genet.* 29:377–387.

Tang, T. H., T. S. Rozhdestvensky, B. C. d'Orval, M.-L. Bortolin, H. Huber, B. Charpentier, C. Branlant, J.-P. Bachellerie, J. Brosius, and A. Hüttenhofer. 2002. RNAomics in Archaea reveals a further link between splicing of archaeal introns and rRNA splicing. *Nucleic Acids Res.* 30:921–930.

Tropp, B. E. 2008. *Molecular Biology, Genes to Proteins*, 3rd ed. Sudbury, MA: Jones & Bartlett Publishers.

Section IV

Multicellularity

14 Multigene Families

INTRODUCTION

Although some of the first genes on the Earth were the products of rare interactions of groups of nucleotides, mutation, drift, and selection, these processes required a great deal of time. Recombination accelerated the generation of new genes by joining the existing coding regions together, but the evolution of completely new genes for each new function still was (and is) a slow process. However, once a few successful genes and gene combinations had evolved, faster processes could occur. One of the most successful was gene duplication followed by mutation and divergence (Figure 14.1). A single gene can serve as a starting material for many other genes that eventually act to diversify the functions in the organism. This process probably is responsible for many (or most) of the gene diversity present today. The presence of similar regions in very different genes and proteins is one piece of evidence for this. Multigene families provide additional direct evidence that this process has occurred often and continues to occur. Many genes within genomes, and especially eukaryotic genomes, are members of multigene families (Table 14.1). These consist of two or more genes that were duplications of a progenitor gene. The duplicated genes then mutated such that they had different characteristics. This was advantageous in most cases, because entirely new genes did not have to evolve, but copies of the existing genes would be free to mutate, and then drift and selection acted to retain some versions, whereas other versions became pseudogenes (nonfuntional copies or partial copies of a functional gene) or they disappeared. Many multigene families have been described, including ribosomal RNA (rRNA) genes (discussed in Chapters 4 and 5), immunoglobulin genes, histones, and others (Table 14.1).

The first step in the establishment of a multigene family is duplication of the original gene (Figure 14.1). This can occur by recombination or during replication by polymerase slippage. There are several possible outcomes. The first is that the presence of two copies is deleterious or lethal to the cell, and therefore, the cell with the duplication dies. The second outcome is that the duplication is advantageous, because it increases the fitness of the organism. The third outcome is that the increase is neutral, such that one of the genes can retain its original function in the organism, whereas the other copy is free to change, so long as it does not produce a protein that is deleterious to the organism. Therefore, this has become a mechanism for relatively rapid change and evolution. Complex genes can be duplicated without reinventing another version. The copy (or copies) changes gradually (or sometimes rapidly) to perform slightly different or sometimes completely new functions. Once an additional copy is made and begins to diverge, there is then a higher likelihood of additional expansion of the family, because there are more homologous regions for crossovers. If two copies exist on a chromosome, misalignment and crossover can then lead to a chromosome with three copies, whereas the other is left with only one copy. There is also a small possibility of generating a chromosome with four copies, with a complete deletion of both copies on the other chromosome. This process can then proceed to make more copies per chromosome. Each copy can change slightly from the previous version until many different versions are present in the genome. Occasionally, some of the copies move to other chromosomes by transposition, splitting of the chromosomes, or chromosomal translocations. Some of the genes may become pseudogenes (Figure 14.2). They are of two types: One type is never expressed, whereas the other type is transcribed and translated, but the proteins are nonfunctional. Some multigene families have evolved into hundreds of different members (Table 14.1).

Multigene families have been crucial to the evolution of life on the Earth. They are responsible for the evolution of motility of cells (flagella and cilia), but also the movement of internal components of cells (actins and myosins). Members of multigene families are responsible for many

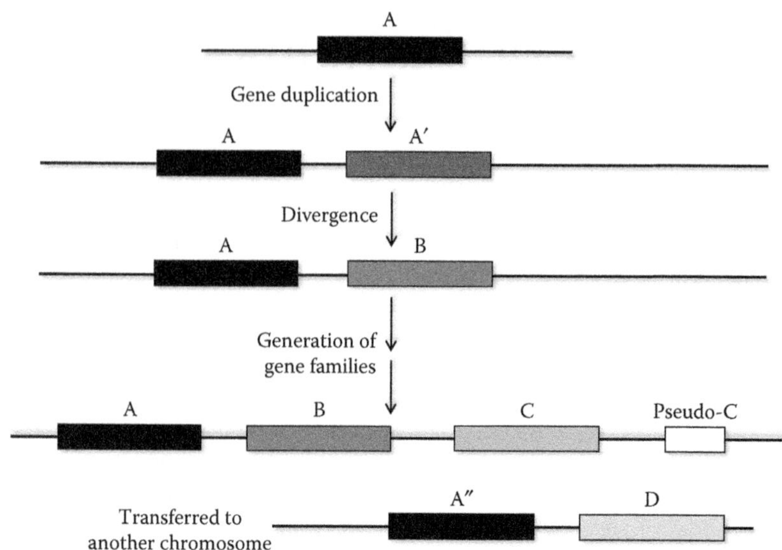

FIGURE 14.1 Generation of members of a multigene family. The first step in the formation of a multigene family is the duplication of a single-copy gene. Normally, this occurs via recombination. The copy has more freedom to mutate, because the original copy still functions normally. Therefore, the duplicate gene can diverge into a gene whose product has slightly to greatly altered functions. In this process, some of the duplicated genes may change to a degree where they are no longer functional (i.e., either not expressed, or the protein produced is nonfunctional), thus becoming pseudogenes. Eventually, some of the gene family may be transferred to other chromosomes, by transposition, chromosomal translocation, or chromosome fragmentation.

TABLE 14.1
Multigene Families Examples

Family	Number of Versions
Actins	61
Cell surface antigens	50–100
Egg shell proteins (insect)	50
α Globins	1–5
β Globins	5
Heat-shock proteins	>10
Histones (H2A, H2B, H3, and H4)	75
Immunoglobins (variable region)	381
Keratins	>20
Laccases (fungi)	3–5
Major histocompatability genes	30
Myosin (heavy chain)	5–10
Protein kinases	10 to >100
Ribosomal RNA genes	1 to dozens (1–23,000 copies)
T-cell receptors	33
Transfer RNA genes	Dozens (1–100 copies each)
Transcription factors	>100
Tubulins (α and β)	3–15
Visual pigment protein (human)	4
Vitellogenin (frog and chicken)	5
Zinc fingers	700

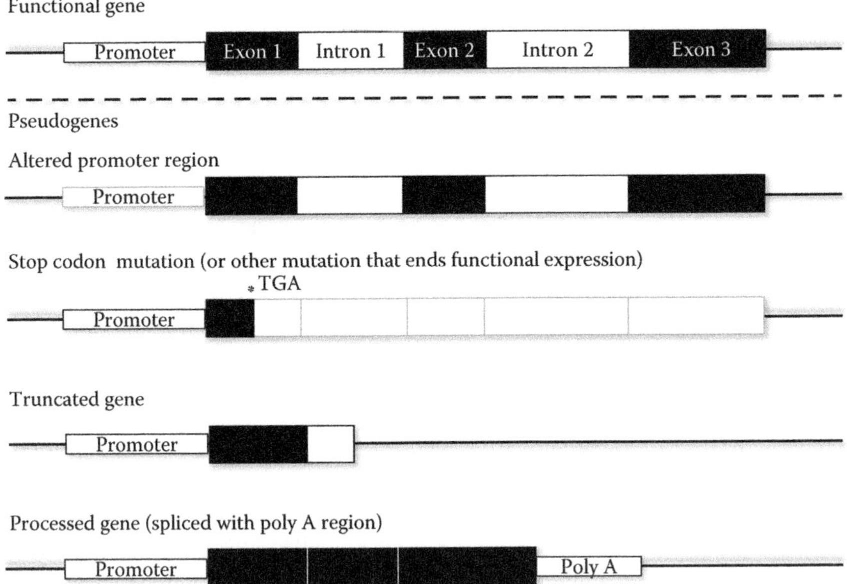

FIGURE 14.2 Types of pseudogenes. Several different types of pseudogenes have been identified in various multigene families. The gene illustrated at the top is functional. It has a functional promoter, as well as several exons and introns (although the presence of introns is not necessary for the generation of pseudogenes). The simplest pseudogene is one in which the coding region is unaltered, but the promoter has mutated such that transcription is never initiated. Another common mutation is one in which a codon has changed to a stop. The RNA is transcribed, but only part of the protein is produced during translation. A third mutation is where the gene itself has been truncated, usually by a recombination event. The fourth type appears to be an inserted DNA copy of the processed messenger RNA (mRNA) for the gene. The exons are present, but all of the introns are missing, as if they have been spliced out. Additionally, a number of adenines are present on the 5′ end of the gene, just as would be present as a poly A tail in the mature mRNA. This last type of pseudogene probably is generated by transposition or reverse transcription, rather than by recombination.

phases of the cell cycle (e.g., chromosome separation and cytokinesis). Several multigene families are responsible for cell-to-cell recognition, response to antigens, and other proteins involved in environmental sensing by cells (e.g., immunoglobulins, T-cell receptors, major histocompatability proteins, and visual pigments). Some of the proteins encoded by members of multigene families act to package DNA, or control gene expression, and have major influences on cell division and organism development (e.g., histones, transcription factors, and homeobox genes).

RIBOSOMAL RNA GENE FAMILY

One multigene family was discussed in Chapter 5. rRNA genes are in a special class of multigene family because their numbers are constantly changing by recombination, although most of the copies must retain their original function, primarily caused by a process called *gene conversion*, which acts to keep all of the copies homologous (or nearly so). If they were not kept nearly identical, there might be excessive mutations, such that fewer ribosomes would be produced and the organism would be weakened or might die. Therefore, there is heavy selection to retain rRNA function. However, this multigene family can experience significant changes in the intergenic regions, some of which control expression of the genes downstream. The copy number of rDNA can change rapidly in plants, amphibians, and other organisms, and the cells must be able to control the expression of these genes. They do this by having a fixed number of factors and RNA polymerase molecules that are able to initiate transcription. Therefore, although the total number of rRNA genes may vary significantly (up to 95-fold in some plants), the number of rRNAs and ribosomes that are produced remains relatively constant per cell. Nonetheless, the mechanism for producing additional copies, and in this case many different copies (at least in the intergenic regions), appears to be operating constantly.

GLOBIN GENE FAMILY

One of the best examples of a multigene family that has led to some interesting and significant evolutionary changes and events is the globin gene family, the genes that encode hemoglobins that carry oxygen in the blood and myoglobin that carries oxygen in other tissues. In humans, the globin gene family is split between chromosomes 11 and 16 (Figure 14.3). The β (beta) globin genes are on chromosome 11, consisting of five functional genes (ε, expressed in embryos; Gγ and Aγ, expressed in fetuses; and δ and β, expressed in adults), and one pseudogene (ψ_β) that is not expressed. The α (alpha) globin genes are on chromosome 16 and include three functional genes (ζ, expressed in embryos, and α_1 and α_2, expressed in fetuses and adults), one expressed pseudogene (θ_2, expressed as a dysfunctional protein), and three other pseudogenes (ψ_1, $\psi_{\alpha 1}$, and $\psi_{\alpha 2}$) that are not expressed. These genes each have specific functions that are related to the transport and delivery of oxygen to cells during different parts of development. The physiology of this transport and delivery system is vital to placental mammals, such as humans. In order for cells to pick up oxygen from the blood, the hemoglobin in the blood must not only pick up oxygen in the lungs but also release the oxygen to the cells that need the oxygen. In cells with high oxygen requirements, such as muscles, another related protein called myoglobin increases the transfer of oxygen from the blood to the cells. Each one of the globins has different affinities for oxygen (Figure 14.4), and therefore at different oxygen concentrations, one type of globin can be releasing oxygen, whereas another type is binding oxygen. In this way, oxygen can be transferred from one type of tissue to another.

These different globins are vital for humans and other placental mammals, because the mothers must be able to pass oxygen to their embryos and fetuses during pregnancy. The pathway for the oxygen is rather long and begins by moving from the mother's lungs into the hemoglobin in the mother's red blood cells (Figure 14.5). From there, the oxygen must pass into the mother's tissues, but it must also be passed through the placenta and into the hemoglobin of her developing offspring.

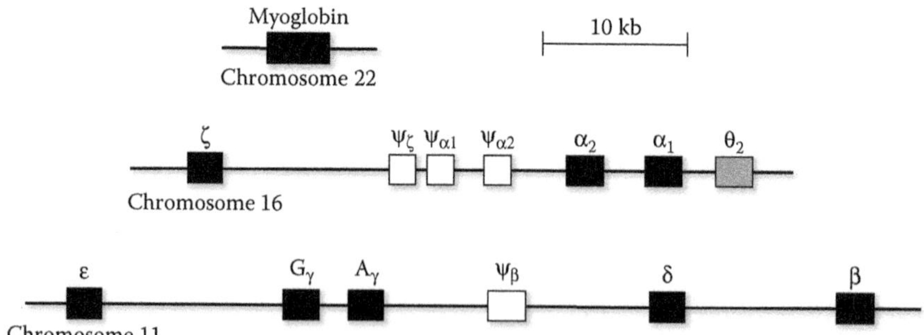

FIGURE 14.3 Arrangement of genes in the human globin gene family. The myoglobin gene is on chromosome 22 and is isolated from all other members of the gene family. The α-type globin genes are all located on chromosome 16. This cluster consists of one gene (ζ) that is expressed only in the embryo, as well as two genes (α_1 and α_2) that are expressed in the fetus and adult. Four pseudogenes are also present, including three (ψ_1, $\psi_{\alpha 1}$, and $\psi_{\alpha 2}$) that are not expressed and one (θ_2) that is expressed, but the protein is nonfunctional. The β-like globins exist as a cluster of five genes and one pseudogene (ψ_β) located on chromosome 11. One gene (ε) is expressed only in the embryo, whereas two genes (G_γ and A_γ) are primarily expressed in the fetus. The two fetal genes continue to be expressed for a few months after birth, but are gradually replaced primarily by adult β globin and a small percentage of δ globin. The genes are expressed developmentally in the order that they appear on the chromosomes. Black boxes indicate the expressed gene. Gray and white boxes indicate pseudogenes that are expressed (but nonfunctional) or not expressed, respectively.

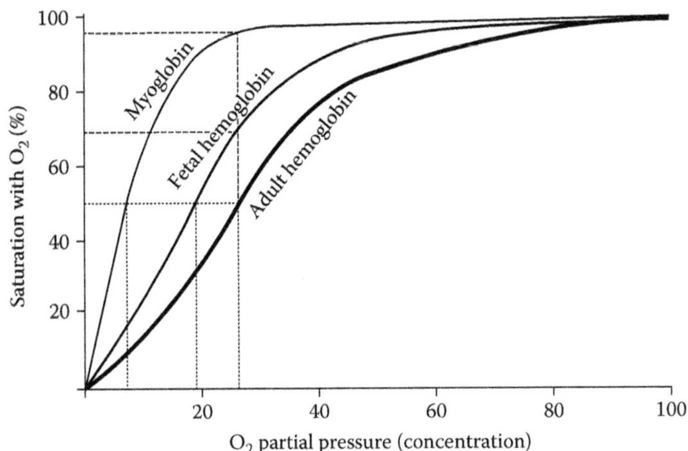

FIGURE 14.4 Oxygen saturation curves for myoglobin and two hemoglobins. The partial pressure of oxygen (a measure of O_2 concentration) is plotted against the amount of oxygen (in percentage of saturation) bound to myoglobin, fetal hemoglobin ($\alpha_2\gamma_2$), and adult hemoglobin ($\alpha_2\beta_2$). Each of the globins binds oxygen at higher partial pressures and releases the oxygen at lower partial pressures, but each of the affinities for oxygen differs, such that the partial pressure where there is 50% saturation of the globins differs for each (dotted lines). For example, when adult hemoglobin is 50% saturated, approximately half of the globins are binding oxygen, whereas the other globins are releasing oxygen. At the same partial pressure (dashed lines), approximately 70% of the fetal hemoglobin molecules are binding oxygen and only 30% are releasing it. Therefore, there is a net movement of oxygen onto the fetal hemoglobin compared to the adult hemoglobin molecules. Again, at the same partial pressure, approximately 95% of the myoglobin molecules would be binding oxygen, whereas only 5% would be releasing it. Therefore, again there would be a net movement of oxygen to myoglobin compared to any of the hemoglobins.

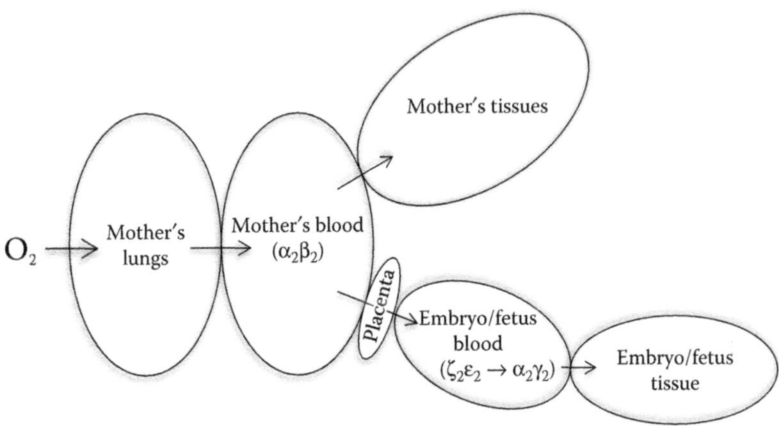

FIGURE 14.5 Diagram of pathway for oxygen in mother and fetus. Oxygen enters the mother's lungs and is transferred into the blood through alveoli. The hemoglobin in the blood is initially depleted of oxygen, and thus, the hemoglobin readily attaches oxygen (one oxygen molecule onto each of the four globins in the hemoglobin tetramer, $\alpha_2\beta_2$). In the mother's tissues, the oxygen pressure is low, and thus, the hemoglobin releases the oxygen more readily, especially if myoglobin is present (as in muscle cells). Also, because the affinity for oxygen is higher for embryonic ($\zeta_2\varepsilon_2$) and fetal ($\alpha_1\gamma_1$) hemoglobins, the mother's hemoglobins, there is a net flow of oxygen across the placenta from the mother's blood to the blood of the developing fetus. Finally, the embryonic or fetal hemoglobins release oxygen to the tissues of the fetus.

Furthermore, the hemoglobin in the developing baby must also be able to release the oxygen into its tissues. If each of the globin proteins had the same affinity for oxygen, this process would be slow, and the amount of oxygen that reached the developing baby would be insufficient, which would lead to its death early in the process. However, each of the versions of globin has a different affinity for oxygen such that there is an efficient flow of oxygen from the mother's hemoglobin to the developing child's hemoglobin, and ultimately to the cells that need the oxygen (Figure 14.4). Of course, this means that the mother is more easily fatigued, not only because of the added weight of the child, but also because that child is using some of the oxygen that would normally be transferred to the mother's tissues. This is especially true late in pregnancy when the oxygen requirements are at their maxima.

The profile of expression of each of the globin genes changes as the child develops (Figure 14.6). Active hemoglobin consists of two globins from the α group and two globins from the β group. In early embryogenesis, the hemoglobin consists of ζ (zeta) globin (an α-type globin) and ε (eta) globin (a β-type globin). Several weeks later, the expression levels of ζ and ε globins have begun to drop, whereas the expression of α and γ (gamma, which is β-like) globins are rising. By the time the embryo has developed into a fetus, levels of ζ and ε globins have dropped to zero, whereas levels of α (the adult version) and γ globins are high. Additionally, the expression of β (the adult version) globin has begun to slowly increase. Although the adult version of α globin continues to be expressed at high levels from then on, the β-like globin expression continues to change. About 1 month prior to birth, the expression of the adult β globin begins to increase more rapidly, whereas the expression of the fetal γ globin begins to decrease. Immediately after birth, the γ globin levels continue decrease, whereas the adult β globin levels continue to rise, and another minor globin, δ (delta) globin starts to appear. After 6 months, the infant has adult levels of α, β, and δ globins.

The evolution of these genes has coincided with a number of very important evolutionary events in large groups of animals (Figure 14.7). The first duplication appears to have occurred more than a billion years ago and was a necessary step in the evolution of multicellularity in eukaryotes. This is because they had begun to develop with cells on their exteriors and cells within the interiors.

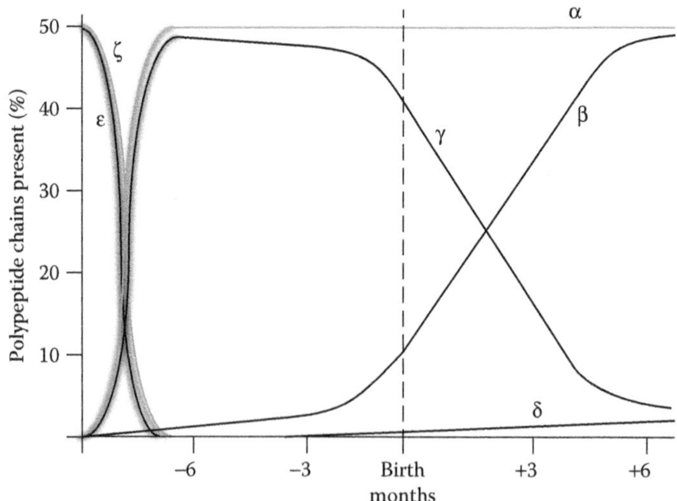

FIGURE 14.6 Expression time course for human globin genes. Initially, only the ζ and ε genes are expressed, as the ζ and ε polypeptides make up 100% of the hemoglobin in the early embryo. By the time the embryo has developed into an early fetus (about 2 months after conception), the hemoglobin of the fetus consists of about equal amounts of α, γ, ε, and ζ globins. By 3 months after conception, the ζ and ε globins are absent, and nearly 100% of the hemoglobin consists of α and γ globins, although β globin is present at a low level. At birth, four globins are present, α, β, γ, and δ, in proportions of approximately 50:9:40:1, respectively. After birth, the globin expression continues to change, and at 6 months, the infant has adult levels of the globins, which are primarily α and β globins ($\alpha_2\beta_2$ hemoglobin) and a small percentage of δ globin.

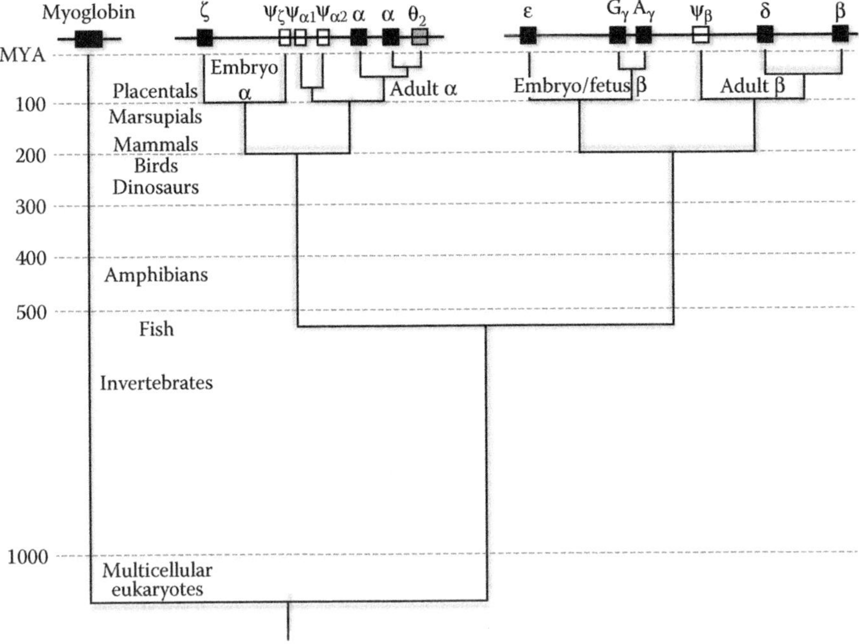

FIGURE 14.7 Evolution of globin genes. When globin genes have been examined in other organisms, a phylogenetic reconstruction can be produced from the sequence information. All globins appear to have evolved from a heme-carrying protein that existed at least 1.1 billion years ago. At about that time, there was a gene duplication, and the two versions began to diverge such that the transfer of oxygen from one globin to the other version would be possible at specific concentrations (partial pressures) of oxygen. This coincided with the appearance of the first multicellular eukaryotes. Thus, the cells on the outside of the organisms could pick up oxygen from the outside environment (water, in this case) and transfer it to the cells on the inside of the organism. This process continued for approximately 500 million years until the next change, which coincided with the appearance of the first fish. The first fish were small, but eventually larger fish evolved with higher demands for energy and oxygen. There was another round of duplications of the globin genes at this time, eventually producing several α- and β-type genes, and these formed tetramers to produce hemoglobin. Therefore, each hemoglobin unit could carry four oxygen molecules, and furthermore, they bound in a cooperative fashion, such that each bound more easily than the previous one. This change ushered in the amphibians. Also, it was around this time the α- and β-type genes became unlinked, that is, they were on separate chromosomes. The change in the globins and their ability to bind and release oxygen in different concentrations of oxygen also meant that the organisms could live outside of an aqueous environment. In other words, they could live on land, at least part of the time, extracting oxygen from the atmosphere instead of the water. Over the next 100 million years, land animals diversified, and eventually, dinosaurs and birds appeared. As with amphibians, they laid eggs, but they laid them on land. This presented yet another environment where oxygen concentrations differed from the previous environments. By this time, there were at least two versions of each α- and β-type globin gene. This allowed the uptake of oxygen from the atmosphere into the egg, as well as from the atmosphere into the adult lungs and blood, while assuring adequate transfer of oxygen into the tissues of the fetuses and the adults. Next, the mammals appeared as another part of the trend toward diversification in development and globin gene family evolution. The earliest mammals laid eggs, but part of embryo development occurred within the mother. Therefore, oxygen had to be delivered from the mother's blood to the blood of the embryo. At this time, a concomitant diversification of both α- and β-type globins is also seen. The embryo versions of these genes appear at this time (200 million years). Further duplication and diversification led to the ability of certain mammal mothers to carry their offspring longer during development. In marsupial mammals, the embryo develops into a fetus before exiting the mother and moving into the pouch for additional growth and development. Finally, the placental mammals, such as humans, represent another stage in this expansion and diversification in globin genes. The most recent globin versions (G_γ and A_γ globins) were necessary to allow an increased transfer of oxygen into a large fetus, although it was still inside of its mother.

Therefore, oxygen that was in the outside environment (i.e., water) had to be transported to the interior cells that also required oxygen, whereas the exterior cells could simply import oxygen from the outside environment. One of the copies of the gene became myoglobin, whereas the other became the oxygen transporting protein, hemoglobin. By the time of jawed fish, about 550 million years ago, the globin genes had duplicated, and mutated, such that there were at least three copies of α globin and three copies of β globin, but they still were physically linked on the same chromosome. This hemoglobin was capable of carrying oxygen from the gills to the tissues, where it would be released to the cells and myoglobin. In organisms that appeared after about 450 million years ago, the α and β globin genes are on different chromosomes. About 425 million years ago, the first amphibians appeared. Instead of gills, they had lungs, and yet spent most of their time in the water, and their eggs obtained oxygen from the water. Amphibians still had only the basic α and β globins, but they had more gene copies, which allowed more flexibility in expression and evolution of these genes. Dinosaurs and birds appeared first between 350 and 330 million years ago. Most were land animals and laid their eggs on land. Some had high metabolic rates, which required large amounts of oxygen. These animals had more than one form of α globin and more than one form of β globin. The reason is that the adults needed one type of hemoglobin, whereas the developing young in the eggs needed different hemoglobin. In bird eggs, oxygen passes through the eggshell and membrane, and then into the fluids surrounding the developing chick, then into the yolk sack and into the blood, and finally into the tissues of the chick. The transfer has to be efficient and sufficient to allow complete development of the animal. And, unlike amphibians, the oxygen was obtained directly from the atmosphere, rather than from a watery environment. Another factor that allowed development of the larger dinosaurs was that the atmospheric oxygen level was about 35% at that time (whereas it is 21% presently).

The next animals to appear on the Earth were the mammals. In early mammals, eggs still were laid, but substantial development occurred within the uterus of the mother. This step requires the delivery of oxygen to the blood of the embryo by the blood of the mother, and therefore, the release and binding of oxygen from the mother to the embryo must be such that sufficient oxygen is transferred to the embryo. Once the egg is laid, different oxygen pressures are present, and therefore, the globins that are expressed must match the requirements of the developing animal. The globin genes found in the platypus are similar to those found in humans, but a few differ significantly, and this may reflect differences in the oxygen-carrying capacity for those globins. Marsupial and placental animals have nearly identical globin genes. This is because for both much of the development of the embryo and fetus occurs in *utero*, and thus the oxygen delivery systems are nearly identical. In both, there is embryonic hemoglobin, fetal hemoglobin, and adult hemoglobin, each consisting of two α-type globins and two β-type globins. Therefore, evolution of the globin gene family was closely synchronized with the major evolutionary events of several large groups of animals.

BACTERIAL FLAGELLA GENE FAMILY

Another example of a multigene family that started from a single gene (or from a small set of genes) is one whose products constitute bacterial flagella. These are long whiplike structures, fueled by adenosine triphosphatase (ATPase) motors, which propel bacteria through aqueous environments. The multigene family often consists of more than 50 genes. Upon initial examination, flagella appear complex and difficult to reconcile given the relative simplicity of the bacterial cell. However, it appears that there was a sequential progression in evolution that proceeded from the inside of the cell to the outside. The genes within the inner membrane have similarities to membrane-bound ATPase complexes, which are common to all cells, and some organelles. Furthermore, the initial structure may have been for secretion, and not for locomotion. Following further duplication and divergence of some of the genes, the portion that spanned the membrane, as well as the proximal and distal rod portions, appeared in succession. Phylogenetic reconstructions support this sequence of evolutionary events. Additional support comes from the fact that when flagella are constructed in bacteria, they are built in a succession from the inner sections to the outermost portions. The final parts of the flagellum, namely, the hook and the filament, are the last portions to be added, and they also appear as the last set of gene duplications and mutations during their evolution. Therefore, rather than depending on the fortuitous combination of a large set of different genes to form a functional flagellum, this structure arose from only a few (possibly only one) ancestral genes that were originally not involved in motility, but were involved in secretion of molecules out of the cell. However, the gene products relied on close coupling with ATPases, and thus, an energy source already existed. Gene duplications and divergence apparently allowed for movements of some of the proteins, which is seen in other proteins (e.g., the TolR and TolQ proteins) that also show similarities to the flagellar proteins. Once movement was established in this system, additions to the multigene family increased the size and efficiency of the apparatus. Presumably, the additional motility greatly increased the fitness of the organisms by allowing them to escape from predators and stressful environments, while allowing them to move toward food and energy sources.

LACCASE GENE FAMILY

Laccases are enzymes that are present in some plants, fungi, and microorganisms. They are component enzymes that can polymerize or depolymerize phenolic compounds. Some have been used in furniture production to achieve a high shine on some types of furniture. However, some fungi have at least two to four distinct forms of laccase and use them in very distinct ways. Most of these fungi degrade wood as a part of the natural recycling of nutrients in forests. Laccases are able to degrade lignins and other compounds, which are polyphenolic compounds that are part of the structural component of plant wood. In fact, they have very wide ranges in their reactions with phenolic compounds. Thus, laccase is needed to aid in degrading the wood of trees, normally dead trees. These are extremely important processes in forest nutrient recycling. Plants also produce a number of phenolic compounds that act as phytoalexins to kill or repel fungal pathogens. Phenolic compounds are toxic to fungi (and most other organisms), and upon infection, plants increase the production and secretion of a class of phytoalexins to protect themselves from the fungi. However, in response, the fungi secrete laccases that can polymerize the phenolics (or alter them in other ways), which renders them nontoxic. The different fungal laccases are very similar in amino acid sequences and are located on the adjacent regions of the same chromosome, and therefore are members of the same multigene family. However, they have different pH requirements. One set of laccases has a pH optimum of approximately 4, which would function best in rotting wood, which is an acidic environment. The other has a pH optimum closer to 7, which would function best at the plant–fungus interface, where the pH is close to neutral. This is the laccase that is being produced and is functioning when the plant is secreting the antifungal phenolic phytoalexins. The two sets of genes have evolved to perform different functions under two different conditions, even though they originated from the same gene.

HISTONE GENE FAMILY

Another example of a multigene family is the histone gene family. Histones are basic (positively charged) proteins that are one of the major proteins that interact with DNA. They form an octamer of two molecules each of H2A, H2B, H3, and H4 (see Figure 7.13). The DNA makes two turns around this octamer, which serves to condense the DNA but also determines whether or not the DNA is transcribed in that region. When histones are present, normally RNA polymerase cannot access the DNA, and therefore, the gene is turned off (although this is not always the case). The DNA can be further condensed by the actions of histone H1. Archaea contain proteins very much like histones, called histone-like proteins, and therefore this was the likely origin of these proteins in eukaryotes (Figure 14.8). Bacteria and some Archaea do not have histones, but they do have basic proteins that are similar in function to histones, and may be distantly related. There are specific members of the family that only attach to centromere regions (a variant of H3), others that attach to nicked DNA (a variant of H2A), some that mark actively transcribing regions (a variant of H3), and other specific regions. In total,

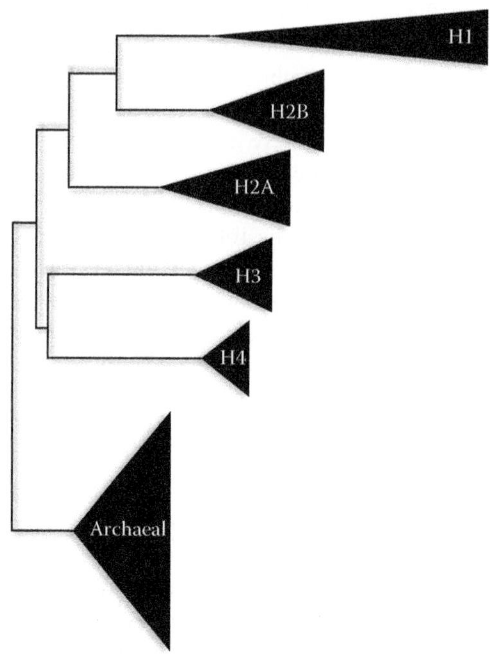

FIGURE 14.8 Evolution of histone genes. All of the histone genes in eukaryotes evolved from histone-like proteins that are found in members of the Euryarchaeota (archaeal, lower triangle). Initially, there was a duplication and diversification event that led to a set of genes related to H3 and H4, and another related to H1 and both H2 types. Further duplication and diversification occurred, probably over a period of at least one billion years that resulted in the five types of histones (H1 through H4). The triangles indicate a summary of the diversity of each of the histone gene types. The lengths of the triangles are indications of the diversity in each group, which shows that H1 is the most variable and H4 is the most conserved of the histone types.

there are approximately 75 different versions of these genes, representing the four classes (H2A, H2B, H3, and H4). H4 exhibits the highest sequence conservation, whereas H2A and H2B are intermediately variable, and H1 is the most variable of the histones (Figure 14.8). All of the histones originated from a single version that is also related to the histone-like proteins present in the Euryarchaeota.

ORTHOLOGS AND PARALOGS

Another part of the evolution of multigene families is that orthologous and paralogous version of the genes can be formed. That is, if the gene evolves in many different species and its protein product retains the same function in all, then each of those genes in each of the species is orthologous. However, if a copy of the gene begins to diverge significantly, and the gene products begin to function differently than (and evolve independently from) the original set of genes, then they are paralogous. This may seem like a minor point, but it is a major part of the evolution of gene and protein function diversity, and when performing phylogenetic analysis, one must be aware of orthologs and paralogs in the analyses. Otherwise, false associations and conclusions are possible. This is because they have different rates of mutation once they have different functions.

POLYPLOIDIZATION AND MULTIGENE FAMILY EVOLUTION

Another mechanism that leads to the expansion and evolution of multigene families is the duplication of chromosomes to produce polyploidy species. These events have occurred many times throughout evolution in broadly divergent phylogentic groups (Figure 14.9). Genome duplications

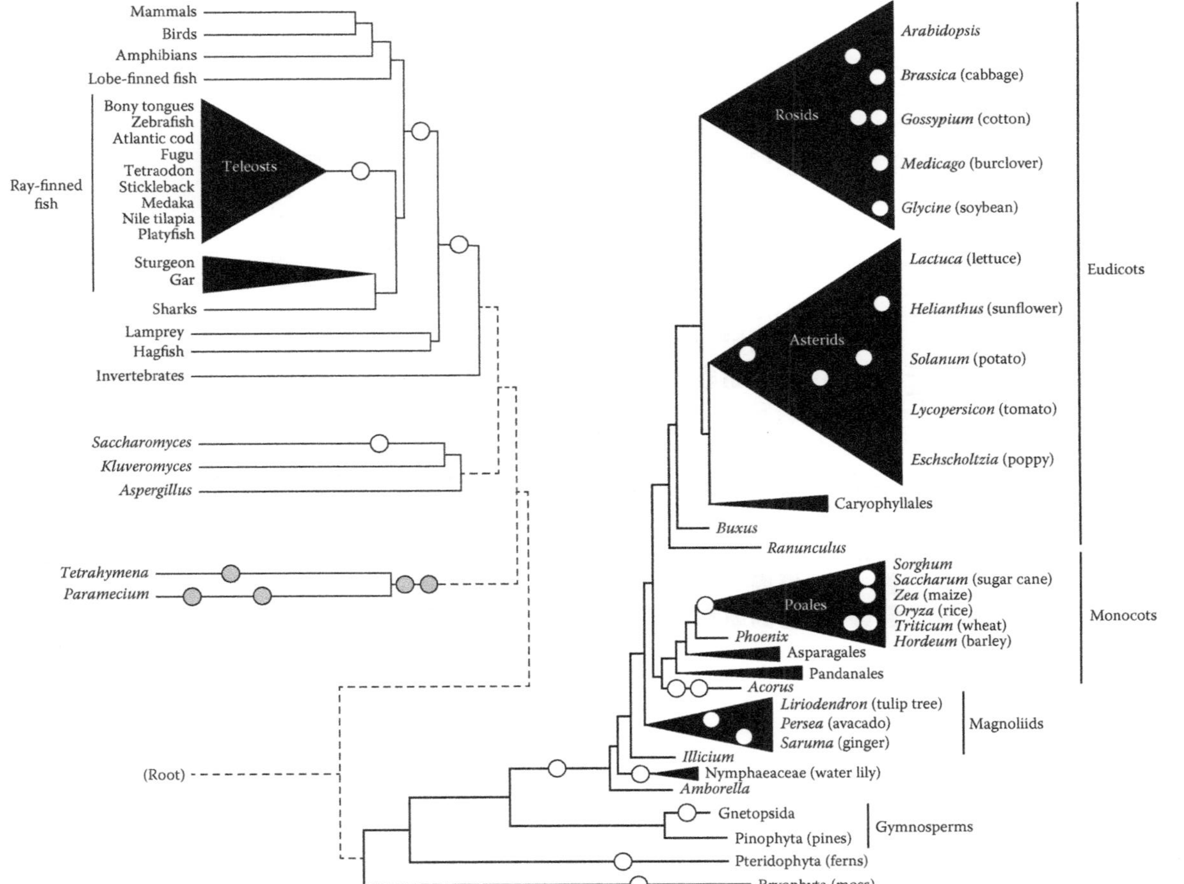

FIGURE 14.9 Genome duplications (indicated by circles) along phylogenetic lines. Polyploid species have been produced within many groups of eukaryotes, including animals, fungi, plants, protists, and amebae. The entire genomes have been duplicated to form tetraploids and species of higher ploidy levels. The duplications have created duplicates of each gene and gene family, which has provided the raw materials for rapid evolution of many of the genes and species.

have occurred several times among vertebrates, fungi, protists, and plants. Some protists and plants are highly polyploidy indicating multiple duplication events during their evolution. Once a duplication has occurred, rendering the organism initially tetraploid, multiple copies of the genes tend to each change their expression patterns, and some disappear from the genome (Figure 14.10). This has led to diversity in these species such that the tissues and organs also have become more diverse. The *Hox* locus of vertebrates is a good example of this. Most invertebrates and ancient vertebrates have one chromosomal locus of *Hox* genes, consisting of one to just over a dozen *Hox* genes. One of the simplest invertebrates, sponges have a single *Hox* gene, whereas more complex invertebrates, such as some echinoderms, have up to 14 *Hox* genes, all formed by gene duplication along

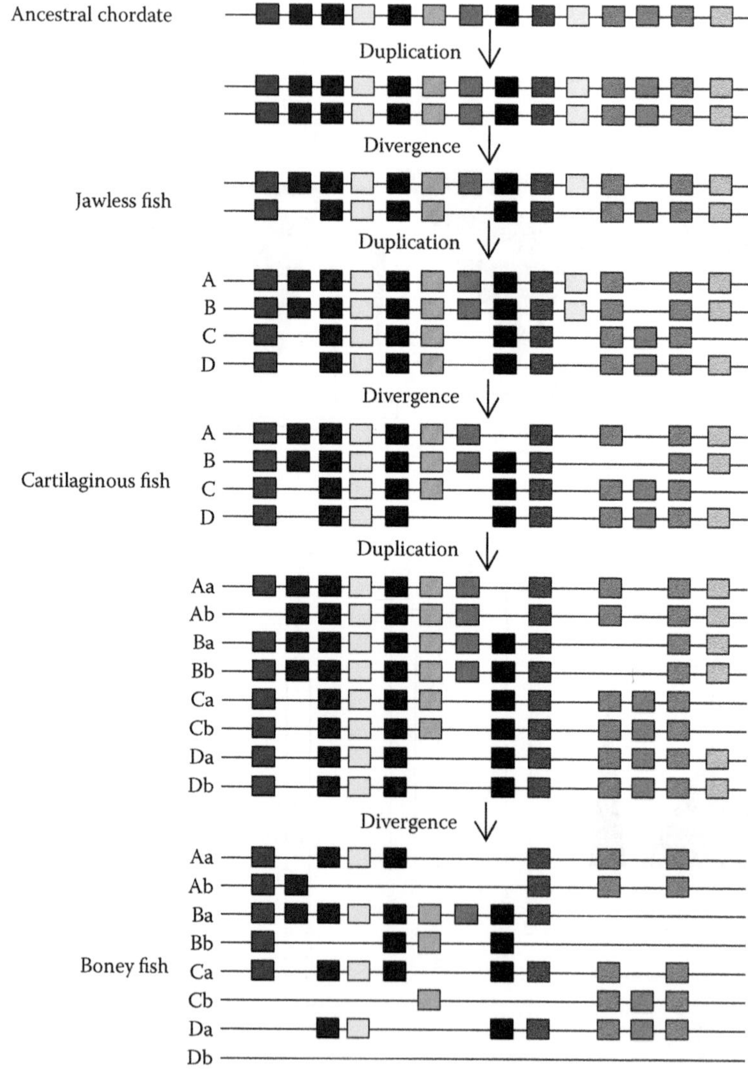

FIGURE 14.10 Genome duplication and divergence in the *Hox* genetic locus. Although invertebrates possess *Hox* loci, they usually have fewer than 14 genes and occur only on one chromosome. Vertebrates usually have 14 versions of *Hox*, and throughout the evolutionary history, there have been polypoloidization events that have duplicated all of the genes, including *Hox* gene loci. Once the duplications occurred, there was divergence, shown here by the loss of selected versions of the *Hox* genes in the evolution of jawless fish, then cartilaginous fish, and finally bony fish. In each case, some gene losses occurred after duplication of the loci.

a single chromosomal region. Ancestral chordates also had 14 *Hox* genes. However, in jawless fish, there was a genome duplication event that produced organisms with additional copies of each chromosome. Instead of having one set of 14 *Hox* genes, these individuals had two sets of 14 *Hox* genes (Figure 14.10). These diverged such that the two loci were no longer identical. Specifically, in one set *Hox* gene 12 was lost, and in the other set, three different *Hox* genes were lost (numbers 2, 7, and 10). The two loci were also expressed differently in different tissues, such that cell and tissue diversity increased. Further duplication and divergence events occurred, which coincided with the evolution of the cartilaginous fish. In these species, there are four distinct *Hox* gene loci, all of which were derived from the original chordate locus, but each of which has diverged, and each of which is expressed in unique ways in the organisms. Another duplication to produce eight versions of the *Hox* locus occurred in the bony fish, and one of the sets has completely disappeared. Because these genes control the expression of many other genes in these organisms, the duplications and changes have resulted in large changes in the cells, tissues, organs, and body plans of each of the organisms.

KEY POINTS

1. Multigene families are generated by gene duplication events followed by mutation and divergence of the duplicated gene copies.
2. Gene duplication and divergence allows much more rapid evolution than from the combinations of extant versions of disparate genes.
3. Multigene families are found in Bacteria, Archaea, and Eukarya.
4. Many important evolutionary processes may be attributed to multigene families.
5. Some of the complex structures and systems in organisms are the results of the evolution of multigene families.
6. Major body plan changes have been caused by the evolution of multigene families (e.g., *Hox* genes) that control large numbers of other genes.

ADDITIONAL READINGS

Adams, K. L. and J. F. Wendel. 2005. Polyploidy and genome evolution in plants. *Curr. Opin. Plant Biol.* 8:135–141.

Alberts, B., D. Bray, K. Hopkin, A. Johnson, J. Lewis, M. Raff, K. Roberts, and P. Walter. 2013. *Essential Cell Biology,* 4th ed. New York: Garland Publishing.

Alberts, B., D. Bray, A. Johnson, J. Lewis, M. Raff, K. Roberts, and J. D. Watson. 1994. *Molecular Biology of the Cell.* New York: Garland Publishing.

Becker, W. M., J. B. Reece, and M. F. Poenie. 1996. *The World of the Cell,* 3rd ed. New York: The Benjamin/Cummings Publishing.

Darnell, J., H. Lodish, and D. Baltimore. 2007. *Molecular Cell Biology*, 6th ed. New York: W. H. Freeman and Company.

Davis, M.C., R. D. Dahn, and N. H. Shubin. 2007. An autopodial-like pattern of Hox expression in the fins of a basal actinopterygian fish. *Nature* 447:473–476.

Duboule, D. 2007. The rise and fall of Hox gene clusters. *Development* 143:2549–2560.

Gregory, T. R. 2005. *The Evolution of the Genome*. San Diego, CA: Elsevier Academic Press.

Liu, R. and H. Ochman. 2007. Stepwise formation of the bacterial flagellar system. *Proc. Natl. Acad. Sci. USA* 104:7116–7121.

Piontkivska, H., A. P. Rooney, and M. Nei. 2002. Purifying selection and birth-and-death evolution in the histone H4 gene family. *Mol. Biol. Evol.* 19:689–697.

Sheth, R., L. Marcon, M. F. Bastida, M. Junco, L. Quintant, R. Dahn, M. Kmits, J. Sharpe, and M. A. Ros. 2012. Hox genes regulate digit patterning by controlling the wavelength of a Turing-type mechanism. *Science* 338:1476–1480.

Swalla, B. J. 2006. Building divergent body plans with similar genetic pathways. *Heredity* 97:235–243.

Tropp, B. E. 2008. *Molecular Biology: Genes to Proteins*, 3rd ed. Sudbury, MA: Jones & Bartlett Publishers.

Valderrama, B., P. Oliver, A. Medrano-Soto, and R. Vazquez-Duhalt. 2003. Evolutionary and structural diversity of fungal laccases. *Antonie van Leeuwenhoek* 84:289–299.

Wolf, K. H. 2001. Yesterday's polyploids are the mystery of diploidization. *Nat. Rev. Genet.* 2:331–341.

Wolf, K. H. and D. Shields. 1997. Molecular evidence for an ancient duplication of the entire yeast genome. *Nature* 387:708–713.

15 Horizontal Gene Transfer

INTRODUCTION

From the first phylogenetic tree, drawn by Darwin in 1837, until the late twentieth century, all of the phylogenetic trees that were constructed were composed of branches that only intersected with one another at a node or bifurcation, assuming that all genes were inherited vertically from parents to progeny, or mother cells to daughter cells. Thus, the trees resembled ordinary physical trees (Figure 15.1). However, beginning in the 1940s, examples of pieces of DNA that could move from one organism to another were discovered, generally known as horizontal gene transfer (HGT). Some of the pieces remained separate from the chromosomes (e.g., plasmids), whereas others integrated into the chromosomes (e.g., prophage and human immunodeficiency virus). These pieces often had patterns of inheritance that differed from standard patterns of inheritance. Additionally, many species import fragments of environmental DNA from dead organisms and integrate the fragments into their genomes (Figure 15.2). Although some of the fragments may have originated in individuals of the same species, most often the DNA fragments are from dissimilar species. For example, species of Archaea and Thermotogales often are found living together in hot pools. When the genomes of species of *Thermotoga* have been analyzed, from 9% to 11% of the genes are from species of Archaea that have been imported into the *Thermotoga* genomes. Up to 90% of the genomes of some dinoflagellates consist of genes from foreign sources.

Often, organisms transfer sections of DNA to other cells directly, or viruses may move portions of genomes from one organism to another. Cells living in close proximity to one another, either because of nutrient availability or because the cells are dependent on one another, often exchange bits of their genomes with one another. Another form of gene transfer is via the formation of viable organisms from genetic combinations of dissimilar species, strains, or varieties (e.g., to form hybrids) to create new combinations of genes within the genome. Each of these phenomena may lead to gene transfers that cannot adequately be depicted using bifurcating trees. A few decades ago, these processes were thought to be minor novelties in evolution, and therefore, bifurcating trees were the norm. However, because the advent of genome sequencing methods and comparative genomic approaches, it has become apparent that these processes occur much more frequently than originally thought, and that HGT and chimeric genomes must be considered when evaluating the evolution of any organism or sets of organisms, especially when phylogenetic trees are being generated (Figure 15.3). Recent studies indicate that HGTs have occurred (and are occurring) recently, in the past, and among ancient organisms (Figure 15.4), such that it appears the processes have been ongoing for billions of years, perhaps from the very beginning of life of the Earth.

These changes can occur very rapidly and repeatedly. Transfer of genetic materials can occur from an individual of one species to another, or can occur between disparate species. The movement of these pieces of genetic material can be as small as a few nucleotides up to movements of entire genomes, as is the case when entire cells have been engulfed by other cells, ultimately ending up as organelles, where most or all of their genomes have been transferred into the nucleus of the host cell. Infectious changes occur throughout the life of an organism because of transfer and integration of viral and other sequences. Most of these produce no symptoms and may be virtually invisible. Others can cause severe symptoms, and some are fatal. Likewise, transposons and mobile introns can cause rapid changes in the cells in which they reside. Some changes are relatively benign, whereas others can be disruptive or fatal. However, the rearrangements to the

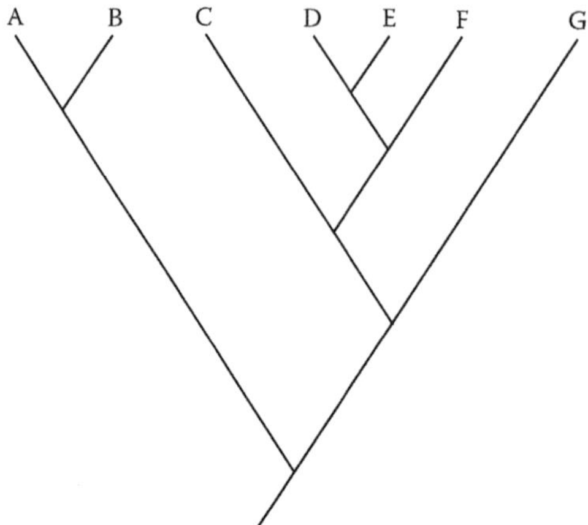

FIGURE 15.1 Phylogenetic tree where the genes are exclusively inherited vertically. Each branch begins at a bifurcation from the ancestral genetic line splitting into only two branches, indicating the divergence of the two resultant genes or taxa.

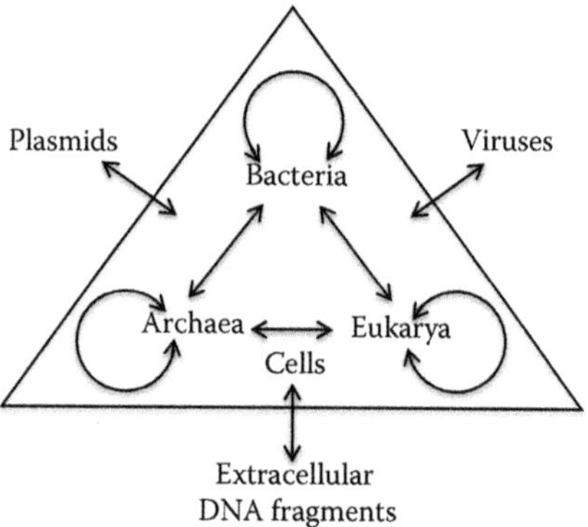

FIGURE 15.2 Diagram of the types of HGTs that have been demonstrated. Genes can be transferred among and between many types of cellular organisms, including Bacteria, Archaea, and Eukarya. They can also be imported as DNA fragments or plasmids from the environment, or enter the cells via virus infections.

genome for all of these genetic movements and insertions can provide opportunities for new combinations of genes and gene products, some of which can provide definite benefits to the organism, and may cause significant increases in the rates of change for specific genes or the species itself. This is clearly the case for eukaryotes that gained genes to help them detoxify and utilize oxygen when they engulfed the proteobacterium that eventually became mitochondria. It is true for plants that gained genes to harness solar energy when they engulfed cyanobacteria that became chloroplasts. Also, it is true for eukaryotes that gained introns that made possible new combinations of gene domains and encoding of multiple polypeptides in a single gene. Often, these rapid changes are difficult to spot, and they can confuse researchers if they are unaware of these frequent movements of genetic materials. After all, these movements are very different than the standard slow methodical changes and vertical inheritance patterns that are normally discussed in genetics and related fields. However, they provide yet another layer of complexity to studies in genetics and evolutionary biology.

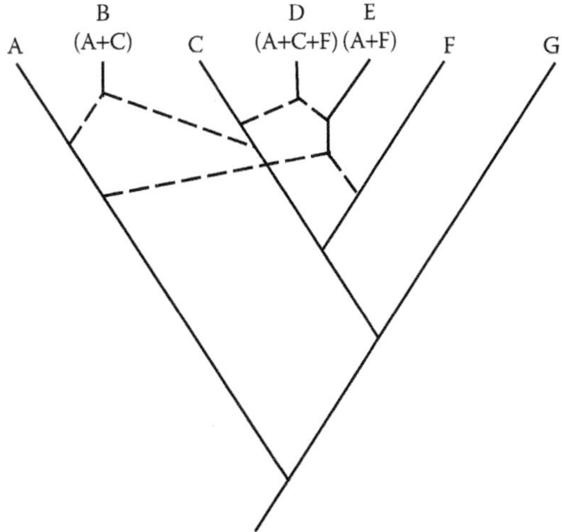

FIGURE 15.3 Phylogenetic tree illustrating both vertical transmission of genetic material, as well as HGTs. The genes or genotypes are indicated by the letters (A, B, C, etc.). When HGTs occur, the resulting organism contains a blend of the genes coming from each of the organisms (i.e., the receiving organism and the donating organism). The proportion of the genetic contribution depends of the type of HGT (e.g., endosymbiont, plasmid, and virus) and the size of the genome of the receiving organism.

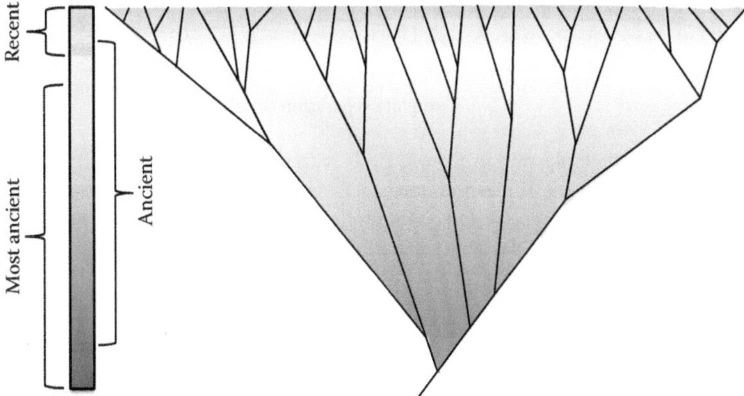

FIGURE 15.4 Phylogram of HGTs over evolutionary time. The transfer of specific genes has been occurring since very early in evolution, such that very ancient transfers (blue), ancient transfers (gray), and recent transfers (orange) have been traced using genomes from a large group of extant species. The result is an amalgamation of a large set of HGTs, showing that these transfers have been frequent and continuous throughout evolution.

PLASMIDS

Plasmids represent a varied group of mobile genetic elements (also discussed in Chapter 13). Most are circular, although some are linear. They are capable of transferring genes from one cell to another, including those that transfer genes between cells of the same species and those that can transfer genes to cells of different species. They are responsible for most of the transfers of antibiotic resistance among bacteria, as well as the transfer of many other genes among and between bacteria, archaea, and eukaryotes. Plasmids range in size from about a kilobase up to large plasmids that are hundreds of kilobases in size and carry dozens of genes. In some cases, they are large enough to be considered chromosomes.

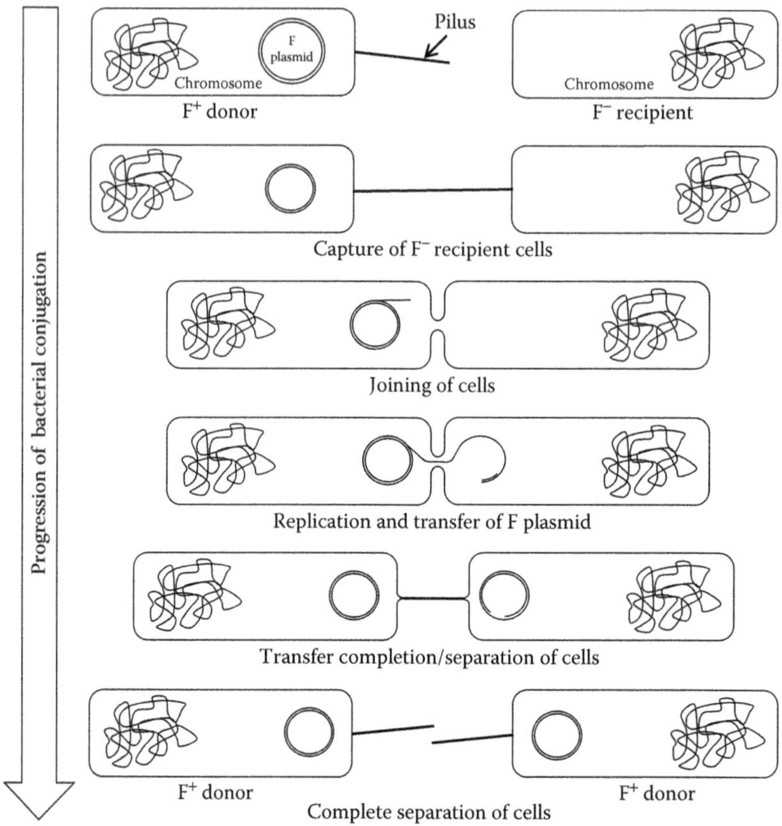

FIGURE 15.5 Transfer of the F plasmid in *E. coli*. Cells with an F plasmid (F^+ cells) use their pilus to seek out recipient cells (those without the F plasmid, termed F^- cells). Once found, they attach to the cells with their pilus. Then, they join with the F^- cells and begin replication and transfer the copy of the F plasmid to the F^- cells. Once the transfer is complete, the cells separate, and the F plasmid begins to express genes to form a pilus (for both cells) and to seek F^- cells.

The first demonstrated example of gene transfer was that of the F plasmid (or F factor) in *Escherichia coli* (Figure 15.5), reported in the 1940s. The F plasmid can move from one *E. coli* cell to another, while being replicated, such that each cell has a copy by the end of the process, although if the process is interrupted, only some of the genes are transferred. It can either remain as a plasmid or integrate into the *E. coli* chromosome, and integrates more often in high frequency of recombination (Hfr) strains of *E. coli*. The F^+ cells (i.e., those with the F plasmid) use a long extension of their cell, called a pilus, to connect with an F^- cell (i.e., one without the F plasmid). It then draws the cell close to itself and fuses with the F^- cell, which initiates the transfer of the F plasmid into the F^- cell. The process (called conjugation) is highly ordered, such that the replicating F plasmid is transferred as a linear molecule starting from a specific region of the plasmid and proceeding along the entire plasmid ending up with the section of the plasmid adjacent to the starting point. It is so precise and organized that the movement can be timed according to which genes have been transferred. Following transfer, the plasmid then circularizes. In Hfr *E. coli* strains, portions of the plasmid can be transferred to the chromosome. Also, integrated pieces of F plasmid, as well as adjacent pieces of the *E. coli* chromosome, can recombine with the F plasmid. This can lead to the production of variants of the F plasmids that carry different combinations of genes, which can then be transferred to other *E. coli* cells.

ColE1 is the native circular plasmid that exists in many strains of *E. coli*. It contains an origin of replication that allows it to replicate autonomously (i.e., extrachromosomally). One of the genes that exists on the plasmid encodes the protein that produces a toxin, called colicin. The toxin inhibits the growth of competing bacteria, thereby increasing the chances that the bacterium possessing the plasmid will survive. Other genes on the plasmid encode proteins responsible for the ability of the plasmid to move from one cell to another (i.e., mobility genes). ColE1 was one of the first plasmids to be used in the construction of plasmid vectors for recombinant DNA technology. Antibiotic resistance genes, as well as other genes, have been used to molecularly engineer the plasmid vectors. Antibiotic resistance genes, as well as genes responsible for resistance to heavy metals, are often naturally shuttled between bacterial cells (of the same species or different species), which can lead to rapid resistance to these toxins in bacterial populations.

The large tumor-inducing (Ti) plasmids that exist within the bacterium, *Agrobacterium tumefaciens*, which infects plant cells, have an interesting and complex mode of genetic transfer (Figure 15.6). They have also been used in research laboratories to transfer genes into plant genomes, something that the bacteria have been doing naturally for millions of years. The large plasmids (200–800 kb) contain a section, called the T-region, or T-DNA (some have more than one T-region), which is the only part of the genome that is physically transferred into the plant cell, and then into the plant cell nucleus, where it integrates into the chromosome. Other genes on the Ti plasmids are needed for replication and mobility. Only a single strand of the T-DNA is coated with specific proteins (encoded on the Ti plasmid) and then transferred from the bacterial cell to the host plant

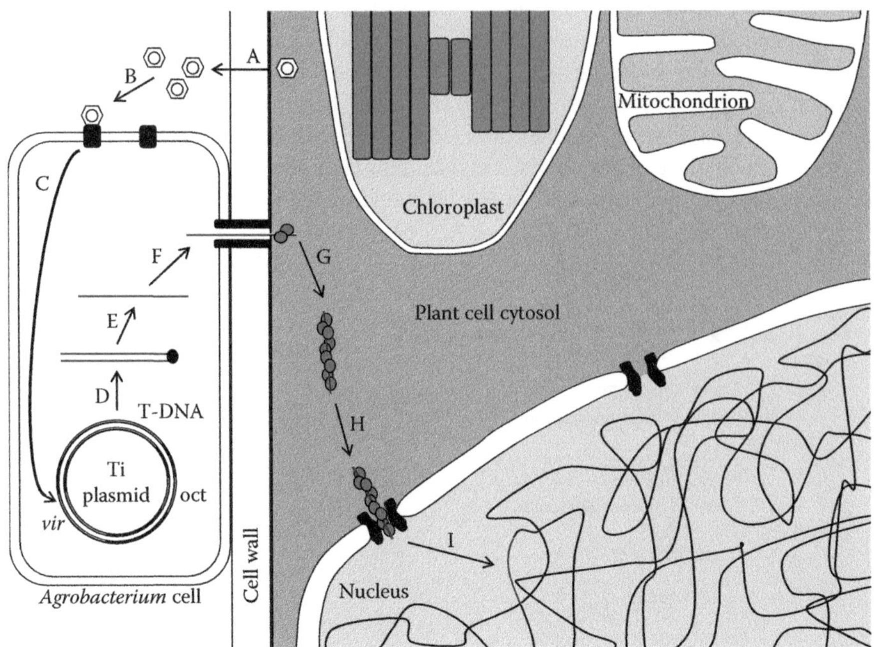

FIGURE 15.6 Transformation of a plant cell through transfer of T-DNA from *Agrobacterium tumefaciens.* (A) Plants exude phenolic compounds to inhibit the invasion and growth of pathogenic organisms. However, the phenolic compounds also signal *A. tumefaciens* to begin the transformation process. (B) Phenolic compounds bind to the receptors. (C) The signal is transduced into the cell, which signals upregulation of the *vir* (virulence) genes on the Ti plasmid. (D) Some of the virulence genes encode proteins that cut out the T-DNA region of the plasmid. (E) Other genes attach to the T-DNA and separate the strands. (F) Proteins encoded by other *vir* genes form a passage between the bacterial cell and the plant cell. (G) As the T-DNA is transferred into the plant cell, DNA-binding proteins (encoded by other *vir* genes) coat the T-DNA to protect it. (H) The coated T-DNA is then transported into the nucleus through the nuclear pore complexes. (I) The T-DNA integrates into the host chromosome, whereas the complementary strand is completed. In native *A. tumefaciens*, octopines are also produced by genes from another part of the Ti plasmid. These cause abnormal cell growth and division, which causes tumorous growth on the plant. Although the T-DNA usually integrates with nuclear DNA, it rarely integrates into chloroplast or mitochondrial genomes.

cell through a channel embedded in the membranes, which is produced from proteins encoded on the Ti plasmid. The coating and transfer into the host cell, followed by translocation to the host cell nucleus, is similar to how some filamentous viruses infect and transform their host cells. Thus, this plasmid has some characteristics of a plasmid and some characteristics similar to those of a virus.

VIRUSES

One of the well-known modes of HGT is that of integration of virus or bacterial sequences into the host genome, in a process called transduction. Many types of bacteriophage integrate into the host bacterial chromosomes, termed temperate bacteriophage (see Figure 13.4). This creates a prophage, which is a virus genome integrated into the bacterial host genome. The cell remains intact, and the expression of certain bacteriophage genes maintains the condition without lysing the cell. In most cases, the integration and excision of the bacteriophage genome is precise. However, this is not always the case, such that parts of the bacterial chromosome may be carried to another bacterial cell within the bacteriophage genome. Additionally, antibiotic resistance genes can sometimes be spread from one cell to another by bacteriophage. Bacterial virus genomes range in size from a few kilobases, containing a few genes, to those that have genomes of several hundred kilobases containing dozens of genes. Although temperate viruses frequently integrate into the host chromosomes, transfers of genes from nontemperate viruses to the host genome also occasionally occur.

Analogous transfers of virus sequences can occur in eukaryotes as well. This is one way that viral oncogenes (i.e., cancer-inducing genes) can be transferred from one cell to another, or from one individual/organism/species to another. They can cause cells to become malignant by integrating next to a cellular oncogene (i.e., a gene that when mutated can lead to the initiation of cancer), by expressing genes that interfere with the expression of normal cellular oncogenes, or by carrying mutant versions of cellular oncogenes, termed viral oncogenes, or v-oncogenes (Figures 15.7 and 15.8). The first

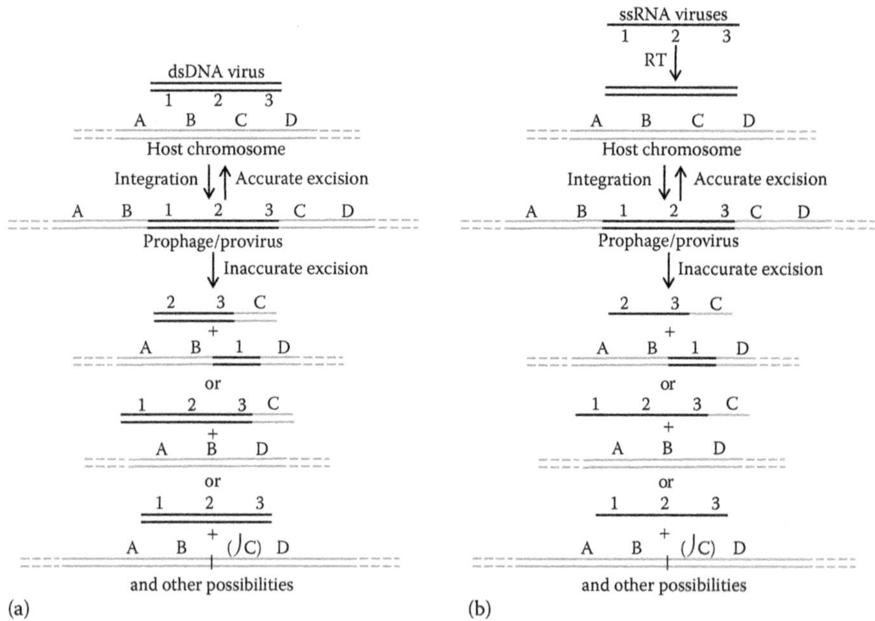

FIGURE 15.7 Illustration of some of the ways that DNA (a) and RNA (b) viruses (black) can affect the host genome (gray). Depending on where they integrate, they can affect the expression of the adjacent genes. For those viruses that can excise themselves from the host genome, the excision process can be imperfect, such that they excise, while leaving a portion of the virus genome in the host chromosome, and/or they can carry a portion of the adjacent host genes, which is then packaged into the new virus particles. Some excision events result in small local mutations at the point of excision that affect the associated gene, sometimes causing it to become a pseudogene.

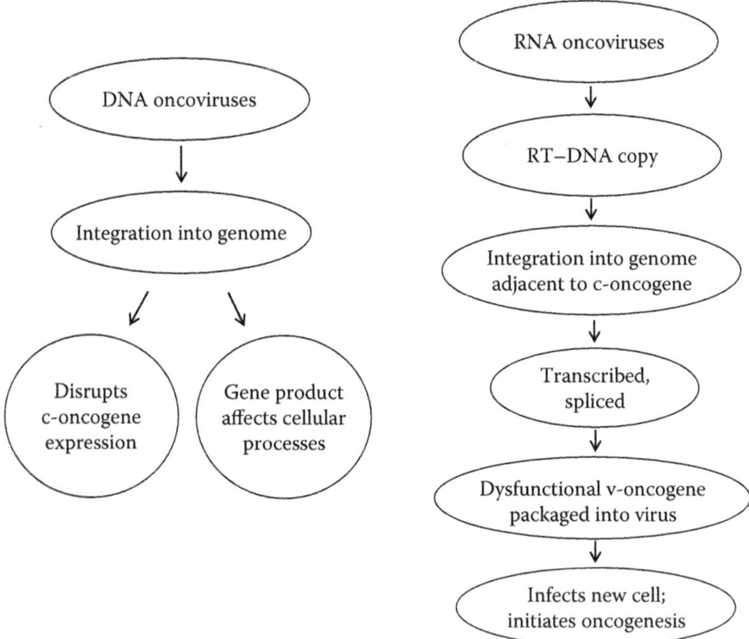

FIGURE 15.8 Effects of infection and integration by DNA and RNA oncoviruses. DNA oncoviruses either produce gene products that interfere with normal cellular control processes or integrate adjacent to cellular oncogenes (c-oncogenes), which affect their proper expression. RNA oncoviruses either integrate near a cellular oncogene, which interferes with their expression, or carry mutated versions of cellular oncogenes. These viral oncogenes (v-oncogenes) are processed versions of the c-oncogenes, having no introns. Additionally, they often are missing large sections of the c-oncogene or contain damaging mutations that encode faulty proteins. RT, reverse transcriptase.

oncoviruses discovered were retroviruses that carried altered versions of host genes. However, some DNA viruses can also cause malignant transformations of cells. About 12% of human cancers are caused by viruses. Another 6% are caused by other infectious agents (e.g., bacteria). However, it is more common to have genes that are not cancer inducing transferred among cells/organism/species. This occurs continuously throughout the lives of all or most organisms, including humans. Although most of these cause no changes in gene expression or cell function, some may integrate into regions of the genome that do lead to changes in the cells, resulting in changes in cell function that range from relatively benign to life threatening. Many of these integration events have had from small to large effects on the evolution of many species, including humans. Human genomes consist of approximately 8% virus sequences. Similar proportions are found in the genomes of other mammals, indicating a long evolutionary history of HGT events by viruses.

SYMBIONTS AND ORGANELLES

In the 1960s, the endosymbiotic theory for the origins of mitochondria and plastids was proposed (Figure 15.9). The theory was heavily challenged at the time, and acceptance spread slowly, and at times not very smoothly. After the development of DNA sequencing methods in the late 1970s, organellar sequence analyses indicated that the photosynthetic apparatus and genes from chloroplasts were similar to those in cyanobacteria, and the oxidative phosphorylation mechanism and genes in mitochondria were similar to the members of α-proteobacteria. However, the numbers of genes in the organelles were much lower than the numbers of genes in the free-living proposed ancestors. Subsequently, large numbers of genes common to the ancestors were found in the nuclear

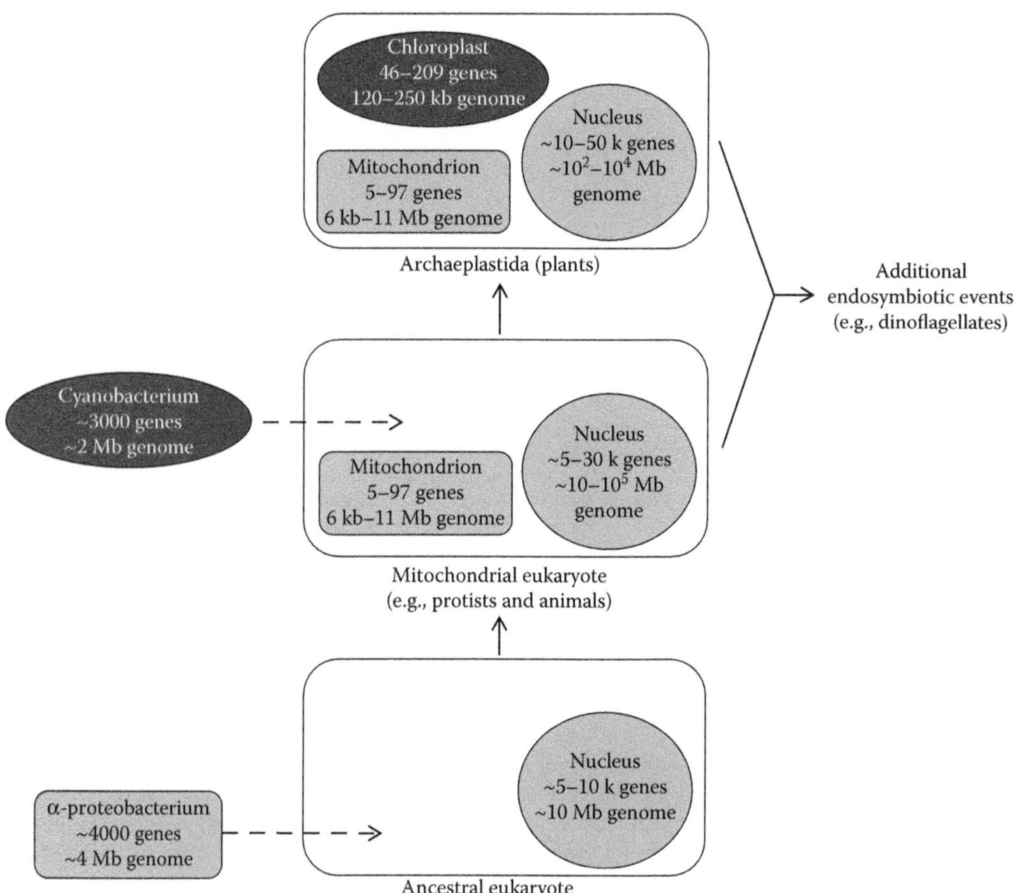

FIGURE 15.9 Major steps in the conversion of endosymbionts (α-proteobacterium and cyanobacterium) to organelles (mitochondrion and chloroplast, respectively). Approximately 2.2–2.4 billion years ago, an α-proteobacterium, with a DNA genome of several megabases with several thousand genes, started on its conversion from an endosymbiont of one of the first eukaryotes into a mitochondrion (lower two diagrams). That cell became the first mitochondrial eukaryote. Since that time, genes from the original α-proteobacterium have been transferred into the nucleus of the host cell. Approximately 1.9 billion years ago, a second event occurred in which a cyanobacterium began its transition from an endosymbiont of a mitochondrial eukaryote into a plastid (top diagram). This cell became the first member of the Archaeplastida. As with mitochondrial evolution, genes from the cyanobacterial endosymbiont were transferred into the nucleus of the cell. Over time, most of the genomes of the original bacterial endosymbionts were transferred into the nucleus. Those genes are now expressed from the nucleus, and the proteins produced from those genes are targeted to be imported into the mitochondria and plastids.

genomes of eukaryotes. By the end of the 1980s, it was widely accepted that the plastids and mitochondria had evolved from bacterial endosymbionts, and that there had been movement of most of their genes into the nuclei of the hosts during the evolution of Eukarya. Additional genomic sequencing studies began a progression of discoveries that demonstrated that horizontal transfers of genes and genomes were more frequent that initially believed, and that they have had huge influences on evolution. The discovery of mitochondrial, plastid, virus, bacterial, and foreign nuclear genes in the nuclei of almost all of the genomes sequenced as well as incongruencies in phylogenetic trees based on certain genes are two indicators that fragments (small and large) have moved from one place to another between and within organisms and cells. These movements often complicate the interpretation of comparative studies of sequencing results.

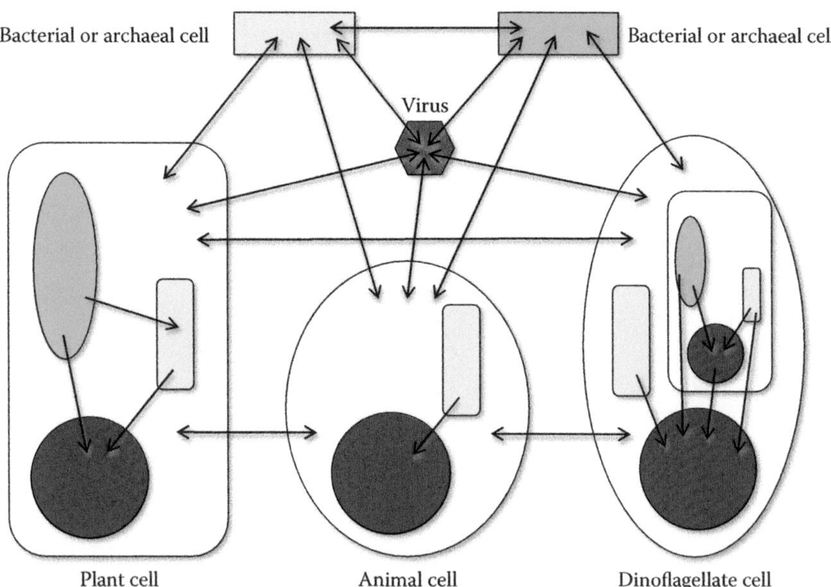

FIGURE 15.10 Summary of movements of genes among organsisms. Pieces of DNA (and RNA in viruses) are constantly moving from one cell compartment to another. Thus, there are movements of small and large pieces of DNA (including one or more genes) between Bacteria and Archaea of the same species (in the process of transformation) or different species (in the process of HGT), as well as between Archaea, Bacteria, and Eukarya. This transfer is sometimes mediated by viruses, which can also transfer some of their genes to the cells that they invade. In Eukarya, other transfers have occurred. Most of the genes that once existed in the ancestor of mitochondria have moved into the nucleus, where many of them are expressed. For plants, most of the genes from the original cyanobacterial genome have moved into the nucleus, many of which are expressed from the nuclear genome. Also, there have been a few transfers of genes between the chloroplasts and the mitochondria, but these have been rare. Among the members of the Chromalveolata, and especially in dinoflagellates, there have been additional gene transfers. These organisms contain organelles whose ancestors entered the cells as complete photosynthetic eukaryotic organisms (red and green algae). Therefore, originally they contained a nucleus, mitochondria, and chloroplasts. There have been movements within each of the organelles, as well as movements of DNA into the main nucleus of the cell.

All cells interact with other cells, whether of the same or different species. Just as humans depend on many other organisms for their fixed carbon, fixed nitrogen, and other nutrients, all organisms are in a constant search for their carbon, nitrogen, and other nutrients. Many organisms develop long-term close associations with neighboring species. In the process, the cells often exchange DNA fragments (Figure 15.10). Some of the exchanges result in permanent integration of genes into the foreign genomes. For mutualistic and endosymbiotic associations, an additional change is that each genome tends to lose genes as well, as the two species of cells become more dependent on one another as they share in the production of what both need to survive. This is because some of the gene functions are supplied by one of the organisms, whereas other gene functions are supplied by the other. The genes that are redundant are free to mutate and eventually disappear. Mitochondria and plastids originated from α-proteobacterial and cyanobacterial ancestors, respectively. In each case, genes from the original endosymbiont genomes have been transferred to the nuclear genomes of the hosts. In the case of the plant *Arabidopsis thaliana*, approximately 19% of the nuclear genome consists of genes that were imported from the cyanobacterial endosymbiont, which became the chloroplast. Based on the amount of DNA that this represents, there are approximately 3 times as many cyanobacterial genes in the nucleus as are in a cyanobacterial genome, although most are not expressed. For chloroplasts and mitochondrial genome, only five to a few hundreds of the original free-living bacterial genes remain in the organellar genomes. Presumably, this transfer allowed more streamlined and coordinated control of

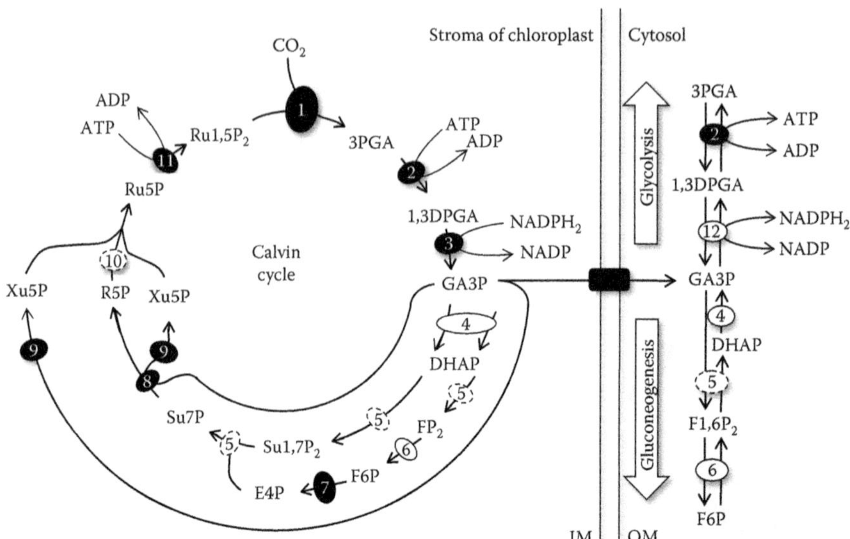

FIGURE 15.11 Origins of the enzymes used in the Calvin cycle and glycolysis in spinach (*Spinacia oleracea*). Enzymes are shown as elipses: Black indicates a cyanobacterial origin, white with solid lines indicates a mitochondrial (α-proteobacterial) origin, and white with dashed lines indicates an unknown origin (although most are probably from the genome of the original host cell). As expected, most of the genes for the Calvin cycle originated in the cyanobacteria. However, a few of the enzymes (e.g., triosphosphate isomerase, enzyme number 4; fructose bisphosphatase, enzyme number 6) originated in the mitochondria, as well as a few with unknown origins (likely to be from the nuclear genome). Glycolysis in the cytosol also exhibits a complex evolutionary pattern, where some enzymes that originated in the mitochondrial lineage, some from the nuclear lineage, and at least one (phosphoglycerate kinase, enzyme number 2) from the cyanobacterial lineage have produced the glycolysis pathway. This indicates that enzymes from different cell compartments and origins have undergone natural selection within the cell to yield chimeric pathways that include enzymes of mixed origins. Enzyme designations: 1, ribulose bisphosphate carboxylase/oxygenase; 2, phosphoglycerate kinase; 3, glyceraldehyde phosphate dehydrogenase; 4, triosphosphate isomerase; 5, fructose bisphosphate aldolase; 6, fructose bisphosphatase; 7, transketolase; 8, sedoheptulose bisphosphatase; 9, ribose phosphate epimerase; 10, ribose phosphate isomerase; 11, ribulose phosphokinase; and 12, glyceraldehyde phosphate dehydrogenase. Chemical abbreviations: $Ru1,5P_2$, ribulose-1,5-bisphosphate; 3PGA, 3-phosphoglycerate; 1,3DPGA, 1,3-diphosphoglycerate; GA3P, glyceraldehyde-3-phosphate; DHAP, dihydroacetone phosphate; $F1,6P_2$, fructose-1,6-diphosphate; F6P, fructose-6-phosphate; E4P, erythrose-4-phosphate; Xu5P, xylulose-5-phosphate; $Su1,7P_2$, sedoheptulose-1,7-bisphosphate; Su7P, sedoheptulose-7-phosphate; R5P, ribose-5-phosphate; and Ru5P, ribulose-5-phosphate.

gene expression in these cells. However, one of the other benefits has been in the evolution of chimeric or hybrid metabolic pathways (Figure 15.11). For example, in plants, the reductive pentose phosphate (Calvin) cycle is a hybrid cycle, in that it uses some enzymes common to the cyanobacterial ancestors, some that are from the host genome, and some enzymes that were originally within the mitochondrial (i.e., α-proteobacterial) genome. Similarly, glycolysis and gluconeogenesis utilize enzymes that originated from the three ancestral genomes. Glycolysis in the cytosol of plants utilizes enzymes encoded by genes that are from the original host genome (prior to the endosymbiotic events), those originating from the mitochondrial genome, and those originating from the plastid genome (Figure 15.12).

Glycolysis in animals utilizes enzymes from the host genome and the mitochondrial genome. Therefore, during the evolution of these organisms, selection favored utilization of only one of the isozymes in each of the biochemical processes (Figure 15.13). Dinoflagellates and other members of the Chromalveolata and related phyla contain other organelles that were originally red algae or green algae (depending on the species), which are the result of secondary and tertiary endosymbiotic events. These organisms have even more complex chimeric biochemical pathways.

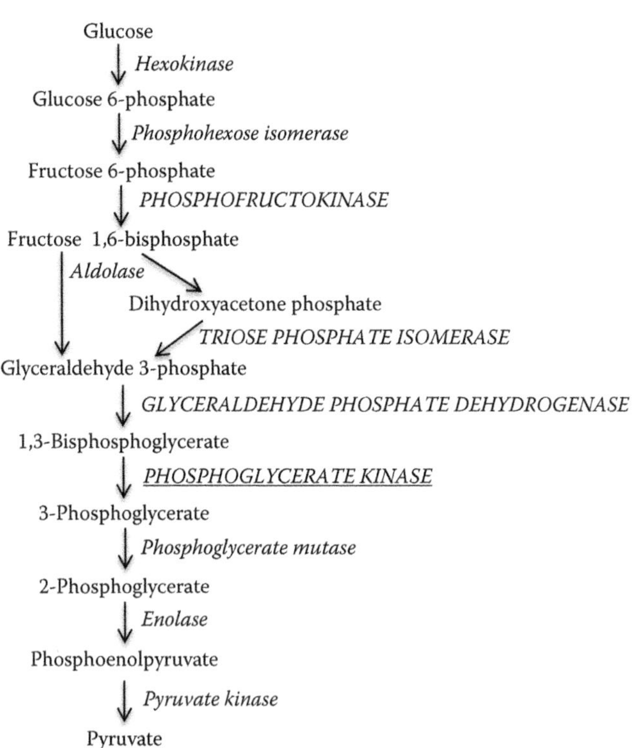

FIGURE 15.12 Glycolysis in spinach cells, showing the chimeric nature of the pathway. Reactants and products are in standard font. Enzymes encoded from the original eukaryotic nuclear genes (with archaeal ancestry) are in italics, whereas enzymes whose genes originated in the mitochondrial ancestor are in italic capital letters, and those that have a cyanobacterial origin are in underlined italic capital letters. Animals have a similar chimeric pathway, except the pathways have no cyanobacterial components.

PARASITES AND PATHOGENS

Organisms that live on other organisms can cause mild to life threatening disease effects on the host organisms. They can also transfer some of their genes to the host, and some have been shown to acquire and integrate host genes into their genomes, some of which allow the organisms to evade some of the host defenses. Oomycetes (members of the kingdom Chromalveolata) contain genes in their genomes that originated in their plant hosts. These genes are used by the pathogen to evade the host defenses, thus allowing it access to the plant cells. They also have acquired transporters that cause the plant cells to release some of the cell constituents which the pathogens then take into their cells. In other words, they transfer nutrients out of the host cell and import those nutrients into their cells using plant gene products that have integrated into their chromosomes.

Chagas disease is caused by *Trypanosoma cruzi* (Figure 15.14), which is a member of the kingdom Excavata (phylum Euglenozoa). This is an organism with an interesting evolutionary history and can be devastating to its hosts. These organisms are capable of transferring some of their DNA to the host organism (humans and other mammals) during infection, which causes permanent genomic changes, some of which can lead to cell and tissue damage in the host. Members of the Excavata originated from the endosymbiotic association of a nonphotosynthetic unicellular eukaryote and a green alga (Figure 15.15). As such, they have mitochondria, and many have membrane-bound organelles and/or genes that are descendant from the green alga endosymbiont. Some, such as species of *Euglena*, contain chloroplasts and have remained photosynthetic. They have an organelle that contains a plastid, which is the descendant of the original green alga, although the algal nucleus

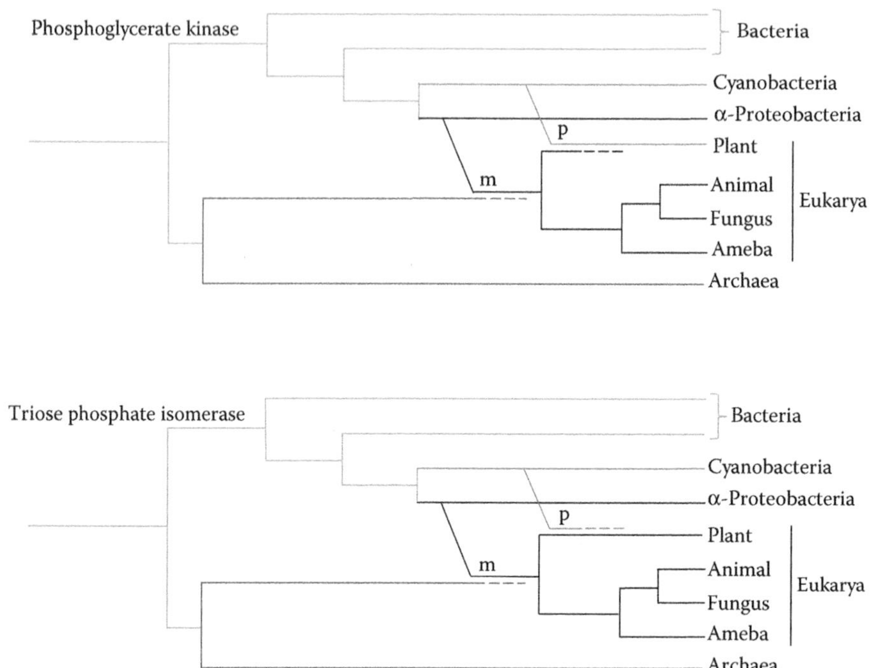

FIGURE 15.13 Phylogenetic models for two enzymes, based on the origins of each of the enzymes in Bacteria, Archaea, and several Eukarya. Diagonal lines indicate HGTs that occurred when the endosymbiosis was established. For both enzymes, the ancestral genes in Bacteria and Archaea diverged and independently mutated into bacterial and archaeal versions. Within the Bacteria, further diversions and independent evolutionary paths occurred for each of the gene versions. When the α-proteobacterium entered into an endosymbiosis with the initial Eukarya, each organism (the archaeal cell and the α-proteobacterial cell) had full complements of genes that encoded the enzymes for glycolysis. As the organelle evolved, selection favored the α-proteobacterial versions of the gene (black lines), which allowed the mutation, and eventual extinction of the archaeal versions (dark gray lines) in the eukaryotic lineage. This is the arrangement in animals, fungi, and amebae. However, in plants, the endosymbiotic cyanobacterial ancestor of the plastids contributed yet another set of genes that encoded the enzymes for glycolysis. Inheritance and evolution of these genes varies with the particular gene. For the phosphoglucokinase (PGK) gene, the cyanobacterial version (light gray line) is the one that survives in plants today, whereas the archaeal and α-proteobacterial versions have gone extinct (dashed lines). However, for the triose phosphate isomerase gene, the mitochondrial version (black line) has survived, whereas the archaeal and cyanobacterial versions have gone extinct in plants.

and mitochondria are absent. However, many of the genes from the algal nucleus, mitochondria, and plastids now exist in the nuclear genome of the organism. Trypanosomes are also members of an Excavata lineage whose members have become parasites and are no longer photosynthetic. Although they have no photosynthetic organelle, the nuclear genome of the organism contains algal chloroplast, mitochondrial, and nuclear genes that were transferred during the evolution of these organisms. They also have different forms of DNA that are enclosed in organelles separate from the nucleus. This DNA is contained within a long mitochondrion that is physically attached to the large flagellum by a protein structure called the tripartite attachment complex (Figure 15.14). Within the mitochondrion and immediately adjacent to the tripartite attachment complex is a region called the kinetoplast. This contains a few dozen DNA molecules called maxicircles (composed of DNA that is from 20 to 40 kb in length) and thousands of smaller minicircles (usually 0.5–1 kb each), bundled together by a protein complex. The maxicircles contain the genes needed by the mitochondrion for parts of its function, whereas the minicircles include sequences that aid in expression of the maxicircle genes.

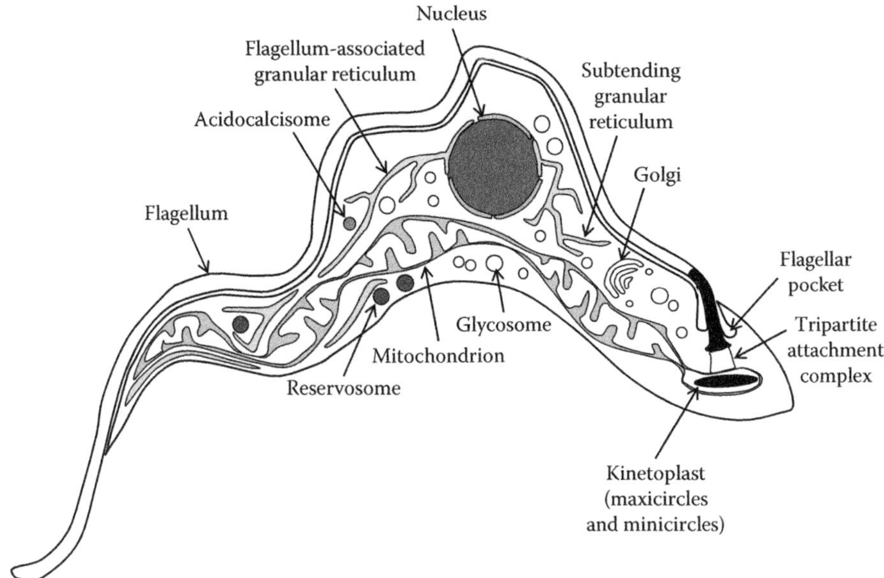

FIGURE 15.14 Diagram of a *T. cruzi* cell. The cell has one flagellum that is attached along the length of the cell and protrudes from the cell at the flagellar pocket. It is attached to the kinetochore region of the mitochondrion through the tripartite attachment complex. In addition to the nucleus and mitochondrion, there are numerous membrane-bound compartments, including glycosomes, reservosomes, acidocalcisomes, Golgi, and an extensive granular reticulum that is physically continuous with the nuclear membrane. The *T. cruzi* cells infect humans through insect bites by kissing bugs, *Triatoma infestans* and *Rhodnius prolixus*. Once released into the bloodstream, the parasites localize in muscle tissue and cause local damage. DNA from the kinetoplast of the *T. cruzi* cells can then be physically transferred into the host cell nuclei, where the kinetoplast DNA (kDNA) integrates into the host genome. The kDNA consists of maxicircles and minicircles, but only the minicircle kDNA becomes integrated into the host genome. From there, the disease becomes chronic, even when all of the *T. cruzi* cells are eliminated from the patient.

Transmission of *T. cruzi* to a human host is by about 12 species of triatomine insect vectors, called kissing bugs, which require blood meals. They usually bite near the eye of the host, and then defacate, which transfers the trypanosome. Redness and swelling occur around the bite, and the trypanosome usually localizes to muscle cells, including cardiac muscle cells, and causes cell damage. Initially, the organisms reproduce and continue to cause disease symptoms, including malaise, fever, chills, weakness, and fatigue. At this point, often drugs are administered to rid the patient of the parasite. The drugs can rid the host of the parasite. However, if treatment is not administered early in this process, kinetoplast DNAs, primarily from the minicicles of the kinetoplast, are transferred and integrated into the genome of the host. These are permanent transfers and no treatments have been developed that rid the patient of the kinetoplast DNA. Integration of the kinetoplast DNA often changes gene expression in the affected cells, which can cause disease symptoms to persist for decades. These changes often cause muscle cells to die, such that often the patients slowly deteriorate as their muscles, include cardiac muscles, and are destroyed. Also, the kinetoplast DNA can be transferred to sperm and egg cells, as well as to a developing embryo or fetus. Children can be born with Chagas disease symptoms without ever having been infected with trypanosomes and can pass the disease onto their progeny, again without ever having been infected with the trypanosome. The major site of integration is within the retrotransposable element long interspersed nuclear element-1 regions. These regions can make duplicate copies of themselves, and adjacent regions, and therefore, the integrated minicircle kinetoplast DNA copies can increase in number over time.

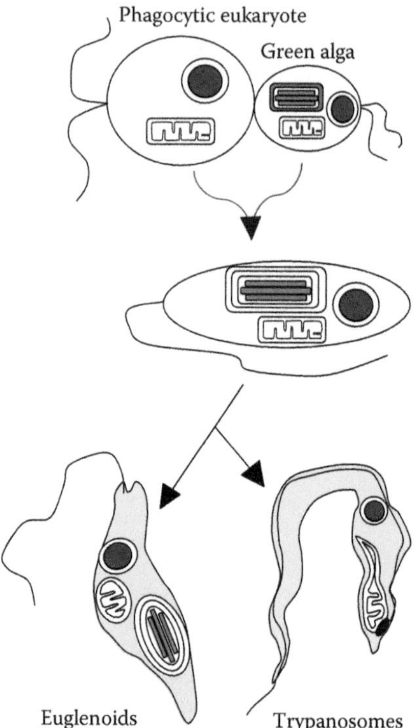

FIGURE 15.15 Evolutionary history of members of the Excavata. The first members of this kingdom were formed when a green algal cell became an endosymbiont with a nonphotosynthetic phagocytic eukaryote. Since that time, several distinct taxonomic groups have emerged. One group remains photosynthetic (Euglenoids), an example of which is *Euglena* sp. Another group consists of species that have become nonphotosynthetic, but parasitic (e.g., trypanosomes). They obtain their nutrients by living off the cells of other species, such as humans and other animals. However, they still have genes that have cyanobacterial/plastid origins. Most of the plastid genes are no longer expressed.

ORIGIN OF GRAM-NEGATIVE BACTERIA

Bacteria can be divided into two distinct groups: those with a single membrane, called monoderms, and those that have two membranes, called diderms. Most monoderms are Gram positive, because they have a thick peptidoglycan layer that binds to the Gram stain (crystal violet). However, most diderms are Gram negative, because the thin peptidoglycan layer is within the periplasmic space between the two membranes, and therefore, the dye cannot reach the peptidoglycan. From an evolutionary standpoint, it would make sense for the monoderms to form a coherent grouping and for the diderms to form another major grouping. Each group would be said to be monophyletic, having a single unified phylogenetic group. However, when conserved genes, such as the ribosomal RNA (rRNA) small subunit genes are used for bacteria, the monoderms and diderms each appear in various locations on the phylogenetic tree (Figure 15.16a). They are said to be polyphyletic, meaning that the bacterial phylogenetic tree is composed of a mixture of the two types of organisms. When other conserved genes are used, the trees often show dramatic changes in the positions of some of the taxonomic groups of bacteria. For decades, this has been ignored, left as unexplained, or thought to be a consequence of the fact that bacteria have evolved over a period of more than 3.5 billion years.

Recently, comparative genomics studies (i.e., simultaneous comparison of dozens or hundreds of genes from the genomes of many species) have shed some light on the possible solution to the apparent polyphyletic nature of monderms and diderms. Analysis of dozens of genes from the major bacterial phyla, including monderms and diderms, has been performed. The genes that were compared

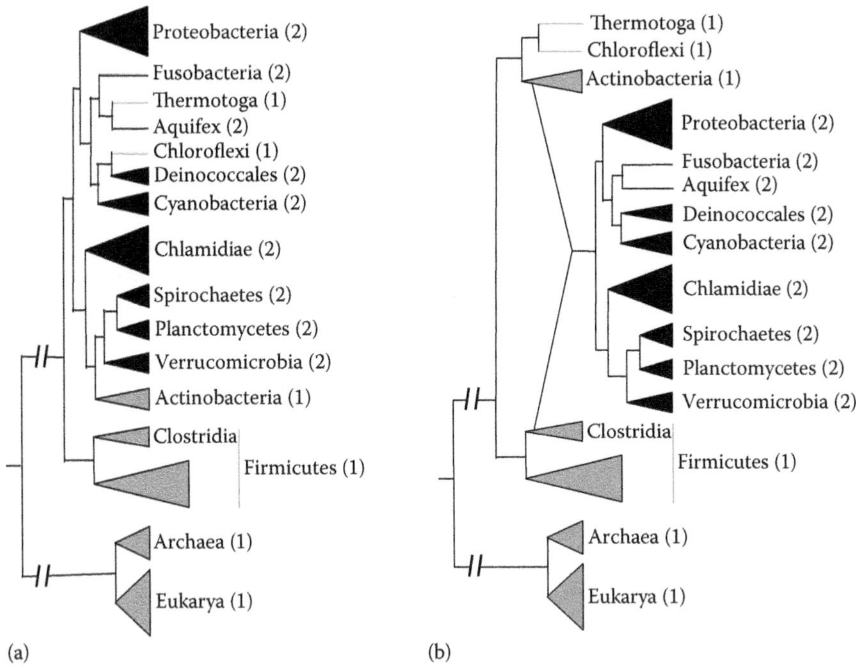

FIGURE 15.16 Possible origin of diderms (Gram-negative bacteria; designated by a2). If a phylogram of bacteria is constructed based on rRNA genes (or other genes), monoderms that have one cell membrane (primarily Gram positive species) and diderms that have two cell membranes (primarily Gram-negative species; designated by a1) are polyphyletic (phylogram[a]). This means that on the tree the two groups fail to form coherent groups, but are instead interdigitated on the tree. Also, when individual gene trees are constructed, very different trees result. Genome comparisons have shown statistically that diderms appear to contain a blend of genes from two groups of monoderms, specifically Actinobacteria and Clostridia (members of the Firmicutes). When a phylogenetic tree is constructed with this in mind, and using sets of gene sequences from these two groups of organisms (as well as several others), the diderms form a monophyletic clade (i.e., a coherent branch), as do the monoderms (b).

were those that existed in all of the genomes that were examined, and phylogenetic distances were calculated for each of the sets of orthologous genes. The most highly supported model was one in which two monoderms had combined to form all of the diderms. Put more simply, one species of monoderm was engulfed by another species of monoderm, and this evolved into the first diderm, which then led to a diversification of this lineage to become all of the diderms. The two monoderms appear to be one member of the Actinobacteria and one member of the Firmicutes (specifically, a Clostridia). When a phylogenetic tree is constructed based on this scenario, the monoderms form a monophyletic group, and the diderms form a monophyletic group (Figure 15.16b). This endosymbiotic event likely occurred at least three billion years ago, because some members of the diderms were already present by that time. This is yet another example showing that evolution cannot always be depicted as a bifurcating tree.

SIGNS OF HGT

Although HGT is common, most inheritance of genes and genomes is via vertical means. How does one determine whether HGT has occurred? The initial indications of HGT were thought at first to be problems with sequence data, the analytical tools, or simply the lack of resolution in the data set. That is, alignments of some genes yielded phylogenetic trees of one topology, and alignments of other genes from the same set of species yielded a different topology. In some cases, the differences were small, but in other cases there were major incongruencies. For example, the rRNA gene alignments almost always yield trees that appear to be consistent with morphological, physiological, ecological, and other information about the taxa begin compared. Humans are close to gorillas, with dogs further separated, mushrooms further than that, petunias more distant than that, and so on. Another gene set might place humans and fungi closer together, possibly indicating that there might have been an HGT of that gene transferred from a fungus into humans. Another indication is when the percentages of nucleotide types differ from one section of the genome to another. Most genomes are somewhat consistent in the percentages of AT and GC pairs that they have. When a piece of DNA is transferred from one species to another, many times the GC percentages of the organism receiving the DNA differ from those of the transferred DNA. These can be located simply by plotting the GC percentages along the DNA. Inserted foreign DNA becomes apparent as islands of DNA with GC percentages that differ from the surrounding DNA.

Another sign of HGT events is the relatively rapid acquisition of complex traits. Fungi are osmotrophic. They secrete chemicals and enzymes that break down complex organic compounds, and then import the resulting needed nutrients. In the case of pathogens, they also secrete proteins that cause the cells of the host organism to leak cellular constituents into the space between the fungus and the host cell. The fungal cells then import the resulting nutrients. However, most protists are phagotrophs. This means that they obtain their nutrients by enveloping particles and cells containing organic molecules with their plasma membrane. Once inside the cell, they digest the enveloped constituents (whether they are bits of organic materials or whole cells), and then use the resultant organic molecules for their metabolic needs. This method of obtaining food works well when the food is in small bits or small cells. However, it fails if the food is enclosed in a multicellular organism. In that case, osmotrophy is effective, which is why so many species are pathogens of multicellular organisms (e.g., they infect plants and animals). One group of protists, called oomycetes, consists of species that are osmotrophs, and many are pathogens of plants or animals. The switch from phagotrophy to osmotrophy requires the addition of at least a few dozen genes. Therefore, this type of change requires coordination of many evolutionary steps, which would not be expected to occur. *Phytophthora* species, primarily, *P. infestans*, *P. sojae*, and *P. ramorum*, cause millions of dollars of damage to crops and trees each year, primarily infecting potatoes, soybeans, and oaks, respectively. Species of *Phytophthora* and related genera (some of which cause monetary losses in aquaculture) have all acquired genes for osmotrophy from an ancestor that acquired these from a fungus in an ancient HGT event. This allowed each of the descendants to exploit new ecological niches, specifically obtaining nutrients from large multicellular host organisms.

Plants have also acquired genes via HGT during their more than 400 million years of evolution. The first important event was that of the successful addition of chloroplast genes from the cyanobacterial endosymbiont that occurred approximately 1.9 billion years ago. This allowed the previously nonphotosynthetic cells to become photosynthetic autotrophs. Much later, as they became more complex multicellular organisms, they acquired genes from several HGT events from bacterial and fungal sources. This includes auxin biosynthesis, cuticle and epidermis development, vascular development, lateral root development, herbivore resistance, and stress tolerance. More recently, they have acquired additional genes for DNA replication and repair, as well as those for pathogen resistance. In addition to these HGT events that have occurred in multicellular organisms, species of bacteria have been incorporating foreign DNA sequences frequently for billions of years. These have allowed them to detoxify toxins, to grow in areas of high metal concentrations, to fix nitrogen and carbon, and to occupy many environmental niches that would be impossible without the ability to incorporate novel sets of genes into their genomes.

INTRONS

Introns are found in a broad range of organisms (Figure 15.17). They have been inherited both vertically and horizontally. Group I introns are found in almost all of the major branches of organisms, including bacteria, archaea, eukarya, and viruses. They appear in some nuclear genes, with the

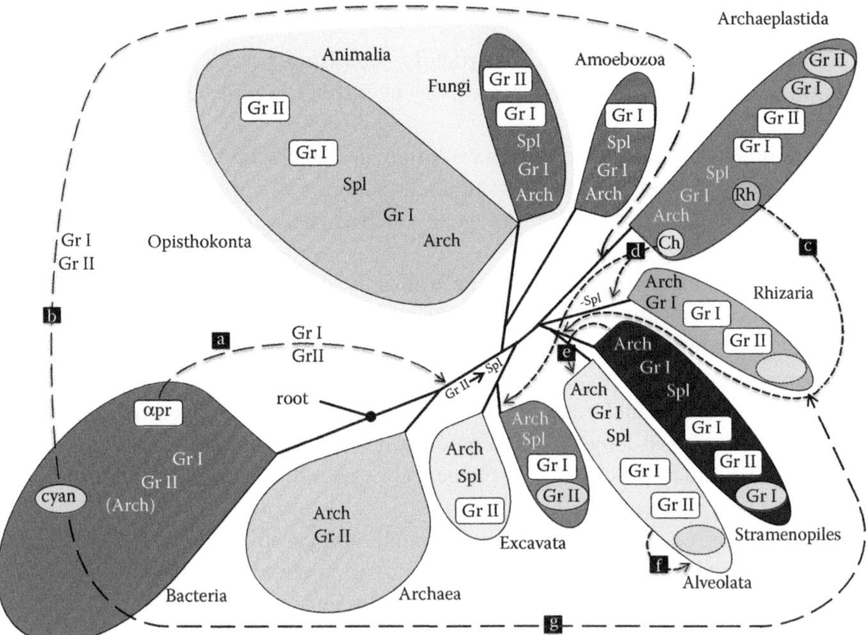

FIGURE 15.17 Phylogenetic tree of organisms overlaid with the presence of introns. Archaeal (Arch) and group I introns (Gr I) appear to have predated group II (Gr II) and spliceosomal (Spl). Group II introns may have been derived from group I introns, and spliceosomal introns appear to have been derived from group II introns. In addition to vertical inheritance, several rounds of HGTs (dashed lines) occurred during the endosymbiotic events that led to the evolution of various groups of eukarya. Labels in white rectangles indicate introns within mitochondrial genomes. Labels in green ovals indicate introns within plastid genomes. All other labels indicate introns in the genomes of bacteria, archaea, or nuclear eukarya. HGTs, primary endosymbiosis of an Archaea and an α-proteobacteria; b, primary endosymbiosis of a protist and a cyanobacterium; c, secondary endosymbiosis of a protist and a red alga (Rh); d, Secondary endosymbiosis of protists and green algae (Ch); e, secondary endosymbiosis of a stramenopile and an Alveolata; f, tertiary endosymbiosis of two dinoflagellates; g, primary endosymbiosis of a protist and a cyanobacterium.

exception of nuclear genes of members of Excavata. They have also been found in mitochondrial genes and those of some chloroplasts. The distribution of group II introns is limited primarily to bacteria, archaea, and organelles. They have not been found in nuclear genes. However, spliceosomal introns are common in the nuclear genes of eukarya. The routes for each of the introns into each of the genomes are partly vertical and partly horizontal. Group II introns may have evolved from group I introns in the progenote of bacteria and archaea, although group I introns have not been found in archaea. They then moved into eukarya via HGT through the endosymbiosis of an α-proteopacterium that became the mitochondrion. They may have been transferred secondarily to the Chromalveolates and Excavata via additional endosymbiotic events. Similarly, group I and group II introns were imported into plant genomes via endosymbiosis of cyanobacteria that became the plastids. Some of these introns were transferred into the host nuclei as well, although some of the introns may also have been inherited vertically into the major eukaryotic groups. Spliceosomal introns most likely evolved from group II introns, but became dependent on spliceosomes. It is unclear whether they all were inherited vertically, of whether some moved into the nucleus as group II introns. The evolution of archaeal introns is less clear. They are most common in transfer RNA genes and splice by a different pathway from all other introns. However, their splicing pathway is very similar to that of rRNA processing, and therefore, rRNA processing in all organisms might be a remnant of archaeal introns in those regions. If so, then they likely evolved in the progenote of all organisms and have been inherited primarily (or completely) vertically. Nonetheless, inheritance of introns has been complex.

KEY POINTS

1. Although most genes are inherited vertically (i.e., from parents or mother cells), HGT (gene transfers between different species) has occurred often throughout the evolutionary history.
2. A large number of important events in evolution have occurred because of HGTs.
3. Cells can integrate foreign sequences into their genomes by importing them from the environment around them, from infectious agents (such as viruses), as plasmids, and from other mechanisms.
4. Eukaryotes have experienced large gene transfers several times during evolution when endosymbiotic organisms became organelles within the eukaryotic cells. Most of the DNA sequences of the organelles have been transferred onto nuclear chromosomes.
5. Some parasites, such as *T. cruzi*, transfer parts of their genomes into the host cells, which causes a permanent transformation of the host cells, such that disease symptoms (e.g., Chagas disease) persist long after the parasite has been eliminated from the host. If the genes integrate into the nuclear DNA of sperm or egg cells, the disease can be inherited by the progeny.
6. Bacterial diderms (i.e., those having two cell membranes, as in most Gram-negative bacteria) may have been the result of an ancient symbiotic even between two monoderms (i.e., those with a single cell membrane, as in most Gram positive bacteria), specifically an ancient member of the Actinobacteria and an ancient member of the Clostridia (Firmicutes).
7. HGTs are suspected when phylogenetic trees for certain genes are incongruous with trees for other genes or with other types of taxonomic classifications. Another characteristic is where islands of genes within a genome have very different percentages of GC than do the surrounding regions.
8. Inheritance of introns has included both vertical and horizontal mechanisms, and their evolution has been long and complex.

ADDITIONAL READINGS

Bhattacharya, D., D. Simon, J. Huang, J. J. Cannone, and R. R. Gutell. 2003. The exon context and distribution of Euascomycetes rRNA spliceosomal introns. *BMC Evol. Biol.* 3:7.

Dagan, T. and W. Martin. 2009. Getting a better picture of microbial evolution en route to a network of genomes. *Philos Trans R Soc Lond B Biol Sci.* 364:2187–2196.

Gregory, T. R. 2005. *The Evolution of the Genome.* San Diego, CA: Elsevier Academic Press.

Hecht, M. M., N. Nitz, P. F. Araujo, A. O. Sousa, A. de C. Rosa, D. A. Gomes, E. Leonardecz, and A. R. L. Teixiera. 2010. Inheritance of DNA transferred from American trypanosomes to human hosts. *PLoS ONE* 5(2):e9181. doi:10/1371/journal.pone.0009181.

Lake, J. A. 2009. Evidence for an early prokaryotic endosymbiosis. *Nature* 460:967–971.

Lylle-Andersen, J., C. Aagaard, M. Semionenkov, and R. A. Garrett. 1997. Archaeal introns: Splicing, intercellular mobility and evolution. *TIBS* 22:326–331.

Martin, W. and C. Schnarrenberger. 1997. The evolution of the Calvin cycle from prokaryotic to eukaryotic chromosomes: A case study of functional redundancy in ancient pathways through endosymbiosis. *Curr. Genet.* 32:1–18.

Moreira, S., S. Breton, and G. Burger. 2012. Unscrambling genetic information at the RNA level. *WIREs RNA.* doi:10.1002/wrna.1106.

Nitz, N., C. Gomes, A. de Cássia Rosa, M. R. D'Souza-Ault, F. Moreno, L. Lauria-Pires, R. J. Nasciemento, and A. R. L. Teixiera. 2004. Heritable integration of kDNA minicircle sequences from *Trypanosoma cruzi* into the avian genome: Insights into human Chagas disease. *Cell* 118:175–186.

Rogers, S. O., Z. H. Yan, K. F. LoBuglio, M. Shinohara, and C. J. K. Wang. 1993. Messenger RNA intron in the nuclear 18S ribosomal DNA gene of deuteromycetes. *Curr. Genet.* 23:338–342.

Rogozin, I. B., L. Carmel, M. Csuros, and E. V. Koonin. 2012. Origin and evolution of spliceosomal introns. *Biol. Direct* 7:11.

Sagan, L. 1967. On the origin of mitosing cells. *J. Theor. Biol.* 14:225–274.

16 Development
Part I—Cooperation among Cells

INTRODUCTION

All cells have programmed limits on their size and how they react to things surrounding them. Unicellular organisms are able to sense certain chemicals and signals from other organisms in their vicinity. When cells form communities, they also communicate in order to maintain the ordered community. Cells often divide and the daughter cells are different. When they divide but do not separate, this is the beginning of multicellularity, a characteristic that is polyphyletic (Figure 16.1), that is, it evolved independently in many lineages. Because multicelluarity involves complex cell communication, it is likely that the basic genes and proteins that led to multicellularity existed long before organisms became multicellular. Also, it is possible that the characteristic has been horizontally transferred one or more times during evolution, or alternatively that it has been lost on many branches. However, for acquisition via horizontal gene transfers, there would have been multiple transfers of large gene clusters, which is unlikely. Also, there is a great deal of evidence that the basic mechanisms that led to multicellular development existed in bacteria and archaea very early in the evolution of life on the Earth. Bacteria and archaea live their lives surrounded by other cells of the same, as well as different, species. Some form filaments where many cells are joined together in multicellular assemblages, and some produce more than one cell type. Molecular mechanisms have evolved in these organisms that allow them to chemically recognize when there are other cells like themselves in the same area, when there are different cells in the area, where cells are signaling that there is food present, where the food is, and where the dangers are (e.g., toxins, predators, heat, cold). These chemical sensory mechanisms are the basis for interacting with their environment and other cells, and they provide the basis for interactions within a multicellular organism. Throughout evolution, the cells that possessed the genes that express these functions were favored in competitions for nutrients and for avoidance of dangers. In fact, some bacteria form multicellular bodies and different types of cells based on their interactions with the environment and other cells. Some bacteria form long chains of cells (e.g., *Streptococcus* spp.) and some form spores, whereas others form more than one cell type attached to each other (e.g., *Caulobacter crescentus*, Figure 16.2). The multicellular forms have been important to the survival and propagation of these species, and cell-to-cell signaling is vital to these processes.

Many types of bacteria form linear chains of cells. Some types of bacteria form branching chains of cells that can appear similar to fungal hyphae. These forms of growth are modulated by chemical signals that signal the timing and direction of cell division. If this were not the case, the cells might divide at random times and in random directions, which would form a disorganized lump of cells resembling a tumor. Some types of bacteria are capable of forming more than one cell type under different conditions, and a few species are able to produce more than one cell type, although joined to a different cell type, thus appearing to be a fundamental type of multicellular organism. One of the most well-studied species is *C. crescentus* (Figure 16.2). One cell type (stalked cell) produces an appendage that glues the cell to a substrate in a lake or stream. It then begins a process of segregation of its duplicated chromosome and gene products, producing a different type of cell (called a swarmer cell) at the end of the stalked cell. The new cell also produces a flagellum at the end of the cell opposite of the region where the two cells will divide. When cell division is complete, the

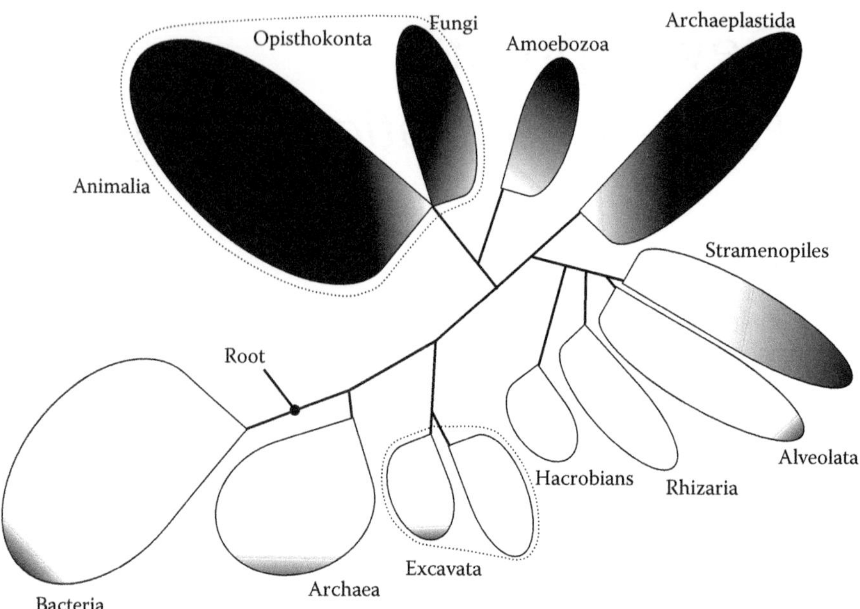

FIGURE 16.1 Phlyogram indicating the proportions of multicellular species (shaded) and single-celled species (unshaded) within each of the major taxonomic groups of organisms.

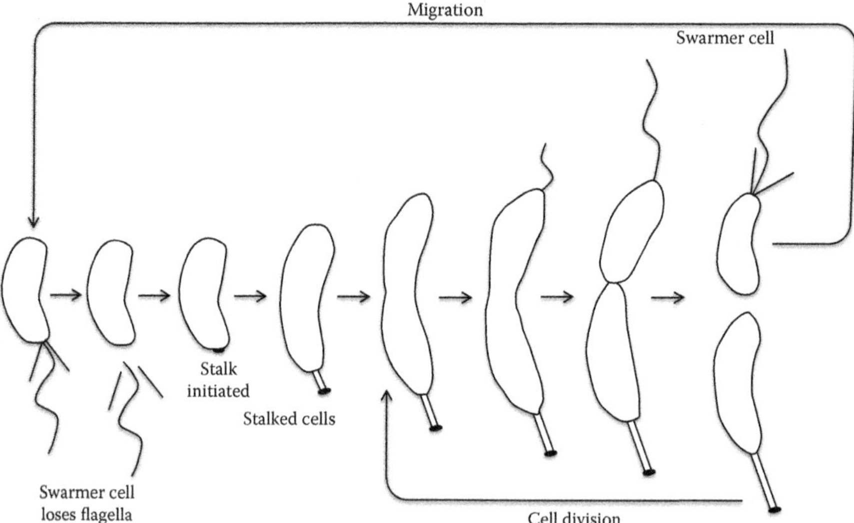

FIGURE 16.2 Cell cycle of *C. crescentus*. Two distinct cell types, stalked cells and swarmer cells, are indicated. Stalked cells are cemented to a substrate on a lakebed or streambed. As they divide, the daughter cells differentiate into swarmer cells that develop flagella at the ends opposite to the mother cells. The swarmer cells then swim away. Once they find a suitable substrate for attachment, they discard their flagella and develop a stalk that becomes cemented to the substrate. Stalked cells can divide many times, but always produce swarmer cells.

flagellated cell (swarmer cell) swims off until it finds a spot suitable for attachment. Then, it drops its flagellum, pushes out a stalk, and glues itself to a suitable substrate, differentiating into a stalk cell. The stalked cells can produce many swarmer cells but can never detach to become swarmer cells themselves. Thus, there is a definite developmental pathway that leads to production of the two cell types.

QUORUM SENSING

Certain species of marine (and other) bacteria fluoresce when the cells are in high concentrations. The fluorescent patches in the oceans had been noted by sailors for centuries, but only recently has been the mechanism studied and described. The phenomenon has been termed quorum sensing (Figures 16.3 and 16.4). It was determined that the bacteria were secreting specific chemicals (the first one discovered was acyl-homoserine lactone [AHL]) into the water, which attach to the receptors on the cell surface or inside the cells, which initiate a set of processes that lead to changes in gene expression of specific genes. In some cases, this causes flashes of light to be emitted by the bacteria, but only when the concentration of the stimulating molecule is sufficient to trigger the receptor molecules. Subsequent to the initial discovery, other similar signaling processes were discovered in other bacteria, some that excrete peptides through adenosine triphosphate (ATP)-binding

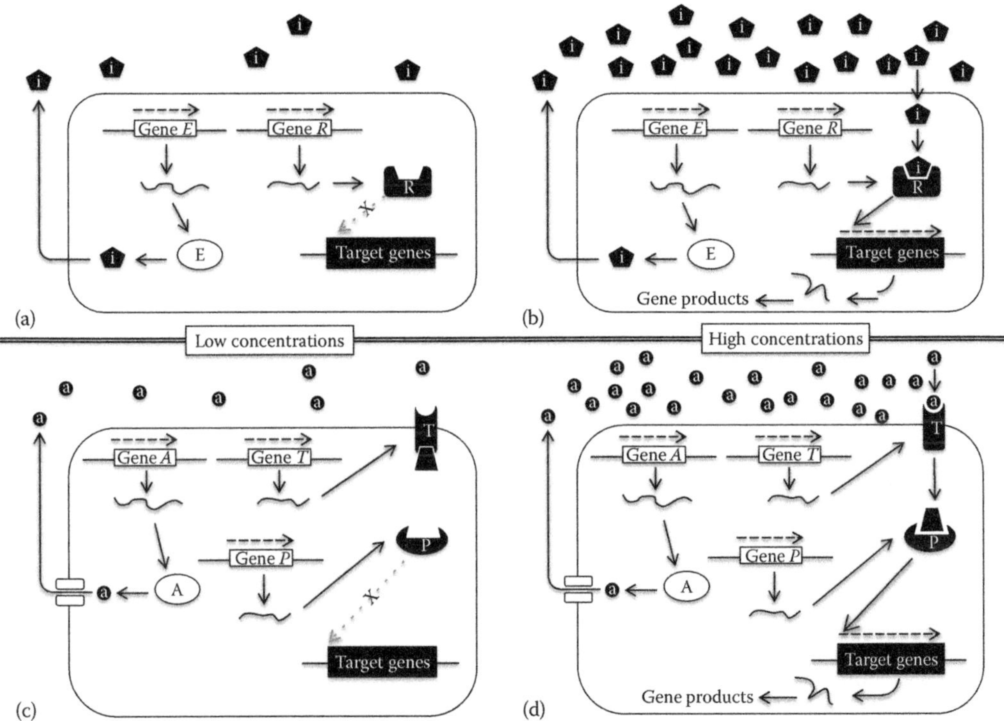

FIGURE 16.3 General features of quorum sensing in bacteria. In the (a), one gene encodes an enzyme (E) that produces an inducer molecule (i), which is exported out of the cell. The molecule can enter the same cell or other cells in the vicinity. A second gene encodes a receptor protein (R) that is capable of turning on the expression of other genes, but only when bound to the inducer molecule. If the inducer is diluted in the surrounding environment, then little of it is imported in the cell. However, if the concentration of the inducer increases in the surrounding environment (b), often in times when the concentration of cells increases, more of the inducer is imported into the cells and the target response genes are expressed. In other cases (c), the inducer is a peptide (a) or protein that is produced by cleavage from a larger protein (A) inside the cell, produced by a specific gene (gene *A*). The peptide (a) is exported from the cell, usually by ABC transporters. It then attaches to a transmembrane receptor (T) on the cell surface that is produced by a second gene (gene *T*). The receptor responds by releasing a factor that binds to an intracellular transcription factor (P) that promotes transcription of the target genes whose products generate the response. When the concentration of the peptide is low (c), insufficient amounts bind to the receptors. However, when the concentrations increase (d), the peptides bind to the receptors, the receptors release the factors the bind to the transcription factors, and the target response genes are expressed.

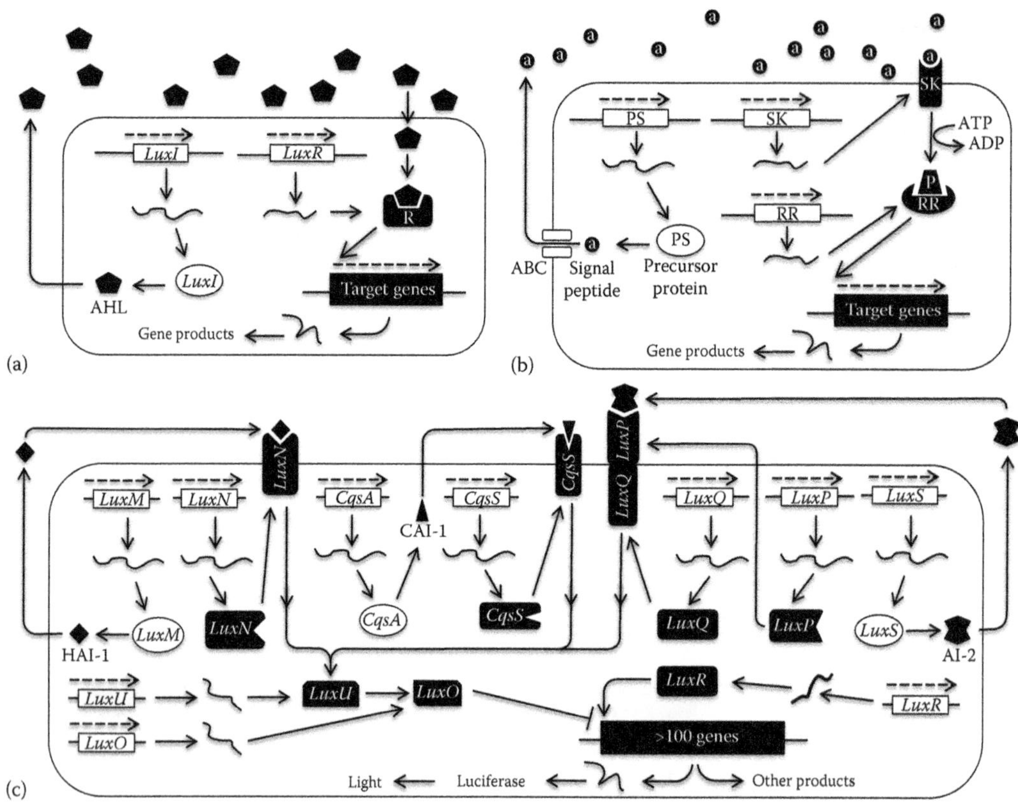

FIGURE 16.4 Three examples of quorum sensing in bacteria. In the first example (a), the *LuxI* gene encodes an enzyme that produces AHL, which is exported out of the cell. When sufficient amounts diffuse into the cells, they bind with a receptor encoded by the *LuxR* gene. This complex then promotes transcription of target genes (additional *Lux* genes), which ultimately produce flashes of light. In the second example (b), gene *PS* produces a precursor protein that is processed, part becoming a peptide (a) that is exported out of the cell by an ABC transporter. The peptide binds to a specific transmembrane serine kinase (SK) encoded by the *SK* gene. This activates the SK, which uses ATP to phosphorylate a protein (P) that binds to a response element (RR). This forms an active complex that increases the expression of specific target genes (*Lux* genes). The third example (c) is more complex. Several genes are involved, including *LuxM* (encoding an enzyme that produces HAI-1, an autoinducer), *LuxN* (encoding the transmembrane receptor for HAI-1), *CqsA* (encoding an enzyme that produces CAI-1, another autoinducer), *CqsS* (encoding a transmembrane receptor for CAI-1), *LuxS* (encoding an enzyme that produces AI-2, another autoinducer), *LuxQ* (encoding the inner portion of the transmembrane protein complex that binds AI-2), and *LuxP* (encoding the outer portion of the transmembrane protein complex that binds AI-2). When all three inducers are in sufficient concentrations, they bind with their respective receptors, which activate the LuxU protein (encoded by the *LuxU* gene), which activates the LuxO protein (encoded by the *LuxO* gene). The LuxO protein is a repressor of the target genes (many of which are additional *Lux* genes). When one of more of the concentrations is lower, the LuxR protein (encoded by the *LuxR* gene) increases the expression of more than 100 genes. Some encode proteins that produce luciferase, and ultimately light.

cassette (ABC) transporters in the cell membrane, and others that use alternative excretion systems. Some enter the cells through transporters, whereas others attach to the receptors embedded in the cell membranes, which trigger a signal transduction pathway inside the cell that leads to changes in gene expression. The cells can signal themselves, as well as other cells. There exist gene expression controls based on the response to environmental conditions but also on what other cells are in the vicinity, and whether they are of the same species/strain. Therefore, the systems responsible for sensing the type and quantity of other cells in the area or indirect contact with the sensing cell appear to have evolved long before multicellular forms appeared. This makes a great deal of sense, because the developmental systems in multicellular eukaryotes are complex and would have taken many evolutionary steps to reach such a level of sophistication and sensitivity that is found in those organisms (e.g., humans).

DEVELOPMENT IN ANIMALS

Although bacteria, archaea, and protists consist primarily of unicellular species, multicellular species predominate in Archaeplastida, Amoebozoa, Fungi, and Animalia. Within Animalia, only choanoflagellates, one of the most ancient groups of Animalia, are unicellular, although some form organized aggregations of cells (Figure 16.5). All other species within the Animalia are multicellular. Several developmental and genetic models (animals: *Caenorhabditis elegans*, *Drosophila melanogaster*, *Mus musculus*; fungi: *Saccharomyces cerevisiae*; plants: *Arabidopsis thaliana*, *Zea mays*, *Oryza sativa*, and others) have resulted in the determination of many facets of the molecular and evolutionary mechanisms of developmental processes. Many of these have been confirmed in *Homo sapiens* and other species, indicating the conservation and complexity of developmental processes in multicellular eukaryotes. Comparative genomics have also helped to elucidate the progression of various diseases, including cancer. Although there are similarities in development in organisms as diverse as plants, fungi, and animals, each group also has unique and specific aspects that lead to the very different forms that result from the developmental processes.

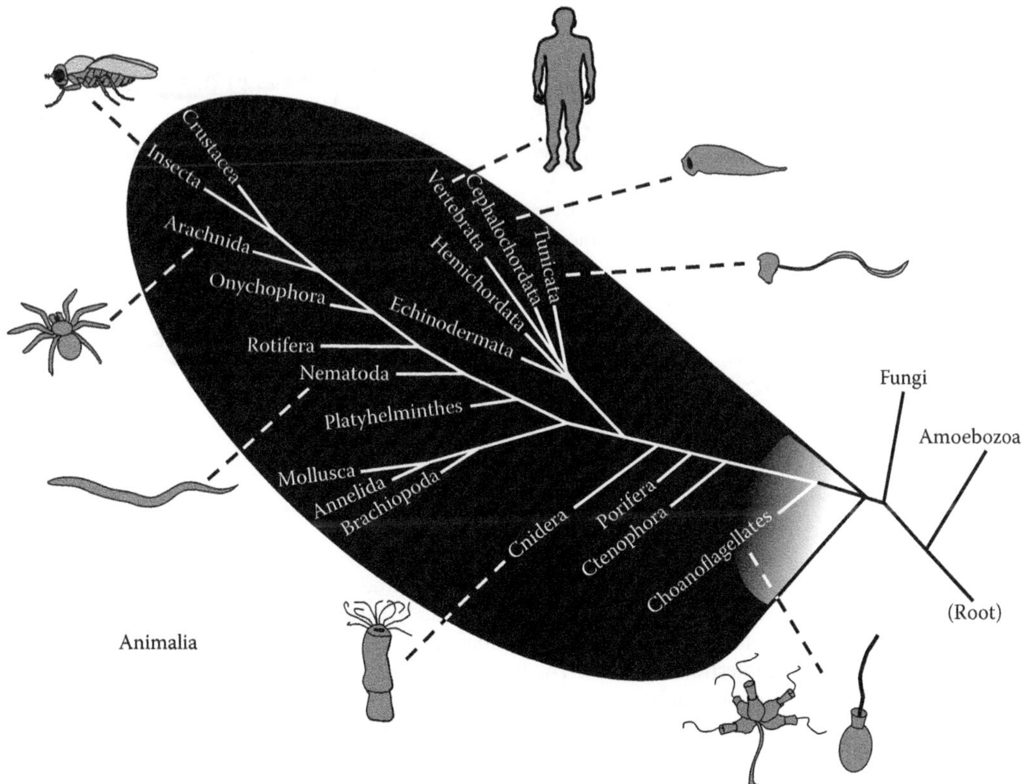

FIGURE 16.5 Detail of the groups of organisms within the Animalia, showing the extent of unicellular (white) versus multicellular (black) taxa. Gray shading indicates the formation of cellular aggreagates.

NEMATODE DEVELOPMENT

C. elegans is a small (about 1.0–1.2 mm in length) free-living nematode that is widespread in the soils of temperate climates. In the 1970s, it began to be used as a developmental and genetic laboratory model organism because of its multicellular structure, small size, and rapid life cycle, and it could be easily cultivated in the laboratory. It has been intensively studied and its relatively small genome has been sequenced. Most of the individuals are hermaphrodites, although about 1/1000 individuals is a male. This combination is not unusual among animals. Some arthropods produce hermaphrodites, females and males, and often the males are haploids. In *C. elegans*, the hermaphrodites consist of exactly 959 somatic cells, including 302 neurons, as well as a variable number of egg and sperm cells (Figure 16.6). The adult male consists of 1031 cells, as well as a variable number of sperm cells. The fate of every cell during development has been mapped, and gene expression during the developmental states has been studied in detail. One of the initially surprising findings in the research was that some cells were destined to die following cell division, whereas the other cells of the division had a specific developmental fate. This was one of the first indications that programmed cell death (termed apoptosis) was a normal part of development. One example of the apoptosis pathway is in neuronal development during sexual differentiation (Figure 16.7). During differentiation of males, cells that are important for hermaphrodites undergo apoptosis, whereas cells that are important for male development, for example, male-specific chemotaxis cells (CEM cells) that are needed so that the male will orient toward hermaphrodites for mating, remain alive

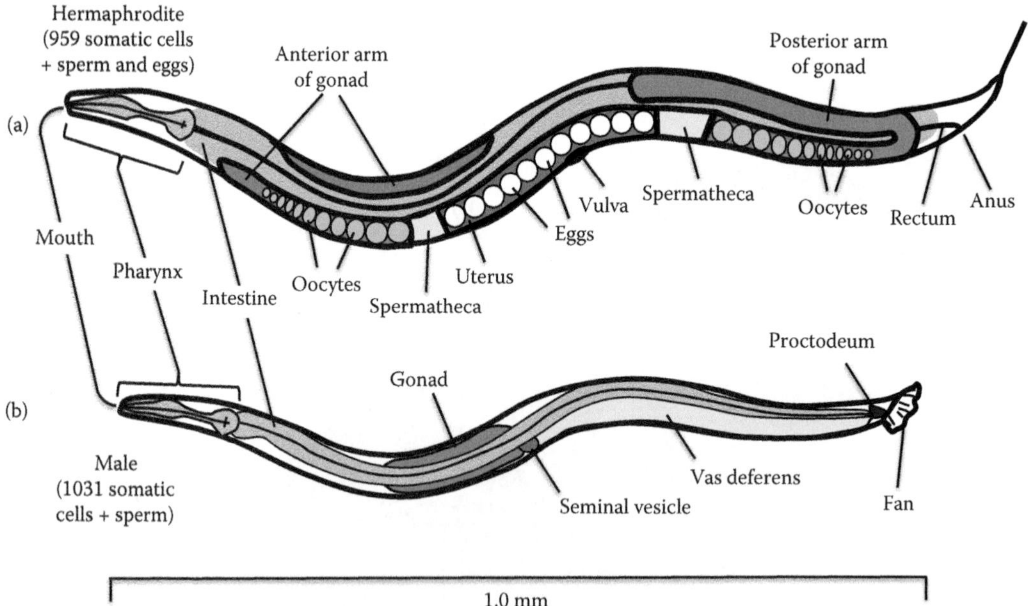

FIGURE 16.6 Anatomy of *C. elegans* (nematode). The hermaphrodites (a) are approximately 1.2 mm in length when fully mature, whereas the males (b) are approximately 1.0 mm long. Generally, males comprise approximately 0.1% of a population. All others are hermaphrodites. Both sexes have a digestive system composed of a mouth, pharynx, intestine, and anus. The hermaphrodite consists of 959 somatic cells (including 302 neurons), as well as a variable number of oocytes, sperm, and fertilized eggs. It has two large gonads that produce the oocytes and smaller organs that produce sperm, termed spermathecae. The mature oocytes are fertilized inside the uterus, and the fertilized eggs are expelled through the vulva. Although smaller than the hermaphrodites, the males each consist of 1031 somatic cells, as well as a variable number of sperm. Each male has a gonad that produces the sperm, a vas deferens that produces seminal fluids, and a seminal vesicle the delivers the sperm to the hermaphrodite. Additionally, both sexes have nervous systems and musculature that facilitate locating food, mobility, mating behaviors, and egg laying (in hermaphrodites).

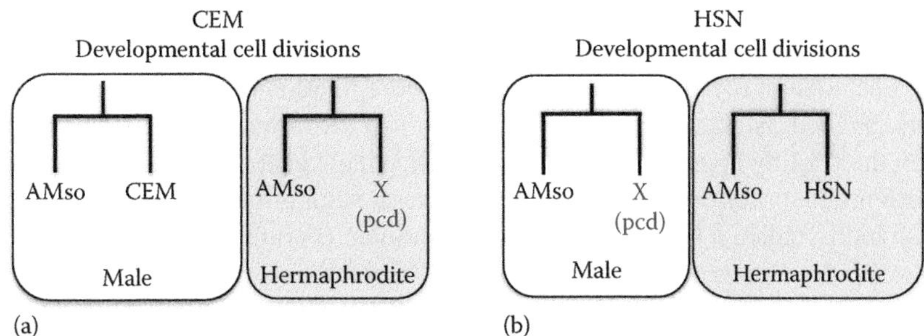

FIGURE 16.7 Example of apoptosis in *C. elegans* neuronal cells during development. Within specific positions of *C. elegans,* developmental pathways differ in the males and hermaphrodites. In males (a, white panel), a specific cell division among neuronal cells leads to one daughter that will become an amphid socket cell (AMso), whereas the other cell leads to neurons that are responsible for the males to orient toward a hermaphrodite for mating (termed CEM cells). However, in a different cell division (a, gray panel), one daughter cell division leads to an amphid socket cell and a cell that undergoes apoptosis (pcd = programmed cell death). In hermaphrodites, that same cell division (b, gray panel) leads to another amphid socket cell and a cell that will lead to the ability of the hermaphrodite to lay eggs. These are termed the HSN cells. In males, the division leads to an amphid socket cell and a cell that undergoes apoptosis.

and differentiate into mature neurons. Conversely, in hermaphrodites, the cells that are important for hermaphrodites, for example, hermaphrodite-specific egg-laying neuron (HSN) cells, grow and differentiate into mature neurons, whereas the cells that would become male CEM cells undergo apoptosis soon after cell division. Therefore, there is a program for each cell division to retain cells that have undergone positive selection in particular versions of the species, whereas cells that have been under negative selection pressures undergo programmed cell death.

Apoptosis had been observed previously in egg cell development. For many animals, meiosis in the formation of the oocyte proceeds via unequal cell divisions. In the first division of meiosis, one large 2C cell and one small 2C cell are produced. The small cell does not divide again, and eventually dies, becoming a polar body. The large cell divides again, producing a small 1C cell and a large 1C oocyte. The small cell dies and becomes a second polar body. As was mentioned in Chapter 5, this form of meiosis has evolved to produce a large oocyte that is capable of rapid cleavage upon fertilization. This process likely evolved in order to produce an oocyte that is fully prepared to begin dividing and differentiating once a sperm nucleus has fused with its nucleus. Another example of apoptosis occurs in cells that have sustained DNA damage. This is a process in which a set of proteins, including one called p53, sets into motion a series of molecular events that lead to the death of the cell. As will be discussed in Chapter 18, on cancer, apoptosis is one of the pathways that become defective in the progression of malignancies. The cells become immortal. Apoptosis triggered by p53 has evolved to eliminate a cell that might become a threat to the survival and reproductive capabilities of the organism. When p53 loses function, the defective cells continue to grow and divide, which eventually leads to more DNA damage and additional losses of function, and ultimately results in damage to the organism as a whole, and sometimes leads to the death of the organism.

HOMEOTIC GENES AND PROTEINS

Fruit flies (*D. melanogaster*, in particular) have been the subject of countless genetic studies that began in the early parts of the twentieth century. *D. melanogaster* became a popular model for genetic studies because they are easily cultured, they exhibit obvious morphological mutations, they have a few large chromosomes, and they have a rapid life cycle. Development has been studied intensively, and many aspects of the complexity of the expression of genes involved in this process have been elucidated (Figure 16.8). One of the surprising aspects of this work in the initial years was that mutations of certain genes caused the misplacement of entire organs. A mutation in the gene called antennapedia caused legs to be formed where antennae normally form. A mutation in the bithorax gene caused an extra thorax, with additional wings, to be formed adjacent to the usual thorax. Each one of the structures requires the coordination of a large number of genes and gene products, as well as accurate developmental timing. However, it was discovered that their positioning on the fruit fly appeared to be controlled by single genes. It was found that these types of genes whose products control the expression of a large suite of other genes were numerous and had a common motif, called a homeobox or a homeodomain (Figure 16.9). Sets of analogous genes in mammals, including humans, act in a coordinated fashion to form various structures on those animals (Figures 16.10 through 16.12). For example, *HOXD9*, *HOXD11*, *HOXD12*, and *HOXD13* are expressed in their order along the chromosome to control gene expression of a large set of genes that lead to the formation of a forelimb/arm, with the upper arm, forearm, wrist, foot/hand, and properly placed toes/fingers (Figures 16.13 and 16.14). When these genes are mutated, the limbs fail to develop correctly, and in some cases, many more fingers and toes result (Figure 16.15), but in most cases they are improperly formed and improperly placed on the hand or foot.

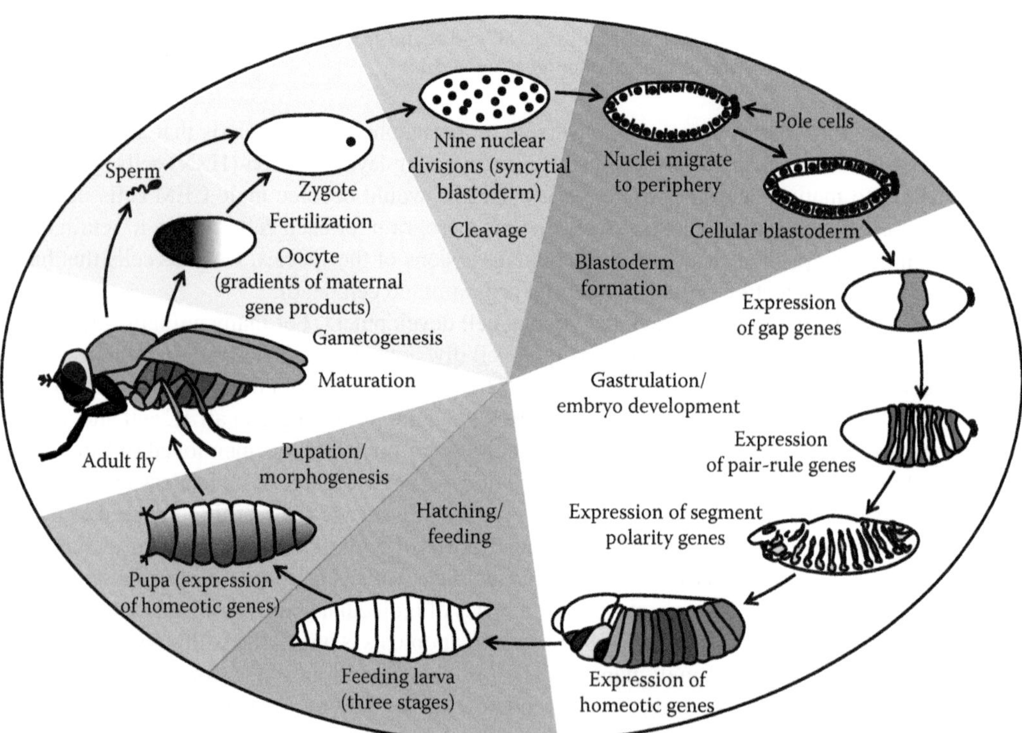

FIGURE 16.8 Life cycle of *D. melanogaster*, showing some of the important features of developmentally regulated genes. Gene product gradients are already present in the oocyte (yellow background). After fertilization to form a zygote (green background), a syncytial blastoderm is the result of nine nuclear divisions without concomitant cell divisions (blue background). The nuclei migrate to the periphery and cell membranes form to produce a cellular blastoderm (pink background). By this time, the gap genes have become activated, followed by the pair-rule genes, and then the segment polarity genes, many of which control expression of the homeotic genes (white background). The larva hatches and feeds (violet background), and proceeds through three phases (called instars), increasing in size during each stage. The larva contains imaginal discs, each of which determines specific structures in the adult during pupation (orange background). During this stage, the animal metamorphoses into the adult form, where the imaginal discs form structures in the adult animal, and the larval tissues are essentially used as food for this process. The adult matures, a male and female mate, and the cycle begins again.

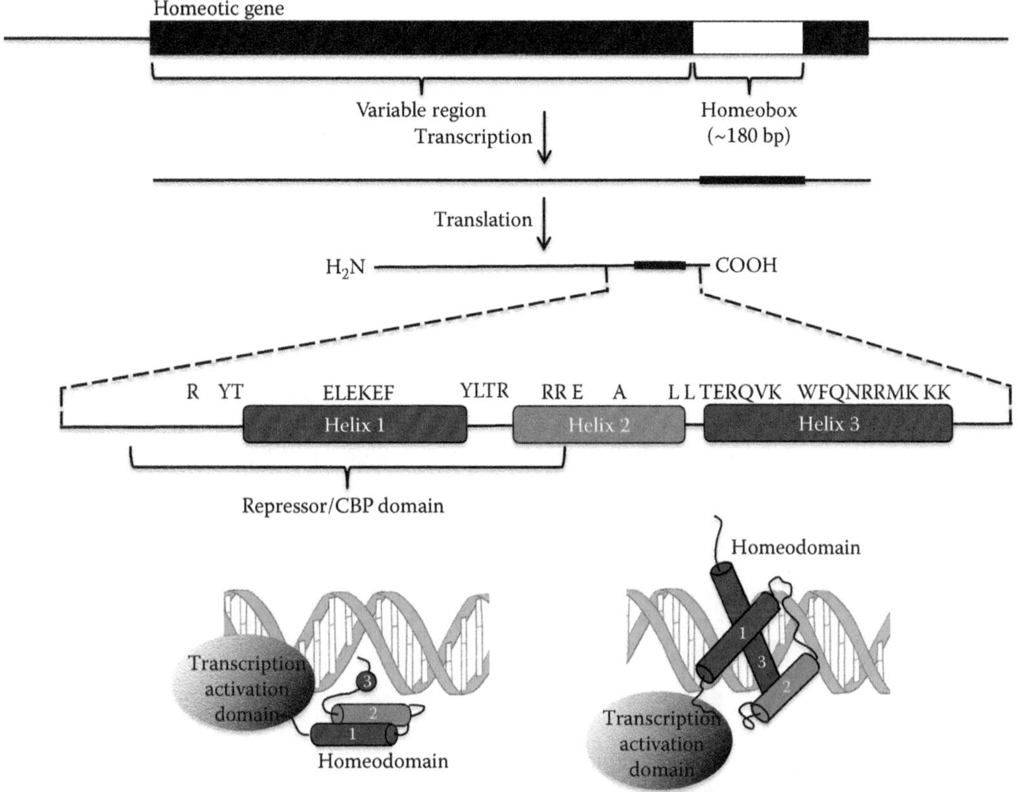

FIGURE 16.9 General structure of a homeotic gene and homeobox domain. The genes all have long variable regions, with a homeobox domain consisting of approximately 180 bp near their 3′ ends. The homeodomain of the protein consists of three α-helices (helix 1, helix 2, and helix 3). Conserved amino acids are listed above the middle diagram. Helix 3 interacts with the major groove of DNA and is the most conserved in amino acid sequences (see Figure 6.1 for one-letter codes for amino acids). Helices 1 and 2 also interact with the DNA, but to a smaller extent. Also, helix 1 interacts with repressors that change the conformation of the homeodomain such that the protein will not attach to DNA. The variable portions that are upstream of the homeodomain interact with other factors to affect the transcription of specific genes and gene sets.

Hox gene expression has had a large impact on the evolution of structures within organisms. For example, as mentioned above, the hands and feet of mammals are primarily determined by HoxD13 expression. This is also true for the fins of fish, except the expression pattern for fish differs from that of mammals (Figure 16.16). The original ancestor of all chordates likely had a single set of *Hox* genes, including *Hox1* through *Hox14*. Prior to divergence of agnathostomes (jawless fish) from the lineage that eventually became all of the other fishes, reptiles, and mammals, the *Hox* locus duplicated, either by unequal crossover or by genome duplication (i.e., polyploidization). This led to an *AB* locus and a *CD* locus that allowed for mutations in the duplicated genes, thus affecting the evolutionary rates in those organisms. Some of the alleles mutated and became inactive, which allowed differential expression of *Hox* genes in different parts of the organism. A second round of duplication produced four loci, A, B, C, and D, and occurred during the divergence of gnathostomes (jawed fish). This allowed yet additional discrimination of expression among the tissues of the organism, leading to more complexity of tissues within the animals. Members of the Chondrichthyes (sharks, skates, and rays) share this pattern. Studies indicate that the distal genes (*Hox9* through *Hox14*) in the *HoxA* locus exhibit low expression levels during development of the fins, whereas those in the *HoxD* locus are expressed to a higher degree (Figure 16.16, upper

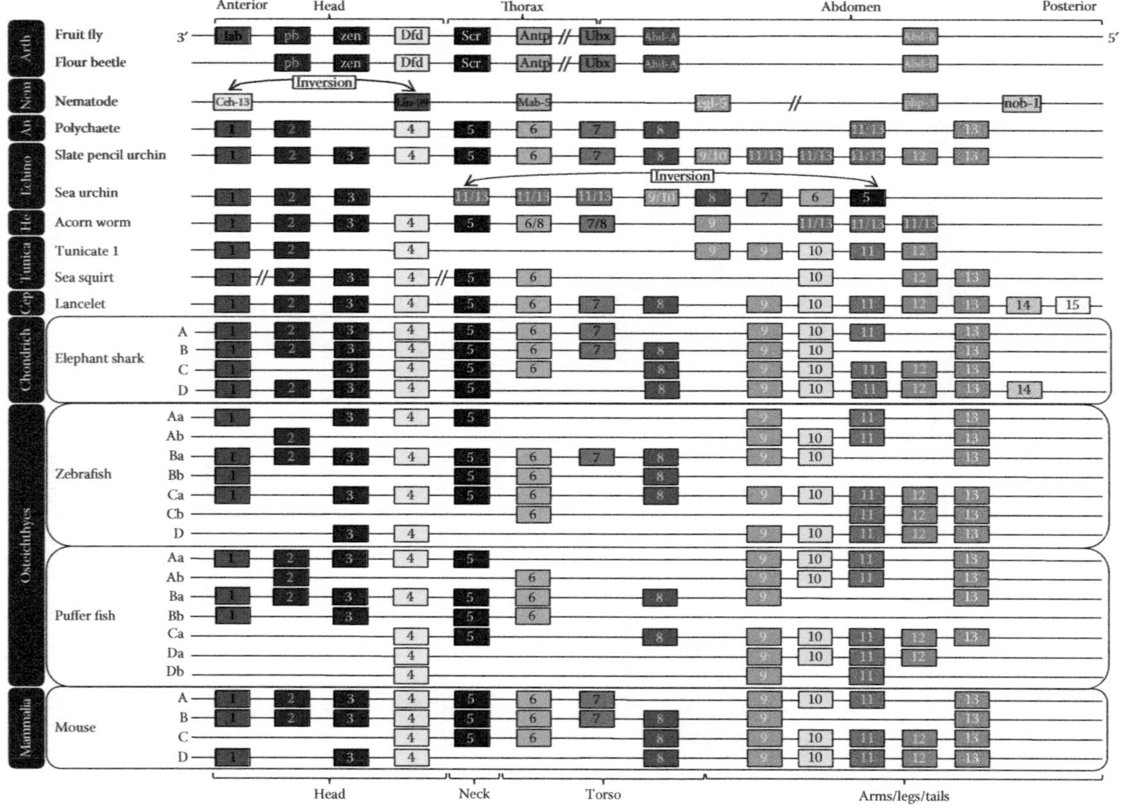

FIGURE 16.10 Chromosomal arrangements of *Hox* genes among a diverse set of animal taxa. Genes are arranged from *Hox1* (or homologs) to *Hox15*, which is the order of expression from the anterior to posterior ends, although the genes are in the 3′–5′ orientation. Abbreviations and species names are as follows: ARTH, Arthropoda, fruit fly (*Drosophila melanogaster*), flour beetle (*Tribolium castaneum*); NEM, Nematoda, nematode (*Caenorhabditis elegans*); AN, Annelida, polychaete (*Capitella teleta*); ECHINO, Echinodermata, slate pencil urchin (*Eucidaris tribuloides*), sea urchin (*Strongylocentrotus purpuratus*); HE, Hemichordata, acorn worm (*Ptychodera flava*); TUNICA, Tunicata, tunicate 1 (*Oikopleura dioica*), sea squirt (*Ciona intestinalis*); CEP, Cephalochordata, lancelet (*Branchiostoma floridae*); CHONDRICH, Chondrichthyes, elephant shark (*Callorhinchus milii*); Osteichthyes, zerbrafish (*Danio rerio*), puffer fish (*Takifugu rubripes*); and Mammalia (also, Aves have the same pattern), mouse (*Mus musculus*).

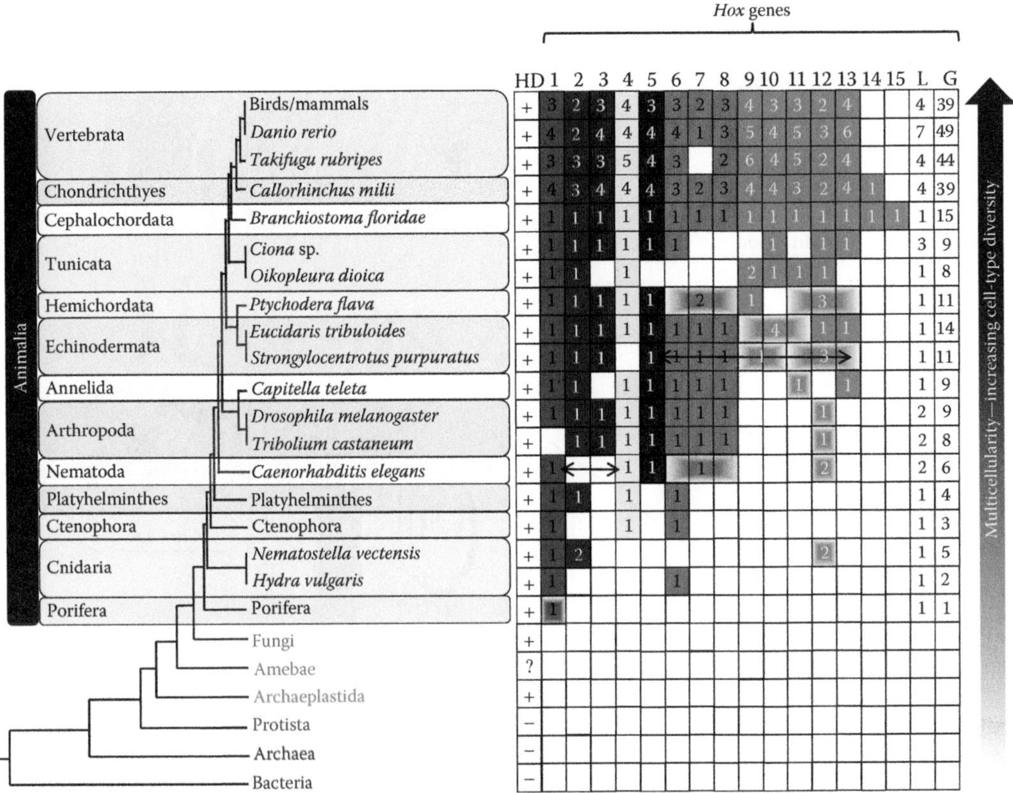

FIGURE 16.11 Phylogram and chart of homeotic genes and the number of *Hox* genes among phylogenetic groups. The number of genetic loci (L) and the total number of *Hox* genes (G) are on the right. Although there are many homeotic genes that produce proteins with homeodomains (HD), *Hox* genes are specific to species in the Animalia. As the number of *Hox* genes increases, organism complexity also increases. The number of loci also has led to the elaboration of specific structures in organisms with multiple versions of the *Hox* locus that are activated in specific periods of development. The evolutionarily earliest genes (*Hox1* and *Hox2*) are those that cause elaboration of the head end of more complex organisms. Tails and appendages appear in organisms with *Hox 9* through *Hox15* genes. Lighter shading indicates genes that have lower degrees of similarity to the genes of comparison. Double arrows indicate an inversion. Homeotic genes are present in the Opisthokonta (Fungi and Animalia), as well as in the Archaeplastida. Although they have not been described in Amoebozoa, because of the many multicellular species, homeotic genes or analogous genes are suspected to be present.

graphs). However, expression of HoxD13, specifically, is only expressed during early limb development, which leads to a large number of cartilaginous elements to be formed, resulting in a large fin (Figure 16.16, left middle). Also, *Hox14* has been lost from all but the *HoxD* locus.

An additional duplication of the *Hox* loci occurred in the ray-finned fish (Actinopterygii), producing loci *Aa*, *Ab*, *Ba*, *Bb*, *Ca*, *Cb*, *Da*, and *Db* (Figure 16.16, lower right). The most likely cause for this is genome duplication where all four loci (*A*, *B*, *C*, and *D*) were duplicated concurrently, again during a genome duplication event, in which the species all became more polyploid. Since that time (approximately 420 million years ago), many of the duplicated *Hox* genes have mutated, such that there is a diversity of patterns among ray-finned species. For example, locus *Db* has been completely lost in zebrafish (*Danio rerio),* whereas locus *Cb* has been completely lost in puffer fish (*Takifugu rubripes*; Figure 16.16, middle). Also, the expression of *Hox* genes differs in the ray-finned fish compared with sharks. Expression of both the *HoxA* and *HoxD* loci is higher in the ray-finned fish. This led to a decrease in the number of solid elements in the fins (in this case, bones). The *Hox* gene arrangement in sarcopterygian (lobe-finned fish) is similar to that in other chordates, in that it

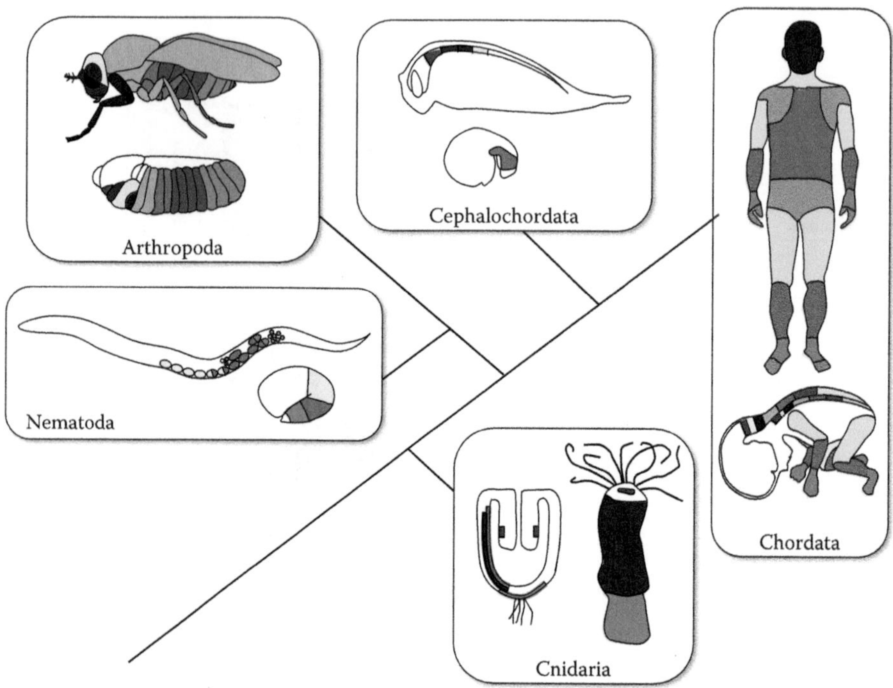

FIGURE 16.12 *Hox* gene expression among species in five phyla of Animalia plotted on a rudimentary phylogenetic tree. Early development and adult stages are shown for each, with *Hox* gene expression indicated for each. Colors correspond to those in Figures 16.10 and 16.11.

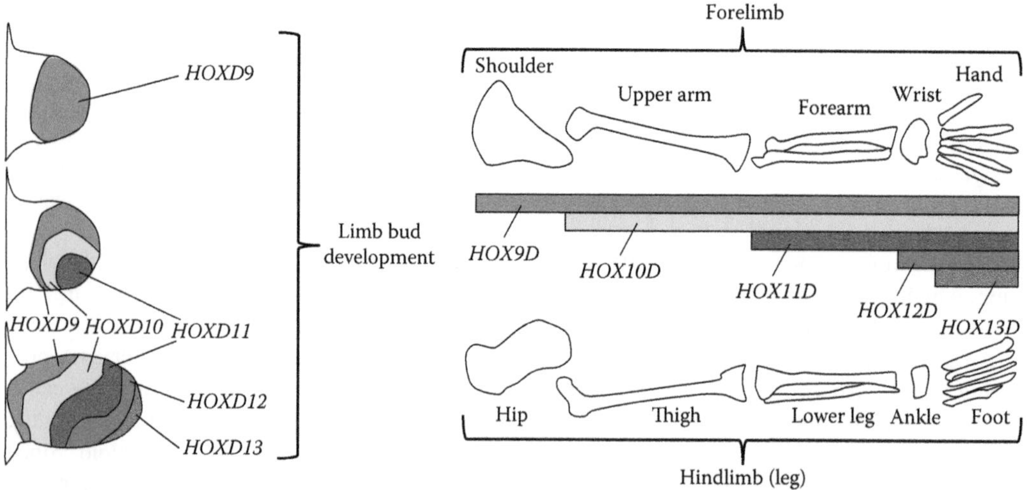

FIGURE 16.13 Expression of *Hox9D* through *Hox13D* in limb bud and limb development. Although *Hox9D* alone is responsible for increasing the expression of genes that lead to the formation of the shoulder and scapula, the pattern of *Hox* gene expression is more complex along the length of the appendages (arms and legs). Although *Hox13D* has a large effect on the formation of fingers and toes, the expression of *Hox9D* through *Hox12D* are also necessary for the development of the distal regions of the appendages. Colors are as in Figures 16.10 and 16.11.

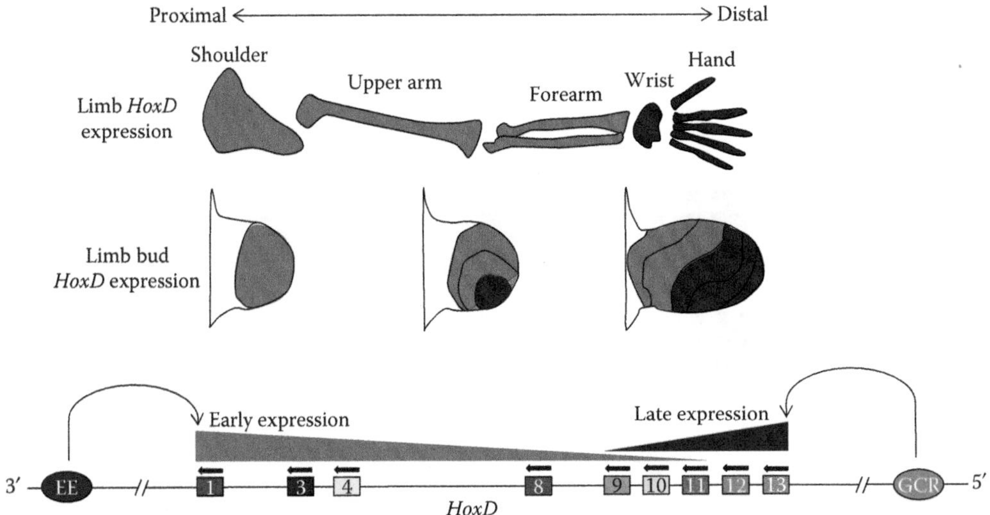

FIGURE 16.14 Regulation of the *HoxD* locus (*Mus musculus*). During early expression in the limb bud, only some of the *HoxD* genes are activated. This pattern sets up the development of the upper (proximal) portions of the limb. The early pattern of expression is controlled by a downstream enhancer element (EE). During the later stages of limb development, a global control region (GCR) activates the *HoxD* genes (9–13), which leads to the development of the distal regions of the limb. *HoxD13* is especially crucial in the development of digits. Arrows over genes indicate direction of transcription.

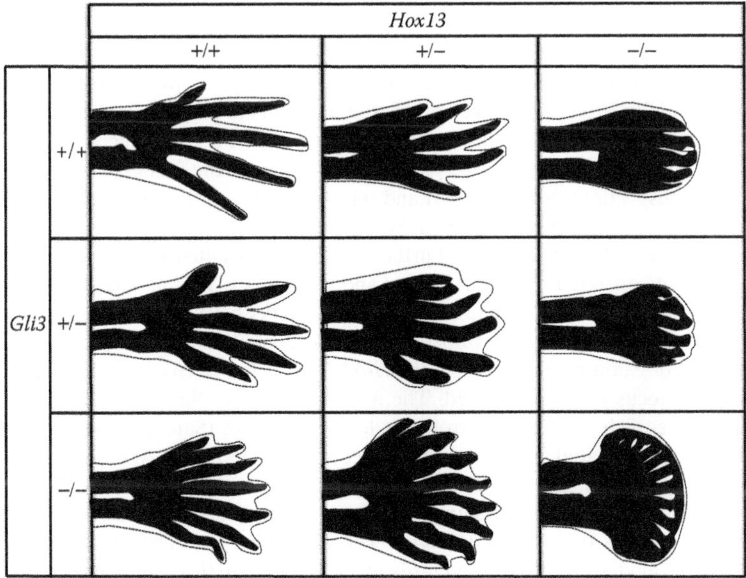

FIGURE 16.15 *Hox13* and *Gli3* (a gene encoding a zinc-finger DNA-binding protein) in relation to digit formation in *Mus musculus* (mouse). Plus symbols indicate functional alleles, whereas minus symbols indicate mutant alleles. Dark areas represent locations where bone will form. Dashed lines indicate the extent of skin. When expression of *Hox13* is at normal level (upper left), five long digits form. When the gene dosage is half (upper middle panel), five digits form, but they are shorter, and the skin forms webbing between the digits. When *Hox13* is not expressed, five digits form, but the foot becomes a lobed structure (reminiscent of those in lobe-finned fish) rather than a foot with toes. Expression of *Gli3* determines the number of digits formed on the foot. When *Gli3* is at normal level, five digits form (left column). However, when the expression of *Gli3* is reduced, the number of digits increases, up to 13 in some cases.

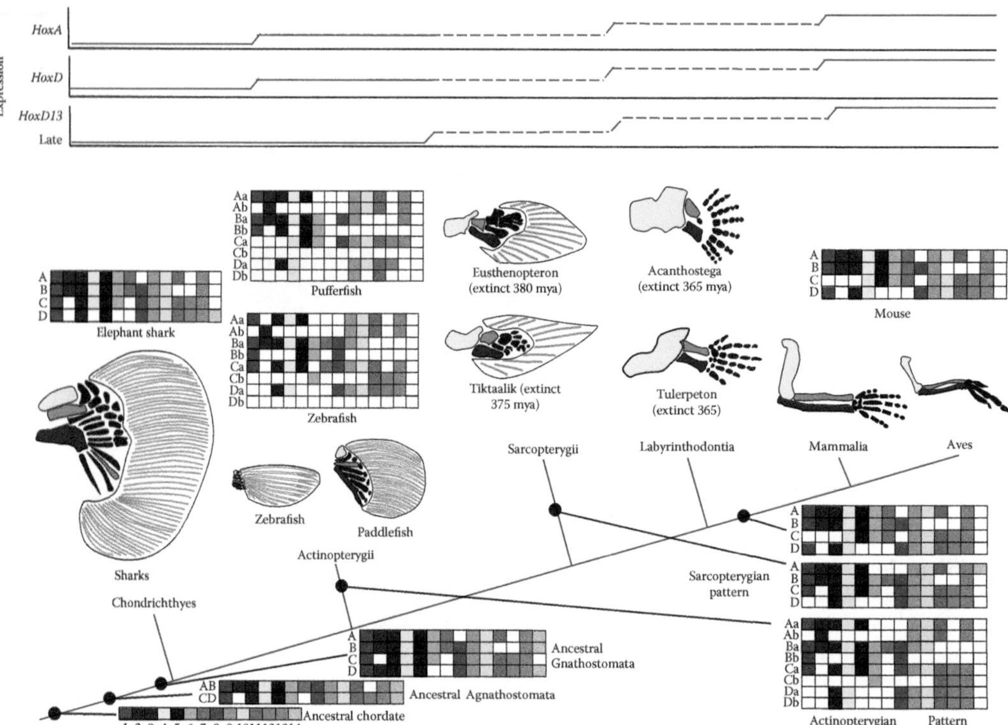

FIGURE 16.16 Evolution of limbs in relation to *Hox* gene number and expression. During the evolution of fish, reptiles, birds, and mammals, the bones in the appendages (fins, feet, hands, and wings) underwent successive morphological changes. These changes can be traced to changes in the number and expression levels of *Hox* genes, and especially *Hox13*. The ancestral chordate appears to have had one copy of each of the *Hox* genes (1–14) at a single locus. During evolution, some of the organisms experienced either a duplication of the locus or a duplication of the entire genome, such that two copies of the *Hox* genes were present (although on separate chromosomes). Through genetic drift and selection, some of the genes mutated to become inactive, pseudogenes, or were deleted (white squares). Some of these patterns were probably present in the first members of the Agnathstomata (jawless fish). Further duplication resulted in four *Hox* loci, designated *A*, *B*, *C*, and *D*, having been generated from an *AB* locus and a *CD* locus. This pattern was probably present in the ancestral members of the Gnathostomata (jawed fishes). Sharks have retained this *Hox* gene pattern. In Actinopterygii (ray-finned fishes), there have been additional duplications of the *Hox* gene loci, and some genes have been lost. For example, in zebrafish, the *HoxDb* locus has been completely deleted, whereas in puffer fish, the *HoxCb* locus has been deleted. The next change in evolution occurred as organisms that were transitional between fish and reptiles, from the lobe finned fish (Sarcopterygii) to primitive reptiles (Labyrinthodontia) which appeared and then went extinct between 350 and 420 million years ago and most of which are extinct. Fossils indicate that the bones of their limbs had become longer, and the numbers of bones were reduced, such that instead of having fins with large numbers of closely spaced bones, they had defined legs and multiple toes (usually six to eight). The exact *Hox* gene patterns are unknown, but they likely resemble those of modern mammals and birds, which have four *Hox* loci that contain *A*, *B*, *C*, and *D* versions. Another variable in the effects of *Hox* genes is the level of expression of the genes. Experiments have shown that the *Hox* genes are controlled by upstream and downstream elements, and that they coordinate the expression of the *Hox* genes during different points in development. During evolution, changes in *Hox* gene expression appears to have led to the changes in limb development (upper graphs). Specifically, increases in expression of the *HoxA* and *HoxD* loci, and ultimately increases in expression of *HoxD13* during the later stages of limb development, led to the transition from large fins to ray fins, to lobed fins, and finally to feet, hands, and wings.

has four loci, *A*, *B*, *C*, and *D*. Additionally, some of the *Hox* alleles have been lost, including one or more versions of *Hox1*, *Hox4*, *Hox6*, *Hox7*, *Hox8*, and *Hox14*. Additionally, the bones analogous to the ulna, radius, and humerus are consolidated in the fin lobe (Figure 16.16, middle). Although the patterns of *Hox* gene expression are unknown at present, it has been hypothesized that the expression of HoxD13 late in limb development is higher in this group than in ray-finned fish (Figure 16.16, upper graphs, dashed lines). This leads to the formation of a lobed fin rather than the flatter fin present in ray-finned fish.

Additional changes in *Hox* gene expression have occurred in mammals (Mammalia) and birds (Aves) that have led to extension of the bones in the arms, legs, wings, fingers, wings, and/or toes. In particular, the expression of HoxD13 late in development causes a reduction in the number of digits, as well as a lengthening of the digits (Figure 16.15). It is likely that this pattern began in the labyrinthodonts, extinct reptile-like organisms that appear to be transitional between fish and mammals (Figure 16.16, middle). Species of labyrinthodonts had from six to eight toes on each leg. The number of digits has decreased in mammals and birds, which have three to five digits on each appendage. Dinosaurs also had three toes, similar to those in birds, which is a sister group to dinosaurs. Although the exact mechanisms remain unknown, it is suspected that the late expression of the distal *Hox* genes, and primarily HoxD13, causes the reduction of digits in these organisms.

Although homeotic genes have been found in most groups of multicellular organisms, the subset of *Hox* genes are only found in animals, and appear to have originated as a single gene in the Porifera (sponges; Figure 16.11). In members of the Porifera, the gene function controls sets of genes to become activated, which leads to differentiation of cell types within the organism. Members of the Cnidaria have from two to five *Hox* genes. Their expression activates genes that determine the anterior tissues and the posterior tissues, as well as control the extent of nerve and muscle cells within the organism. Although the number of genes and genetic loci increases with increasing organism complexity (Figures 16.10 and 16.11), the pattern of expression also becomes increasingly complex (Figures 16.17 and 16.18a and b). The genes are not turned on one at a time. Instead, in many cases several genes are activated in individual cells and tissues, such that two, three, or more genes are expressed in the same cells and tissues. Furthermore, the patterns of expression for each of the *Hox* genes differ from one tissue type to another. One example is shown for the Ascidian, *Oikopleura dioica*, within the Tunicata (Figure 16.19). In the larval stage, it has a body and a tail, and some resemble small tadpoles. *Hox* gene expression has been studied in the various tissues of the organism, including the epidermis, muscle, nerve cord, and notocord. Only *Hox1*, *Hox10*, *Hox12*, and *Hox13* are expressed in the epidermis. *Hox1* is expressed only in a short portion of the anterior portion of the animal, *Hox10* and *Hox13* are expressed in the anterior and posterior half, respectively, and *Hox12* is expressed along roughly 80% of the animal. The expression patterns of *Hox* genes are completely unique for each of the other tissue types, but all lead to the formation of the complete animal.

ARTHROPOD DEVELOPMENT

One of the major reasons for studying *D. melanogaster* development is that the genes responsible for development (both normal development and abnormal development, such as in cancer) have analogs in the human (and other animals) genome (Figures 16.10 and 16.11). Therefore, by studying the developmental pathways in fruit fly, much can be learned about analogous pathways in humans, as well as other animals. Development in fruit flies is initially set up in the oocyte (Figures 16.8 and 16.20). Gene expression and partitioning of the products of gene expression are already in place by the time the cell is fertilized by the sperm. This establishes a pattern within the cell that is set up by the maternal genes. A gradient of several proteins is formed. The proteins expressed by the bicoid (*bcd*) and hunchback (*hb*) genes are highest in the anterior half of the oocyte (where the head will

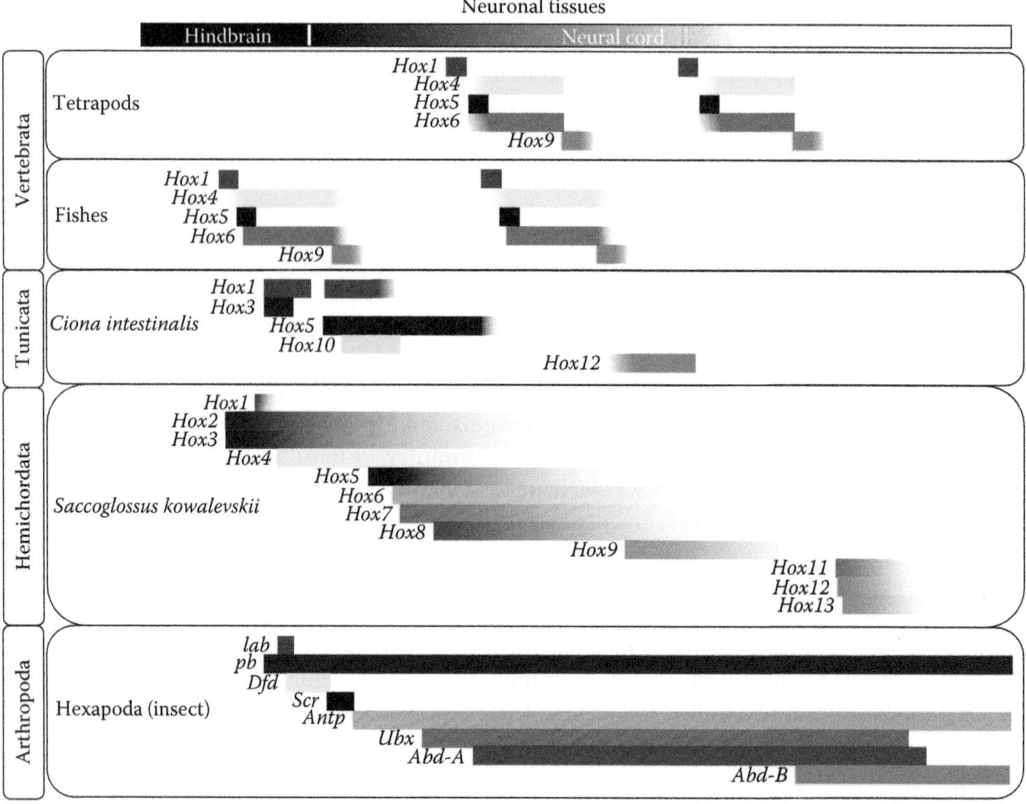

FIGURE 16.17 *Hox* gene expression patterns that produce nerve neuronal along the longitudinal axis of several organisms. The patterns for the Vertebrata show only the positions of enervation of limbs. This indicates the relative positions of neurons for limbs in fish versus mammals. The patterns for the other species are complete summaries of *Hox* expression. Tunicates have relatively rudimentary nervous systems, whereas hemichordates and arthropods have more elaborate nervous systems, which are reflected in the *Hox* expression patterns. Colors are as in Figures 16.10 and 16.11.

eventually form) and the proteins expressed by the nanos (*nos*) and the caudal (*cad*) genes are highest in the posterior half (the tail end) of the oocyte (Figure 16.20).

Once fertilization occurs, there is a rapid succession of replication and nuclear division events to form a syncytium (i.e., a multinucleate cell; Figure 16.8). Cells form around each nucleus, and the embryo begins its program of development. These maternal proteins are regulators of other genes, and depending on their concentrations in the embryo, they influence another set of genes called *gap* genes (Figures 16.8 and 16.20). The expression of these genes is such that it forms a central region, as well as defined anterior and posterior ends. The gradients of concentrations of the *gap* genes then affect the expression of another set of genes called the primary pair-rule genes, which begin to define the segments of the animal. The products of the primary pair-rule genes lead to differential expression of secondary pair-rule genes (also called segment polarity genes) that bisect each of the larger segments set up by the primary pair-rule genes. Many of the gene products from each of these categories influence the expression of the homeotic genes. The products produced by these genes affect gene expression of large sets of genes that set up the body plan of the organism. For example, they control where legs, wings, antennae, major organs, and other structures form (Figure 16.8).

As stated above, each of these sets of genes has analogs in other organisms, although the *Hox* genes have been studied in the greatest detail. When development in fruit fly is compared to other arthropods, the genes are nearly identical in all of them, but their expression in the developing embryos differs. For example, nearly the same set of maternal, gap, pair-rule, segment polarity, and homeotic genes that are found in fruit flies is also found in wasps. However, they differ with respect to exactly where in the embryo the genes turn on and off, and where the protein products are at higher or lower concentrations (Figure 16.21). These subtle differences produce two different body plans, and therefore two different organisms. However, the conservation of the genes and their expression patterns points strongly to a common origin for these genes and developmental patters.

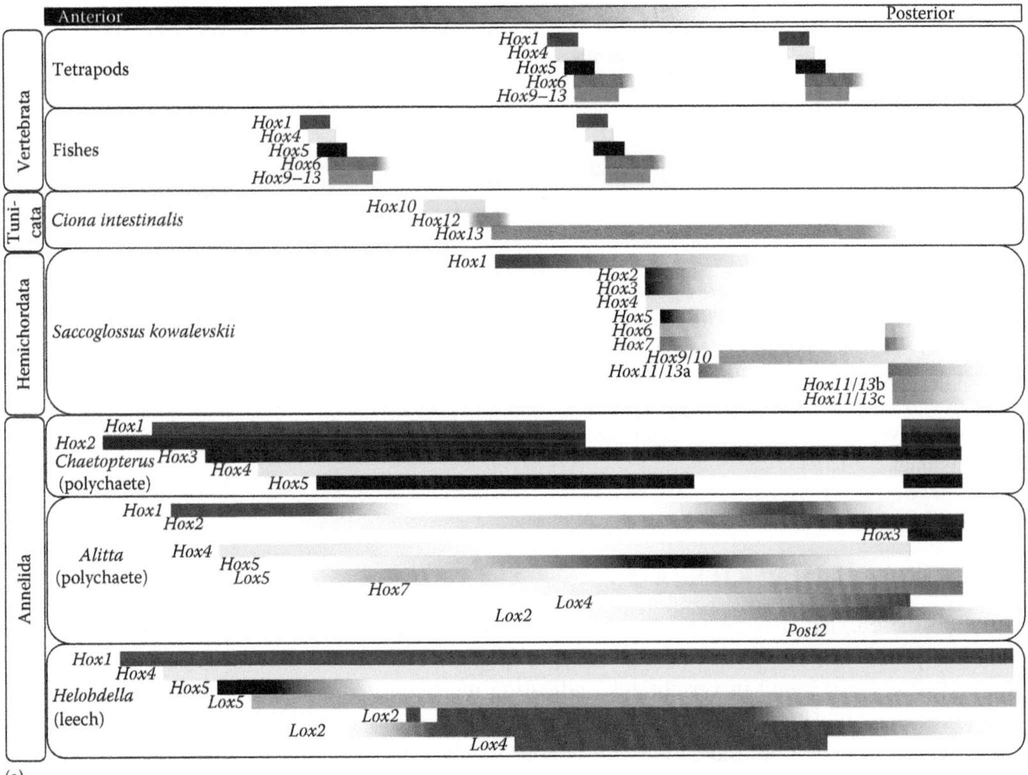

FIGURE 16.18 *Hox* gene expression patterns that produce somatic tissues along the longitudinal axis of several organisms. (a) The patterns for the Vertebrata show only the positions of the limbs. This indicates the relative positions of fins in fish versus hands and feet in mammals. Fins are initiated more toward the anterior end of fish than hands and feet in tetrapods. In general, the expression of each of the *Hox* genes is more strictly regulated in fish and tetrapods than in other animals. *Hox* expression in somatic cells is limited in tunicates, although this is not so for neuronal tissues (see Figure 18.17). Although *Hox* expression is spread over large portions of the organisms in Annelida, the patterns for Vertebrata, Tunicata, and Hemichrodata are more limited and defined, which leads to a more complex body plan. Color shading indicates partial to full expression of the *Hox* genes.

(Continued)

FIGURE 16.18 (Continued) *Hox* gene expression patterns that produce somatic tissues along the longitudinal axis of several organisms. (b) *Hox* gene expression patterns in somatic tissues of several members of Arthropoda and one member of Cnidaria. Although the *Hox* gene nomenclature varies in these organisms, the colors correspond to the *Hox* gene types indicated in (a). The comparisons of the different arthropods shows how body plan varies according to the differences in positional expression of each of the *Hox* genes. Cnidarians have fewer *Hox* genes, with relatively simple expression patterns, which leads to a more rudimentary body plan.

DEVELOPMENT IN VERTEBRATES

Although there exists a wide variety of vertebrates, their developmental patterns and genes are remarkably similar. Also, some of the same types of genes that are found in arthropods, such as fruit flies and wasps, are found in vertebrates, indicating their conservation during evolution. Also, the gene order along the chromosomes has been mostly conserved (Figure 16.10) over hundreds of millions of years of evolution, indicating that they probably were primarily vertically inherited as the animals expanded over the past 500 million years. However, the expression patterns of these genes in vertebrates and arthropods vary greatly, even from the earliest of stages, which lead to the different body plans observed for each group. For example, in arthropods, the *Abd* and *Ubx* genes are responsible for setting up the abdomen and the joining points to the thorax. The analogous homeotic genes for mammals (*HOX7* through *HOX13*) set up the torso, legs/hindlimbs, and arms/forelimbs. On the anterior end, genes *lab*, *pb*, *zen*, and *Dfd* determine the formation of the head, whereas the analogous genes in mammals (*HOX1* through *HOX4*) also set up the formation of the head. These genes are mostly syntenous along the chromosomal loci.

Varying amounts of the protein products cause various cell types, tissues, and organs to form. As was described earlier, in order to form vertebrate arms and hands, several *HOX* genes are turned on sequentially to form each part of the arm/forelimb or leg/hindlimb. In the bud that will form the limb, the *HoxD* locus becomes active. *HoxD9* (in the forelimb) or *HoxD10* (in the hindlimb) is first expressed. Then, *HoxD11*, *HoxD12*, and *HoxD13* are turned on sequentially,

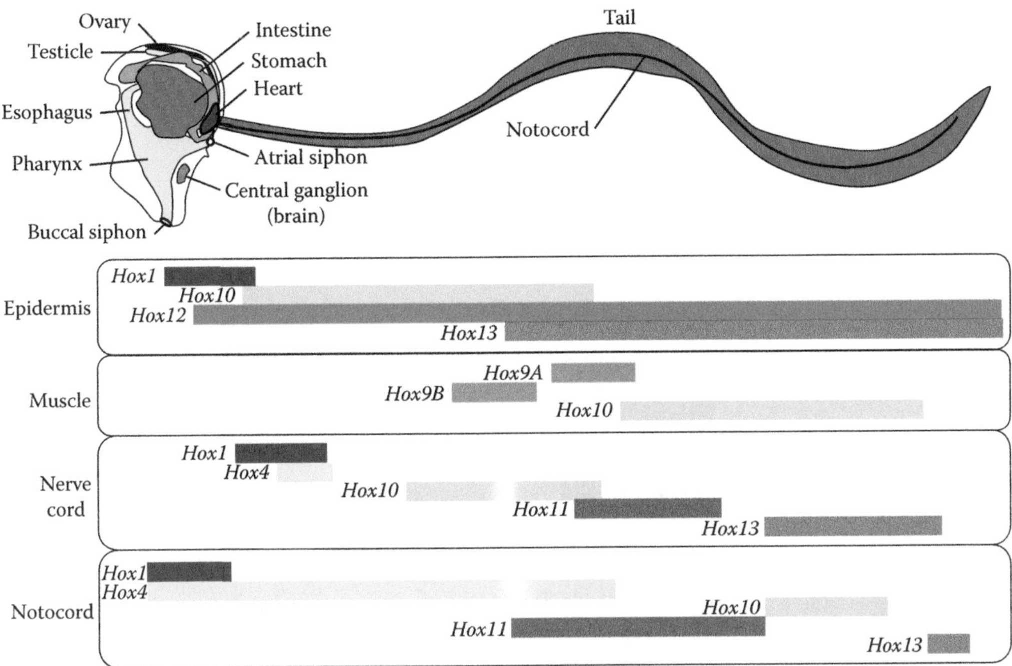

FIGURE 16.19 Anatomy and *Hox* gene expression in four tissues (epidermis, muscle, neuronal, and notochord) of the tunicate *Oikopleura dioica.* The animal consists mainly of a main body and a tail, in which the notochord persists. Expression of the *Hox* genes varies with tissue type, which leads to the characteristics of each of the tissues. Although *O. dioica* is a relatively simple animal, it shares many features of more complex animals, and the basic *Hox* gene expression patterns are similar in many ways to more complex animals. The order and orientation of the genes are consistent with all animals.

such that the combination of the HoxD proteins sets up the developmental expression of genes needed to form the shoulder/hip, upper limb, lower limb, and hand/foot, respectively. Actually, it is not this simple, in that other regulatory elements and genes affect the process as well. These act in conjunction with the *Hox* genes in ways that produce hands or feet with more or fewer fingers or toes, and cause either separation of the fingers and toes or causes them to form together as webbed feet or fins. Therefore, by varying the amounts, the timing of expression, and the differing functions of alleles and mutants of these genes, appendages from a hand with four fingers and a thumb, to a fin with a dozen or more bones can be produced, even when all other genes are the same in both instances.

Although tetrapods (such as humans) and fish have similar *Hox* gene loci, the expression differences lead to very different, although analogous, body plans. In fish, the fin buds originate closer to the head than the limb buds in tetrapods (Figures 16.17 and 16.18a). Not only do the positions of the fin and limbs differ, but the musculature and neuronal tissues must match the placement of the appendages. Much of this is controlled by the *Hox4*, *Hox6*, *Hox9*, *Hox10*, *Hox11*, *Hox12*, and *Hox13* genes. In fish, genes *Hox4*, *Hox6*, and *Hox9* are expressed in the hindbrain and the upper region of the spinal column to form motor neurons that control the fins. Additionally, *Hox6*, as well as *Hox9* through *Hox13* are expressed in the upper portion of the subtending tissues to signal the formation of the somatic tissues of the fin bud. In tetrapods, neuronal tissues that will develop into the motor neurons are initiated by the expression of *Hox4* and *Hox6* further down in the spinal column, and *Hox9* is expressed below that region. The somatic

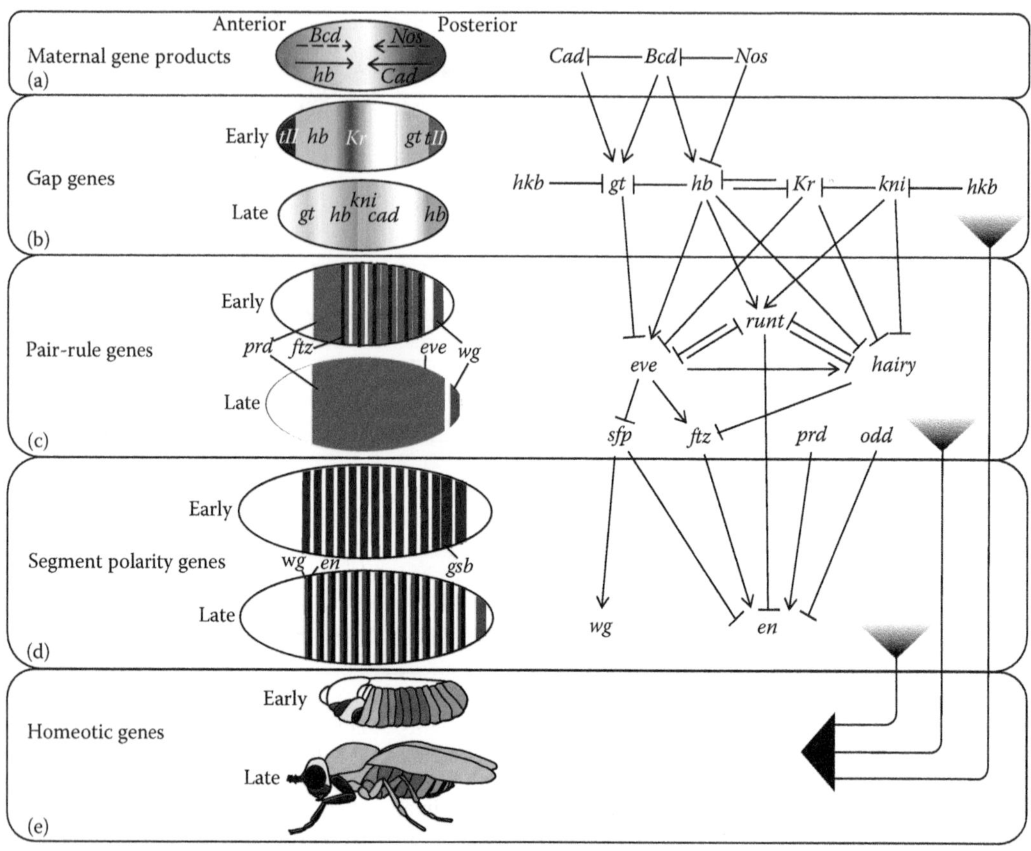

FIGURE 16.20 Sequential gene activation and inactivation during development in *Drosophila melanogaster*. The egg already contains gradients of developmental determinants by the time it is fertilized. Thus, the anterior and posterior ends have been defined. Upon fertilization, the maternal factors (a) regulate the expression of another set of genes, termed gap genes (b). These further define the ends of the animal, and they set up the expression of genes in the middle portion of the embryo. They control expression of another set of genes, called pair-rule genes (c). These are turned on or off in segments along the axis of the body of the embryo and act to control expression in yet another set of genes, called segment polarity genes (d), which essentially bisect the segments set up by the pair-rule genes. The products of the gap genes, pair-rule genes, and segment polarity genes all have effects on expression of the homeotic genes (e). Colors for homeotic genes (e) correspond to those in Figures 16.10 and 16.11.

tissues of the limb bud are induced by the expression of *Hox6*, as well as *Hox9* through *Hox13* in the same regions. Therefore, with a slight change in position, expression of analogous genes can cause large differences in the ultimate body form.

HIERARCHY AND EVOLUTION OF HOMEOTIC GENES

At one time, it was thought that there was a strict hierarchy of gene regulation, such that there were master regulatory genes that controlled a cascade of other hierarchically arranged genes. This has turned out to be only partially true, as many of the genes act at several levels in the hypothesized hierarchy, so if there are master regulatory genes, they are not truly masters over the entire process, and there are many protein and gene interactions that can change the expression patterns during the developmental process. However, promoters, enhancer elements (EEs), control regions, and global

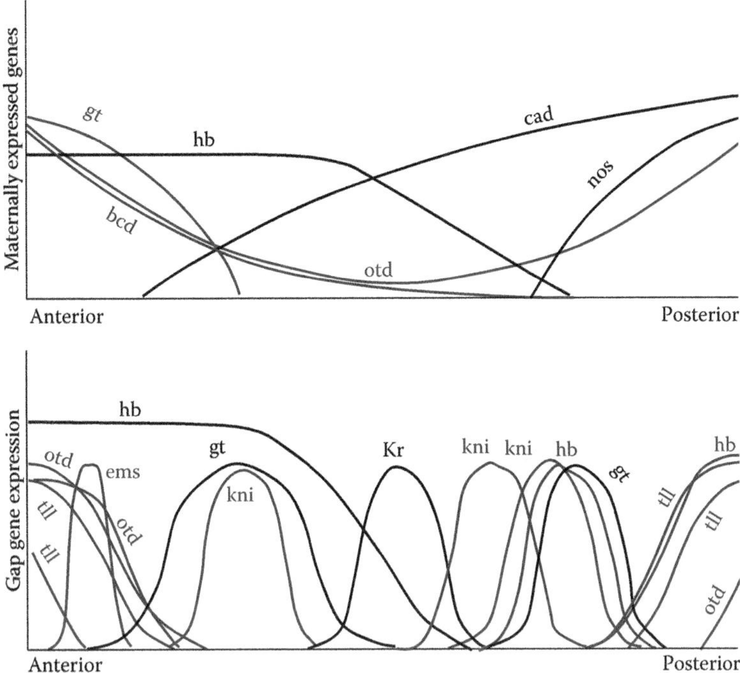

FIGURE 16.21 Relative expression of selected maternal gene products and gap genes in fruit fly (*Drosophila melanogaster*—in blue) versus wasp (*Nasonia vitripennis*—in red). For some genes, expression is nearly identical in both species (in black). Although *hb*, *nos*, *cad*, *gt*, and *Kr* exhibit very similar expression in both species, all other expression characteristics differ between the two species, which leads to the characteristic body shapes for the two species.

control regions (GCRs) appear to control the homeotic genes. These appear to be mostly *cis*-acting control regions that are able to switch on one or more of the *Hox* genes (usually in a sequential manner) that can be either upstream or downstream of the controlling elements. They do this in a coordinated fashion, such that the *Hox* genes for particular functions are tightly regulated in the timing and amount of their expression. In doing so, strict control of each part of the developmental process from the beginning to the end is maintained.

Evolution of homoetic gene clusters have been hypothesized to proceed through a process of local control of gene expression, gene consolidation, distant control, and locus duplication (Figure 16.22). Initially, one homeotic gene was present (A in Figure 16.22). Local duplication of the gene (B and C in Figure 16.22) was followed by control of the genes exerted by the effects of a region of the chromosome in the same region or by a protein or RNA factor acting to control the genes (D in Figure 16.22). This was followed by control by a downstream EE that controlled expression of many of the homeotic genes and consolidation of control by loss of function for some of the local controlling factors (E in Figure 16.22). Eventually, other upstream elements, termed GCRs, evolved which also coordinated sequential control of the expression of the homeotic genes, but the expression pattern differed from those controlled by the downstream EEs (F in Figure 16.22). This allowed differential expression of the genes in the locus at different times during development. For example, differences in early and late expression of the *HoxD* locus are thought to have led to the transition from fins to lobed fins, and lobed fins to feet, hands, and wings. Finally, duplication of the homeotic loci allowed changes in the alleles in different tissues within the organisms, such that a greater variety of tissue types and structures could be formed (G in Figure 16.22).

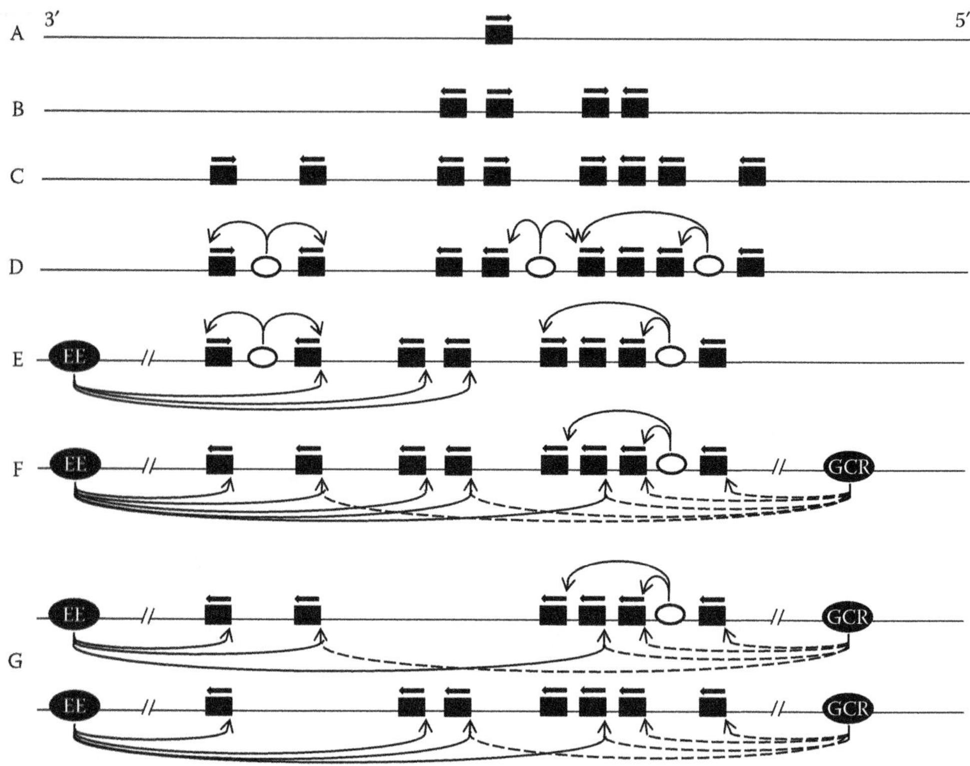

FIGURE 16.22 Model for the evolution of *Hox* gene clusters and coordinated developmental coordination of the *Hox* genes. (A–C) The locus began as a single gene that underwent several rounds of duplication and divergence of each copy. (D) Each copy was under the control of one or more controlling regions and/or proteins (white ovals). (E,F) Eventually, groups of the genes became controlled by downstream EEs and upstream GCRs. (G) Duplication of the loci by recombination of genome duplication led to changes in each of the loci that allowed more developmental elaboration.

KEY POINTS

1. Differentiation of two or more cell types exists in almost all kingdoms of organisms.
2. Bacterial cells communicate with other bacterial cells using quorum sensing mechanisms.
3. The cells in multicellular organisms must communicate with adjacent and distant cells of the organism in order to maintain the shape and functions of the organism.
4. Programmed cell death (or apoptosis) is a normal part of development in animals.
5. Homeotic proteins control the expression of many other genes and are responsible for the correct placement, structure, and function of major multicellular parts of multicellular organisms.
6. *Hox* genes and Hox proteins are a subset of the homebox class of genes and proteins, and are only found in animals.
7. Development in animals is controlled by locational and temporal expression of genes whose proteins and sometimes RNAs control sets of genes that determine the timing and location of development of structures in the organism.

8. During evolution, *Hox* genes (as well as other homeotic genes) have changed in number, timing of expression, and location of expression to produce the wide variety of extant and extinct body types.
9. Although homeotic genes and proteins (as well as other genes, proteins, and RNAs) control gene expression during development in a hierarchical manner, there are many checks and balances in the hierarchy, such that a network of control exists.

ADDITIONAL READINGS

Alberts, B., D. Bray, K. Hopkin, A. Johnson, J. Lewis, M. Raff, K. Roberts, and P. Walter. 2013. *Essential Cell Biology,* 4th ed. New York: Garland Publishing.

Alberts, B., D. Bray, A. Johnson, J. Lewis, M. Raff, K. Roberts, and J. D. Watson. 1994. *Molecular Biology of the Cell.* New York: Garland Publishing.

Freitas, R., G. J. Zhang, and M. J. Cohn. 2004. Biphasic Hoxd gene expression in shark paired fins reveals an ancient origin of distal limb domain. *PLoS ONE* 2:e754.

Lemons, D. and W. McGinnis. 2006. Genomic evolution of Hox gene clusters. *Science* 313:1918–1922.

Monteiro, A. S. and D. E. K. Ferrier. 2006. Hox genes are not always collinear. *Int. J. Biol. Sci.* 2:95–103.

Ravi, V., K. Lam, B.-H. Tay, A. Tay, S. Brenner, and B. Venkatesh. 2009. Elephant shark (*Callorhinchus milii*) provides insights into the evolution of Hox gene clusters in gnathostomes. *Proc. Natl. Acad. Sci. USA* 106:16327–16332.

Sheth, R., L. Marcon, M. F. Bastida, M. Junco, L. Quintana, R. Dahn, M. Kmita, J. Sharpe, and M. A. Ros. 2012. Hox genes regulate digit patterning by controlling the wavelength of a Turing-type mechanism. *Science* 338:1476–1480.

Shubin, N. H., E. B. Daeschler, and F. A. Jenkins. 2006. The pectoral fin of *Tiktaalik roseae* and the origin of the tetrapod limb. *Nature* 440:764–771.

Swalla, B. J. 2006. Building divergent body plans with similar genetic pathways. *Heredity* 97:235–243.

17 Development
Part II—Plants

INTRODUCTION

As mentioned in Chapter 16, *Hox* genes are not the only genes that control development in animals. The maternal factors, gap genes, pair-rule genes, segment polarity, and many other homeotic genes exist in multicellular organisms. Each of the aspects of the development is tightly controlled by the products of these genes. They include proteins and RNAs. Some of them show a degree of similarity to some of the genes that control development in plants. The products of these genes, as well as many others, control the organization of various tissues, as well as the placement of organs on the plants. This indicates either a very ancient origin for these genes that predated the separation of the Archaeplastida (plants) from the Opisthokonta (animals and fungi), or that there was an ancient horizontal transfer of one or more versions of these genes that occurred after the divergence of the two groups.

Approximately 1.9–2.0 billion years ago, a nonphotosynthetic protist formed an endosymbiotic association with an ancestral cyanobacterium, thus benefitting the host organism by gaining the ability to obtain nutrients by photosynthesis, and presumably benefitting the cynaobacterial endosymbiont, as well. This amalgamation of two different organisms, one a eukaryote with a nucleus and mitochondria and the other a photosynthetic bacterium, became the initiating member of the Archaeplastida. By 1.9 billion years ago, two distinct lineages had diverged (Figure 17.1), glaucophytes and rhodophytes (red algae). These diversified, and while glaucophytes consisted mainly of unicellular species (and a few species that formed small clusters of cells), as the rhodophytes diverged, multicellular forms evolved. By 1.2 billion years ago, some members of one line of the Archaeplastida had undergone changes in their photosynthetic and other cellular processes, becoming unicellular chlorophytes (green algae). Initially, most were unicellular, but soon thereafter several multicellular forms appeared. Some were essentially collections of identical or very similar cells, while in others more than one cell type developed (Figure 17.2). As the organisms evolved, the cell types became more differentiated from one another, taking on different roles in the body of the organism. For example, in some of the multicellular organisms, there were cells specialized in motility and orientation of the organism to optimize the amount of photosynthesis, while other cells performed the majority of the photosynthetic processes for the organism. One branch of the Chlorophyta diverged and led to the evolution of all other groups of the Archaeplastida, including the unicellular Prasinophyta (marine and freshwater planktonic algae), and the multicellular taxa from the Charophyta (freshwater green algae) to the Spermatopsida (seed plants, such as gymnosperms and angiosperms). Some unicellular rhodophytes and chlorophytes became endosymbionts, and later organelles, of several protist groups, including Chromalveolata (Chromista, Stramenopiles, and Hacrobians) and Excavata. Some of these groups diversified to produce unicellular and multicellular forms, including the brown algae, golden algae, yellow-green algae, diatoms, dinoflagellates, and others. Thus, the innovation of photosynthesis was transferred via horizontal gene transfer of entire genomes to other kingdoms of organisms.

As the evolution of these lineages continued, the male and female gametes became more specialized and had different morphologies. Thus, the basal members of the Archaeplastida primarily exhibit isogamy (i.e., the gametes all have the same morphology), while the multicellular species all exhibit anisogamy (i.e., the male and female gametes are morphologically distinct from one

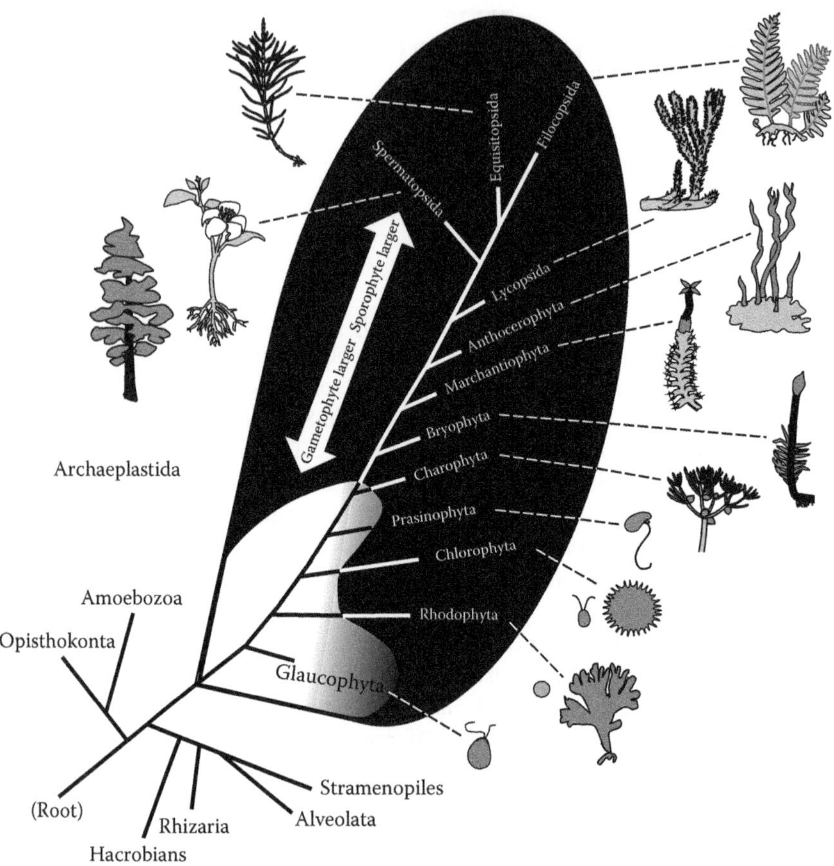

FIGURE 17.1 Phylogram of the major groups in the Archaeplastida based on rRNA genes. Rooting and relationships to other Kingdoms are indicated. The extent of multicellular taxa is indicated with black background, while unicellular taxa are indicated with a white background. Most taxa are multicellular. Diagrams of representatives of each group are shown. Gametophytes dominate the more primitive multicellular taxa, while sporophytes dominate in the more recently derived, and more complex, groups.

another; Figure 17.2). In addition to a shift from isogamy to anisogamy, there was a shift from a gametophyte-dominant lifestyle (haploid organisms), to a sporophyte-dominant lifestyle (diploid organisms). In members of the Bryophyta, Marchantophyta, and Anthocerophyta, the gametophyte (the haploid phase) is the dominant form of the organism. The sporophytes (the diploid form) are short lived and generally small (often microscopic). Up until about 450–500 million years ago, plants were small and primarily marine or aquatic, and exhibited a gametophyte-dominant lifestyle. However, during the Devonian period (420–360 million years ago), a large increase in the diversity of terrestrial plants (as well as animals) occurred. All of the major groups of land plants first appeared during that time, including the Lycopsida (lycopods), Filocopsida (ferns), Equisitopsida (horsetails), and Spermatopsida (gymnosperms and angiosperms). In each of these, the sporophytes are the dominant forms, while the gametophytes are generally inconspicuous and short lived. For example, the sporophytes of some members of the Spermatophyta are some of the largest organisms on Earth (e.g., giant sequoia trees), while the male gametophytes consist of four cells (including two sperm nuclei), and the female gametophytes consist of a few hundred cells (including one egg cell).

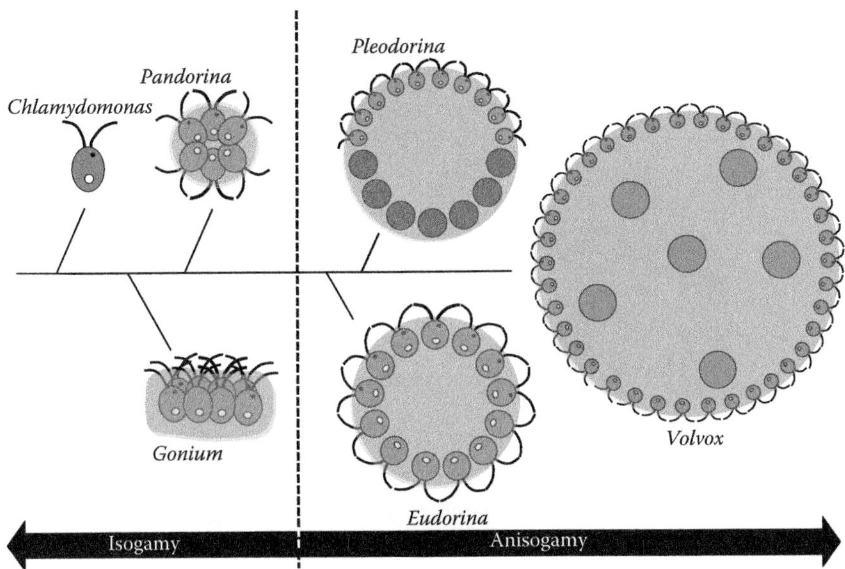

FIGURE 17.2 Cladogram including several species of green algae demonstrating the increasing complexity of multicellular forms. *Chlamydomonas* species are unicellular motile green algae. They have a single large chloroplast, as well as a light sensitive organelle (black dot inside cell) that is used to orient the cell toward the light. While cells of two different mating types join to form a diploid zygote, individuals of this species are almost always found as single cells. *Gonium* species are found in small groups of cells of identical morphology embedded in a gelatinous matrix extruded by the cells. The cells are oriented with their flagella pointing outward so that their movement will move the ball of cells. Only some cells have retained the photosensor organelle, indicating a degree of differentiation among the cells of the colony. In *Pandorina* species, two cell types also exist, but they are tightly organized in a ball and they are embedded in a gelatinous matrix, similar to that of *Gonium* species. In each of these genera, the gametes of each species are indistinguishable from one another, and are thus termed isogamous. Further elaboration is found among species of the genus *Eudorina*. The ball of cells consists of a larger number of cells, and the cells with the photoreceptors are on one side of the ball. Additionally, the male and female gametes are morphologically distinct, and thus are termed anisogamous. Species of *Pleodorina* develop into balls of cells, some of which form other small ball colonies that eventually separate from the main colony to produce a new colony. Species of *Volvox* form large colonies with small motile photosynthetic cells on the outer surfaces and new smaller colonies on the inside, which are released to form new large colonies. As with species of *Pleodorina* and *Eudorina*, species of *Volvox* are anisogamous.

PLANT MORPHOLOGY

Almost all plant species form rigid cellulose cell walls that protect the cells from damage and desiccation. However, the walls also restrict changes in cell size and present obstacles to cell-to-cell communication and cell division (Figure 17.3). Plant cells communicate with other via two main routes. The cell walls and the areas between the cell walls are filled with aqueous fluids that allow the passage of small molecules. Some signaling and other molecules are present in this fluid. This is similar to some types of signaling in animal and other types of cells, because some of the molecules are recognized or transported into the cells via receptors and transporters in the plasma membrane. The second mode of communication is via plasmodesmata. These are pores where the membranes of the adjacent cells are physically joined, thus creating direct connections between the cytosol of the two cells. These passageways are controlled by specific sets of proteins, as well as via part of the endoplasmic reticulum. Signaling proteins, as well as small RNAs, are transferred from cell to cell via these plasmodesmata.

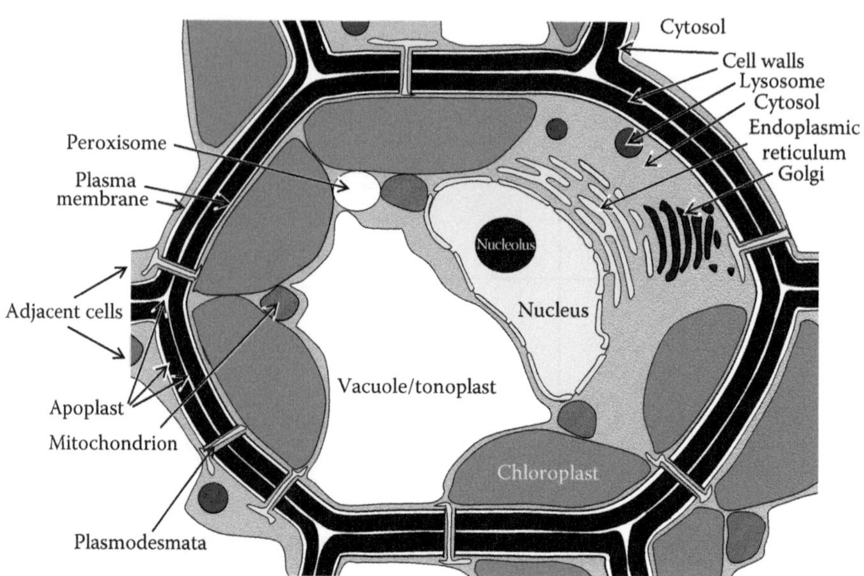

FIGURE 17.3 A typical plant cell surrounded by six adjacent cells. The major organelles and structures are indicated.

Because plant cells can change shape only if the cell wall is softened and new wall material is synthesized and assembled, the volume of the cell and the pressure inside the cell must be controlled. Plant cells have a large vacuole, called the tonoplast, which is partly responsible for this purpose. However, it is also used to maintain concentrations of specific cellular constituents in the cells, and these are varied in different cell types. Another problem arose during the initial stages of evolution of the Archaeplastida, as occurs with every endosymbiotic association. The original protist cell had nuclear genes and their products (RNAs and proteins), as well as mitochondrial genes and their products. The products controlled the internal cellular characteristics, as well as transport of molecules across the membranes. But, in the beginning, the cyanobacteria that would become the plastids carried with them a full set of their own sets of genes, RNAs, proteins, and other molecules. Compatibility among the gene products of the host cell and the cynobacterial cell probably required some time to rectify. However, in all of the contemporary members of the Archaeplastida, all of the genes and gene products interact cooperatively in the cells. In fact, in many cases the gene products are shuttled directly from one organelle to another. In many cells, the membranes of the nuclei, mitochondria, and chloroplasts are physically in contact with one another in the plant cells. Often, another organelle, the peroxisome (which contains no DNA), is also in contact with the plastid and mitochondrial membranes (Figure 17.3). The reason for this is that various molecules are being shuttled across these membranes during photosynthesis (and other cellular processes). Apparently, this is more efficient than having any of those products traverse the cytosol.

DEVELOPMENT IN PLANTS

The process of development in plants has taken a very different path from that in animals, although homeodomain/homeobox genes and proteins are also used in the process. As with animal development, the homeobox proteins contain a conserved homeodomain near the carboxyl end of the protein that is well conserved among these genes (and is similar to the homeobox proteins in animals), while the majority of the protein is unique to each family or set of proteins (see Figure 16.9). Also, as with all homeobox proteins, they are DNA binding proteins that interact with other transcription factors to activate specific sets of genes that then cause a specific cellular process to occur that leads to formation of specific tissues and structures in the organism.

In gymnosperms, the female gametophyte (called an archegonium) is found in the female cones, and consists of a few hundred to a thousand haploid cells, including a single egg cell. In angiosperms, the female gametophyte, which exists within the ovule of the flower, consists of from six to a few dozen haploid cells, including a single egg cell. In a typical angiosperm, the single large egg cell is located at one end of the gametophyte surrounded by two smaller cells, called synergids (Figure 17.4). Three cells at the opposite end of the gametophyte are called antipodals. Two nuclei (called polar nuclei) are located near the equator of the cell. During fertilization, two sperm nuclei are delivered to the gametophyte through a pollen tube that grows out of the pollen grain. One of the sperm nuclei enters the egg cell and joins the egg nucleus to form the diploid zygote (Figure 17.4). The other sperm nucleus fuses with the two polar nuclei to initially produce a triploid cell that

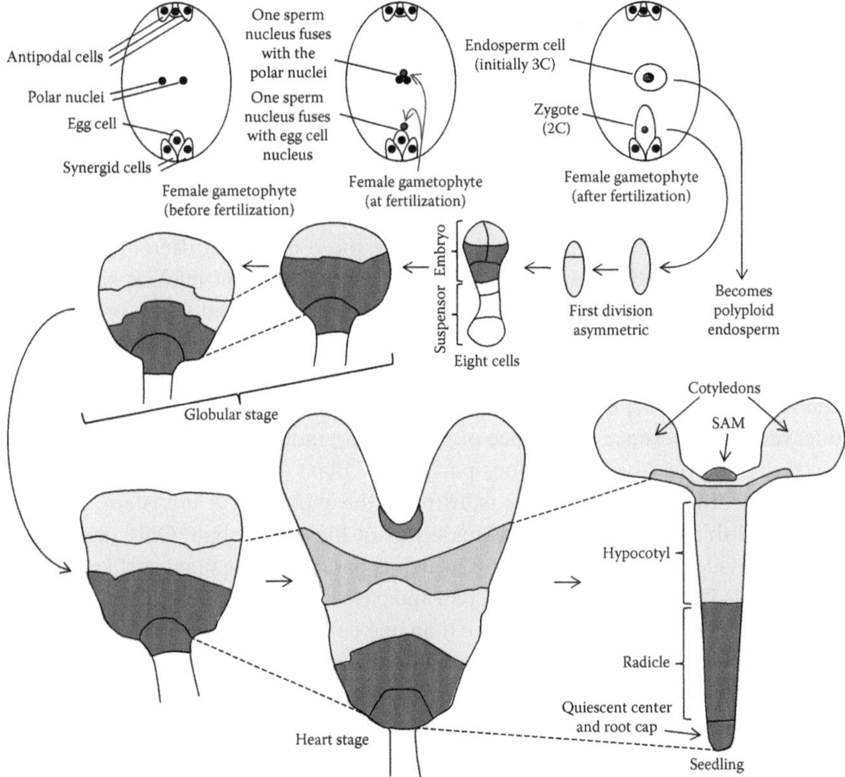

FIGURE 17.4 Fertilization and development of an angiosperm (eudicot). The female gametophyte develops in the ovule that is formed within the ovary of the sporophyte plant. In most eudicots, there is a single egg cell, two synergid cells, three antipodal cells, and two polar nuclei (other patterns of cells exist among angiosperms). When the pollen tube reaches the ovule, two sperm nuclei are released. One fertilizes the egg, while the other fuses with the two polar nuclei to form a triploid nucleus that will produce the nutritive endosperm. The zygote undergoes an asymmetric division. The larger cell becomes the suspensor that connects the embryo to the ovary wall, while the small cell develops into the embryo. By the eight-cell stage, the shoot and root axes have been established. The shoot, hypocotyl, and radicle have been defined by the globular stage, and the shoot apical meristem has been set up by the heart stage, consisting of thousands of cells. As the embryo grows further, the cotyledons develop and lengthen, while the shoot apical meristem begins to produce the first leaf primordial that will eventually become the true leaves of the seedling. The following colors indicate each of the various regions of the embryo: dark green = shoot apical meristem; light green = shoot tissues; blue = root tip tissues; yellow = hypocotyl (stem) tissues; orange = radicle (root) tissues.

will become nutritive tissue called the endosperm. Although the endosperm is triploid at first, the cells become polyploid in many species. It becomes a source of amino acids, nucleic acids, carbon, etc. for the developing embryo in most species (e.g., in most dicots, including *Pisum sativum*), but becomes a nutrition source for the seedling in other species (e.g., in grasses, including *Zea mays*). When you eat corn, the sweet and starchy parts are the endosperm, whereas when you eat peas, the sweet and starchy parts are the cotyledons (seed leaves). In those plants, the endosperm provides nutrition to the embryo and is completely gone by the time the seed is mature.

The fate of each portion of the zygote is determined by the time of fertilization. The first division is asymmetric, producing a small cell destined to become the embryo, and a larger cell that will become the suspensor (Figure 17.4). This cell continues to divide primarily along its long axis, and the cells become polyploid and polytene. The other cell divides in all three planes to become a ball of cells, organized to become the various parts of the plant. At the late globular stage, groups of cells differentiate into the upper portions of the shoot, lower portions of the shoot, hypocotyl (i.e., embryonic stem), radicle (embryonic root), and quiescent center/root cap region. Once the embryo has progressed to the heart stage, a true shoot apical meristem has developed. This will become the region from which all of the cells of the shoot will originate in the plant. While the embryo is still in the plant ovary, the shoot apical meristem will begin to form the first true leaves of the plant that will be of the same type as those that will be produced by the adult plant.

Other patterns of organization exist in the plant. Layers of cells (called histogenic layers) that are genetically identical have been described in a large number of plants. These histogenic layers were discovered when plants were mutagenized during the early stages of development. Sometimes, a chimeric mutant would be formed. The first ones were those that had different ploidy levels (some layers were diploid, while others were tetraploid leading to different nucleus sizes) in the layers. Diploid layers would continue to produce diploid cells, while tetraploid layers produced only tetraploid cells. As the nuclei were observed progressing from the shoot apical meristem to the basal portions of the shoot, two, three, four, and sometimes five layers of genetically related cells were found. Other chimeras consisted of plants that had particular types of variegation (color variants in the shoots and leaves). The presence or absence of green color, indicating chloroplasts and chlorophyll, was used to trace the layers through various parts of the plant (Figures 17.5 and 17.6a, b). Shells of tissues were being formed in these plants, starting at the shoot apical meristem. The first layer is always a single cell thick and becomes the epidermis of the entire plant. Only one layer is formed because the cells divide in only one direction, which is parallel to the plant surface. Also, this layer only has chloroplasts in the guard cells that open and close to allow gases to pass into and out of the leaves. The next histogenic layer often is more than one cell thick in the shoot apical meristem, and becomes thicker further down in the shoot and in the leaves, because the cells divide in more than one direction. The third layer is several cells thick in the shoot apical meristem. It produces cells that essentially become the entire middle portion of the plant body. Vascular tissues (e.g., vessels and tracheids) form independent of these layers, producing yet another pattern within the plant.

Additional patterns are found in plants, all of which are caused by differential expression of genes that control transcription of other sets of genes, whose products are responsible for the structured patterns (Figure 17.6c and d). Some of the genes and their protein products contain homeodomains, analogous to those in animals. Several models of the shoot apex have been proposed to explain the structure and the developmental events that occur in the shoot apex. The tunica-corpus model (Figure 17.6c) is similar to the histogenic layer view of the shoot, although each histogenic layer may be comprised of more than one tunica layer in the shoot apical meristem region. The tunica cells form layers that cover the corpus. The corpus consists of cells that do not necessarily form precise layers, but instead form a ball of cells that is distinct from other regions of the shoot apical meristem. The zonation model of the shoot (Figure 17.6d) was first described cytologically, but has been since confirmed in molecular studies of gene expression. The cells in the central zone are large with diverse cytological staining patterns. The cells in this region divide infrequently. In the center of this zone exists a small number of cells often referred to as apical initials, which

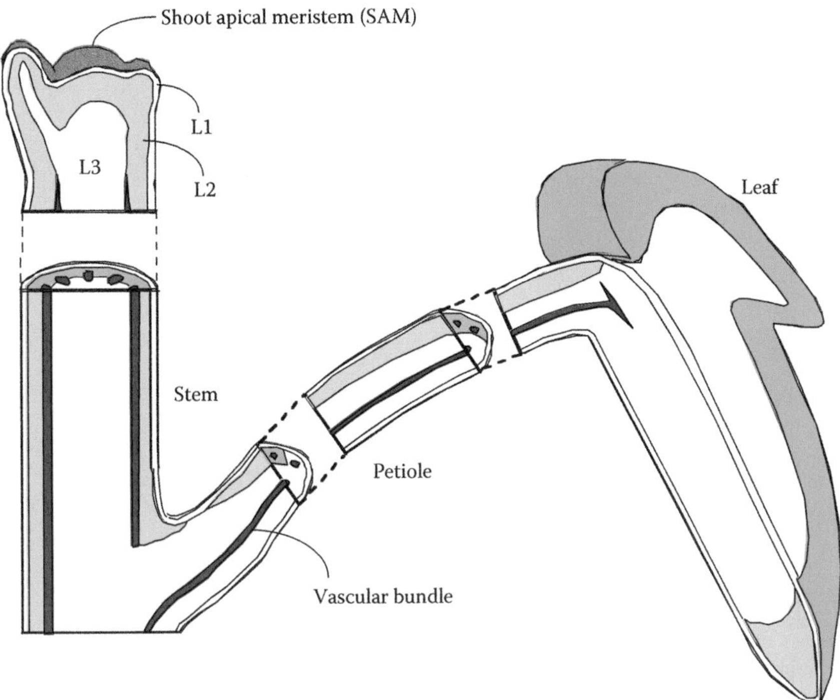

FIGURE 17.5 Cross section of a chloroplast chimera (variagation) of *Hedera helix* (English ivy) showing the histogenic layers in various parts of the plant (each layer uniquely colored). All cells in each layer are genetically derived from cells in the shoot apical meristem (SAM) that are continuous with that layer. Histogenic layer L1 is the epidermis, which is one cell thick throughout the plant. Only the guard cells have functional chloroplasts. All other cells are phenotypically white. Layer L2 (green) is 2–4 cells thick at the SAM, although in some plants this ranges from 1 to 5 cells thick. This layer forms a shell of cells that becomes the outer portions of the stems and leaves. Layer L3 (white) consists of a ball of cells in the SAM that is approximately 5 cells deep and 3–5 cells wide. Cells from this layer are the source of all of the inner portions of the stems and leaves.

have sometimes been compared with stem cells in animals. They are large cells with very large nuclei and one cell on each side of the zone divides about once per leaf initiation (called a plastochron). This can be from once per day to once per several weeks (depending on the species). Genetically, all of the cells of the plant are derived from cells in this zone. The cells in the peripheral zone are very different from the cells in the central zone. Cytologically, they stain more densely, the cells are much smaller, and they undergo frequent mitosis and cytokinesis events. This is the zone where leaf primordia are initiated. Essentially, control of growth and the positions of leaves exist within the cells of the central zone. But once that control has been exercised, the cells of the peripheral zone respond to those signals and begin the genetic and molecular programs to form leaves, petioles, stems, flowers, or other organs.

GENE EXPRESSION DURING DEVELOPMENT

Developmental regulation begins soon after fertilization, and polarity of the embryo is established. Development of the embryo proceeds through several crucial stages: the 8-cell, 16-cell, globular, heart-shape, early embryo, and late embryo stages. A gene called WUS (shortened from WUSCHEL) is expressed by the eight-cell stage in the cells that will become the shoot end of the embryo (Figure 17.7). It continues to be expressed throughout the stages of development and eventually establishes the central zone of the shoot apical meristem. Three other gene products

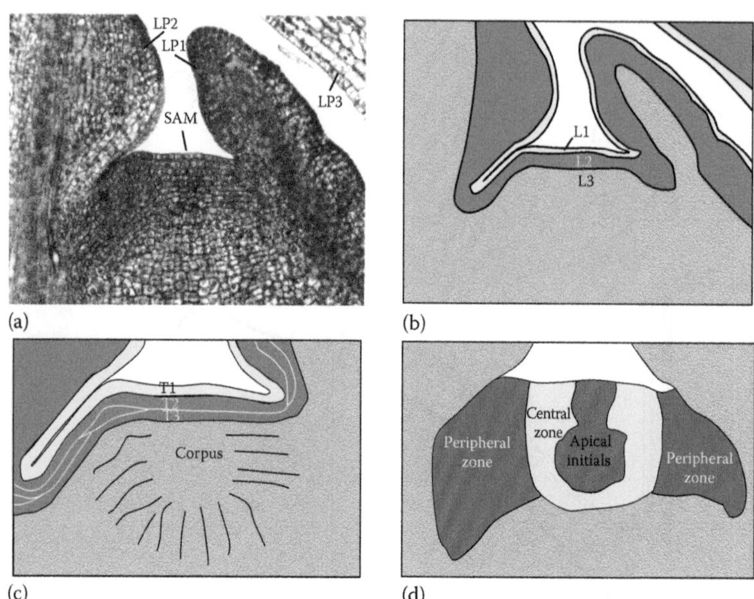

FIGURE 17.6 Photomicrograph and cell patterns of a *Hedera helix* (English ivy) shoot apex. (a) Photomicrograph showing the cellular arrangement of the SAM and three leaf primordial (lp1, lp2, and lp3). This plant has leaves that are arranged alternately on the stem. One leaf is initiated on one side, and then the next leaf is initiated on the opposite side of the apex above the previous leaf primordium. A new leaf primordium has just been initiated in this shoot apex, as indicated by the slightly raised area on the left side of the SAM. (b) The same shoot apex outlining the three histogenic layers (L1, L2, and L3) present in this plant. (c) The same shoot apex using the tunica-corpus model of shoot apical structure. Note that histogenic layers L2 and L3 consist of more than single layers of cells. (d) The same apex showing the zonation model of shoot apex structure. These patterns were derived from cytological staining and morphological characteristics of the cells in the apex. Cells in the peripheral zone stain darkly and the cells and nuclei are small. Cells in the central zone stain lightly and the cells and nuclei are larger. The apical initials stain lightly, and the cells and nuclei are large, and are similar to those in the central zone. However, they also exhibit a higher degree of extracellular staining, indicating that their cell walls are less developed, and may be more porous.

then begin to be expressed: CLV1 (CLAVATA1), a cell surface receptor only expressed in the central zone; CLV2, a cell surface receptor that interacts with the CLV1 protein and is expressed throughout the shoot apical meristem; and CLV3, a ligand protein only expressed by cells in the upper portion of the apical initials zone, which activates the CLV1/CLV2 receptor once it attaches to the receptor. WUS is secreted by cells in the globular portion of the apical initials zone and activates cells in the upper region of the apical initials zone to express CLV3. Once produced and secreted, CLV3 finds its way to the apical initials in the globular region and inhibits further expression of WUS. This feedback mechanism tightly controls the cells in this region that causes them to express other sets of genes that control the structure and mitotic activities of cells in the immediate vicinity.

By the middle of the globular stage another gene called WOX (WUSCHEL-related homeobox gene) is expressed at the radicle end of the embryo (Figure 17.7), which begins to set up the root structures of the embryo. WOX is secreted and influences cells in the vicinity that maintains them as stem cells. When the root is mature, the CRN protein expressed in cells deeper in the root apical meristem stimulates WOX in the cells of the quiescent center. Cells lower in the root meristem express ACR4, which inhibits WOX and stem cells, thus allowing the affected cells to begin to differentiate in the root tip. Additionally, CLE40, expressed by the root cap cells, activates the ACR4 gene. In mature plants, if the root is still making progress through the soil, CLE40 is

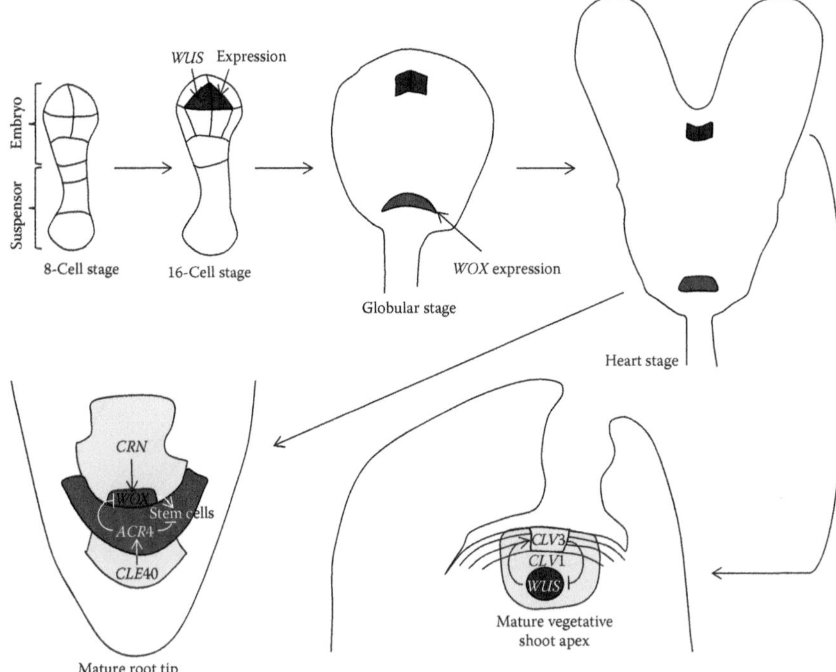

FIGURE 17.7 The role of the homeotic proteins WUS and WOX in plant development. By the 16-cell stage, WUS is expressed in what will become the shoot end of the plant. It continues to be expressed throughout the life of the plant in the shoot apex. Its expression is downregulated by the effects of CLV1, CLV2, and CLV3. CLV1 (expressed only in cells within the central zone) and CLV2 (expressed throughout the shoot apex) are cell surface receptors, while CLV3 is a ligand that attaches to the CLV1/CLV2 cell surface complex. This creates a balance system that determines the structure and function of the shoot apical meristem. Similarly, WOX begins to be expressed in the radicle end of the embryo by the early globular stage. It continues to be expressed in the root in the root apical meristem throughout the life of the plant, and its expression is controlled from the cells immediately above and immediately below the quiescent center of the root. This region is termed the quiescent center because the cells in this region rarely divide. This is analogous to the cells in apical initials zone of the shoot apical meristem. Control by the other gene products (e.g., CRN, ACR4, and CLE40) maintains the structure and function of root apical meristem. Arrowheads in the root tip and shoot apex indicate stimulation, while bars indicate suppression of affected cells.

expressed, which activates the expression of ACR4, which inhibits WOX, which causes the stem cells to begin to grow and differentiate, thus causing the elongation of the root tip.

FORMATION OF LEAVES AND FLORAL ORGANS

Leaves come in many forms and their positions and orientations along the lengths of the shoots are characteristic for each species. The positions of the leaves and determination of the upper and lower surfaces of the leaves are caused by the expression of a set of genes that control other larger sets of genes, in a manner analogous to developmental processes in animals (Figure 17.8).

The primary genes that are expressed in the center of the shoot apical meristem are STM (SHOOT MERISTEMLESS) and KNAT1 (knotted-1-like 1 gene) first identified in *Arabidopsis thaliana*. These gene products inhibit the expression of AS1 and AS2 (ASYMMETRIC LEAVES 1 and 2), two gene products necessary for the initiation and growth of leaf primordia. Once cells are far enough away from the sources of the STM and KNAT1 gene products, the cells are no longer blocked from growth and division and leaf primordia are initiated, beginning with increases in mitotic activity. Once initiated the primordia become influenced by PHB (PHABULOSA) and

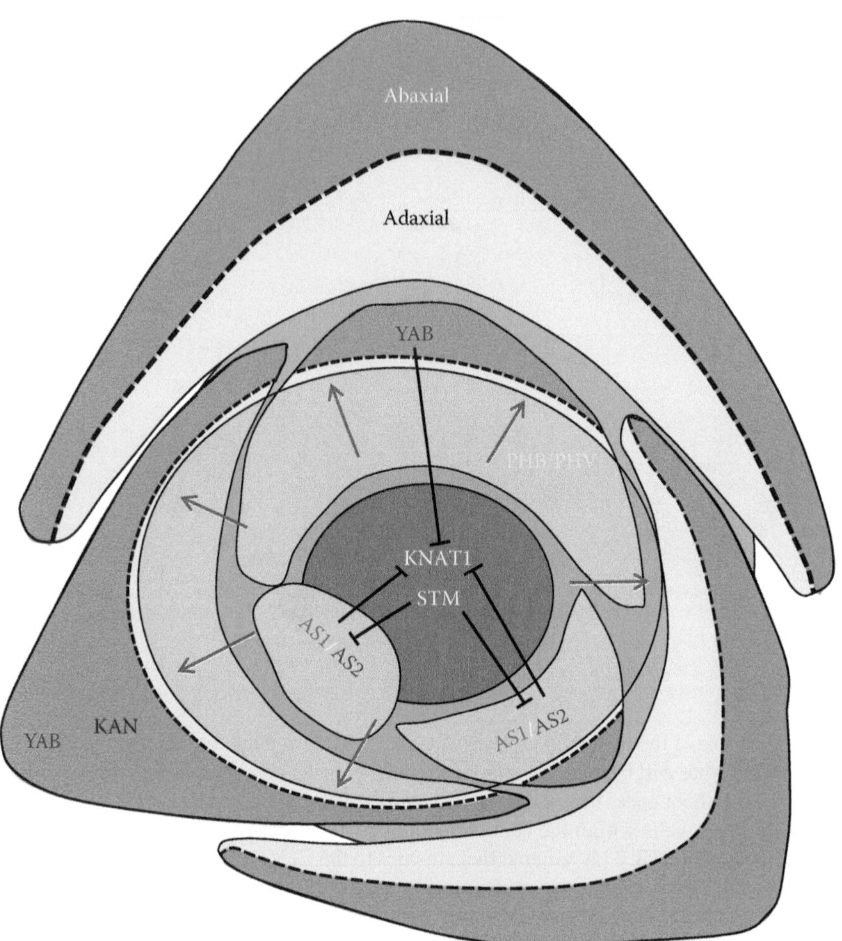

FIGURE 17.8 Control of phyllotaxy (arrangement of leaves on the stem) by specific genes and gene products in a cross section through the shoot apex. The homeodomain protein STM (STEMLESS) is central to the control of leaf primordium initiation, in that it inhibits the process. However, its effects are diluted out away from the central portions of the SAM (brown). KNAT1 has similar effects, although it is inhibited by other proteins (e.g., AS1, AS2, and YAB) emanating from the leaf primordia (gray, green). The PHB and PHV proteins are present throughout the apex, and influence the orientation of the leaf primordial by defining the adaxial sides of the leaves (those initially facing the SAM, but later becoming the upper surfaces of the expanded mature leaves). Conversely, the YAB and KAN proteins inhibit the influences of KNAT1, and thus allow growth and differentiation of the leaf primordia, and influencing formation of the abaxial side of the leaf primordia. The relative concentrations of each of the proteins define where the next leaf primordium will initiate and determine the structure of the adaxial and abaxial sides of the leaves. Homeodomain genes are in white font; Zn finger transcription factors are in violet font; homeodomain leucine zipper proteins are in yellow font; repressor of Zn finger transcription factor is in brown font; MYB-domain transcription factors are in red; other transcription factors are in light blue font. Arrows indicate stimulatory effects, while blocked lines indicate inhibitory effects.

PHV (PHAVOLUTA), two homeodomain leucine zipper proteins that promote transcription of a number of genes. In doing so, the adaxial side (the side closest to the meristem, eventually becoming the upper side of the leaf) is defined. Next, the YAB (YABBY, transcription factor containing Zn finger and high mobility group domains) and KAN (KANADI, transcription factor containing a GARP domain, related to MYB proteins that control cell proliferation and differentiation) genes are activated, which are transcription factors for a set of genes that define the abaxial side of the leaves (the side away from the meristem, eventually becoming the lower side of the leaf). Part of the function of YAB is to inhibit signaling molecules being expressed in the meristem and in the adaxial portion of the leaf.

Control of the position of the next leaf primordium is dependent on the genes being expressed and the signals being sent by the cells in the center of the shoot apical meristem. It also is dependent on where the last primordium was initiated, because that primordium is expressing different genes, which are producing a different set of signals. The size of the shoot apical meristem, the genes being expressed, and the signals being sent and received all combine to set the position for initiation of the next leaf primordium, which is characteristic for each species. The gene expression patterns have evolved to form a wide variety of leaf ontogenies, ranging from simple to complex (Figures 17.9 and 17.10). Some produce two leaf primordial synchronously (or nearly so) on the opposite sides of the meristem, producing a leaf pattern simply termed opposite. Others produce leaves that alternate sides of the meristem with each plastochron (the time/distance between the appearance of each new leaf). A decussate pattern is produced when primordial are produced in an opposite pattern that rotates 90° with each plastochron. Some produce more than two leaves that

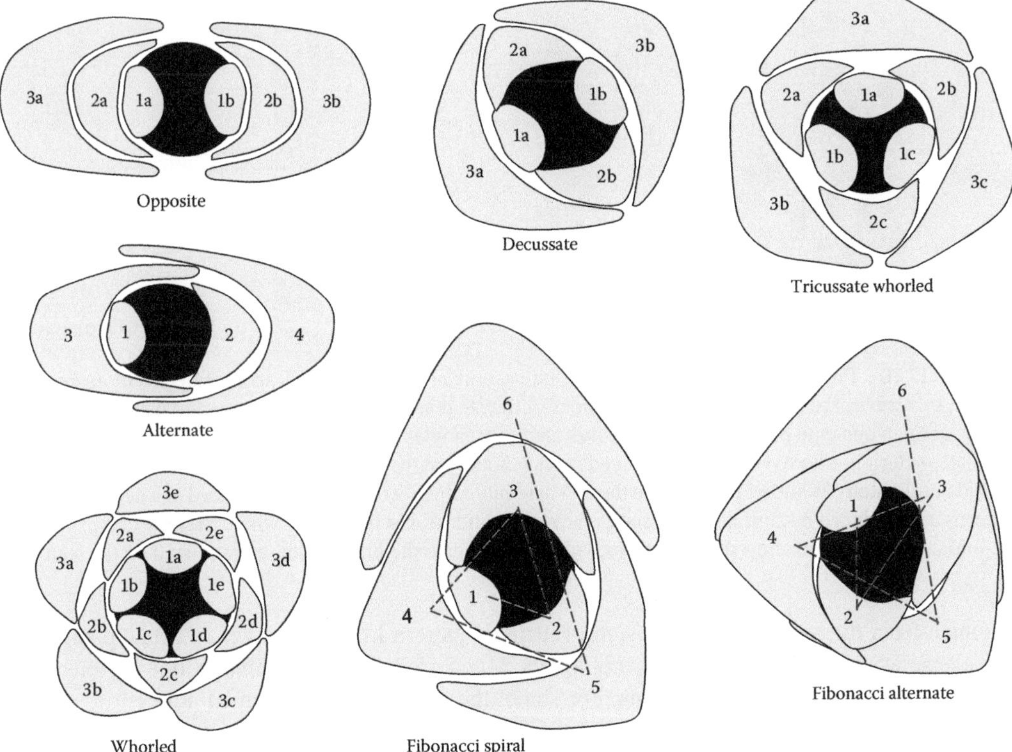

FIGURE 17.9 Leaf arrangements (phyllotaxy) exhibited by a variety of angiosperm species, as viewed from the top of the shoot. Some species produce two leaves 180° from one another each node (the position on the stem where leaves appear), in an opposite pattern. Others produce one leaf per node, but they alternate sides of the stem at each node. A decussate pattern is similar to an opposite pattern, except the leaves are rotated 90° around the shoot axis every node. Tricussate is similar, except that three leaves are produced at each node, and the leaves at the next node are rotated by 120° around the shoot axis. Three or more leaves per node are termed a whorl, and some plants have many leaves per whorl. Many plants produce leaves and flowers in spiral patterns, known as Fibonacci spirals or Fibonacci alternating arrangements. Both are similar. In one, the next primordium initiated appears close to the last one initiated, such that they are adjacent to one another on the stem. In the other, the next primordim initiated is across the shoot by a specific number of degrees around the stem from the last formed leaf primordium.

FIGURE 17.10 Photographs of plants that exhibit several of the phyllotaxy arrangements. In an opposite pattern, two leaves are produced on opposite sides of the stem at each node. In a decussate pattern, leaves are produced in an opposite pattern, although leaves are at 90° angles to the leaves of the adjacent two nodes. An alternate pattern is where one leaf is produced at each node, but the leaves of adjacent nodes are on the opposite side of the stem. Whorled patterns are those where more than two leaves are produced at a node. Fibonacci patterns are produced by similar processes as the other patterns, but the angles where the new leaf primordial are initiated differ, producing spiral patterns. Numbers indicate the order of leaves produced on the stem.

originate from the same spot on the stem, causing a pattern known as a whorl. Some plants have very dense whorls consisting of dozens of leaves. Many produce patterns that exhibit various spiral patterns, known as Fibonacci patterns. For some, the new leaves are formed adjacent to the last formed leaf, creating a Fibonacci spiral pattern. For others, the new leaves are spaced away from the last formed leaf, producing a Fibonacci alternate pattern. Nonetheless, all of the patterns are set up by the expression patterns and the concentrations of signaling molecules and receptors on the cells within the shoot apical meristem and subtending leaf primordia.

Most plants are constrained geographically, such that once they are rooted in the ground they cannot physically move. They are then subjected to heat, cold, flooding, drought, herbivore damage, infectious organisms, and other physical challenges. They must be able to sense when to grow and when not to grow. They must develop reproductive structures when the conditions are correct for doing so. For example, if they produce reproductive structures when the pollinators are not present, they may go extinct. Because of this, plants have evolved to have elaborate molecular mechanisms to determine the environmental conditions around them that signal growth and dormancy, as well as when to grow vegetative structures (e.g., leaves) versus when to grow reproductive structures (e.g., cones or flowers). Plants can respond to light, temperature, nutrient availability, and other environmental factors. The mechanisms and some of the genes involved have been identified (Figure 17.11). Several receptor molecules react when exposed to specific wavelengths of light, primarily in the blue range and in the red/far-red ranges. These signal circadian rhythm proteins. This allows the plants (including some seeds) to measure day length, thus signaling other proteins and genes in the plant (or seed) whether to initiate growth (or germination) or not. These mechanisms are not as important in the tropics, and are less prevalent in tropical plants, but they are crucial in temperate and colder regions. Evolution has favored these mechanisms and genes in plants native to temperate and polar climates. Similar situations are found in proteins and genes important in sensing temperature. Warm temperatures stimulate growth in most plants, while cold temperatures produce the opposite effect. However, some plants and seeds require a period of cold prior to germination, growth, or flowering. This is called vernalization. This is usually found in cold weather plants that need mechanisms that allow them to flower and germinate during the proper seasons. Specific proteins inhibit growth, germination, and/or flowering in certain plants. Cold temperatures over a period of time are needed. During this time, the inhibitor effects degrade, such that when the

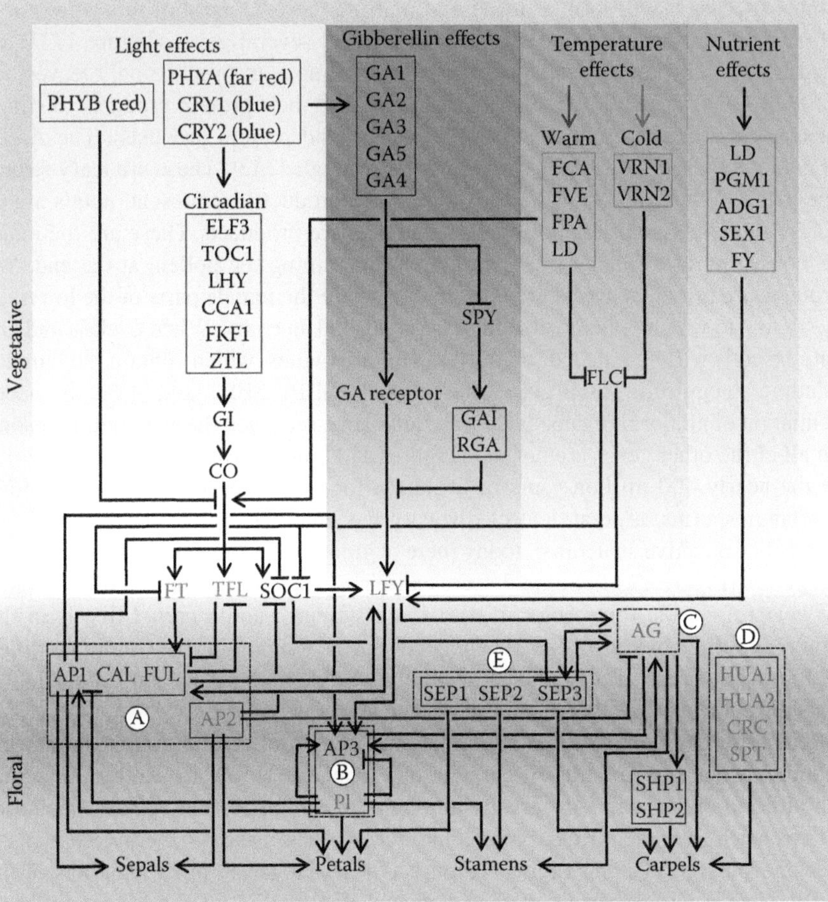

FIGURE 17.11 Model of some of the genes/proteins/factors that lead to the formation of shoots, leaves, and floral organs in angiosperms. Plants respond to light, temperature, nutrient levels, and hormones (in this case gibberellins). Each pathway leads into a set of genes that either stimulate vegetative growth or signal a transformation from vegetative growth to floral development. MADS-box transcription factors are in violet font; kinase regulation proteins are in light blue; other transcription factors are in light green font; and all others are in black font. Arrows indicate stimulation, while bars indicate inhibition. Gene classes A–E correspond to those in Figure 17.13.

temperatures warm the plant will begin to grow, germinate, or flower (depending on the species). These effects are most common in plants that grow in climates with cold winters. Nutrient levels also affect growth and development of plants, which is self-evident. For example, if a plant could not sense that particular nutrients were in low concentrations, it might grow at a rate that exceeds the amount of nutrients available, which could be fatal. All of these mechanisms have led to the evolution of plants that can grow almost everywhere on Earth, including areas of extreme heat, extreme cold, high light, low light, high moisture, low moisture, salty soils, low nutrient soils, and so on.

While most plants need to produce vegetative structures for photosynthesis, they also need to reproduce. While some of the reproductive structures in some plants are capable of photosynthesis (after all, they are modified leaves), they are less efficient than the vegetative structures. Strict molecular controls have evolved to produce reproductive structures. Most research has centered on angiosperms, *Arabidopsis thaliana* specifically, although more studies on other species have expanded the view of how flowers develop. It is important that a plant form flowers only when needed, and only when the conditions favor successful reproduction. Vegetative shoot apices are mostly indeterminate organs, given that under the proper conditions, they will continue to form leaves indefinitely. However, most floral apices are determinate. Once an apex becomes a floral apex, for most species, it will form a flower and nothing more. Control of this conversion is often dependent on light, temperature, and nutrients that signal several genes (Figure 17.11), especially LFY (called LEAFY, because the mutant causes a floral apex to produce only leaves) and SOC1 (SUPPRESSOR OF OVEREXPRESSION OF CO 1), and their protein products to stimulate sets of other genes, classified as A, B, C, D, and E genes (and protein products). The A-class genes alone will induce sepals to be formed (Figures 17.11 through 17.13). These are leafy structures that subtend the petals. When the A-class and B-class gene products are present, petals are produced. When B-class and C-class protein are present, stamens are produced. These are the male parts of the flower, consisting of a filament with an anther (containing the pollen) at the end. When only C-class proteins are present, carpels are formed. These are the female parts of the lower, consisting of a sticky stigma (to which pollen with adhere), a style, and an ovary. When C-class and D- proteins are present, the cells will be induced to differentiate into ovules, which contain the female gametophyte, including an egg cell. The E-class gene products, SEP1, SEP2, and SEP3, are necessary for the differentiation of all floral organs. If they are not expressed, then the apex will form only leaves, even when all of the other genes are normal (Figures 17.11 and 17.13).

During the nearly 400 million years of evolution for gymnosperms and 200 million years of evolution of angiosperms, a great deal of diversity has emerged. While the first such plants had inconspicuous reproductive structures, today there is great diversity in size, color, and form among the extant species (Figure 17.12). Some flowers exhibit a simple form, such as lilies (Figure 17.12b) with three colorful sepals, three colorful petals, six stamens, and a carpel that is in three parts (three stigmas and three ovaries). Another monocot group, the orchids have more modified flowers (Figure 17.12c) with six colorful tepals (organs that resemble both sepals and petals). However, one of the tepals differs from the others, and usually the stamens and the styles are physically fused to the tepal. This form has evolved to accommodate the insect pollinators that move into the tubular tepal to obtain nectar, and in doing so carry pollen from one flower to another. The number and orientation of floral organs can vary greatly among species. For *Prunus* species (including important fruit trees, such as apple, prune, plumb, and peach), they usually have five stamens, to match the number of sepals, petals, and stigmas (Figure 17.12d). However, other members of the Family Rosaceae can have much higher numbers of stamens (Figure 17.12e), while the number of petals remains at five. In some species, not only are there larger numbers of stamens, but also the filaments are fused, forming a tube that surrounds the style (Figure 17.12f).

During the past 200 million years, the arrangement of floral organs has proceeded through some interesting steps. In *Amborella* sp., a species that is basal to all other extant angiosperms, the boundaries of the expression of the A, B, C, D, and E genes are less distinct than in other groups. This is exhibited by the fact that while the flowers have distinct carpels and stamens, they have tepals rather than distinct

FIGURE 17.12 Diagram of an idealized flower and examples of simple flowers. All are perfect flowers, in that they have both male and female organs. (a) The majority of flowers have from 3 to 5 sepals (s), 3 to 5 petals (p), variable numbers of stamens (consisting of filaments and pollen-containing anthers), and 3 to 5 carpels (consisting of stigmas, styles, ovaries, and ovules). (b) Lily flower, with 3 sepals, 3 petals, 6 stamens, and 3 carpels. (c) Orchids belong to a group of monocots that lack sepals and petals, but have tepals (t) that are more ancestral organs that have not differentiated into petals and sepals. One of the tepals (t*) is modified into a tube structure, with the stamens and styles fused to the modified tepal. (d) *Prunus* sp. flower, with 5 sepals (that are behind the petals in this view so that they cannot be seen), 5 petals, 5 stamens, and 5 carpels. (e) Rose flower, with 5 sepals (not visible in this view), 5 petals, numerous (approximately 80) stamens, and 5 fused carpels. (f) *Hibiscus* sp. flower with 5 sepals (not visible in this view), 5 petals, numerous stamens (approximately 100) with fused filaments that form a tube surrounding the style, and 5 carpels.

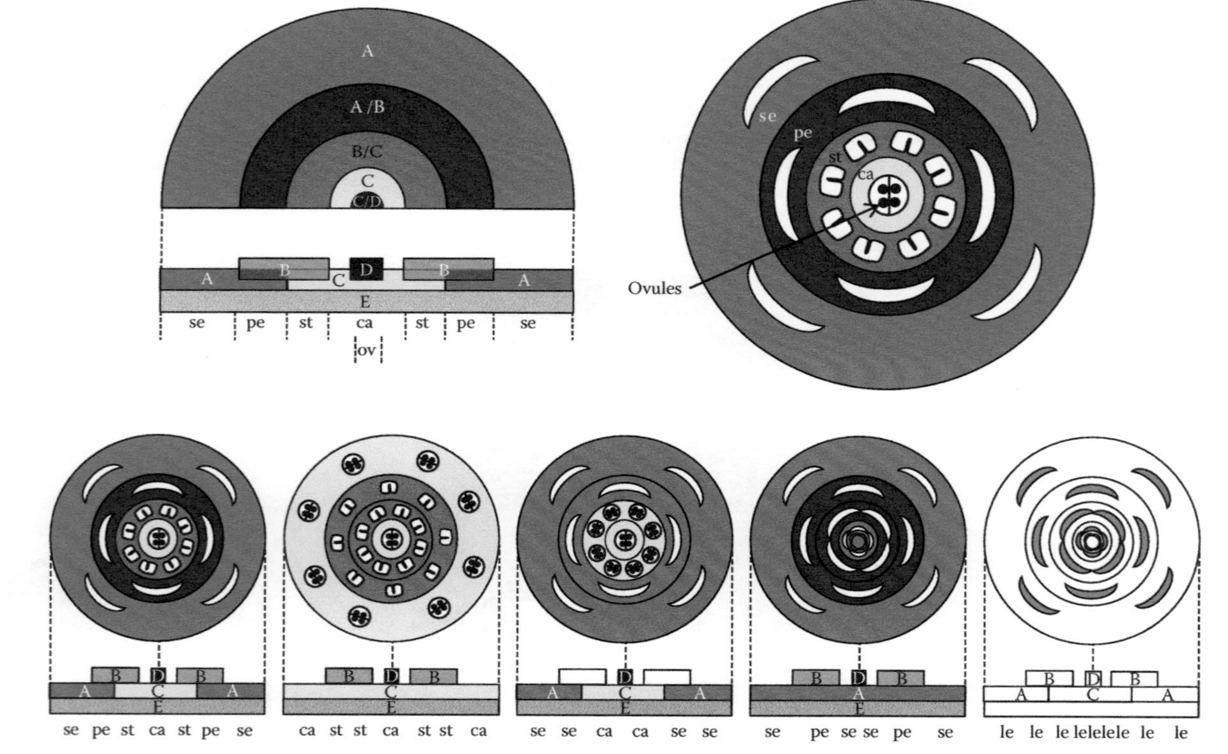

FIGURE 17.13 Floral development model based on A-, B-, C-, D-, and E-class genes/proteins (details of the genes/proteins are included in Figure 17.11), based on experiments with *Arabidopsis thaliana* mutants. Wild type *A. thaliana* plants have flowers with four of each organ type (se = sepals; pe = petals; st = stamens; ca = carpels; ov = ovaries). These genes are expressed in concentric rings in the floral apex, which influence the expression of sets of genes that produce the floral organs. When the A-class genes are expressed, sepals are produced, but when the A-class and B-class genes are expressed, petals are produced. However, when B-class and C-class genes are expressed, stamens are formed, and when only C is expressed, carpels are produced. The ovules within the carpels are formed where the D-class genes are also expressed. Results from mutants of specific genes are illustrated at the bottom. The wild type is on the left. A mutant where the A genes have been mutated produces a flower that has carpels where sepals are normally formed, and stamens where petals are usually formed. Mutants lacking B-class gene expression produce flowers that consist of only sepals and carpels. Mutants lacking C-class gene expression produce flowers consisting of only sepals and petals. If the E-class genes are mutated, no floral organs are formed, and only leaves are formed. This is because the E-class functions are needed to form any of the floral organs (see Figure 17.11).

sepals or petals. Tepals have characteristics of both sepals and petals, and represent an ancient floral organ that predates the appearance of sepals and petals. Among species of Magnoliids (e.g., *Magnolia* sp.), the tepals have become more differentiated, but they are not quite sepals or petals. The subtending organs are termed sepal-like tepals, and the organs above them are petal-like tepals. Evolutionarily, they are on the way to sepals and petals (respectively), but they have not fully differentiated into sepals and petals. Among monocots, several unique forms have evolved. Within the Order Pandanales (a rare plant group), the positions of the anthers and carpels have switched places, indicating a shift in expression of the B-class genes, and probably the D-class genes as well. Within the Order Asparagales (including orchids), the expression of the B-class genes is expanded such that only tepals are produced, and no true sepals or petals are present. In the Poales (grasses and related taxa), there has been additional divergence and specialization, due primarily to most members being wind pollinated. While the gene expression patterns with respect to the A-, B-, C-, D-, and E-class genes are similar to that in all eudicots, the structures of the flowers differ. Instead of sepals, grass flowers have lemma and palea, which are bracts surrounding the other floral organs, and lodicules, which are classified as scales that are just outside of the stamen and carpels. Among the eudicots, the pattern is remarkably similar, indicating the evolutionary success of this pattern of gene expression and floral arrangement. Almost all have a similar arrangement of sepals, petals, stamen, and carpels (Figure 17.14).

While there is a diversity of floral types, there also exists a great diversity in the arrangements of floral apices and flowers among plant species (Figures 17.15 and 17.16). The first flowers were likely solitary and inconspicuous, and were almost indistinguishable from vegetative organs. The current diversity of floral arrangements runs from single small flowers to multiple inflorescences to packed floral heads (Figures 17.15 through 17.17). As with leaf ontogeny, some of the simplest floral arrangements are flowers in an alternate pattern (Figure 17.15), including inflorescences called spikes, spikelets, racemes, and catkins. More complex patterns have also been described, including chymes (where a branch diverges to form a flower and another branch, followed by another such divergence), umbels (where many florets diverge from a single stem position), and corymbs (where multiple florets emerge from different parts of the stem). Other inflorescences are formed from multiple flowers or florets emerging from one of more positions along the stem. Some flowers form on heads, some of which become involuted, where the flowers are inside of a ball-shaped inflorescence. Figs (*Ficus* sp.) have this type of inflorescence, called a syconium. Species of wasps pollinate the flowers by entering through a small opening on one end of the inflorescence. The female wasps also lay their eggs inside the fig inflorescences, which is why they enter in the first place. Many species of plants, especially asterids, produce flowers on various sized and shaped heads (Figure 17.16). In these inflorescences, the individual flowers often differ according to their position on the head. The outer flowers, called ray flowers, often have a small appendage called pappus, instead of sepals, and usually have only one large colorful petal (Figure 17.17). Often, they have no stamens, but do contain a carpel. To differentiate in this way, the influence or expression of the B-class genes would have to be altered (as in Figure 17.17). All of the remaining flowers, called disk flowers, are perfect flowers (having both male and female organs), although they have pappus that is more feathery in appearance, instead of sepals. They are usually very small, but numerous. Most are pollinated by arthropods or birds. The pattern of expression of the A-, B-, C-, D-, and E-class genes appears to be similar to the pattern found in most eudicots.

PLANTS VERSUS ANIMALS

The lineage that led to the evolution of plants and the lineage that led to evolution of animals diverged at least 1.9 billion years ago. Nonetheless, the genes and proteins that are used in the developmental processes of both have some striking similarities. Developmental processes are highly controlled and coordinated primarily by sets of transcription factors and repressors. This includes proteins with homeodomains, Zn fingers, leucine zippers, high mobility group domains, MADS boxes, MYB domains, and others. Some of the regions are so well conserved between plants and animals that protein sequences can be aligned. Because of the large numbers of these genes and gene families, and the

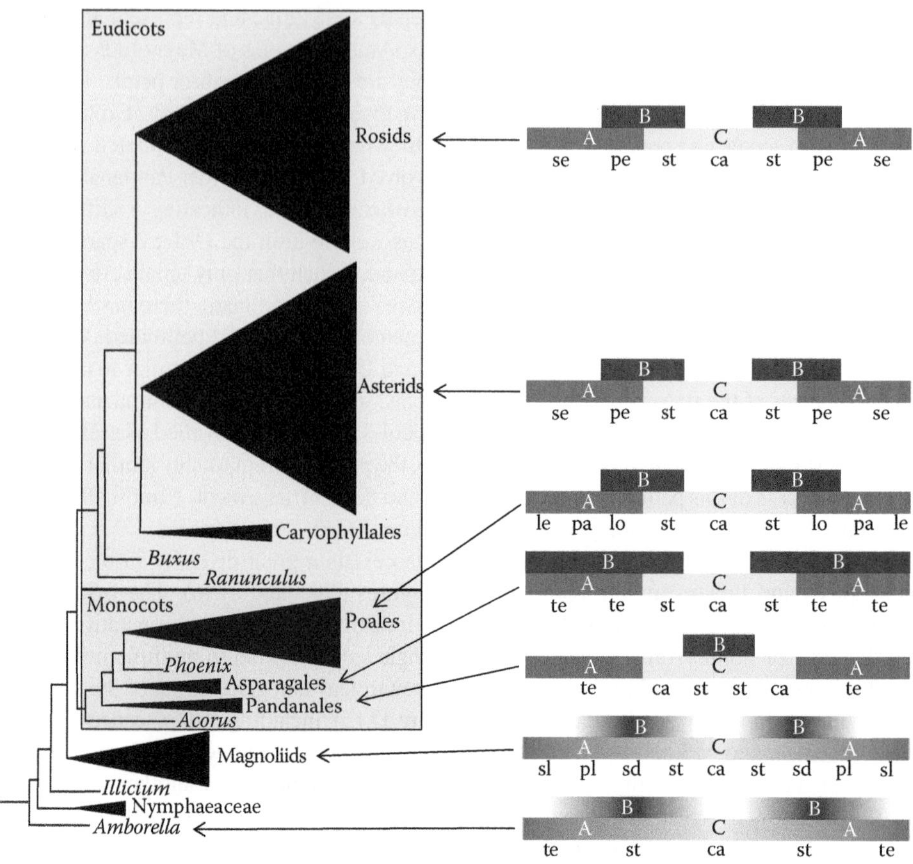

FIGURE 17.14 Phylogeny of plants, superimposed with the patterns of expression of the A-, B-, and C-class genes. The class D and E genes are presumed to be constant throughout evolution, and are not shown. Among the basal groups (*Amborella*, Nymphaeaceae, and Magnoliids), expression of the genes extends over wide portions of the shoot apex, such that the organs are less morphologically distinct from one another. For example, *Amborella* flowers lack sepals (se) and petals (pe), but instead have organs that resemble both sepals and petals, which are termed tepals (te). In Magnoliids, there is more distinction in the organs, but true sepals and petals are not present. Instead, the flowers have sepal-like tepals (sl) and petal-like tepals (pl). They also have stamenoids (sd), as well as true stamens (st). Stamenoids are similar to stamens, but they contain no pollen. Monocots have several types of flowers, including those in the Pandanales where the positions of the carpels and stamens are reversed compared with other angiosperm flowers. Apparently, this is caused by a shift in the expression of the B-class genes. Also, these flowers have tepals, with no evidence of true sepals or petals. Plants in the Asparagales (which includes orchids) also have only tepals, primarily because of an extended area of expression of B-class genes in the floral apex. Members of the Poales (grasses and related plants) have highly modified flowers, partly because most are wind pollinated. Instead of having showy colorful sepals and petals, they have a few small bracts and scales, called the lemma (le), palea (pa), and lodicules (lo). While the organs differ from other angiosperms, the expression of the A, B, C, D, and E genes is similar to that in the eudicots, most of which exhibit a similar pattern of expression, and produce similar floral organs, including distinct sepals, petals, stamens, and carpels. All angiosperm flowers possess true carpels and stamens (except in the male and female flowers of dioecious plants—those with separate male and female plants—where the plants each bear only stamens or only carpels, as well as sepals and petals).

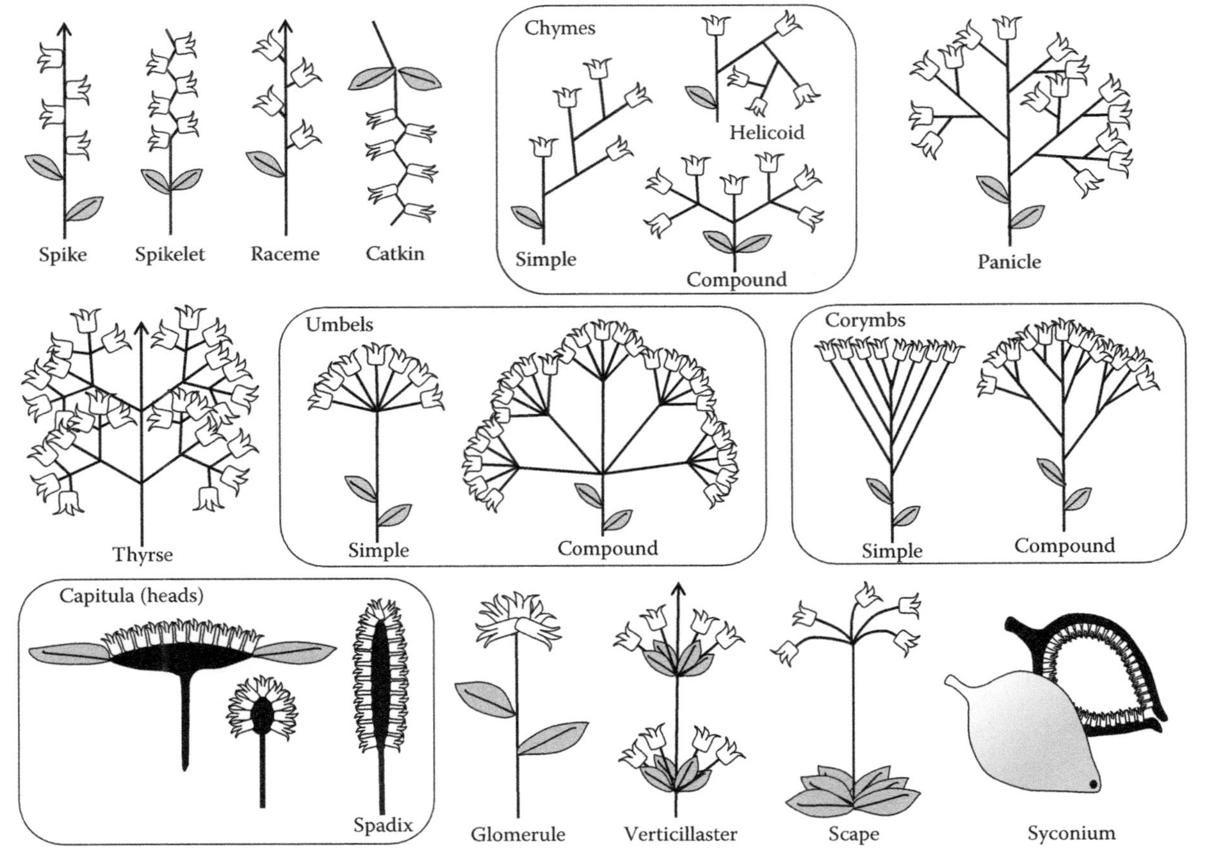

FIGURE 17.15 Types of angiosperm inflorescences. In unbranched inflorescences, the individual flowers may be produced in an alternate pattern to form a spike or spikelet raceme or catkin. Spikes and racemes may possess indeterminate apices (indicated by arrow heads), leading to the formation of long inflorescences. Flowers can also be produced in branching patterns on the inflorescence, to form a chyme, panicle, thyrse, or corymb. When the flowers appear in whorls, they appear as umbels, glomurules, verticillasters, or scapes. The flowers can also be arranged on heads or receptacles (including heads or a spadix), including some that involute to form a hollow ball with a single opening (e.g., a syconium).

FIGURE 17.16 Examples of inflorescence types. (a): *Allium cepa* (onion), a scape; *Robinia pseudoacacia* (black locust tree), a raceme; *Daucus carota* (wild carrot, or Queen Anne's lace), a complex umbel. (b): *Euphorbia millii* (crown of thorns), a chime (*Note*: the flowers of this species are very small; the two red organs are bracts [B], or modified leaves, which form below the sepals [se] and petals [pe]); *Trifolium* sp. (clover), a head; *Echinacea purpurea* (echinacea), a head. (c): *Taraxacum officiale* (dandelion), a head, flower and mature head with seeds (inset); *Dipsacus fullonum* (teasel), a head, flower, and mature head (inset); *Helianthus annuus* (sunflower), a head.

overlap between the plant and animal versions, it is most likely that these genes and gene families existed prior to the divergence of the two groups. This implies that these genes and proteins probably were used as means for the unicellular organisms to communicate, organize, and cooperate to some extent. This brings the discussion of the origin and evolution of these genes back to the discussion of quorum sensing in bacteria that was introduced in the previous chapter. This illustrates a fundamental aspect of evolutionary processes. Many begin by being beneficial to the organism for one aspect of its life, but they may evolve slowly to lead to new unanticipated functions that are beneficial to its descendants. Thus, a transcription factor that is activated in response to an external stimulus in a unicellular organism may evolve into a transcription factor that is activated in response to a stimulus being emitted by an adjacent or nearby cell within a multicellular organism. Both mechanisms are the same, but the stimulus and the response might be different.

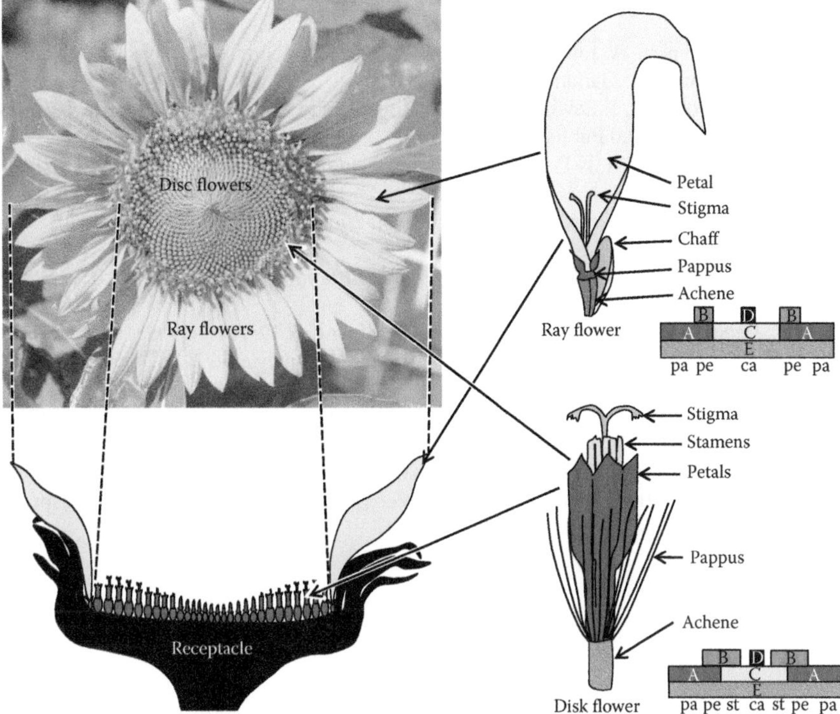

FIGURE 17.17 Structure of a sunflower inflorescence (head). Most sunflower heads have hundreds of individual flowers. Two different types of flowers are present, ray flowers and disk flowers. Ray flowers have small sepal-like organs, called papuas (pa), and at least one large petal, in addition to a carpel (ca), with style and stigma. No stamens are present. Each flower is subtended by an achene and chaff. Reduction of the expression of the B-class genes might cause the lack of stamens in these flowers. The disk flowers are perfect, having both stamens (st) and carpels (ca). They also produce pappi, rather than sepals, but their pappi are feathery in morphology. The petals are small and fused together, as are the stamens and styles. The pattern of A-, B-, C-, D-, and E-class genes is similar to those in other eudicots, except that it is repeated in each of the hundreds of flowers.

KEY POINTS

1. All members of the Archaeplastida were derived from an endosymbiotic event between a unicellular protest and a cyanobacterium approximately 1.9–2.0 billion years ago.
2. The earliest plants were unicellular, but many groups became multicellular early in their evolution.
3. Gametophytes are the dominant life form in lower plants, while sporophytes predominate in higher plants.
4. Multicelluarity in plants is controlled by a number of genes and their products that alter the expression of genes to produce each of the organs.
5. Plants are organized in cell layers that respond to developmental signals to form organs in pattern characteristic for each species.
6. Many of the genes that control development in plants have some similarities to analogous genes in animals.
7. Position and timing of vegetative organs is controlled by signals emanating from the central zone of the shoot apical meristem
8. Floral organ (as well as vegetative organ) development is controlled by a large number of genes that respond to changes in temperature, light, hormones, and nutrients.
9. Floral development includes organ production, as well as flower arrangement on the plants.

ADDITIONAL READINGS

Alberts, B., D. Bray, K. Hopkin, A. Johnson, J. Lewis, M. Raff, K. Roberts, and P. Walter. 2013. *Essential Cell Biology,* 4th ed. New York: Garland Publishing.

Alberts, B., D. Bray, A. Johnson, J. Lewis, M. Raff, K. Roberts, and J. D. Watson. 1994. *Molecular Biology of the Cell.* New York: Garland Publishing.

Becker, W. M., J. B. Reece, and M. F. Poenie. 1996. *The World of the Cell,* 3rd ed. New York: The Benjamin/Cummings Publishing.

Bidlack, J. and S. Jansky. 2010. *Stern's Introductory Plant Biology*, 12th ed. Columbus, OH: McGraw-Hill Education.

Darnell, J., H. Lodish, and D. Baltimore. 2007. *Molecular Cell Biology*, 6th ed. New York: W.H. Freeman and Company.

Levy, Y. Y. and C. Dean. 1998. The transition to flowering. *Plant Cell* 12:1973–1989.

Reece, J. B., L. A. Urry, M. L. Cain, S. A. Wasserman, P. V. Minorsky, and R. B. Jackson. 2010. *Campbell Biology*, 9th ed. New York: Benjamin/Cummings Publishing Company.

Rogers, S. O. and H. T. Bonnett. 1989. Evidence for apical initials in the vegetative shoot apex of *Hedera helix* cv. Goldheart. *Am. J. Bot.* 76:539–545.

Tropp, B. E. 2008. *Molecular Biology, Genes to Proteins*, 3rd ed. Sudbury, MA: Jones & Bartlett Publishers.

Wolters, H. and G. Jürgens. 2009. Survival of the flexible: Hormonal growth control and adaptation in plant development. *Nat. Rev. Genet.* 10:305–317.

18 Cancer

INTRODUCTION

Cancer is not a single disease, but a collection of diseases related by the fact that the affected genes change the ways in which cells respond to various signals that normally control cell division, growth, and response to surrounding cells and tissues. Cancers vary in their incidence and virulence. For example, while breast cancer has the highest rate of incidence (Figure 18.1), lung cancer has the greatest mortality rate (Figure 18.2). Cancer can occur in virtually any multicellular organism, and in some ways can be conceptualized as an evolutionary disease. That is, the same mechanisms responsible for evolution, such as mutation, recombination, gene duplication, gene amplification, chromosomal translocations, and other phenomena, also can lead to initiation of a cancer cell. Because cancer cells begin as normal cells, they are not recognized as foreign or dangerous cells by the immune system of the organism. The cells proliferate to form small foci, and then larger tumors, many of which attract blood supplies, and then migrate to other parts of the organism to set up colonies in other locations. This often makes treatments and cures difficult.

If a unicellular organism mutates into a form that grows faster and uses nutrients at higher levels than normal, it might reproduce faster, but it is more likely to die because it will outgrow the available resources. Whether it grows faster or dies may have little effect on the population of cells. However, in a multicellular organism, each of the cells is dependent on other cells within the organism, and growth and nutrient acquisition are highly regulated. When a cell in a multicellular organism grows and divides uncontrollably the entire organism is jeopardized. Initially, this might not cause many problems. However, as the number of cells increases, the tumorous cells begin to encroach on and invade normal tissues, affecting their normal functions. Also, the tumor cells utilize increasing amounts of nutrients and produce large amounts of waste products, some of which are toxic to the normal cells. Cancer is as old as multicellularity (i.e., at least 1 billion years old). Humans and fruit flies both have many of the same genes that are implicated in many types of cancers. For this reason, *Drosophila melanogaster* has been used to study the causes and progression of cancer that may be useful in treating human malignancies. Cancer and cancer genes have been demonstrated in animals as simple as hydra (Cnidaria). Yes, they can get cancer.

Cancer cells exhibit one or more of the following aberrant characteristics:

1. Uncontrolled growth
2. Mutations in signal transduction pathways
3. No apoptosis
4. Loss of contact inhibition
5. Loss of anchoring to adjacent cells
6. Reduced need for growth factors
7. Ability to invade other tissues
8. Ability to attract blood supply (angiogenesis)
9. Ability to move to distant sites (metastasis)

The last three of these greatly increase the severity of the disease, and often once the cancers have metastasized, it is difficult to control the tumor and the survival of the organism is

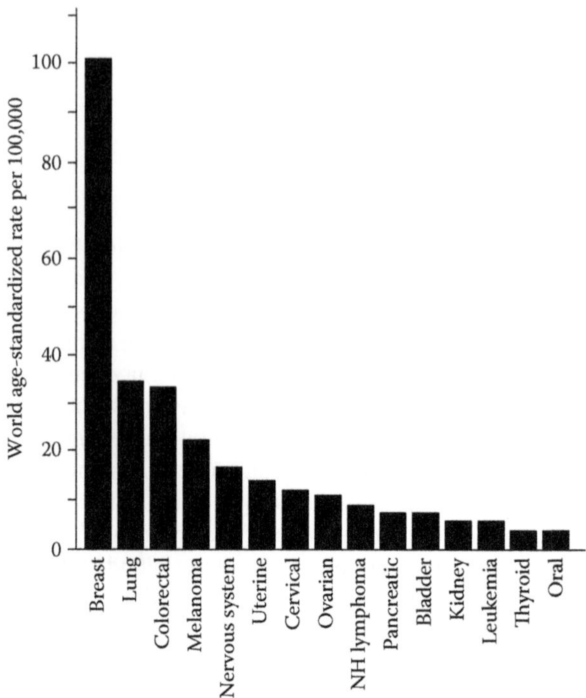

FIGURE 18.1 Graph of the world age-standardized rate per 100,000 people of the top 15 most prevalent cancers (2008 statistics).

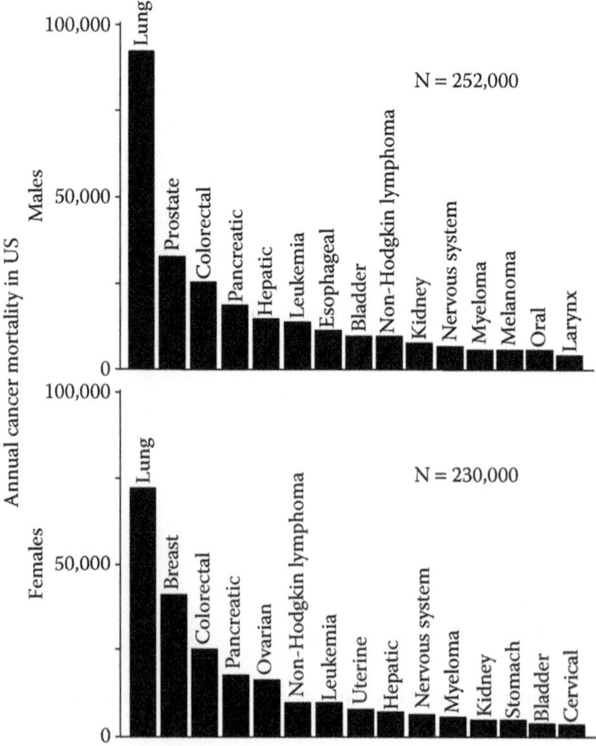

FIGURE 18.2 Annual US mortality for males and females by cancer type (2008 statistics).

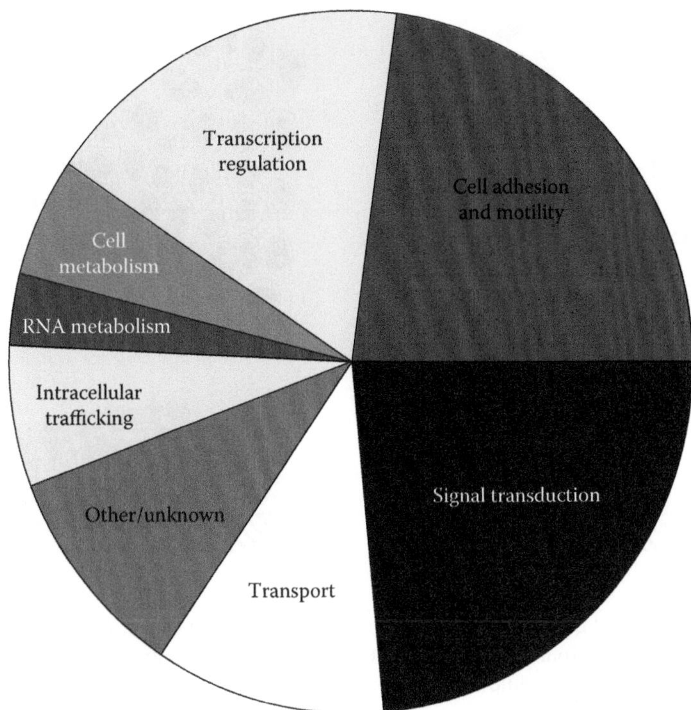

FIGURE 18.3 Proportions of gene categories that are found in a wide range of cancers. Genes that encode proteins involved in signal transduction, cell motility/adhesion, and transcription regulation comprise more than 60% of the genes that mutate in cancer cells.

jeopardized. The normal characteristics of cells are strictly regulated by specific types of gene products that act to control cells in relation to other cells in the organism. But these mutate during the process of carcinogenesis. In other words, the cells must be controlled to react appropriately to other cells in the organism so that order is maintained in the entire organism. The majority of the cancer genes are involved in cell adhesion, motility, signal transduction, and control of transcription (Figure 18.3). However, a small number of the genes are responsible for transport into and out of the cell, intracellular trafficking of proteins, RNA metabolism, and cell metabolism.

PROGRESSION OF CANCER

Normal cells begin as undifferentiated stem cells, but go through a series of molecular and cellular changes dictated by specific programmed gene expression patterns. As the cells differentiate, they take on characteristic shapes. Each cell type has a specific shape based on the evolutionary history of the species and its component tissues and cells. As the cells differentiate they are in communication with the surrounding cells, such that they are inhibited from growing larger than their placement in the organism dictates. When grown in cell culture, they will normally differentiate and form a monolayer of cells (Figure 18.4). Cancer cells lose their ability to communicate or detect the presence of other cells. Also, they tend to be more rounded in shape, similar to the morphology of a stem cell. When grown in culture, they continue to grow uncontrolled by contact with other cells. Some cultures of cancer grow beyond the confines of the culture plates.

Time and several mutations are required to transform a normal cell into a cancer cell (Figures 18.5 and 18.6). These changes do not happen overnight. Early mutations tend to cause the cells to grow larger than normal, because they are losing the signals that inhibit their growth. In this

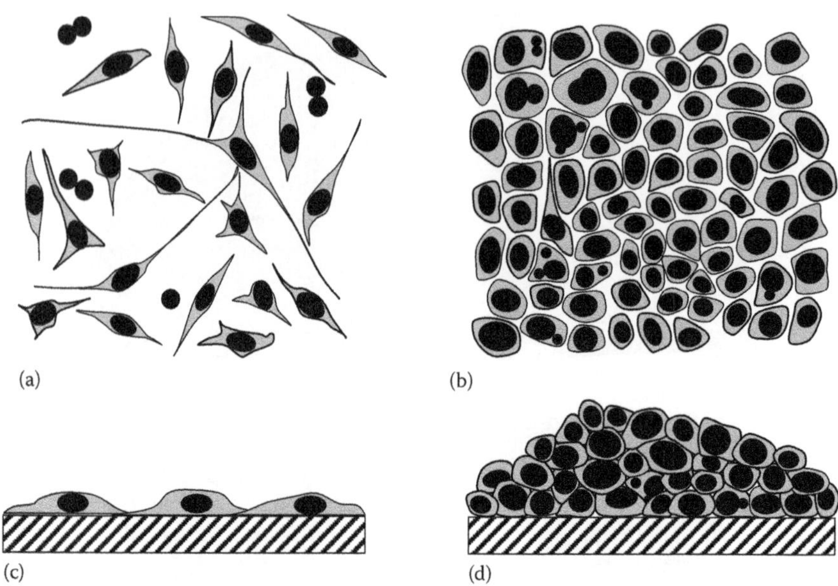

FIGURE 18.4 Generalized diagrams comparing normal cells (a) with cancer cells (b). Normal cells spread out, while cancer cells tend to round up, similar to cells that are preparing for cell division. In culture, normal cells tend to form a monolayer (c), while cancer cells pile up in the culture plate (d).

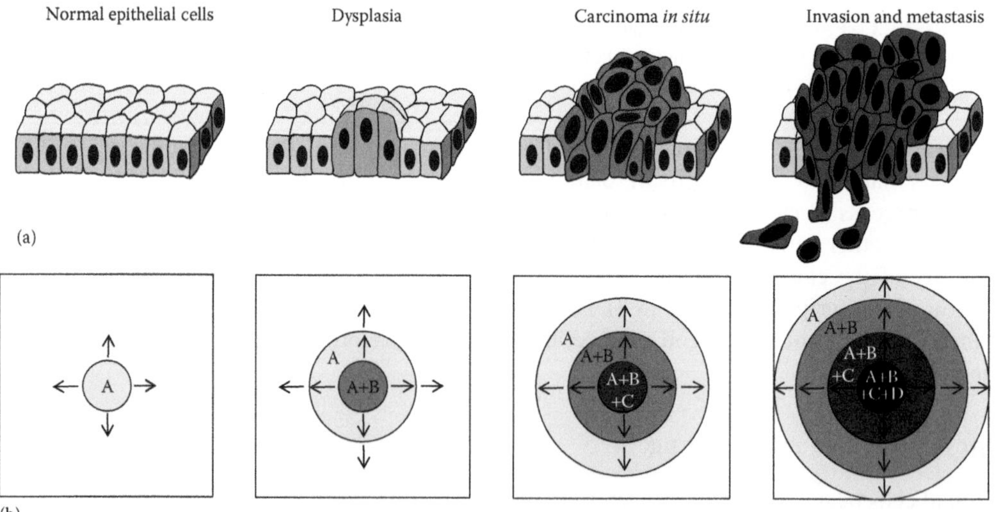

FIGURE 18.5 Cellular progression of cancer (a). Initially, the cells begin to change in size and shape, exhibiting abnormal growth called dysplasia. After further mutations, the cells become more misshapen, and the nuclei are abnormal in size, position, and shape. At this point the tumor is termed a cancer *in situ*, indicating that it has not yet invaded surrounding tissues. In the later stages, the cells become more rounded and abnormal in shape, size, and position, and the nuclei are also highly abnormal. Often, micronuclei (small pieces of chromosomes that have broken away from the nucleus) are present. Also, they begin to invade other tissues and eventually migrate to other areas. (b) Demonstrates the progressive mutations that lead to the development of increasingly abnormal cells. A cancer is initiated when a mutation in a cancer gene (A) first occurs (yellow). This cell divides producing a ring of cells containing the same mutation (A). Additional mutations (B) occur, such that some of the cells in the tumor have A + B mutations. These cells then divide forming more cells with the same genotype. This process continues until the cells in the center of the tumor contain many more mutations (A + B + C + D), becoming a more dangerous population of cells.

Cancer	Feature	Normal → Dysplasia → Adenoma/neoplasm → Carcinoma *in situ* → Invasion → Metastasis
Colorectal	Karyotype: LOH	5q 18q 17p 15q 4p 5p 8p 10q 15q 16p 22q, 4q 9p 10q 12q
	Karyotype: Gains	Xp Xq 4q 5q 5p 7p 7q 8q 12q 13q 20q 6p
	Methylation	*HPP1 hMLH1SFRP1*
	Gene mutations	*APC p16 DCC GAS AMACR COX2 KRAS CTNNB hMSH2 CCKBR SMAD4 ECAD TGFb p53 BCL2* *BAX*
Prostate	Karyotype: LOH	8p 16q 10q 13q 17p
	Karyotype: Gains	7 8q Xq
	Methylation	*GSTP1* *CD44*
	Gene mutations	*AMACR CDKN1B telomerase C-CAM FLT1 MYC KAI1 MXI1 PTEN p16 TMSB15 AR C-MET ECAD FGF8 FLK1 VEGF PSCA NKX3.1 EZH2 p53 BCL2 PAI TGFb RAF*
Pancreatic	Karyotype: LOH	1p 3p 5q 6q 8p 9p 10q 17p 18q 21q
	Karyotype: Gains	3q 5p 7p 8q 20q
	Methylation	*BNC1 ADAMTS1*
	Gene mutations	*PSCA KRAS Fascin Mucin5 Mucin1 p16 DPC4 Cyclin D1 14-3-3s p53 BRCA2 Ki-67 TOPO IIa* *Mesothelin*

FIGURE 18.6 Changes in chromosomes and mutated genes during carcinogenesis in three types of cancer. Loss of heterozygosity indicates changes in the chromosomes, such that one of the alleles for a particular gene is deleted from the homologs. Gains indicate additional copies of genes that are included in the nucleus. Both indicate gross changes that occur in these cancer cells. A *p* indicates changes to the short arm of the chromosome, while a *q* indicates changes to the long chromosomal arm. Genes that undergo methylation changes are indicated, and genes that are mutated also are indicated. Font colors indicate gene classification as follows: blue—tumor suppressor, orange—DNA repair; green—growth regulator; red—enzyme; violet—signal transduction; black—transcription regulation; tan—cell proliferation; olive—controls apoptosis; gray—detoxification; aqua—adhesion/migration; brown—*Hox* gene; light green—cell cycle regulator; hazel—chromosome integrity; dark blue—cell organization; lavender—tissue integrity.

phase the cells exhibit dysplasia. Their sizes, shapes, and positions vary from those of normal cells of the same type. Further mutations and cell changes lead first to a benign tumor that may undergo additional mutations and changes to eventually become a carcinoma *in situ*. This is a true cancer that has not yet invaded the surrounding tissues. However, many carcinomas advance beyond this stage to invade the surrounding tissues, many attract blood vessels to grow toward them by producing chemicals that cause the growth of blood vessels, and some of these metastasize to colonize distant sites. Cancers tend to become more damaging and dangerous with time. This is because the cells at the center of the tumor tend to accumulate more mutations. The initial mutations may produce cells that still appear normal. These cells replicate and divide to form tissues that possess the same mutations, which were copied during replication (Figure 18.5). However, as they replicate and divide, additional mutations occur, partly due to the loss of control of some cellular functions. This produces cell populations with additional mutations and additional changes in cellular functions. If the apoptosis pathways are not initiated, then additional mutations may occur, causing additional losses in cellular control. Because of this, cells surrounding the tumor may contain some of the initial mutations that led to the formation of the tumor. This is the reason, in many cases, surgeons will not only remove the cancerous tumors, but they will also remove some of the surrounding tissues, because often they are prone to forming new tumors. As cancers progress, the cells and their nuclei become more abnormal in shape (Figure 18.5). Often, the cells contain micronuclei, which contain pieces of chromosomes and/or amplified copies of various genes. Additionally, some epigenetic effects are implicated in various types of cancer. Usually, methylation of specific genes is increased, causing a decrease in expression of those genes.

Each type of cancer has distinctive mutational patterns and chromosomal abnormalities (Figure 18.6). For example, one of the initial genes to mutate in the most common type of colorectal cancer is the APC (adenomatous polyposis coli) gene. This gene encodes a protein that is a tumor suppressor. Two additional tumor suppressors, p16 (a cyclin-dependent kinase that controls part of the cell cycle, encoded by the CDKN2A gene) and DCC (deleted in colorectal carcinoma, is a cell surface receptor capable of regulating apoptosis), also mutate in the progression to colorectal cancer, which causes the cells to continue to grow when they otherwise would be inhibited from growing. Other genes that encode proteins that control cell proliferation, cell division, gene transcription, cell adhesion, and apoptosis also become mutated. Usually, proteins that inhibit these processes lose their functions. This causes a loss of inhibition processes. Loss of p53, a regulator of apoptosis (programmed cell death), is common in many types of cancer. It is normally activated when DNA damage is sensed by a specific set of proteins. The activated p53 then causes a set of changes in the cell leading to its death. When p53 is nonfunctional, the apoptotic pathway fails to initiate, and the cell essentially becomes immortal.

GENES INVOLVED IN CANCER

Several classes of genes lead to the transformation of a normal cell into a cancerous one. As mentioned earlier, the cell loses its ability to communicate and respond to its environment. This occurs when a cell surface receptor fails to function properly (Figure 18.7). These receptors bind molecules on the outer surfaces of the cells, and respond by sending a molecular signal to other proteins inside the cell. In this way, they are part of the signal transduction pathways of the cell. Within the cell, other signal transduction pathways exist that eventually transmit the signals to perform specific tasks within the cell. Some signal changes in transcription, while others control steps in the cell cycle or initiate events leading to apoptosis. Other sets of proteins constantly traverse the genomic DNA to identify and/or repair mutations. However, if the damage is too great, then these proteins will signal p53, which will initiate apoptosis. Therefore, although the cell has undergone multiple mutations, it will not progress into a cancerous cell because it will undergo programmed cell death.

Many of the genes involved in carcinogenesis are normally expressed at low levels in specific cells during specific stages of development and growth of cells. However, during the transformation of the cells into cancer cells, loss of control of these genes occurs causing large increases in expression of their protein products. Some of these are kinases, which normally phosphorylate sets of proteins at specific times to send molecular signals throughout the cell (Figure 18.7). Increases in transcription or changes to the proteins often cause increased rates of phosphorylation of sets of proteins, which usually causes them to be constantly active. For example, when phosphorylated, some proteins cause the cell to grow, while others cause the cell to proceed through the cell cycle. The protein encoded by the APC gene is an important regulator of the amount of β-catenin produced in the cell, and β-catenin is a regulator of transcription (Figures 18.7 and 18.8). As such, the APC protein is a tumor suppressor. During fetal development, GSK3-β (glycogen synthase kinase 3-β) and CK1 (casein kinase 1) phosphorylate a protein called axin, which causes it to associate with the transmembrane LRP5/6 (low-density lipoprotein receptor) protein that is associated with the transmembrane receptor Fz (frizzled). In this form, APC associates with GSK3-β and CK1, it has no effect on β-catenin, and therefore, β-catenin stimulates gene transcription, cell growth, and cell division. In adults, the axin is not phosphorylated, and it complexes with the APC, GSK3-β, and CK1 proteins. In this form the GSK3-β phosphorylates the β-catenin, which marks it for destruction by proteasomes. Therefore, transcription is not initiated, which inhibits cell growth and division. However, during initiation of colorectal cancer, the APC gene is mutated. Often, the gene produces an aberrant APC protein or fails to produce an APC protein. In these cases, the APC/GSK3-β/CK1/axin/β-catenin complex fails to form, and therefore the β-catenin is not phosphorylated, allowing it to stimulate transcription of the genes that it normally only stimulates during fetal development. This initiates tumorous growth.

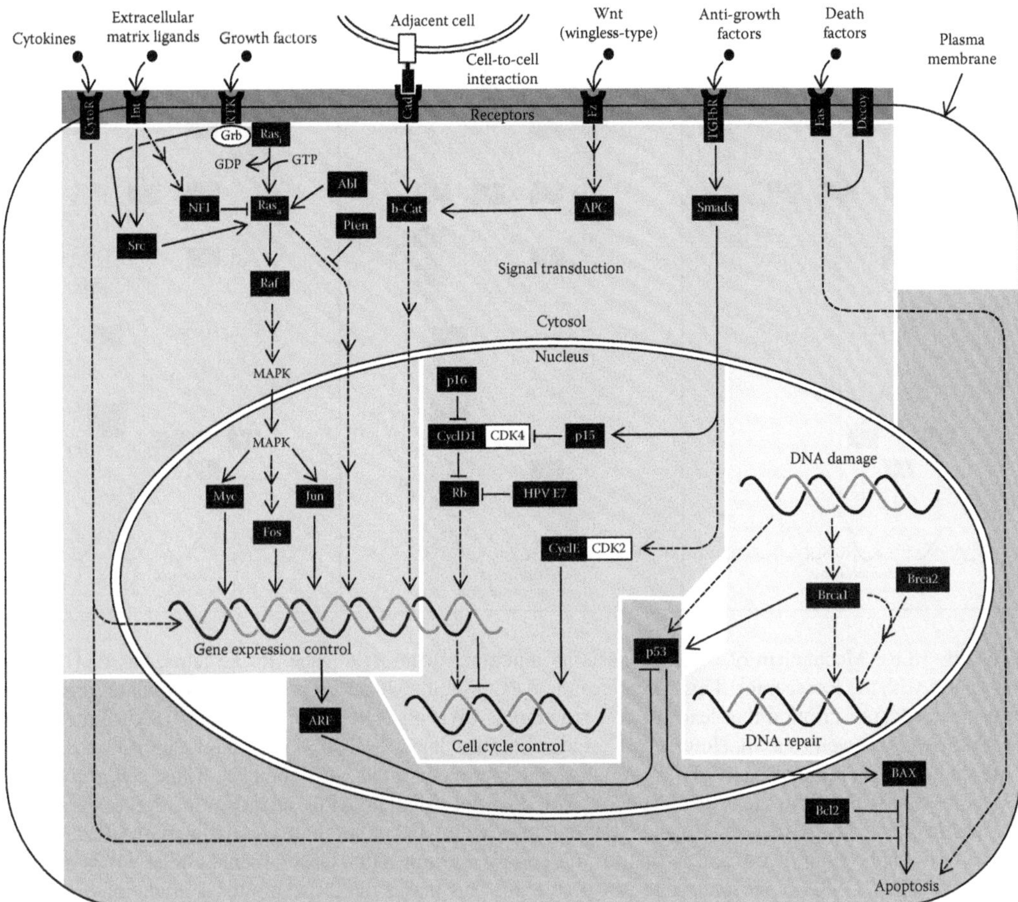

FIGURE 18.7 Summary of some of the proteins involved in cancers. Cancer genes are black with white lettering, while accessory genes/proteins are white with black lettering. Arrowheads indicate stimulatory effects, while bars indicate inhibitory effects. Solid lines represent single steps, while dashed lines indicate pathways with multiple steps, where all the steps are not shown (for simplicity). Proteins with a red background are cell surface receptors that bind a ligand externally and produce a signal inside the cell. Proteins shaded in blue are part of the signal transduction pathways that control the response to various internal signals (or those received from cell-surface receptors) through the cytosol. Many of them eventually send signals into the nucleus to influence transcription (shaded in orange) and/or the cell cycle (shaded in green). Also, within the nucleus are proteins that are able to sense DNA damage (yellow). Some of these signal DNA repair (yellow) and/or apoptosis (lavender), which is controlled by proteins from the nucleus, as well as from the cytosol and mitochondria (e.g., BAX).

Most cells react to external stimuli through specific cell surface receptor proteins. These receptors bind specific ligands (hormones, small molecules, or other proteins), and on binding change conformation that affects portions of the receptors that are inside the cell. These changes then send signals to internal molecules (small internal messenger molecules or other proteins) that signal other molecules inside the cell. These can cause one or many changes in the cell, including changes in gene transcription. This is known as signal transduction. Some cancer cells lose their ability to sense these external signals, such that the cell begins to act as if no other cells were around. This often leads to cell growth and division in cells that normally would be inhibited from growing and dividing by contacts with other cells.

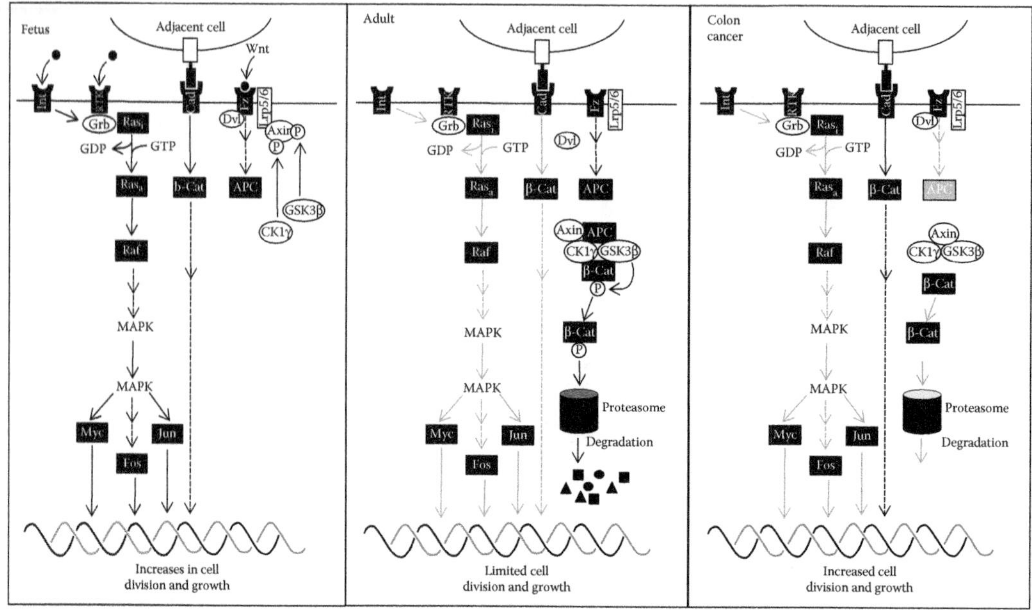

FIGURE 18.8 Mechanism of carcinogenesis by mutation of the *APC* gene. In the fetus, the Wnt protein associates with the Fz receptor. This stimulates the APC protein, which is a protein that stimulates a pathway that increases transcription that leads to cell growth and division. APC complexes with GSK3-β and CK1γ (both kinases), as well as axin. However, if axin is phosphorylated by GSK3-β and CK1γ, it forms a complex with the transmembrane protein LRP5/6, and cannot interact with the APC complex. Thus, cell growth and division are stimulated. In the cells of an adult, Wnt does not stimulate APC, such that the level of expression is lower in adult cells. Also, GSK3-β and CK1γ-complex with APC rather than associating with LRP5/6. This leaves axin unphosphorylated, and in this form it complexes with APC, GSK3-β, and CK1γ, which binds to β-catenin. Then, GSK3-β phosphorylates the β-catenin. This marks the β-catenin for degradation by proteasomes. Therefore, the β-catenin concentration is reduced, such that growth and division are not stimulated. In APC cancer cells, the APC gene is mutated, and the APC protein is either not produced or it is a nonfunctional protein. In both cases, the APC does not complex with the GSK3-β, CK1γ, or the axin, and thus the β-catenin is not phosphorylated, leaving it in the active form that stimulates growth and cell division. Gray diagrams indicate nonfunctioning pathways.

Another characteristic of cancer cells is that the cells continue to divide and become immortal. In healthy cells, the cell cycle is tightly controlled at several points in the cycle. In cancer cells, one or more of the checkpoints in the cell cycle are lost, and the cell continues indefinitely through the cell cycle and division. Additionally, cancer cells become immortal. Because all cells eventually contain many mutations, specific fail-safe mechanisms have evolved to cause cells with many mutations to die. This is called apoptosis, or programmed cell death. It occurs naturally during development (as already mentioned in Chapter 16), but it is also initiated to eliminate cells that are dangerous to the organism. The protein known as p53 is central to this process. When a great deal of DNA damage is detected in the nucleus, signals are produced that reach p53, and once activated it signals other proteins that begin the process of apoptosis to kill the cell. When the p53 gene is mutated, this pathway is lost, and whether or not a cell has numerous mutations, it continues to grow, divide, and spread.

Normal cells do not invade other tissues. Their cell surface receptors sense where they are and where they should and should not grow. Thus, cell types are confined to specific regions of tissues, organs, and body locations. Once they fail to recognize those signals, they can begin to grow uncontrollably to the point where they begin to encroach and invade other tissues, causing disruption of those normal tissues and organs. As they grow, they may also secrete chemicals that cause blood vessels to grow toward them, which increases their supplies of nutrients and oxygen, which causes them to grow faster, becoming more dangerous. Eventually, for many types of cancer, some of the cells can break off of the main tumor and enter the blood vessels or lymphatic system, and they may be capable of colonizing other parts of the organism. This is known as metastasis. This is why there can be a lung cancer tumor in the brain or bones. It starts in the lungs, but then metastasizes into other locations in the body. This is usually one of the last stages of carcinogenesis, and the cancer becomes the most dangerous at this point, because multiple tumors begin to grow in disseminated locations of the body, and controlling the tumors becomes increasingly difficult or impossible.

TYPES OF CANCER

Cancers are identified by the cell origins. These origins relate to the characteristics of the cells during the early stages of development, and often provide some prediction as to their behaviors as cancer cells. Carcinomas are primarily cancers of epithelial cell layers that form the skin, gut lining, and other tissue/tube lining structures, as well as neural tissues. These are primarily cancers of the skin, breast, and colon, and include melanomas. Adenocarcinomas are cancers that originate in glandular tissues, such as those that initiate in breast ductal glands, esophagus, exocrine glands, and endocrine tissues. Sarcomas are cancers that begin in mesodermal tissues, including cancers of connective tissues, bone, cartilage, fat, and muscle. Blood cells are also formed from mesodermal cells, but cancers of these specialized cells are called lymphomas, which originate in lymph tissues, and leukemias, which originate in blood cells. Myelomas are related to these, but are cancers specific to bone marrow. Germ cell tumors originate in testicular or ovarian tissues. Blastomas originate during embryonic development from any cell type. During cancer diagnosis, it is important to identify the type of cancer. Some cancers, such as melanomas metastasize early in the disease, and can rapidly spread throughout the body. This propensity to be mobile is a characteristic of the melanocytes during development. They migrate and colonize the skin early in development. When they become cancerous, they again assume this embryonic characteristic and begin to migrate. On the other hand seminomas (a type of testicular cancer) rarely metastasize, and thus are easily curable when treated early. Another reason to determine the origin of a tumor is to detect metastasis. If a tumor is found in the bone, it might not have originated in that tissue. If it has lung cancer characteristics, it is likely that the primary tumor is in the lung, but parts of the tumor have metastasized to the bone. Once metastasis has occurred, treatment and/or cure are much more challenging. Cancers are classified according to cellular characteristics, genetic markers, and or tumor size and spread. Cells are termed grade 1, when the cells are slightly abnormal, but are well differentiated. Grade 2 cells are those that are more abnormal and only partially differentiated. Grade 3 cells are highly abnormal and only slightly differentiated. Grade 4 cells appear immature and undifferentiated. Tumors are ranked by selected characteristics: tumor size is ranked from T0 (no tumor) to T4 (large tumor); lymph node involvement is ranked from N0 (no nodes involved) to N4 (many lymph nodes contain cancer cells); and disease spread is ranked from M0 (no metastasis) to M1 (metastasis to distant sites). Therefore, a finding of T1, N0, M0 would be a newly initiated cancer that would be easily treated, while a finding of T4, N4, M1 would be extremely dangerous, with limited options for treatment. Alternatively, cancer can be classified as stage I (cancer *in situ*), stage II (large cancer *in situ*), stage III (limited spread of cancer to local tissues), stage IV (extensive local and regional spread), and stage V (distant metastasis). Again, stage I would present treatment options that could eliminate the cancer, while stage V is extremely difficult to treat and eliminate.

CAUSES OF MUTATIONS IN CARCINOGENESIS

As was mentioned earlier, some of the same mechanisms that cause mutations in evolutionary processes also cause mutations that lead to the development of cancer. Included are point mutations, recombination, insertions, deletions, amplification, defective DNA repair, translocations/transpositions, and some viruses.

Point Mutations

Several cancer genes have been mapped with respect to point mutations among a population of tumors. Breast cancer genes *Brca1* and *Brca2* were mapped in detail (Figure 18.9). While nonsense and frameshift mutations were spread throughout both genes, fewer total mutations were found in *Brca2* than in *Brca1*, possibly because *Brca1* is more central to the detection of DNA damage in the cell, and directly signals p53 if the DNA damage is excessive. Therefore, mutations in the *Brca2* gene are more likely to lead to carcinogenesis. Interestingly, no splice-site mutations were found in the Brca1 gene. The major conclusion from these results was that few common point mutations were found within the population of tumor mutations, indicating that no overall predictions could be made that correlate specific mutations in *Brca1* or *Brca2* to initiation of breast cancer. The only conclusion possible was that many mutations in those genes could result in a dysfunctional protein that could initiate a breast tumor. However, some common mutated cancer genes have been found within some families, and these often are predictive of cancer within an individual.

Recombination

Recombination events have been documented in several types of cancer. Normally, recombination occurs when homologs or sister chromatids exchange homologous pieces of the two chromosomes/chromatids (Figure 18.10a). Occasionally, the homologs or sister chromatids misalign in such a way to reverse the order of genes in a portion of the homologs/chromatids. In this case, inversions are formed (Figure 18.10b). Duplications, deletions, amplifications, and chromosomal translocations are more common in cancer cells. Duplications and deletions often occur during a single recombination event through misalignment of parts of the homologs or sister chromatids (Figure 18.10c). One homolog/chromatid gains a section of DNA, while the other homolog/chromatid loses the same section of DNA, as it is transferred to the other homolog/chromatid. Amplification can occur either through frequent exchanges, by additional rounds of replication in a section of the chromosome, or via extrachromosomal replication of regions of the chromosome. Chromosomal translocations are recombination events between nonhomologs, and these are common in some forms of cancer (Figure 18.10d). These exchanges bring portions of the chromosomes together that

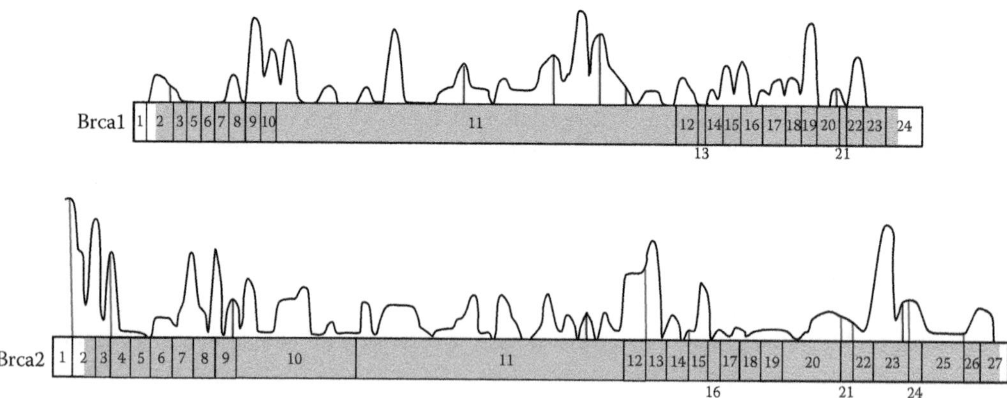

FIGURE 18.9 Map of the Brca1 and Brca2 mutations, based on data from a large number of tumors. The black line indicates the relative frequencies of all mutations (synonymous, nonsynonymous, missense, frameshift, and intron borders) in both genes. Red bars are positions of missense mutations. Blue bars indicate mutations in intron borders.

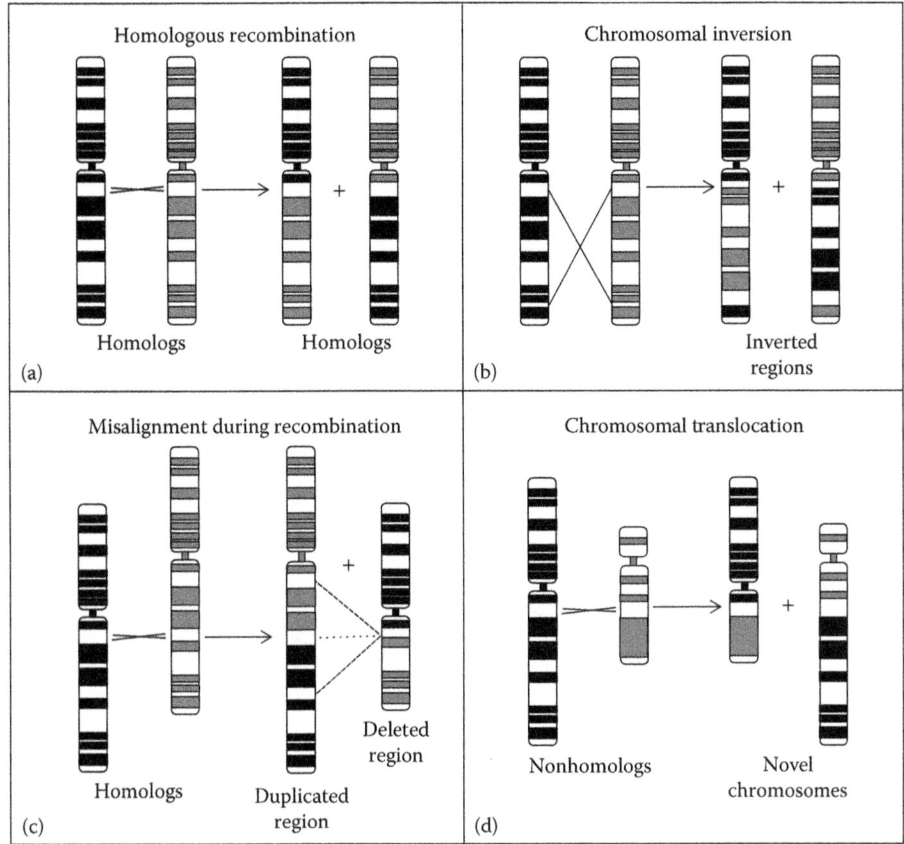

FIGURE 18.10 Chromosomal recombination events. Most recombination between chromosomes occurs between homologs and sister chromatids during meiosis. However, mitotic crossovers also occur between homologs (a). Occasionally, parts of the chromosomes misalign, such that inversions of regions on the homologs occur (b). Some of the inversions result in changes in expression of the genes, especially those in the regions of the crossovers. When the homologs misalign during a recombination event, some regions are duplicated on one chromosome, but deleted from the other in a nonhomologous exchange (c). These conditions usually produce physiological effects in the cells. Losses and gains of large pieces of chromosomes have been described in cancer cells. Most prominent among cancer cells are chromosomal translocations (d). In this form of recombination, nonhomologous chromosomes recombine, creating two new chromosomes with each event. These have been described in many different diseases, including some types of cancer, Down's syndrome, and schizophrenia.

are normally nowhere close to each other. This almost always either changes the levels of expression of the genes in the region of the crossover, or creates chimeric genes and gene products, many of which are nonfunctional. A large number of chromosomal translocations have been identified in many different human diseases, including some cancers. A map of some of these is presented in Figure 18.11.

After the crossover, several general types of products have been found in certain cancers. One type is the result of a crossover between two alleles of the same gene, where they are misaligned, usually within an intron (Figure 18.12). One of the alleles gains one or more exons, while the other loses the same number of exons. Often, both alleles produce proteins that have altered function or no function at all. When the crossover produces a chromosomal translocation, new

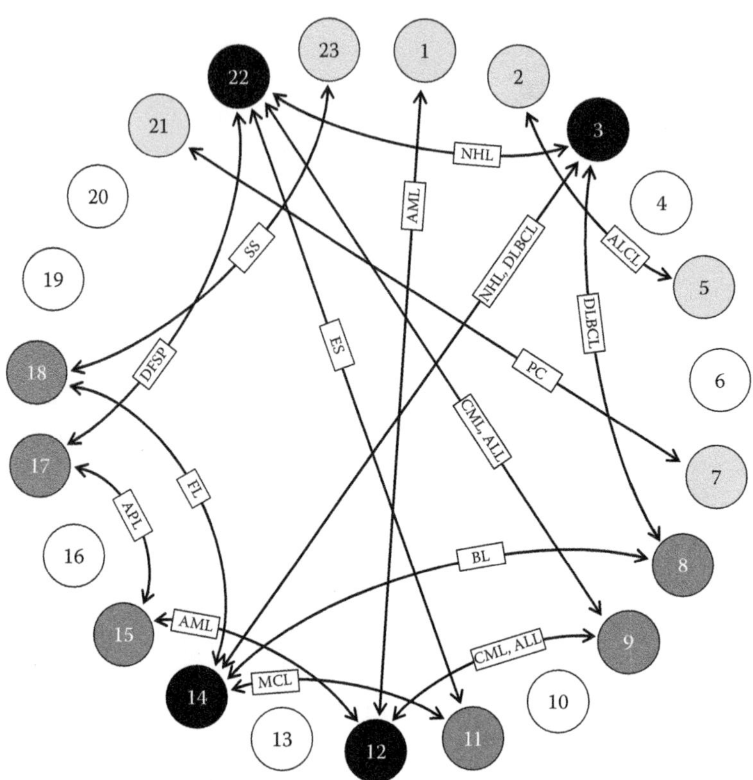

FIGURE 18.11 Some of the translocations that have been described for various forms of cancer. Each of the 23 human chromosomes is represented by a circle. Translocations are indicated by arrows between each of the two chromosomes involved in the translocation. The types of cancer are in boxes along the translocation lines. ALCL, anaplastic large cell lymphoma; ALL, acute lymphoblastic leukemia; AML, acute myelogenous lymphoma; APL, acute promyelocytic lymphoma; BL, Burkitt's lymphoma; CML, chronic myelogenous leukemia; DFSP, dermatofibrosarcoma protuberans; DLBCL, diffuse large B-cell lymphoma; ES, Ewing's sarcoma; FL, follicular lymphoma; MCL, mantle cell lymphoma; NHL, non-Hodgkin's lymphoma; PC, pancreatic cancer; SS, synovial sarcoma. The degree of shading of each chromosome indicates the frequency that the translocations are found in cancer cells.

alleles often are produced (Figure 18.12). Additionally, promoters and enhancers are often are moved close to genes that normally are controlled by very different promoters and enhancers. Several examples of these have been detailed in specific tumors (Figure 18.13). In human follicular lymphoma, a chromosomal translocation between chromosomes 14 and 18 produces a chimeric gene that has part of the *Bcl-2* gene from chromosome 18 (including the promoter) and the J_H region of the immunoglobulin locus (including the enhancer). The Bcl-2 protein regulates apoptosis, but the chimeric gene produces a nonfunctional protein, thus causing a loss in the regulation of apoptosis. A similar result is found in some diffuse large B-cell lymphomas, where a chimeric gene consisting of part of the J_H region of immunoglobulin locus (with the promoter) from chromosome 14 has recombines with part of the *Bcl-6* gene from chromosome 3. Again, a loss of apoptosis regulation is the result. In Burkitt's lymphoma, the *Myc* gene is translocated next to a constitutive promoter, which causes over expression of the Myc protein, which causes an increase in gene transcription that leads to cell growth and division. In many types of prostrate cancer, a translocation between the *TMPRSS2* gene (which encodes a transmembrane protease) on chromosome 21 and the *ERG* gene (which encodes an ETS [erythrocyte transformation

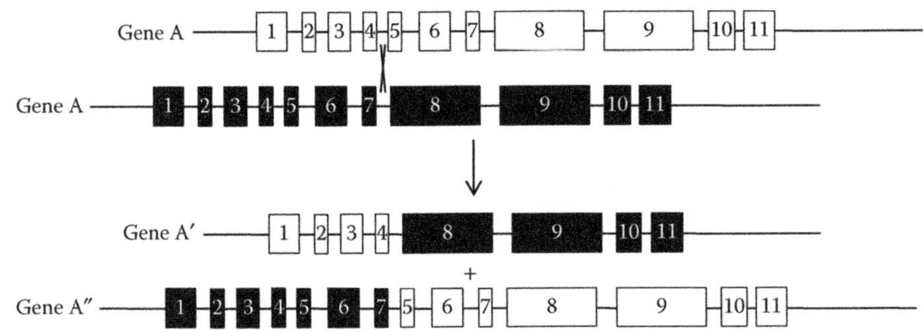

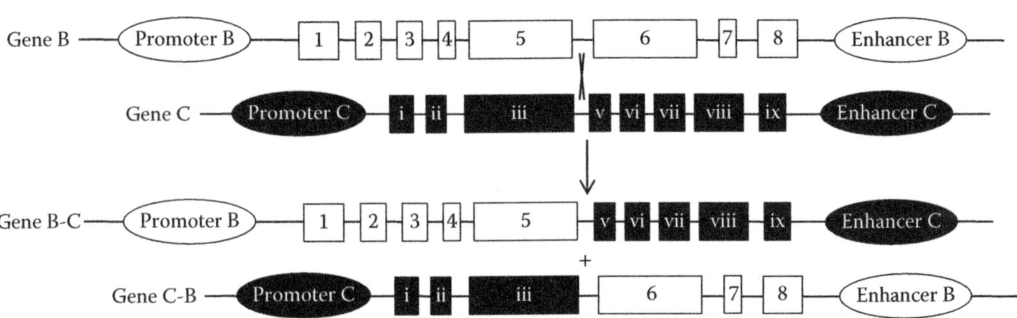

FIGURE 18.12 Generalized recombination events characteristic of those in cancer cells. In the event at the top, a crossover occurs between two alleles of the same gene when they are misaligned. This creates two new versions of the gene, one with a deleted section and the other with a duplicated portion. Often, both produce nonfunctional proteins. In the event at the bottom, two genes from different chromosomal regions recombine to form chimeric genes. Often promoters and enhancers are involved in this process, causing large changes in expression of the genes.

specific]-related protein) on chromosome 7 causes an increase in the transcription of genes that lead to increased cell growth and division. Chronic myelogenous leukemia is often the result of a translocation between the *BCR* gene on chromosome 22 and the *Abl* gene on chromosome 9. The normal Abl protein is a membrane-associated tyrosine kinase, whose expression is normally tightly controlled. However, the BCR-Abl chimeric protein is produced in higher concentrations and constantly phorphorylates tyrosine residues, which activates entire sets of proteins, some of which lead to increases in cell growth and division.

Amplification

Amplification is known to cause some cancers, as well as increase the aggressiveness of some cancers, including some breast, colon, lung, ovarian, and pancreatic cancers. While an increase in the copy number of specific genes has been measured in some cancers, the number of proteins produced may far exceed the increase in gene copy number (Figure 18.14). For example, the SAPR (somatostatin–angiotensin-like receptor protein) gene copy number is often measured at eight copies per haploid chromosome set in colon cancer cells (vs. a single copy in normal cells). However, the amount of protein produced can be over 1000 times that of normal cells. Amplification of certain genes can sometimes lead to more aggressive tumors that decrease the survival rate of patients

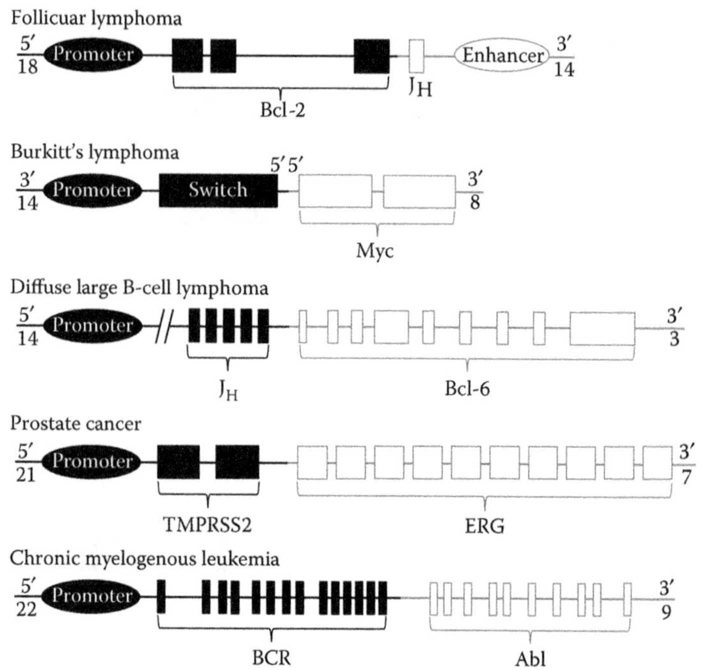

FIGURE 18.13 Examples of documented chimeric genes found in cancer cells. Direction of each chromosome (above) and the chromosome number (below) are indicated on each end of the chimeric gene map. In follicular lymphoma, an active enhancer originally on chromosome 14 increases the expression of a mutated *Bcl-2* gene from chromosome 18 that encodes a protein that inhibits apoptosis. Chromosomal translocations found in Burkitt's lymphoma and other B-cell lymphomas cause a chimeric gene that includes a portion of the *Myc* gene from chromosome 8 attached to a constitutive promoter from chromosome 14. The Myc protein stimulates transcription, which causes cell growth and division. Similar to follicular lymphoma, diffuse large B-cell lymphoma cancer cells contain chimeric genes that join portions of the immunoglobulin gene locus from chromosome 14 to the *Bcl-6* gene on chromosome 3. Normally, the Bcl-6 protein controls transcription in B cells, but the chimeric gene produces a nonfunctional protein. In some prostate cancers, the *TMPRSS2* gene on chromosome 21 is joined to the *ERG* gene on chromosome 9. The TMPRSS2 portion of the gene causes overexpression of *ERG*, which associates with other proteins to increase transcription, leading to cell proliferation. One of the most studied chimeric genes is responsible for causing chronic myelogenous leukemia. A portion of the *BCR* gene on chromosome 22 is fused to a portion of the Abl gene on chromosome 9. The Abl protein is a transmembrane tyrosine kinase that is usually controlled by other proteins, such that its activity is tightly controlled. However, the kinase activity of the chimeric BCR-Abl is always on, and the phosphorylation of other proteins leads to uncontrolled cell growth.

(Figure 18.15). For example, when five or more copies of the *HER2* gene are present in breast cancer tumors, the rate of survival is lower than when the cells have a single copy of the *HER2* gene.

Viruses

Some RNA and DNA viruses are known to cause cancers of various types. DNA oncoviruses include hepatitis B and C viruses (HBV and HCV, respectively), herpesviruses, and papilloma viruses. Although the precise mechanisms are unclear, some carry genes that affect cell cycle control and hormonal responses, and most may affect cellular regulation through integration near a cellular oncogene. RNA oncoviruses include feline, murine, avian, and human leukemogenic retroviruses, and Simian type-D retroviruses. RNA viruses utilize RNA as their genetic material, and carry an RNA-dependent DNA polymerase (i.e., a reverse transcriptase) inside their viral particles, such that they can create a DNA copy of their genomes. Also, the nucleic acids often undergo recombination events,

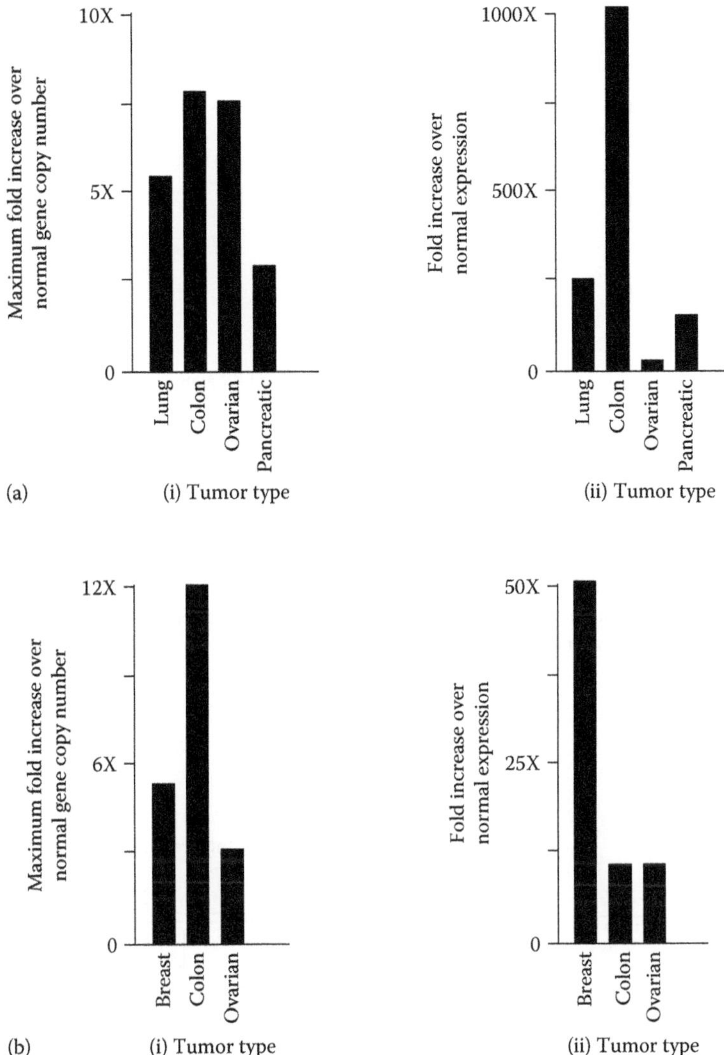

FIGURE 18.14 Gene amplification levels for several cancers. (a) Somatostain- and angiotensin-like receptors control cell proliferation. The genes for these receptors are amplified up to eight times in several types of cancer, including lung, colon, ovarian, and pancreatic (i). However, the number of proteins expressed from those genes can exceed 1000 times normal levels (ii). (b) Cathepsin Z is a peptidase that will degrade various proteins. The genes for this protein are amplified in most types of cancer cells. In breast, colon, and ovarian cancers they are amplified up to 12 times, although expression of the proteins can be as high as 50 times normal.

and are capable of integration into the host chromosomes. In some cases, integration alone can trigger the initiation of a cancer cell through the interruption of a cellular oncogene or by placing an active viral promoter next to a cellular oncogene. As mentioned in Chapters 12, 13, and 15 these exchanges of genetic material have actually led to many innovations in the evolution of cellular organisms with large genomes, because they result in horizontal transfers of DNA regions, including genes, promoters, and other genetic elements. However, it also allows a virus to integrate a cellular gene into its genome and control that gene using viral promoters. It also allows integration of versions of cellular genes into the virus genomes, which can then transform other host cells with those genes (Figure 18.16). The DNA repair mechanisms in these viruses are less accurate than the analogous cellular processes,

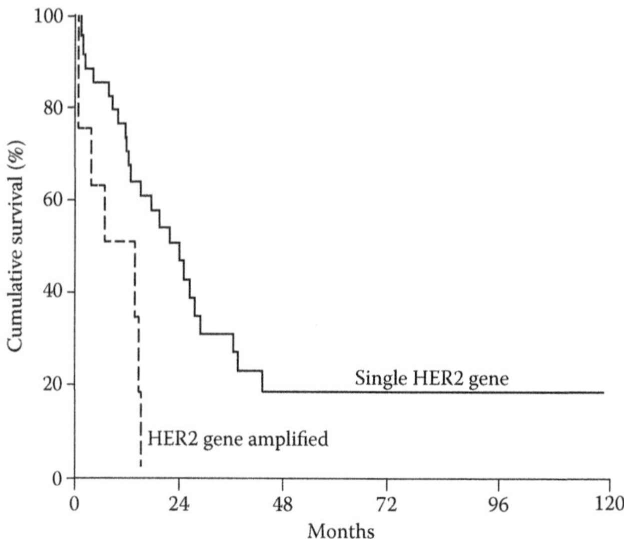

FIGURE 18.15 Effects of amplification on survival. When the HER2 gene is present in multiple copies in certain breast tumors, cumulative survival of patients is lower than in patients where the tumors have a single copy of HER2 (per haploid genome).

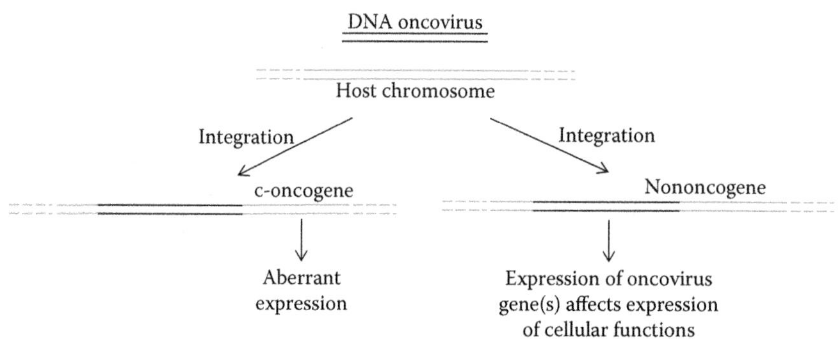

FIGURE 18.16 Integration of a DNA virus into a c-oncogene region, as well as into a nononcogene region. In both cases, the presence of the DNA virus genome can cause changes in transcription in the surrounding host genes. This can cause tumor initiation or other disruptions in normal cell functions.

resulting in higher rates of mutations in the viral version of the cellular genes. The genes are termed proto-oncogenes, indicating that they are capable of causing a cancer when mutated. The cellular versions are termed c-oncogenes (cellular oncogenes), while the viral versions of the same genes are called v-oncogenes (viral oncogenes; Figure 18.17). The primary differences between the c-oncogenes and the v-oncogenes are that the v-oncogenes contain only exons, they have major mutations, and they are controlled by a viral promoter. Removal of the introns is probably due to integration and reverse transcription of a mature c-oncogene mRNA. This has become integrated into the virus genome, and packaged into virus particles. Then, they infect other host cells, carrying the v-oncogene version. All of these lead to the production (usually an over production) of a mutant oncogenic protein within the infected cells. The result is aberrant gene expression leading to oncogenesis.

Avian leucosis virus (ALV) is a typical RNA retrovirus (Figure 18.18), which contains a gag gene (produces a polyprotein that forms the interior structure of the virion), a pol gene (produces a polyprotein that consists of a reverse transcriptase, an RNase, an integrase, and a protease), and an env gene

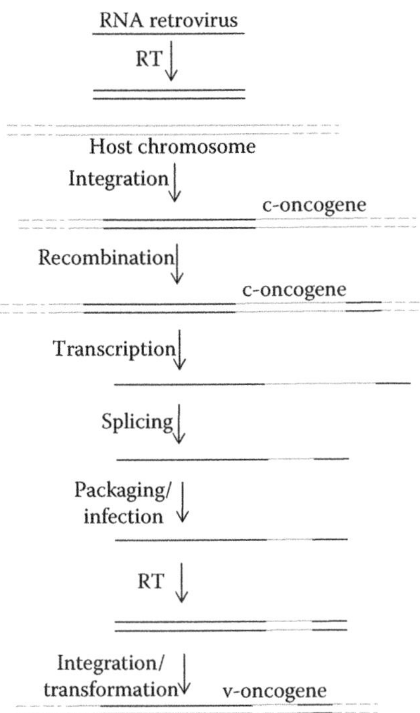

FIGURE 18.17 Conversion and integration of a c-oncogene into a v-oncogene by retroviruses. A standard retrovirus integrates into a c-oncogene region. Transcription and splicing produce a v-oncogene that includes both virus genes and a processed version of the c-oncogene. This is then incorporated into a newly forming virus particle, which can then infect a new host cell.

(a polyprotein consisting of a few glycoproteins). It has the ability to integrate a DNA copy of its genome into avian (and mammalian) host genomes. In some cases this leads to development of sarcomas. In the past, three exons of a host *Myb* gene (which encodes a transcriptional repressor that is important in controlling blood cell proliferation) have become incorporated into the virus genome, which has created a novel virus, called the avian myeloblastosis virus (AMV). The three exons of the *Myb* gene are integrated in the middle of the env gene. The Myb protein expressed from this gene is only partially functional, leading to carcinogenesis in the infected avian hosts. Some viruses have lost much of their genome, but continue to infect and cause cancers. For example, the Ableson murine leukemia virus (A-MuLV) carries a virus version of the *Abl* gene (a tyrosine kinase, involved in control of cell growth and division), but the insertion of the gene has caused a deletion of much of the *gag* gene, and all of the *pol* and *env* genes that are present in MuLV (Figure 18.18). Infection by this virus causes cancers in mice and other mammals. Similarly, Rous sarcoma virus has originated from the insertion of a virus version of the *Src* gene (which encodes a tyrosine-protein kinase that is important in controlling cell growth) into a MuLV. However, in this case, the *gag*, *pol*, and *env* genes are still fully intact. Infectious changes caused by this virus have been found in colon, lung, pancreatic, liver, and breast cancers.

DNA Viruses

Some DNA viruses also can cause cancers (as well as other health problems from warts to heart disease). Human papilloma virus most often causes cancers of the cervix, but also can cause cancers of the vulva, vagina, penis, anus, and oral/pharyngeal tissues. Its genome encodes nine proteins, one of which binds to p53, and marks it for degradation by proteasomes. This effectively shuts off the apoptosis pathway, and the cells continue to grow and divide to produce small warts and tumors, and in some

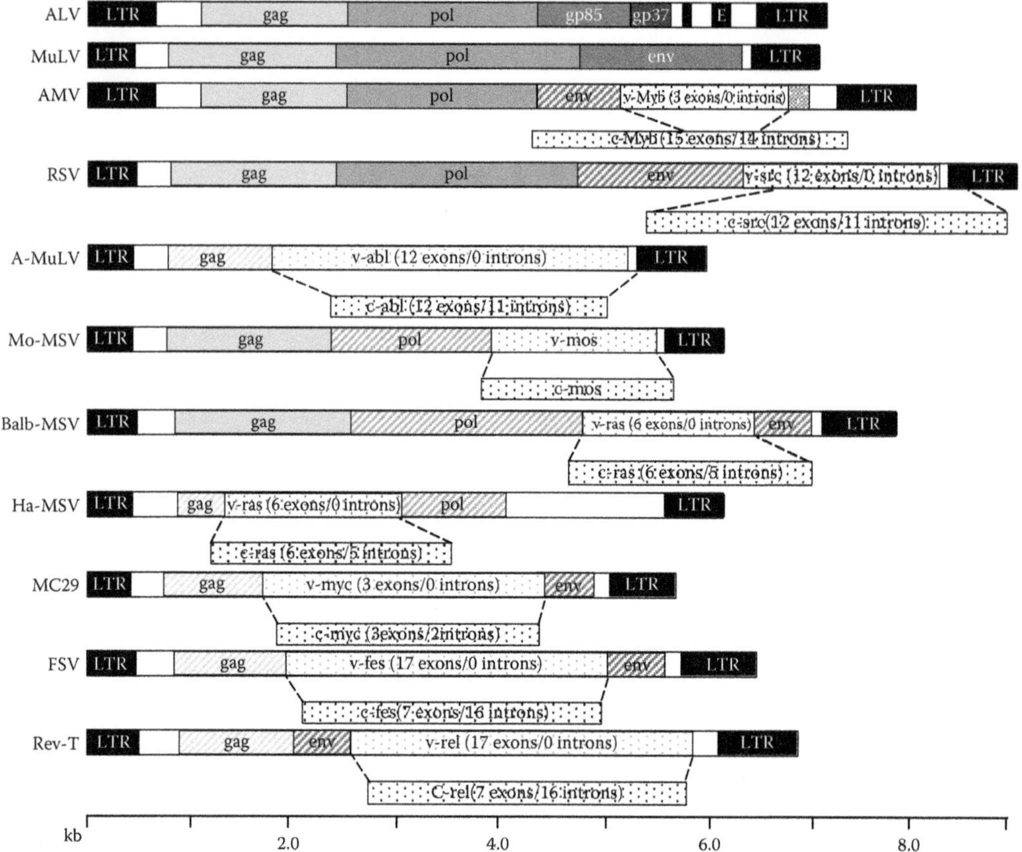

FIGURE 18.18 A selection of retroviruses that can cause cancers. Typical retroviruses have at least three genes: gag, pol, and env (top). These encode proteins that produce the structural internal proteins, polymerase/reverse transcriptase/RNase/integrase, and external glycoproteins, respectively. These can cause cancer by integrating within or near cancer genes. Alternatively, retroviruses can incorporate mutated versions of cellular proto-oncogenes (i.e., cancer genes) that cause aberrant expression patterns of genes that control cell proliferation and division. For example, avian leukosis virus (ALV) can cause cancer. Avian myeloblastosis virus (AMV) also causes cancer, but in a different way. AMV is a version of ALV, where part of a cellular Myb gene (without introns) has inserted into the env gene. On viral infection, the viral version of Myb (a viral oncogene, v-oncogene), which is a repressor of transcription that controls cell proliferation, is expressed and causes uncontrolled cell proliferation, due to loss of control of transcription of the repressor. Other viruses cause cancers by disturbing controls on cell proliferation, cell division, or apoptosis.

cases cancers. Epstein–Barr virus (EBV), a member of the herpesvirus group, is known to cause several types of lymphomas, as well as nasopharyngeal cancers. It can also trigger certain types of autoimmune diseases, and is the cause of infectious mononucleosis. It infects both B cells of the immune system, and epithelial cells. Once infected, the virus genome moves into the nucleus and circularizes, remaining there as an episome for the entire life of the host. It can be reactivated to once again produce virions and lyse the cells. Its genome is about 192 kilobases in length, which encodes 85 proteins. Although the exact mechanism whereby EBV causes cancer are unknown, it is suspected that some of the genes that are expressed by the virus during the latent parts of infection stimulate the host B cells to proliferate, thereby increasing the number of cells containing EBV, but at the same time causing damaging growth of the host cells. Kaposi sarcoma-associated herpesviruses (KSHV) are capable of causing skin cancers. While they were originally rare cancers, patients with AIDS (autoimmune deficiency syndrome, caused by infection and chromosomal integration of human immunodeficiency virus—HIV) have a higher incidence of these cancers due to their lowered immunity to viruses, such as KSHV. The virus has a genome of approximately 165 kilobases, that includes several genes that it has gained from the human genome. These include some genes that encode proteins involved in the immune system, cell cycle regulation, signal transduction, nucleotide synthesis, and DNA polymerase. Although the entire mechanism for its ability to induce Kaposi sarcomas in cells is unknown, it is known that it causes inactivation of some of the normal tumor suppression pathways in the cells.

HORMONES

Steroid hormones are chemicals that can cause widespread changes in cells either by interacting with receptors on the cell surfaces, intracellular receptors, or through other interactions. For example, estrogens can attach to intracellular estrogen receptors to increase transcription of specific genes. It can also attach to other receptors that associate with other factors to increase transcription of a number of genes. In a third mode, the estrogen can bind with cell surface and/or intracellular receptors to signal other molecules in the cell to perform specific physiological functions. Some of the estrogen receptors can act without binding estrogen by becoming phosphorylated by kinases that are activated by the cell surface receptors once they are bound to their specific growth factors. Steroid hormones all are produced in animals utilizing cholesterol, which consists of three linked six-carbon rings, and a linked five-carbon ring, with an alcohol on one end and a saturated side chain at the other end (Figure 18.19). A total of 11 enzymes are required to produce all of the steroid

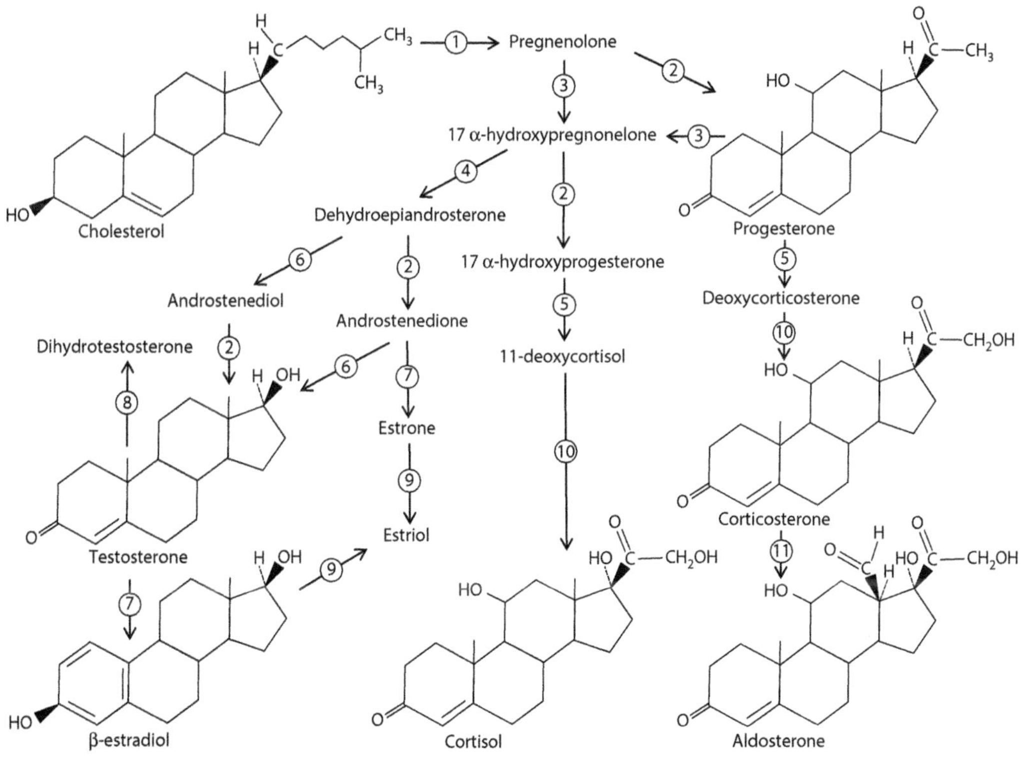

FIGURE 18.19 Biosynthetic pathways for steroid hormone synthesis. Starting with cholesterol, a series of 11 additional enzymes produce progesterone, testosterone, β-estradiol, cortisol, corticosterone, and aldosterone, as well as a number of other related intermediates and products. The enzymes are as follows: 1 = cytochrome P455scc (mitochondrial); 2 = 3β-hydroxysteroid dehydrogenase; 3 = 17α-hydroxylase; 4 = 17,20 lyase; 5 = 21-hydroxylase; 6 = 17α-hydroxysteroid dehydrogenase; 7 = aromatase; 8 = 5α-reductase; 9 = liver/placenta enzymes; 10 = 5β-hydroxylase (mitochondrial); aldosterone synthase (mitochondrial).

hormones, including progesterone, testosterone, cortisol, β-estradiol, aldosterone, and corticosterone, as well as 11 other related compounds and intermediates in the pathways. While hormones themselves do not cause cancers, they can stimulate the growth of cancers that already have been initiated. Also, they can stimulate cell growth and division, which can increase the possibility of mutation that may initiate a cancer. Therefore, steroid hormones can influence the initiation of a cancer, and then stimulate its growth, or they can stimulate an existing tumor to change it from a relatively small growth into a large life-threatening tumor. Both sometimes occur in people taking steroid hormones (for sports or for disease therapy) and in pregnant women, where tissues are being subjected to large amounts of these steroid hormones.

KEY POINTS

1. Cancer is a collection of diseases characterized by mutations in genes that control cell growth and division.
2. Cancers begin as mutations in specific genes that cause uncontrolled growth. Most eventually invade surrounding tissues, and some are able to attract blood vessels and metastasize to other parts of the body.
3. Several classes of genes can cause cancerous growths, including those that encode cell surface receptors, proteins involved in signal transduction pathways, kinases, hormone receptors, proteins that control apoptosis, cell adhesion proteins, cell cycle regulators, and transcription factors.
4. Cancers are classified based on their origin. Carcinomas begin from epithelial cells, sarcomas originate from endodermal cells, such as those in muscle, fat, connective tissues, bone, and cartilage. Lymphomas and leukemias begin from lymph and blood cells, respectively. Myelomas are cancers of bone marrow, and blastomas are cancers that originate in embryonic tissues during early development.
5. Cancers can be initiated by changes in specific genes caused by point mutations, recombination events (including unequal crossovers and chromosomal translocations), gene amplification, and some viruses (both RNA and DNA viruses).
6. Hormones can intensify and/or speed the progression of some cancers.

ADDITIONAL READINGS

Alberts, B., D. Bray, K. Hopkin, A. Johnson, J. Lewis, M. Raff, K. Roberts, and P. Walter. 2013. *Essential Cell Biology,* 4th ed. New York: Garland Publishing.

Alberts, B., D. Bray, A. Johnson, J. Lewis, M. Raff, K. Roberts, and J. D. Watson. 1994. *Molecular Biology of the Cell.* New York: Garland Publishing.

Andersen, P., H. Uosaki, L. T. Sheje, and C. Kwon. 2012. Non-canonical Notch signaling: Emerging role and mechanism. *Trends Cell Biol.* 22:257–265.

Darnell, J., H. Lodish, and D. Baltimore. 2007. *Molecular Cell Biology*, 6th ed. New York: W.H. Freeman and Company.

Domazet-Lošo, T., A. Klimovich, B. Anokhin, F. Anton-Erxleben, M. J. Hamm, C. Lange, and T. C. G. Bosch. 2014. Naturally occurring tumours in the basal metazoan Hydra. *Nat. Commun.* 5:4222. DOI: 10.1038/ncomms5222.

Freiberg, E. C., G. C. Walker, W. Siede, R. D. Wood, R. A. Schultz, and T. Ellenberger. 2005. *DNA Repair and Mutagenesis*, 2nd ed. Washington, DC: ASM Press.

Karp, G. 2013. *Cell and Molecular Biology*, 7th ed. New York: John Wiley & Sons.

Lewin, B. 2008. *Genes IX.* Sudbury, MA: Jones & Bartlett Publishers.

Rosenbluh, J., X. Wang, and W. C. Hahn. 2014. Genomic insights into WNT/β-catenin signaling. *Trends Pharmacol. Sci.* 35:103–109.

Section V

Molecular Biology and Bioinformatic Methods

19 Extraction and Quantification of Biological Molecules

INTRODUCTION

Most molecular biology projects begin by extracting DNA, RNA, and/or proteins, followed by quantification and characterization of the biological molecules (Figure 19.1). The tissues can be fresh, frozen, or those that have been preserved either artificially or naturally (Figure 19.2). However, tissues are preserved best if they are kept frozen or dried (Figure 19.3). DNA and RNA extractions can also be performed in the field in cases where laboratories are distant, but extraction of high-quality DNA and RNA is needed from field-collected specimens. Projects range from characterization of the biological molecules from a species or tissue to characterization of those molecules following experimental manipulations of the tissues or cells. DNAs have been recovered from specimens up to millions of years old. The Neanderthal genome was determined from the bone marrow of an individual that had died 50,000 years ago. That led to the discovery that the human population that first moved into Europe more than 30,000 years ago hybridized with the Neanderthals, and that the Neanderthal alleles for some genes, up to several percent, can be found in modern humans with European and East Asian ancestry.

DNAs play a central role in molecular biology and genetics, RNAs are used to produce proteins and control gene expression, and proteins produce the characteristics of the cells, including control of gene expression. Members of each category can be extracted, studied, and manipulated to understand more about particular cellular processes or to transform organisms to produce specific products or functions. To purify high-molecular-weight DNA (>50–100 kb) so that it can be used in molecular biology procedures and experiments requires specialized equipment, supplies, and chemicals, but for most samples, it is not especially difficult to extract (Table 19.1). On the other hand, additional chemicals and equipment are needed to purify high-quality RNA, and greater care must be taken to eliminate as much RNase activity as possible. RNAs are generally more reactive than are DNAs, and they are degraded much more readily than are DNAs. During the process of extracting nucleic acids of all types, proteins (including nucleases that degrade DNA and RNA) as well as polysaccharides (some of which can inhibit enzymes used in molecular biology procedures), and other cellular debris are removed. For extraction of RNA, RNase inhibitors are also used.

In general, nucleic-acid-extraction procedures must accomplish several outcomes (Table 19.1). First, if there are cell walls present (e.g., in plants, fungi, many protists, bacteria, and archaea), they must be broken or digested to release the cellular constituents (Figures 19.4 and 19.5). This is usually accomplished by grinding the tissue in dry ice, liquid nitrogen, or in a buffered solution, or by adding enzymes that will break or soften the cell walls. A mortar and pestle, food grinder (e.g., coffee or spice mill), or homogenizers can be used for grinding and tissue dissociation. The tissues can be fresh, frozen, or dried (as with herbarium, lyophilized, or mummified tissues). Second, the cell, organelle, and nuclear membranes must be disrupted so that the nucleic acids are released into the extraction buffer. This is accomplished using a detergent, such as SDS (sodium dodecyl sulfate) or CTAB (cetyltrimethylammonium bromide, also known as hexadecyltrimethylammonium bromide). Third, the nucleic acids must be protected from endogenous nucleases. Detergents can be somewhat effective for this purpose, because some will partly denature these enzymes. However, a common additive is EDTA (ethylenediaminetetraacetic acid), a chelating agent that binds magnesium ions (Mg^{2+}), which are cofactors for many of the nucleases. Therefore, removal of magnesium ions inactivates most nucleases. However, if RNA is sought, specific RNase inhibitors need to be added (e.g., guanidinium thiocyanate,

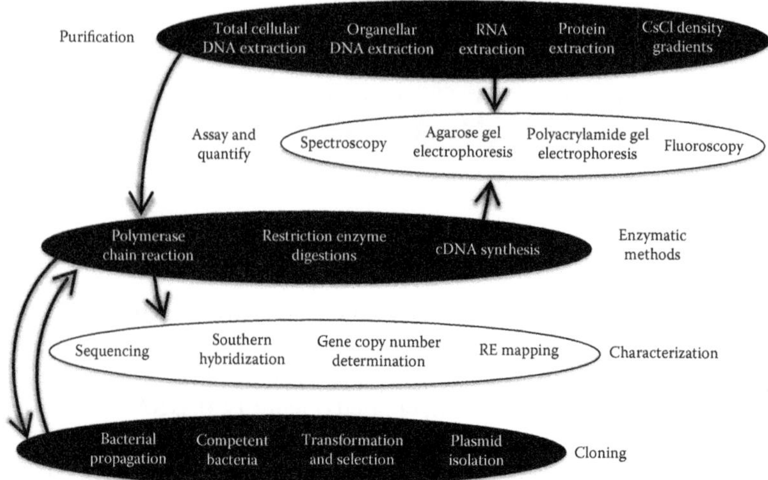

FIGURE 19.1 Flow chart of procedures, assays, and other methods used in molecular biological research. Initially, the molecules being studied are extracted and purified. Next, assays are performed to determine the quantity and quality of each extract. Additional procedures are performed to increase the utility of the extracted molecules. Characterizations of the molecules are then performed. Alternatively, portions of the nucleic acid extracts are cloned, and the recombinant molecules are studied and characterized.

FIGURE 19.2 Examples of preserved materials, many of which have contained usable biological molecules. Specimens A, B, C, D, and E are fossilized organisms where minerals have replaced most of the biological materials. For most fossils of this type, extraction of nucleic acids and proteins has been unsuccessful. The F samples are plant and arthropod tissues encased in amber. Both are over 1 million years old, but they likely contain usable biological molecules, as DNA has been extracted from amber-encased tissues more than 100 million years old. Desiccated tissues, such as those in herbaria (specimens G and H—fungi), as well as those in dry tombs (I—cedar wood from King Midas' tomb), and packrat middens (J—plant materials) often contain extractable nucleic acids. Specimens K and L are meltwater samples from glacial ice from Antarctica that were 10,000 and 1.5 million years old, respectively. Biological molecules can also be extracted from cultures grown from ancient materials. Specimen M is a fungus culture that was grown from a 140,000 year old sample of glacial ice from Greenland.

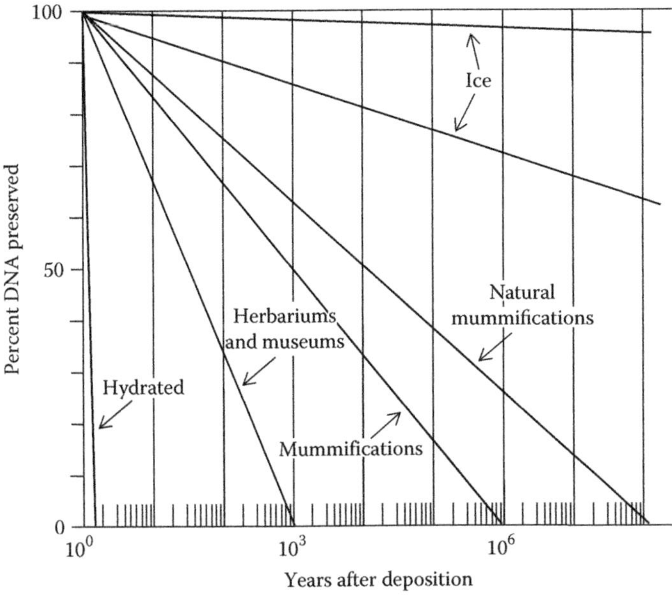

FIGURE 19.3 Graph showing the relative degree of DNA preservation under various conditions. Hydrated DNA is rapidly degraded, such that very little intact DNA remains after days or weeks when continuously hydrated at room temperature. DNA extracted from hundreds of herbarium and museum specimens indicate that the dry conditions preserve the DNA to a degree. Measurable DNA that is in pieces of 100 bp or more can be extracted from specimens that are several hundred years old. In some mummified humans, animals, and plants, preservation is better, such that usable DNA has been extracted from specimens that were several thousand years old. Natural mummifications of humans, animals, and plants have yielded DNA from specimens that were tens of thousands to hundreds of thousands of years old. Ice appears to yield the best preservation (although reports of successful extraction and sequencing from 100+ million year old tissue in amber have been reported). Not only have DNA and RNA been recovered from ice as old as 2 million years old, but viable bacteria and fungi have been cultured from ice up to 8 million years old.

TABLE 19.1
Considerations When Extracting Nucleic Acids

Property/Consideration	Countermeasure/Process
Cell walls—surround the cells in plants, fungi, bacteria, and others	Digest or abrade cell walls with enzymes and/or by grinding
Cell membranes—surround cells and organelles (including nuclei, plastids, mitochondria, lysosomes, peroxisomes, etc.)	Disrupt/solubilize membranes with detergents, such as SDS and/or CTAB
High pH (basic, alkaline)—denatures DNA, degrades RNA	Buffer to pH 6.5–8.5 (e.g., Tris, phosphate)
Low pH (acidic)—degrades both DNA and RNA, primarily by depurination and sugar/phosphate backbone breakage	Buffer to pH 6.5–8.5 (e.g., Tris, phosphate)
Mg^{2+} present in all cells—activated nucleases (DNases and RNases)	Add chelating agent, such as EDTA to sequester Mg^{2+}
RNases present in all cells (and in environment, on hands, etc.)—rapidly degrade RNA	Treat with DEPC or RNase inhibitors
Proteins—including nucleases and other proteins that inhibit or interfere with the use of the extracted nucleic acids	Denature and/or remove proteins with chloroform and/or phenol
Polysaccharides—many inhibit enzymes (including those used in molecular biology protocols)	Precipitate with CTAB and/or alcohol
Purify nucleic acids	Precipitate with CTAB and/or alcohol

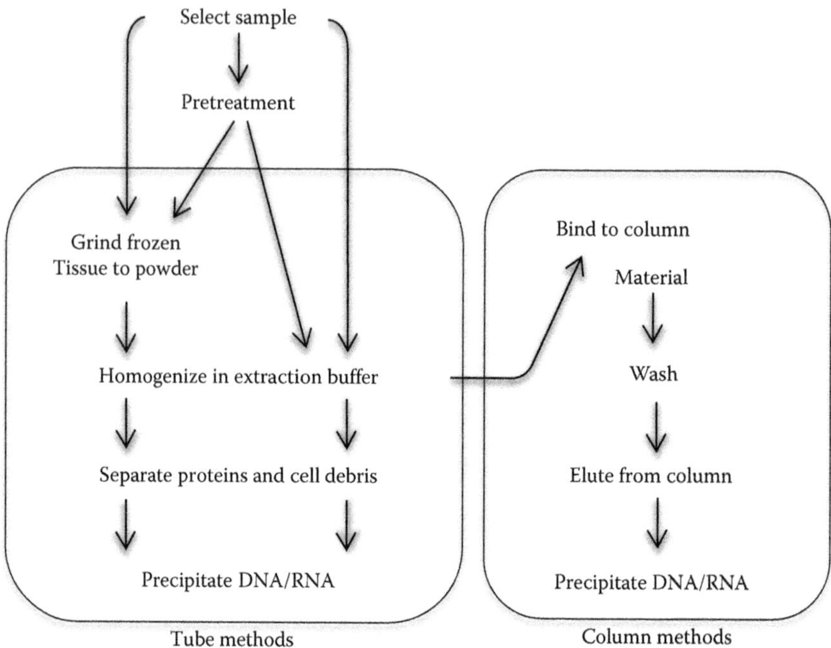

FIGURE 19.4 General steps that are used in extracting nucleic acids using tubes (microfuge, centrifuge, or test) or columns containing substances that bind nucleic acids. Many types of samples need pretreatment, including tissues with cell walls (bacteria, archaea, fungi, plants, and some protists), as well as mixed tissues (such as stool, water, and soil samples).

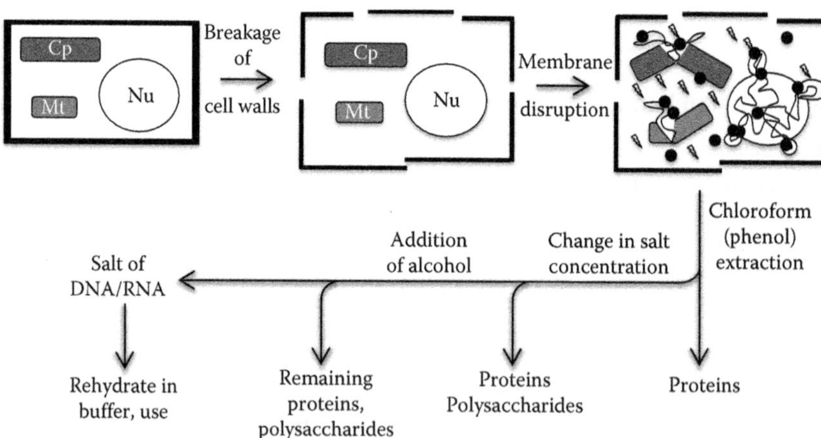

FIGURE 19.5 Critical steps necessary to extract high-purity DNA and RNA. If the cells have cell walls, they must first be broken or digested with enzymes. Next, the membranes must be disrupted (usually with a detergent), while components in the extraction solution inactivate degradative enzymes and reactive molecules. Next, the proteins are removed using chloroform and/or phenol. The DNA and RNA can then be separated from polysaccharides and the remaining proteins by adding salt and alcohol, which causes precipitation of the nucleic acids. After drying the nucleic acids, they are rehydrated in a buffer (usually Tris, as pH 8.0, with a low concentration of EDTA to inactivate any remaining nucleases).

RNasin, or analogous inhibitors). Additionally, the extraction solution must contain a buffer to control extremes in pH, because DNA and RNA can be damaged both by acidic and basic conditions. Most often, a phosphate buffer or a Tris (Tris[hydroxymethyl]aminomethane) buffer is used to maintain the pH between 7.0 and 8.5. The buffer/tissue homogenate is usually emulsified either with chloroform or phenol (or both) to denature and remove many of the proteins from the nucleic acids. Fourth, shearing of the DNA should be minimized while the nucleic acids are in aqueous solutions. The long strands of DNA can be fragmented by high levels of turbulence (e.g., being drawn through a small orifice, such as a pipette tip; or by vortexing). Typically, DNA that is from 30 to 150 kb in length can be obtained without great care being taken. If lower molecular weight molecules are sufficient, then shearing the DNA is less of a concern. Fifth, if frozen tissues are used, at no point in time should the tissue be thawed prior to addition of the extraction buffer. Thawing may cause breakage of cell and internal membranes, which will lead to the exposure of the nucleic acids to nucleases and other components that may degrade the nucleic acids. The time between the addition of the extraction buffer and precipitation of the nucleic acids should be minimized. DNA and RNA are vulnerable to nuclease activity when in aqueous solutions, but are protected if they are precipitated as salts in ethanol, or when dried. In fact, DNA can be dried and sent by regular mail, and it will be intact when the recipient receives the letter or package. On the other hand, if an aqueous solution of DNA or RNA is sent through regular mail (at ambient temperatures), it will be highly degraded once it arrives at its destination.

Once the sample is homogenized in the extraction buffer, various parts (e.g., proteins and polysaccharides) of the homogenate must be rapidly separated from the nucleic acids. There are three common ways to accomplish this. One is to send the sample through a column containing a matrix of material that differentially attaches to one or more components in the homogenate (Figure 19.5). One of the most common is a type of silica. The silica polymers in the column have a net positive charge, while nucleic acids have a negative charge (due to the phosphate groups on the backbones on the molecules), and thus the nucleic acids bind to the silica in the column, while most other components will flow directly through the column. Larger insoluble components are trapped at the top of the column. A solution containing monovalent cations (e.g., NaCl) is then sent through the column that strips the DNA away from the silicate by binding to the phosphate groups on the nucleic acids (while the Cl^- ions replace the nucleic acids bound to the silica), and the DNA flows out of the column. The DNA is then precipitated as a salt (most often as a sodium salt) using ethanol (nucleic acids have low solubilities in alcohols). A second mode of extraction is to centrifuge the cell debris, and then precipitate the DNA salt (usually a sodium salt) with an alcohol (usually ethanol or isopropanol). However, for samples that contain high concentrations of proteins, polysaccharides, or other compounds, these other molecules frequently coprecipitate with the nucleic acids, rendering the nucleic acid solutions useless for molecular biology procedures. A third way that has been useful for the extraction of DNA from plants and fungi (and other organisms with high concentrations of polysaccharides) is to use the differential solubilities of nucleic acids, proteins, and polysaccharides when they are in the presence of CTAB and salt (usually NaCl), which is described in the following.

EXTRACTION OF NUCLEIC ACIDS USING CTAB

While standardized extraction methods are successful for many tissues and cell types, there are some differences that must be considered when choosing an extraction method. Most commercial kits are optimized for general purposes, and not for specific tissues. For example, bacteria, archaea, plants, fungi, and many protists have cell walls in addition to cell membranes, and many animals have hard exoskeletons. In each of the groups, the cell wall or exoskeleton composition is unique. In bacteria, peptidoglycans are the major wall component, while in plants, it is cellulose; for fungi and many animal exoskeletons, it is chitin, and for protists, a variety of polysaccharides may be present. In addition to the rigid polysaccharide structures, most organisms have constituents that inhibit attack by molecules from foreign organisms. If these are copurified with the nucleic acids, they will often inhibit the enzymes that are used for molecular biology manipulations and assays.

The inhibitors can be enzymes, polysaccaharides, peptides, or other compounds. CTAB has been used successfully in an extensive list of tissues and cells of plants, fungi, animals, and bacteria for DNA and RNA extraction to remove these inhibitors during the extraction process (Figure 19.6). It is an ionic detergent, which dissociates in water into a cetyltrimethylammonium cation, CTA^+, and a bromide anion, Br^-. The CTA^+ cation complexes with nucleic acids, polysaccharides, and proteins (as long as they have negative charges on their surfaces). However, the solubility of these complexes differs, especially when in the presence of varying concentrations of another cation, such as sodium cations (Na^+). Therefore, by altering the Na^+ concentration, proteins, polysaccharides, and nucleic acids can be separated, usually by precipitation as a CTA^+ salt by using different concentrations of Na^+ ions. For nucleic acids, precipitation occurs when the concentration of sodium ions falls below about 0.5 M. The initial extraction buffer usually contains 0.7 M NaCl. The extraction buffer/tissue mix is emulsified with chloroform to denature and remove most of the proteins, and then the nucleic acids are precipitated either by reducing the salt concentration to approximately 0.3–0.4 M or by adding isopropanol. The precipitate is then rehydrated (i.e., dissolved) in a buffer containing 1 M NaCl, and precipitated again by adding an alcohol (usually, ethanol or isopropanol). In these last steps, Na^+ replaces the CTA^+ on the nucleic acids, so that the final product is a sodium salt of DNA and RNA. In this form, the purified nucleic acids can be used in a wide range of molecular biological techniques. In addition to the extraction of DNA from fresh tissues, CTAB methods have been used to extract nucleic acids from milligram amounts of dried herbarium and mummified tissues,

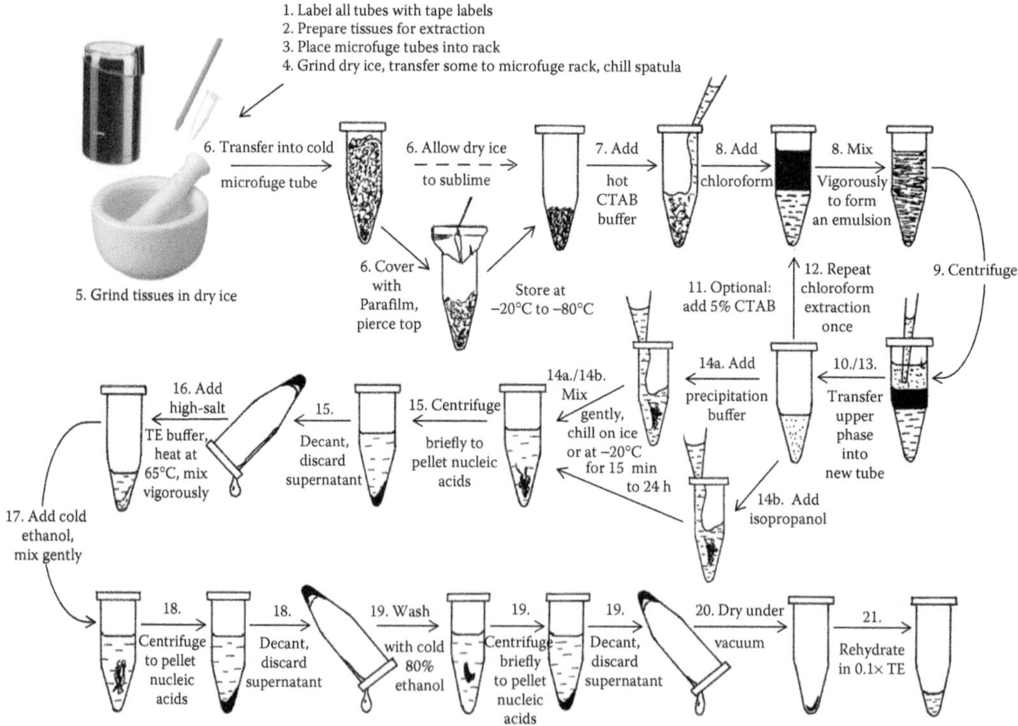

FIGURE 19.6 Extraction of nucleic acids from plants and fungi using CTAB. Cell walls are first broken by grinding in dry ice or liquid nitrogen. Following sublimation, hot CTAB extraction buffer is added and thoroughly mixed with the tissue. Two chloroform extractions are performed to separate proteins from the nucleic acids. The nucleic acids are precipitated either by reducing the salt concentration or by adding isopropanol. Following centrifugation, the pellets are rehydrated in high salt TE (Tris, EDTA). The nucleic acids are then precipitated as sodium salts. They are then pelleted, dried, and rehydrated in 0.1× TE (diluted TE). They may then be used in molecular biology protocols.

some that were more than 100,000 years old, and from ice samples, some of which were more than 1 million years old. In addition to extraction of nucleic acids, CTAB can be used to extract polysaccharides and proteins by varying the NaCl concentration.

PURIFICATION OF ORGANELLAR DNA

The study of organellar DNA often requires that DNA samples be obtained that are highly enriched in the organellar DNA. In the case of animals, fungi, and so on mitochondria are the only other cellular components that contain DNA (other than the nucleus). On the other hand, plant cells contain mitochondria, as well as plastids (chloroplasts, amyloplasts, etc.) that contain their own DNAs. While all three components (chloroplast, mitochondria, and nuclei) contain different DNA sequences, much of the DNA from the original bacteria that became the plastids and mitochondria has moved to the nucleus during evolution. Additionally, some mitochondrial sequences have been transferred into plastid genomes. Transfer of sequences into mitochondria is rare. Most cells generally contain more than one of each organelle. Each organelle often contains multiple (often hundreds of) copies of its genome. Because of the multiplicity of the organelles and their genomes, the proportion of DNA in a cell that is organellar DNA often exceeds 20% of the total DNA per cell.

Organellar genome sizes are small compared with nuclear genomes. Plastid genomes are between approximately 35 and 250 kb, while mitochondrial genomes range from approximately 6 to 11,000 kb. Nuclear genomes range from approximately 6×10^6 to 6×10^{11} bp. Because of this, bands can clearly be seen when restriction digests of the organellar DNA are separated by electrophoresis on agarose gels. This is in contrast to the restriction enzyme patterns exhibited by nuclear DNAs of higher eukaryotes that are so large and diverse in sequence that only a few bands can be seen (corresponding to long highly repetitive and middle repetitive sequences) over a background smear of most of the DNA (low and single copy number sequences). Although the organellar DNA can sometimes be observed on restriction digests of total DNA preparations, to accurately study these genomes, enrichment of the organellar DNA must be accomplished. All methods of enrichment employ low- and high-speed centrifugation, and many employ density gradients (Figure 19.7) of one type or another, used to separate cellular debris from the organelles. Initially, a low-speed centrifugation step is used to pellet the nuclei and starch grains (in plants). Next, the supernatant is loaded onto a density gradient, or the organelles are pelleted using a high-speed centrifugation step. After centrifugation, the organelles are located in the gradient or in the pellet and are removed for study and DNA/RNA extraction. The two most frequently used density gradients utilize either sucrose or Percoll (colloidal silica). Sucrose is less expensive and easier to use, but separation and resolution are generally better with Percoll. After the organelles have been isolated by differential centrifugation or by density gradients, a high amount of nuclear (and other) DNA is usually present in the solution and adhering to the surfaces of the organelles. Further purification can be achieved by treating the solution with a low concentration of DNaseI. The enzyme cannot pass through intact membranes, and therefore will digest only the DNA outside of the intact organelles. However, the DNA within any organelle with a compromised membrane will be digested. Therefore, this method is also a test for the condition of the organelles. Before lysing the organelles for DNA extraction, the DNaseI must be inactivated, usually by using a high concentration of EDTA. The DNA extracted in this way is highly enriched in organellar DNA. Of course, in the case of plants, both mitochondria and plastids will be included among the sections of the gradient. Depending on the gradient conditions, the two types of organelles may be separated or they may be found together. In some species, cesium chloride (CsCl) isopycnic gradients can be used to further purify the DNA. Bisbenzimide (also known as Hoechst 33258) is used in CsCl gradients to accentuate the base composition differences and to visualize the DNA. This dye associates with AT-rich portions of the DNA and causes AT-rich DNA to be buoyed up in the gradient (lowers the density of the AT-rich DNA). For higher plants, plastid DNA has nearly the same base composition as most of the nuclear DNA, such that the plastid DNA comigrates with the main nuclear DNA. However, for some algae, the chloroplast DNA

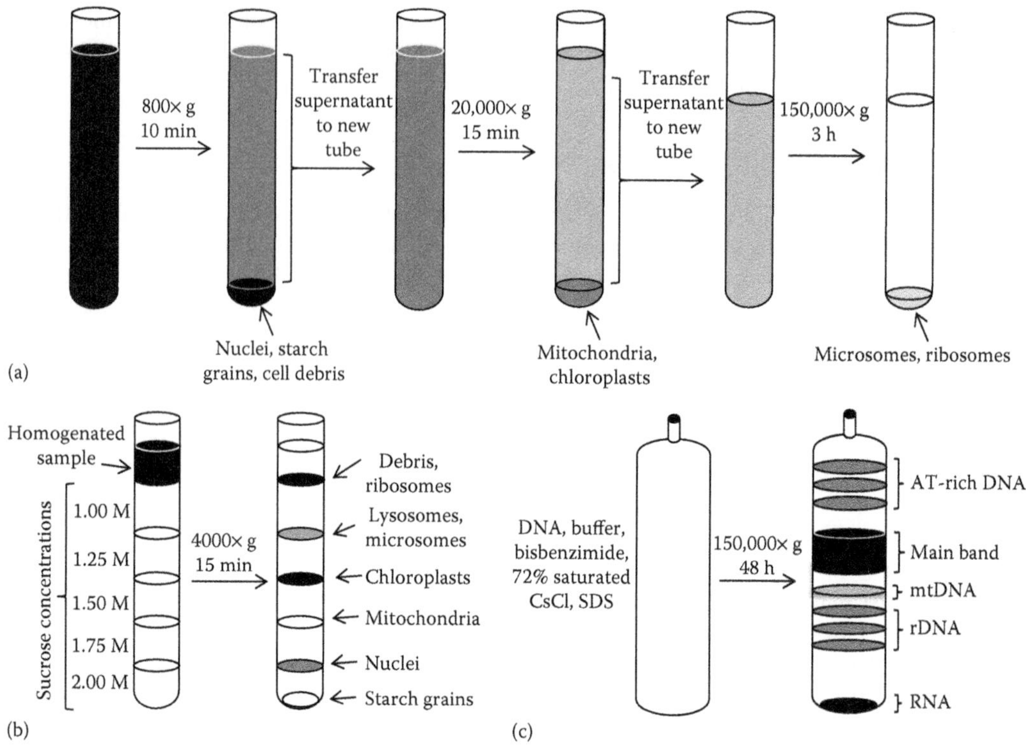

FIGURE 19.7 Separation of cellular constituents and biological molecules using centrifugation. (a) Differential centrifugation entails low-speed spins progressing to higher speeds. After each centrifugation, pelleted material is collected, and the supernatants are used in higher speed centrifugation runs. (b) Rate-zonal centrifugation or density gradient centrifugation involves pouring a gradient of concentrations of solutions from high density to low density, bottom to top. During centrifugation, the cellular constituents migrate to the densities that are closest to their own densities. (c) Isopycnic gradients employ ultracentrifugation. These can be used to separate species of DNA and RNA (as well as other large biological molecules). Often, chemicals (dyes) are used that attach to the nucleic acids, and cause density changes based on the regions of DNA and RNA where they attach. For example, bisbenzimide attaches to AT-rich regions of DNA, and therefore those regions are buoyed up in the gradient, thus accentuating differences in density among the species of DNAs in the gradient.

can be separated from the main nuclear DNA due to the different base compositions for each of the DNA components. Mitochondrial DNA from most higher plants contains a higher proportion of GC than does nuclear DNA or plastid DNA, and therefore does not incorporate as much of the dye as other components of DNA. Because of this, plant mitochondrial DNA appears below the main and plastid bands, and can often be separated from nuclear DNA using this method.

EXTRACTION OF RNA

It is more difficult to extract intact RNA than DNA. There are several reasons for this. First, RNases are everywhere, and they are difficult to inactivate. They exist in every cell in every organism, on your skin, on your hair, in your breath, in the air, and almost anywhere else you might look. Some of them can be boiled for long periods of time, and they lose little of their activity. Additionally, there are many different RNases. Some are highly active and can destroy an RNA molecule in seconds. Also, RNA itself is more reactive than DNA, primarily because of the hydroxyl on the 2′ carbon of the ribose sugar. The hydroxyl is a reactive group that can attack other portions of the RNA molecules. Finally, while DNA is merely denatured in alkaline solutions (above about pH 9.0), RNA is degraded. Thus, additional measures must be employed while working with RNA.

The amount of RNA in each eukaryotic cell averages approximately 10–50 pg. About 80% of the RNA is rRNA that is the primary component of ribosomes (discussed in Chapter 5). However, the sequence diversity and complexity of rRNA is low because there are only three (in bacteria and archaea) or four (in eukarya) different rRNAs (aside from the precursor molecules). About 15% is tRNA, while only 1%–5% is mRNA. However, the mRNAs are diverse, consisting of thousands or tens of thousands of different sequences that encode at least the same number of proteins. For many purposes, mRNA is sought. Because most mRNAs have poly-A tails, a poly-dT column, or beads that have covalently bound poly-dT attached to them, can be used to concentrate mRNA. Alternatively, RT-PCR (reverse transcriptase PCR) can be used, in conjunction with a poly-dT primer for the 3′ end of the RNA, to amplify only the mRNA to produce cDNA. If all types of RNAs are sought, then random hexamer primers can be used to produce cDNA copies of the RNAs, most of which will consist of copies of regions of the rRNAs (due to the abundance of the rRNAs).

Much more care is needed when extracting RNA from cells. Because of its instability and the ubiquity of RNases, measures must be taken to obtain intact RNA. One of the primary concerns is to treat the solutions, tubes, and other equipment with chemicals that will destroy or inhibit RNases. One common method is to treat tubes and instruments with a solution of DEPC (diethylpyrocarbonate), which denatures proteins. After treatment, the tubes and instruments can be autoclaved, which sterilizes the instruments and causes the DEPC to decompose into carbon dioxide and ethanol, thus rendering it inactive. Because the DEPC is used in low concentrations, the CO_2 and ethanol present no problems with subsequent procedures. Several other RNase inhibitors are also effective and available commercially. When working with RNA, gloves should be used at all times, and care should be taken not to exhale directly into any solution or tube. RNases are on skin, hair, clothing, in exhaled air, in dust, and elsewhere.

A sequence can be obtained directly from isolated pure RNA of one type. However, it is difficult and time-consuming. Most researchers synthesize cDNA (i.e., a complementary copy of the RNA), and then either sequence it directly as DNA or clone it into a bacterial vector first and then sequence it. Next-generation sequencing methods produce a cDNA prior to sequence determination. Alternatively, a cDNA library can be created on synthetic beads using RT-PCR methods in preparation for high throughput sequencing. To produce cDNA, a first strand is produced that is complementary to the RNA. This is accomplished using a poly-dT primer (for mRNA) in conjunction with a reverse transcriptase to produce the complementary first strand. To produce cDNA from total RNA, random hexamer primers often are used, again in conjunction with a reverse transcriptase. RNase H is then used to eliminate the RNA, and second strand synthesis (using the first cDNA strand as the template) is initiated using a DNA polymerase. At this point, the double-stranded cDNA can be cloned or sequenced directly, or small known sequences, called adapters or linkers, can be ligated to the ends of the cDNAs. Subsequently, the cDNAs can be amplified (using the adapter/linker sequences as the primers) or cloned into a vector for characterization, sequencing, and/or to study expression *in vivo* or *in vitro*.

There are many methods available for isolating intact RNA. Columns can be used, which are effective for most bacteria. However, eukarya can be more challenging, and more variable. There are many methods based on the use of guanidinium thiocyanate that denatures proteins (including RNases). It is generally used in conjunction with phenol and/or chloroform. A popular and effective method uses all three in a reagent known as TRIZOL. It is relatively simple, and yields high-quality RNA. Alternatively, various types of columns (some that can be used for RNA and DNA) can be used to extract high-quality RNA.

QUANTIFICATION OF NUCLEIC ACIDS

DNA and RNA can be quantified in several ways, including methods that can detect less than a picogram (10^{-12} g) of each, although most have sensitivities well above a picogram (usually in the nanogram or microgram range). A common method of quantification is to use a UV spectrophotometer (Figure 19.8). The purine and pyrimidine rings of nucleic acids maximally absorb UV irradiation between 250 and 280 nm. By measuring the amount of absorbance at 260 nm, the concentration of both DNA and RNA can be estimated. Specifically, for pure DNA, an absorbance of 1.0 O.D. (optical density) at 260 nm (i.e., A_{260}) equates to a concentration of 50 μg/mL, while 1.0 O.D. at A_{260} for RNA is 40 μg/mL, and for ssDNA, it is 33 μg/mL. The reason for the difference is that RNA and ssDNA have unpaired bases, many of which are protruding from the molecule. Therefore, there are more bases exposed to the UV irradiation, which leads to additional absorbance as compared with DNA, which contains bases that are paired within the central axis of the molecule, thus limiting their exposure to the UV irradiation.

Another quantification method for DNA and RNA is fluorimetry (Figure 19.9). For this method, a fluorescent dye is used that fluoresces at a specific wavelength when associated with DNA or RNA (some dyes are specific for either DNA or RNA, while others fluoresce when associated with either nucleic acid). Usually, UV wavelengths are used for excitation, and the fluorescence emission is in the visible range of wavelengths. Fluorimeters can be used for this specific purpose, although simple UV light and camera combinations can also be used. DNA or RNA of known concentrations are used to calibrate the device using a specific dye and wavelength, and then unknowns (DNA and/or RNA mixed with the same fluorescent dye) are measured and compared with the standards to determine concentrations. Sensitivities into the picogram range are possible. Alternatively, DNA and RNA can be quantified using agarose gels (described in the following), by spotting the nucleic acids onto paper or aluminum foil, or by encapsulating the nucleic acids inside glass capillaries. Again, standards of known concentrations must be used to determine the concentrations of the unknowns. UV irradiation is used in each case to excite the fluorescent dyes, and then the amount of emitted light is measured. Quantification of the emission may be accomplished using photography, digital analyses, a light sensor, or a combination.

AGAROSE GEL ELECTROPHORESIS

One of the most common methods of separating nucleic acids by size is by agarose gel electrophoresis (Figure 19.10). Agarose is an inert polysaccharide derived from marine macroalgae (Figure 19.11). It differs from agar in that the agaropectin has been removed. Several different agaroses are commercially available that vary in the side groups on the sugars, as well as in the melting and gelling temperatures. Several agaroses are produced that can be used for *in-gel* procedures, wherein enzymes can be used on the DNA and RNA while they are still in the gel. Some side groups on other

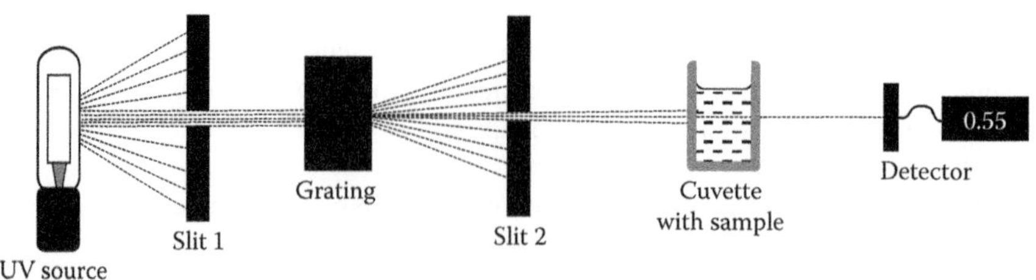

FIGURE 19.8 General structure of a UV spectrophotometer for quantifying nucleic acids. The UV light is sent through a slit so that it forms a coherent column of illumination. It is then split by a grating to split the UV into discrete wavelengths. A movable slit allows only the wavelength selected by the user, which then is passed through the solution of nucleic acids. The absorbance, or OD, is measured by the detector, based on a blank control that contains only the buffer solution without the nucleic acids.

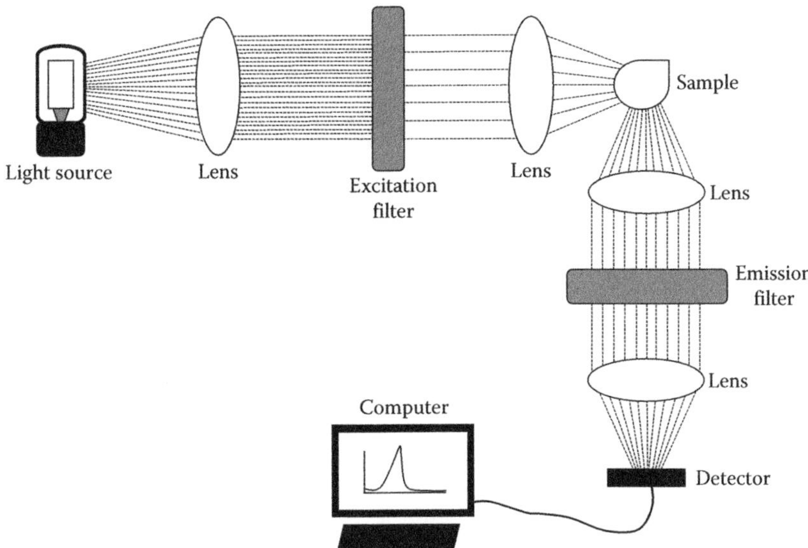

FIGURE 19.9 General structure of a fluorimeter for quantification of nucleic acids. The illumination from the light source is filtered so that it is primarily light of the wavelength that will excite the flourophore that associates with the nucleic acids. The light beam is focused with a lens on the sample, which fluoresces in all directions. At 90° to the beam path, another lens disperses the light through an emission filter that allows primarily the wavelength of light that is the maximal emission of the fluorophore associate with the nucleic acids. This is then focused with another lens onto the detector that measures the amount of light and transmits the amount to software in a computer.

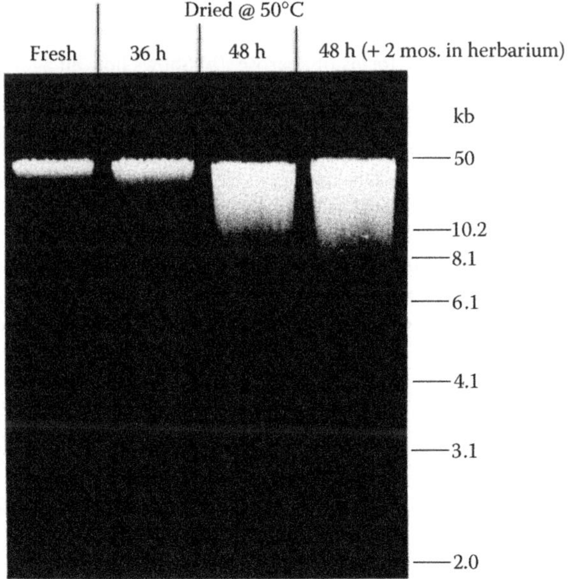

FIGURE 19.10 DNA separated by agaorse gel electrophoresis. The DNA in the first lane (left) was extracted from a fresh fungus sample. The average molecular weight of the DNA is just below 50 kb. In the second lane, the same fungus sample had been in a 50°C drying oven for 36 h prior to extraction. A small amount of degradation is evident by the light fluorescence below the main band. After 48 h of drying (third lane), additional degradation is observed in the extracted DNA as a smear of fluorescence between approximately 10 kb and the main band. Following 48 h of drying and subsequently 2 months in a herbarium, the extracted DNA exhibits additional degradation, with DNA fluorescence extending down to approximately 9 kb.

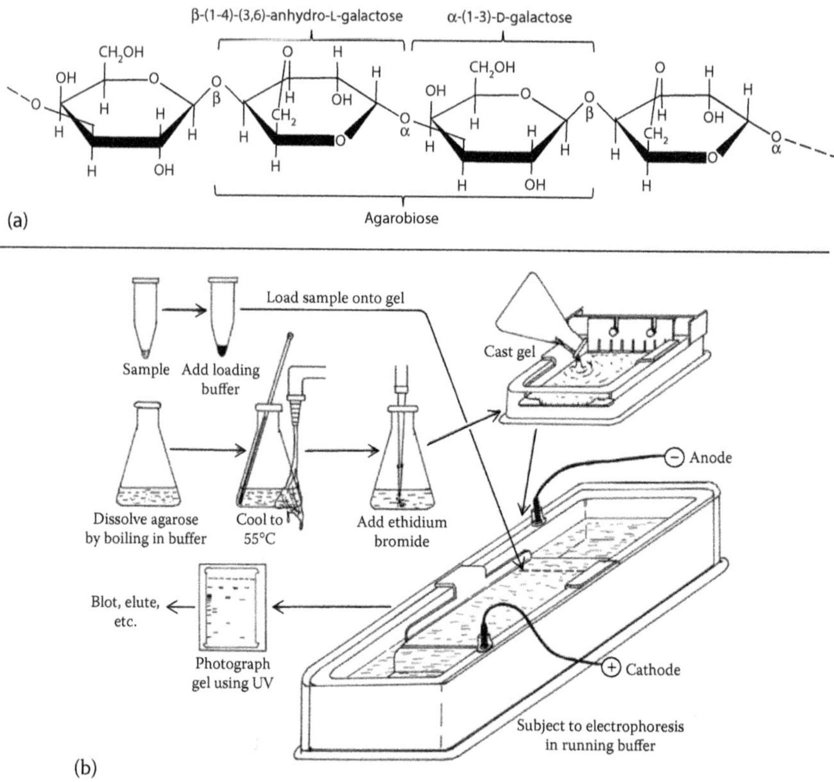

FIGURE 19.11 Fundamentals of agarose gel electrophoresis. Agarose is a polymer of alternating D-galactose and 1,3-anhydro-L-galactose moieties (a). Some agaroses have side groups (not shown), and these determine some of the special properties of different forms of agarose. (b) Shows how an agarose gel is set up. The agarose is measured and mixed with a buffer (usually a Tris, borate, EDTA [TBE]; or Tris, acetate, EDTA [TAE]). The buffer is boiled to dissolve the agarose. The solution is cooled somewhat (so that the plastic trays are not melted), mixed with the appropriate amount of ethidium bromide, and then poured into a casting tray with a comb that will form the wells into which the DNA samples will be loaded. Once the gel solidifies (30–90 min), the comb is removed and the gel is placed into the electrophoresis unit that has been filled with the same buffer (with ethidium bromide) sufficient to completely cover the gel and fill the wells with the buffer. The DNA samples are first mixed with a loading buffer solution that contains a densifier (such as Ficoll or SDS) so that the sample will flow downward into the well, and a dye so that it is visible when loading and the dye can be observed when the gel is running so that electrophoresis can be gauged to stop before the DNA is electrophoresed of the end of the gel. The gel is then photographed using a UV light source. Ethidium bromide fluoresces red-orange when complexed with nucleic acids.

forms of agarose will inhibit the enzymes, and therefore cannot be used for in-gel applications. Others are used for the isolation and purification of specific DNA fragments from the gels, following gel electrophoresis to separate the DNA from other DNA fragments in a mixture. In aqueous solutions, agarose forms a gel through the formation of hydrogen bonds between adjacent polymers. The gel is a matrix that acts as a molecular sieve. When an electric current is passed through the matrix, charged molecules, such as nucleic acids, can be separated by molecular mass based on how rapidly they traverse the polymer matrix. Smaller molecules generally move faster through the matrix than larger ones, although the total conformation of the molecule also has some influence on the speed at which the molecules move through the matrix. Additionally, when dyes, such as ethidium bromide, are used during electrophoresis, the apparent molecular mass of the DNA may change due to the conformation of the DNA. For example, for plasmid DNA that is supercoiled, the amount of ethidium bromide that can intercalate between the bases is smaller than for nicked

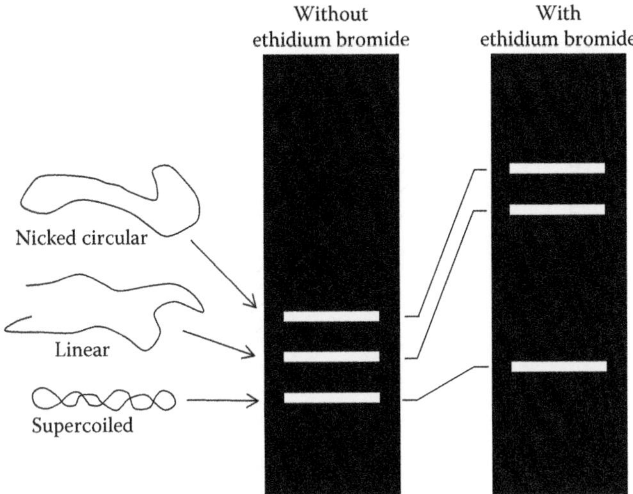

FIGURE 19.12 Effect of ethidium bromide on topologically constrained DNAs. Plasmid DNAs and other DNAs often are covalently closed and supercoiled. This conformation restricts the amount of ethidium bromide that they can incorporate, because ethidium bromide intercalates between adjacent base pairs. When plasmids are nicked on one strand, they lose supercoiling and become relaxed circles. If both strands are nicked in the same area, the plasmid becomes linear. If a gel is run with supercoiled, circular, and linear versions of the same plasmid without ethidium bromide, the three appear at slightly different positions after electrophoresis. However, if the same plasmids are separated on an agarose gel with ethidium bromide, the circular and linear molecules incorporate much more ethidium bromide than does the supercoiled molecule. Therefore, while all three molecules run slower, the circular and linear molecules are slowed to a greater extent.

circular or linear plasmids of the same type (Figure 19.12), due to the torsional constraints of a covalently closed circular supercoiled molecule. Therefore, the supercoiled plasmids migrate faster through the gel than nicked circular or linear molecules of the same size because their molecular masses have increased less than the circular and linear molecules (i.e., additional molecular mass of the ethidium bromide has added to the molecular mass of the DNAs).

The linear range of separation depends on the percentage of agarose in the gel (usually from 0.4% to 2.0%). In the lower portions of the gel, the nucleic acids are separated as a linear distance to the starting point (the well) as the log of molecular mass of the DNA/RNA molecules (Figure 19.13). In the upper portions, this relationship ends, and all of the larger fragments run together as a single band. For a 0.4% standard agarose gel, this occurs when the fragments are from about 50 kb and larger (Figure 19.14). On the lower end, fragments smaller than about 50 bp are at the limit of resolution for a 2.0% standard agarose gel. The range of 50 bp to 100 kb can be extended to approximately 10 bp to 150 kb by using specialty agaroses. These characteristics hold true for standard gel electrophoresis, which is where a constant single voltage is applied from one end to the opposite end of the gel. However, methods that vary the angle, direction, strength, and/or duration of the electric current can raise the upper end of separation to several megabases (Mb; Figure 19.15). Pulsed-field gel electrophoresis (PFGE, where the electrical field is pulsed forward and then reversed for a shorter time), Orthogonal-field-alternation gel electrophoresis (OFAGE, where the electrical field is switched between various angles across the gel), contour-clamped homogeneous electric field gel electrophoresis (CHEF, also involving the switching of the electrical field and field strength through various angles through the gel), and other techniques allow the separation up to the sizes of bacterial chromosomes and small eukaryotic chromosomes. Large and small DNA fragments have been observed and recorded traveling through agarose by electrophoresis. While it was once thought that they travel in ways similar to snakes (e.g., a process called reptation), they do not (Figure 19.16). Different parts move through the gel at different rates, until one of the leading edges finds a path

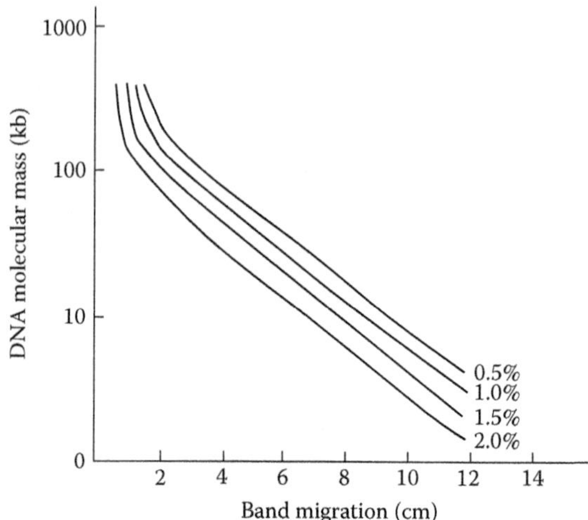

FIGURE 19.13 Migration of DNA molecules separated on various percentages of agarose gels. DNA produces a characteristic linear separation (x-axis) as a log of the molecular weight (y-axis). Each has a characteristic linear range of separation. All of them vary from a linear pattern as the DNA fragments become larger (left portions of each line).

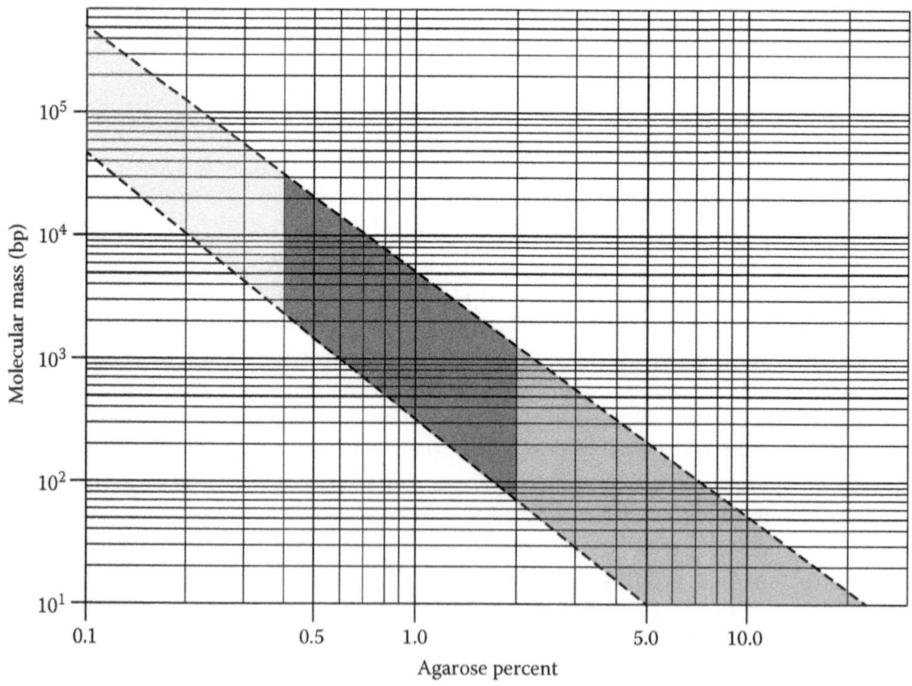

FIGURE 19.14 Effective separation ranges for various concentrations of three classes of agarose. The graph shows the upper range (upper dashed line) and lower range (lower dashed line) for percentages of agaroses. Standard agaroses (dark gray) perform best from about 0.4% to about 2.0%. The gels begin to become too soft below 0.4%, while above 2.0% they begin to solidify as they are being poured. Specialty agaroses, or standard agaroses can be cast below 0.4%, but they must be supported around all sides with a gel of higher concentration (light gray). Specialty agaroses must be used when preparing a gel greater than 2.0%–2.5% (medium gray).

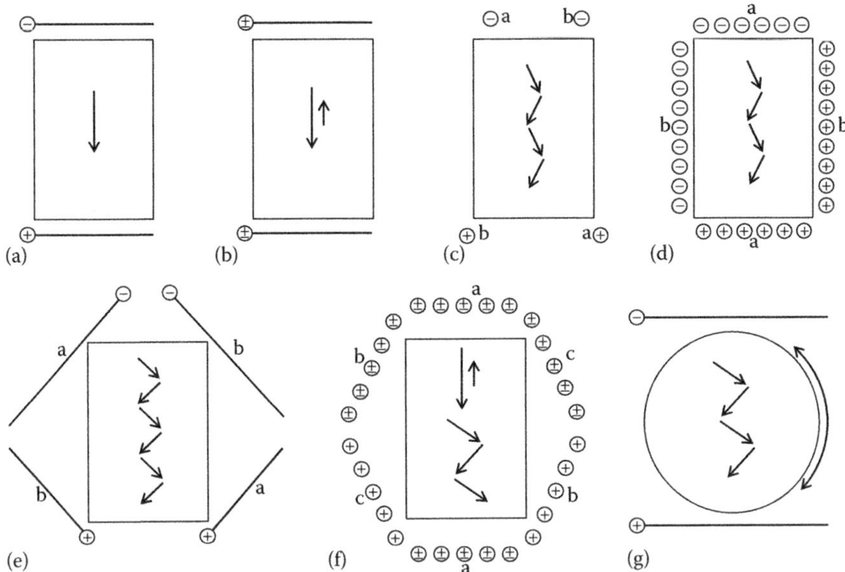

FIGURE 19.15 Alternative methods of agarose gel electrophoresis. All of the methods, except standard electrophoresis (a) can be used to separate large molecules up to several Mb. (a) Standard electrophoresis, with one long anode wire and one long cathode wire. (b) FIGE—field inversion gel electrophoresis. The electric field is primarily run in one direction, with periodic short reversals of the electric field. (c) TAFE—transverse alternating gel electrophoresis. The electric field is switched between two sets of discrete poles, such that the DNA moves through the gel in a zigzag pattern. (d) PFGE. This is similar to TAFE, except the poles are sets of discrete electrodes, rather than single long electrodes. This allows additional variation in voltages across the poles. (e) OFAGE—orthogonal field agarose gel electrophoresis. This is similar to TAFE, except the electrodes are long, rather than discrete short poles. (f) CHEF—contour-clamped homogeneous electric field electrophoresis. This is a combination between PFGE and OFAGE, although it can be run as a FIGE system, as well. (g) RGE—rotogel electrophoresis. In this process, the round gel is on a platform that is rotated by a motor, such that the electrical field is traveling through the gel at different angles.

of least resistance. It then proceeds through that region, pulling all other portions of the molecule with it. Eventually, another part of the molecule begins moving faster, and then it pulls the remainder of the molecule through that path. This repeats many times during the electrophoresis of each molecule, or set of molecules.

The molecular mass of the DNA (or RNA) can be estimated by loading molecular mass DNA (or RNA) standards to one of the lanes on the gel. These can be obtained commercially for many size ranges, but most often they are in multiples of 100 bp or 1000 bp. These can be graphed on semilog paper (molecular mass on the log axis, and distance migrated on the linear axis), and the molecular masses of the sample DNAs can be estimated by where they fall on the graph (Figure 19.13). Often, the molecular mass standards also include the amount of DNA in each band (usually in nanograms), and therefore the amount of DNA in the sample bands can also be estimated, based on the size and fluorescence intensity of the band. The minimum amount of DNA that can be visualized on an agarose gel is approximately 1 ng for most fluorescent dyes.

EXTRACTION OF PROTEINS

While protein sequences are often used in evolutionary studies, the sequences are usually deduced from the DNA sequences, rather than from the proteins themselves. Nonetheless, proteins are often extracted and studied separately from the DNA and RNA, especially when investigating differences in the expression of those proteins. As with nucleic acids, there are many different

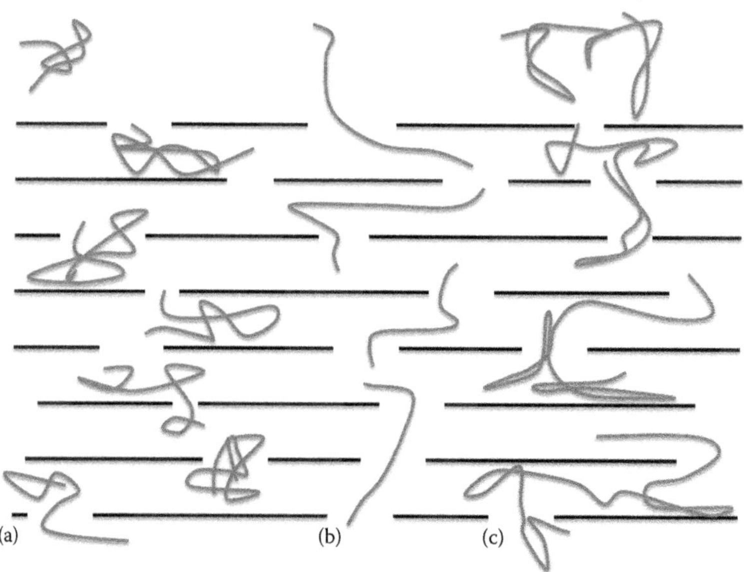

FIGURE 19.16 Three models of the movement of DNAs during agarose gel electrophoresis. The black lines are parts of the agarose, with holes indicated by gaps. The DNA molecules are indicated as gray random shapes. (a) Random coils. This was an early model based on Brownian motion of the molecules. As they change shapes randomly, the electric current moves them in one direction. When they are in a favorable conformation, they move through the pores in the agarose. (b) Biased reptation. In this model, the DNAs stretch into long fibers and thread their way through the pores by leading with one end. (c) Multiple leads. The other two models can explain electrophoresis of molecules smaller than 50–100 kb. However, they do not explain larger molecules adequately. In experiments that were able to visualize the DNA microscopically while undergoing agarose gel electrophoresis, it was found that the DNA formed multiple leads, one of which finds a way through a portion of the gel, and then pulls the remainder of the DNA in through that path. The molecules then are stopped again, form multiple leads, and repeat the process repeats.

methods for isolating proteins. The buffers are somewhat similar to those used for nucleic acids, in that Tris (or similar buffer) is used to maintain the pH of the solution close to 7.0. The extraction solutions also contain a detergent (usually SDS) to solubilize membranes, protease inhibitors, and salts. CTAB can also be used, and the proteins can be precipitated by altering the salt concentration. The proteins are often recovered by using centrifugation, gradients, or columns. Proteins can be fractionated using various concentrations of ammonium persulfate. In this way, different *cuts* or groups of proteins can be separated from the other proteins for more detailed studies of particular proteins. They can also be separated on one-dimensional and two-dimensional polyacrylamide gels.

QUANTIFICATION OF PROTEINS

Proteins can be quantified in several ways. Similar to nucleic acids, spectrophotometry can be used. The tryptophan and tyrosine residues in polypeptides absorb in the UV range between 270 and 300 nm. The concentration can be calculated by using the equation $(1.55 \times A_{280}) - (0.76 \times A_{260})$ = mg/mL total proteins. A common chemical assay is the Bradford method. This method uses a blue dye (Coomassie Brilliant Blue G-250) to determine the concentration of proteins in a solution. Following the addition of the dye and heating, the color density is assayed in a standard spectrophotometer. A version of the dye (coomassie or silver stains) can also be used to stain proteins in polyacrylamide gels (see Polyacrylamide Gel Electrophoresis).

POLYACRYLAMIDE GEL ELECTROPHORESIS

Although large proteins can be adequately separated on high-percentage agarose gels, polyacrylamide gels are superior in their ability to finely separate smaller molecules, such as proteins, as well as short nucleic acids (up to approximately 1500 bp, depending on the gel length). Polyacrylamide gel electrophoresis (PAGE) is used for DNA and RNA sequencing, because it can separate fragments that vary by single nucleotides. Rather than being used horizontally, as are agarose gels, polyacrylamide gels are usually cast between two glass plates and electrophoresed vertically between two buffer trays (Figure 19.17). The gels are usually very thin, usually from 0.75 to 1.5 mm. Lengths of gels can vary. For proteins, the gels are usually short or moderate in length, approximately 10 cm; while sequencing gels are approximately 60 cm in length. Proteins can be electrophoresed in a native gel, meaning that the proteins are folded in their native conformations (or something close to their native conformations). However, more often, gels are cast using a buffer with SDS (a detergent). This neutralizes the charges on the proteins and causes them to unfold, or relax. When these molecules are separated on gels, their positions on the gels at the end of electrophoresis are more closely related to their molecular masses. Molecular mass standards are usually loaded into one lane of the gel so that molecular masses of the protein samples that were loaded in the other lanes can be estimated.

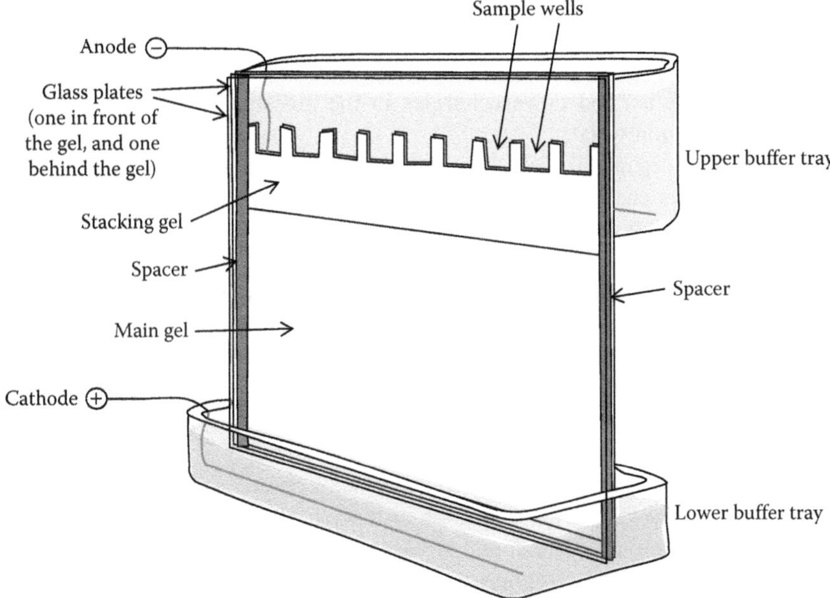

FIGURE 19.17 General PAGE equipment. The components of the gel are mixed and immediately poured between two glass plates separated by thin spacers, which determine the thickness of the gel. A comb of the same thickness is inserted at the top of the gel, and a spacer is placed at the bottom of the gel to prevent the unpolymerized gel solution from leaking. For some gels, the main gel is poured first, and then a *stacking gel* of a lower polyacrylamide concentration is poured at the top after the main gel has polymerized. The stacking gel aids in separating the proteins prior to entering the main gel, and thus increase the resolution of the gel. Once the gel and stacking gel have completely polymerized, the comb and bottom spacer are removed, and the gel (still between the glass plates) is loaded into a vertical gel electrophoresis unit. The top tray and bottom tray are filled with running buffer (which is also in the gel), and the samples are carefully loaded into the wells of the gel. Then, the power is applied and electrophoresis proceeds. Once electrophoresis is completed, the gel is removed from the glass plates and it is stained (usually with Coomassie blue or a silver stain) to visualize the proteins.

Two other PAGE methods that are commonly used are isoelectric focusing and two-dimensional gel electrophoresis. In isoelectric focusing gels, the gels are cast with ampholytes that produce a pH gradient within the gel. When the proteins are sent through the gel by electrophoresis, they migrate to their isoelectric point. That is, they stop moving when the charges on their surfaces are at equilibrium, such that there is no net charge on the molecule. Two-dimensional gels are produced when a sample is loaded into only one well on one side of the gel. Then, the gel is electrophoresed in that direction. The first direction often is based on isoelectric focusing, or another parameter. Following completion of that electrophoresis procedure, electrophoresis is performed in a direction that is 90° to the first electrophoresis run. This is usually standard electrophoresis. Therefore, the proteins spread out according to their characteristics based on the two forms of electrophoresis. Rather than appearing as bands, the proteins are usually spread out in spots across the entire gel. Using this method, particular proteins, can be characterized based on their locations on the gels. Additionally, comparisons of the proteins within tissues or populations of cells can be compared.

KEY POINTS

1. Success in molecular biology experiments and characterizations depends on the extraction of purified DNAs, RNAs, and/or proteins.
2. Biological molecules are best preserved when frozen and/or dried, although temperatures well below freezing are best for long-term preservation.
3. DNAs have been successfully recovered from ancient samples up to millions of years old.
4. The basis of extraction of nucleic acids using CTAB is the differential solubilities of nucleic acids, proteins, and polysaccharides in the presence of CTAB and monovalent cations (most commonly Na^+).
5. To obtain purified organellar DNA (e.g., from mitochondria and chloroplasts), the organelles and the DNAs can be enriched using gradients in conjunction with centrifugation and/or ultracentrifugation.
6. RNAs are more sensitive to degradation than are DNAs. Additional care must be taken when extracting RNA, including the use of RNase inhibitors. RNAs often are converted to cDNAs prior to full characterization and study.
7. DNA and RNA can be quantified using UV spectrophotometry, by measuring absorbance at 260 nm; by fluorimetry, using dyes that fluoresce when in association with DNA and/or RNA; by gel electrophoresis, in conjunction with a fluorescent dye (e.g., ethidium bromide); or by other similar methods.
8. Agarose gel electrophoresis can separate nucleic acids based on molecular mass. The molecules are separated as the linear migration distance to the log of their molecular weight.
9. The range of separation for standard agarose gel electrophoresis is from approximately 50 bp to 100 kb. Other electrophoresis methods (e.g., FIGE, OFAGE, PFGE, CHEF) that switch the electric field during electrophoresis can separate lager molecules, up to several Mb.
10. The methods to purify proteins are similar to those to purify nucleic acids, although the specific concentrations of the solutions used for extraction, and the precise steps differ. Also, protease inhibitors are used for most procedures.
11. Proteins can be quantified using a UV spectrophotometer, keying in on the amino acids with rings, such as tryptophan and tyrosine, which have absorbance peaks in the 270–300 nm range. Additionally, quantification can be accomplished using a Bradford reagent, which is primarily Coomassie blue stain, or by PAGE.
12. Separation of proteins is usually accomplished using PAGE. Following PAGE, the proteins are stained with Coomassie blue or silver stains. Several methods of DNA and RNA sequencing also employ PAGE.

ADDITIONAL READINGS

Abu Almakarem, A. S., K. L. Heilmann, H. L. Conger, Y. M. Shtarkman, and S. O. Rogers. 2012. Extraction of DNA in the Field. *BMC Res. Notes* 5:266. DOI:10.1186/1756-0500-5-266.

Alberts, B., D. Bray, K. Hopkin, A. Johnson, J. Lewis, M. Raff, K. Roberts, and P. Walter. 2013. *Essential Cell Biology,* 4th ed. New York: Garland Publishing.

Alberts, B., D. Bray, A. Johnson, J. Lewis, M. Raff, K. Roberts, and J. D. Watson. 1994. *Molecular Biology of the Cell.* New York: Garland Publishing.

Castello, J. D., S. O. Rogers, W. T. Starmer, C. Catranis, L. Ma, G. Bachand, Y. Zhao, and J. E. Smith. 1999. Detection of tomato mosaic tobamovirus RNA in ancient glacial ice. *Polar Biol.* 22:207–212.

D'Elia, T., R. Veerapaneni, and S. O. Rogers 2008. Isolation of microbes from Lake Vostok accretion ice. *Appl. Environ. Microbiol.* 74:4962–4965.

Douglas, M. P. and S. O. Rogers. 1998. DNA damage caused by common cytological fixatives. *Mut. Res.* 401:77–88.

Doyle, J. J. and J. L. Doyle. 1987. A rapid DNA isolation procedure for small quantities of fresh leaf tissue. *Phytochem. Bull.* 19:11–15.

Knowlton, C., R. Veerapaneni, T. D'Elia, and S. O. Rogers. 2013. Microbial analysis of ancient ice core sections from Greenland and Antarctica. *Biology* 2:206–232.

Murray, M. G. and W. F. Thompson. 1980. Rapid isolation of high molecular weight plant DNA. *Nucleic Acids Res.* 8:4321–4325.

Rogers, S. O. 1994. Phylogenetic and taxonomic information from herbarium and mummified DNA. In *Conservation of Plant Genes II: Utilization of Ancient and Modern DNA*. eds. R. P. Adams, J. Miller, E. Golenberg, and J. E. Adams. St. Louis, MO: Missouri Botanical Gardens Press, pp. 47–67.

Rogers, S. O. and A. J. Bendich. 1985. Extraction of DNA from milligram amounts of fresh, herbarium and mummified plant tissues. *Plant Mol. Biol.* 5:69–76.

Rogers, S. O. and A. J. Bendich, 1988. Extraction of total cellular DNA from plants, algae and fungi. In *Plant Molecular Biology Manual*. eds. S. B. Gelvin and R. A. Schilperoort. Boston, MA: Kluwer Academic Press, A6:1–10.

Rogers, S. O. and A. J. Bendich. 1994. Extraction of total cellular DNA from plants, algae and fungi. In *Plant Molecular Biology Manual,* 2nd ed. eds. S. B. Gelvin and R. A. Schilperoort. Dordrecht, the Netherlands: Kluwer Academic Press, D1:1–8.

Rogers, S. O. and Z. Kaya. 2006. DNA from ancient cedar wood from King Midas' Tomb, Turkey, and Al-Aksa Mosque, Israel. *Silvae Genet.* 55:54–62.

Rogers, S. O., K. Langenegger, and O. Holdenrieder. 2000. DNA changes in tissues entrapped in plant resins (the precursors of amber). *Naturwissenschaften* 87:70–75.

Rogers, S. O., L. Ma, Y. Zhao, C. M. Catranis, W. T. Starmer, and J. D. Castello. 2005. Recommendations for elimination of contaminants and authentication of isolates in ancient ice cores. In *Life in Ancient Ice*. eds. J. D. Castello and S. O. Rogers. Princeton, NJ: Princeton University Press, pp. 5–21.

Rogers, S. O., S. Rehner, C. Bledsoe, G. J. Mueller, and J. F. Ammirati. 1989. Extraction of DNA from Basidiomycetes for ribosomal DNA hybridizations. *Can. J. Bot.* 67:1235–1243.

Rogers, S. O., Y. M. Shtarkman, Z. A. Koçer, R. Edgar, R. Veerapaneni, and T. D'Elia. 2013. Ecology of subglacial Lake Vostok (Antarctica), based on metagenomic/metatranscriptomic analyses of accretion ice. *Biology* 2:629–650.

Shivji, M. S., S. O. Rogers, and M. J. Stanhope. 1992. Rapid isolation of high molecular weight DNA from marine macroalgae. *Marine Ecol. Prog. Ser.* 84:197–203.

Shtarkman Y. M., Z. A. Koçer, R. Edgar, R. S. Veerapaneni, T. D'Elia, P. F. Morris, and S. O. Rogers. 2013. Subglacial Lake Vostok (Antarctica) accretion ice contains a diverse set of sequences from aquatic, marine and sediment-inhabiting bacteria and eukarya. *PLoS ONE* 8(7):e67221. DOI:10.1371/journal.pone.0067221.

ADDITIONAL READINGS

20 Recombinant DNA and Characterization of Biological Molecules

INTRODUCTION

After DNA has been extracted and quantified, it can be manipulated and studied in several ways. One of the initial characterizations that can be performed is analysis using restriction enzymes. The enzymes are specific to each bacterial species, have specific sequences that they recognize and cut, and have evolved to digest primarily foreign DNA. The bacterium that produces the enzyme protects its own DNA by adding protecting groups onto the recognition sequences. The enzyme will then digest the foreign DNA, but the DNA from the organism itself will be protected by the protecting group on the DNA. The protecting group is usually a methyl attached to one or more of the bases. The names of the enzymes are usually based on the bacterial species from which they were isolated. For example, *Eco*RI was the first restriction endonuclease isolated from *Escherichia coli*. Some make staggered cuts that leave the nucleotides on the 5′ ends exposed (e.g., *Eco*RI, *Bgl*I, and *Bam*HI; Figure 20.1), some leave the 3′ nucleotides exposed (e.g., *Hha*I, *Kpn*I, and *Pst*I), and some make blunt-end cuts (e.g., *Alu*I, *Eco*RV, and *Hae*III). When one or more of the bases are methylated, many will fail to cut at those spots. However, a few cut only when one or more bases are methylated. Most require magnesium ions, specific salt concentration ranges, physiological pH ranges, and narrow temperature ranges to function properly. Many can be stopped by adding EDTA (to chelate the magnesium) and/or by heating to more than 65°C. Most are inhibited by certain polysaccharides (found in plants and fungi), as well as by some other molecules. If the proper environment is not provided for the enzymes, they may cut inefficiently causing partial digestion of the DNA, or they may completely fail to cut the DNA. For some, altered conditions may cause the enzyme to recognize and cut slightly different sequences (often called *star* or *asterisk* activity). Because the enzymes cut at specific sites and much of the DNA changes little, these enzymes can cut the DNA into fragments of reproducible lengths. When separated on agarose gels and hybridized to labeled gene probes, maps of the DNA regions of interest can be produced, based on their restriction enzyme sites.

POLYMERASE CHAIN REACTION

Another method for analyzing DNA is to amplify regions of the extracted DNA using the polymerase chain reaction (PCR). This method was developed in the early 1980s by Kary Mullis, a nucleic acids chemist (then at Perkin–Elmer–Cetus), based on a procedure that was published by H. Gobind Khorana and colleagues in 1971. Mullis was awarded a Nobel Prize in 1993 for his work. Since that time, tens of thousands of papers have been published in journals and books using PCR, or describing other applications for this technology. It is a powerful molecular biology method, and because of its simplicity and sensitivity, it has a multitude of uses. While initially it was developed to amplify from double-stranded DNA samples, methods have also been developed to amplify from single-stranded DNA (ssDNA) in a process called asymmetric PCR, as well as methods to amplify cDNAs that have been produced by reverse transcription of RNAs (called RT-PCR, or reverse transcriptase PCR—not to be confused with another process that is also termed RT-PCR, which is real-time PCR). Through cycles of heating and cooling, it is possible to amplify several thousand to

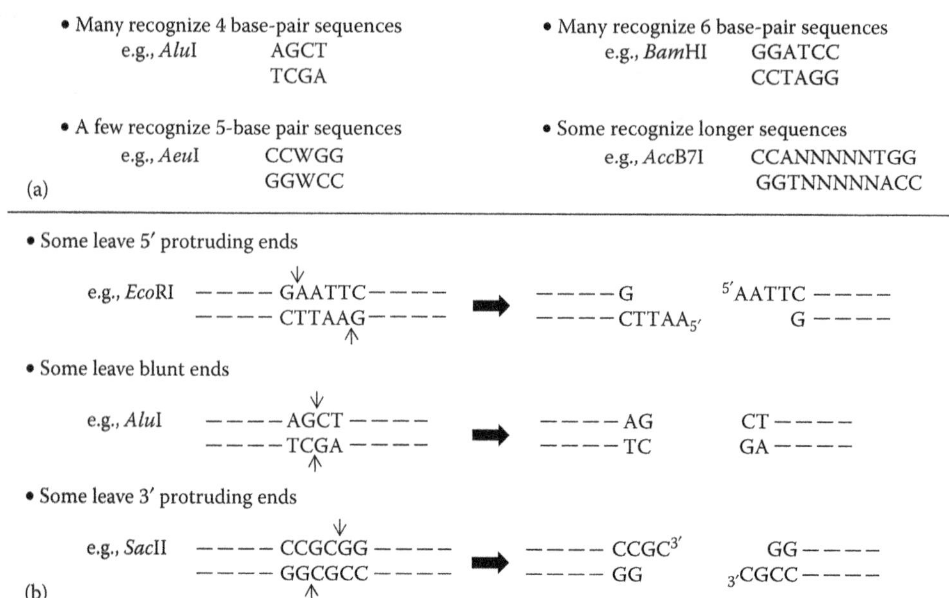

FIGURE 20.1 Examples of restriction enzymes (endonucleases), illustrating some of the varieties of their characteristics. Restriction enzymes usually recognize pseudopalindromic sequences; that is, the sequence in one direction is the same as the reverse complement in the opposite direction. Most recognize four, five, or six base pair sequences (an example of each is presented in [a]). However, some recognize longer sequences, and some recognize a sequence, but cut further down on the DNA. (b) provides examples of enzymes that produce staggered cuts, either leaving 5′ protruding ends (e.g., *Eco*RI) or 3′ protruding ends (e.g., *Sac*II); as well as those that produce blunt ends (e.g., *Alu*I).

several million amplicons in a few hours (Figures 1.4 and 20.2). When first developed, the method was labor intensive. During each cycle of heating and cooling, more enzyme (DNA polymerase) had to be added because the enzyme was inactivated during the heating (usually to 94°C or 95°C) necessary to denature the double-stranded DNA during each cycle. Eventually, a heat-stable DNA polymerase (the first one being *Taq* DNA polymerase isolated from the thermophilic bacterium, *Thermus aquaticus*) was isolated for use with PCR. This polymerase has an optimal temperature of 72°C, and is not denatured by temperatures as high as 95°C. Subsequently, automated thermal cyclers were developed that were essentially programmable heating blocks (Figure 20.2). The amount of time needed to set up a set of PCR reactions requires about an hour.

While one of the biggest advantages of PCR is its ability to amplify down to very small amounts of DNA (theoretically to a single DNA molecule), this is also a weakness in some ways. Because the method does not necessarily discriminate between the DNA of interest and contaminating DNA molecules, either or both types of molecules may be amplified. This means that contamination must be kept to an absolute minimum (preferably at or near zero); otherwise, contaminating DNAs will be amplified, either producing invalid or confusing results. For a standard PCR reaction mixture, the following components are needed: a heat-stable DNA polymerase (e.g., *Taq* DNA polymerase), two synthesized ssDNA primers (usually around 20 bp in length, one homologous to the left end of the region to be amplified, and one homologous to the opposite DNA strand on the right end of the region to be amplified), deoxynucleotides (dATP, dGTP, dCTP, and dTTP), a buffer (usually with a pH of 8.3–9.3, containing Mg^{2+} ions, necessary cofactors for the enzyme), and the DNA (approximately 1–10 ng) to be tested/amplified. The mixture is then taken through 25–45 cycles of heating (to 94°C to denature the DNA), cooling (usually between 45°C and 60°C, depending on the GC content and the lengths of the primers, to anneal the primers to the test DNA, and to begin polymerization of new DNA strands), and heating (to 72°C, the optimal temperature for the polymerase, to

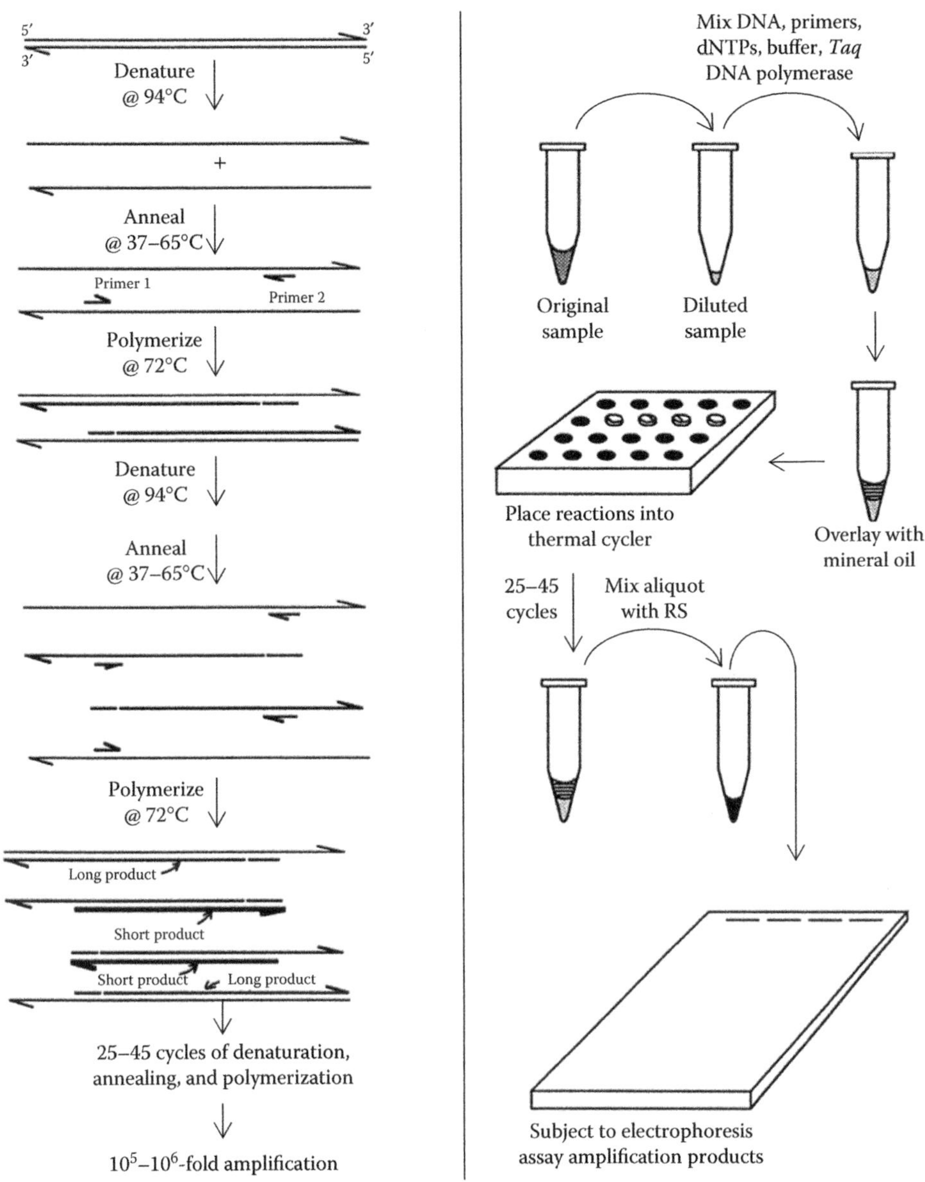

FIGURE 20.2 Theory (left) and practice (right) of PCR. Following addition of a buffer containing Tris (pH 8.3–9.3), $MgCl_2$ (1.0–2.5 M), primers, template DNA, and *Taq* DNA polymerase, the solution is first heated to 94°C to denature the template DNA. Then, the temperature is lowered to 37°C–65°C (depending on the primer melting/annealing temperature) to anneal the primers to the template DNA. Polymerization begins at this time. The temperature is slowly increased to 72°C, which is the optimal temperature for *Taq* DNA polymerase. The mixture is taken through 25–45 cycles of 94°C–annealing temperature (37°C–65°C)–72°C. In each cycle, the long product increases linearly, while the short products (defined by the two primer) increase exponentially. The process is also presented, on the right. The DNA sample is chosen, diluted so that 1–10 ng of DNA is added to the reaction. All of the components are added and mixed thoroughly, without introducing bubbles. Light mineral oil is layered on top of the reaction mixture (optional). Then, the tubes are placed into a thermal cycler with the appropriate program, and allowed to go to completion (4–6 h). Aliquots of each reaction are assayed by agarose gel electrophoresis to determine whether the PCR reaction was successful. Additional details are provided in Figure 1.4.

complete the polymerization of the new DNA strands). Initially, longer strands of DNA predominate in each cycle, but after a few cycles, the majority of amplification products are ones that span from the first primer to the second primer (Figures 1.4 and 20.2). The DNA can be amplified from a few copies to 10^5–10^6 copies in from 1 to 6 h.

Several other methods have PCR at their basis, including quantitative PCR (qPCR) using real-time PCR (RT-PCR, which is unfortunately the same abbreviation for reverse transcriptase PCR). Fluorescence-tagged primers are used in this method, as well as internal standards to indicate how many copies of a particular sequence have been produced. One class of PCR reporter primers contains three parts to the primer. One section contains the specific sequences that will hybridize with the template DNA. It is in the middle portion of the molecule. The two ends contain sequences that are complementary to one another, such that a hairpin structure is formed. On one end of the primer, a reporter molecule has been covalently attached to the primer. This will fluoresce at a specific wavelength when exposed to UV light. The primer has a quencher molecule covalently attached to the opposite end. This absorbs the UV energy, such that when it is in close proximity to the reporter molecule, it absorbs the energy needed for the reporter to fluoresce, thus emitting no light. However, once PCR begins, the primer unfolds as its center hybridizes to the template DNA, thus separating the reporter from the quencher. As DNA polymerization proceeds, it also cuts off the reporter from the primer and this allows the reporter to absorb the UV light and fluoresce. The amount of fluorescence is proportional to the number of amplicons that are produced, and thus, PCR can be assayed in real time. When compared to standards, in which the exact number of molecules is known for a particular amount of fluorescence, accurate determinations of the number of amplicons can be made.

RECOMBINANT DNA METHODS

Recombinant DNA technologies rely heavily on the ability to propagate the desired DNA molecules in bacteria and other vectors. This includes insertion of DNA fragments into plasmids, phage, or artificial chromosomes so that they can be used to transform bacteria, yeasts, plant cells, or other types of cells, including human cells. Transformation is simply the introduction of a genetic element into a cell, which is replicated multiple times in the cell and distributed to daughter cells during cell division. The following are the general steps of a genetic engineering project:

1. Determine length and type of insert desired.
2. Will it be a restriction enzyme fragment, a PCR amplicon, or fragments with random ends?
3. Choose appropriate vector (plasmid, lambda, cosmid, bacterial artificial chromosome [BAC], yeast artificial chromosome [YAC], etc.) based on insert and proposed protocols after cloning.
4. Choose matching vector organism (bacterium, yeast, etc.).
5. Make growth media, selective media.
6. Grow organism.
7. Construct recombinant molecules (e.g., ligate inserts into plasmids).
8. Transform vector organism.
9. Select, based on antibiotic resistance, α-complementation, etc.
10. Grow clones individually in liquid culture.
11. Isolate recombinant molecules.
12. Characterize the recombinant molecules.
13. Perform other experiments, as needed.

After carefully designing the cloning project, and selection of the DNA vector, it is crucial to select and grow the strain of bacteria (usually a strain of *E. coli*) that is appropriate to the research that will be performed. These include strains of *E. coli* in which plasmid or bacteriophage lambda (or other vector DNAs) are used as the DNA vector to contain the desired recombinant sequences.

The foreign DNA can be inserted into a restriction enzyme site using DNA ligase, into other plasmids that will integrate a PCR-amplified piece of DNA using a topoisomerase, or integrated using host bacterial functions. The vector cells usually have many mutations so that they would have a difficult time surviving outside of a laboratory. Concurrent with the first step in the recombinant process is to select an appropriate plasmid (or other) vector. For small inserts (up to approximately 10 kb), plasmids are generally the first choice (Figure 20.3). However, for inserts that are above 5 kb, cloning can be difficult in plasmids, because of instability. For pieces between 5 kb and approximately 30 kb, bacteriophage (usually lambda, λ) can be used. Cosmids (hybrids that are partly plasmids and partly bacteriophage) can also be used to clone these larger fragments. Above 30 kb, artificial chromosomes can be constructed, some containing more than a megabase of DNA, although these can be difficult to use. BACs are one example of this type of vector, which are suitable for propagation in bacterial. These have been used extensively in bacterial genomic sequencing projects. YACs are a second type, which can be propagated in yeast cells (most often in *Saccharomyces cerevisiae*).

When DNA sequencing first became a routine lab procedure, cloning was most often performed using bacteriophage M13. This is because it had a double-stranded growth phase and a single-stranded phase. It could be readily propagated to form double-stranded molecules for cloning and characterization, and could also be grown so that it produced primarily single-stranded versions. The single-stranded phase was used by the virus to infect other cells, but this part of the life cycle was mutated by scientists, so that the phage still produced single-stranded molecules that could be used as templates for sequencing reactions, but the phage was incapable of making viral coat proteins for infection of other cells. Some of the M13 sequences still are found in many cloning vectors that are used for sequencing.

All of the plasmid, phage, and other vectors have one or more selectable markers (Figure 20.4). These must be present to be able to distinguish between the bacteria (or other vector organism) that have the recombinant molecules, and those that do not. Also, often the vector organisms may lose the recombinant molecules, such that constant selection must be used to maintain the vector molecules in those cells. Many plasmids have a gene that produces a protein that degrades the antibiotic ampicillin. A few also have a gene that degrades the antibiotic tetracycline, and a few have a gene that encodes a protein that degrades chloramphenicol. (Ampicillin inhibits bacterial cell wall assembly, tetracycline binds to the 30S bacterial ribosome, and chloramphenicol binds to the 50S ribosomal

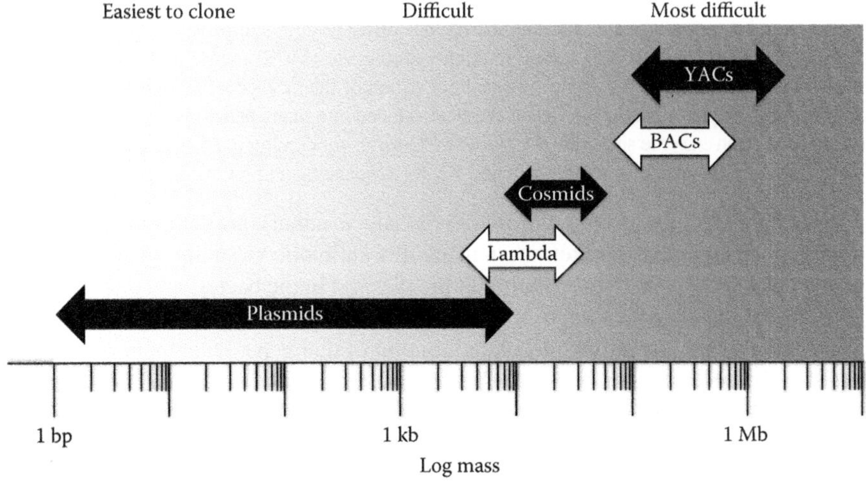

FIGURE 20.3 Comparison of vectors and the inserts that can be accommodated by those vectors. Plasmids can be used for inserts up to approximately 10 kb, although the efficiency of cloning is much lower than for inserts that are 1 kb. Lambda can be used for inserts up to approximately 30 kb, while cosmids can accommodate inserts up to nearly 50 kb. For larger inserts, BACs and YACs can be used. YACs have been used to clone inserts that are over 1 Mb.

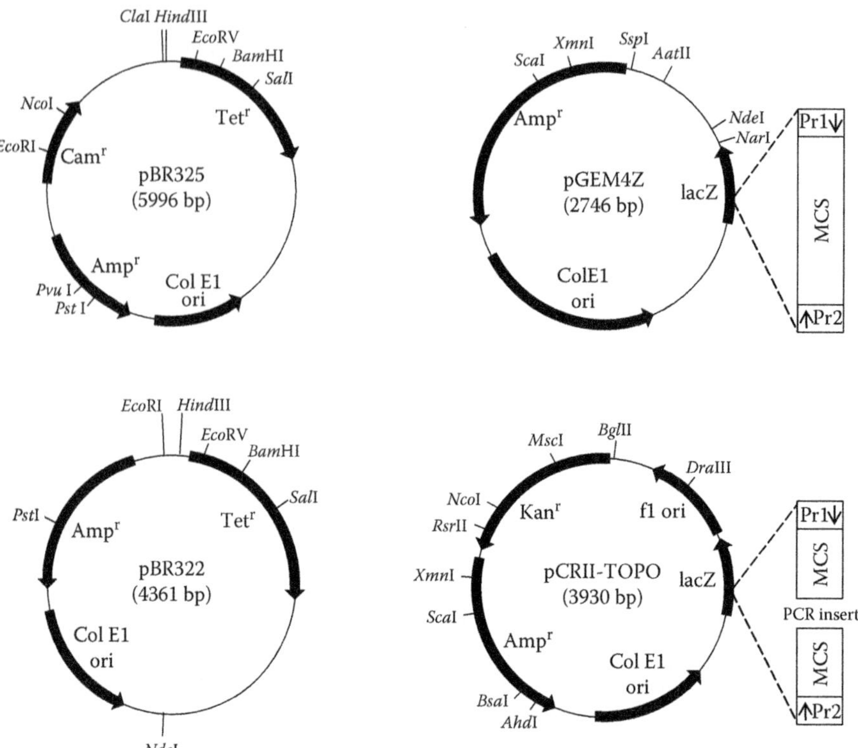

FIGURE 20.4 Examples of a variety of plasmids. Plasmids pBR325 and pBR322 were used extensively during the initial decades of recombinant DNA technology. They share a common laboratory lineage, in that they were derived from a common set of plasmids that were constructed using the ColE1 origin of replication (from the wild-type *E. coli* ColE1 plasmid). The tetracycline resistance and ampicillin resistance genes (Tet^r from plasmid pSC1201, and Amp^r from plasmid pMB9, respectively) were engineered into the plasmid, eventually producing pBR313. Additional engineering produced pRB322, which has several unique restriction enzyme recognition sites, the ColE1 origin of replication, a tetracycline resistance gene, and an ampicillin resistance gene. A chloramphenicol resistance gene from bacteriophage P1 was engineered into pBR322 to create pBR325. Bacteriophage M13 was engineered as a sequencing vector to include a multiple cloning site (MCS) region (containing many restriction enzyme sites), which is a small section near the promoter region of the lacZ gene, to produce pUC plasmids. The multiple cloning sites, coupled with the lacZ gene, were introduced into many plasmids, including the pGEM series for cloning into restriction enzyme sites, and the pCR-TOPO series for cloning of PCR amplicons. The PGEM and PCR-TOPO vectors include promotor regions (P1 and P2) for sequencing and for transcription studies.

subunit. Chloramphenicol has been implicated in damage to eukaryotic cells, and induction of some tumors; therefore, care should be taken when using this antibiotic.) Usually, each vector has at least two selectable markers. One is used to maintain the plasmid in the bacteria, and the other is used as a cloning site. Once the gene is disrupted, it will no longer function, and therefore the bacteria that contain the recombinant plasmid with lose the function of that gene. One popular gene that is used in a wide range of plasmid vectors is the β-galactosidase gene (called lacZ), usually used in conjunction with an ampicillin resistance gene. In wild-type bacteria, the lacZ gene produces β-galactosidase, which breaks down lactose. The lacZ gene can be split to produce two polypeptides, both of which are necessary for activity of the enzyme. Bacterial vectors using this enzyme have the gene that encodes the alpha (α) portion on the plasmid, and the other portion (called the omega, Ω, portion) of the lacZ gene is on their chromosome. If both genes are present in the bacterium, then the enzyme is produced, in a process called α-complementation. Normally, it would be difficult to assay for enzyme activity. However, the enzyme also acts on a dye that has a galactoside moiety that is a substrate for the β-galactosidase. The dye is 5-bromo-4-chloro-3-indoyl-β-D-galactoside, often called X-gal. In solution, and when added to bacterial growth media, it is either colorless, or at high concentrations, it has a slightly yellow color. However, when cut by β-galactosidase, a dark blue precipitate is produced. Therefore, if a bacterium (carrying only lacZΩ – the distal portion of the β-galactosidase gene) has been transformed with a plasmid carrying ampicillin and the lacZα gene (the proximal portion of the gene), then a functional β-galactosidase will be produced, which will form blue colonies when grown on agar growth plates containing LB (lysogeny broth, also called Luria–Bertani) medium with ampicillin and X-gal. There are cloning sites within the lacZα gene, and when it is interrupted by an inserted piece, the gene is inactive. In this case, the bacteria still grow on the culture plates because they contain intact ampicillin resistance genes on their plasmids, but they form white- or cream-colored colonies because they cannot produce β-galactosidase.

There are many different media for growing bacteria. One of the most common is LB medium. Bacteria can be grown in liquid or solid (with agar) media. When grown in liquid media, the containers must be stirred so that the bacteria are aerated and grow aerobically. If they are not stirred, they settle to the bottom where they rapidly exhaust the oxygen and then switch to anaerobic growth, rendering them unsuitable for use in recombinant DNA methods. There are two crucial steps in molecular cloning: (1) engineering/ligating the DNA fragment of interest into a vector, and (2) transforming the bacterial vector with the engineered molecule. Ligation of the fragment is normally accomplished in one of two ways. The first is by using DNA ligase to covalently join the fragment into the plasmid (or other) vector (Figure 20.5). This is most often used when the DNA insert and the plasmid have been digested with the same (or compatible) restriction endonucleases.

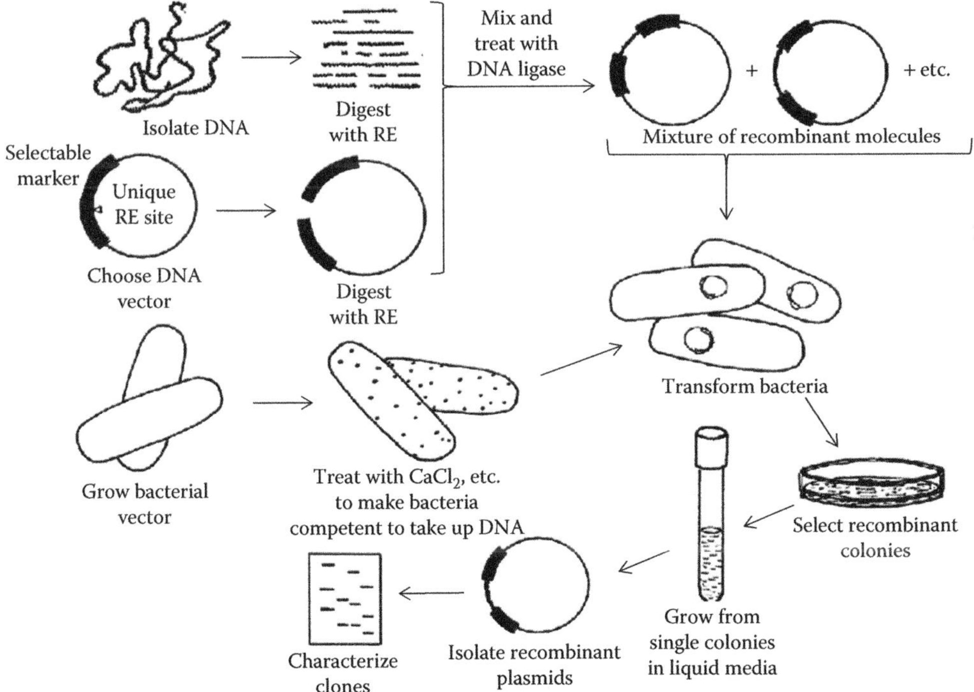

FIGURE 20.5 Cloning using ligation into restriction site cut sites. The DNA to be inserted is digested with a restriction endonuclease, while at the same time, the plasmid (or other) vector is digested with the same enzyme, or one that generates compatible ends. DNA ligase is used to join the two together. This mixture is used to transform competent bacteria (produced using $CaCl_2$ and other chemicals), and the bacteria are placed onto media that will allow transformants carrying the plasmids to survive, but all other cells will die. The potential transformants are selected and grown in liquid culture, followed by isolation of the recombinant molecules, and characterization of those molecules.

Ligation of molecules with complementary protruding bases (called compatible ends) are joined together with the highest efficiency. Blunt-ended molecules also can be joined by DNA ligase, but the efficiency is 10–20 times lower. Usually, other chemicals must be added, and the time of the ligation reaction must be extended. Ligase will only join the fragments together when the 5′ end has the phosphorus group still attached, and a 3′ hydroxyl. The second way to ligate the insert into the plasmid is to use vectors specifically engineered to accept PCR products. PCR adds an adenine onto the 3′ ends of the amplicons (Figure 20.6). The plasmid vector is engineered with 5′ overhanging thymines, and therefore the As and Ts on both ends can form hydrogen bonds, and then an enzyme is used to covalently bond the insert into the vector. The plasmids are usually called TA vectors, or TOPO TA vectors, because the enzyme used with this system is a topoisomerase.

Cloning PCR products with the TOPO TA vectors is relatively simple, and inhibition of the topoisomerase is not often observed. While cloning with restriction enzyme digests and DNA ligase is more efficient, the ligase is sensitive to various conditions. It utilizes adenosine triphosphate (ATP), and is inhibited by ADP and pyrophosphate. Usually, manufacturers supply the buffer that already contains ATP. However, if the buffer is made separately, only the highest grades of ATP should be used, because lower grades often contain significant amounts of ADP. Some contaminants from DNA eluted from agarose gels can inhibit DNA ligase. Use of spermidine (at 3–5 mM) can often alleviate this inhibition. Also, this problem can be totally avoided if an agarose rated for in-gel applications is used. These contain no inhibitors of the enzymes. Monovalent cations (e.g., Na^+, K^+) can also inhibit ligase at moderate concentrations (100–200 mM). However, when added in conjunction with polyethylene glycol (PEG), these cations actually cause increases in ligation rates. PEG comes in various molecular weights, but most often molecular weights of between 3000 and 8000 are most effective

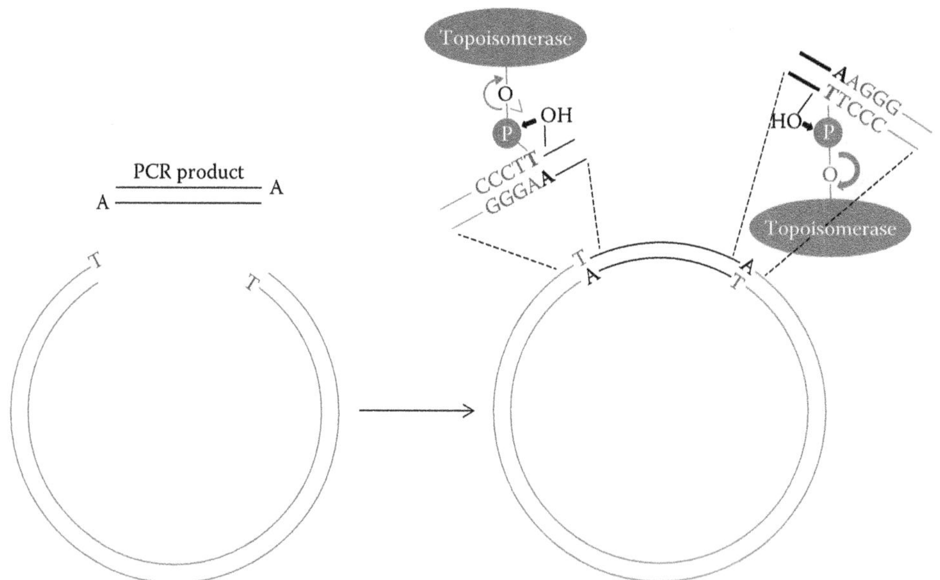

FIGURE 20.6 Cloning using PCR amplicons and topoisomerase (TOPO cloning). Many of the steps with TOPO cloning are similar to those of standard cloning (Figure 20.5). However, the insert is a PCR amplicon, and the vector begins as a linear molecule with topoisomerase covalently bonded to the 3′ ends (T residues) of the vector through a phosphate and oxygen attached to a tyrosine residue in the topoisomerase. PCR usually adds A residues to the 3′ ends of the amplicons, which hydrogen bond to the overhanging T residues of the vector. Topoisomerase then joins the ends of the strands together through the hydroxyl of the residue on the 5′ ends of the amplicons. The double-stranded recombinant molecules are used to transform competent bacteria that have an ampicillin resistance gene and the lacZ gene for α-complementation. The PCR product is black, while the parts of the vector are gray.

for ligation. While ligation increases at PEG concentrations between 5% and 15%, for unknown reasons, the ligated molecules resulting from PEG concentrations of 10%–15% fail to transform the bacterial cells efficiently. Therefore, 5% PEG is most often used.

The ligated plasmids with inserts are then introduced into bacterial cells in a process know as transformation. The first part of the process is to make the bacterial cells competent to take up the DNA (competent cells are also available commercially). This is usually accomplished by the treatment of cells in mid-log growth phase with calcium chloride, as well as additional chemicals. It is thought that this introduces holes in the membranes, but it also affects some of the membrane proteins that are active during normal transformation processes of these cells. Using this method, from 10^5 to 10^7 transformed cells can be obtained from a microgram of introduced recombinant DNA. In 1983, Doug Hanahan carried out a detailed analysis of conditions for optimizing transformation efficiency of bacteria. He was able to increase transformation efficiencies up to 10^8–10^9 transformants per microgram of introduced DNA. He estimated this as being equivalent to one in every 400 plasmid molecules being effectively taken up and replicated by the *E. coli* cells. His final conclusions stated that a solution containing magnesium ions, calcium ions, and rubidium ions and/or potassium ions, as well as dimethyl sulfoxide (DMSO), dithiothreitol (DTT), and hexamine cobalt (III) ions yielded the best results. His methods are still used by companies that are selling competent bacterial cells.

Following transformation, the cells are spread on media that will allow the detection of those cells carrying the recombinant plasmids (most often carrying an ampicillin resistance gene and the α portion of lacZ gene. Once the putative recombinant colonies are identified, they are grown in liquid medium with the same antibiotics, and the plasmids are isolated from the broth using one of a number of methods. Lysis and purification of the plasmids by alkaline lysis or detergent lysis can be used. Alternatively, various columns can be used to separate the long and viscous chromosomal DNA from the relatively small and compact plasmid DNAs. These recombinant plasmids are then subjected to analyses to determine whether the desired insert is present in the plasmid.

SOUTHERN HYBRIDIZATION

In 1975, Edwin M. Southern first described a method for the transfer of DNA restriction fragments from agarose gels (for his initial reports, he was using tube gels, not slab gels) onto cellulose nitrate (nitrocellulose) membranes. Then, he hybridized the resulting blot-immobilized DNA to radioactively labeled RNAs (ribosomal RNA, in his case), and exposed the blot to X-ray film. Once developed, the film indicated where the ribosomal genes were located on the blot, and therefore where they were in the original gel used for blotting. Since that time, most researchers have used labeled DNA probes rather than RNA probes (Figure 20.7). They are more easily labeled and are more stable. However, whether the probe is DNA or RNA, blotting of a DNA gel onto a membrane is referred to as *Southern blotting*, or *Southern hybridization*. Northern blotting/hybridization refers to transfer of RNA from a gel onto a blotting material, and hybridizing with either a DNA or an RNA probe. Western blotting/hybridization refers to transfer of proteins from a gel onto a membrane, with hybridization to a labeled antibody.

The original and subsequent methods are based on movement of the DNA by capillary action of the solution (usually containing sodium chloride, sodium citrate, and/or sodium hydroxide), which carries the DNA to the membrane. The solution is able to travel through the gel and the membrane, while the DNA is attracted to, and trapped by, the charged membrane. Because the membranes are more reactive toward ssDNA than double-stranded DNA, the DNA must first be denatured by treatment with a basic solution (most often sodium hydroxide). Other methods have been developed for the transfer of the DNA onto the membranes. Several vacuum blotting systems are available commercially. They speed up the blotting process, but sometimes are inferior to simple blotting, because high vacuum pressure forces some of the DNA completely through the membrane without binding to the membrane. Electroblotting methods and systems have also been developed. While these have

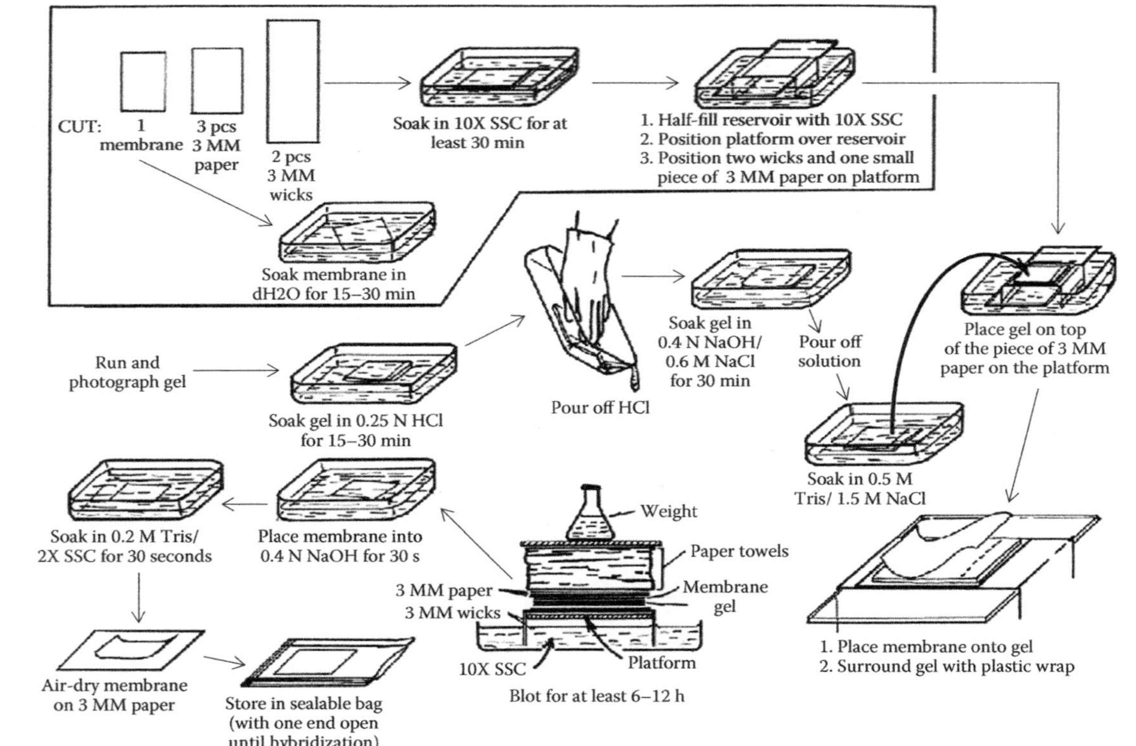

FIGURE 20.7 Basic steps in Southern hybridization. The method can be used with blots of gels, or dot or slot blots. For gel blots (left middle), the gel is run, and after it has been photographed (or otherwise recorded), the blotting membrane and 3 MM paper pieces (top left) are cut. The membrane is cut to the size of the gel, while three pieces of 3 MM paper is cut about 1 cm longer and wider than the blot. Two additional pieces of 3 MM paper are cut that are 1 cm wider than the gel, and several cm longer than the gel, which will form wicks that dip into the blotting solution (SSC—sodium chloride, sodium citrate). Once the two wicks and one of the smaller pieces of 3 MM paper are placed on a glass or plastic platform over the SSC reservoir, the gel is treated with HCl to nick the DNA (so that it will flow through the gel onto the membrane), then NaOH to denature the DNA (only ssDNA will stick to the membrane), and finally a buffer solution to neutralize the gel. The gel is placed (right side up) onto the 3 MM paper on the platform, and then the blotting membrane is placed onto the gel. The two pieces of 3 MM paper are placed on top, followed by paper towels. After 12–24 h, the paper towels and 3 MM paper are removed, and the blot is removed. It is placed into a NaOH solution to fully denature the DNA, then into a buffer solution, and finally it is dried. It is ready for hybridization (see Figures 20.8 through 20.17).

been effective for transfer of proteins onto membranes, they have produced mixed results for DNA and RNA blotting.

Nitrocellulose membranes were the first used for this process, and many researchers still use them. However, cellulose nitrate is easily oxidized, and once oxidized, it will not bind DNA. Therefore, it needs to be carefully stored, preferably under vacuum. Nylon-based membranes are available from several manufacturers and suppliers. These have a backing of nylon with either a cellulose nitrate layer, or other chemical layer that will otherwise bind the DNA. The DNA binding capacities are much higher for these nylon-based membranes (Table 20.1), but it is sometimes difficult to remove DNA (e.g., probe DNA) if multiple probes are to be used for the blot.

A radioactively labeled probe yields the greatest sensitivity, as amounts of DNA down to 0.1–1.0 pg often can be determined. These probes can also be used to detect up to several micrograms of DNA, so the sensitivity and range are excellent (Table 20.1). However, care must be taken not to add excessive amounts of radioactivity to the probe DNA, because the radioactivity can cause damage to the probe DNA, causing fragmentation of the probe (Table 20.2). The DNA to be used as a probe is subjected to a polymerization reaction using the radioactive nucleoside triphosphates, which become incorporated into the newly synthesized strands (Figure 20.8). When the 5′ end of the gene in the DNA or the transcript in RNA needs to be identified, γ-labeled dNTPs or rNTPs are used in conjunction with polynucleotide kinase (Figure 20.9). While the probe is radioactively

TABLE 20.1
Characteristics of Three Methods of Labeling and Detection for Blot Hybridizations

Characteristic	Radiolabeling	Colorimetric Labeling	Chemiluminescent Labeling
Sensitivity	0.2 pg	5 pg	1–5 pg
Specificity	Good	Good	Good
Useful range	5 pg to 1 μg	20 pg to 1 μg	5 pg to 1 μg
Background	Low	Low–high	Low–moderate
Suggested membrane	Any	Nytran Nitrocellulose Biodyne BioBond	Any
Detection of multicopy genes	Best	Good	Good
Detection of single copy genes	Best	For small genomes	Good
Reprobing with additional probes	Best	No	Good

TABLE 20.2
Incorporation of Radionuclide and Specific Activity, Starting with 50 μCi of ^{32}P-Labeled dNTP

Amount of ^{32}P-Labeled dNTP[a] (μCi)	Amount of DNA Labeled	Specific Activity of Probe[b] (dpm)	Percent Incorporation of ^{32}P (%)	Stability of Probe
50	1 μg	10^7	80	High
50	100 ng	10^8	60	Moderate
50	10 ng	10^9	40	Low
50	1 ng	10^9	10	Low

[a] 100 μCi—3,700,000 becquerels.
[b] dpm = disintegrations per minute.

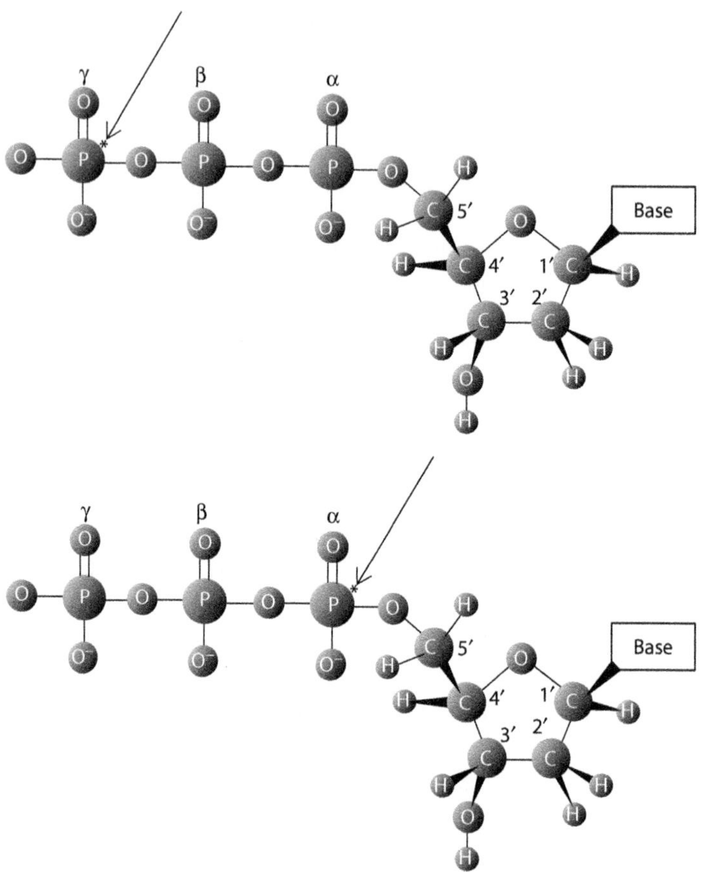

FIGURE 20.8 Labeling of nucleotides for probes in hybridizations. Most often, ^{32}P or ^{33}P isotopes are used. The former is more energetic, and therefore usually provides greater sensitivity. However, the sharpness of the bands is better with ^{33}P because of the narrower field of radioactive particles surrounding the radioactive phosphorus atoms. The nucleoside triphosphates are labeled wither on the γ (gamma) or the α (alpha) phosphorus. A γ label is used for end labeling of DNA or RNA, to determine the end of a particular molecules or nucleic acid region. An α label is used for uniformly labeling a DNA or RNA.

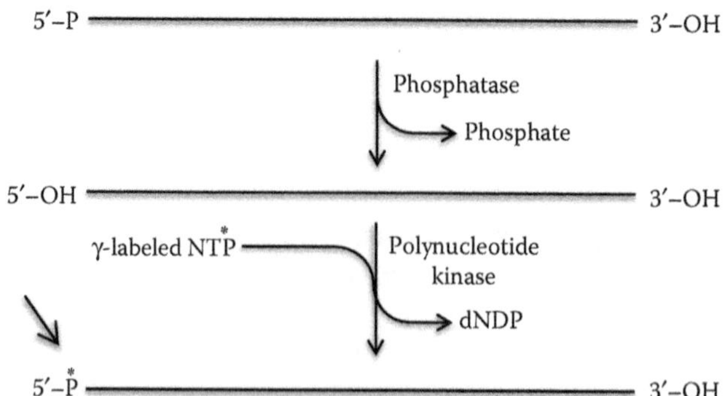

FIGURE 20.9 End-labeling of DNA using ^{32}P nucleoside triphosphates. This method uses a γ-labeled deoxynucleotide triphosphate and polynucleotide kinase to transfer the ^{32}P to the 5′ nucleotide on a DNA molecule. If the 5′ end of the DNA has a phosphate group prior to polynucleotide kinase treatment, the phosphate group must first be removed, using a phosphatase. Asterisk indicates ^{32}P-labeled phosphate.

labeled at a low specific activity, this is a useful method for defining the 5′ end of a gene or transcript. Several methods can be used to label the probe uniformly throughout its length. One is nick translation (Figure 20.10). The unlabeled probe DNA is mixed with DNA polymerase, dNTPs (one of which is an α-^{32}P dNTP), buffer, and a low concentration of DNaseI. The DNaseI produces single-stranded nicks on the DNA, and polymerase binds to those regions and polymerizes new strands (displacing the old strands), including incorporating the radioactive nucleotides. Another method is by random primer labeling (Figure 20.11). DNA polymerase, dNTPs (one of which is an α-^{32}P dNTP), a buffer, and a random mixture of hexanucleotide primers (including various combinations of the four nucleotides) are mixed with the unlabeled denatured (by heating to 95°C) probe DNA. After incubation at 37°C, the new strands of DNA produced by the DNA polymerase are uniformly labeled. PCR can also be used to label probe DNA (Figure 20.12). A standard PCR reaction is set up with one of the dNTPs substituted with an α-^{32}P dNTP. All three of these methods produce probes with high specific activities. The probes are hybridized to the DNA immobilized on blots and detected using autoradiography with X-ray film (Figure 20.13).

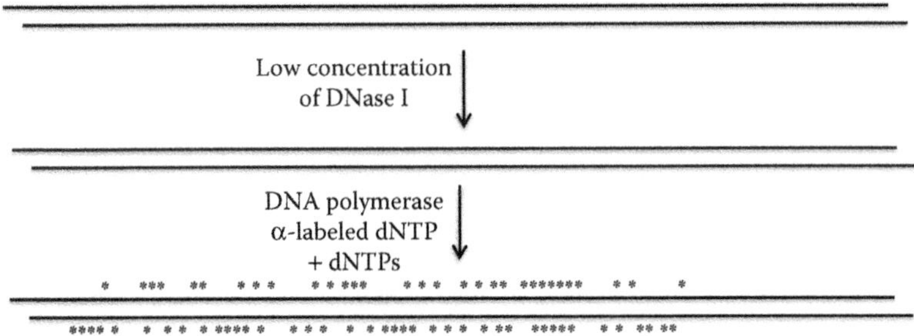

FIGURE 20.10 Nick translation for labeling DNA. This method can be used to label DNA with radioactive or nonradioactive dNTPs (deoxyribonucleotide triphosphates). The DNA is mixes with a low concentration of DNaseI, DNA polymerase, buffer, and dNTPs (one of which is labeled). The DNaseI forms single-stranded nicks in the DNA, which are recognized by DNA polymerase, which then begins the polymerization of new strands to repair the gap, as well as regions downstream of the gap. The term *translation* simply means to move down the molecules, hence the name *nick translation*. Radioactive probes produced by this method are usually of high activity, and therefore are highly sensitive.

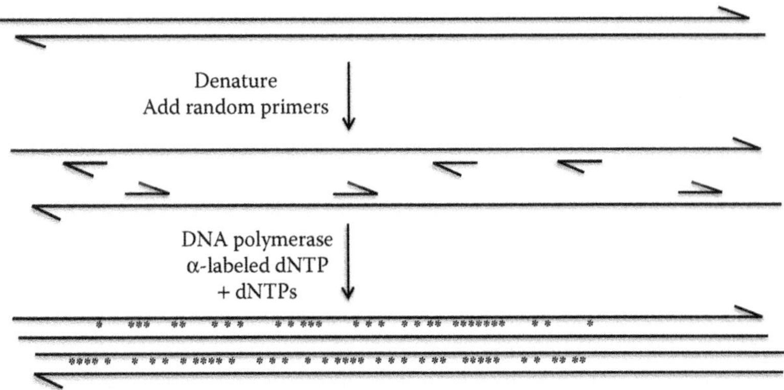

FIGURE 20.11 Random primer labeling of DNA. This method is similar to PCR methods. The template DNA is denatured by heating to 95°C. Then, a set of random hexanucleotide fragments was added in the presence of DNA polymerase, a buffer, and dNTPs (one of which is labeled). The primers hybridize to many places along the single strands of DNA, and DNA polymerase produces complementary strands starting at each of the primers. This method also produces sensitive probes.

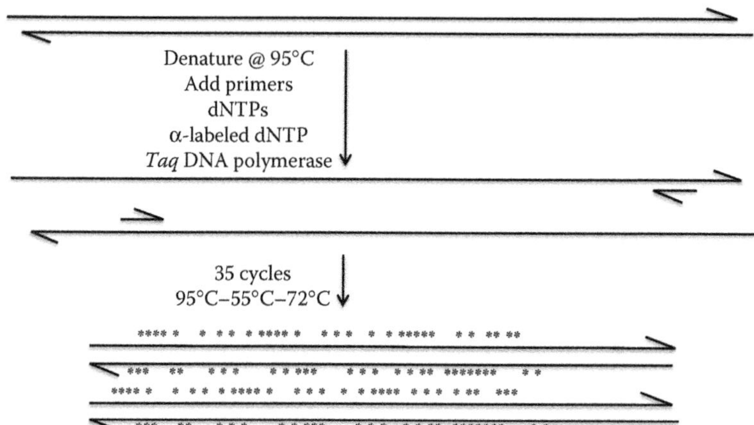

FIGURE 20.12 Labeling of probes using PCR. This method consists of a standard PCR reaction, where one of the dNTPs is replaced with one (or more) labeled dNTP. PCR is explained in Figure 20.2.

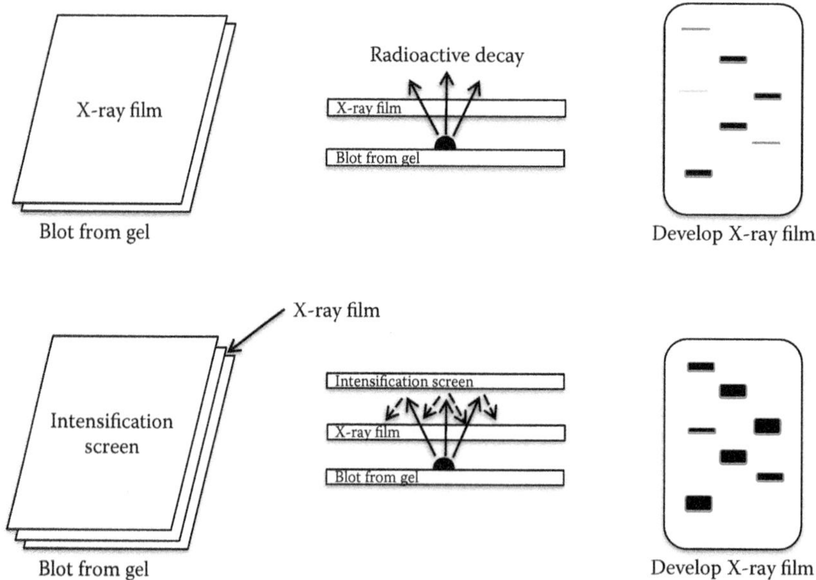

FIGURE 20.13 Detection of radioactivity on blots. The blot is first wrapped in plastic wrap to keep moisture away from the X-ray film. In the dark, a piece of X-ray film is placed in direct contact with the plastic-wrapped blot, and sealed in an X-ray film cassette (or similar light-tight container). The cassette is placed in the dark for hours, days, or a few weeks, as needed to achieve the best exposure. The film is then developed after the exposure period. Another piece of film can be used for another exposure. Intensification screens can be used when the radioactive signal is low. Intensification screens increase the signal by approximately seven times. This occurs because radioactive decay excites molecules in the intensification screen, which produce flashes of light when excited. The screens only function when at temperatures below −55°C, because the reaction is a two-step process. The first step excites the molecule, but this is rapidly reversible above −55°C. However, below −55°C, this step is stabilized long enough for the second portion of the reaction, which produces a flash of light. Because the radioactivity decays in all directions, as do the light flashes from the intensification screen, the exposure on the film forms a cloud of exposure.

Nonradioactive methods have been developed, some of which are almost as sensitive as radioactive methods. Colorimetric probes (those that form a colored precipitate on the blot) also can be used, although the sensitivities are up in the high picogram to low nanogram ranges, so the sensitivity is about 20–100 times less than with radioactive probes. These probes are labeled with a nucleoside triphosphate, where the base contains a side group (Figures 20.14 and 20.15) ending in either a digoxigenin or a biotin moiety. When incorporated into a probe DNA by a polymerization reaction, the additional side group is protruding from the DNA (from the uridines, which are used as a substitute for thymidines). An antibody to this side group is used to attach to this (Figure 20.16), and then a second antibody with an attached enzyme is attached to that (this increases the total signal, which increases the sensitivity). The enzyme cleaves a substrate that yields a blue precipitate (with the proper substrate), which indicates where the DNA is on the blot. As with radioactive probes, amounts of DNA into the microgram range can also be easily detected. Other labeling and detection methods can be used, some that are more sensitive than colorimetric methods, such as chemiluminescence methods, but not as sensitive as radioactive probe methods (Table 20.1). Chemiluminescent methods utilize the same digoxigenin label and antibodies, but the substrate produces light when acted upon by the enzyme.

DETERMINATION OF GENE COPY NUMBER

Hybridization methods can be used to accurately determine gene copy number. Many genes occur in multiple copies, such as rRNA genes. Gene copy numbers can be determined using various hybridization methods, but internal standards must be used so that a comparison with known

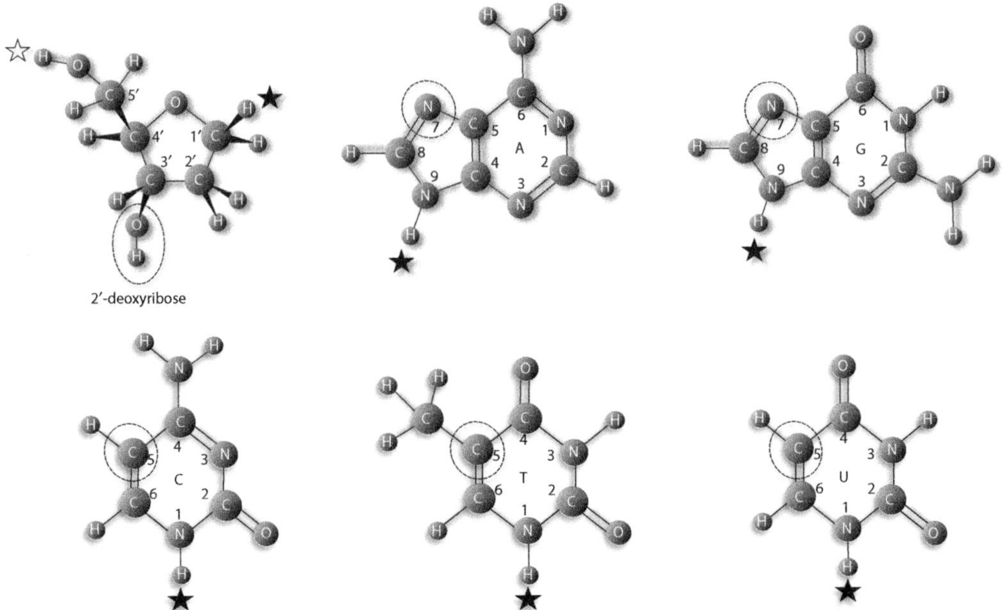

FIGURE 20.14 Labeling of nucleotides for probes in nonradioactive hybridizations. Most methods utilize either 2′-deoxyribose nucleoside triphosphates, or 2′,3′-dideoxynucleotides (where the 3′ carbon has only a hydrogen without an oxygen attached). The open star indicates where the phosphates are attached, while the filled star indicates where the bases are attached. The five bases (A, G, C, T, and U) are shown. Additional molecular sidegroups are attached to each in labeled nucleotides. Dashed circles around atoms indicate where the side groups are attached. Attachment is to N-7 on the purines, and to C-5 on the pyrimidines. Because thymidine has a methyl group attached to C-5, uridine is used to take the place of thymidines for nonradioactive labeling. Side groups are shown in Figure 20.15.

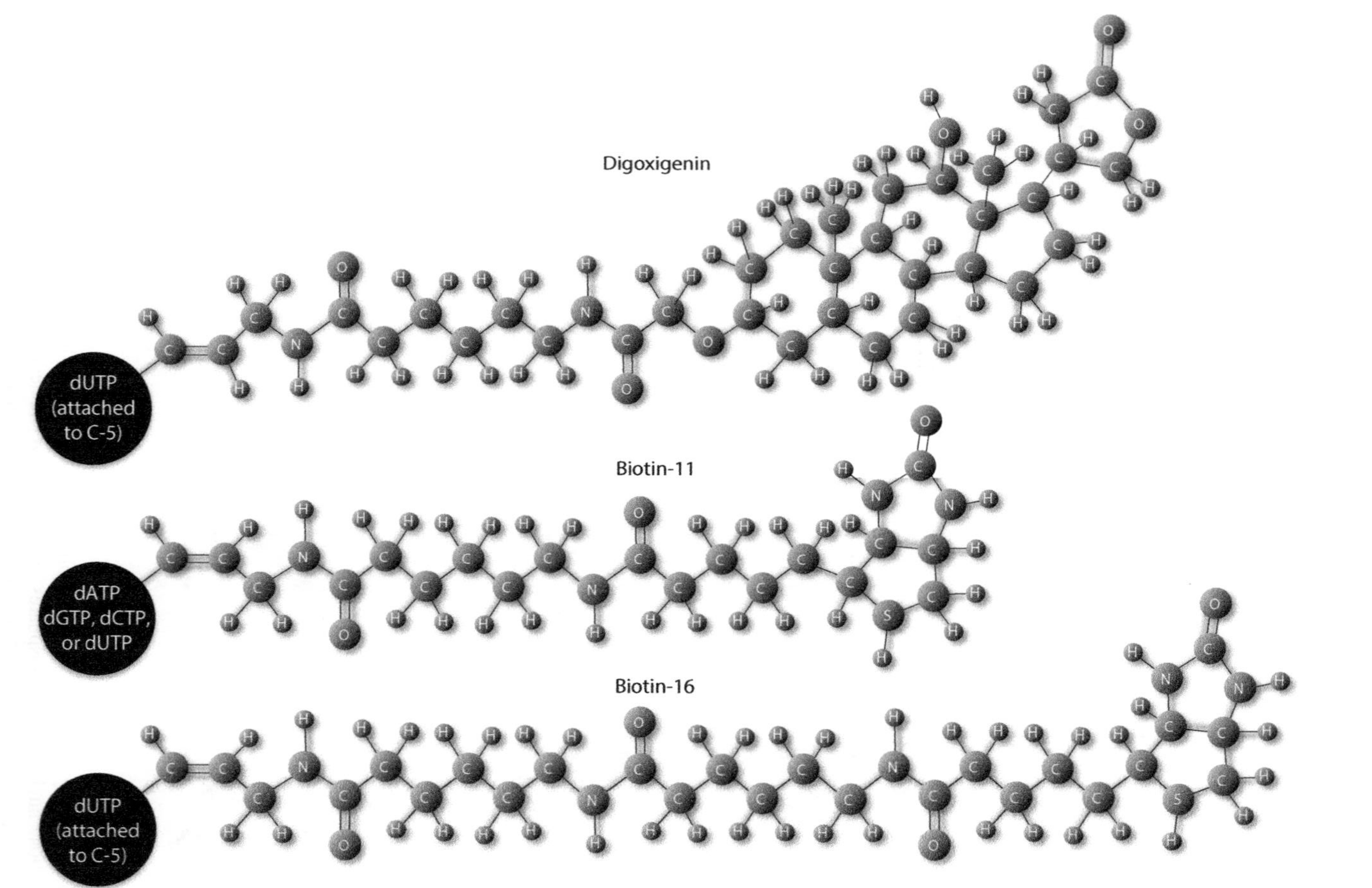

FIGURE 20.15 Side groups used in nonradioactive labeling for hybridizations, and similar protocols. Each is attached to the base at C-5 for each of the pyrimidines, or N-7 for the purines. In each case, there are long hydrocarbon linkers that end in a series of rings that serve as epitopes for antibody recognition (which is the basis of detection for nonradioactive detection methods). Digoxigenin provides the most sensitive nonradioactive detection method. Biotin-11 was one of the first nonradioactive labeling/detection systems, followed by biotin-16 labels that were more sensitive.

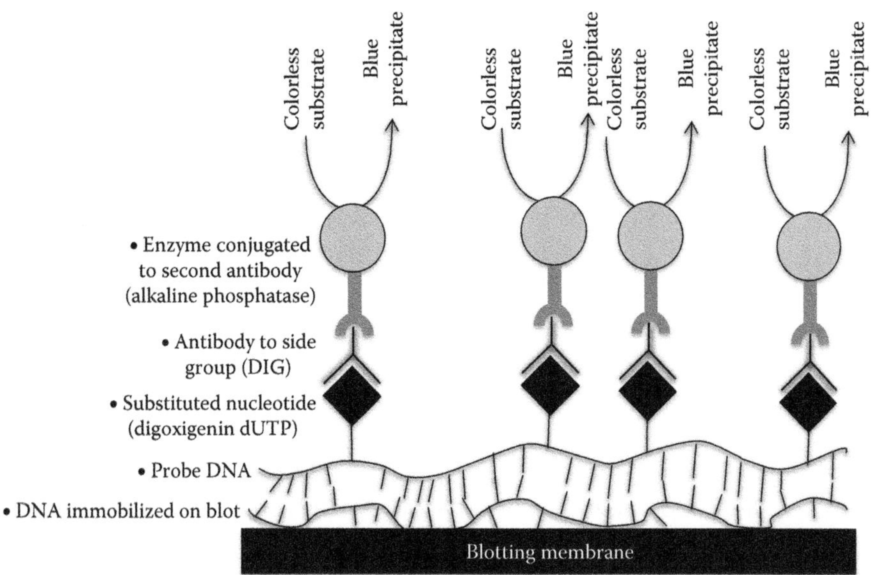

FIGURE 20.16 Detection of probe using nonradioactive detection methods. Antibodies are raised to the side group on the labeled nucleotide incorporated in the probe. These antibodies attach to the blot where the probe is hybridized. A second antibody that recognizes the first antibody is then added. This antibody has conjugated to it an enzyme that reacts with a substrate. In some cases, the substrate is a dye that is converted to a blue pigment that precipitates onto the blot. Alternatively, other substrates produce flashes of light when acted upon by the enzyme. These can be used in conjunction with X-ray film to produce a film that resembles autoradiography with radioactive probes.

amounts of DNA can be made. Similar methods can be used to determine RNA copy numbers for studies of gene expression. The nucleic acids can be studied on blotted gels, or by dot or slot-blotting methods. Slot-blotting is a process whereby a DNA sample is loaded onto one small location on the blot (Figure 20.17). The original methods were called *dot blots* because the DNA was loaded from a pipet to form a small dot. However, it was difficult to quantify the amount of hybridization in a dot because the concentration of hybridized probe decreased exponentially out from the center of the dot. Therefore, only the very center of the dot yielded an accurate reading. In slot-blotting, the DNA is spread out in a line. Therefore, when a detection apparatus, such as a densitometer passes over the line, there is a uniform distribution of hybridized molecules along the entire center of the line of DNA. This improves the accuracy and repeatability. The best results are obtained when the DNA is first digested with a restriction enzyme or is sheared to a uniform length. Then, it is denatured using an alkaline solution (usually NaOH), neutralized, and then blotted onto the membrane. Subsequently, labeled probe DNA is hybridized to the membrane, and the amount of hybridization is quantified. A standard curve is made from the probe standards that are loaded onto the same blot, and then the amounts of hybridization for each of the samples are determined based on the standard curve.

MICROSCOPY

DNA, RNA, and some protein complexes can be visualized by light, fluorescence, phase contrast, electron, and other types of microscopy. For light microscopy, the tissues are usually fixed and then sectioned using a microtome (Figure 20.18). Then, various stains are used that are specific for nucleic acids, and for various types of proteins (as well as for lipids and polysaccharides). Some of these can be used for quantification of DNA and RNA levels in cells. Genome sizes for organisms

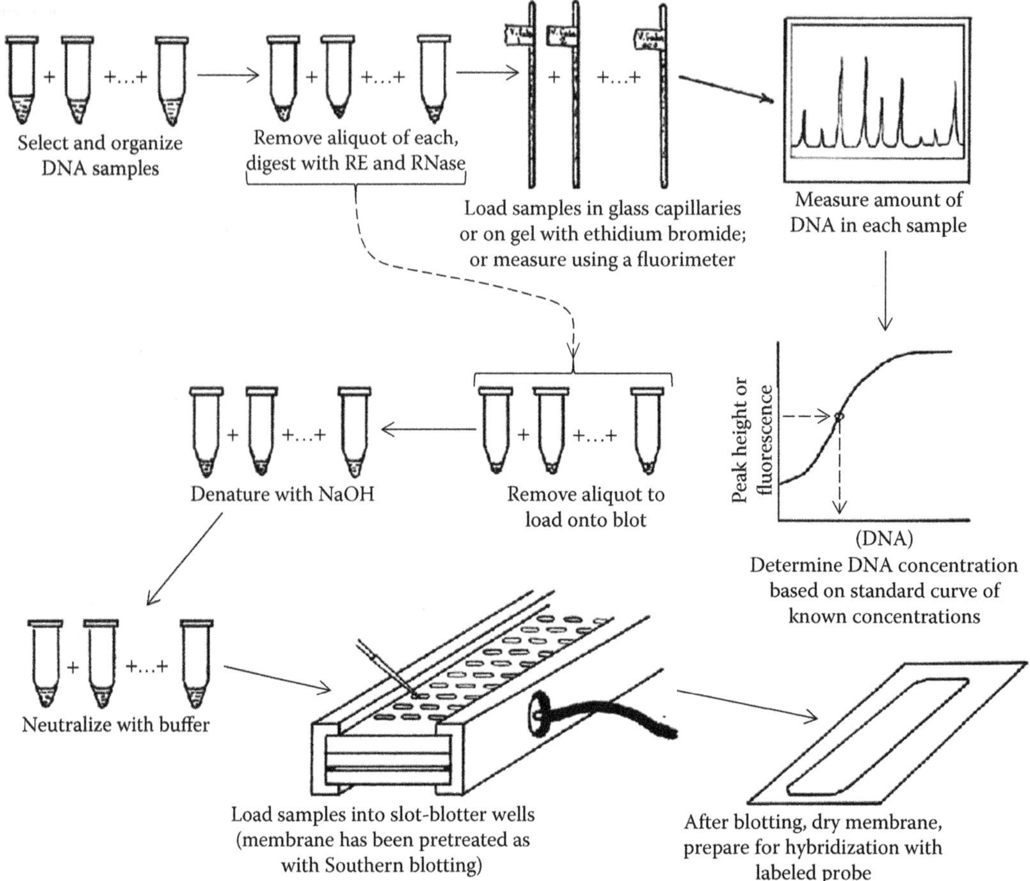

FIGURE 20.17 Slot-blotting to determine gene copy numbers. The DNA samples to be assayed must first be carefully organized and diluted (or concentrated) so that they will be in the correct range for detection of the genes to be assayed for gene copy number. Their concentrations must be accurately measured either on gels (with ethidium bromide), in a fluorimeter (with fluorescent dye), or in glass capillaries (with ethidium bromide). Usually, they are pretreated with a restriction enzyme and RNase to produce a homogeneous solution. An appropriate amount of each of the samples is then processed for loading on the blot in the slot blotter. As with Southern hybridization, the blot is pretreated to prepare for blotting. The samples are loaded and slowly drawn onto the blot. Then, the blot is removed, treated with NaOH, buffer, and then dried. It is then hybridized to a radioactive or nonradioactive probe for detection, quantification, and eventual calculation of gene copy number. The calculation is dependent on the amount of DNA loaded, the genome size of the organism, the length of DNA that hybridizes, and the quantity of DNA as indicated by the hybridized probe.

with large genomes were first estimated using staining methods for light microscopy. Subsequently, methods using fluorescent stains coupled with fluorescence microscopy provided more accurate measurements of large genomes, and allowed estimation of genome sizes for small genomes and organellar genomes, as well. Ribosomes were first observed using staining methods for light microscopy. They also were observed by transmission electron microscopy (TEM), and much of their structures were first determined using TEM. DNA and RNA can also be observed by TEM, as well as by phase-contrast microscopy. For TEM, the nucleic acids can be viewed in sectioned materials (sectioned on an ultramicrotome) or as nucleic acids spread on TEM grids (Figure 20.19). The stains that are used for TEM are heavy metals (such as uranium, usually in the form of uranyl acetate) that are electron dense so that they deflect the electron beam to increase the contrast of the molecules. For DNA and RNA spreads, the nucleic acids are mixed with proteins (e.g., cytochrome c) that

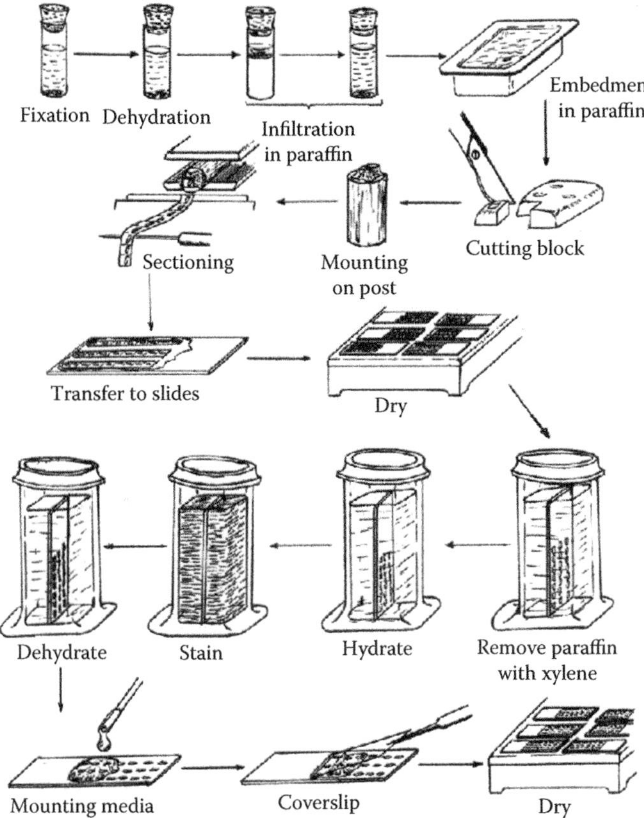

FIGURE 20.18 Processing of tissue samples for sectioning and viewing by light microscopy. Tissues degrade rapidly unless treated with a fixative that crosslinks molecules within the tissues and stops degradative processes. Following fixation, they are dehydrated in alcohol and other chemicals, and then they are embedded in a paraffin/plastic mixture. The fixed tissues are sectioned (usually between 5 and 8 μm thick) using a microtome. The sections are affixed to slides, the paraffin/plastic is removed by dissolving, and the sections are stained. Stains have been developed that will accentuate certain parts of the cell, such as proteins, DNA, RNA, membranes, and so on.

attach to DNA, and then they are spread on an aqueous phase and picked up on grids that are coated with a thin layer of nitrocellulose. The DNA and/or RNA are then shadowed by evaporating palladium/platinum wire in a vacuum. Spreads such as these led to discoveries such as DNA replication in bacteria, intron splicing, and other fundamental aspects of nucleic acids.

PROTEIN ANALYSIS

While protein sequences are often used in evolutionary studies, the sequences are usually deduced from the DNA sequences, rather than from the proteins themselves. Nonetheless, proteins are sometimes extracted and studied separately from the DNA and RNA, especially when investigating differences in the expression of those proteins. Proteins can be extracted, purified, and used to study various aspects of cell functions, and are often used in molecular biological procedures. Proteins can be sequenced using various proteases that cut polypeptides next to specific amino acids. The fragments can then be separated and digested with other enzymes, such that the entire protein sequence can be determined. There also are several other methods for sequencing proteins.

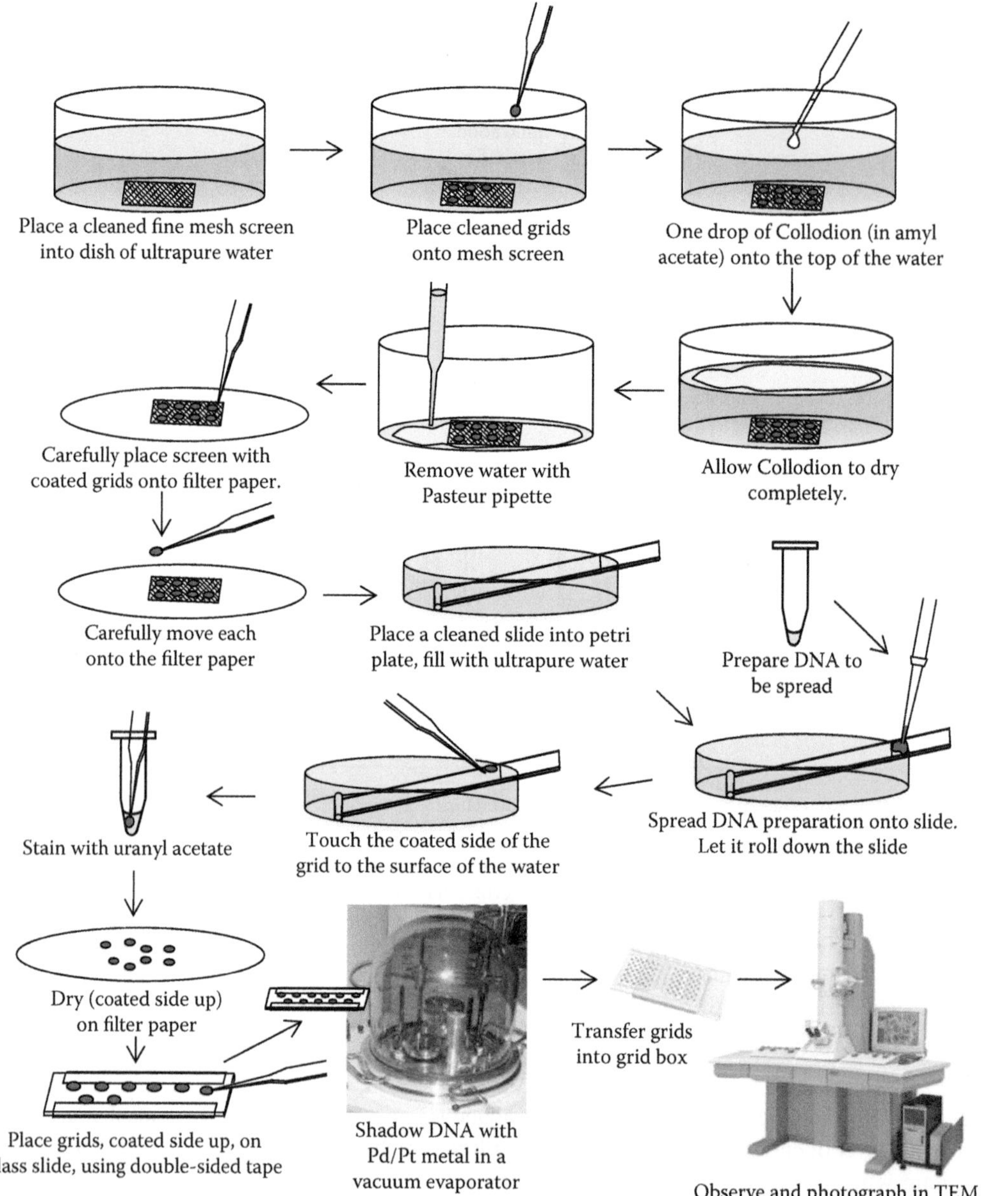
Place a cleaned fine mesh screen into dish of ultrapure water
Place cleaned grids onto mesh screen
One drop of Collodion (in amyl acetate) onto the top of the water
Allow Collodion to dry completely.
Remove water with Pasteur pipette
Carefully place screen with coated grids onto filter paper.
Carefully move each onto the filter paper
Place a cleaned slide into petri plate, fill with ultrapure water
Prepare DNA to be spread
Spread DNA preparation onto slide. Let it roll down the slide
Touch the coated side of the grid to the surface of the water
Stain with uranyl acetate
Dry (coated side up) on filter paper
Place grids, coated side up, on glass slide, using double-sided tape
Shadow DNA with Pd/Pt metal in a vacuum evaporator
Transfer grids into grid box
Observe and photograph in TEM

FIGURE 20.19 Preparation of DNA for viewing in a transmission electron microscope (TEM). DNA and RNA can be spread out on a water surface and transferred to electron microscope grids with a nitrocellulose coating. The grids are coated with Collodion (a nitrocellulose solution in amyl acetate) by placing one drop of Collodion on a water surface, and then drawing down the water over grids placed at the bottom of the container. Because nitrocellulose is rapidly oxidized, the coated grids must be used within a few hours of preparation. The DNA (and/or RNA) to be viewed is mixed with cytochrome C (or similar protein that will adhere to nucleic acids to increase their apparent diameters), and then allowed to flow down a glass slide ramp into a dish of water. The DNA (or DNA) is picked up on the grids by touching the coated side of the grid to the water surface. Excess water is removed, and the grids are stained with uranyl acetate, and coated with palladium/platinum in a vacuum evaporator. The grids are then viewed in a TEM.

KEY POINTS

1. DNAs can be characterized and studied using various methods. One of the simplest assays is to digest the DNA with restriction enzymes, and examine the resulting fragments by agarose gel electrophoresis.
2. PCR can be used to amplify specific regions of DNA, and can be used to amplify from extremely small quantities of DNA.
3. Recombinant DNA methods can be used to isolate, amplify, and study specific genes or genomes in detail. In recombinant methods, the DNA is isolated, digested with an enzyme or amplified by PCR, inserted into a DNA vector, and used to transform a bacterial vector. The transformants are then isolated and characterized using molecular biological methods.
4. DNAs can be examined using hybridization methods, such as Southern hybridization, to identify specific genes and sequences, and investigate the functions and evolution of the genes. Northern hybridizations are used to investigate RNA species, while western hybridizations are used to investigate proteins.
5. Gene copy number can be determined using hybridization methods, either using dot blots, slot blots, or on blots of gels.
6. Nucleic acids and proteins can be studied by light microscopy (with stains), fluorescence microscopy (with fluorescent stains), electron microscopy (with heavy metal stains), or by other methods.
7. Extracted proteins can be studied by many methods, including staining, immunological procedures, functional assays, etc.

ADDITIONAL READINGS

Alberts, B., D. Bray, K. Hopkin, A. Johnson, J. Lewis, M. Raff, K. Roberts, and P. Walter. 2013. *Essential Cell Biology,* 4th ed. New York: Garland Publishing.

Davis, L. G., M. D. Dibner, and J. F. Battey. 1986. *Basic Methods in Molecular Biology*. New York: Elsevier.

Glover, D. M. 1980. *Genetic Engineering—DNA Cloning.* New York: Chapman and Hall.

Green, M. R. and J. Sambrook. 2012. *Molecular Cloning: A Laboratory Manual*, 4th ed. Cold Spring Harbor, NY: Cold Spring Harbor Laboratory Press.

Hanahan, D. 1983. Studies on transformation of *Escherichia coli* with plasmids. *J. Mol. Biol.* 166:557–580.

Innis, M. A., D. H. Gelfand, and J. J. Sninsky, eds. 1995. *PCR Strategies.* San Diego, CA: Academic Press.

Innis, M. A., D. H. Gelfand, J. J. Sninsky, and T. J. White, eds. 1990. *PCR Protocols*. San Diego, CA: Academic Press.

LoBuglio, K. F., S. O. Rogers, and C. J. K. Wang. 1991. Variation in ribosomal DNA among isolates of the mycorrhizal fungus *Cenococcum geophilum*. *Can. J. Bot.* 69:2331–2343.

Rogers, S. O., G. C. Beaulieu, and A. J. Bendich. 1993. Comparative studies of gene copy numbers. *Meth. Enzymol.* 224:243–251.

Rogers, S. O. and A. J. Bendich. 1987. Heritability and variability in ribosomal RNA genes of *Vicia faba*. *Genetics* 117:285–295.

Rogers, S. O. and A. J. Bendich. 1987. Ribosomal RNA genes in plants: Variability in copy number and in the intergenic spacer. *Plant Mol. Biol.* 9:509–520.

Rogers, S. O. and A. J. Bendich. 1992. Variability and heritability of histone genes H3 and H4 in *Vicia faba*. *Theor. Appl. Genet.* 84:617–623.

Rogers, S. O., S. Honda, and A. J. Bendich. 1986. Variation in ribosomal RNA genes among individuals of *Vicia faba*. *Plant Mol. Biol.* 6:339–345.

Rogers, S. O., S. Rehner, C. Bledsoe, G. J. Mueller, and J. F. Ammirati. 1989. Extraction of DNA from Basidiomycetes for ribosomal DNA hybridizations. *Can. J. Bot.* 67:1235–1243.

Southern, E. M. 1975. Detection of specific sequences among DNA fragments separated by gel electrophoresis. *J Mol. Biol.* 98:503–517.

REFERENCES

[illegible]

ADDITIONAL REFERENCES

[illegible]

Rogers, S.O., [illegible] [illegible] rDNA [illegible]

Rogers, S.O., [illegible] [illegible] DNA [illegible] *Can. J. Bot.* [illegible]

Southern, E.M. 1975. Detection of specific sequences among DNA fragments separated by gel electrophoresis. *J. Mol. Biol.* 98:503–517.

21 Sequencing and Alignment Methods

INTRODUCTION

Sequencing of nucleic acids began in the 1960s, when the sequences of transfer RNAs (tRNAs) were determined. It was accomplished by methods that partially digested different portions of the RNA, and then the individual pieces were joined together on paper manually by the researchers. This was a painstaking process, so much so that the first RNA virus genomes were not fully determined until 1976 (for MS2, which is approximately 3.6 kb in length; Figure 21.1). Because most RNAs are single-stranded molecules, obtaining their sequences was limited to degradative methods at that time. However, at about the same time, DNA sequencing methods were developed, one using a degradative process and one using a polymerization reaction. The latter gained greater acceptance and was used initially for virus and organellar DNA sequencing projects, and eventually for projects on larger genomes, including those for Bacteria, Archaea, and Eukarya (including the initial human genome projects).

DEVELOPMENT OF DNA SEQUENCING METHODS

In the year after the RNA sequence of MSZ was published, two different methods for DNA sequencing were published in *PNAS* (*Proceedings of the National Academy of Sciences*). One method, published by Allan Maxam and Walter Gilbert, was somewhat analogous to the RNA methods for sequencing, in that it used a set of controlled degradative reactions to create a set of variously sized DNA fragments (Figure 21.2).

One reaction contained chemicals that would cut the DNA at purines, another would cut the DNA only at adenine residues, another would cut only at pyrimidines, and the final reaction would cut only at cytosine residues. The resulting products were each loaded into individual lanes of a long polyacrylamide gel, and then subjected to electrophoresis, such that fragments varying by single nucleotides could be distinguished from one another. Furthermore, the starting DNA molecule was labeled with ^{32}P at their 5′ ends, so that only the fragments containing the intact 5′ end would show up by autoradiography of the gel. This allowed reconstruction of the DNA sequence reading from the 5′ end. If a band was present in both the purine and adenine gel lanes, then the nucleotide at that position was an adenine. However, if there was a band in the purine lane, but not in the adenine lane, then the nucleotide at that position was a guanine. Similarly, for pyrimidines, a band in the pyrimidine and cytosine lanes indicated a cytosine, whereas a band in the pyrimidine lane, but not in the thymidine lane, indicated a thymidine at that position in the sequence.

The second method, which became more widely used, was developed by Frederick Sanger and colleagues (Figure 21.3). Instead of being a degradative reaction, this method used DNA polymerase and a set of modified mononucleosides to polymerize DNA fragments of varying lengths. Standard nucleoside triphosphates for DNA synthesis have 2′deoxyribose sugars, and it is the 3′ hydroxyl on the 2′deoxyribose on the growing nucleotide chain that joins with the 5′ phosphate group of the incoming nucleoside triphosphate during DNA polymerization. For the Sanger method, a small amount of 2′,3′-dideoxyribose triphosphates are added to each of four reactions (one for each of the four nucleotides). When one of the dideoxyribose nucleotides is added to a growing DNA chain, no additional nucleotides can be added to the chain, because the end contains no 3′-hydroxyl on the

```
   1 gggtgggacc cctttcgggg tcctgctcaa cttcctgtcg agctaatgcc atttttaatg
  61 tctttagcga gacgctacca tggctatcgc tgtaggtagc cggaattcca ttcctaggag
 121 gtttgacctg tgcgagcttt tagtaccctt gatagggaga acgagacctt cgtcccctcc
 181 gttcgcgttt acgcggacgg tgagactgaa gataactcat tctctttaaa atatcgttcg
 241 aactggactc ccggtcgttt taactcgact ggggccaaaa cgaaacagtg gcactacccc
 301 tctccgtatt cacgggggc gttaagtgtc acatcgatag atcaaggtgc ctacaagcga
 361 agtgggtcat cgtggggtcg cccgtacgag gagaaagccg gtttcggctt ctccctcgac
 421 gcacgctcct gctacagcct cttccctgta agccaaaact tgacttacat cgaagtgccg
 481 cagaacgttg cgaaccgggc gtcgaccgaa gtcctgcaaa aggtcaccca gggtaatttt
 541 aaccttggtg ttgctttagc agaggccagg tcgacagcct cacaactcgc gacgcaaacc
 601 attgcgctcg tgaaggcgta cactgccgct cgtcgcggta attggcgcca ggcgctccgc
 661 taccttgccc taaacgaaga tcgaaagttt cgatcaaaac acgtggccgg caggtggttg
 721 gagttgcagt tcggttggtt accactaatg agtgatatcc agggtgcata tgagatgctt
 781 acgaaggttc accttcaaga gtttcttcct atgagagccg tacgtcaggt cggtactaac
 841 atcaagttag atggccgtct gtcgtatcca gctgcaaact tccagacaac gtgcaacata
 901 tcgcgacgta tcgtgatatg gttttacata aacgatgcac gtttggcatg gttgtcgtct
 961 ctaggtatct tgaacccact aggtatagtg tgggaaaagg tgcctttctc attcgttgtc
1021 gactggctcc tacctgtagg taacatgctc gagggcctta cggcccccgt gggatgctcc
1081 tacatgtcag gaacagttac tgacgtaata acgggtgagt ccatcataag cgttgacgct
1141 ccctacgggt ggactgtgga gagacagggc actgctaagg cccaaatctc agccatgcat
1201 cgaggggtac aatccgtatg gccaacaact ggcgcgtacg taaagtctcc tttctcgatg
1261 gtccatacct tagatgcgtt agcattaatc aggcaacggc tctctagata gagccctcaa
1321 ccggagtttg aagcatggct tctaacttta ctcagttcgt tctcgtcgac aatggcggaa
1381 ctggcgacgt gactgtcgcc ccaagcaact tcgctaacgg ggtcgctgaa tggatcagct
1441 ctaactcgcg ttcacaggct tacaaagtaa cctgtagcgt tcgtcagagc tctgcgcaga
1501 atcgcaaata caccatcaaa gtcgaggtgc ctaaagtggc aacccagact gttggtggtg
1561 tagagcttcc tgtagccgca tggcgttcgt acttaaatat ggaactaacc attccaattt
1621 tcgctacgaa ttccgactgc gagcttattg ttaaggcaat gcaaggtctc ctaaaagatg
1681 gaaacccgat tccctcagca atcgcagcaa actccggcat ctactaatag acgccggcca
1741 ttcaaacatg aggattaccc atgtcgaaga caacaaagaa gttcaactct ttatgtattg
1801 atcttcctcg cgatctttct ctcgaaattt accaatcaat tgcttctgtc gctactggaa
1861 gcggtgatcc gcacagtgac gactttacag caattgctta cttaagggac gaattgctca
1921 caaagcatcc gaccttaggt tctggtaatg acgaggcgac ccgtcgtacc ttagctatcg
1981 ctaagctacg ggaggcgaat ggtgatcgcg gtcagataaa tagagaaggt ttcttacatg
2041 acaaatcctt gtcatgggat ccggatgttt tacaaaccag catccgtagc cttattggca
2101 acctcctctc tggctaccga tcgtcgttgt ttgggcaatg cacgttctcc aacggtgctc
2161 ctatggggca caagttgcag gatgcagcgc cttacaagaa gttcgctgaa caagcaaccg
2221 ttaccccccg cgctctgaga gcggctctat tggtccgaga ccaatgtgcg ccgtggatca
2281 gacacgcggt ccgctataac gagtcatatg aatttaggct cgttgtaggg aacggagtgt
2341 ttacagttcc gaagaataat aaaatagatc gggctgcctg taaggagcct gatatgaata
2401 tgtacctcca gaaaggggtc ggtgctttca tcagacgccg gctcaaatcc gttggtatag
2461 acctgaatga tcaatcgatc aaccagcgtc tggctcagca gggcagcgta gatggttcgc
2521 ttgcgacgat agacttatcg tctgcatccg attccatctc cgatcgcctg gtgtggagtt
2581 ttctcccacc agagctatat tcatatctcg atcgtatccg ctcacactac ggaatcgtag
2641 atggcgagac gatacgatgg gaactatttt ccacaatggg aaatgggttc acatttgagc
2701 tagagtccat gatattctgg gcaatagtca aagcgaccca aatccatttt ggtaacgccg
2761 gaaccatagg catctacggg gacgatatta tatgtcccag tgagattgca ccccgtgtgc
2821 tagaggcact tgcctactac ggttttaaac cgaatcttcg taaaacgttc gtgtccgggc
2881 tctttcgcga gagctgcggc gcgcactttt accgtggtgt cgatgtcaaa ccgtttttaca
2941 tcaagaaacc tgttgacaat ctcttcgccc tgatgctgat attaaatcgg ctacggggtt
3001 ggggagttgt cggaggtatg tcagatccac gcctctataa ggtgtgggta cggctctcct
3061 cccaggtgcc ttcgatgttc ttcggtggga cggacctcgc tgccgactac tacgtagtca
3121 gcccgcctac ggcagtctcg gtatacacca agactccgta cgggcggctg ctcgcggata
3181 cccgtacctc gggtttccgt cttgctcgta tcgctcgaga acgcaagttc ttcagcgaaa
3241 agcacgacag tggtcgctac atagcgtggt tccatactgg aggtgaaatc accgacagca
3301 tgaagtccgc cggcgtgcgc gttatacgca cttcggagtg gctaacgccg gttcccacat
3361 tccctcagga gtgtgggcca gcgagctctc ctcggtagct gaccgaggga cccccgtaaa
3421 cggggtgggt gtgctcgaaa gagcacgggt gcgaaagcgg tccggctcca ccgaaaggtg
3481 ggcgggcttc ggcccaggga cctcccccta aagagaggac ccgggattct cccgatttgg
3541 taactagctg cttggctagt taccaccca
```

FIGURE 21.1 Complete genomic sequence of bacteriophage MSZ. This was the first DNA genome that was sequenced. It is 3569 nucleotides in length.

ribose to react with the 5′-phosphate on the incoming nucleoside triphosphate. Thus, the chain is terminated when a dideoxynucleotide triphosphate is added to the DNA. This method is also called the *chain termination method*. This method has three advantages over the Maxam and Gilbert method: (1) Because it uses a set of synthesis reactions rather than a set of degradative reactions, much less DNA is needed for sequencing; (2) because synthesis starts from the 5′ end, the region being sequenced can be radioactively labeled (or labeled using nonradioactive methods) throughout its length, which also meant that less starting DNA was needed for sequencing; and (3) the degradative reactions in the Maxam and Gilbert method have to each be carefully controlled in order to only partially digest the DNA to be sequenced. The reactions in the Sanger method were more easily controlled and more consistent for a variety of template DNAs.

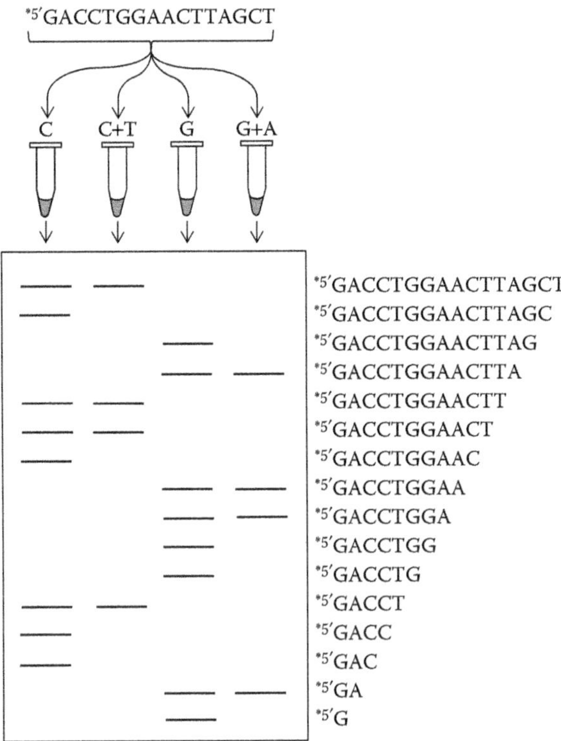

FIGURE 21.2 Sequence determination using the Maxam and Gilbert method. This method uses degradative reactions to cut the DNA at specific bases. Four reactions are set up, which cut the DNA at cytosines, pyrimidines (cytosines and thymidines), guanosines, or purines (adenosines and guanosines). The concentrations of the DNA and the degradative chemicals must be carefully calibrated, such that the degradation does not cut at every site but cuts the DNA into a series of fragments that represent each of the nucleotides in the DNA template to be sequenced. Prior to the degradative reactions, the DNA is labeled on the 5′ end (usually with radioactive phosphorus, ^{32}P or ^{33}P), so that all of the digested fragments represent distances to the 5′ end.

The original Sanger method used radioactive labeling of one or more of the nucleoside triphosphates, followed by gel electrophoresis through thin polyacrylamide gels, and then autoradiography to visualize the bands (described in Chapter 19). By the late 1980s and the early 1990s, several methods using nonradioactive labeling methods were developed, one of which was used extensively in semiautomated sequencing methods. The labeled nucleotides had side groups attached to them that would fluoresce when excited by specific wavelengths of light, emitted from lasers. The first methods had only one such chromophore, so that four lanes still had to be run for each sequence, corresponding to the adenine, guanine, cytosine, and thymidine reactions. The gel was loaded, the fragments were separated in the gel during electrophoresis, and at the bottom of the gel a laser would move across the gel, along with a light detector, which would send a signal to the attendant computer. Software in the computer would then interpret the light signals as nucleotides and their positions in each of the lanes, and a sequence would be generated from the information. Subsequently, Applied Biosystems developed a system with four chromophores attached to the dideoxynucleoside triphosphates (Figure 21.4), one for each type of nucleotide (A, G, C, and T), each of which fluoresced a different color (i.e., wavelength). The advantage of this system was that all four reactions could be mixed into a single reaction and loaded into one lane, as opposed to four lanes for the other systems. Once a fragment reached the bottom of the gel and was struck by the laser light, only one color would fluoresce, corresponding to the specific type of nucleotide. The signal sent to the computer indicated the nucleotide and that position, and the computer built up the sequence results.

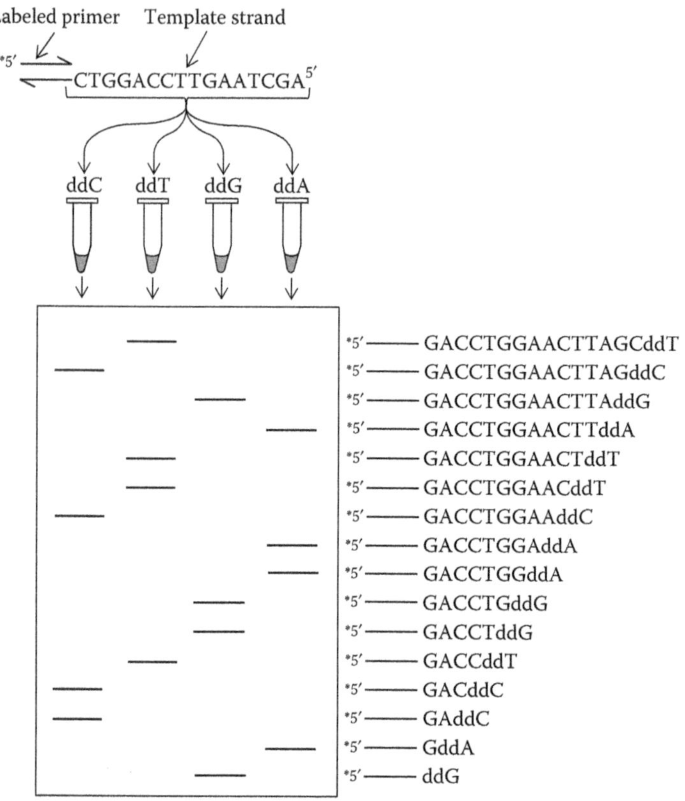

FIGURE 21.3 Sequence determination using the Sanger et al. method. This method uses DNA polymerase to synthesize DNA molecules of varying sizes. It is also called the *dideoxynucleotide*, or *chain termination* method, because of its reliance on a set of dideoxynucleotides, which when incorporated into a DNA molecule do not allow additional bases to be added to the DNA chain. As with Maxam and Gilbert sequencing, the DNA is first labeled with radioactive phosphorus so that all of the fragments vary in size relative to the 5′ end. Alternatively, a small amount of radioactively labeled deoxynucleotide triphosphate (dNTP; usually one of the four) is added. Because sequencing is initiated using a known primer, the 5′ end is already defined for each of the synthesized new DNA molecules. Four reactions are prepared, all of which include DNA polymerase, all four dNTPs, and buffer. In addition, 2′,3′-dideoxynucleotide triphosphates (ddNTPs) are added specifically to each reaction. For the C reaction, only a small amount of ddCTP is added; for the A reaction, a small amount of ddATP is added; and ddGTP and ddTTP are similarly added to the G and T reactions, respectively. During polymerization, most often a dNTP is added to the growing polypeptide chain. However, because there are low concentrations of ddNTPs, occasionally a ddNTP is added to the chain. Once one of the ddNTPs is added, no additional dNTPs can be added to that chain, because the 3′ hydroxyl is missing. When the concentrations of ddNTP are correct for each of the four reactions, the bands can be clearly seen in each lane, and up to 1500 bp can be read per sequencing set.

HIGH-THROUGHPUT TECHNOLOGIES

Each step in the development of sequencing technology led to an increase in the capabilities of the DNA sequencers, and thus an increase in the number of nucleotides that could be determined in each facility. The next improvement eliminated gels, which often were time consuming to make and to load. Polymers were developed that were loaded into capillary tubes by the automated DNA sequencer that allowed separation of the DNA fragments when electric currents were applied through the polymers (Figure 21.5). The advantage was that the loading and electrophoresis could be automated. A robotic system loaded the capillaries with the polymer from a reservoir of polymer. Next, it loaded the sample to be sequenced from a tray of prepared DNA samples loaded by the researchers. Initially, the sequencers

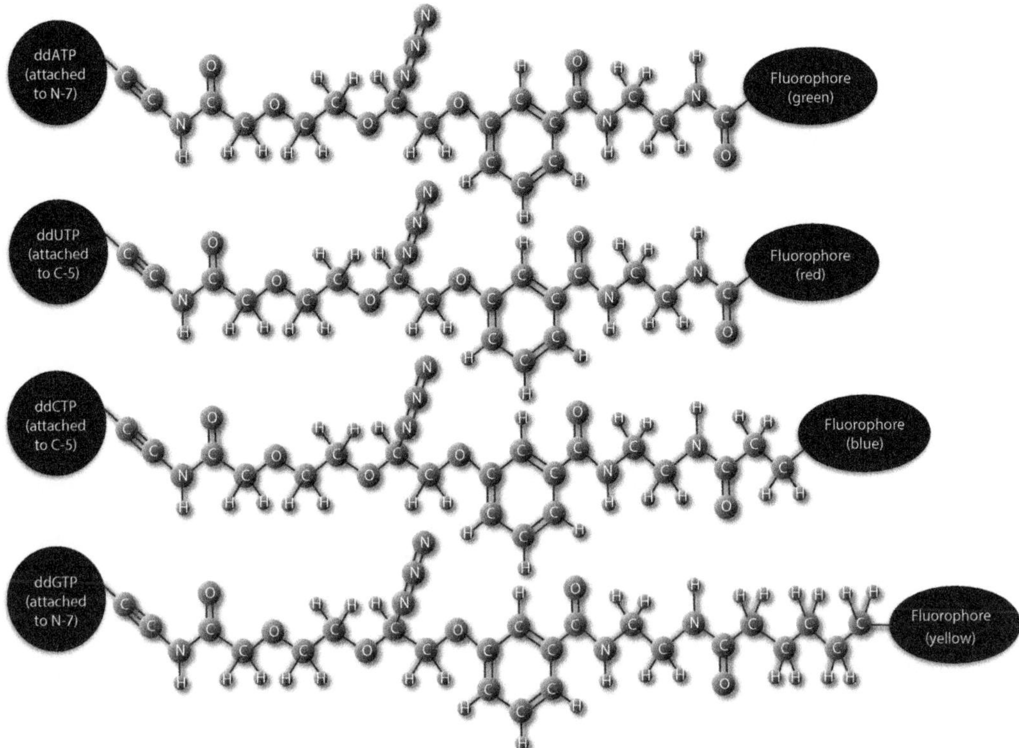

FIGURE 21.4 Fluorescently labeled 2′,3′-dideoxynucleotide triphosphates (ddNTPs) used in automated DNA sequencing. In order to speed up sequencing, each of the four ddNTPs was synthesized with a unique fluorophore that emitted light of a characteristic wavelength when excited by light from a laser. This reduced preparation time, because one reaction could be used rather than four. It also sped up electrophoresis of the fragments, because only one lane was used instead of four for each fragment to be sequenced. Eventually, capillaries and then manifolds were used, which further sped up the sequencing process for Sanger-type sequencing protocols.

had single capillaries, but eventually machines having up to 16 or more were produced. Ultimately, manifolds replaced the capillaries, and machines with 96 position manifolds were produced. These were the machines that were used to determine much of the initial human genome sequences. For large genome projects like this, rooms full of DNA sequencers churned out the data for many years.

NEXT-GENERATION SEQUENCING

By the end of the 1990s and the start of the 2000s, it was clear that standard sequencing methods were inadequate, too expensive, and too time consuming for the personnel to efficiently sequence entire genomes, especially eukaryotic genomes. For example, it took many years and several laboratories collaborating to sequence the first bacterial genomes, which were accomplished by creating clone libraries spanning all or most of the genome, sequencing each of the cloned fragments, and then reassembling the genomes using computer analyses. It took years and hundreds of Sanger-based sequencers to determine the human genome. Several researchers and companies began developing other methods and machines, collectively now called next-generation sequencing (NGS). NGS methods rely on performing many DNA synthesis/sequencing reactions synchronously, using sensors and computers to record and analyze the data from each of the individual sequencing reactions. The sequencing reactions are performed at the molecular level, such that a large number of sequencing reactions occur simultaneously in small wells, on slides, or on computer chips.

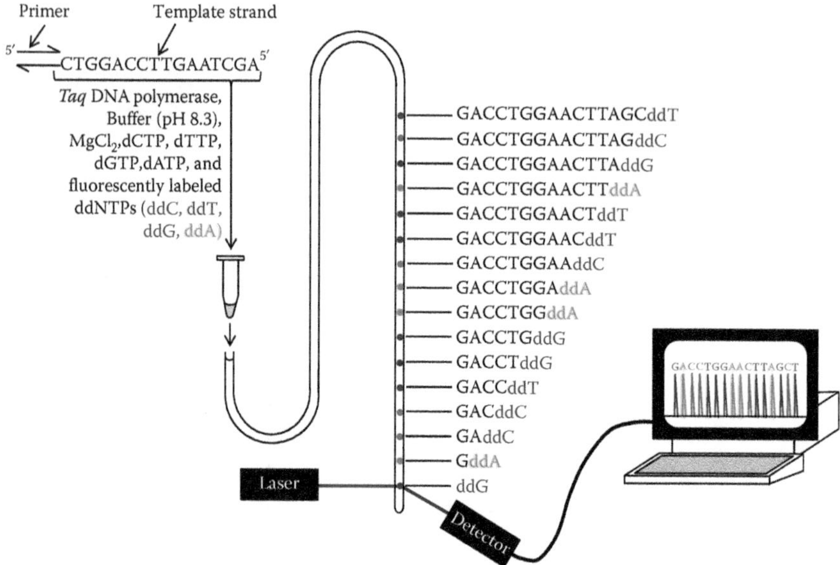

FIGURE 21.5 Automated sequencing using a capillary, a laser, and fluorphores. Following the sequencing reaction, using DNA polymerase, a primer, a buffer, deoxynucleotide triphosphates (dNTPs), and small amounts of the fluorescently labeled 2′,3′-dideoxynucleotide triphosphates (ddNTPs), the reactions are loaded into the automated DNA sequencer. The sequencer fills the capillary with polymer, and then loads a sample. If the system has multiple capillaries or a manifold, then more samples are loaded for each sequencing run. Electrophoresis is then carried out. The fragments separate based on the size, with the smallest proceeding before the larger molecules. As they pass the laser, they fluoresce based on the ddNTP at their ends. The color is interpreted as a particular base in that position of the sequence. The sequence of bases is recorded by the computer, and later retrieved by the user.

Among the first NGS methods to become commercially successful was simply called 454 pyrosequencing (Figure 21.6). As with Sanger sequencing, the DNA is sequenced using DNA polymerase, but polymerization occurs while the template DNAs are immobilized on microbeads. The microbeads have a single-stranded DNA of known sequence covalently bonded to them. The DNA fragments to be sequenced have had the complementary sequence attached to them, such that they hydrogen bond firmly to the single strand on the beads. A second sequence serves as the primer for the sequencing reaction. The DNA fragments to be sequenced each have the complementary sequence attached to their opposite ends. To begin the process, the DNA to be sequenced is diluted such that on average only a single DNA fragment is hybridized to a single bead. These are then subjected to emulsion polymerase chain reaction (emPCR), where the beads and the surrounding aqueous reaction mixture amplify the DNAs, which then adhere to the covalently attached DNA on the beads. Each of these small reaction regions is surrounded by mineral oil, which is the other part of the emulsion. The beads are then centrifuged into a microplate that has tiny wells that are only large enough so that one nanobead will fit inside of each. Polymerization (again, a PCR reaction) is then performed to increase the number of identical DNA fragments on each bead in each well. Finally, Sanger-type sequencing is performed, washing a single type of nucleoside triphosphate over the microplate for each cycle. Also, the enzyme luciferase, the chromogenic compound luciferin, and adenosine 5′-phosphosulfate (APS) are also present in the microplate. When a nucleotide is incorporated into a DNA chain, pyrophosphate is produced. Pyrophosphate reacts with the APS to produce adenosine triphosphate (ATP), which activates luciferase, which then acts on luciferase to produce a flash of light. Light sensors and a computer record the positions of each

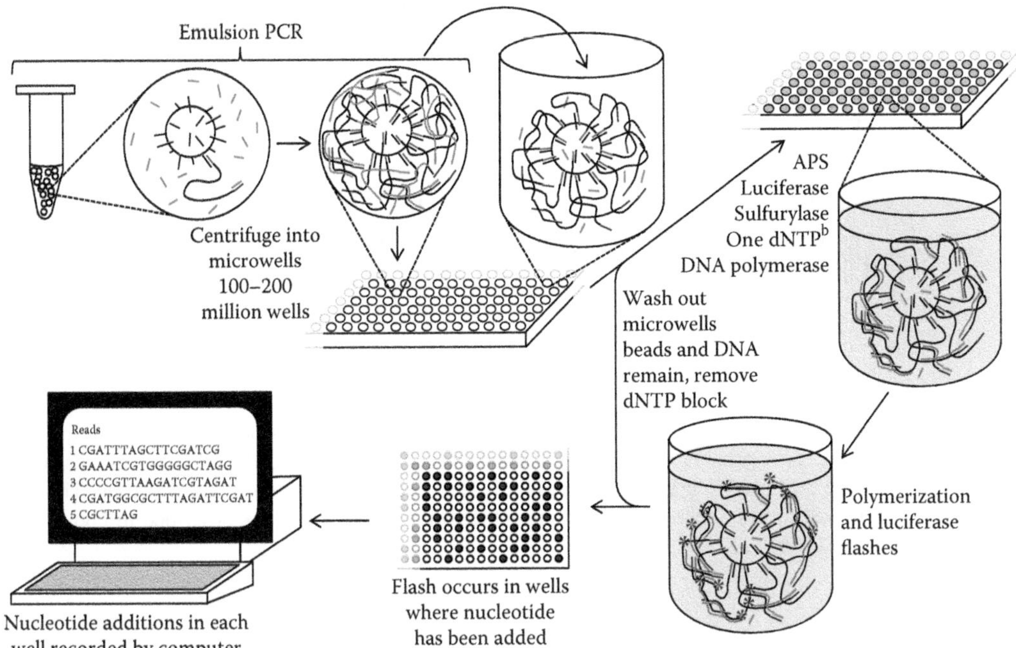

FIGURE 21.6 NGS using 454 pyrosequencing. This method uses microbeads that have short single-stranded DNA primers attached to them. Before sequencing, specific primer sequences (sequences 1 and 2, blue and red, respectively) are added to each end of the templates (black lines) to be sequenced. One of them is complementary to the primer sequences on the beads (violet), and the other (green) is used to prime the DNA sequencing reactions. Next, emPCR is performed. The beads are mixed with PCR reagents with the DNAs to be sequenced, such that, on average, only one DNA binds to one bead. This is also mixed with mineral oil and emulsified. The aqueous portions of the emulsion form around each bead, creating microreactors, and PCR cycling is performed. At the end, each bead is covered with amplified DNAs that all are of the same sequence. The emulsion is then broken, and the beads are centrifuged into a plate that contains 100–200 million microwells. The wells are of the size that allows only one bead to enter each well. Next, a solution containing the sequencing primer, DNA polymerase, sulfurylase, luciferase, adenosine 5′-phosphosulfate (APS), and one of the modified deoxynucleotide triphosphates (dNTPs) such that only one nucleotide can be added to any of the growing DNA chains. Upon incubation, if a nucleotide is incorporated in the DNAs in one of the wells, pyrophosphate (PP_i) is released in the reaction. The sulfurylase catalyzes the reaction between PP_i and APS to produce ATP. The ATP is used by luciferase to cause a flash of light, which is sensed and recorded by a light sensor connected to a computer. After the reaction, the wells are washed and the blocking groups are removed from the DNA chains, followed by addition of the reaction mix, including a different blocked dNTP. The flashes are recorded, and the process is repeated. The computer keeps track of the positions of each of the flashes and compiles all of the flashes into sequences, knowing which dNTPs were added in each cycle. Millions of unique reads, up to 800 bp in length, have been obtained using this method.

of the flashes and correlate this with the nucleoside triphosphate that was added during each cycle. The computer then produces a set of DNA sequence fragments, called *reads*. Read lengths for each microwell average about 200–400 nucleotides, although reads of nearly 1000 nucleotides sometimes result. Often more than one million such high-quality reads can result from a single sample of DNA. Because there is a level of error involved in this method, usually, it is best if multiple reads of the same region are produced for each region. During the assembly phase, which is where all of the short reads are joined into longer contiguous sequences (called contigs), when some nucleotide positions exhibit variation from one sequence to another, a consensus base is chosen to report in

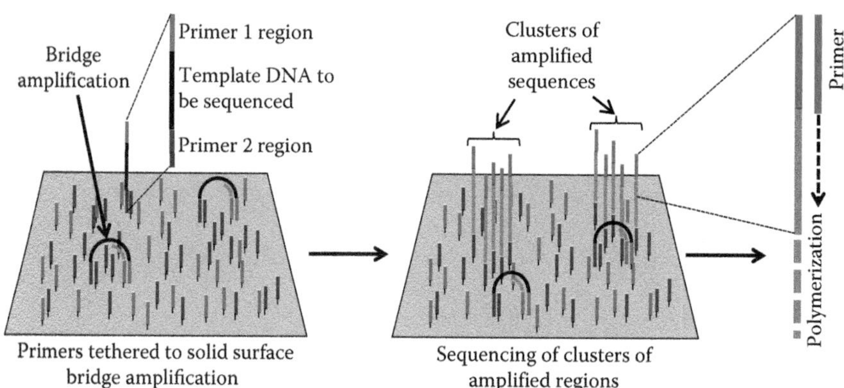

FIGURE 21.7 NGS using Illumina, which uses two types of primers that are bound to a substrate. Sequences complementary to the two primers are covalently attached (i.e., tethered) to the DNA fragments to be sequenced. These are then used in bridge PCR to amplify the islands of templates on the substrate. The second primer is then used for the sequencing reactions, which are analogous to those described in Figure 21.6 (although the details differ).

the final fully assembled sequence. Unfortunately, because of the high cost of the method, in 2014 it was announced that 454 pyrosequencing would no longer be supported, and thus has been discontinued in many facilities.

Other methods, such as Illumina (Solexa), also use specific sequences on each end but tether the DNA fragments to a slide or chip (similar to a computer chip; Figure 21.7). Again, the DNA fragments are first amplified in number using polymerization reactions, such that each position on the chip contains a high concentration of identical templates. Then the chips are flooded with solutions of each nucleoside triphosphate, which have been modified with a fluorophore and blocking group on the nucleoside triphosphate, such that only one nucleotide can be added at a time. The patterns of fluorescence are recorded by a high-resolution digital camera after each cycle, and if there is a light signal for a specific spot on the chip, then that nucleotide is added to the data set (the *read*) for that spot. Once the data have been recorded, both the fluorophore and the blocking group are chemically removed, and the chip is flooded with the next modified nucleoside triphosphate. Many cycles are repeated, which results in individual reads of 50–300 nucleotides. These can be *assembled* into longer sequences in the computer, which examines the reads for overlaps and then joins them together in longer contigs. Other similar NGS methods include those in which a single primer is bound to a substrate, which becomes the initiation site for polymerization and detection (Figure 21.8a); a method in which the templates are bound to the substrate, and then primed on the opposite end for sequencing (21.8b); and those that utilize DNA polymerase bound to the substrate (Figure 21.8c).

Another sequencing technology is called the Ion Torrent system. This takes advantage of the fact that a hydrogen ion is liberated during the incorporation of a nucleotide into a DNA strand during polymerization. Standard chemistry and nucleoside triphosphates are used in this system, and thus, costs can be lower. The other important part of the system is the use of ion-sensitive field-effect transistors (ISFETs), which detect when the hydrogen ions are released during the reaction. The chip containing the ISFET detectors is directly below the microwells into which each DNA template molecule has been deposited. One nucleotide at a time is flooded into the plate. When the detector sends a signal to the computer, the position is recorded by the computer, thus determining the nucleotide added for each position, and eventually leading to a determination of the sequence in each microwell. When more than one nucleotide is added in a particular microwell, the signal sent by the detector is proportional to the number of ions detected, and hence the number of the particular nucleotide that has been added to the chain. Therefore, it is simpler in some ways than other methods, and because

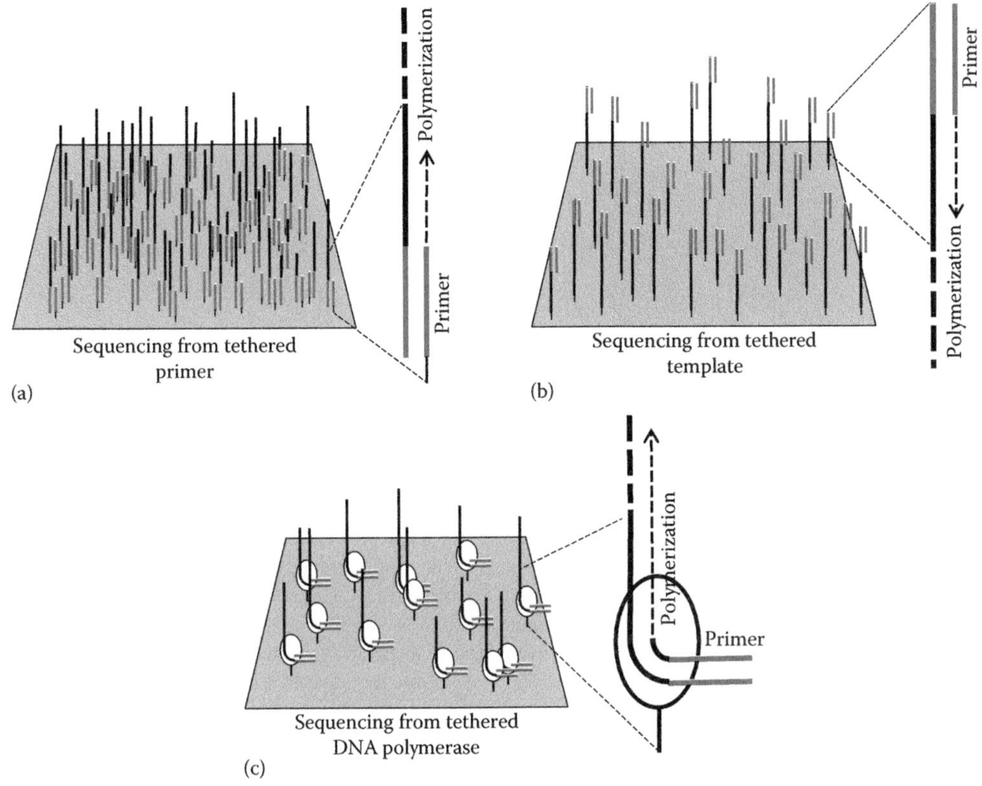

FIGURE 21.8 Three additional NGS methods. In the first method (a), the primers are tethered to the substrate, and then the templates that have a sequence complementary to the primer, and sequencing commences from there. In the second method (b), the templates (with a sequence complementary to the primer) are tethered to the substrate. In the third method (c), the DNA polymerase is tethered to the substrate, and the templates move through the polymerase molecules as they are sequenced. After adding the primer, polymerization begins. Again, the sequencing and detection methods are analogous to those described in Figure 21.6.

it uses standard DNA polymerase chemistry, it is also less expensive than some of the other methods. When first developed, the reads were short, but several improvements have increased the average read lengths to 200–400 nucleotides, and reads of 800 nucleotides are sometimes produced.

Another step in the technology uses specific pore proteins that are immobilized on a membrane (Figure 21.9). The pore is of the size that only one nucleotide at a time of a single-stranded DNA molecule can fit through the pore. Furthermore, because of the size of each of the nucleotides, the time that it takes for each nucleotide to move through the pore differs. Also, ions that are used in the system can pass through the pore in addition to the nucleotides. An electrical potential is created by the flow of ions, such that it can be determined when a nucleotide is passing through the pore. Therefore, by recording the electrical potential, and deducing the flow of ions through the pore, as well as the time it takes for the electrical potential to change, the computer can determine the sequence of nucleotides that are passing through the pore.

PROTEIN SEQUENCING

It is possible to determine amino acid sequences of proteins by sequencing the DNA for specific genes, and because DNA sequencing methods have become simpler and less expensive, many proteins are sequenced using translation of the DNA sequence and comparison of the estimated

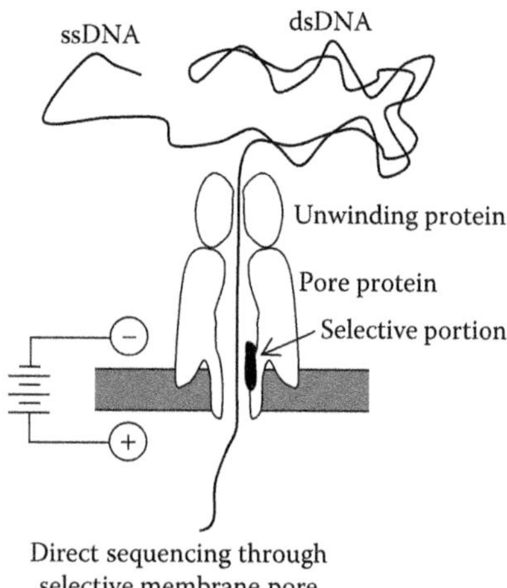

FIGURE 21.9 A membrane-based method for sequencing. The DNA moves through a protein pore through a membrane, which has a charge difference between the two sides of the membrane. The pore has a restriction at the end, which allows only one nucleotide to move through the pore at a time. As the nucleotides are moving through the pore, they interrupt the flow of ions through the pore. Because of the size differences among the four bases, each base takes slightly different amounts of time to pass through the pore, which interrupts the ion movements differentially. Therefore, the sequence of the DNA can be read by recording the changes in the electrical potential between the two sides of the membrane. Although the DNA that passes through the pore must be single stranded, it must begin as a double-stranded molecule that is unwound by another protein that is associated with the pore protein.

molecular weight based on gel electrophoresis of the protein. However, in some cases, it is necessary to determine the amino acid sequence from the purified protein. First, the protein is hydrolyzed using acid, and then the individual amino acids are separated on columns, usually ion exchange, hydrophobic interaction, or silica columns. Each is then quantified using a ninhydrin reaction, which turns purple with all of the amino acids, except proline, which yields a yellow color. The intensity of color is proportional to the amount of amino acid in the solution. The N-terminal amino acid can be determined using a chemical that binds to the amine group on the N-terminus. After the reaction is completed, the protein is hydrolyzed and the amino acids are separated. Based on the chemical used, a specific color indicates the N-terminal amino acid. The C-terminal amino acid can be determined by digestion using carboxypeptidase, which will cleave the protein from the C-terminus inward. Samples taken at intervals will have higher concentrations of the amino acid at the C-terminal end. The protein can then be sequenced, but first the disulfide bridges must be cut to separate any pieces of the proteins that are covalently linked. Using a combination of digestions with selected proteases that cut next to specific amino acid, the proteins are cut into pieces that are less than 50 amino acids in length. For short proteins, each of the smaller fragments can also be digested into smaller pieces using other proteases, eventually separating them and digesting them into individual amino acids. The data can then be combined and compared to determine all or most of the sequence of amino acids. Alternatively, the pieces that are 50 amino acids in length can be sequenced using mass spectroscopy, coupled with a computer database that contains the characteristics of short peptides and amino acids that have been previously subjected to mass spectroscopic analyses. After sequencing, each of the sequenced fragments can then be joined together into the complete amino acid sequence for the protein being analyzed. This can be confirmed from DNA sequences, but more often the amino acid sequence is determined from the DNA sequence first, and if there are ambiguities, only then are the proteins sequenced.

SEQUENCE HOMOLOGY SEARCHES

Once the DNA, RNA, or protein sequences have been determined, sequence homology searches of databases often are one of the first procedures that are performed. This may be the primary procedure, or it can be a search of similar sequences to be used in phylogenetic, or other, analyses. When sequences were short, and the databases were relatively small, sequence comparison algorithms were inefficient in their comparisons. However, the concepts were similar. That is, the algorithms sought an alignment of a query sequence (the sequence entered by the user) to a set of sequences in a database (most often the National Center for Biotechnology Information GenBank database or the European Molecular Biology Laboratory database). Alignment scores are calculated by applying a model of base pair or amino acid change probabilities, in conjunction with sliding parts of query sequence along the sequences in the database. The best scores, based on statistical analysis of the sequence changes, as well as the lengths of the regions of highest similarity, are returned to be used in a tabular form. The basic information about the taxonomic identifiers of the sequence is included in the table. Basic Local Alignment Search Tool (BLAST) is a set of algorithms that are most often used in homology searches (Figure 21.10). BLASTn searches for the best similarities among DNA (or RNA) sequences in the database. BLASTx first translates the nucleic acid sequence in all three reading frames (for both strands) into an amino acid sequence, and then searches for the best matches among the protein database. BLASTp uses an amino acid sequence entered by the user to find the best matches among the protein sequences in the database. For large query sequences, such as searches using genomic sequences, megablast is used. This compares nucleotide sequences, by first grouping them and then searching for regions that are well aligned. It then focuses on the regions that have the best alignments.

All of the BLAST algorithms perform their searches in a similar manner. They begin by examining the first several nucleotides or amino acids (depending on the query sequence).

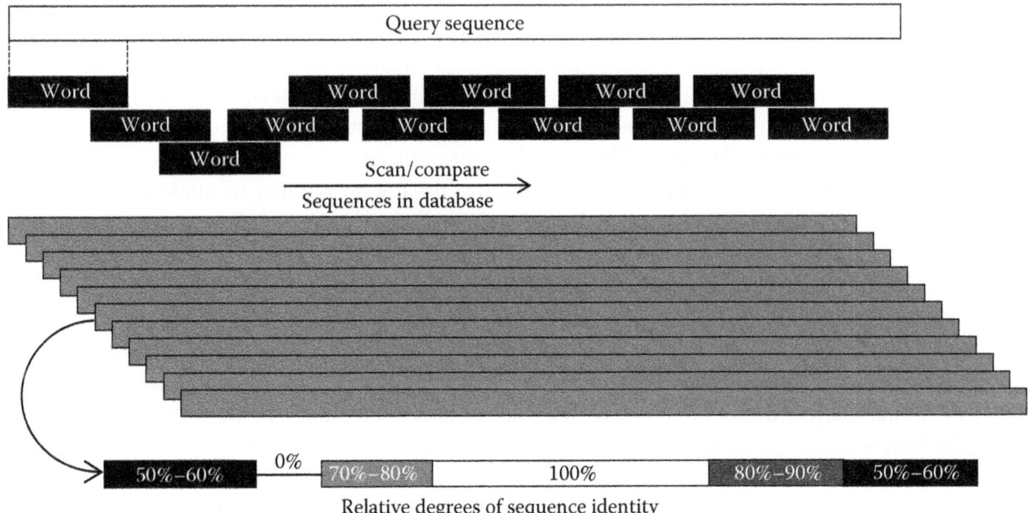

FIGURE 21.10 General protocol for BLAST sequence homology searches. The search begins by comparing a portion of the query sequence (the sequence that the user submits), called a *word*, which is usually set to 11 letters. It compares this word with those of other sequences in the database. It will move along both sequences calculating the number of bases that match. Eventually, it will return sequences from the database that have significant regions of homology to the query sequence and will provide a map of the regions of high and low sequence identity.

This is called a *word*, and usually, the default is set to 11 nucleotides or amino acids, but the length can be changed, if necessary (Figure 21.10). This is compared with the sequences in the database, and calculations are made based on the number of contiguous matching nucleotides or amino acids. When a match is found that is within the bounds of other parameters, the search is moved by one or more nucleotides or amino acids to the next word. This process continues, and sequences within the bounds of the chosen parameters are returned as *hits*, or possible matches with the query sequence. The parameters include percentage identity (calculation of the percentage of identical nucleotides or amino acids), percentage similarity (calculation of the identical, as well as similar nucleotides or amino acids—purines, pyrimidines, polar amino acids, etc.), and e-value. The e-value is the calculation of the probability of the sequence not being like the query sequence. Therefore, an e-value of 10^0 means that the probability of the sequence not matching the query sequence is 1, or 100%. An e-value of 10^{-6} means that the probability of the sequence not matching the query sequence is about 1 in a million. This indicates that there is good agreement between the query sequence and the sequence selected from the database. The user can preselect cutoff values for the computer to return. For conserved DNA and RNA sequences, it is common to select e-values of 10^{-6}–10^{-10}. For less conserved regions, and for protein sequences, often e-values of 10^{-2}–10^{-6} are chosen. Similarly, cutoff values for similarity of conserved nucleic acid sequences are generally in the 95%–98% range, whereas cutoff values for proteins are often in the 30%–70% range.

ALIGNING SEQUENCES

After the sequences have been determined and/or selected by homology searches, often they need to be aligned for further analyses (e.g., phylogenetics). This is necessary so that identical nucleotides or amino acids can be determined, and those that have changed can eventually be quantified. When all of the sequences are identical, or nearly so, alignment is simple. On the other extreme, if all of the sequences are different at every site, then alignment is impossible. Therefore, the sequences must have regions where they are either highly similar or identical to other sequences in the dataset. However, they must differ somewhat from other sequences in the dataset in order to determine the distance between each pair of sequences. There are many possible alignments for each pair of sequences, and some can lead to very different phylogenetic tree arrangements. There is an attempt when aligning sequences to find the alignment that has the fewest number of nucleotide or amino acid differences. Alignment algorithms, such as cluster analysis of pairwise alignments (CLUSTAL) and multiple alignments using fast Fourier transformation (MAFFT) are two common computing packages. CLUSTAL first performs pair-wise alignments for each pair of sequences. Then, it builds a phylogenetic tree (called a guide tree), and then performs alignments of all of the sequences together using the guide tree to decide the alignments for each of the nucleotides. Several versions of CLUSTAL are available. MAFFT calculates regions of high homology by searching for identical and then similar nucleotides or amino acids. Two algorithms are included in the software package, a faster, but less accurate algorithm, and a slower, but more accurate algorithm.

For most alignment algorithms, identical nucleotides (or amino acids) receive the highest scores, followed by similar nucleotides (transitions) or amino acids (charged versus uncharged), less similar nucleic acids (transversions) or amino acids (polar versus nonpolar), and finally gaps are needed in order to accommodate alignments that are separated by one of more nucleotides compared to the other sequence being aligned (Figures 21.11 and 21.12). Although the alignments are performed by computers, in most cases some manual adjustments are needed after the computational alignments are completed. Accurate alignment is crucial to any phylogenetic analysis, especially for analyses in which few differences are present in the dataset. Because all of the phylogenetic calculations are based on the alignment, any inaccurately aligned sections will cause inaccuracies in the resulting phylogeny.

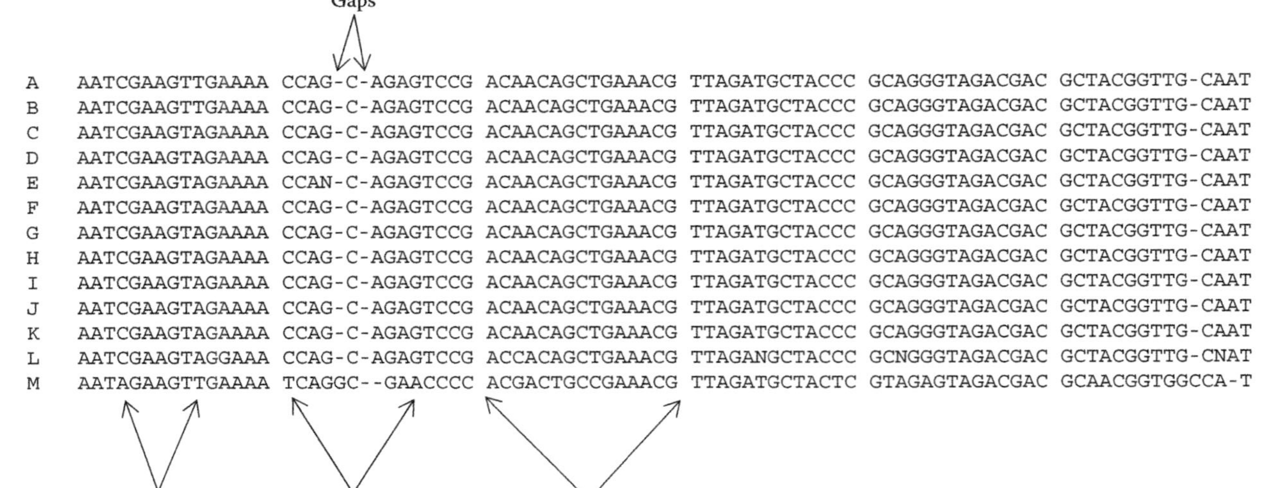

FIGURE 21.11 Example of a DNA sequence alignment. In this case, there are 13 sequences CA-M that were aligned using CLUSTAL. The algorithm first makes a rough phylogenetic tree based on the percentage of similarity between each pair of sequences. It places a higher score on transitions than transversions, and often assesses penalty scores for gaps. In doing so, it produces the alignment of highest probability for that set of sequences.

```
A    MALHQFDYLFMIFAFLDAWNIGANDVANSFSSVSSRSLKYWQAMILAAIMEFLGAVLVGSRVSDTIRNKIV
B    ........I.....A....------........................Q.........A...........
D    .V............A..................I..........V....Q.....................
E    .V......I..........------........I...............Q.........A..........I
F    .V......I..L.......---...........I..........V..G.Q.....A.......G....N.I
G    .V......I..L......F...........A..I..........V..GLQ.....A...A...G....N.I
H    .V......I..L..A...F..........WA..IA...T.......GGL.....SA...A..AG....N.I
I    .V.......L.L..A...F..........WAT.AA...T.......GGL.....SAG..A..AG....N..
J    .V.......L.L..A.....---......WAT.AA...T.......GGL.....S.G..A..AG...T...
K    .V.......L....A..............WAT.AA..VT.L.....GS....A.S.G..A..A....T.V.
L    .........L.R..A...Q..........WA...A...T.L...------..A.S.G....SA..Q...VV
```

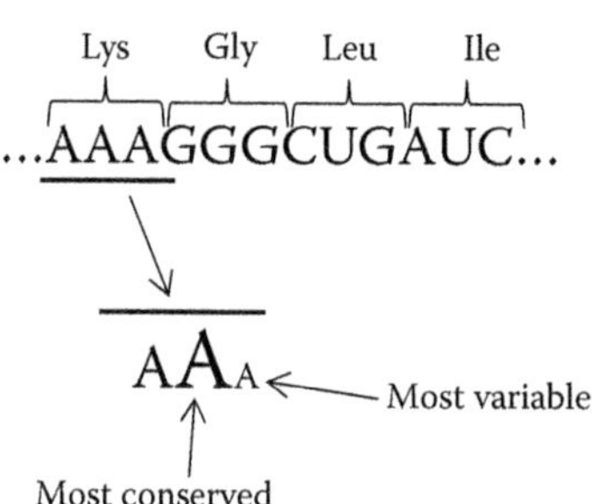

FIGURE 21.12 Example of an amino acid sequence alignment. A matrix of the 12 amino acid sequences (A–L) indicates each of the amino acids, except where they differ from the reference sequences at the top. In those cases, a dot indicates that they have the identical amino acids in those positions. Adjacent amino acids, as well as amino acid changes, are often determined using a correlation matrix (a point accepted mutation matrix) that is based on the examination of hundreds of proteins to assess the relationships among the amino acids and the types of changes that are either common or rare in those proteins. This is used to calculate which amino acid change is the most likely when the amino acids fail to match at certain locations in the sequence. In cases in which the DNA or RNA sequence has been used to determine the amino acid sequence, the first, second, and third nucleotides of the codon can be differentially weighted in sequence and phylogenetic analyses.

KEY POINTS

1. Sequencing of biological molecules was first performed on RNAs and proteins.
2. The first DNA sequencing methods, published in 1977, allowed sequencing to be performed on relatively small molecules.
3. By the 1980s, sequencing was becoming routine, and large numbers of sequences were being published.
4. Automated DNA sequencing allowed large numbers of sequences to be generated in small amounts of time.
5. The initial draft of the human genome was completed in 2003, using sequencing methods developed in the 1980s and 1990s.
6. NGS methods caused an exponential growth in the number of sequences that could be determined and published.
7. Sequence homology searches allow a user to search large databases for sequences that are similar to the one(s) submitted by the user.
8. Accurate sequence alignments must be performed prior to further analyses, including performing phylogenetic analyses.

ADDITIONAL READINGS

Alberts, B., D. Bray, K. Hopkin, A. Johnson, J. Lewis, M. Raff, K. Roberts, and P. Walter. 2013. *Essential Cell Biology*, 4th ed. New York: Garland Publishing.

Claverie, J.-M. and C. Notredame. 2006. *Bioinformatics for Dummies*, 2nd ed. New York: Wiley Publishing.

DeSalle, R. and J. A. Rosenfeld. 2013. *Phylogenomics: A Primer*. New York: Garland Science.

Graur, D. and W.-H. Li. 2000. *Fundamentals of Molecular Evolution*, 2nd ed. Sunderland, MA: Sinauer Associates.

Li, W.-H. 1997. *Molecular Evolution.* Sunderland, MA: Sinauer Associates.

Li, W.-H. and D. Graur. 1991. *Fundamentals of Molecular Evolution.* Sunderland, MA: Sinauer Associates.

Maxam, A. M. and W. Gilbert 1977. A new method for sequencing DNA. *Proc. Natl. Acad. Sci. USA* 74:560–564.

Mount, D. W. 2004. *Bioinformatics: Sequence and Genome Analysis*, 2nd ed. Cold Spring Harbor, NY: Cold Spring Harbor Laboratory Press.

Sanger, F. and A. R. Coulson. 1977. A rapid method for determining sequences in DNA by primed synthesis with DNA polymerase. *J. Mol. Biol.* 94:441–448.

22 Omics
Part I

INTRODUCTION

Large amounts of sequence data are becoming available more rapidly with the advent of next-generation sequencing (NGS) methods. The first genome sequenced was that of the RNA virus MS2 in 1976 (see Figure 21.1). It consists of about 3569 nucleotides and required the work of several researchers several years to obtain the complete sequence. In 1977, the first DNA genome sequence was reported, that of ϕX174, with a genome of approximately 5386 base pairs. Again, several researchers worked for several years to determine this sequence. By the 1990s, bacterial genomes consisting of millions of base pairs and thousands of genes were being determined and published. Again, however, these sequencing projects often required the collaboration of multiple laboratories working for several years in order to completely sequence and annotate these genomes. By the year 2000, a few dozen genomes had been sequences, whereas currently, thousands of bacterial genomes are sequenced, and hundreds of eukaryotic and archaeal genomes are sequenced.

GENOMICS

Genomics is defined as the determination of the entire genomic sequence of an organism or species (Figure 22.1). It usually includes annotation (identification and labeling) of genes, or as many as can be determined. For virus genomes, this usually was done manually. However, computational methods had to be developed for larger genomes, including those of species of bacteria, archaea, and eukarya (Figure 22.2). Eukaryotic genomes required the development of additional strategies, due to their larger sizes (from tens of millions to hundreds of billions of base pairs per genome), their organizations on multiple chromosomes, their large numbers of introns, the large amounts of non-protein-encoding regions, and the organellar genomes that accompanied the nuclear genomes. Some of the smallest genomes were sequenced first: *Saccharomyces cerevisiae* (brewer yeast, 12.5 Mb), *Caenorhabditis elegans* (nematode, 100 Mb), *Drosophila melanogaster* (arthropod, 132 Mb), and *Arabidopsis thaliana* (plant, 157 Mb). Conveniently, eukaryotic genomes are subdivided into chromosomes, and although the sizes of chromosomes vary widely, each is roughly the size of some bacterial genomes. Much of the sequencing of eukaryotic genomes was initially accomplished using bacterial artificial chromosomes (BACs). Essentially, large sections of the eukaryotic genomes were cloned as BACs, which were used to transform bacteria so that large amounts of each would be produced by the transformed bacteria. Then, smaller clones were made from them and each was sequenced. Eventually, the entire sequence for each chromosome was determined, and then the entire genome was assembled. Each of these projects required enormous efforts from many laboratories.

As the human genome project was being planned in the mid-1980s, the scope of the project was beyond the molecular biological and sequencing technologies at the time. However, by the early 1990s, the technology was sufficiently developed to begin the project. However, the project was huge, and the sequencing facilities that were constructed were essentially large warehouses filled with automated DNA sequencers (all using Sanger-based sequencing methods). The project was more than an order of magnitude larger than the largest genome sequenced by that time. At more

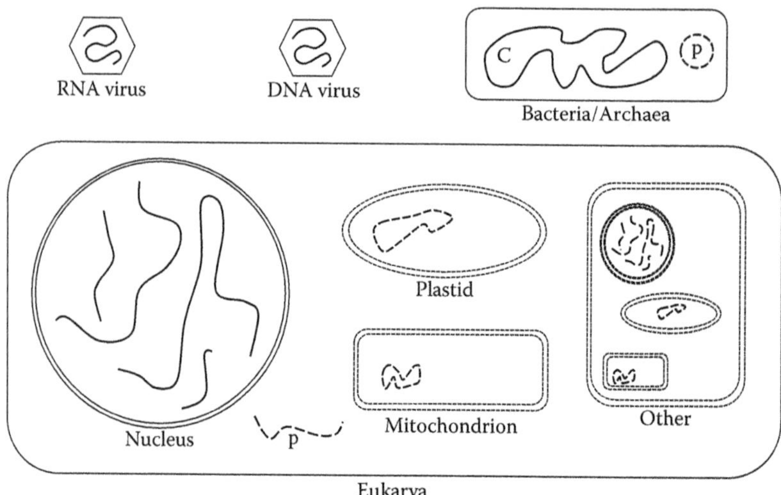

FIGURE 22.1 Locations of genomic and subgenomic genetic material in various organisms. Solid lines indicate nucleic acids, organisms, and organelles that are included in all organisms of each type. Dashed lines indicate DNA and organelles that are found in most or some of the organisms, but not all. Viruses contain either RNA (single or double stranded) or DNA (single or double stranded). The genomes of bacteria and archaea all consist of one or more double-stranded DNA chromosomes (C), and many have extrachromosomal elements, most of which are classified as plasmids (p). Eukarya all contain genomes consisting of one or more chromosomes. Most also contain mitochondria, with a mitochondrial genome. Often, mitochondria are polyploid. A large number contain plastids, with plastid genomes. Often, chloroplasts are polyploid. Members of the Excavata, Chromalveolata, and Rhizaria often have additional organelles that originated from either red or green algal ancestors. Each of the organelles originally contained nuclear, mitochondrial, and plastid genomes; but all have been reduced in size. Some eukaryotes also carry plasmids.

than 3 Gb in total length with 23 chromosomes, it represented an almost impossible task when it was begun. For the National Institutes of Health (NIH) sequencing project, led by Francis Collins, standard methods for the time were used, that is, the use of BACs to selectively clone fragments of each chromosome. The locations of each of the large clones was known, which would aid in later assembly of each of the chromosomes, and eventually the entire genome. Each of the clones was sheared into smaller pieces, cloned, and sequenced. The pieces were assembled *in silico* (by computer) by overlapping sequences that were identical (Figure 22.3). One of the main researchers, Craig Venter, thought that there was a better way to accomplish this, and thus early in the project, he split from the others on the project and began a company to perform the sequencing using his protocol. It consisted of shearing the entire genome into small pieces, cloning and sequencing all of the small pieces, and then reassembling the genome *in silico* (i.e., using computers) by linking up fragments by their overlapping ends. Therefore, he avoided the time-consuming first steps of initial cloning and mapping of large fragments from each of the chromosomes. The computer compared the ends of each of the sequences, and it would link up the sequences if they had identical overlapping sequences. At first, some thought that it was an impossible task, but it worked, and it allowed more rapid sequencing and assembly than did the BAC-based methods. The Venter group began sequencing a year after the NIH group started, and the two groups completed their initial drafts of the human genome at roughly the same time.

The sequencing and partial annotation of the human genome was more or less completed by 2003, to coincide with the 50th anniversary of the publication of the 1953 Watson and Crick paper reporting the structure of DNA. Because of the nature of animal genomes (as well as some other eukaryotic genomes), work is still proceeding on the human genome. For example, based

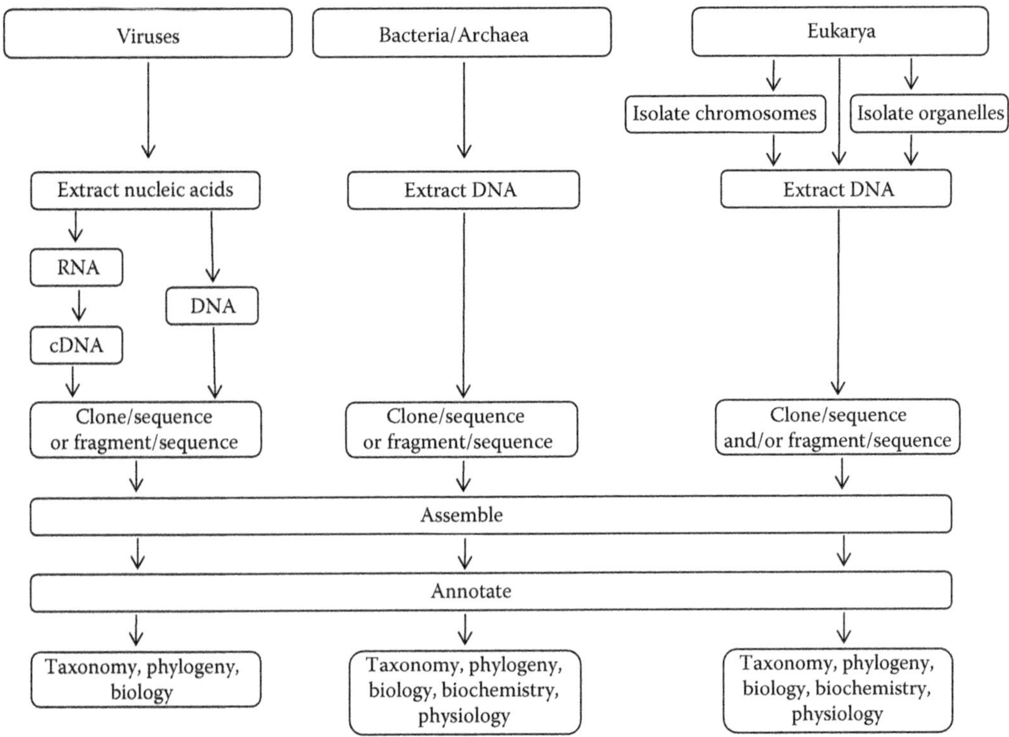

FIGURE 22.2 General scheme for genomic sequencing projects. For RNA viruses, the RNA is extracted, followed by synthesis of cDNA, using reverse transcriptase and DNA polymerase. The resulting cDNA is subjected to cloning, followed by Sanger sequencing; or fragmentation, followed by NGS. The sequences are then assembled and annotated. For most such projects, taxonomy, phylogeny, and some information of the biology of the viruses can be deduced from the sequence information. For DNA viruses, synthesis of cDNA is unnecessary. Otherwise, the steps in determination of the genome are the same as for RNA viruses. For bacteria and archaea, the process is similar to virus genomic sequence determination, but the genome sizes are usually much larger. Although all of the early bacterial and archaeal genomic sequencing project relied on cloning in plasmids, phage, and cosmids, during the past decade, NGS methods have been preferred because of the speed at which the sequences can be determined. Using NGS methods, the computer algorithms are also able to determine the linkage groups that are present, and therefore can determine whether there is more than one chromosome, and whether extrachromosomal elements and plasmids are present. Most eukaryotic genomes are larger and more complex than those for bacteria and archaea. For the first dozen genomes that were determined, each chromosome was isolated and sequenced separately, by cloning overlapping regions of the chromosomes. Later, the chromosomes were separated, but they were fragmented and sequenced using NGS methods. Most recent genome projects simply fragment the genomic DNA, sequence using NGS methods, and join the sequences together *in silico*. The main linkage groups are the chromosomes, whereas the smaller ones generally are the organelles and plasmids.

on some of the initial sequencing results, the first estimates of the number of genes in the human genome were 30,000–40,000. However, the current estimates center around 23,000 genes, based on the extensive study of the completed genome. Only about 1.5%–2.0% of the human genome consists of protein-encoding genes. These genes also often have many introns, and some of the introns are extremely long (some are tens of kilobases long, and a few are much longer). Because of this, annotation of the genome has been painstakingly slow for some regions of the genome, even though many researchers are working on this process. Depending on the research group, Internet database site, and methods of analyses, estimates for the total number of genes in the human genome range from just over 20,000 to more than 27,000. Researchers are also carefully

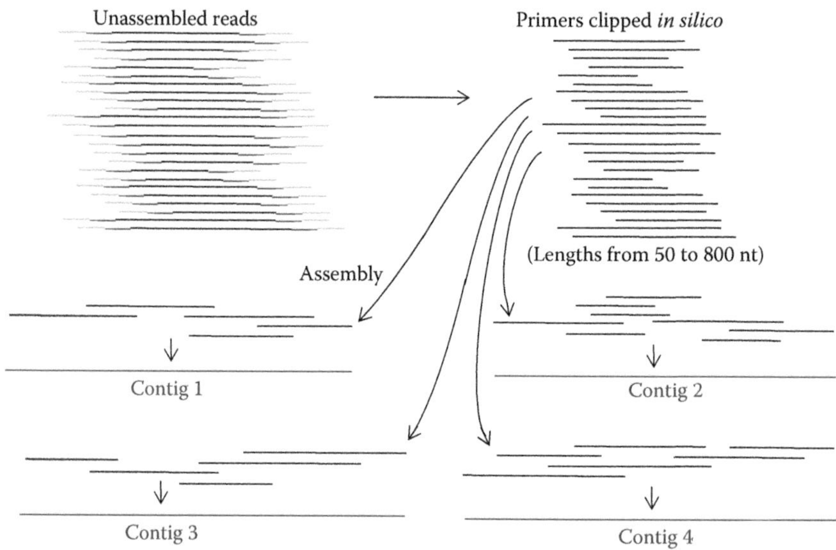

FIGURE 22.3 General process of fragment assembly. The reads usually contain terminal sequence specific to the amplification and sequencing processes, whereas the central portion of the reads contains the sequences from the genome being determined. In the first part of the process, the terminal sequences are clipped from the central portion *in silico* (by the computer algorithm), based on the known terminal sequences that were added during the NGS process. Then, the computer algorithm joins together reads that have overlapping regions of homology, indicating that they are to be joined to form longer fragments, called *contigs*, for contiguous sequences. These may be joined later into longer contigs, or they may be determined to be complete linkage groups (i.e., chromosomes).

examining the other 98% of the human genome, which consist of long and short repetitive elements, some of which are transposons, segments of virus genomes, and a number of genes that produce small RNAs, which are important in controlling gene expression. Specific genomes from viruses to animals are covered in Chapters 27 through 34.

TRANSCRIPTOMICS

Each species is characterized by its genome. This is primarily defined as the set of genes that are expressed. However, each organism is only expressing a portion of its genome at any point in time. They do not continuously express every gene in the genome. Transcriptomics is a field of science that examines the RNAs that are being expressed at different times by the organism, as well as in various cells and various circumstances. For example, the human transcriptome in a skin epithelial cell would differ significantly from the transcriptome of a T lymphocyte. The patterns of gene expression change with the age of the cell, tissue, and organism, as well. Many transcriptomes have been determined in many different types of cells, including human cells. One characteristic of almost all cell types is that there is a set of genes, comprising about 10%–30% of the transcriptomes (depending on the cell type and organism) that are found expressed constitutively. These are the so-called *housekeeping genes*, whose products comprise the basic elements of central metabolism, such as glycolysis, amino acid metabolism, replication, transcription, translation, and lipid synthesis. Another large set of genes is expressed in similar cell types. For example, they are expressed in most epithelial cells, whether they are in the skin, gut, lungs, and so on. Then, there are sets of RNAs that encode proteins that are specific to one or a few cell types. These encode proteins that are characteristic for certain types of cells.

For example, B lymphocytes express specific immunoglobulins in high concentrations, which are destined to be released into bodily fluid to inactivate, tag, and eliminate antigens.

Transcriptomes can consist of total RNA, or only messenger RNAs (mRNAs) (Figure 22.4). In the former case, total RNA is carefully extracted from the sample material, and then reverse transcribed using random DNA primers to produce complementary DNA (cDNA) copies of the RNA. To select only the mRNAs, the RNA is first sent through a column that has poly-deoxyribose-thymidine attached to beads in the column matrix. mRNAs will bind to the column because of their 3′ poly-A tails. Once all of the other RNAs are washed through

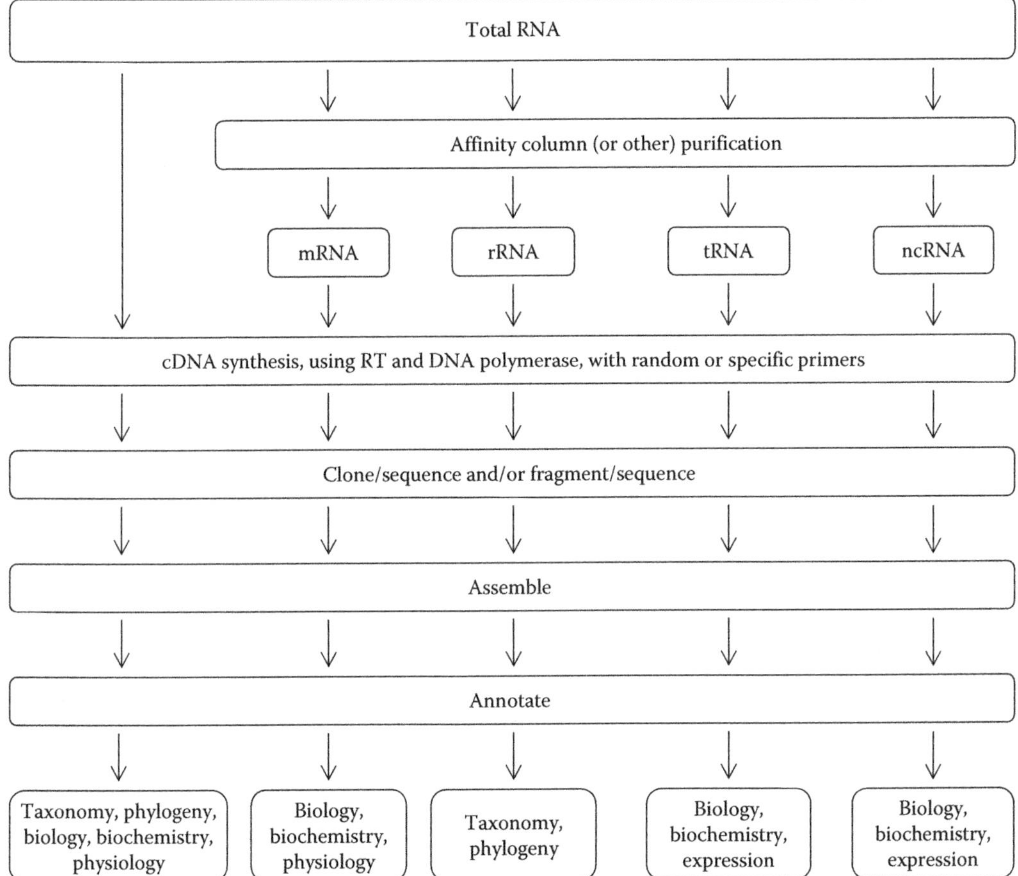

FIGURE 22.4 General scheme for transcriptomic sequencing projects. Initially, total RNA is extracted. This can be used to synthesize cDNA using random hexanucleotide primers, or it can be fractionated using affinity columns, gels, or other methods into mRNA, rRNA, tRNA, and other ncRNA portions. Usually, to purify mRNA, an oligo-dT column is used, which causes mRNAs with poly-A tails to be retained in the column. Subsequently, a salt solution is used to detach the purified mRNA from the column. Similarly, rRNA can be isolated using columns with DNAs complementary to the rRNAs, or they can be purified on gels or by CsCl isopycnic gradient ultracentrifugation, followed by fractionation. The tRNAs and other ncRNAs can be separated by gel electrophoresis or by affinity column methods. Each of the RNAs can be used to synthesize cDNA, followed by either cloning and sequencing, or fragmentation and NGS. Following assembly and annotation, each of the fractions can convey various types of information. Total RNA sequences can be used for taxonomy and phylogenetics (because 80% of the RNA is rRNA), as well as to determine the biology, biochemistry, and physiology of the organism (by analysis of the mRNA sequences). The mRNA, rRNA, tRNA, and other ncRNA sequences can provide various types of information about the organism (as indicated at the bottom). RT, reverse transcriptase.

the column, the mRNA is released from the column using a salt solution. The mRNA can then be used as a template to create cDNA copies using reverse transcriptase, in conjunction with random primers, or an oligo-dT primer, as well as random primers. The cDNAs can either be cloned as a library, and then sequenced, or be directly sequenced using NGS methods. However, in many cases, additional ends must be ligated or attached using polymerase chain reaction (PCR) methods in order to modify the ends so that they are compatible with the primer or tag sequences of the NGS method and software. Because ribosomal RNA (rRNA) comprises about 80% of the total RNA, for total RNA sequencing projects, usually 80% of the sequences are rRNA sequences, and only about 5% are mRNA sequences. The remainder are transfer RNA (tRNA) and noncoding (ncRNA) sequences. For sequencing projects in which mRNA has been purified, roughly 90% of the sequences are those from mRNAs. One of the major uses in human medicine is the study of human cancers, whose transcriptome differs greatly from the transcriptomes of normal tissues of the same type.

METAGENOMICS/METATRANSCRIPTOMICS

Genomic and transcriptomic methods have been applied to ecological and environmental scientific questions aided by NGS methods. These applications are called metagenomics and metatranscriptomics (Figure 22.5). Both involve collection a sample from the environment, extracting the nucleic acids, and then sequencing all of the nucleic acid fragments. These fragments originate from all of the organisms that are present at (or near) the sample site. Metagenomics can determine not only what genes are present in the sample, but it can often lead to a determination of what taxa are present in the sample. This can lead to a scientific reconstruction of the community of organisms at a particular environmental site. Metagenomic studies have been carried out worldwide from hot deserts to frozen deserts, and from oceans, lakes, streams, underground caves and mines, as well as hundreds of other sites. Each of the sites examined to date contains unique assemblages of DNA, and presumably of species. Significantly, usually more than half of the sequences are unlike any sequences that have been deposited in the National Center for Biotechnology Information and European Molecular Biology Laboratory databases. A metagenomic investigation was carried out on the dust in the International Space Station (ISS), which found a large variety of microbes, some of which can be pathogenic on humans. These studies will be important in the planning of long-range space missions, in order to consider the potential for infection of personnel on those missions.

Metatranscriptomics can answer two basic questions: What are the biochemical pathways that are active at the sample site, and what organisms are living at that site? Because RNA degrades rapidly after cells die, most of the RNA in a sample comes from organisms that were living when the sample was collected. Therefore, the metatranscriptome can indicate the metabolic processes that were ongoing at the time. These methods have been successfully used to study biological processes in oceans, in within and on the surface of humans, in subglacial lakes, in glaciers, in deep mines, in caves, in the atmosphere, in the ISS, and in many other environmental sites. A metagenomic/metatranscriptomic study of ice from subglacial Lake Vostok in Antarctica indicated that a complex ecosystem exists in this lake, which has been buried by kilometers of glacial ice for over 14 million years (Figure 22.6). An analogous study of the sediment at the bottom of another subglacial lake, Lake Whillans, came to similar conclusions. Therefore, metagenomic and metatransctiptomic studies can provide strong evidence for the complexity of life in many environments, including those that are extreme.

MICROBIOMICS

During the past decade, other types of studies have been undertaken, which have yielded very interesting results. These are studies of the microbiomes. These are essentially metagenomic and metatranscriptomic investigations to characterize the consortia of microorganisms that are living together

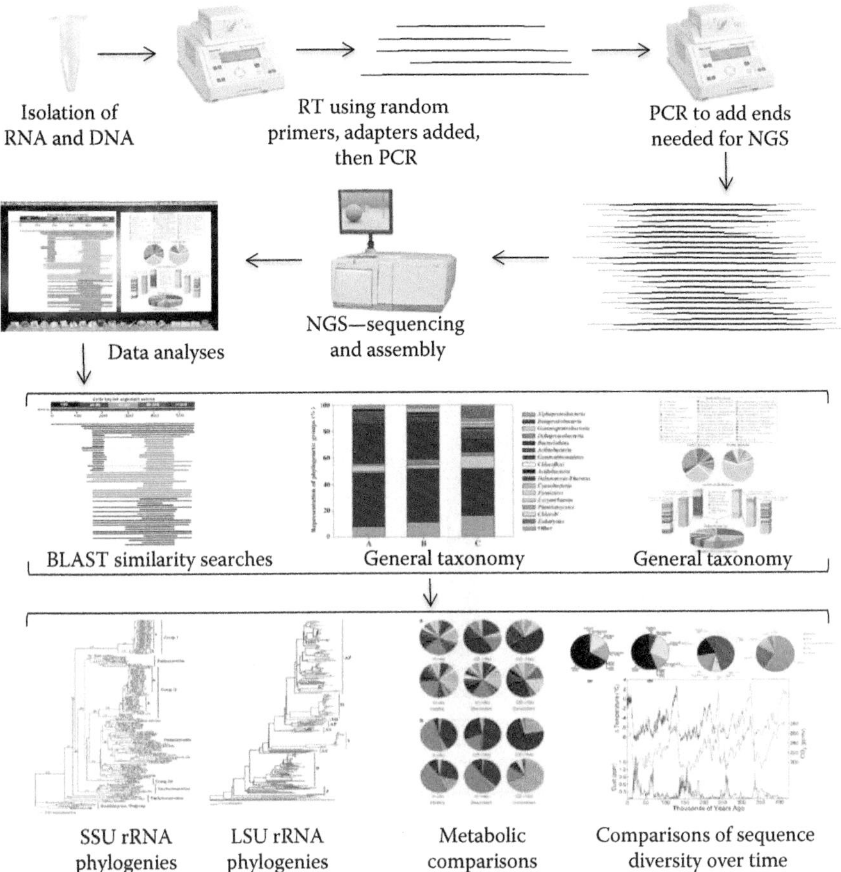

FIGURE 22.5 Overall process for a metagenomic and/or metatranscriptomic project. After careful collection of an environmental (or other) sample, the DNA and/or RNA are carefully and aseptically extracted. For metatranscriptomics, the RNA is used as a template for cDNA synthesis using random hexanucleotide primers. The DNA and/or cDNA must then be amplified, and therefore, terminal ends of known sequence which are complementary to the primers being used for PCR (red regions) are often attached using DNA ligase. The fragments are amplified, and additional sequences are added to the termini (either by incorporating them into PCR primers or by ligating them to the amplified fragments). These additional sequences are need for the specific NGS protocols (see Chapter 21; Figures 21.6 through 21.8). The sequences are determined and assembled using NGS systems, and then the assembled sequences are analyzed. RT, reverse transcriptase.

in specific environments. Microbiomes are studied most often in their relation to various niches on other organisms, including humans (Figure 22.7). For example, many studies have been completed on microbiomes from various parts of the gut from hundreds of healthy people, as well as those who had diseases. From these studies, it has been found that the microbial communities in healthy people of normal weight differ significantly from those who are obese. Also, those with colitis, colon cancers, and other diseases of the gut have very different combinations of organisms inhabiting their intestines. An important finding in the studies of gut microbiota is that the microbes not only aid in digesting the food that we eat, but produce valuable nutrients, including amino acids, important lipids, and cofactors needed for various aspects of our metabolism. Therefore, the gut is not just a place where food is broken down, and then the parts are absorbed for use, but a bioreactor where parts of the food are used by the microbes to produce vital products that are then absorbed and used.

Microbiomes of other parts of the human body have been published. Microbiomes of many parts of the human skin indicate that each portion of the human body has a different microbiome,

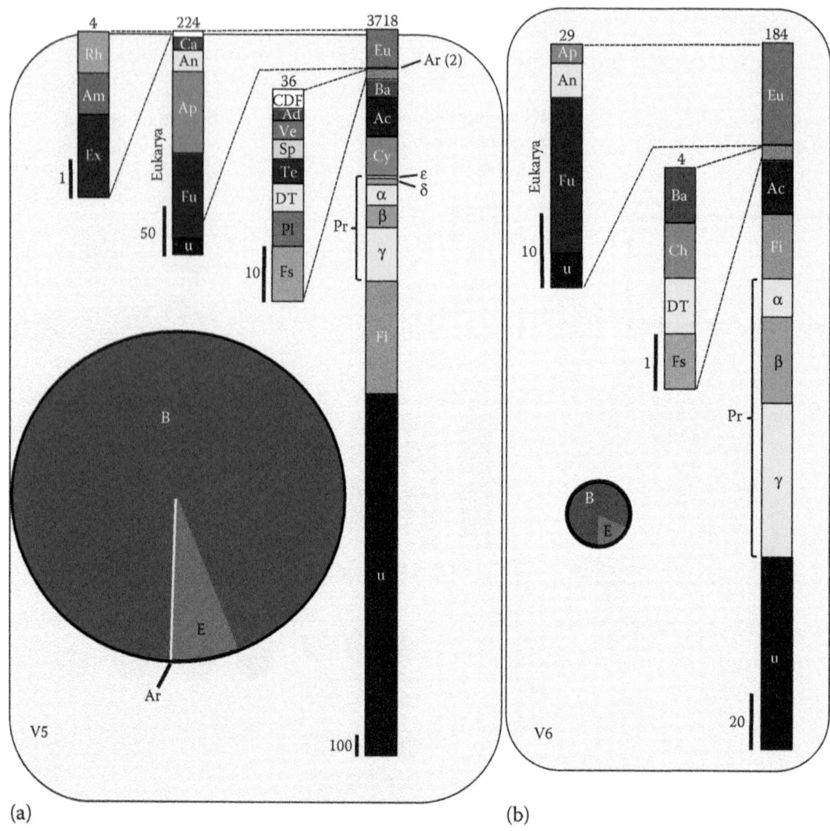

FIGURE 22.6 An example of the taxonomic analysis for two samples. Two ice core sections were obtained from the Lake Vostok (Antarctica) ice core. These two depths represent ice that originated as water from the surface of Lake Vostok, which is a subglacial lake that lies beneath 3.7 km of glacial ice, and has been buried by ice continuously for 14 million years. Although the concentration of microbes is very low, the diversity of organisms in the samples is moderately high. The numbers indicate the number of species identified in each category. V5 (a) is one of the ice samples, whereas V6 (b) is the other. The pie charts indicate the relative numbers of sequences from bacteria, archaea, and eukarya in each of the two samples. Ac, Actinobacteria; Ad, Acidobacteria; Am, Amoebozoa; An, Animalia; Ap, Archaeplastida; Ar, Archaea; α, Alphaproteobacteria; Ar, Archaea; Ba, Bacterioidetes; β, Betaproteobacteria; Ca, Chromalveolata; CDF, Chlorobi/Deferribacteres/Fibrobacteres; Ch, Chloroflexi; Cy, Cyanobacteria; δ, Deltaproteobacteria; DT, Deinococcus/Thermus; ε, Epsilonproteobacteria; Ex, Excavata; Fi, Firmicutes; Fs, Fusobacteria; Fu, Fungi; γ, Gammaproteobacteria; Pl, Planctomyces; Pr, Proteobacteria; Rh, Rhizaria; Sp, Spirochaetes; Te, Tenericutes; u, uncultured/unidentified; Ve, Verrucomicrobia.

and again, these are for healthy individuals. The microbiome on the foot differs from that on the upper arm, that on the face, and so on. Similarly, the mouth, throat, larynx, and so on all have different microbiomes. When individuals are compared, their microbiomes differ with respect to the same body region. However, when the microbiomes are analyzed in a different way, they are amazingly similar. When studies of gut microbiomes were evaluated as to what metabolic functions each of the microbial species was performing, it was found that the overall biochemistry that was occurring in the sample was nearly identical, regardless of the species that were present. Vaginal microbiomes have been studied, primarily to investigate infection patterns, menstrual cycles, and fertility. From those studies, it was clear that the vaginal microbiomes changed during the menstrual cycles of all of the individuals. However, they changed differently among the individuals,

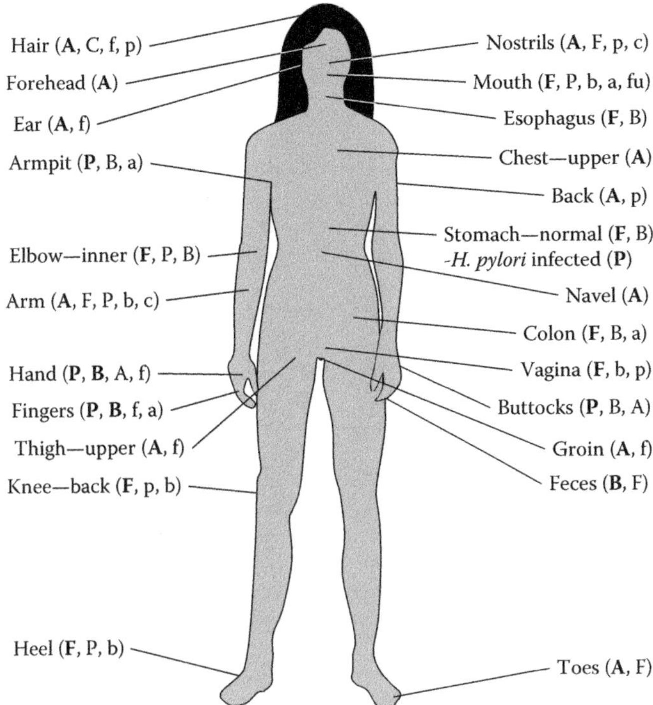

FIGURE 22.7 Summary of human microbiome results. Samples from many different locations on humans have been subjected to NGS methods to determine the types of microorganisms that inhabit those locations. This is a brief summary of some of those results, showing the normal microbial communities that have been found on and in humans. One instance of a pathogen (stomach infected by *Helicobacter pylori*). Although Actinobacteria species predominate on the outer surfaces of humans, Firmicutes are most prevalent within humans. A, Actinobacteria; B, Bacteroidetes; C, Cyanobacteria; F, Firmicutes; Fu, Fusobacteria; P, Proteobacteria. Capital letters in bold type indicate the phylum in the highest concentration (usually >50% of the total); capital letters indicate phyla that occur in high concentrations (usually 15%–30% of the total); lowercase letters indicate phyla that are minor components in the sample (<15% of the total). Other phyla are also present, but were excluded if their amounts were <5% of the total.

such that no firm conclusions could be made to predict what was a normal microbiome versus an abnormal microbiome.

KEY POINTS

1. Genomics is the process of sequencing, assembly, and annotation of complete genomes.
2. Virus genomes were completed with only partial use of computers, whereas the genomic projects with cellular organisms rely heavily on computational methods.
3. Transcriptomics is the study of the expression of genes as the RNAs in order to determine the expression profiles for normal as well as abnormal cells.
4. Transcriptomics has been used extensively in studies of expression patterns in cancer cells.
5. Metagenomic and metatranscriptomic methods sequence all of the DNA or RNA, respectively, of an environmental (or other) sample to determine the community of organisms and metabolic processes in the sample.
6. Microbiomics is the study of microbiomes from various environments, including those found in the human gut and on human skin.

ADDITIONAL READINGS

Alberts, B., D. Bray, K. Hopkin, A. Johnson, J. Lewis, M. Raff, K. Roberts, and P. Walter. 2013. *Essential Cell Biology*, 4th ed. New York: Garland Publishing.

Blackburn, G. M. and M. J. Gait (eds.). 1996. *Nucleic Acids in Chemistry and Biology*. New York: Oxford University Press.

Claverie, J.-M. and C. Notredame. 2006. *Bioinformatics for Dummies*, 2nd ed. New York: Wiley Publishing.

DeSalle, R. and J. A. Rosenfeld. 2013. *Phylogenomics: A Primer*. New York: Garland Science.

D'souza, N. A., Y. Kawarasaki, J. D. Gantz, R. E. Lee, Jr., B. F. N. Beall, Y. M. Shtarkman, Z. A. Koçer, S. O. Rogers, H. Wildschutte, G. S. Bullerjahn, and R. M. L. McKay. 2013. Microbes promote ice formation in large lakes. *ISME J.* 7:1632–1640.

Graur, D. and W.-H. Li. 2000. *Fundamentals of Molecular Evolution*, 2nd ed. Sunderland, MA: Sinauer Associates.

Knowlton, C., R. Veerapaneni, T. D'Elia, and S. O. Rogers. 2013. Microbial analysis of ancient ice core sections from Greenland and Antarctica. *Biology* 2:206–232.

Krebs, J. E., E. S. Goldstein, and S. T. Kirkpatrick. 2012. *Lewin's Genes XI*, 11th ed. Sudbury, MA: Jones & Bartlett Publishers.

Li, W.-H. 1997. *Molecular Evolution*. Sunderland, MA: Sinauer Associates.

Li, W.-H. and D. Graur. 1991. *Fundamentals of Molecular Evolution*. Sunderland, MA: Sinauer Associates.

Mount, D. W. 2004. *Bioinformatics: Sequence and Genome Analysis*, 2nd ed. Cold Spring Harbor, NY: Cold Spring Harbor Laboratory Press.

Rogers, S. O., Y. M. Shtarkman, Z. A. Koçer, R. Edgar, R. Veerapaneni, and T. D'Elia. 2013. Ecology of subglacial Lake Vostok (Antarctica), based on metagenomic/metatranscriptomic analyses of accretion ice. *Biology* 2:629–650.

Shtarkman, Y. M., Z. A. Koçer, R. Edgar, R. S. Veerapaneni, T. D'Elia, P. F. Morris, and S. O. Rogers. 2013. Subglacial Lake Vostok (Antarctica) accretion ice contains a diverse set of sequences from aquatic, marine and sediment-inhabiting Bacteria and Eukarya. *PLoS ONE* 8(7): e67221.

Venkateswaren, K., P. Vaishampayan, J. Cisneros, D. L. Pierson, S. O. Rogers, and J. Perry. 2014. International Space Station Environmental Microbiome—Microbial inventories of ISS filter debris. *Appl. Microbiol. Biotechnol.* 98:6453–6456.

23 Omics
Part II

INTRODUCTION

Genomic and transcriptomic studies, as well as related studies (Figure 23.1), have yielded enormous amounts of data and information regarding biological systems over the past three decades. Progress in the elucidation of the evolution of genes and genomes, including the mixing of genes and genomes, caused by horizontal gene transfers, is proceeding at ever increasing rates. Studies of the many functions of various types of RNAs have expanded greatly. However, although nucleic acids provide the genetic basis, and some of the controls of gene expression, energy transfer, and signal transduction, proteins and some of the products that they produce form the major characters of cells. Although organic chemistry and biochemistry have formed a foundation for much of the study of proteins, lipids, carbohydrates, and other organic molecules, molecular biological and bioinformatics methods have allowed more comprehensive studies to be performed.

PROTEOMICS

High-speed computers and methods to rapidly analyze proteins, their cellular locations, and their functions, sometimes a single molecule at a time, have contributed to large-scale studies of proteins, known as the field of proteomics (Figures 23.2 and 23.3).

Proteins can be extracted and purified using a number of methods, although all methods are not amenable to all proteins. This is because some proteins are in the aqueous parts of the cell, called the soluble proteins, although others are bound to membranes. Usually, the extraction buffer contains a detergent, such as sodium dodecylsulfate (SDS) to solubilize the membranes and the membrane-bound proteins. Following extraction from the cells, the proteins can be further purified by column chromatography, high pressure liquid chromatography, gel electrophoresis, ammonium sulfate precipitation, or other methods. For electrophoresis, one-dimensional gels are used to initially characterize total proteins in a cell. Two general types of gel conditions are used: (1) native conditions, in which the proteins may assume more of their native conformations, or (2) denaturing conditions, using SDS, which stabilizes charges and relaxes the proteins, which causes the proteins to migrate primarily according to their molecular masses. In native gels, the proteins may not necessarily migrate through the gels according to their molecular weight and charges. Secondary structure may cause them to move faster or slower relative to other proteins of similar molecular weights. For SDS gels, the proteins all migrate primarily to their molecular weights, and therefore, specific proteins will migrate to the same extent in each experiment.

Isoelectric focus (IEF) gels are useful in determining the charge on a particular protein. Gels are prepared, which have a gradient of pH levels. When proteins are loaded onto these gels and subjected to electrophoresis, they will migrate in the gel when they have a charge, but stop migrating when they have no charge. As they move through the various regions of pH, they will either gain or lose protons (actually hydronium ions, H_3O^+). When the protons equalize the total charge on the proteins, the protein will stop moving. This is called its isoelectric point, which helps to characterize proteins. Although one-dimensional gels are very useful in initial characterizations of protein preparations, two-dimensional gels can yield much more information about the proteins in a sample. For these gels, the entire sample is loaded into the well on one side of the gel. Electrophoresis is then carried out in one direction until the smallest proteins have reached

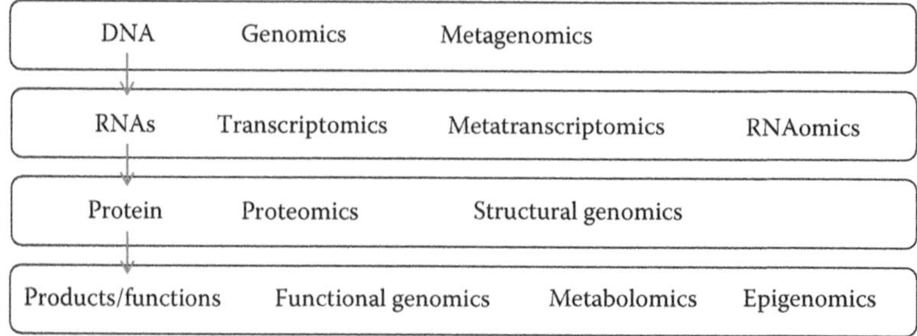

FIGURE 23.1 Categorization of the omics covered in this chapter and Chapter 22. Genomics and metagenomics are the analyses of DNA sequences from individual species or from collectives of species, respectively. Similarly, transcriptomics, metatranscriptomics, and RNAomics use RNA sequence data (usually from synthesized cDNA) to investigate the transcribed RNA in cells and tissues of a single species at one time point or several, the collective RNA in an environmental sample, or the functional structures of the RNA molecules, respectively. Proteomics is the collection and analysis of the protein contents of individual cells, tissues, organs, or other samples, and includes comparative studies to investigate changes in normal and disease states. Structural genomics analyzes the secondary, tertiary, and quaternary structures of proteins and portions of proteins in order to discern function, reactions, interactions, and changes. Functional genomics, metabolomics, epigenomics, and other related fields utilize genomic, transcriptomic, proteomic, and other data to determine the functions and interactions of the molecules, complexes, and systems.

close to the end of the gel. Then, the gel is rotated 90°, and electrophoresis is carried out in that direction. At the end of electrophoresis, the proteins are visualized by staining, or by autoradiography if they were radioactively labeled. The gel can be blotted as well, and western blotting can be carried out with a radioactively labeled antibody that is specific for an epitope of one or more proteins. Often, two-dimension gel electrophoresis is carried out on more than one sample, each of which originated from different tissues, or similar tissues under different conditions. Comparisons are then made to determine the differences in expression between the two samples.

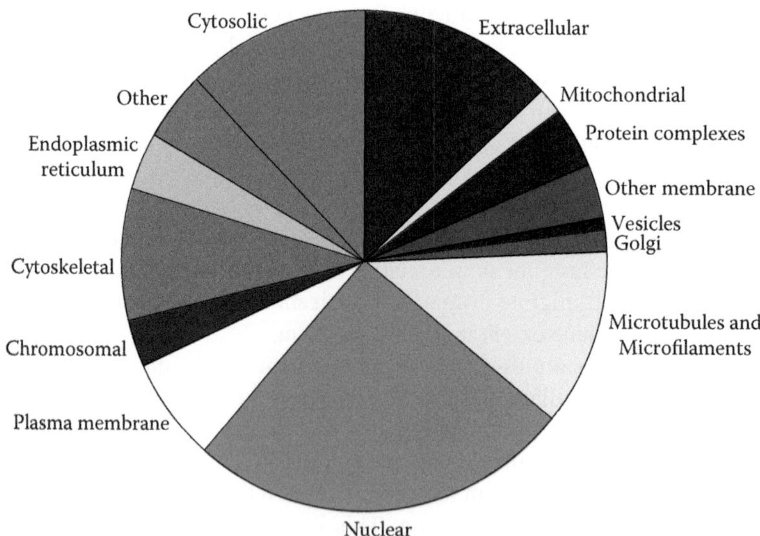

FIGURE 23.2 Example of a proteomic categorization based on the final destinations/functions of proteins in a eukaryotic cell. More than 50% of the proteins are found in the nucleus and cytosol, or are transported to the outer cell surface. Cytoskeletal elements, including microfilaments and microtubules, comprise about 20% of the proteins in a eukaryotic cell. Proteins found in the membrane are about 10% of the total.

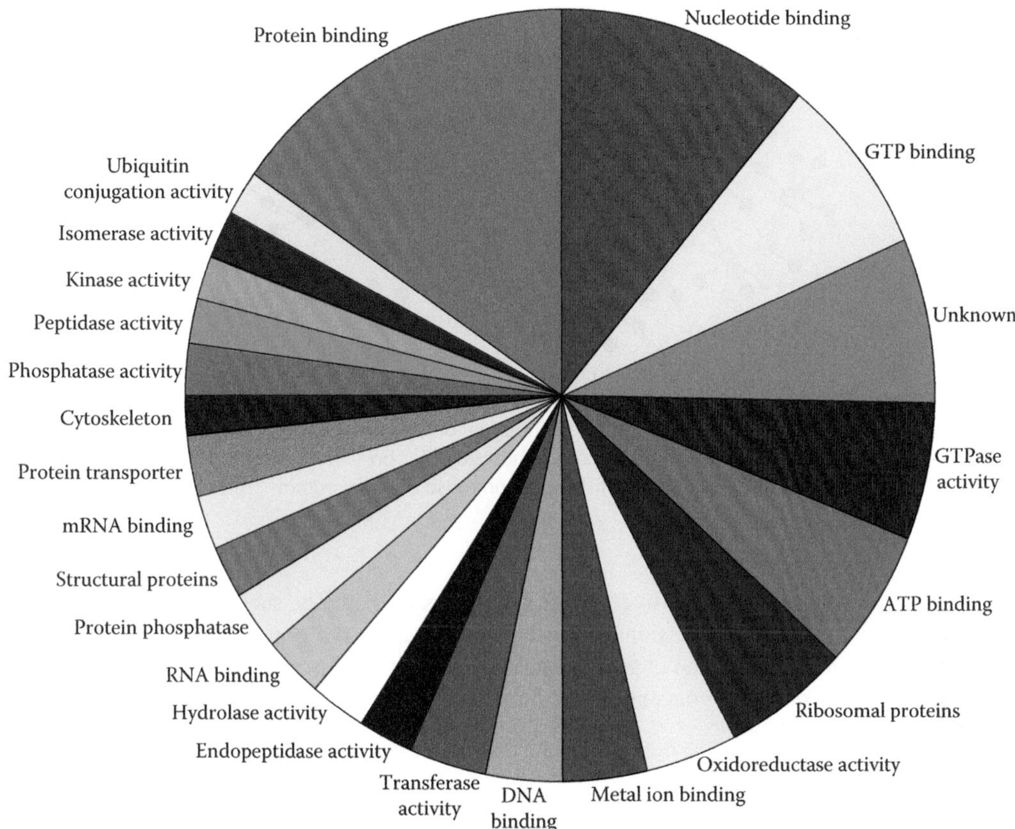

FIGURE 23.3 Example of a proteomic analysis indicating the proportions of proteins in several categories. More than 70% of the proteins are involved in binding to various other molecules in the cell. Approximately 15% are enzymes, whereas less than 10% provide structure and movement capabilities to the cell. GTP, guanosine triphosphate; GTPase, guanosine triphosphatase.

Various types of columns can be used to separate, identify, and/or purify proteins, including those that separate based on molecular mass, solubility, charge, binding to column materials, and so on. The columns can also be coupled to equipment that detects the proteins emerging from the end of the column, such as those that measure UV absorption and those that measure the mass-to-charge ratio (i.e., mass spectrometry). They can also be coupled to a fraction collector, which collects aliquots as they emerge from the column. One of the simplest methods to fractionate a protein sample is to use various concentrations of ammonium sulfate. Various proteins will precipitate from the solution under different ionic strengths of the solutions of ammonium sulfate, such that a range of proteins can be separated in initial characterizations of the proteome.

The proteome of *Homo sapiens* contains about 90,000 proteins, which are expressed from approximately 23,000 genes. The difference in number is due to extensive alternative splicing (as well as multiple promoters for some genes) in humans (and many animals), which yields several proteins from some genes. For example, the gene that encodes the protein dystrophin, found mainly in muscle cells, also produces at least six other proteins, including utrophin, which is only a bit shorter than dystorophin, and five other short proteins, called dystrobrevins. On the other hand, for most Bacteria and Archaea, one gene encodes a single protein. Therefore, for those organisms, a genome provides a good estimation of the proteome, whereas the genomes of mammals and other animals can provide only a partial estimation of the proteome. In those cases, much more protein analysis must be performed to accurately determine the proteomes. Categorization of the proteins

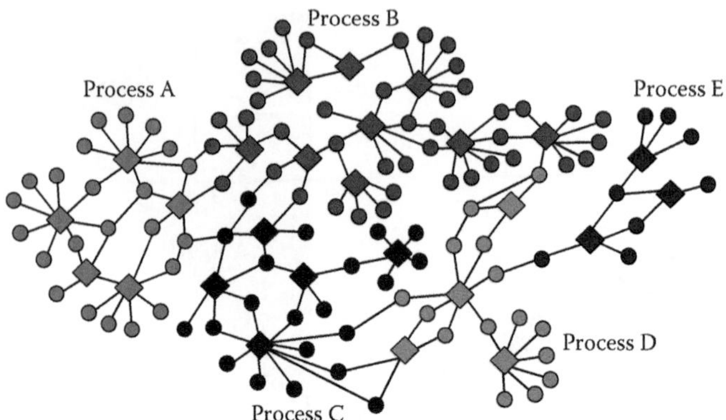

FIGURE 23.4 A generalized map of the interactions among proteins involved in five processes (A–E). Diamonds indicate transcription factors, and circles indicate other proteins. Proteins from each process usually interact with proteins from one or more of the other processes. These protein–protein interactions are vital to all cellular functions.

in the proteome is based on their functions and locations in the cells, such as DNA-binding proteins, cell cycle control proteins, signal transduction proteins, and phosphorylases. Because the proteome determines most of the characteristics of a cell, it is important to understand not only the proteins and their relative concentrations, but also their interactions and functions in the cells (Figures 23.2 through 23.4).

STRUCTURAL GENOMICS

Structural biology attempts to determine the three-dimensional structures of biological molecules. When Watson and Crick reported the structure of DNA, they were engaging in structural biology research. Others, such as Linus Pauling, had already determined that portions of proteins wind up into α-helices, and entire structures of proteins, such as lysozyme, were determined (Figure 23.5). Certain structures in proteins, called motifs, appear in many proteins, and most proteins contain

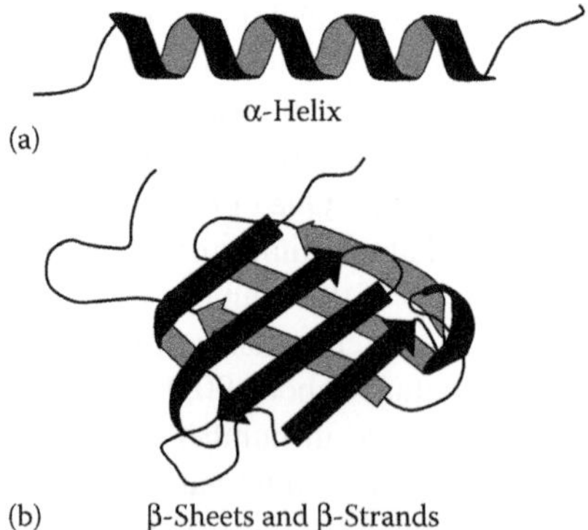

FIGURE 23.5 Two of the main basic structures found in proteins. The α-helices (a) are found in most proteins. It is a right-handed helix in which the amino group of one amino acid in the chain hydrogen bonds with the carboxyl group of four amino acids in the chain. Proteins usually contain from one to many α-helices. For membrane proteins, they often form the portions that span the membrane. The β-sheets and β-strands (b) are another major feature of proteins. They can be found forming wide flat regions of some proteins and barrel-like structures in other proteins.

motifs found in other proteins. Part of structural genomics is to search for common motifs in the amino acid sequences of proteins in order to provide estimates of the complete structure of the proteins. This information can be used to plan experiments, using methods such as X-ray crystallography and nuclear magnetic resonance), which can determine the structure of the proteins at high resolution. X-ray crystallography has been used to construct high-resolution models of small and large proteins, as well as protein complexes. This procedure was used to determine the structure of bacterial ribosomes.

Structural genomics also includes the determination of DNA and RNA structures. By comparing genomic sequence data with structural information from X-ray crystallography studies, structural conclusions can be made regarding the molecular interactions of the sequenced DNAs and RNAs. Unlike DNA, which has only a limited number of conformations, RNAs can form a large number of structures, not limited to double helices, stems, bulges, and loops, and often form long-range interactions between bases in different parts of the sequence.

RNAOMICS

A part of structural genomics is called RNAomics. Although RNAomics might seem superficially similar to transcriptomics, it seeks to use sequence and structural RNA data to determine the important structures and functions of RNA molecules (Figure 23.6).

Many RNAs are structural and/or functional, such as ribosomal RNAs (rRNAs), transfer RNAs, noncoding RNAs (ncRNAs), and introns. Their structures are dictated by the portions of the RNAs

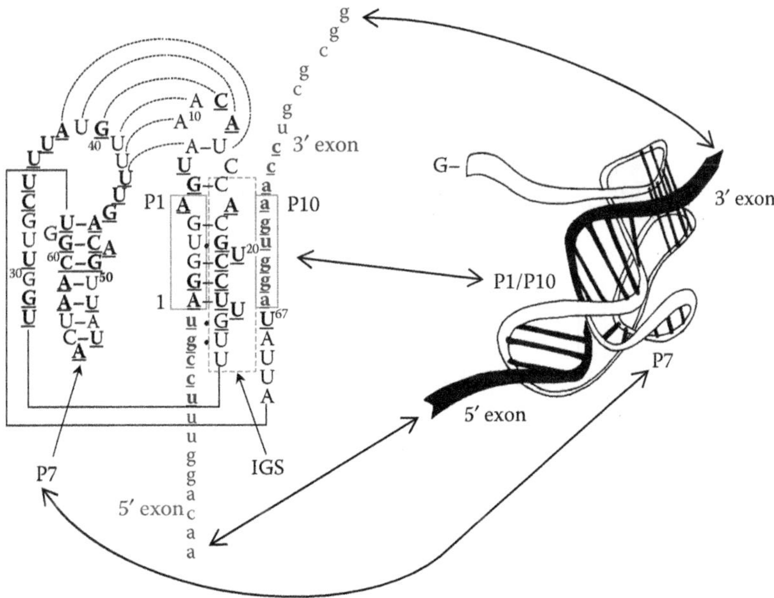

FIGURE 23.6 Structural elements of a small group I self-splicing (RNA) intron. RNAomics analyzes sequence and biochemical data to determine RNA structures and functions. The two-dimensional structure of this intron (left) indicates four regions of pairing in the molecule, P1, P10, P7, and pairing of two loops at the top (indicated by dotted lines). Dashes indicate standard Watson–Crick pairing, whereas dots indicate non-Watson–Crick base pairs. The intron is in black capital letters, whereas the 5′ and 3′ exons are in red lowercase letters. Nucleotides that are conserved among the seven such small introns that have been studied are in bold font. Numbers indicate the nucleotide position beginning with the first nucleotide (1) of the intron. The three-dimensional structure (right) is the same molecule, but it is a model of the structure, based on sequence data, three-dimensional structures of larger group I introns, and biochemical data.

that can fold and hydrogen bond to other parts of the molecules, or to adjacent molecules. Certain structures have been found in many RNAs, such as adenine platforms (consisting of two or more consecutive adenines) and GNRA (each having a G, then any nucleotide, followed by a purine and an A) loops at the ends of double-stranded stems. There are many such motifs in RNA molecules, and some of them are catalytic sites, which convert the RNA into a ribozyme (catalytic RNA). One such site is the peptidyl transferase section of the LSU rRNA molecule, which connects amino acids together to form polypeptides during translation. Scientist use structural data from a variety of species in comparative studies to try to determine the structures in the molecules that are conserved, and which have similar functions because of the conservation of the structures.

EPIGENOMICS

Epigenetics has become an important field of study in the past several decades, due to the major effects on development and tumorigenesis due to modifications of histones and nucleotide bases (usually C; Figure 23.7). Epigenetic effects occur as a result of modifications to the structure of the DNA, rather than changes in the genes themselves. These modifications cause changes in the accessibility of the molecules that control transcription of the affected genes, and therefore, depending on the gene being affected, can cause major changes in gene expression and cellular functions. Methylation, demethylation, acetylation, and deacetylation of histones are the main mechanisms of epigeneic changes. Methylation of DNA bases (especially cytosine) also affects the structure of DNA and its accessibility to the transcription machinery necessary for expression of the genes in the region. Where high amounts of 5-methylcytosine are present in DNA, Z-form DNA can result, which usually renders the DNA unrecognizable to transcription factors that normally attach to the region to initiate transcription. Epigenomics maps changes in these regions and correlates them with

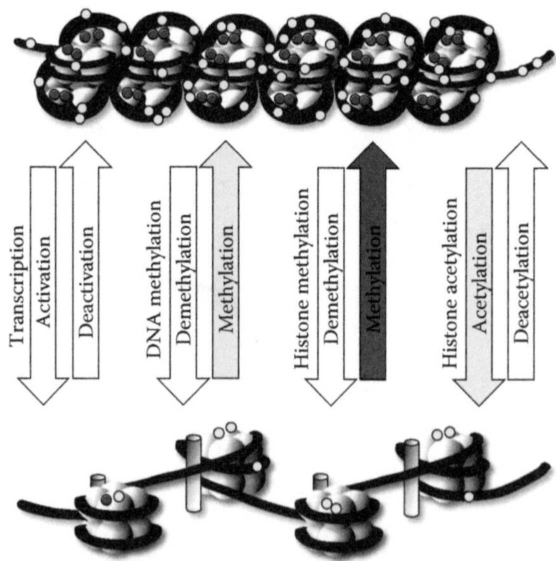

FIGURE 23.7 DNA and histone modifications that change transcriptional activity. When DNA is heavily methylated on C residues (yellow dots), its conformation changes, which usually deactivates transcription. Histone modification is also important in this process. When histone H3 is methylated (on amino acids 4 and 27, which are both lysines; red dots), it causes tightening of the chromatin structure, which excludes the transcription machinery. However, if those sites are demethylated, the DNA is demethylated, and histone H3 is acetylated (on amino acids 9 and 14, which are lysines; green dots), the chromatin structure relaxes, which allows access of the transcription machinery. These changes are the focus of epigenomic studies.

changes in gene expression. The changes and their effects are then studied to determine whether their effects are detrimental, especially in cases in which human health is affected, with hopes to be able to develop a control for the malady.

METABOLOMICS

Metabolites are the relatively small molecules that are the result of the cellular reactions and processes that occur in cells. The specific metabolites and their concentrations are characteristic of cells and tissues, and normal versus diseased cells and tissues can produce very different metabolite patterns. Metabolomics is the process of characterizing and categorizing these metabolite patterns and correlating them with specific functions in cells, tissues, organ, species, environments, and so on. They have been used extensively in medical fields, due to their sensitive detection of small changes when cells begin to change their patterns of gene expression. In some cases, early signs of disease can be detected using these metabolite signals. In environmental applications, they can be used to detect the initial signals from toxic algal blooms that might affect the organisms in a lake, and in some cases may pose a risk to human health. Because there are many different classes of metabolites that are produced by cells, the methods to detect them are also varied. Some can be detected by spectroscopy, whereas others require specific chemical assays, and others require more elaborate tests. Because of this, laboratories must be set up specifically to test a particular subset of metabolites.

Lipidomics is a subset of metabolomics that seeks to identify the lipids, as well as the pathways that produce them. Lipids are extremely important in cells, being the major components of all membranes, as well as being important carbon storage molecules for carbon and in the production of adenosine triphosphate. Lipids can also be important in the identification of organisms. For example, although Bacteria and Eukarya both use D-glycerol and fatty acids joined by ester linkages in their membranes, Archaea use L-glycerol and isoprene units joined by ether linkages in their membranes. Lipids are also found in lipoproteins embedded in membranes. Often, these are characteristic of certain species and/or cell types. The degree of saturation of the lipids in the membranes can indicate the temperature at which the cells normally grow, or at which they have recently been growing. Saturated lipids are often signs of growth in elevated temperatures, whereas unsaturated lipids are normally found in cells growing at lower temperatures, due to their increased fluidity at lower temperatures.

FUNCTIONAL GENOMICS

This branch of genomics utilizes genomic, transcriptomic, proteomic, and metabolomic data to draw conclusions about gene, RNA, and protein functions, as well as their interactions (Figure 23.8). As such, functional genomics is often at the crux of genomics, transctiptomics, and proteomics. Genes and their products (RNAs and proteins) interact either directly or indirectly, which together determine the characteristics and functions of the encompassing cell. For example, many separate proteins are involved in the control of transcription of each gene, and often include repressors and activators, which can suppress or speed up, respectively, transcription of the gene. Additionally, disease states, such as carcinogenesis, involve the interactions between proteins on the cell surfaces, which interact with small molecules, hormones, proteins, and other molecules, and transfer the signal to the inside of the cell. The signal is passed on to other proteins that become activated to produce changes that lead them to pass the signal to other proteins in the process of signal transduction. Many of these signals are passed on to other proteins into the nucleus where they affect gene transcription, most often by activating transcription of sets of genes. Thus, functional genomics has been extremely important in cancer research, and many therapies have been developed from studying these pathways for specific cancers.

Major branches of functional genomics include studies of DNA–protein, RNA–protein, and protein–protein interactions. Biological molecules almost never act isolated from other

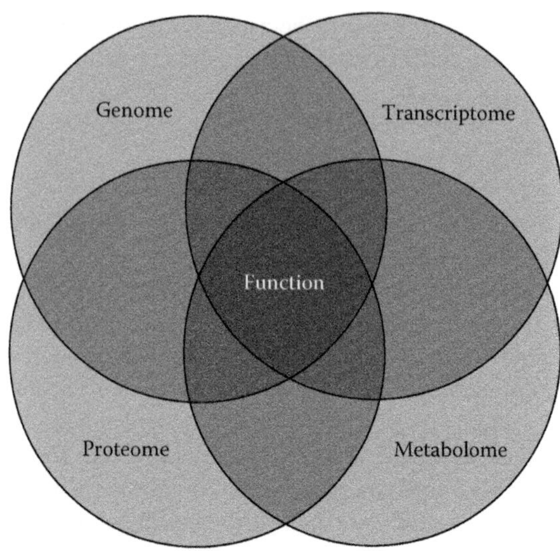

FIGURE 23.8 Overlapping data sets are used for functional genomic research. DNA data from genomic studies, RNA data from transcriptomic studies, protein data from proteomic studies, and/or other chemical analyses from metabolomics studies are used in functional genomic research either to discern the molecules, complexes, and pathways responsible for the functions, or to examine the molecules, complexes, and pathways in order to deduce a particular function.

molecules, and most form DNA–protein, RNA–protein, and/or protein–protein complexes. These interactions are often crucial to the function of the entire complex. For example, transcription of RNA involves more than a dozen proteins in Bacteria, and 2 times that number in Eukarya. In addition to histones, there are hundreds of proteins that associate with nucleic acids. DNA-binding proteins usually affect either transcription or replication, whereas those that associate with RNA include ribosomal proteins, spliceosomal proteins, and those involved in aiding in folding of ncRNAs. These interactions can be studied by polyacrylamide gel methods, assays using antibodies, precipitation experiments, and recombinant DNA methods. Protein–protein interactions have been the focus of many research studies over the past few decades. Most attempt to identify which proteins are involved in specific processes in the cell and how the proteins interact. Some assays involve cloning the genes for the proteins, and then using a reporter gene (e.g., similar to α-complementation used with bacterial cloning) to determine when and how the proteins interact.

Another important aspect of proteomics and functional genomics is determining how and when proteins are modified. These are posttranslational modifications and include the formation of disulfide bridges (between cysteine residues), phosphorylation (of tyrosines, serines, histidines, and other residues), alkylation (of lysines or arginines), acetylation (on the amino end, or on lysines), addition of lipids, addition of sugars, and many others. These modifications greatly increase the number of functions that proteins can carry out. Phosphorylation often activates certain proteins, which are important in signal transduction pathways, and many activate transcription of many genes concurrently during developmental processes. Glycosylated proteins (those with covalently attached sugars) are mostly concentrated in the cell membranes (such as those involved in the ABO blood groups of humans), and are important in cell-to-cell recognition. Lipoproteins also are concentrated in cell membranes and cell walls to provide structural integrity and fluidity to the membranes and walls. Because the proteome is so diverse, with too many interactive components (many of which can be modified in the cell) and too many functions, research groups usually concentrate on one aspect of the proteome in one organism.

KEY POINTS

1. Proteomics is the study of the total diversity of proteins found in single cells, tissues, organs, or other samples.
2. Structural genomics uses sequence data, biochemical data, and other information to determine the structures of molecules, complexes of molecules, or parts of molecules.
3. RNAomics is the process of correlating RNA structures, sequences, and motifs with their functions in the RNA molecules.
4. Epigenomics is the study of the effects of epigenetic changes in expression of the adjacent gene regions, which are caused by changes in methylation of DNA and histones, as well as changes in acetylation of histones.
5. Metabolomics uses chemical analysis data to determine the molecular constituents of cells that are the result of biochemical reactions (usually modulated by proteins).
6. Functional genomics uses genomic, transcriptomic, proteomic, and/or metabolomic information to either determine the molecules, complexes, and pathways responsible for a particular function, or use information about the molecules, complexes, and pathways to deduce a particular function in the cell.

ADDITIONAL READINGS

Alberts, B., D. Bray, K. Hopkin, A. Johnson, J. Lewis, M. Raff, K. Roberts, and P. Walter. 2013. *Essential Cell Biology*, 4th ed. New York: Garland Publishing.

Blackburn, G. M. and M. J. Gait (eds.). 1996. *Nucleic Acids in Chemistry and Biology*. New York: Oxford University Press.

Claverie, J.-M. and C. Notredame. 2006. *Bioinformatics for Dummies*, 2nd ed. New York: Wiley Publishing.

DeSalle, R. and J. A. Rosenfeld. 2013. *Phylogenomics: A Primer*. New York: Garland Science.

Graur, D. and W.-H. Li. 2000. *Fundamentals of Molecular Evolution*, 2nd ed. Sunderland, MA: Sinauer Associates.

Harris, L. and S. O. Rogers. 2008. Splicing by an unusually small group I ribozyme. *Curr. Genet.* 54:213–222.

Harris, L. B. and S. O. Rogers. 2011. Evolution of small putative group I introns in the SSU rRNA gene locus of *Phialophora* species. *BMC Res. Notes* 4:258–262.

Harris, L. B. and S. O. Rogers. 2014. Finding and characterizing small group I introns. In *Genomics III: Methods, Techniques, and Applications*. Hong Kong, China: iConcept Press, pp. 327–341.

Knowlton, C., R. Veerapaneni, T. D'Elia, and S. O. Rogers. 2013. Microbial analysis of ancient ice core sections from Greenland and Antarctica. *Biology* 2:206–232.

Krebs, J. E., E. S. Goldstein, and S. T. Kirkpatrick. 2012. *Lewin's Genes XI*, 11th ed. Sudbury, MA: Jones & Bartlett Publishers.

Li, W.-H. 1997. *Molecular Evolution*. Sunderland, MA: Sinauer Associates.

Li, W.-H. and D. Graur. 1991. *Fundamentals of Molecular Evolution*. Sunderland, MA: Sinauer Associates.

Mount, D. W. 2004. *Bioinformatics, Sequence and Genome Analysis*, 2nd ed. Cold Spring Harbor, NY: Cold Spring Harbor Laboratory Press.

Stombaugh, J., C. L. Zirbel, E. Westof, and N. B. Leontis. 2009. Frequency and isostericity of RNA base pairs. *Nucleic Acids Res.* 37:2294–2312.

Zirbel, C. L., J. E. Sponer, J. Sponer, J. Stombaugh, and N. B. Leontis. 2009. Classification and energetics of the base-phosphate interactions in RNA. *Nucleic Acids Res.* 37:4898–4918.

24 Species Concepts and Phylogenetics

INTRODUCTION

Organisms are classified according to their shared characteristics. Initially, the Greeks classified organisms based on their gross morphological characters, and therefore at that time, there appeared to be only two major groupings, plants and animals. Within each group, there were recognizable entities that could breed to produce more of the same kind of organism. Eventually, these became known as species, but this was not formalized until the eighteenth century by Carl Linnaeus. In the same century, scientists began to use microscopes to add additional characters and groups to taxonomic classifications. Higher order groupings were also formalized to logically classify organisms for study, as well as for commerce. Eventually, in the twentieth century, electron microscopes aided classifications based on microscopic characters. During the past three decades, molecular biology methods have been at the forefront of taxonomy, because often they are more direct measures of genetic relatedness among the taxa. Additionally, they allow accurate phylogenetic reconstructions of the taxa that can yield precise measures of evolutionary events and the timing of those events.

WHAT IS A SPECIES?

Everyone knows the word *species*, but few understand exactly what it means, even the taxonomic experts. One reason for this is that it means different things at different times. For bacteria, species are difficult to define for several reasons. They are microscopic, have few morphological characters, and most cannot be cultured for detailed study. For fungi, there are similar constraints to species identification, but there are others as well. Some fungi are morphologically very different during different times of their life cycles. For most plants and animals, species identification may be easier, but even then, it takes a great deal of knowledge to determine the species, and even more expertise to discover and describe new species. For example, some plants and animals are dioecious, that is, they have separate male and female sexes. When the entire life cycle of the plant or animal is not observed, the males and females of the species might be described as separate species. More specifically, if one knew nothing about humans, and only had a single picture of a male and a female, they might describe the two as being two species, because they differ physically. If one observes two birds at a bird feeder, one bright yellow and one a dull brown color, one might conclude that they are two species, when in reality they are a male and a female goldfinch. However, some bacterial species cannot be distinguished from one another based on physical and growth characteristics. Some have been described as single species, but genomic studies in which the entire genomes have been determined have concluded that some of what originally were thought to be varieties of distinct species are, in fact, different species.

Species are described as entities that can produce cells or individuals of the same type and that are able to interact genetically with others of the same kind. However, there are problems in making this determination. The first was illustrated above, that of sexual dimorphism in plants and animals. Another example of this is the peacock and peahen, the male and female version of the species *Pavo cristatus* or *Pavo muticus*. The peacock has a set of very large and colorful feathers, whereas the female is drab in coloration. Morphologically, they have some major differences. Another problem is illustrated by the genus *Canis*, which includes domestic dogs (Figure 24.1). One part of this problem was noted by Charles Darwin in his book *The Origin of*

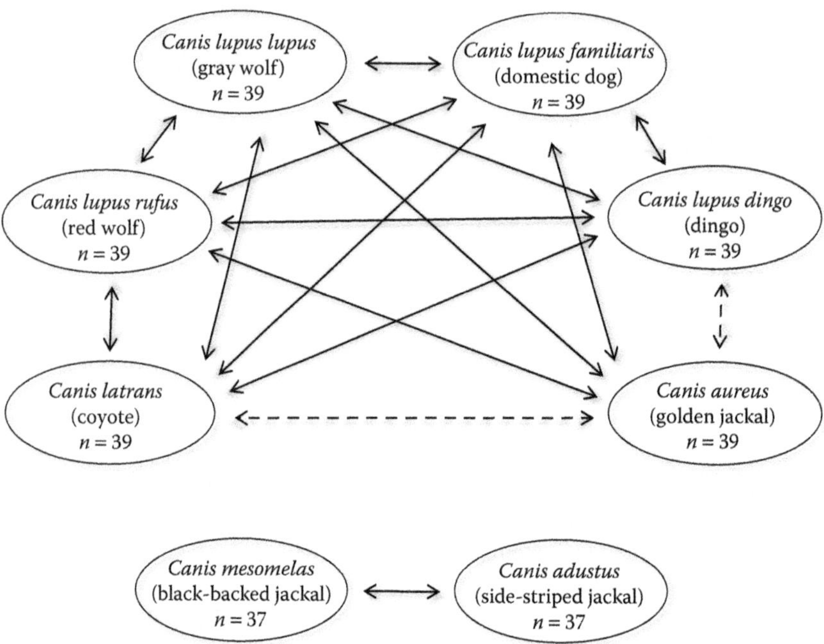

FIGURE 24.1 Production of viable offspring within the genus *Canis*. Members of this genus can interbreed to form hybrids that can also reproduce. This is partly due to their recent separation as species and subspecies, and the fact that most still have the same number of chromosomes and likely the same gene order along each of the chromosomes. Solid lines indicate evidence of successful mating to produce viable hybrids. Dashed lines indicate mating that should be possible but have not been observed. Domestic dogs (*C. lupus familiaris*) can breed with wolves (*C. lupus lupus* and *C. lupus rufus*), dingoes (*C. lupus dingo*), coyotes (*C. latrans*), and golden jackals (*C. aureas*). Additionally, many of the other members of the genus can successfully interbreed. However, two species of jackal (*C. mesomelas* and *C. adjustus*) cannot mate with other members of the genus because they have a different number of chromosomes, although they have the same number of genes. Several million years ago, two of the chromosomes fragmented, such that some of the species within the genus have two more chromosomes than other species, although the genome has remained very similar. *C. mesomelas* and *C. adjustus* have the same number of chromosomes, and they can mate to form viable hybrids.

Species, and clearly had a major influence on his ideas about species variation and natural selection. Dogs exhibit a great deal of phenotypic variation because people have selected for particular characteristics. All of the dogs in the world comprise a single subspecies, *Canis lupus familiaris*. Individuals with very different physical characters can breed and produce viable offspring. For example, very large breeds of dogs can breed with very small breeds to produce viable offspring. Therefore, without the direct observation of dog breeding, the various breeds of dogs might otherwise be described as many different species. At the same time (something of which Darwin was unaware), other species within the genus *Canis* can interbreed to produce viable offspring capable of reproducing. Domestic dogs can breed with coyotes (*C. latrans*), red wolves (*C. lupus rufus*), gray wolves (*C. lupus lupus*), dingoes (*C. lupus dingo*), and golden jackals (*C. aureus*) to produce viable offspring capable of producing offspring (Figure 24.1). Interbreeding among other species in this genus is also possible.

Hybridization events, and other processes, such as parthenogenesis (see Chapter 10), also can cause confusion regarding species designations. One example of hybridization and taxonomic ambiguity is illustrated in the *Cnemidophorus*, a group of whiptail lizards, found in southwestern North America. The little striped whiptail (*C. inornatus*) has both males and females, and when they breed, they produce new individuals of the same species. Morphologically and behaviorally, there are no ambiguities. The tiger whiptail (*C. tigris*) also has both males and females, and when they breed, they also produce new individuals of the same species, which are morphologically and behaviorally similar to their parents. However, these two species can interbreed to form hybrids that are known as New Mexico whiptails that have been given the species designation, *C. neomexicanus*. Morphologically, individuals are intermediate between *C. inornatus* and *C. tigris*, in that they have stripes similar to *C. inornatus*, but are larger, approaching the sizes of *C. tigris* individuals. Also, all of the individuals are female, and they produce new females by parthenogenesis. Similar situations occur among some species of copepods and in many other groups of organisms. Therefore, even the definition of a biological species is sometimes somewhat ambiguous.

The main problem with defining a species as ones that breed successfully with other members of the species is that it is nearly impossible to make these determinations for most organisms. In the vast majority of cases, species are described based only on their morphological characteristics. Descriptions based on their genomes have increased during the past few decades but are still small in number compared to the number of species described by morphology, and both are only small representations of the total number of species on the Earth. Morphological descriptions can be problematic. Often, characters within a species or among closely related species diverge rapidly (as seen in dogs). Darwin observed this firsthand when he visited the Galápagos Islands. As he collected the birds on each island with very different morphologies, he labeled them as being members of different groups of birds (e.g., blackbirds, grosbeaks, and finches). However, when the specimens were sent to the bird expert, John Gould, at London's British Museum, he determined that all of them were finches. This was one of the factors that initiated Darwin to begin writing about slow change and natural selection. Divergent morphology also occurs in other organisms, including sharks and many groups of plants. This can also occur with DNA. Genomes often change rapidly in some plant groups via polyploidization (multiple copies of the entire genome), followed by divergence of individual genes.

Another problem with morphological description and identification of species is illustrated in fungi. Many species of fungi are dimorphic, and some are polymorphic. This means that they can grow as two very different forms (i.e., they are dimorphic) or many different forms (i.e., they are polymorphic or pleomorphic). Many animal pathogenic fungi (e.g., *Coccidioides immitis*, *Penicillium marneffei*, and *Candida albicans*) can grow as filaments (hyphae) in the environment. However, once they enter an animal, they begin to grow as unicellular yeasts, and some have the potential to produce widespread infection in the host organism. Fungi may also exhibit very different morphologies based on their stage of life. As haploids, many form simple branching hyphae. Under some circumstances, they can form reproductive structures (conidiogenous structures) that produce and release haploid spores (conidia). Many species have been described based on the dimensions and appearance of these reproductive structures and spores. However, some species can form two or more different reproductive structures, which has led to some confusion in naming and identifying these organisms. To compound the problem, when the fungi mate (by anastamosis) to form dikaryotic (or multicellular) cells, often the cells, hyphae, and reproductive structures are completely different in appearance than the haploid cells. Because of this, often they have been described as two (or sometimes more) species. However, DNA studies have begun to sort out some of the confusion of fungal taxonomy. In many cases, the asexual forms (mitosporic or anamophic fungi) have been connected with their sexual (perfect or teleomorphic fungi) stages.

Another concern with assigning species based on morphology is essentially the opposite of divergent morphology and evolution—convergent morphology and evolution. Very different species may exhibit some characters that are very similar to those in other species. This is seen in the beaks of birds, as well as in plant leaves. Beak dimensions in birds primarily are related to the types of food that they eat. Many species and groups of birds eat similar foods, and their beaks have evolved to be able to eat those foods. In plants, leaves as well as stems have evolved to maximize photosynthetic productivity. Because of this, there are many plants with leaves that are very similar, even when the plants are unrelated. Just as with divergent evolution, DNA can exhibit convergent evolution as well when similar processes and genes are employed by different organisms.

Therefore, in essence, a species is a collection of cells or individuals with the same (or similar) genome that exists during a finite period of time. Operationally, a species can be defined as a group of organisms that can be identified as a single entity by morphology, heredity, and/or molecular similarities. Morphology is an indirect indicator of the genetic components of the cells and organisms, and in many cases, it is the fastest and easiest method for describing and identifying a species. Heredity is direct measure of species ties for a specific individual or cell, but is often impractical to determine. Molecular methods provide a relatively simple way to directly determine a species and its relation to other species.

CLASSIFICATION OF LIFE

Many different classification schemes have been developed in attempts to make sense of the diversity of life on the Earth (Figure 24.2). The first such system was developed by Aristotle, and later by Linnaeus. Two kingdoms were defined, one consisting of animals and the other of everything else, but primarily plants. Once microscopes were developed, three kingdoms were recognized: plants, animals, and protists (unicellular organisms). Once electron microscopes were developed, four kingdoms were proposed where the protists were split into Monera (bacteria) and Protists (eukaryotic microbes). The placement of fungal species had always been a problem, some classifying them as plants and others within the protists. Finally, a five kingdom scheme was developed splitting off fungi as a separate kingdom. However, in the 1970s, Carl Woese and George Fox started using DNA to study bacteria to attempt to better determine the taxonomy and evolution of this large group. They soon discovered that bacteria in the historical sense actually consisted of two major groups. This split became the domains Bacteria and Archaea, which include the kingdoms Bacteria and Archaea, respectively. The third domain of life is Eukarya (eukaryotes). During the past two decades, two separate schemes for classification have been proposed: The first includes the three domains, namely, Bacteria, Archaea, and Eukarya and the second encompasses six kingdoms, namely, Bacteria, Archaea, Protista, Plantae, Fungi and Animalia (the last four being within the Eukarya). More recently, changes have been suggested based on additional genomic sequencing results. Specifically, the original kingdom Protista has been subdivided into several additional kingdoms, because of the recognition of the high degree of diversity within this designation. Also, Animalia and Fungi have been combined into the Opisthokonta, based on recent phylogenetic studies of ribosomal RNA (rRNA) genes uniting the two, as well as morphological characteristics that are in common between the two. The following kingdoms have been suggested based on recent genomic results (Figure 24.3): Domain Eukaryota would include the kingdoms, Amoebozoa (amoeba and slime molds), Opisthokonta (animals, fungi, and related organisms), Archaeplastida (plants, red algae, green algae, and glaucophytes), Rhizaria (Foraminifera, Radiolaria, and related protozoans), Chromalveolata (stramenopiles, dinoflagellates, ciliates, oomycetes, brown/golden brown algae, and apicomplexans), and Excavata (diplomonas, euglenoids, and kinetoplastids); the domain Bacteria contains one kingdom, Bacteria; and the domain Archaea contains one kingdom, Archaea.

Of course, taxonomic classification does not end at the kingdom level. Groupings based on shared characteristics flow all the way down to genus and species, as well as further down to subspecies, variety, and so on. For example, the taxonomic classifications for humans would be as follows: domain = Eukarya, kingdom = Animalia, Metazoa, or Opisthokonta (depending on the classification scheme), phylum = Chordata, class = Mammalia, order = Primates, family = Hominidae, genus = *Homo*, species = *sapiens*, and subspecies = *sapiens*. Thus, humans are

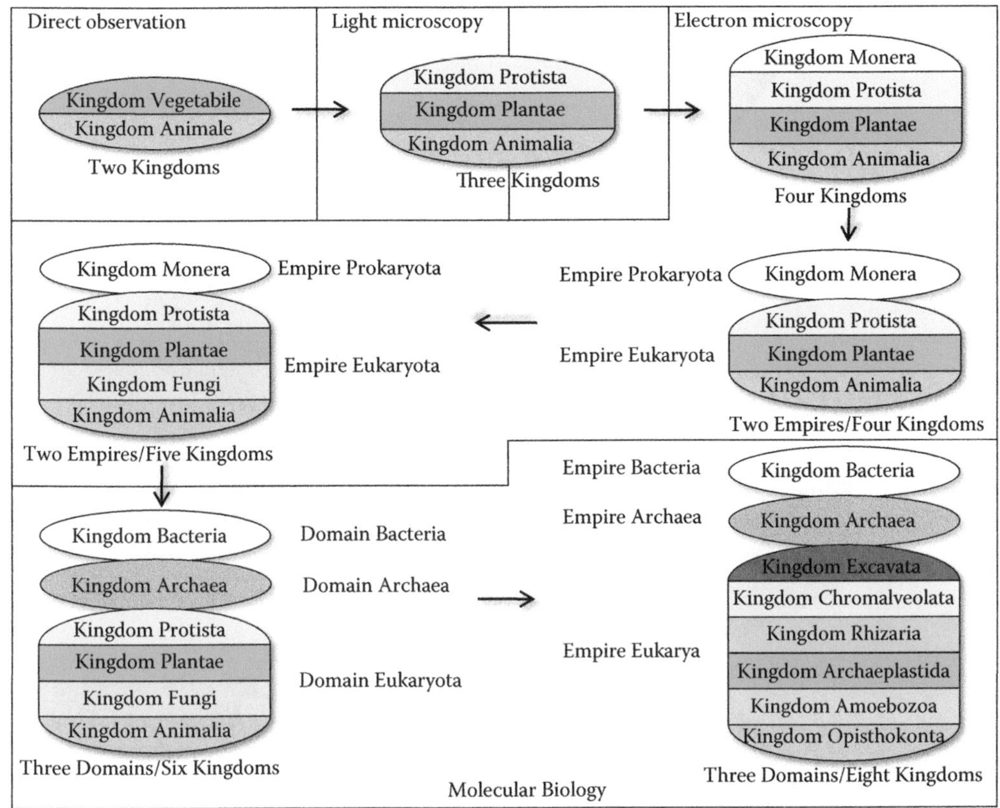

FIGURE 24.2 Short history of the schemes developed to classify life on the Earth. Aristotle and Theophrastus devised the first known classification system (upper left). At that time, they described what they could observe directly, which were the animals and plants. In the seventeenth century, the first observations using a microscope were possible, and by the nineteenth century, it had become clear that the two-kingdom system was inadequate. A new kingdom, the Protista, was proposed (top middle). The advent of the electron microscope led to the discovery that a large group of single-celled organisms had nuclei, whereas others clearly lacked them. An additional kingdom, Monera, was added (top right). Because all (or most) of the multicellular organisms also had nuclei, a system of two empires (Prokaryota and Eukaryota), including four kingdoms (Monera, single-celled prokaryotes; Protista, single-celled Eukaryotes; Plantae; and Animalia), was then established (middle right). Fungi had been problematical from the beginning. At one time, they were classified with plants, and later with animals. By the mid-twentieth century (middle left), they were split off into a separate kingdom. By the 1970s, molecular biology methods were becoming established, and Carl Woese and George Fox started using some of these methods to study particular groups of bacteria. By using molecular characters, they discovered that the original kingdom Monera consisted of two large, but separated, branches of the phylogenetic tree of life, based on the small subunit genes of rRNA. Originally, they were called Archaebacteria, but they have since been renamed Archaea. By the mid-1990s (lower left), the classification system included three domains (Bacteria, Archaea, and Eukarya) and six kingdoms (Bacteria, Archaea, Protista, Plantae, Fungi, and Animalia). Over the past two decades, an enormous amount of sequence data has been determined, leading to further distinctions, especially within the Protista (also known as protozoa). This group has been found to be extremely diverse genetically, morphologically, and developmentally. It also includes many species that have undergone one or more endosymbiotic events. Currently (lower right), there are three accepted domains (Bacteria, Archaea, and Eukarya) that include eight kingdoms (Bacteria, Archaea, Excavata, Chromalveolata (sometimes split into Cryptophyta, Haptophyta, Alveolata, and Stramenopiles), Rhizaria, Archaeplastida, Amoebozoa, and Opisthokonta).

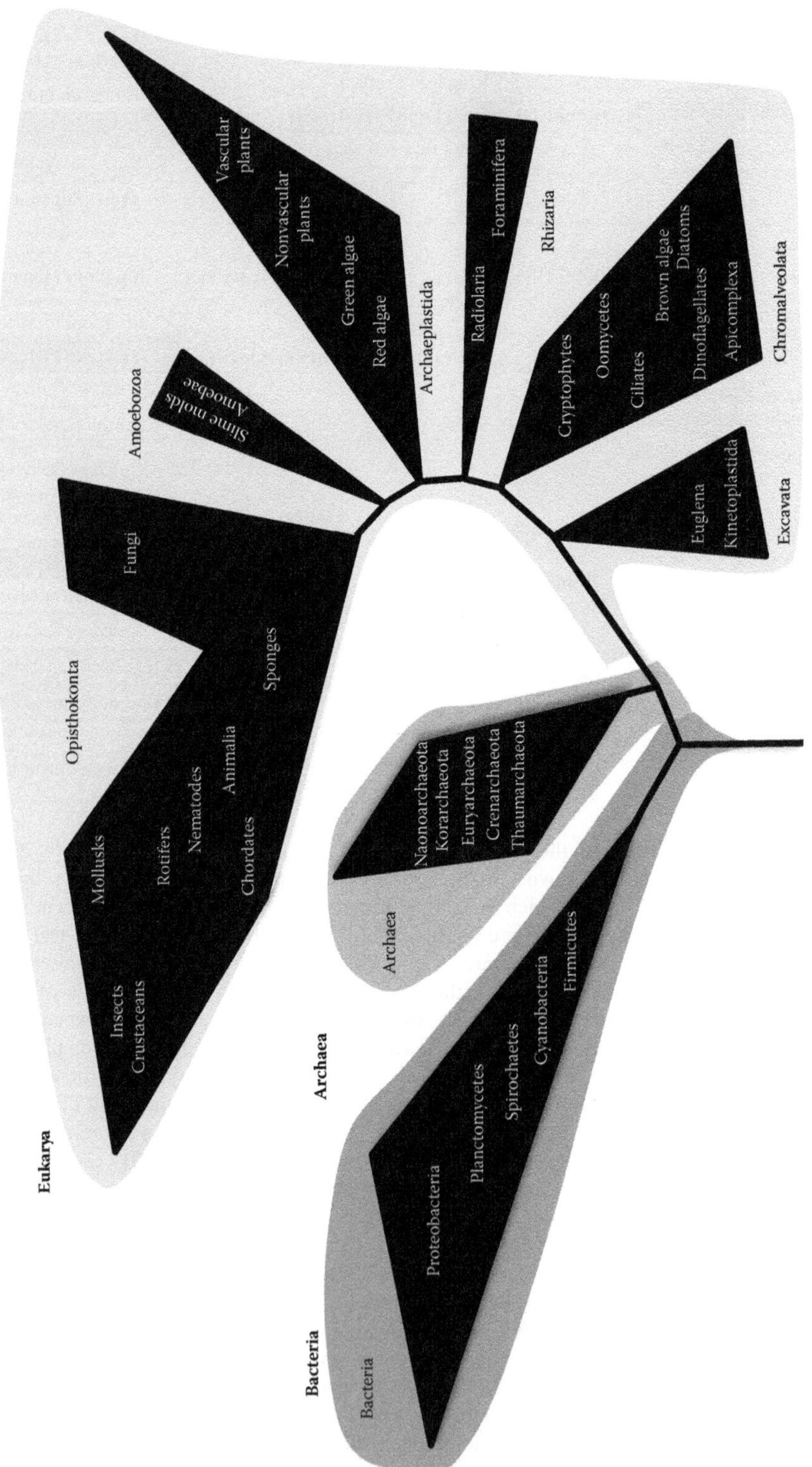

FIGURE 24.3 Phylogram of all domains and kingdoms of life on the Earth, based on the small subunit rRNA genes. The domain Bacteria (bold font) includes the kingdom Bacteria (inside gray area), which includes all phyla of Bacteria (examples in standard font inside black areas). *(Continued)*

referred to as *Homo sapiens sapiens*. The taxonomic classification for *Escherichia coli* would be as follows: domain = Bacteria, kingdom = Bacteria, phylum = Proteobacteria, class = Gamma Proteobacteria, order = Enterobacteriales, family = Enterobacteriaceae, genus = *Escherichia*, and species = *coli*. The first *E. coli* strain to be sequenced was strain K-12, which further details its taxonomic position. Even with all of the knowledge that has been collected regarding species on the Earth, scientists still have only scratched the surface. It has been estimated that at least 13.0 million species exist on the Earth currently (Figure 24.4). However, only about 1.7 million have been described scientifically.

RECONSTRUCTION OF EVOLUTIONARY HISTORY

The evolutionary history of organisms can be traced by using phylogenetics. Gene sequences can be aligned and the number or proportional changes among the sequences can be calculated. Then, a graphical representation of the evolutionary history can be produced. These are the so-called evolutionary trees. They are very much like physical trees in that they are three dimensional, although on paper they are two-dimensional representations. If evolution could be viewed as it actually occurs, each of the lines on the phylogram would actually become a dense network of finer lines joining each of the individuals to its ancestors and progeny (Figure 24.5).

This is the vertical inheritance in these lineages. However, we know that horizontal exchanges have occurred frequently in the past (see Chapter 15). Viruses pass bits of genetic material from one individual to another and from one species to another constantly. Mitochondria are the descendants of proteobacterial endosymbionts, and chloroplasts are the descendants of cyanobacterial endosymbionts that have passed their genetic material into the nuclei and cytosol of their hosts. Therefore, evolution can be viewed as the combination of all of the changes occurring to each individual in a population of organisms of the same species. Thus, there is an averaging of the changes over time in the phylogenetic trees. The relative contributions of each of the gene versions (e.g., alleles) that are vertically inherited and the addition of genes via horizontal exchange of genetic material add complexity to the simple linear inheritance of genes. However, both processes have been powerful driving forces in evolution.

FIGURE 24.3 (Continued) The domain Archaea includes the single kingdom Archaea, which includes all phyla of Archaea (five examples are listed). Many groups of extremophiles are found in this kingdom, including halophites, methanogens, and thermophiles. Note that phylogentically Archaea are closer to Eukaryotes than Bacteria. The domain Eukaryota includes six kingdoms. The kingdom Excavata includes Euglenida, Diplomonads, Kinetoplastida, and other flagellated single-celled protozoans. (*Note*: This is sometimes split into two groups, the Excavata and the Discicristata.) The Chromalveolata is a large kingdom that includes brown algae, diatoms, dinoflagellates, ciliates, apicomplexans, oomycetes (sometimes called water molds, and originally placed together with fungi), and others. Members of this kingdom have undergone one to several rounds of endosymbiotic events (see Figure 15.17), such that they have complex mixtures of genes from various sources, and they sometimes have unusual and unique organelles (having originated as eukaryotic endosymbionts). Some have elaborate cell wall and flagellar structures. The kingdom Rhizaria consists of the Foraminifera, Radiolaria, Euglyphida, and other groups that have elaborate cell structures. The kingdom Plantae (Archaeplastida) encompasses all of the vascular (e.g., ferns, gymnosperms, and angiosperms) and nonvascular plants (mosses, liverworts, hornworts, green algae, red algae, and glaucophytes). The kingdom Amoebozoa includes slime molds, or what is also called the Myxomycetes (originally considered fungi). The kingdom Opisthokonta includes the previous kingdoms Animalia and Fungi. They have been grouped into one kingdom primarily based on their molecular and key morphological similarities. The Fungi consist primarily of the Ascomycota, Basidiomycota, and Zygomyces, but the chytrids and microsporidians are included in this group. The animals include sponges, mollusks, insects, crustaceans, and chordates (including all vertebrates).

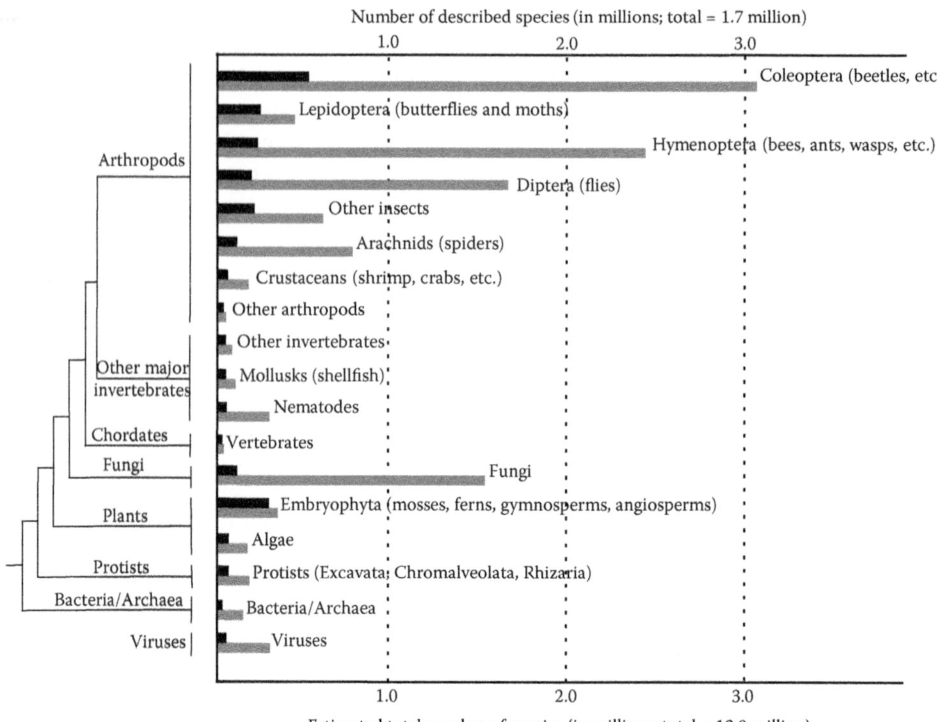

FIGURE 24.4 Comparison of the number of species that have been scientifically described versus the number of species that have been predicted. Approximately 1.7 million species have been described, but it has been estimated that currently about 13.0 million species exist on the Earth. Arthropods dominate on the basis of having the most species. Plant species are the next most numerous, in terms of described species, although fungi may be more numerous, based on estimates of undescribed species. The cladogram on the left indicates relationships among major groups. Branch lengths are not proportional to evolutionary distances.

PHYLOGENETICS

Mutations, gene flow, and selection lead to evolutionary events. Data can be collected on mutation rates, patterns, and directions, which can be used to infer the evolution and evolutionary patterns of genes and taxonomic groups. How does one take sequence (or other) data and convert that into evolutionary trees? The methods to perform these evolutionary reconstructions are known as phylogenetic methods. The mutations among a group of analogous sequences are compared, and calculations are made based on a set of assumptions and models, and a graphical representation of the sequence relationships is constructed. These are the familiar phylogenetic trees, evolutionary trees, or simply trees. The first tree was drawn by Charles Darwin in one of his notebooks while developing his theories of descent by natural selection. Of course, it was based on morphological characters of the organisms rather than on sequences. Since that time, many tree construction methods have been developed, and in the past several decades, DNA, RNA, and amino acid sequences have been used to determine relationships and construct trees based on comparisons among sets of sequences.

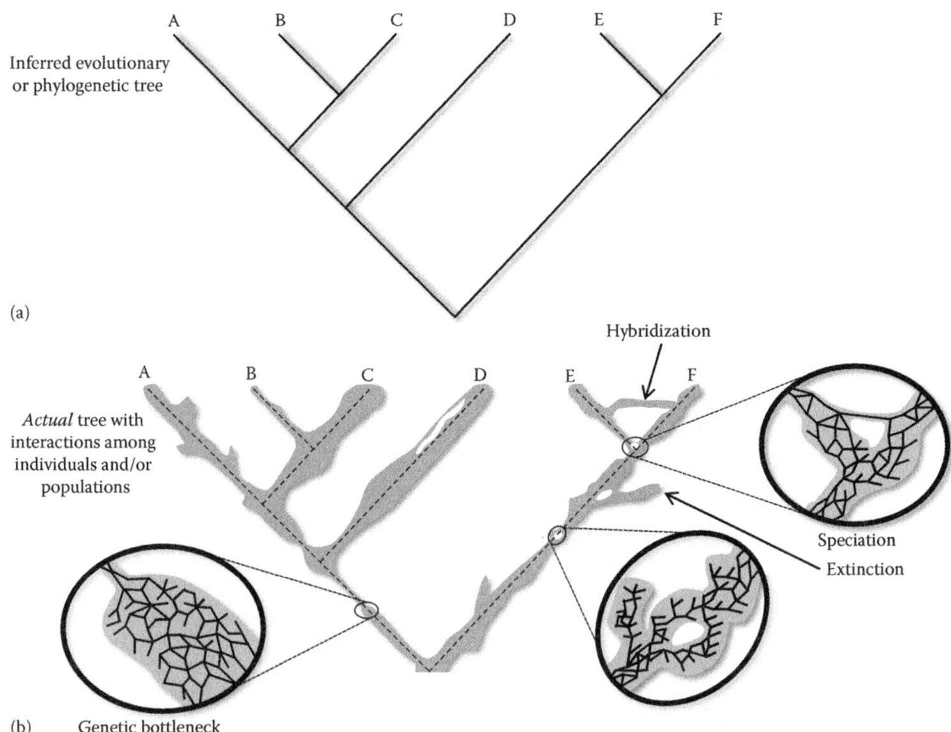

FIGURE 24.5 Evolutionary tree derived from sequence data (a) and a theoretical view of the interactions among individuals along the tree (b). In order to produce a phylogenetic tree, one or a small number of representative sequences are used. Essentially, they are the end points (A, B, C, D, E, and F) of each of the branches. However, the populations for each are the product of a large number of individuals interacting genetically (by cell division, sexual reproduction, etc.). Because of this, at any point in time, the number of individuals and gene versions varies. The population might increase, decrease, or remain static during different time periods. There may be genetic bottlenecks where the population or genetic heterogeneity has greatly declined. There are speciation and frequent extinction events. As was explained in Figure 15.3, some species can form hybrids, and thus, there can be horizontal transfers of genes over short timescales. Although phylogenetic trees are always drawn with lines as the branches, the actual populations in each case rarely form straight uniform lines.

TREE TERMINOLOGY

Phylogenetic trees are graphical representations of reconstructions of evolutionary histories. Each of the branches joins taxa (e.g., species) or sequences in a pattern that describes the relationships and/or distances between each of the taxa or sequences. These are defined as operational taxonomic units (OTUs; Figure 24.6) and appear on the terminal nodes of the tree.

The branches joining them to other parts of the tree are called terminal branches, and all of the other branches are internal branches. Nodes are the points where branches meet. The root is the most ancient node. There are two general types of trees that can be drawn (Figure 24.7). The first is a cladogram. In this tree, the branches are not always proportional to the distances between the OTUs. The branching is meant only to indicate the branching order among the OTUs. The second type of tree is a phylogram. These trees convey both the branching pattern and the distances among the OTUs, which were calculated from the data. Trees can be produced, which are slanted

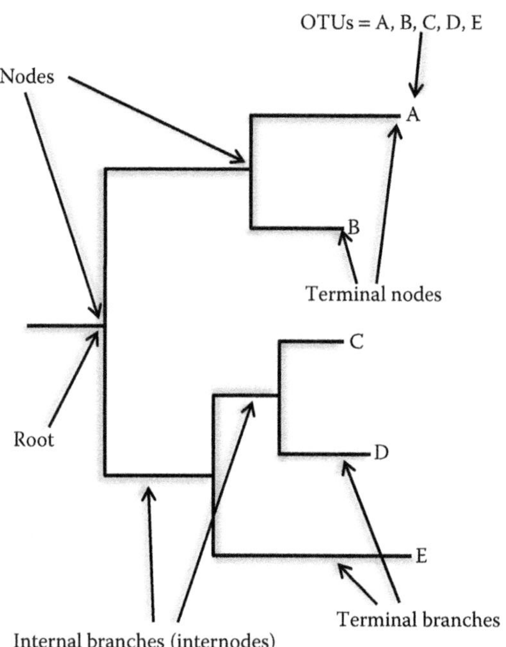

FIGURE 24.6 Parts of a phylogenetic tree. The OTUs are the names of the sequences/genus/species. They are at the ends (terminal nodes) of the terminal branches. The branches are joined to other branches by nodes. Internal branches (also called internodes) join nodes to other nodes within the tree. If the tree is rooted, the root is at the most internal node of the tree.

or rectangular in form. For rectangular phylograms, the horizontal branches convey the distances, whereas the vertical branches are only used to join adjacent branches together. They contain no distance information. Circular trees are simply rectangular trees bent into a circle. All types of trees can be either rooted, when the location of the most ancient branch is known, or unrooted. Changes in rooting can change the evolutionary relationships as well as some of the distances, and therefore, trees should be rooted only when their location is certain (Figure 24.8). Rooting is often accomplished by including a sister group taxon or gene into the analysis. A sister group is defined as an OTU that is closely related but clearly outside of the group being analyzed. In this case, the sister group is termed the *outgroup*, and the group being analyzed is the *ingroup*. Trees can be rotated at any of the nodes, because this does not alter the relationships or the distances among the OTUs (Figure 24.9).

Terminology is important when using trees. Trees based on characters are character trees, and those based on sequences are sequence trees, showing the evolution of the characters or sequences, respectively. This might seem obvious, but most trees have species or other taxonomic names listed at their terminal nodes. Therefore, many times researchers are attempting to produce species trees based on sequences. This is valid in most cases, but it should be remembered that the trees are actually species trees inferred by the sequence data, and that the trees are therefore more accurately described as sequence (or gene) trees.

CHOOSING A GENOMIC REGION FOR PHYLOGENTICS

First of all, it is impossible to examine all of the evolutionary steps that have occurred on the Earth. Also, it is impossible to compare all of the sequence changes that have occurred in humans during the past 200,000 years. The number of sequences that can be examined, quantified, and compared for any study must be limited. Every sequence that is chosen and every base pair change provides a narrow window

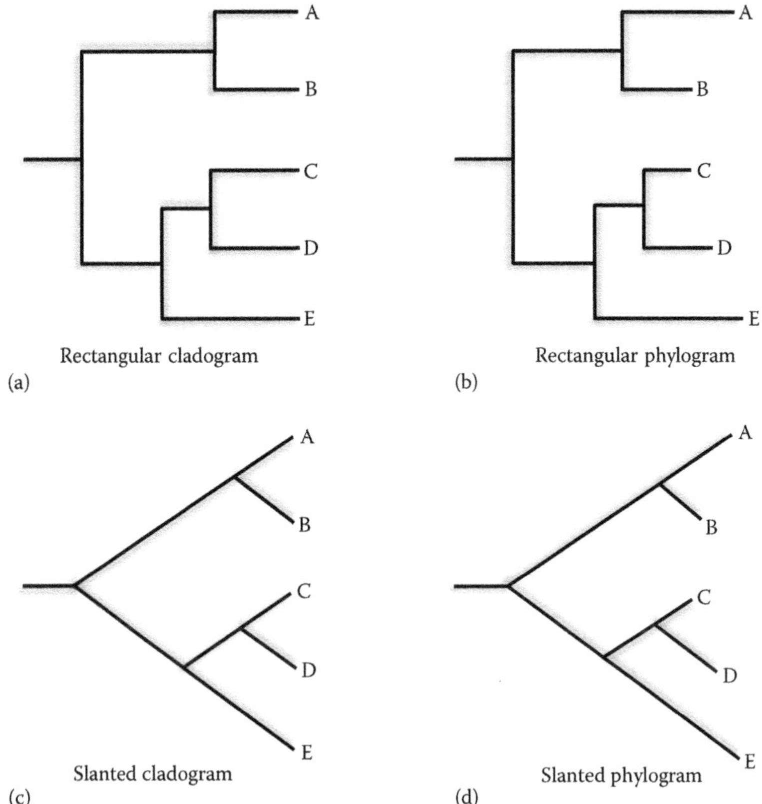

FIGURE 24.7 Types of trees. Rectangular trees are on (a,b) and slanted trees are at (c,d). Cladograms are (a,c). They show branching patterns to indicate evolutionary relationships, but usually do not indicate phylogenetic distances. Phylograms are (b,d), which show evolutionary relationships, as well as phylogenetic distances. For rectangular phylogenetic trees, only the horizontal branches indicate the phylogenetic distance. The vertical branches are included only to join together the horizontal branches.

into the question that is being addressed. Because of this, careful selection of the sequences is an absolute necessity prior to beginning any phylogenetic analysis. Different parts of every genome have different rates of nucleotide substitution (Figure 24.10). Even specific gene loci have regions that may change rapidly and others that are highly conserved (Figure 24.11). If the question deals with the evolution of a particular gene and its origins, specific sequences should be chosen to approach an answer. If the origin of particular organelles is being sought, then different sets of genes from various species should be selected. If the project is to determine the evolutionary relationships within a taxonomic group of organisms, then a different set of genes should be chosen. If the evolutionary question involves a recent event or a narrow taxonomic range, then the genomic region being compared should be one that changes rapidly (Figure 24.12). If the question involves a very ancient event or a broad taxonomic range, then the genomic region being compared must change very little over long periods of time (i.e., it is conserved). For example, if one wanted a comparison of a population of humans, one would not choose the small subunit (SSU) rRNA gene as the sequence of choice, because all humans have exactly the same sequence for this gene. There would be no differences, and thus, a comparison would be meaningless. However, SSU rRNA gene sequences can be used to produce a phylogenetic tree of all cellular organisms on the Earth because it has enough variation to distinguish humans from birds, and humans from *E. coli*, but at the same time, there is enough sequence conservation to allow alignment of those sequences. To make

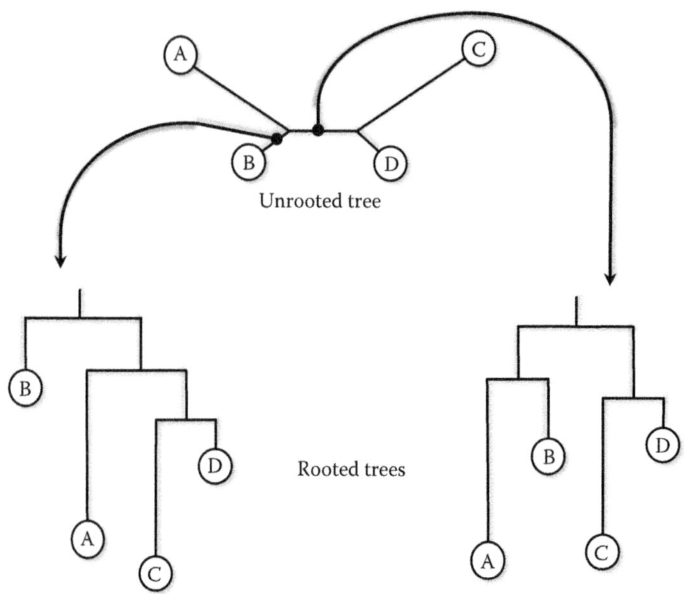

FIGURE 24.8 Rooting a tree. An unrooted tree can be rooted at any chosen point, based on the knowledge of the most ancient spot on the tree. Although the distances between the OTUs remain unchanged, the interpretation of the relationships among the OTUs can change, because once a tree is rooted, an evolutionary direction is implied.

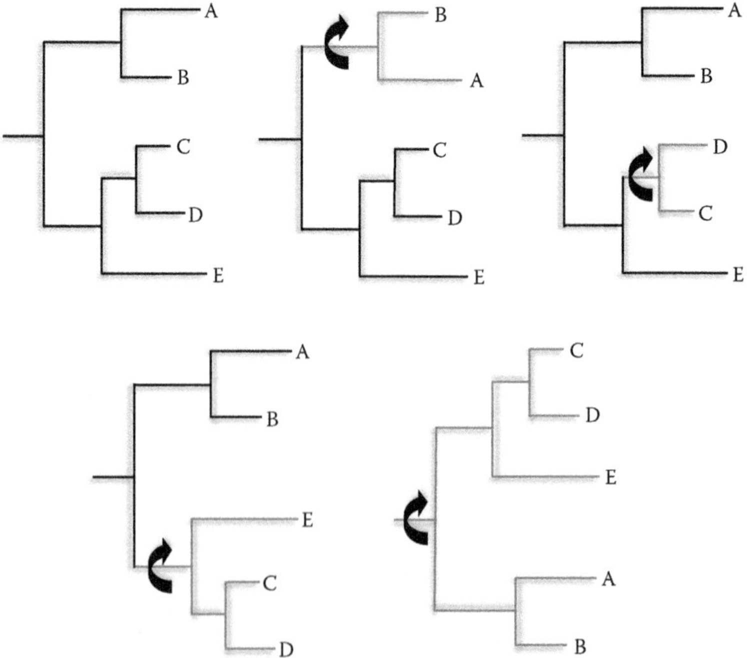

FIGURE 24.9 Rotation of branches. Free rotation is possible around each of the horizontal branches at their nodes. The relationships and distances (branch lengths) remain unchanged by each of these rotations.

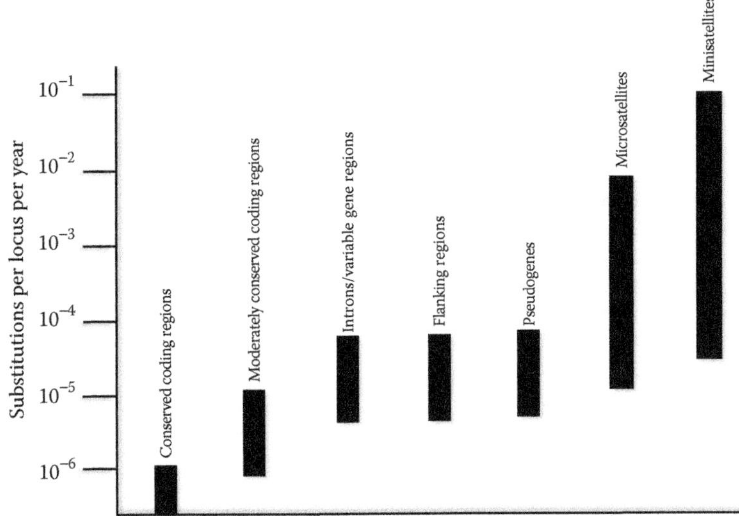

FIGURE 24.10 Rates of nucleotide substitution for various classes of sequences. Genomic regions have different rates of mutation. Conserved genes have substitution rates close to one change every million years. Moderately conserved genes have higher rates of mutation. Introns and intergenic spacers have similar rates of change, and some spacers may have originated from introns. Regions that flank genes, as well as psuedogenes, generally have similar rates of mutation. Microsatellites and minisatellites have the highest substitution rates. Some of these cause tumors in humans, and thus have rates of change that are within the life span of humans.

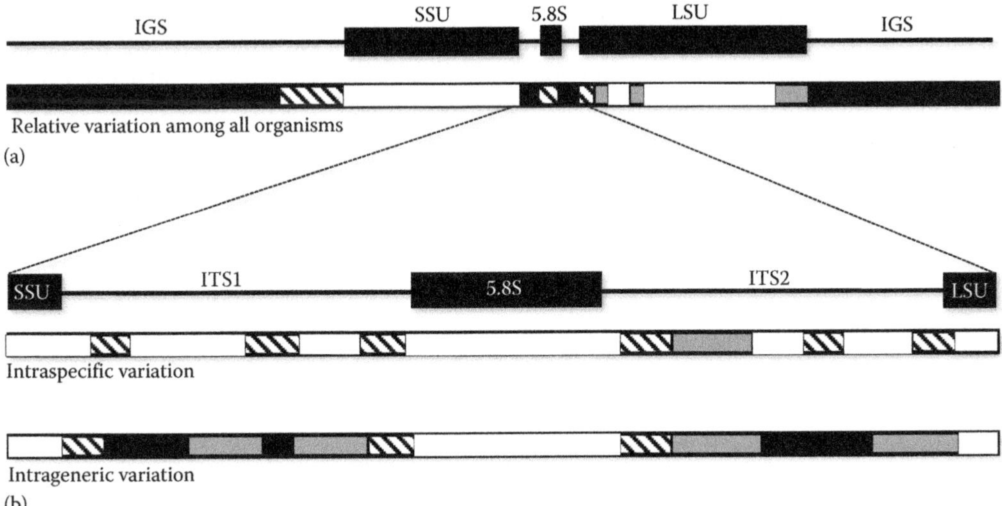

FIGURE 24.11 Sequence variation within the rRNA gene region. White regions indicate high sequence conservation, whereas black regions have the highest degree of sequence variation. Striped and stippled regions indicate regions of moderate conservation and moderate variation, respectively. Throughout eukaryotes, the rRNA gene locus (a) exhibits a great deal of variation in the amount of sequence diversity through the various regions of the locus. The intergenic spacer (IGS) and internal transcribed spacers (ITS1 and ITS2) exhibit the highest variation. The SSU and large subunit (LSU) genes have the highest degrees of sequence conservation. The promoter region, 5.8S, and 5′ end of the LSU are moderately conserved, whereas parts of the LSU exhibit higher degrees of variation. When the genes and the internal transcribed spacers are examined at the species (intraspecific) and genus (intrageneric) levels (b), variable and conserved regions are found. Although most of the gene regions are highly conserved, parts of ITS1 and ITS2 are also well conserved at the species level, although they are more variable at the genus level.

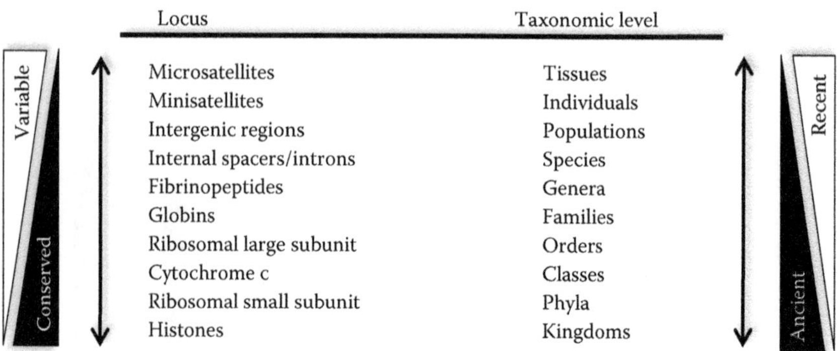

Locus
Taxonomic level
Variable
Conserved
Microsatellites
Minisatellites
Intergenic regions
Internal spacers/introns
Fibrinopeptides
Globins
Ribosomal large subunit
Cytochrome c
Ribosomal small subunit
Histones
Tissues
Individuals
Populations
Species
Genera
Families
Orders
Classes
Phyla
Kingdoms
Recent
Ancient

FIGURE 24.12 Relative rates of change for various gene loci compared to a taxonomic level of study. The genomic region of study must be matched with the timescale for the study being performed. Genomic regions that change slowly can be used to study questions encompassing a broad range of taxonomic groups (or genes), involving ancient evolutionary events. Use of genomic regions that change rapidly are appropriate for studies of narrow taxonomic scope, involving recent evolutionary events.

comparisons among humans, it would be more appropriate to use intergenic regions, microsatellites, or minisatellites, which all change rapidly. Once a region is selected, the exact sequences that will be used must be selected from a database, or sequences have to be determined from DNA samples collected for the project. For many such studies, a set of new sequences is determined, but additional sequences are chosen for comparison by searching for appropriate sequences that are available from databases.

Another consideration is that when a region is sequenced, it almost always comes from an organism that is living today. When phylogenetic trees are constructed, the sequences are contemporary and lines joining taxa or sequences together are inferred based on the differences and similarities among the sequences (Figure 24.13). Short branches indicate more recent changes and divergences, whereas long branches indicate more distant changes and divergences, and therefore changes that occurred earlier in the Earth history. For these trees, the deeper one goes in the tree, the more uncertainty there will be along that branch (Figure 24.14). One way to reduce the uncertainty is to use ancient DNA. This is DNA that is extracted from preserved or fossil specimens (Figure 24.15).

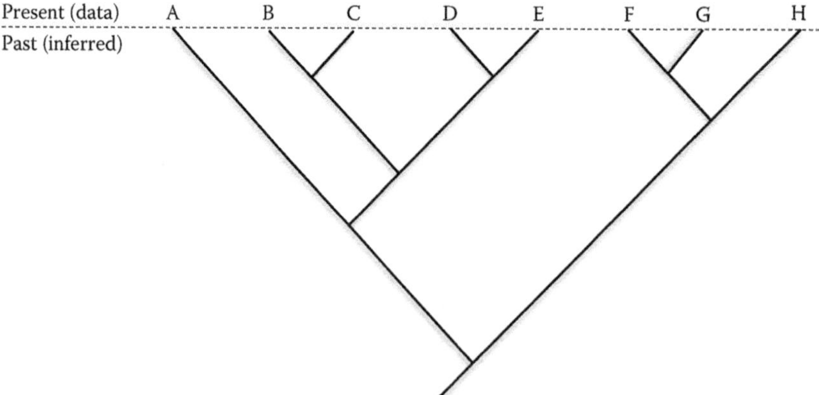

FIGURE 24.13 Relationship between sequences and phylogram. Most sequences used to produce phylogenetic trees are from modern samples. This includes sequences from all groups of organisms from multicellular eukaryotes to primitive bacteria. The branches of the phylogram are graphical representations of evolution based on the dataset that was used for the analysis. As such, the phylogram is an evolutionary reconstruction based on the data.

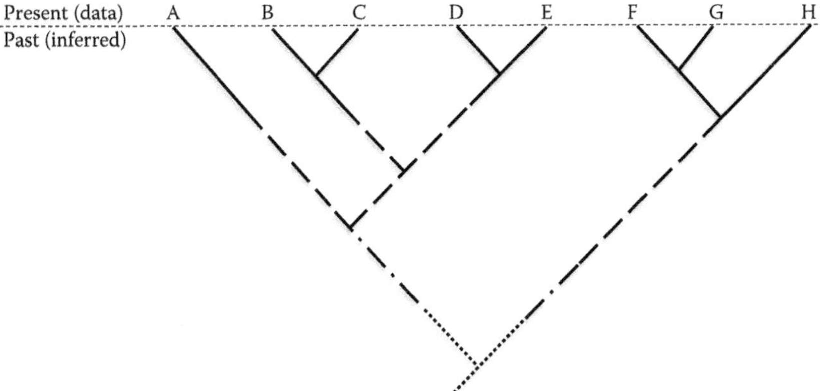

FIGURE 24.14 Phylogenetic resolution decreases with increasing evolutionary time. Depending on the dataset, the resolution of the branches may decrease for times that are more distant in the past. Although the branch points may be correct, distances may be more difficult to determine because more of the mutations may have been missed. Solid lines represent statistically supported branches. Dashed lines represent branches with less support.

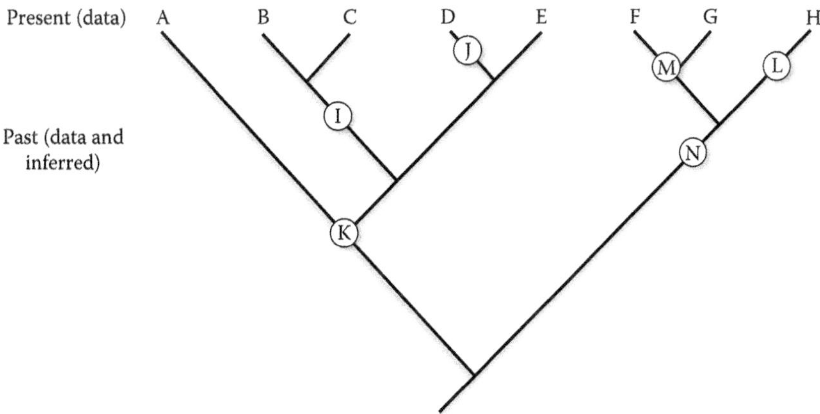

FIGURE 24.15 Use of ancient DNA in phylogenetic studies. If ancient DNA samples are used, then some of the branches can be resolved further back in the past. However, ancient DNA is often damaged from the millennia to millions of years that it spends without being repaired. This can cause the scoring of mutations that did not exist in the living organism at the time it died and was deposited as a fossil. Also, most ancient samples are less than tens of millions of years old, and therefore, its use is limited to studies involving relatively recent evolutionary events.

This includes samples that have been dried or mummified, frozen, encased in amber, or otherwise protected from degradation (see Chapter 19).

Another consideration is whether to use a nucleotide sequence or the deduced amino acid sequence, if the sequenced region encodes a protein (i.e., this is not applicable for rRNAs, transfer RNAs [tRNAs], or other noncoding RNAs [ncRNAs]). When converting a DNA (or RNA) sequence to a protein sequence, it should be noted that 67% of the characters will be eliminated by using the amino acid sequence, because each amino acid is encoded by three nucleotides. However, in this case, selection acts on the proteins and not directly on the DNA, and therefore, the amino acid sequence might have more evolutionary information than would the DNA sequence alone. However, for many phylogenetic algorithms, there are ways to weight each of the nucleotides in the triplet code differently. Therefore, the second nucleotide can be given more weight than the third nucleotide when the distances between sequences are calculated. For genes that only produce RNA (e.g., rRNA, tRNA, and other ncRNA genes), there is no reading frame for translation into proteins. However, many of the nucleotides in RNA hydrogen bond to other nucleotides (sometimes more than one at a time), so that there are certain nucleotides that are conserved, because they are important in maintaining functional structures within the RNAs. More importantly, there are nucleotide interactions that are conserved. Another possibility with nucleotide and amino acid sequence data is that gaps can be scored as additional pieces of data for some of the phylogenetic algorithms. For amino acid data, this is less of a consideration, because essentially it adds a 21st amino acid to the 20th that are already possible. However, for nucleotides, consideration of gaps adds a fifth base possibility to the four that are normally considered. This can be useful for datasets that are of low resolution because it increases the possible number of characters at each nucleotide position by 25%.

OTHER CONSIDERATIONS WHEN PERFORMING PHYLOGENETIC ANALYSES

Evolution once was thought to be s slow steady process. We now know that this is true for some gene regions, but for other gene regions, mutation, genetic drift, and selection can be rapid during some intervals and slower at other times (Figure 24.16). This can lead to branches of unequal lengths on the phylogram because of different rates of change. This sometimes causes problems of inadequate resolution for some sequences and taxa. This is often true for organisms that have become pathogens or endosymbionts. Another consideration when interpreting trees is that not all mutations can be observed (Figure 24.17), and therefore, accurate measurements of actual mutation rates are not always possible. In most cases, mutation rates are underestimated. Again, careful selection of sequences can mitigate this problem somewhat, but cannot eliminate it completely. Multigene families (discussed in Chapter 14) present another challenge. When there is vertical inheritance among members of multigene families, phylogenetic analysis can be performed, and the results are straightforward. The genes are said to be orthologous. However, some gene versions within multigene families can begin to change into versions that differ significantly from the original versions. Some can begin to have somewhat different functions, and sometimes very different functions, and thus, the selective pressures may

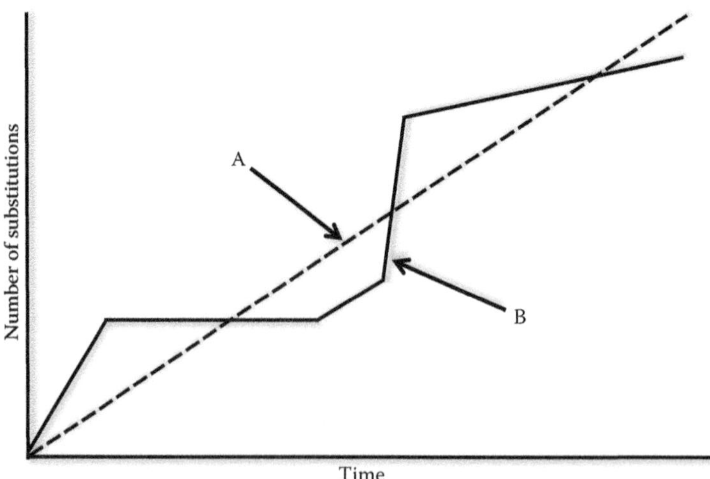

FIGURE 24.16 General rates of substitution. For many genes, the rates of substitution are slow and relatively stable over time (line A). Therefore, graphically, their rates of substitution form a straight line, which makes them useful as rulers of evolutionary time. Other genes appear to experience different rates of substitution at different points in time (line B). This means that they may appear to be conserved genes over one time interval and more variable at other points in time.

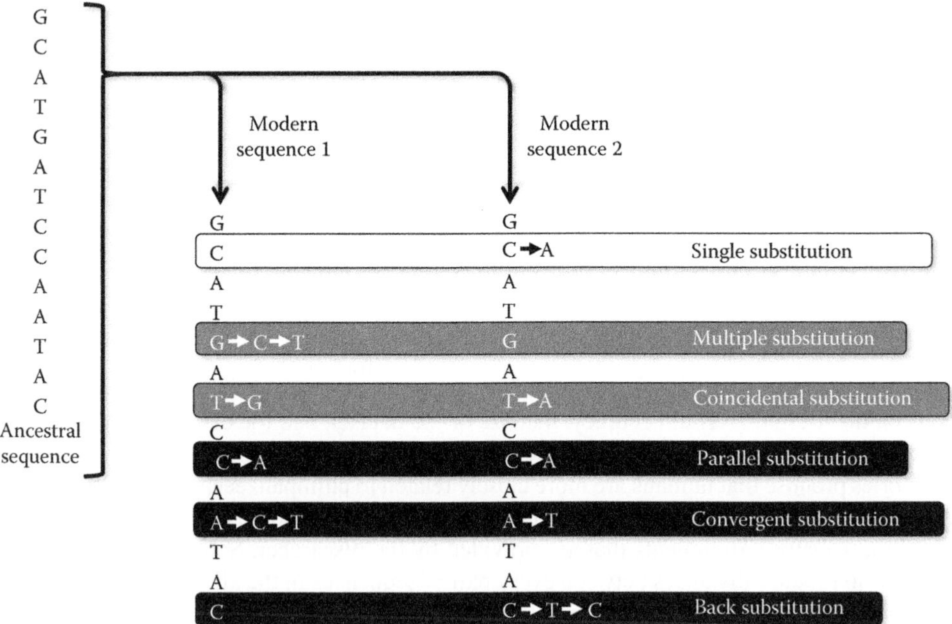

FIGURE 24.17 Observable and nonobservable substitutions. Not all nucleotide substitutions can be observed during evolution. Changes in white boxes are readily observable, those in gray boxes are sometimes observable, whereas those in black boxes often cannot be observed. Some mutations are erased over time, whereas others are masked by other substitutions. Starting with the sequence on the left, which is the ancestral version that is extinct, two modern sequences are presented to illustrate which substitutions may be observed and which ones are essentially invisible, or partly so. When a single substitution occurs (a C-to-A change in sequence 2), it can be observed in all cases. When multiple changes occur, such as the G-to-T-to-C change in sequence 1, only that last substitution is observed, which appears as a G-to-T change. If coincidental changes occur, such as the T-to-G and T-to-A changes in sequence 1 and 2, respectively, it appears to be a single G-to-A change. Parallel substitutions essentially are invisible, such as the two C-to-A changes. Therefore, although there have been two substitutions, neither is scored as a change. Convergent substitutions, such as the A-to-C-to-T change in sequence 1 and the A-to-T change in sequence 2, are also not scored as changes, because they cannot be observed. As indicated, there can be more than two changes that are missed if there are multiple substitutions that result in convergent sequence changes. Back substitutions are also invisible. Of course, with all of these invisible or semi-invisible changes, they may be observed if more sampling is performed, because there are possibilities that some of the intermediate steps in these processes will exist in some individuals and taxa.

differ significantly compared to the unaltered versions. These genes are paralogous. When assembling sequences for alignments and analysis, assessment of whether the sequences are orthologs or paralogs must be made. If all are assumed to be orthologs, inaccurate conclusions may result. However, if orthologs and paralogs are compared in the same analysis, erroneous conclusions may result. However, in some projects, comparisons of orthologs and paralogs might be the purpose of the study.

MODELS OF MUTATION

Considerations regarding the types and rates of mutations affect phylogenetic results. There are several models of mutation that are commonly used, each based on various assumptions. The simplest model is known as the Jukes–Cantor model. It assumes that each of the nucleotide changes is equally likely. That is, a change from an A to a G (a transition) is considered equivalent in cost to a change from an A to a T (a transversion). As was mentioned in Chapter 12, this is usually not true, because in most cases transitions occur more often than transversions. Other models provide alternative calculations that weight transversions more heavily than transitions because of this. Models by Kimura, Tamura, Tajima, Nei, and others include other parameters to their equations to account for the fact that transitions are more common than transversions. In most sequences, the increase of transitions over transversions is from about 1.5 to 3.0 times. Another consideration is that nucleotides at each position of the sequence may not experience the same probability to mutate. Although some of this is due to selective forces, in many cases, it has more to do with the amount of exposure the nucleotide has to mutagens. To account for this variation, other terms can be included in the equations in some of the algorithms that are used to produce the distance values to construct the phylogenetic trees. Often, this can produce more accurate phylogenetic results.

ANALYZING ALIGNED SEQUENCES

The next consideration is in choosing a method to analyze the distances between the sequences. This depends on the desired level of accuracy, the size of the dataset, and the capacity of the computer. It also depends on whether you have specific tree models that you want to compare. For example, some theories propose that humans are more closely related to chimpanzees, whereas other theories propose that gorillas are closest to humans. Certain phylogenetic methods can test the two theories based on the sequence alignments that are provided by the researcher. Some methods use simple calculations to produce distances between every pair of sequences in the alignment, whereas others determine all of the possible ways to join all of the sequences together, and then test each one to determine the one with the fewest number of nucleotide or amino acid changes. Because of the very different calculations, some require very little computing time, whereas others are computationally costly. Therefore, for small datasets (fewer than 30 sequences), all of the methods can be applied; for moderate datasets (30–60 sequences), some of the methods cannot be used; and for large datasets (>60 sequences), only a few of the methods can be used. Of course, these numbers will differ if one is using a super computer versus a small laptop computer.

UNWEIGHTED PAIR GROUP METHOD WITH ARITHMETIC MEAN

This is one of the simplest methods, and also relies on a simple calculation, such that it can be used on very large datasets. The analysis begins by forming a matrix of the aligned sequences. Then, each sequence is compared to every other sequence, and distances are calculated between each pair of sequences, producing a distance matrix (Figure 24.18). The distances can be based on the total number of differences or proportional distances based on the total differences among all sequences. The calculation can be based on a variety of models, including those that weight transversions differently than transitions. Next, the phylogenetic tree is constructed. Tree construction begins by connecting together the two closest OTUs. Next, distances from each of the other sequences and the mean of the two branches connecting the initial two OTUs are calculated. Then the next closest OTU is connected to the tree. This process is continued until all of the OTUs are added to the tree. The final tree produced by the unweighted pair group method with arithmetic mean (UPGMA) is usually a cladogram. Because of the averaging of branch lengths, UPGMA can produce inaccurate trees, especially when different rates of sequence change occur on adjacent branches.

NEIGHBOR JOINING

Another relatively simple, but more powerful method, is neighbor joining (NJ). Although this method also calculates distances based on pairs of sequence differences, it keeps track of the nodes, and therefore does not average the branch lengths, as in UPGMA. Because of this, the trees are more accurate. As with UPGMA, the computation begins by calculating the distances between every possible pair of sequences. Again, various models can be applied, including those that provide

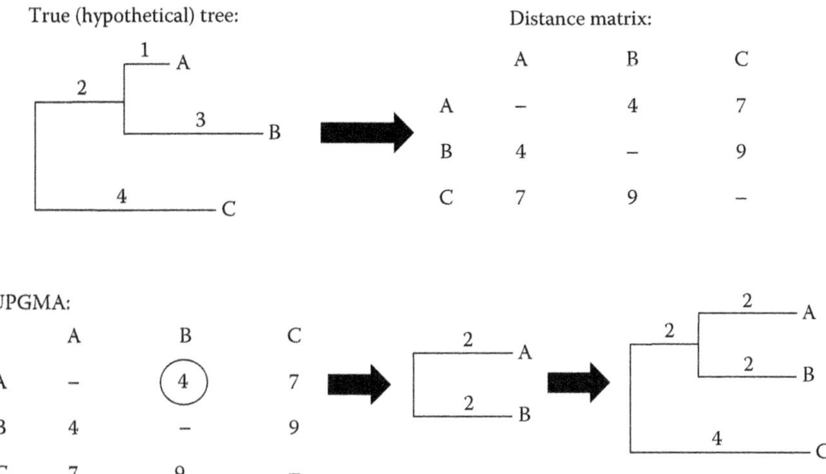

FIGURE 24.18 Construction of tree using UPGMA. Starting with a hypothetical evolutionary pathway (tree at top) with unequal branches, a distance matrix can be constructed. UPGMA averages branch lengths. Therefore, the branch leading to OTU A would be 4/2 = 2, and the branch leading to OTU B would also be 4/2 = 2. The matrix would then be recalculated to determine the average distance from C to the A–B branch, leading to one branch that is 8/2 = 4, and the other would be 2. Therefore, the resultant tree is accurate along some branches and inaccurate along branches that have unequal rates of evolution.

weighting for transitions and transversions. After producing a distance matrix from the sequence matrix, NJ then joins the closest two OTUs together. However, when it joins them, it also considers the distances between these two OTUs and the next closest OTUs. In this way, it can place the node more accurately that does in UPGMA. It then joins the next closest OTU with the initial OTUs based on its original calculation of the distances from those two OTUs (Figure 24.19). Thus, NJ produces a phylogram, rather than a cladogram (although for most software packages, a cladogram option is also provided). NJ is used frequently for moderate and large datasets because it is rapid due to its relatively simple calculations.

MAXIMUM PARSIMONY

Parsimony simply means the simplest explanation. For phylogenetic analysis, it means finding the tree that requires the fewest sequence changes for the entire dataset. Maximum parsimony (MP) differs significantly from UPGMA and NJ, which are based on the initial distance calculations. It begins by creating all of the possible ways to join the OTUs together into bifurcating trees (Figure 24.20). It then searches through the trees to determine which is the shortest tree, based on the sequence changes derived by evaluating the sequence alignment data (Figure 24.21). The number of possible trees increases rapidly as the number of sequences (i.e. OTUs) increases (Figure 24.22), and therefore, this method can be used only for datasets that are small to moderate in size. There are three main search strategies that can be used in MP. The first is an exhaustive search. In this case, the length (number of total nucleotide changes in the tree) of every possible tree is calculated, and the MP tree(s) is reported. Another method is called a heuristic search. In this case, a random number is generated from among the number of possible trees, and the length of the tree corresponding to the selected random number is determined. Adjacent trees are then analyzed until a local minimum length tree is found, which is reported as the MP tree. Although this finds the local MP tree, sometimes it does not find the true MP tree for that dataset. The third method can be termed a semi-exhaustive method, because it does find the MP tree, but does not examine every single tree.

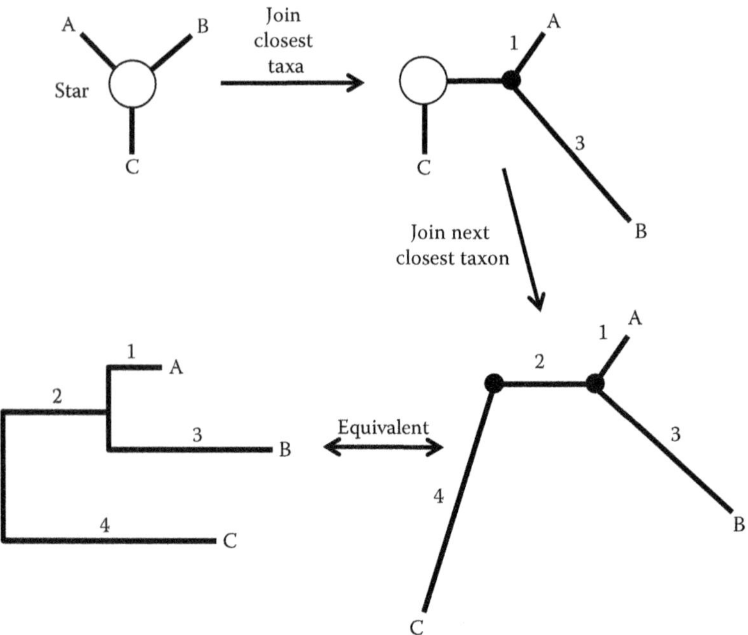

FIGURE 24.19 Construction of trees using NJ. Beginning with the same hypothetical evolutionary scenario and distance matrix as in Figure 24.18 (top), NJ produces branches that not only take into account the distance between the two OTUs being joined but also keep track of the distances to other OTUs. Therefore, the distance between A and B is 4, but the distances between A and C, and B and C differ. Therefore, it places the node at a position that accounts for all of the data. To construct the tree, all of the taxa rejoined into a *star*. Then, the two OTUs that are closest are first joined. The next closest OTU is joined, and the process continues until all of the OTUs are joined. On the bottom left, the tree is converted to a rectangular version, which is comparable to the original hypothetical tree (Figure 24.18, upper left).

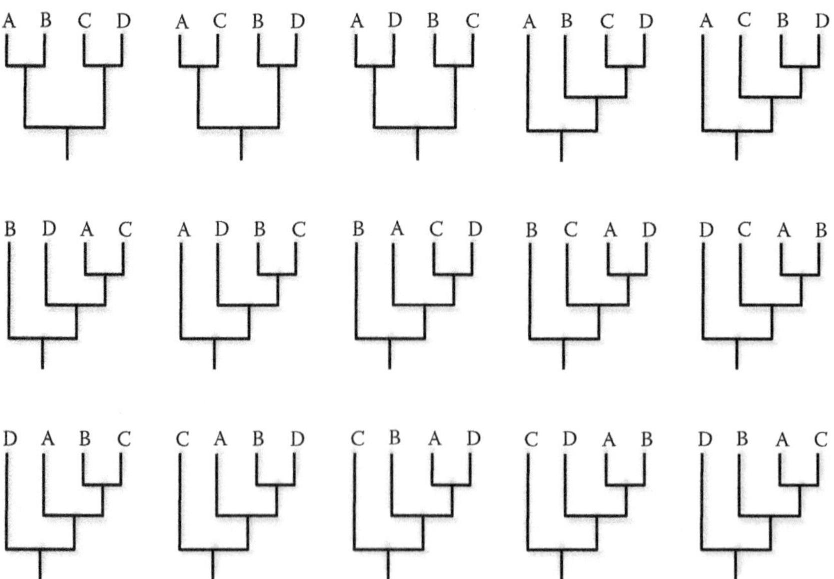

FIGURE 24.20 Joining of four OTUs (A–D). There are 15 possible ways to join four OTUs together in rooted trees. For unrooted trees, there are only three possible trees (see Figure 24.22).

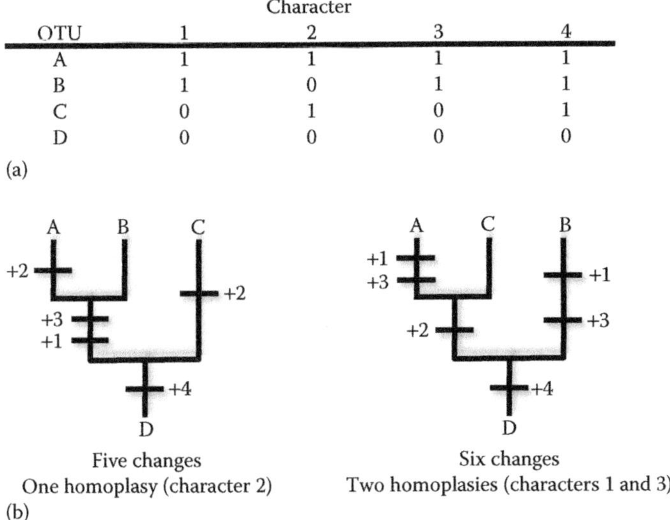

OTU	Character 1	2	3	4
A	1	1	1	1
B	1	0	1	1
C	0	1	0	1
D	0	0	0	0

FIGURE 24.21 Evaluation of trees using parsimony analysis. The matrix at (a) shows the presence or absence of a particular character (e.g., a restriction enzyme site or a nucleotide at a specific location). At (b) are given two of the possible trees that join the four OTUs, with D as the most ancient OTU. Because all of the OTUs, except D, have character 4, that character has to be added early in the tree. For the tree on the left, character 2 must then be added to the branch ending in OTU C, and characters 1 and 3 must be added to the branch joining A and B. However, character 2 must also be added to the branch with A. The addition of a character more than once is called a homoplasy, literally *same change*. Usually, there are some homoplasies in the most parsimonious trees. The tree on the right requires at least six changes and contains two homoplasies. Therefore, in this example, the left tree is the most parsimonious tree, because it has only five changes and one homoplasy.

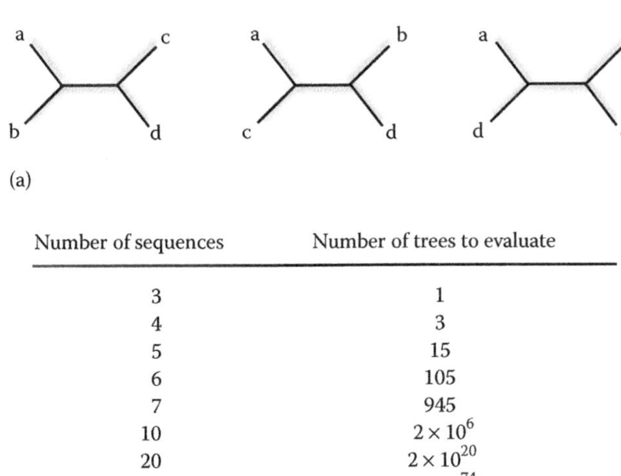

Number of sequences	Number of trees to evaluate
3	1
4	3
5	15
6	105
7	945
10	2×10^6
20	2×10^{20}
50	3×10^{74}

FIGURE 24.22 Number of unrooted trees possible for various numbers of OTUs. The three possible unrooted trees joining four OTUs are shown in (a). The number of possible trees for joining higher numbers of OTUs is listed in (b). The number of possible trees surpasses two million by the time 10 OTUs are being joined and is more than 3×10^{74} by the time 50 OTUs are joined.

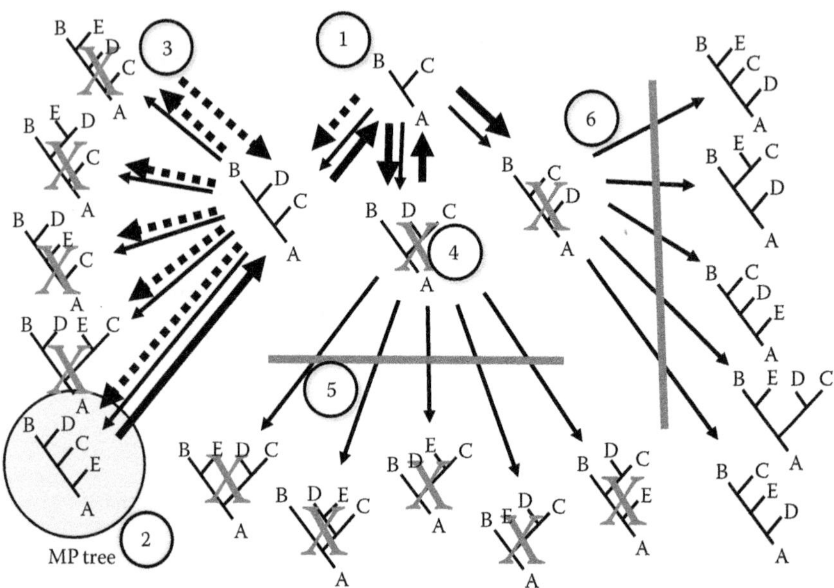

FIGURE 24.23 Branch-and-bound search strategy. This semi-exhaustive search begins by first creating a search tree. This is different from the phylogenetic tree in that it is simply a protocol for performing the search of all possible phylogenetic trees. (1) It begins by joining three OTUs together, and then adding additional OTUs, including all of the possible trees at each addition. (2) It proceeds down one of the branch (dashed arrows) and calculates the length of that tree. It saves that tree as the current MP tree. (3) It evaluates all other trees on that branch (dashed arrows). If they are longer than the current MP tree, then they are eliminated (indicated by an "X"). (4) Next, it proceeds down another branch of the search tree, evaluating each tree as it moves down that branch. (5) If any of them is longer than the current MP tree, then that tree is eliminated, and all other trees along that search branch are eliminated without being evaluated. This is because each one will be longer than the tree that has just been eliminated. (6) Then, it will proceed down other search tree branches until it has eliminated all other possible trees or has found a tree that is shorter than the current MP tree. At the end, it will report the MP tree (or trees, if equally parsimonious trees were found).

However, in a way it is able to evaluate all of the possible trees. It begins by joining each OTU one by one and creating a *search tree* (Figure 24.23). Once it has joined all of the OTUs, it then evaluates one of the resulting trees with all of the OTUs, as well as each of the local trees to determine the one with the shortest length. It then moves back up to the top of the search tree and proceeds down another search branch one node at a time. If one of those trees is longer than the current shortest tree, then that search is stopped at that point, and all of the trees lower on the search tree are eliminated from the search. In this way, large groups of trees are eliminated from the search because they all are longer than the one being evaluated, and therefore they all are longer than the current shortest tree. Although this is a good method for determining the MP tree and is much faster than an exhaustive search, it still requires a great deal of computing time, and therefore, can be used only for small or moderate datasets.

MAXIMUM LIKELIHOOD

This method is based on the likelihood of sequence differences as determined by the frequencies of changes of each of the types of bases (or amino acids). The algorithm will determine this based on the frequencies of each base in the dataset provided or the user can provide a model for base pair change and frequencies. Then, the natural log (ln) of the likelihood (L) is calculated for the branches joining each of the OTUs together, based on the changes in the sequence along those branches, beginning with the two closest OTUs first, and then joining the next closest OTUs sequentially

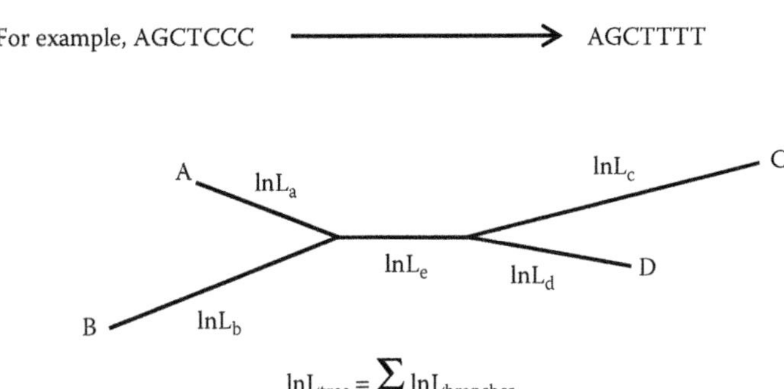
Given the data (which gives the probabilities of
each type of nucleotide change), what is the likelihood
of sequence S1 changing to sequence S2?
For example, AGCTCCC
AGCTTTT
A
lnL$_a$
lnL$_c$
C
lnL$_e$
lnL$_d$
D
B
lnL$_b$
$\ln L_{tree} = \sum \ln L_{branches}$

FIGURE 24.24 ML tree construction. This method begins by calculating the base frequencies in the entire dataset, and then determines the probabilities of change among each of the possible base changes. Alternatively, the user can provide base frequencies. Also, models for nucleotide change can be chosen by the user. Then, ML begins to join the OTUs together, beginning with the two that are closest and proceeding stepwise by adding each successive OTU onto the tree in order of their distances to the other OTUs. Next, it begins to calculate each of the branch lengths based on the likelihood of base changes along that branch. Once it has calculated all of the lnL (natural log of the likelihood) of the branches, it calculates a value of the entire tree by summing all of the lnL branch lengths. This will be the ML tree. The likelihoods for other trees can also be calculated by this method and compared to the ML tree.

(Figure 24.24). The ln likelihood of the entire tree is the sum of the ln likelihoods of each of the branches. One of the other useful capabilities of maximum likelihood (ML) is in testing hypothetical trees. User-supplied trees can be compared to the ML tree to determine whether the hypotheses are supported or rejected, based on likelihood calculations and comparisons. As with the example presented earlier in this chapter, the ML tree can be tested against a tree that has humans closest to chimpanzees as well as a tree that has gorillas closest to humans. Although this is a powerful statistical method, it requires a great deal of computing power. Because of this, it can only be used for small datasets.

BAYESIAN PHYLOGENETIC ANALYSIS

During the past decade, Bayesian inference methodologies have been applied to phylogenetic analyses. Although the method is somewhat conceptually analogous to ML, because it relies on the evaluation of probabilities of sequence change along the branches of the trees, the calculations involved are different than those for ML. In general, this method examines one tree topology based on a previously determined tree that was established previously. The model used to produce the trees can be supplied by the user, or the Bayesian program can produce a model based on the nucleotide frequency and/or a default model. It inches its way through various tree possibilities, testing a current tree with another tree, determining whether or not it is of higher or lower probability, based on the evolutionary model. In doing so, it normally evaluates millions of possible trees. As a statistically based method, it can be used to determine the reliability of the final tree, and it can be used to evaluate and compare more than one tree.

BOOTSTRAPPING

Bootstrapping is a statistical method to measure the strength (confidence) of the branches resulting from the phylogenetic analysis. It is often used in NJ and MP analyses but can be used for other phylogenetic methods as well. It is a multiple resampling method. It first analyzes the original

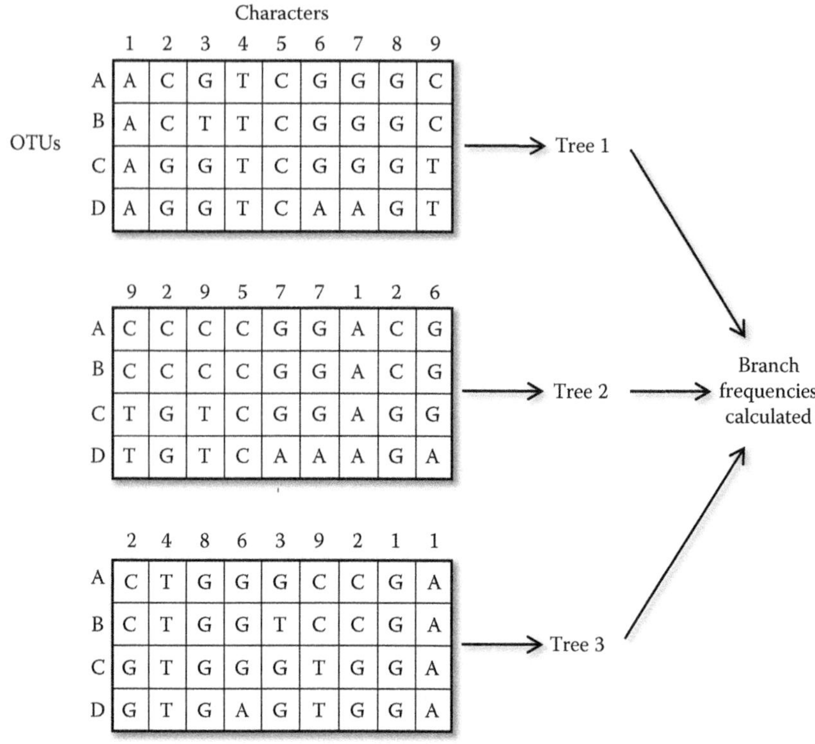

FIGURE 24.25 Bootstrap analysis. This method begins by finding the best tree for a particular dataset. It then resamples the data by building a new data matrix of the same size. Some characters may be selected more than once, and other characters might not be sampled at all. It builds a tree (tree 2) based on this new data matrix. Then, this process is repeated a number of times (usually 500–1000 times). The frequencies of each of the branches are then calculated. Some branches will occur in every tree, and therefore will have a value of 100 (i.e., occurring in 100% of the trees). Other branches may occur less often.

data matrix and creates a phylogenetic tree. Then, it randomly samples characters from the matrix repeatedly, until it has constructed a matrix of the same size as the original data matrix (Figure 24.25). The sampling allows repeated sampling of individual characters, which is called sampling with replacement. This means that some characters (nucleotide or amino acid positions) may be sampled more than once, whereas some characters may not be sampled at all. Once the resampled data matrix is constructed, a new tree is determined and each branch is scored. This process is repeated, usually 500–1000 times (determined by the user). At the end, the frequencies of each of the branches are calculated. If a branch was present in all 1000 resamplings (replications), then the branch is scored at 100 (for 100%), indicating the confidence level for that branch. Usually, branches that are at 90% or above are considered well supported, those at 70%–90% are considered moderately supported, those that are 50%–70% are weakly supported, and those that are below 50% are unsupported by the data. Usually, the unsupported branches are based on only a few nucleotide changes, whereas well-supported branches are based on many nucleotide changes.

VERTICAL VERSUS HORIZONTAL EVOLUTIONARY EVENTS

Phylogenetic analyses are based only on the nucleotide (or amino acid) homologies and differences. No matter which sequences are aligned and analyzed, a tree can usually be produced. Normally, they are used to indicate instances of vertical inheritance and evolution of the sequences (and by inferences, the taxa). However, they can also indicate when horizontal transfers have occurred (see Chapter 15). This is usually indicated when a phylogram produced using one gene region that is vertically inherited (such are rRNA genes) differs from a phylogram derived from another gene region. This usually means either that there are different rates of sequence change or that there have been one or more horizontal gene transfers.

KEY POINTS

1. Species are classified based on shared characters.
2. Although there are many problems with taxonomical classification schemes, and the criteria for determining species vary among the taxa, the systems allow determination and study of species for many purposes.
3. Currently, all life on the Earth is classified into three empires that are divided into eight kingdoms, based on morphology as well as molecular characters.
4. Phylogenetic trees are simplified reconstructions of evolutionary history.
5. Phylogenetic trees are reconstructions of evolutionary history, based on data collected from sampled organisms.
6. Most contemporary phylogenetic reconstructions are based on molecular data, primarily, DNA, RNA, and/or amino acid sequence data.
7. The genomic regions to use in phylogenetics must be carefully chosen to reflect the time of the evolutionary events and/or the taxonomic group being evaluated.
8. Accurate sequence alignment is vital to determining an accurate phylogenetic analysis.
9. All of the mutations that occur during evolution cannot be observed, and therefore, rates of evolution are underestimates of the actual rates.
10. When multigene families are involved in the analysis, a distinction must be made between orthologs and paralogs.
11. The major methods that are used are UPGMA, NJ, MP, ML, and Bayesian phylogenetic analyses.
12. Bootstrapping can be used to test the strengths of each of the branches on a phylogenetic tree.
13. Phylogenetic analysis is useful for evaluating possible instances of horizontal gene transfer, including those involving endosymbiotic events.

ADDITIONAL READINGS

Adams, R. P., J. Miller, E. Golenberg, and J. E. Adams (eds.). 1994. *Conservation of Plant Genes II: Utilization of Ancient and Modern DNA.* St. Louis, MO: Missouri Botanical Garden Press.

Alberts, B., D. Bray, K. Hopkin, A. Johnson, J. Lewis, M. Raff, K. Roberts, and P. Walter. 2013. *Essential Cell Biology*, 4th ed. New York: Garland Publishing.

Darnell, J., H. Lodish, and D. Baltimore. 1990. *Molecular Cell Biology*. New York: W. H. Freeman and Company.

Graur, D. and W.-H. Li. 2000. *Fundamentals of Molecular Evolution*, 2nd ed. Sunderland, MA: Sinauer Associates.

Higgs, P. H. and T. K. Attwood. 2005. *Bioinformatics and Molecular Evolution*. Carleton, Victoria, Australia: Blackwell Publishing.

Karp, G. 2002. *Cell and Molecular Biology*. New York: John Wiley & Sons.

Kumar, S., K. Tamura, and M. Nei. 1993. *Molecular Evolutionary Genetics Analysis*, version 1.01. University Park, PA: The Pennsylvania State University.

Li, W.-H. 1997. *Molecular Evolution.* Sunderland, MA: Sinauer Associates.

Li, W.-H. and D. Graur. 1991. *Fundamentals of Molecular Evolution*. Sunderland, MA: Sinauer Associates.

Liu, Y., J. F. Ammirati, and S. O. Rogers. 1997. Phylogenetic relationships of *Dermocybe* and *Cortinarius* species based on nuclear ribosomal DNA internal transcribed spaces. *Canad. J. Bot.* 75:519–532.

Mount, D. W. 2004. *Bioinformatics: Sequence and Genome Analysis*, 2nd ed. Cold Spring Harbor, NY: Cold Spring Harbor Laboratory Press.
Smith, J. M. and E. Szathmary. 1995. The *Major Transitions in Evolution.* New York: Oxford University Press.
Swofford, D. 1993. *Phylogenetic Analysis Using Parsimony*, version 3.1. Washington, DC: Smithsonian Institution Press.
Wallace, R. A., J. L. King, and G. P. Sanders. 1981. *Biology: The Science of Life*. Santa Monica, CA: Goodyear Publishing Company.
Wilson, E. O. 1992. *The Diversity of Life*. New York: W. W. Norton & Company.

25 Phylogenetic Networks and Reticulate Evolution

INTRODUCTION

Phylogenetic trees present evolutionary patterns based on unaltered vertical inheritance. They are well suited when analyzing ribosomal RNA (rRNA) gene sequences, because these genes are almost always vertically inherited, and they usually have stable rates of change. Most of the conserved and moderately variable regions exhibit stable rates of change over billions of years. It is only the highly variable regions (e.g., the intergenic spacers) that often exhibit variable rates of change over short periods of time. Some other genes, such as histone genes, also have similar patterns of inheritance and genetic change. Because of this, phylogenetic trees based on these genes produce relatively well-resolved linear branches. However, for many other genes, the evolutionary patterns may often be nonlinear, and the genes may not be inherited in a strictly vertical fashion. Nonetheless, the sequence data can be used in phylogenetic analyses, and a tree will most often be produced. However, the tree may represent a model that is far from the actual evolutionary steps that resulted in the sequences. Even in analyses where rRNA genes are used, erroneous conclusions may result when the resolution in the sequence data is below that necessary to resolve the evolutionary patterns. However, lack of resolution may indicate other processes that have occurred, which are unresolvable using standard phylogenetic analyses. Utilization of phylogenetic network analyses can sometimes resolve these issues and lead to a more accurate conclusion regarding the patterns of evolution.

Up until the second half of the twentieth century, evolution on the Earth was thought to be a series of genetic bifurcations, many of which led to speciation events. It was thought to be a progression of the separation and isolation of two populations of a single species that each eventually diverged genetically such that they would never interact genetically after that point. Therefore, the evolution of organisms could be represented as a bifurcating tree, and an especially bushy one at that, with each of the branches being a single straight line. With the advent of molecular methods from the 1960s onward, molecular data began to be used to determine those bifurcations using phenetic and phylogenetic methods. In other words, the nucleic acid or amino acid sequences could be aligned, and the similarities and/or differences could be used to determine the molecular relationships among the sequences.

By the mid-1980s, problems began to surface with respect to the determination of purely bifurcating phylogenetic trees. Some sequences evolved rapidly, some evolved very slowly, and many evolved at moderate rates. When comparing the gene trees with species separations, the trees determined using different gene sequence sets sometimes resulted in trees that were incongruent with one another. Although some were moderately different, some trees contained particular species that were placed within completely different taxonomic groups than their usual positions. Although this was sometimes due to lack of resolution in the data, in many cases it resulted from genetic changes within a genome, exchanges among members of the same species, and occasionally from genetic exchanges between different species. In several cases, the incongruencies resulted from genetic exchanges among hosts, pathogens, symbionts, and/or endosymbionts. These horizontal gene transfers (HGTs) were problematic, but at the same time they were evolutionarily interesting and important. Rather than producing bifurcating branches on the phylogenetic trees, they formed branches that interacted and fused with other branches, indicating the mixing of genomes. That is, they formed networks, rather than simple trees. It was discovered that reticulate evolution was common, and often produced very different results than would stable vertical evolutionary processes. Other processes also occur that complicate the analyses of sequence data. Species hybridization, recombination (both meiotic and mitotic), transposition, gene conversion, and reassortment all are known to occur, and each complicates phylogenetic analysis. These are collectively categorized as non-Mendelian mechanisms.

PHYLOGENETIC ANALYSES OF RETICULATE EVENTS

When a phylogenetic analysis yields incongruous results, or unresolved branches, the problems might not be with the data or the alignment. In some cases, the analytic tools might be applying inappropriate calculations for the evolutionary processes recorded in the sequences. Phylogenetic network analysis can often be used to tease out these evolutionary processes and yield a more accurate result and conclusion regarding the evolutionary history of the sequence dataset. Phylogenetic trees and phylogenetic networks are both graphs of the evolutionary steps that are indicated by the dataset. However, although phylogenetic trees utilize algorithms that yield bifurcating trees, phylogenetic network algorithms examine each of the bifurcations to determine whether a more complex graph is more compatible with the changes indicated in the dataset. Rather than being a simple bifurcating tree, a phylogenetic network has multiple branches emanating from some nodes (called splits). Each branch in a split represents a slightly different path (i.e., an edge in the graph) to each of the adjacent nodes. As noted in Chapter 24 (see Figure 24.4), this more closely represents an accurate model of the processes that constantly contribute to evolution on the Earth.

In phylogenetic analyses, trees are built using algorithms that either join branches based on calculations of distances to the next closest taxon or join the sequences based on the minimum number of nucleotide or amino acid steps from one sequence to the next. In cases in which there are very few differences between two sequences, or the total dataset indicates that there are equivalent, but incongruous solutions, a tree with weak statistical support results, or a tree with one or more unresolved branches is produced. In phylogenetic network analyses, the algorithm analyzes each of the incongruous or unresolved areas in the data and produces a graph that includes two or more possible routes for the branches (Figure 25.1). Because the phylogenetic network graph provides a model that considers more of the data than a standard phylogenetic tree, it provides clues regarding the alternative pathways that have led to the patterns in the sequence data.

ADVANTAGES OF PHYLOGENETIC NETWORKS

Transfer of DNA and rapid genomic changes have been attributed to several mechanisms, including HGTs, species hybridization, recombination, gene conversion, transposition, reassortment, gene duplication, and gene loss. During the past few decades, the frequency and importance of these mechanisms have become more apparent. This has presented problems for those attempting to perform phylogenetic reconstructions of genes and taxa because the gene trees and the species trees are often incongruent. At the same time, these problems have opened up new areas of interesting research and have spawned additional tools for determining the evolutionary pathways for large sets of organisms. These reticulate events result in an evolutionary pathway that can be best represented as a network rather than as simple lines. In fact, phylogenetic trees are the simplest type of phylogenetic network. However, instead of determining detailed bifurcating trees, many of which would be incorrect when these types of changes occur, phylogenetic networks attempt to reconstruct the true pathways of specific gene changes regardless of taxonomic assessments.

The most obvious example of the need for phylogenetic networks came from results based on the genes from organelles that have moved into the nuclei of their hosts (e.g., mitochondrial and plastid genes). Based on small subunit rRNA (rRNA SSU) gene sequences, bacteria, archaea, and eukarya all form distinct monophyletic clades. However, during past evolutionary events, many of the genes (e.g., most adenosine triphosphate synthase subunit genes, most of the tricarboxylic acid cycle genes, most of the genes involved in photosynthesis) have moved from the organellar genome

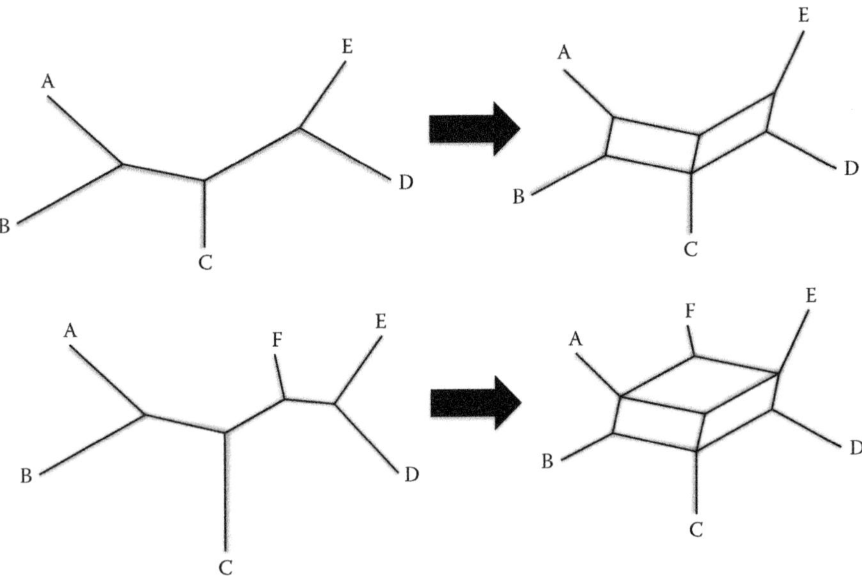

FIGURE 25.1 Hypothetical phylogenetic trees and possible conversions to phylogenetic networks. The upper conversion demonstrates how splits are expanded into networks that better fit the data. Although the relationships are essentially the same, additional refinement of the splits are indicated in the phylogenetic network. The lower conversion shows how the networks can become more complex and that some relationships (pathways to sequence F) may differ from those in the simple phylogenetic tree.

into the host nuclear genome. When some of these nuclear-encoded genes are used to construct phylograms, the trees position the eukaryotes closest to either proteobacterial or cyanobacterial species. However, when the nuclear rRNA SSU genes from eukaryotes are used, the eukaryotic clade is clearly separated from both the bacterial and archaeal clades. Clearly, the rRNA SSU genes and the genes that originated from the organelles indicate two separate parts of the evolution of these organisms. They indicate that the rRNA genes have been inherited vertically, whereas the organellar genes entered the eukaryotic hosts via HGT. In the case of mitochondria and plastids, this occurred by importation of the entire ancestral bacterial genome into the eukaryotic host. After the transfer, much of the ancestral bacterial genomes were integrated into the nuclear genomes. When genomes are compared, concatenating the two types of genes in a single analysis leads to erroneous conclusions, unless algorithms are used that specifically identify these incongruencies. Some algorithms to generate phylogenetic networks address these issues and can lead to the discovery of genetic changes that would otherwise be missed.

As mentioned above, the simplest phylogenetic network is a phylogram (Figure 25.1). However, the algorithms for determining networks are capable of examining connections in greater detail. Depending on the evolutionary model used and the sequence variation, the algorithm may find a solution to a particular branch of a phylogram that incorporates more of the sequence data by building a network rather than a line. It may result in a highly branched network (Figure 25.2), with some of the splits being difficult to interpret, but which account for more of the data than do simpler phylogenetic methods. Most build simpler networks that incorporate as much of the data as possible. For some algorithms, only the data that indicate sequence differences are used, whereas the parts of the sequence that are identical throughout the dataset are not considered. This is done to reduce the amount of computer time needed for the analyses. The amount of complexity in the network can also be controlled. However, the amount of complexity might not fully correspond to the actual biological processes that have led to the sequences being examined. This statement is essentially true for all phylogenetic reconstruction methods. They all computationally deduce a particular

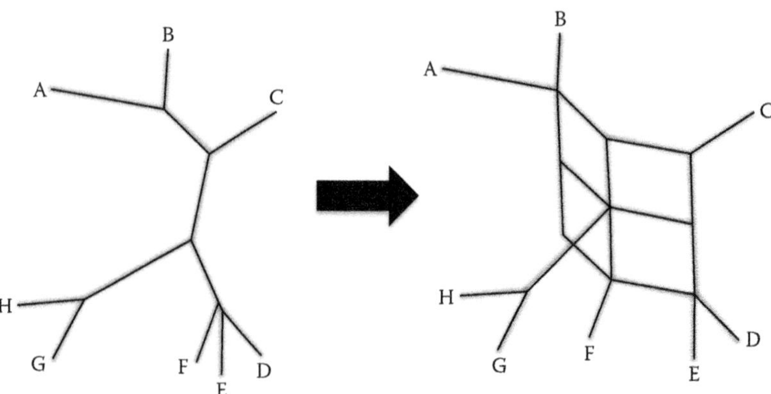

FIGURE 25.2 Hypothetical phylogenetic tree that contains an unresolved branch {D,E,F} converted to a phylogenetic network in which the taxa on the branch are resolved. Also, note that the terminal branches can attach to the network at any intersection (node) of network edges (branches), including external and internal intersections, (e.g., the branch joining {H,G}).

evolutionary pattern based on the data that are being evaluated. The dataset is always a small fraction of the total number of changes and events that have occurred in comparisons of two or more sequences, and thus, the results are always under representations of the actual events. Nonetheless, a phylogenetic network is a working model used to examine a dataset or to test a hypothesis based on what is known about the evolution of the sequences.

HORIZONTAL GENE TRANSFERS

HGTs (which are described in more detail in Chapter 15) have occurred often in the evolution of organisms on the Earth. Either small or large pieces of genetic material are transferred from one genome into another (Figure 25.3). HGTs may be missed by phylogenetic algorithms, especially if they were small or very ancient HGTs. Often phylogenetic network results can clarify some of the HGTs that may be missed by standard phylogenetic methods. HGTs are primarily determined when more than one gene set is being compared among a group of taxa, or when the phylogenetic relationships among the group of organisms or genes already are known. The phylogenetic trees of each set are compared, and if one is grossly incongruous, then HGT may be suspected. A phylogenetic network can be determined based on the incongruencies, and statistical tests (e.g., maximum likelihood) can be performed to determine whether the phylogenetic network can explain the data

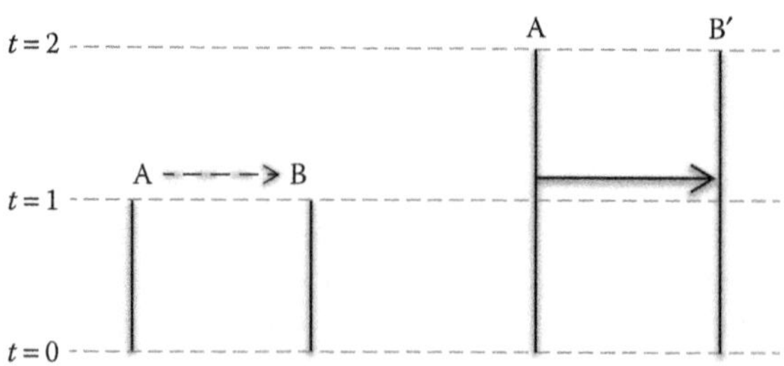

FIGURE 25.3 An evolutionary pathway indicating horizontal transfer of one or more genetic elements. Taxa A and B can be either members of the same species or members of different species. Note that after transfer taxon B has been changed to become taxon B′. If the genetic changes are few, it may retain all of the characteristics of taxon B. However, if there are many changes, the taxon may be changed significantly, sometimes becoming a new species.

better than the phylogenetic trees alone. If a genomic dataset is used, which is a concatenation of a large number of genes, then the HGT signal might be lost for any single gene. In this case, the search could be performed on subsets of the dataset to determine whether any signs of HGTs might be present. Utilization of phylogenetic networks is useful for such searches as they are more sensitive to reticulation than standard phylogenetic methods. As mentioned before, HGTs have been demonstrated in the evolution of mitochondria and plastids, but they have also been described as frequent events in the evolution of a number of other organelles present in members of the Chromalveolata, Excavata, and Rhizaria. In some of these, there have been secondary and tertiary endosymbiotic events, with the concomitant mixing of genomes (discussed in Figure 31.2). In most cases, these have been photosynthetic eukaryotes (red or green algae) that have become endosymbionts (and eventually organelles) of nonphotosynthetic eukaryotic hosts. Thus, there have been entire genomes (including nuclear, mitochondrial, and plastid genomes) introduced into eukaryotes that have their own nuclear and mitochondrial genomes. Over evolutionary timescales, the genes from each compartment have moved primarily into the nucleus of the host organism to the extent that for some dinoflagellates, up to 90% of the nuclear genome has been imported by HGTs. Phylogenetic network algorithms can be useful in elucidating the ancestry and movements of the genes in these complex genomes.

SPECIES HYBRIDIZATION

Another mechanism that creates reticulate evolution of genes is species or varietal hybridization (Figure 25.4). Individuals and populations always contain mixtures of alleles, and therefore, it is often difficult to determine whether an individual from a population or two individuals from two different populations are genetically representative of the species or population. Thus, selection of the individuals and genes to study is crucially important to the outcome of the research. As populations of a species become isolated, or in some cases in which speciation has already occurred, the genomes of each have already diverged to some extent, either due to mutation and selection or simply from genetic drift. If the two are able to hybridize, that is, mate to form viable individuals, then the genomes will mix, and the allelic versions for each of the genetic loci will sort out in the progeny. Although standard phylogenetic methods may reflect these events, they may be extremely complex, and each of the loci may provide only a partial answer regarding the hybridization event. Analysis may simply result in fully resolved phylograms that fail to indicate hybridization, unresolved phylograms, due to the inability of the algorithms to discern hybridization events, or phylograms that

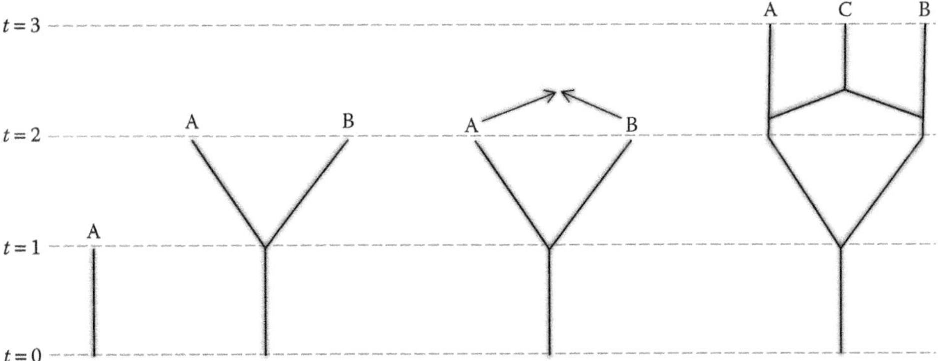

FIGURE 25.4 An evolutionary pathway leading to species hybridization. At $t = 1$, a single lineage is present that splits into two species (A and B) beginning at $t = 1$. At $t = 2$, members of species A and B mate to form a hybrid species that then becomes established as a separate species at $t = 3$. The final network accurately represents the evolutionary pathway of the three species.

appear accurate, but do not reflect the actual evolutionary events that led to the sequences in the dataset. Phylogenetic networks may be capable of accurately reflecting some of these hybridization events, as long as the sequences that are selected have sufficient resolution to indicate the changes, and that sampling of individuals in the population is sufficient to provide enough sequence data to reflect the event(s).

RECOMBINATION

Homologous and nonhomologous recombination events can occur at almost any time within a genome. Although they occur in bacterial, archaeal, and eukaryotic somatic diploid (and haploid) cells, they are more frequent during meiosis in eukaryotes and during DNA repair processes. Recombination can lead to a number of different outcomes, including gene duplication, gene loss, exchange of alleles, and gene conversion (Figure 25.5). Because these events occur often, rearrangements in the genome, especially for tandemly repetitive sequences (e.g., rRNA loci), are present throughout populations and multicellular organisms, and can be observed because of the non-Mendelian inheritance in the population, and within the cells of individuals. Phylogenetic networks can be useful in spotting these changes, although adequate sampling and gene locus are crucial to determining whether recombination is active and what types of changes are exhibited. Depending on the gene locus being analyzed, they may be detectable within an individual, within a tissue (or tumor), among individuals of a population, or at higher taxonomic levels. One note of caution with genes that have undergone duplication events is that the separate genes might still be orthologous, but in many cases, they may have evolved

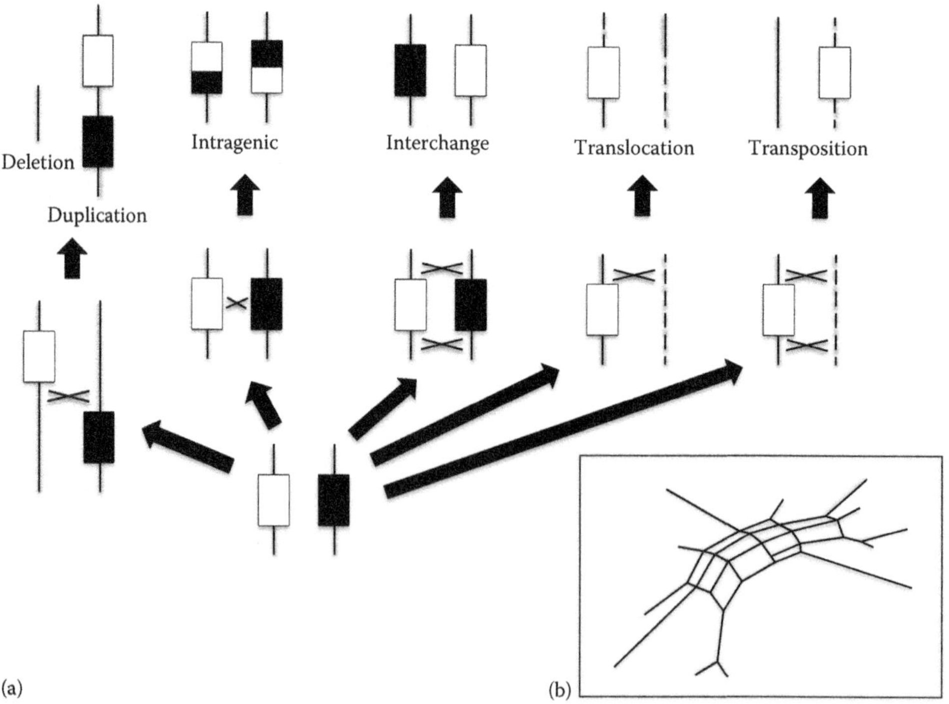

FIGURE 25.5 Potential products of recombination processes. (a) From *left* to *right*: An unequal crossover resulting in a deletion of the gene from one chromosome and a duplication of the gene on the reciprocal chromosome; a crossover within the gene, resulting in chimeric genes on both chromosomes; a double crossover on either side of the alleles resulting in an exchange of the two alleles on their respective chromosomes; a chromosomal translocation resulting in novel chromosomes and genetic linkages; and a transposition of a gene into a novel location. (b) Hypothetical network resulting from recombination processes.

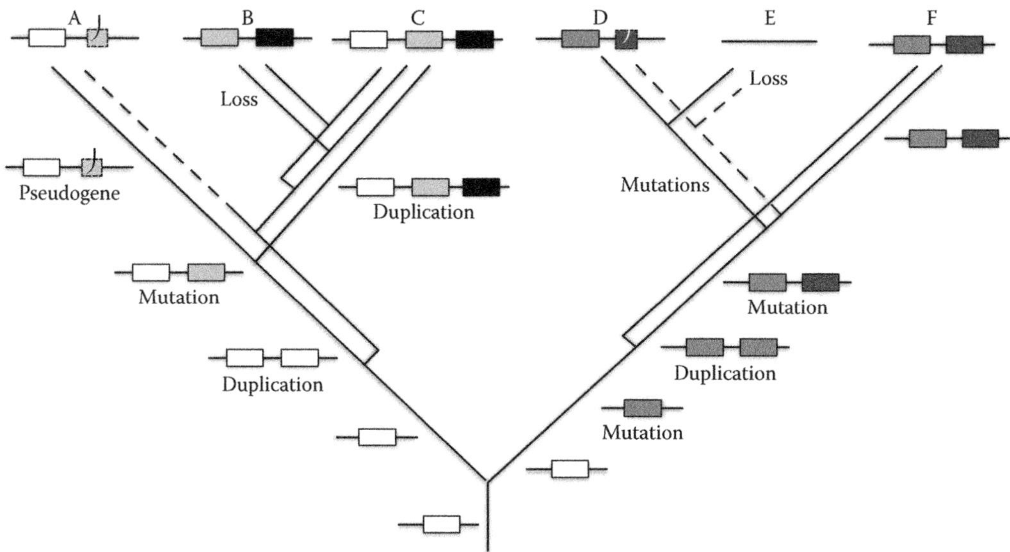

FIGURE 25.6 Phylogram with gene duplications, changes, and losses mapped onto the phylogram. The single copy gene (root at the bottom) is initially split through speciation of two taxa. It remains unchanged on the left branch, but there is a duplication of the gene. Following duplication, additional mutations occur such that the two copies differ from one another, and in some cases one of the genes loses its function, thus becoming a pseudogene (ψ). Additional duplications and deletions occur such that each version (A–F) eventually differs from the other versions. Phylogenetic network analyses can help to elucidate some of these complex patterns.

to become paralogs (Figure 25.6). Phylogenetic network analyses can sometimes yield information that indicates the separation of orthologs from their paralogs. Standard phylogenetic methods can sometimes indicate these separations, but can also sometimes lead to erroneous conclusions regarding the evolutionary relationships of certain genes or taxa.

TRANSPOSITION

Transposition can occur within a genome or between genomes, and in some organisms (e.g., *Zea mays*) can affect a significant portion of those genomes. As with recombination, transposition may occur during one or more of the phases of the cell cycle or the life cycle of the organism, often occurring during DNA replication (Figure 25.5a, upper right). Many types of transposons (the main agents of transposition; discussed in Chapter 13) can move themselves, move other transposons, can disrupt genes, and can cause chromosome breakage, some of which may be repaired incorrectly. Although phylogenetic networks have not been specifically used to investigate transposition events, they may be useful for spotting signs of transposition within an organism or among a group of organisms.

REASSORTMENT

Reassortment has been documented in influenza A viruses, but the process is likely more widespread than this and is similar to sorting out in the genomes of eukaryotes with more than one chromosome. Influenza A viruses have genomes consisting of eight RNA chromosomes, each carrying at least one gene (see Chapter 13). These viruses also infect a large number of host species of birds and mammals. Some genotypes can infect many different species. All 16 hemagglutinin (H1–H16) and all 9 neuraminidase (H1–H9) subtypes have been found in birds. However, only H1–H3 are transmissible among humans, although H5, H7, and H9 can be transmitted from birds to humans, but human-to-human transmission does not occur. Reassortment of the chromosomes can occur when an animal is concurrently infected by more than one subtype. This has been documented in several cases, including some reassortant genotypes that have been virulent in the human population. In these cases, mixing occurs in some of the cells, such that some of the chromosomes that originated from an avian strain, some that originated from a human source, and some that originated from a swine source are packaged together in individual virions, which infect other cells, and are eventually transmitted to another host organism. In some ways, this is similar to species hybridization (described earlier).

EXAMPLES OF RETICULATE EVOLUTION EVENTS

As mentioned earlier in this chapter, mitochondria and plastids originated as endosymbionts, and phylogenetic analyses and networks support this conclusion. Additional endosymbiotic events have also been indicated and/or confirmed using phylogenetic and phylogenetic network analyses (see Figure 31.2). Moreover, alternative types of evolutionary trees are being used to more accurately visualize evolutionary patterns and events that have occurred. For example, the standard tree of life is composed of straight bisecting branches (Figure 25.7a), with the inclusion of intersecting lines representing the horizontal transfer events of the endosymbiotic events that led to the origin of mitochondria and chloroplasts. However, a stronger statement, and more accurate tree of life, is one that directly shows the endosymbiotic nature of the origination of the two organelles (Figure 25.7b). More such trees are appearing in the scientific literature to more accurately indicate the modes of evolution. One version of a network tree was used to illustrate the possible origin of the two membranes of diderms (i.e., Bacteria that have a double membrane, often referred to as Gram-negative bacteria). When a phylogenetic tree of Bacteria is constructed, species having one membrane (mostly Gram-positive bacteria) and those having two membranes are polyphyletic (Figure 25.8a). That is, there appear to be multiple origins of diderm bacteria. However, the most ancient bacteria appear to be monoderms (Gram positive). One hypothesis is that the double membrane was the result of an endosymbiotic event that

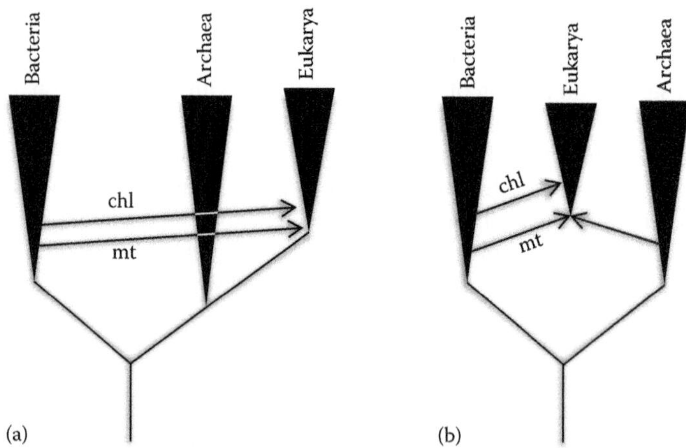

FIGURE 25.7 Two versions of the evolutionary history of the origination of eukarya that show some of the HGTs that have led to the eukarya. (a) A standard phylogram with HGTs superimposed on the phylogram. (b) A phylogenetic network indicating a more accurate depiction of the origination of the eukarya, which joins an archaeal genome with a bacterial genome to form the first eukaryotic cell. It is unclear whether the mitochondria were present in the first eukaryotic cells or if they were introduced shortly after (within approximately 200 million years) the first eukaryotic cells were formed mt, mitochondrion; chl, chloroplast.

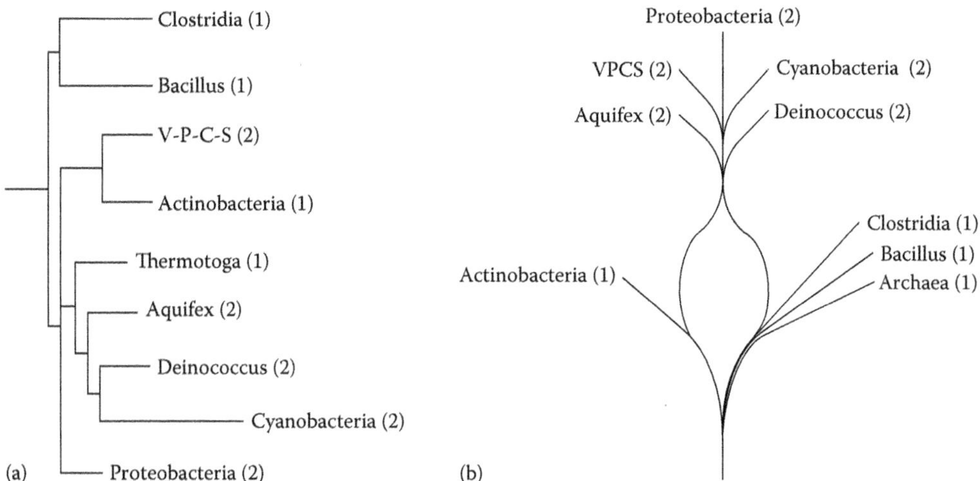

FIGURE 25.8 Phylogram (a) and circular phylogram (b; a type of phylogram network) of Bacteria to examine whether double-membrane bacteria originated from an endosymbiotic event between two single-membrane bacteria. (a) A phylogram based on rRNA SSU genes from selected bacterial groups. The numbers in parentheses indicate the number of cell membranes for each taxonomic group (1 = monoderm; 2 = diderm). The number of membranes is polyphyletic, because the monoderms and diderms each fail to form monophyletic clades. The circular phylogram (b) is based on a phylogenetic reconstruction using a dataset containing conserved ancient genes from the same bacterial groups. The circular tree was the best tree in a maximum likelihood analysis, which indicated that the diderm (double-membrane) bacteria may have originated from a symbiotic event between a member of the Actinobacteria (a monoderm) and a member of the Clostridia (Firmicutes; a monoderm). In this phylogenetic network, monoderms and diderms each form monophyletic clades.

occurred at least three billion years ago between two species of bacteria, each with a single membrane. Because the event occurred so long ago, the phylogenetic signal coming from individual genes has been insufficient to test this hypothesis with standard phylogenetic methods. However, through examination of sets of genes from genomic projects, and using a phylogenetic network approach, trees have been constructed that account for more of the sequence data and yield trees in which monoderms form a monophyletic group and diderms form a monophyletic group. This was done by examining a set of genes that are known to be very ancient using maximum likelihood methods. The tree (Figure 25.8b) results indicated the possibility that a symbiotic event between a member of the Clostridia (a Firmicute) and a member of the Actinobacteria had occurred. This tree had a statistical likelihood score that was significantly better than the rectangular phylogenetic tree (Figure 25.7a), which included the same organisms. In addition to being favored by maximum likelihood analysis, diderms are included on a large monophyletic clade. The conclusion is that a single endosymbiotic event may have led to the formation of Bacteria with double membranes (diderms; or Gram-negative bacteria).

KEY POINTS

1. Often, evolution cannot be represented as simple bifurcating phylogenetic trees.
2. Reticulate evolution is defined as the partial merging of different genetic lineages.
3. Processes that cause reticulate evolution include HGT, species hybridization, recombination, transposition, gene conversion, reassortment, gene duplication, and gene loss.
4. Phylogenetic network analyses can be used to indicate and study reticulate evolutionary events.
5. Endosymbiotic events, including the mixing of the individual genomes, are more accurately resolved and described using phylogenetic network analyses.

ADDITIONAL READINGS

Higgs, P. H. and T. K. Attwood. 2005. *Bioinformatics and Molecular Evolution.* Carleton, Victoria, Australia: Blackwell Publishing.

Huson, D. H., R. Rupp, and C. Scornavacca. 2010. *Phylogenetic Networks: Concepts, Algorithms and Applications.* Cambridge: Cambridge University Press.

Li, W.-H. 1997. *Molecular Evolution.* Sunderland, MA: Sinauer Associates.

Li, W.-H. and D. Graur. 1991. *Fundamentals of Molecular Evolution.* Sunderland, MA: Sinauer Associates.

Mount, D. W. 2004. *Bioinformatics: Sequence and Genome Analysis*, 2nd ed. Cold Spring Harbor, NY: Cold Spring Harbor Laboratory Press.

26 Phylogenomics and Comparative Genomics

INTRODUCTION

Other than virus genomes, the first genomes of cellular organisms began to be published about two decades ago. However, the initial genomes were primarily descriptive, covering the contents of a single genome. Once a few bacterial genomes were completed, comparisons were made among the genomes outlining the similarities and differences between the genomes. Once the first archaeal genome (*Methanocaldococcus jannaschii*) and the first eukaryotic genomes were published (*Saccharomyces cerevisiae* and *Caenorhabditis elegans*), the genomes of the three large domains of life were compared for the first time, which initiated the field of comparative genomics. However, the number of genomes was still too small for phylogenetic reconstructions, and therefore, for a few more years, phylogenetics consisted of the analyses of one or a few gene regions from sets of organisms.

Once the number of genomic sequences increased sufficiently (by about a decade ago), it became possible to compare sequences from many genetic loci among a number of taxa. One of the first uses for the comparisons was in gene discovery and gene prediction. When a gene was identified in one genome, a search for orthologs and paralogs was performed on a second genome. A large proportion of genes have been determined using genomic and phylogenetic tools using this strategy. Early in this process, not only were there too few genomes to perform valid phylogenetic analyses, but the phylogenetic tools were insufficient to analyze the enormous amounts of data being generated from genomic sequencing projects. The computational capabilities suffered from a lack of computing power and a lack of algorithms that could speed up the process of analysis. Eventually, both improved to the point where phylogenetic reconstructions based on sequence alignments from multiple concatenated genes from multiple species could be analyzed. However, not every sequence from each genomic dataset could be used. First, the data had to be screened for those genes and other regions that were in common to all of the genomes being analyzed (Figure 26.1). These had to be further screened to determine whether the sequences could be accurately concatenated (joined together *in silico*) and aligned for phylogenetic analysis (Figures 26.2 and 26.3). The first phylogenomic studies found that the branching patterns matched those based on ribosomal RNA (rRNA) sequences, as well as many of those based on morphology and physiology. However, as more genomes were sequenced, in some cases there were incongruencies. When individual gene sequences were compared, often signs of horizontal gene transfer (HGT) were indicated for some of them. Thus, an additional use of phylogenomics became the search for HGTs, which have been found in a large number of genomes, including some in which most of the genome appears to have been gained through HGTs.

IMPROVEMENTS IN SEQUENCING AND PHYLOGENOMICS

After the entire human genome was sequenced using Sanger-based (dideoxynucleotide) methods, many researchers, engineers, and biotechnology companies realized that methods faster than Sanger sequencing would have to be developed to speed up progress in genomic sequencing, especially for organisms with large genomes (megabase lengths and larger). Next-generation sequencing (NGS) methods (e.g., 454, Illumina, Ion Torrent; see Chapter 21) began to be developed in the 1990s and were employed in large sequencing projects by the mid-2000s. Using these methods, bacterial genomes can be sequenced in a few days, whereas only a decade before they required

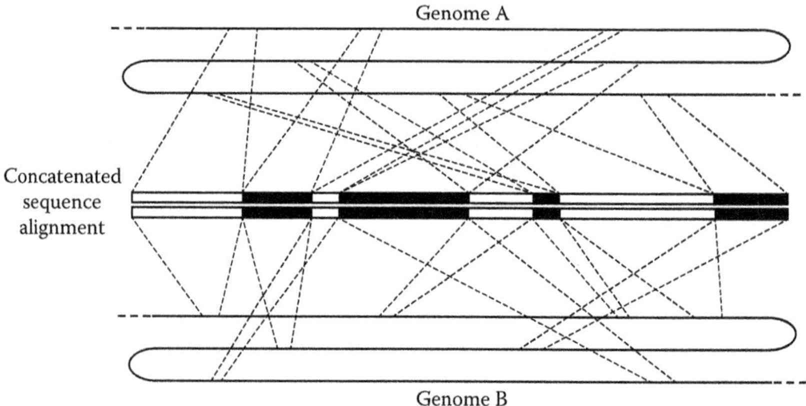

FIGURE 26.1 Alignment of sequences for phylogenomic analyses. Orthologous genes and other regions are located in each genome (in this case, genomes A and B). The regions are concatenated in the same order for each genome, and then they are aligned. Any nonhomologous regions or regions that cannot be unambiguously aligned are excluded from the phylogenetic analyses.

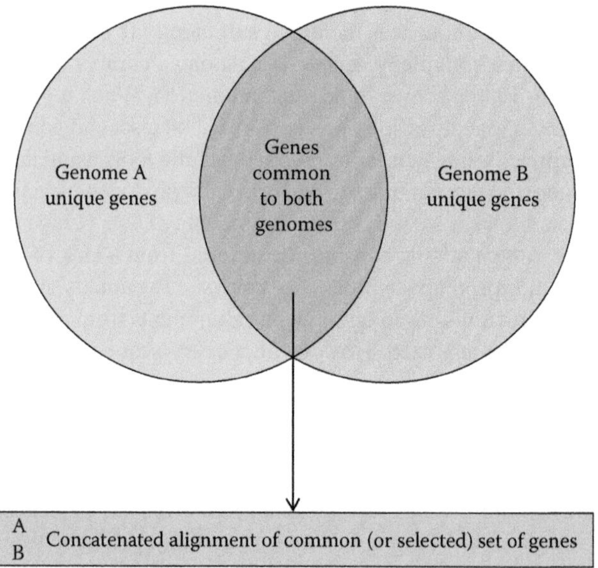

FIGURE 26.2 Venn diagram of the genes common to two genomes. A proportion of the genes in each genome are in common and can be aligned, compared, and used in phylogenetic analyses. The number of genes that can be compared is dependent on the genome sizes, as well as the phylogenetic distance between the two genomes.

months or years of effort to sequence. Currently, sequencing information is being added to databases at unprecedented rates. The problem of genome analysis has shifted from the limitations of sequencing to the limitations of the analytical software and computers that are used to elucidate and describe the genomic sequences. That is, once a genomic sequence is generated, the genes, controlling regions, introns, and other key features, must be located and compared with analogous features in other genomes. This annotation is needed prior to phylogenomic analysis, because the correct regions must be properly and accurately aligned prior to any phylogenetic analysis. Although the problems are not insurmountable, they are formidable.

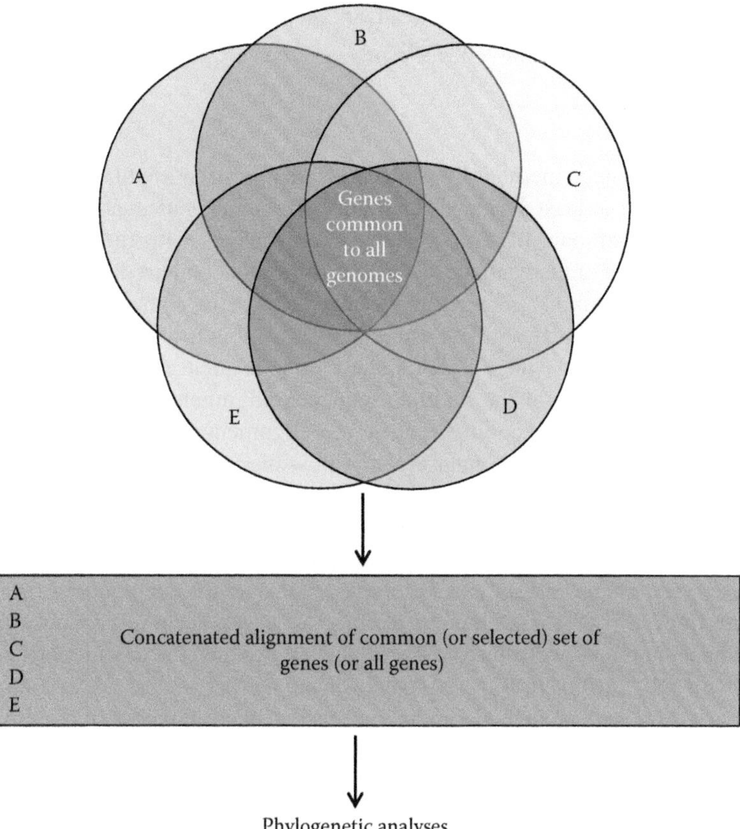

FIGURE 26.3 Venn diagram of overlap of genes for five genomes. In this case, genes in common to any of each pair of genomes, as well as genes unique to a single species, are often used in a complete analysis of all of the genomes. For the phylogenetic analyses, unique genes would increase the length of a branch for a particular species, whereas genes that are common to only two of the genomes would cause those to be more closely grouped on the phylogram. In general, if the proportion of genes common to all genomes is high, all of the branches in the phylogram would tend to be shorter, whereas the branch lengths would be longer when the proportion of common genes is small.

For standard phylogenetic analyses, generally a single gene locus is being compared among dozens, or sometimes hundreds of taxa (usually tens of thousands or hundreds of thousands of nucleotides per data set). In phylogenomics, dozens or hundreds of gene loci are being compared among dozens of taxa (millions of nucleotides in the dataset). This represents a change of 1–3 orders of magnitude in the amount of data to analyze. However, many of the analytical bioinformatics tools used to perform phylogenetic analyses use calculations that increase exponentially as the dataset increases in size. Therefore, many of the phylogenetic methods that have been successfully applied to relatively small phylogenetic studies cannot be applied to genome-sized comparisons. For example, maximum parsimony (MP) and maximum likelihood (ML) methods (see Chapter 24) become difficult or impossible to perform on datasets larger than about 10,000–50,000 nucleotides in size. Although these are accurate phylogenetic methods for smaller datasets, they cannot be easily applied to genomic alignments. However, neighbor joining (NJ) has been successfully used in many phylogenomic projects. It requires much less computer processing time, because its algorithms are purely computational, and it does not use any searching functions. Bayesian methods have also been used in phylogenomic research, but it does use search functions, which are iterative in nature. The advantage of these methods is that it is statistically robust, which is similar to ML in this respect.

WHAT TO COMPARE

As with standard phylogenetic methods, phylogenomic analyses require that the regions being compared are orthologous, or at least paralogous, and that they can be adequately aligned for analysis. However, the enormous amount of data creates some problems with the analyses. For genomic comparisons, only a subset of the genomic sequences is usually compared. These are the sections of the genomes that are in common among all of the sequences being compared, and those that can be aligned unambiguously (or nearly so). The usual method for this is to concatenate the gene regions to be used for each genomic sequence into a single long sequence (Figures 26.1 through 26.3), and then align them using one of the multiple sequence alignment applications (e.g., CLUSTAL or MAFFT; see Chapter 21). For quality control, the alignments are usually examined manually (i.e., *by eye*), although for very large datasets, it is possible only to examine a few short regions for a small number of sequences. Once the alignment is finalized, phylogenetic analyses can be attempted. As stated above, because of the large datasets, methods such as MP and ML become almost impossible to perform, even when a supercomputer is used. However, distance methods, such as NJ, which uses very simple calculations to compute pair-wise distances among the sequences, are able to generate trees in reasonable amounts of time. However, accuracy is not assured. Bayesian methods have been employed with increasing frequency because the calculations are statistically supported (as with ML). Although Bayesian phylogenetic methods require less computational time than ML, they still require a great deal of computer processing time. Therefore, for rapid comparisons, NJ is often used first, and the Bayesian methods are used for more detailed, rigorous, and statistically based phylogenomic analyses.

The problems of dataset size are limited somewhat because of three factors: repetitive regions, nonhomologous regions, and variations in gene mutation. Repetitive regions are difficult to resolve by NGS, because many of the regions consist of dozens to thousands of identical copies of two to hundreds of base pair units. The reads in NGS results are often too short to encompass the entire region, and although the computer can easily locate many overlaps, often it cannot determine how many repeats constitute the entire region. When the computer attempts to assemble the sequenced fragments into longer contiguous sequences (i.e., contigs), the precise beginnings, middles, and ends of the regions cannot be determined because all of the pieces have regions that are identical with those of other pieces. Also, because of frequent recombination events, the number of repeats often changes from cell to cell, such that the DNA that is being sequenced contains many different versions of the repetitive regions. They may also change as a result of DNA polymerase slippage during replication. Therefore, these regions often are left unfinished in the final genomes, and for comparative genomic and phylogenomic studies, they are excluded from the dataset.

Size of the dataset is also limited because some regions cannot be aligned. This often is dependent on relatedness of the sequences/taxa. That is, closely related organisms will have highly similar genomes, and therefore, more of the genomes will be amenable to alignment. However, when the organisms are from different domains, or when regions of the genomes have rapid rates of evolution, the sequences have much less in common, and therefore, many regions of the genomes must be excluded from analyses, because they cannot be accurately or reliably aligned. A caveat to this is that the most ancient genes are common to most organisms, and therefore, for broad-range phylogenomic studies, primarily ancient genes are being analyzed. This necessarily limits the analyses to highly conserved genes, and therefore, the results can only be indicative of evolutionary events that have occurred over billions of years.

The datasets that can ultimately be used for genomic comparisons are reduced somewhat when it contains only bacterial genomes, for many of the reasons outlined earlier. However, when comparing bacterial and archaeal genomes, the datasets are further reduced, due to their evolutionary distance. The data sets are reduced considerably more for eukaryotic genome comparisons, where only from 1% to 10% of the genome is composed of mRNA genes, and for comparisons among bacteria, archaea, and eukarya, where the number of genes in common to all is less than the number in each genome. Nonhomologous regions, including introns, are also often disregarded in comparative and phylogenomic studies, unless they are specifically aimed at the study of the introns, splicing, alternative splicing, and/or products (i.e., RNAs) from the introns.

When examining genomic sequences from different organisms, many regions cannot be aligned because either one or more genes are not present in one of the genomes, or the sequences are so different that they cannot be unambiguously aligned. Part of the reason for this is that not all genes, or all regions of each gene, mutate at the same rates, and some organisms may experience large genomic changes compared to other organisms. Punctuated evolutionary events are one of the causes of these differences. Because of this, phylogenomic analyses tend to result in averages of the overall rates of change in each of the sequences and may fail to indicate some changes that have occurred in specific loci in a particular genome. However, although phylogenomic methods usually cannot utilize complete genomes, and they may miss some changes, due to the sheer volume of nucleotides in each sequence, they provide yet another tool to analyze the evolutionary events that have led to the diversity of life on the Earth. Additionally, a phylogenomic analysis analyzes tens or hundreds of genes at a time and can indicate changes in a number of genes that would require much more time to evaluate were the genes to be analyzed individually using phylogenetic analyses.

SINGLE-NUCLEOTIDE POLYMORPHISMS

Within any population, and within any multicellular organism, mutations constantly occur. In most instances, they are single-nucleotide changes and they cause no deleterious effects. Throughout a population, these changes can be mapped as polymorphisms in the population and can be compared to sets of polymorphisms in other populations (or among individuals or cells within an individual). Maps of single-nucleotide polymorphisms (SNPs) have been constructed for the human population (and this work goes on) using DNAs from individuals around the world, such that millions of such SNPs have been recorded and stored in databases. One of the popular uses of SNPs is to determine ancestry of individuals. One can submit a saliva sample to a company or research laboratory and the SNP profile can be determined. From the profile, it can be determined which SNPs are in the samples that correlate with those from specific populations around the world. For example, a typical sample might indicate that the DNA sample is 40% similar to profiles from Northern Europeans, 30% is similar to profiles from Scandinavians, 20% is similar to profiles from Central Africans, and the remaining 10% is a mixture from various locations in Europe and Asia. SNP profiles are also being examined to determine whether they can be used in the diagnoses of specific diseases, including malignancies. SNPs have not yet been used extensively for studies of other types of organisms, primarily because extensive SNP libraries and databases have not yet been produced and/or constructed. However, SNP libraries are potentially useful for a wide range of taxa, and probably will be developed in other groups of organisms in time.

MICROSATELLITES AND MINISATELLITES

Like SNPs, microsatellites and minisatellites are widespread among eukaryotes and have already been used to categorize individuals in populations, as well as to characterize populations. Microsatellites consist of short (2–6 bp) tandem repeats, flanked by unique sequences, whereas minisatellites are tandem copies of 10 bp to over 100 bp, which are also flanked by unique sequences. The number of tandem repeats varies from a few to hundreds in a single locus. Because of their repetitive nature, they are subject to frequent recombination events and DNA polymerase slippage, both of which have the potential to produce variants of the original locus. Although they were first classified based on hybridization methods, more recently, molecular methods, including polymerase chain reaction, cloning, and sequencing, have been used for various types of studies. In this sense, they can be used in phylogenomic studies to identify the origin of individuals within populations and to infer genetic and geographical origins of individuals. They have also been used to correlate blood samples with suspects in criminal cases. Some regions change so rapidly that the probability of two humans having identical repetitive regions is less than one in several billion. Analogous studies can identify individual trees (and other plants) from microsatellite and minisatellite sequences in their leaves and other plant parts. These have also been used in judicial forensics cases to determine whether the suspects were at a crime scene.

HOW TO COMPARE

Some of the same basic models are used for comparing genomes as those used for comparing single genes. For example, the Jukes–Cantor model can be used in which the bases all have the same probability of mutating into any other base. More accurate models use different modes of change, weighting transversions more heavily than transitions, because transitions occur more often than transversions, and C-to-T changes occur with the highest frequencies. The same considerations that need to be addressed in phylogenetic analyses apply to phylogenomic analyses. That is, homologous regions must be properly aligned, orthologs must be differentiated from paralogs, regions of sequence ambiguity must be removed from consideration, and so on. The differences primarily relate to the sizes of the databases. As mentioned earlier, this creates some issues regarding the types of analyses that can be performed. Although the models and algorithms utilized in ML and MP are rigorous, they are difficult to perform on very large datasets, due to the number of calculations involved. The primary method of analysis with very large datasets is NJ or NJ coupled with bootstrapping.

TESTING FOR SELECTION

Genomic sequences can be used to test for signs of selection on one or more regions of the genomes of several sequenced genomes. This can be important in the gene discovery process, as well as identifying any genes and portions of genes that are being subjected to positive or negative selective pressures. Once the reading frame is deduced, calculations can be made of the synonymous versus the nonsynonymous mutations. If the number of synonymous and nonsynonymous mutations is comparable (after considering the number of redundant codons for particular amino acids), the region is probably not under selective pressures. However, if the numbers are skewed, then the region might be under selective pressure. Furthermore, specific regions of the genes that are under the heaviest selective pressures can be identified, indicating that they may serve an important functional role in mature proteins.

INCONGRUENT TREES

rRNA genes are well conserved over much of their lengths, and they have been inherited vertically in virtually all organisms that have been examined. Because of this, they have been used to construct various versions of the trees of life. In effect, the evolution of bacteria and archaea can be traced using rRNA genes, whereas the evolution of nuclear genomes and organellar genomes in eukarya can be traced separately using rRNA genes. Some gene trees are incongruent with the rRNA trees from the same species (Figure 26.4). When this occurs with genomic gene sets, HGTs are suspected, and the selected genes can be studied in detail after locating them using phylogenomic analyses.

Darwinian evolution and Mendelian genetics are linear processes. That is, species arise from ancestors by passing on their genetic components to their progeny. This yields a very linear process

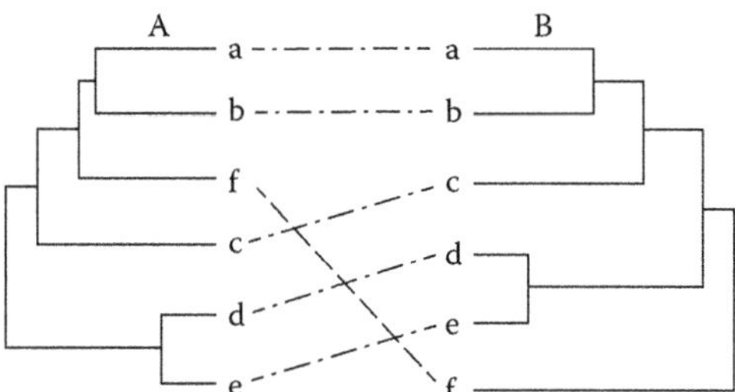

FIGURE 26.4 Example of incongruent phylogenetic trees. The hypothetical phylogenetic tree on the left (A) is based on one set of aligned sequences (e.g. genomic sequences), whereas the hypothetical phylogenetic tree on the right (B) is based on another set of aligned sequences (e.g., rRNA gene sequences). Although most of the taxa (a, b, c, d, and e) appear on the two trees with similar topology, taxon f appears on branches of different topology for the two trees. The incongruency between the trees indicates a possible HGT in taxon f, which might have involved a taxon similar to either taxon a or b.

upon which a phylogenetic tree can be produced. This works remarkably well for highly conserved genes involved in complex processes, such as rRNA genes. However, it was clear very early in phylogenetic studies that more variable genes often produced trees that were incongruous with phylogenetic trees based on rRNA gene sequences and phenetic trees based on morphological characters. This created some confusion regarding the original supposition that gene trees should be coincident with species trees. In fact, in many cases, the gene trees were dissimilar to the inferred species trees that were based on other characters and sequences from other genes. Once entire genomes were available for comparisons, the inconsistencies became even more apparent. Some of the inconsistencies were caused by horizontal transfers of genes from the organellar genomes into the nuclear genomes. However, this explanation was insufficient for many gene comparisons. Instead, the only possible explanation was that there had been other HGT events that had occurred in the ancestors of the species being examined. In some dinoflagellates, this process has resulted in genomes that are composed of 90% horizontally acquired genes and only 10% through direct vertical inheritance from the ancestral lines. Thus, phylogenomics has become a powerful tool to study the evolution of species and genes by investigating both the vertical and horizontal events that have led to evolutionary events among species.

COMPARATIVE GENOMICS

Phylogenomic analyses often include one or more aspects of comparative genomics. These projects usually compare the numbers of genes and/or alleles that are in common among the genomes being compared, as well as those that are shared by two or more of the taxa, and those that are unique for each taxon/sample/strain. Some of the first studies compared pathogenic bacteria with nonpathogenic free-living bacteria. Most of the initial genomes that were sequenced were those from bacteria with small genomes. Several of these were parasites of humans and primates. For example, *Mycoplasma genitalium* (genome size = 5.8×10^5 bp) and *Haemophilus influenza* (genome size 1.8×10^6 bp) were some of the first genomes to be determined. When they were compared with *Escherichia coli* K-12 after its sequence was completed, the comparison showed that most of the genes in the two parasites were also found in *E. coli* K-12. However, the *E. coli* K-12 genome also contained a large number of genes that were absent from the genomes of the parasitic bacteria, indicating that their genomes had decreased as they had evolved into a parasitic lifestyle.

When the first eukaryotic genome was sequenced (that of *S. cerevisiae*), the genomes of several Bacteria (including *E. coli* K-12) and the genome of an Archaea (that of *M. jannaschii*) had been completed. This allowed a comparative genomic analysis of the three domains: eukarya, bacteria, and archaea. The analysis supported the relative relationships among the three domains that had been established in several phylogenetic analyses. That is, archaea first diverged from a group of early bacteria, and later an archaeal branch diverged to found the eukarya. Another conclusion of the study was that the majority of genes were in common to taxa from all three domains. In fact, the archaeal genome was more similar to the eukaryotic genome than to the bacterial one, and there were very few genes unique to the archaeal genome that were missing from the eukaryotic genome (Figure 26.5). When multiple eukaryotic genomes became available, the genomes were compared. One of the first such studies compared the genomes of three animal species (*C. elegans*, *Homo sapiens*, and *Drosophila melanogaster*), with a fungus genome (*S. cerevisiae*), all within the Opisthokonta (Figure 26.6). The animals all had high degrees of genes in common (>40%), whereas approximately 25% of their genes were found in the *S. cerevisiae* genome. Based only on the proportions of genes that are in common, *D. melanogaster* (an arthropod) is slightly more closely related to *H. sapiens* than *C. elegans* (a nematode). Additionally, the number of transcription factors increased dramatically as the organisms became more complex, in terms of development, size, tissue types, and cell types (Figure 26.7). This is especially true for the Cys2 His2 zinc zipper transcription factors, of which there are more than 850 in *H. sapiens*, approximately 360 in *D. melanogaster*, about 100 in *C. elegans*, and less than 50 in *S. cerevisiae*. Other transcription factors are also higher in the *H. sapiens* genome than in the other species, and *S. cerevisiae* generally had the lowest numbers of each of the transcription factors (Figure 26.7).

Genomic studies of bacteria have surpassed those of archaea and eukarya, partly because of the availability of bacterial cultures and partly because of the relatively small genome sizes of Bacteria. The genomes of mutualistic and endosymbiotic bacteria have also been determined even when culturing has not been possible. This has been accomplished in several ways, including whole genome sequencing from single cells. Comparative genomic research has been carried out on many of these organisms, and the results indicate that various gene losses have occurred in lineages of these organisms over long periods of time (Figure 26.8). These studies have shown that most of the genes have been retained as functional orthologs in each of the members of a taxonomic group. However, they also show that frequently in one strain or species, pseudogenes are generated in one to many

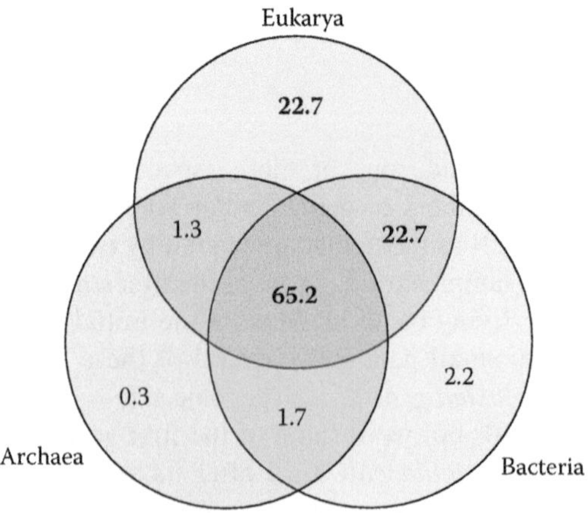

FIGURE 26.5 Venn diagram comparing members of the bacteria, archaea, and eukarya. More than 65% of the total number of genes among all of these genomes are common to all three genomes. Only a small percentage of the genes are unique to archaea and bacteria. This is primarily due to the fact that those genomes are generally smaller than those of eukarya, and that most of the genes in eukarya originated in archaea and bacteria. The total number of genes analyzed was 776.

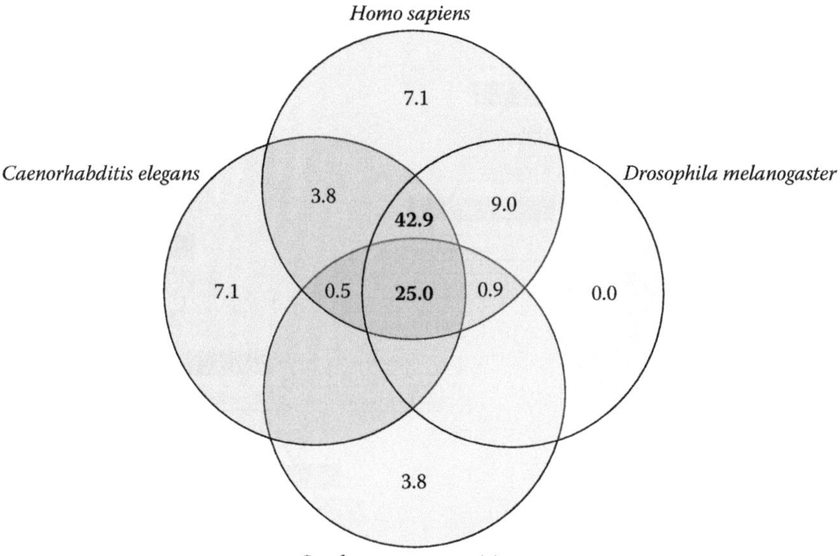

FIGURE 26.6 Venn diagram comparing protein kinase gene content for three animals and one fungus. As expected, the proportion of genes that is common (42%) among the animal genomes (*Caenorhabdits elegans*, *Homo sapiens*, and *Drosophila melanogaster*) is larger than the proportion common to all of the organisms, including the fungus (*Saccharomyces cerevisiae*), which is 25%. The total number of protein kinase genes analyzed was 212.

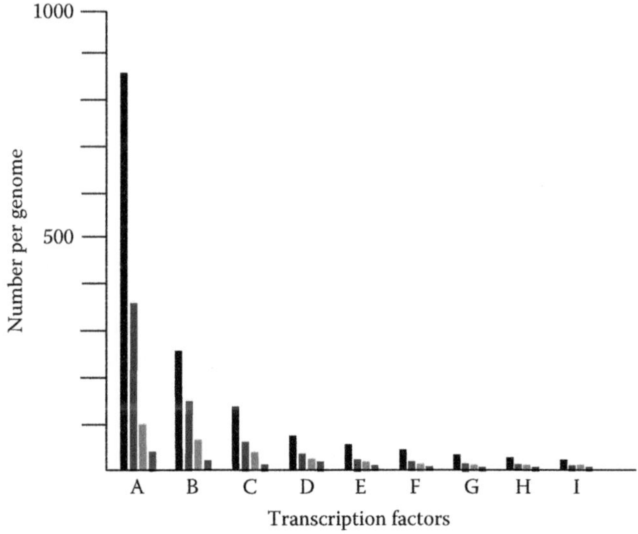

FIGURE 26.7 Relative numbers of transcription factors among four eukaryotes. For each set of bars (A–I), the bars indicate the number of transcription factors of each type for each of the four genomes: black for the *Homo sapiens* genome, red for the *Drosophila melanogaster* genome, green for the *Caenorhabditis elegans* genome, and blue for the *Saccharomyces cerevisiae* genome. The transcription factor types are as follows: A, Cys2 His2 zinc finger; B, Homeobox; C, helix–loop–helix binding domain; D, basic leucine zipper domain; E, forkhead box transcription factors; F, E26 transformation-specific domain; G, POU domain; H, retinoic acid receptor; and I, methyl-CpG domain.

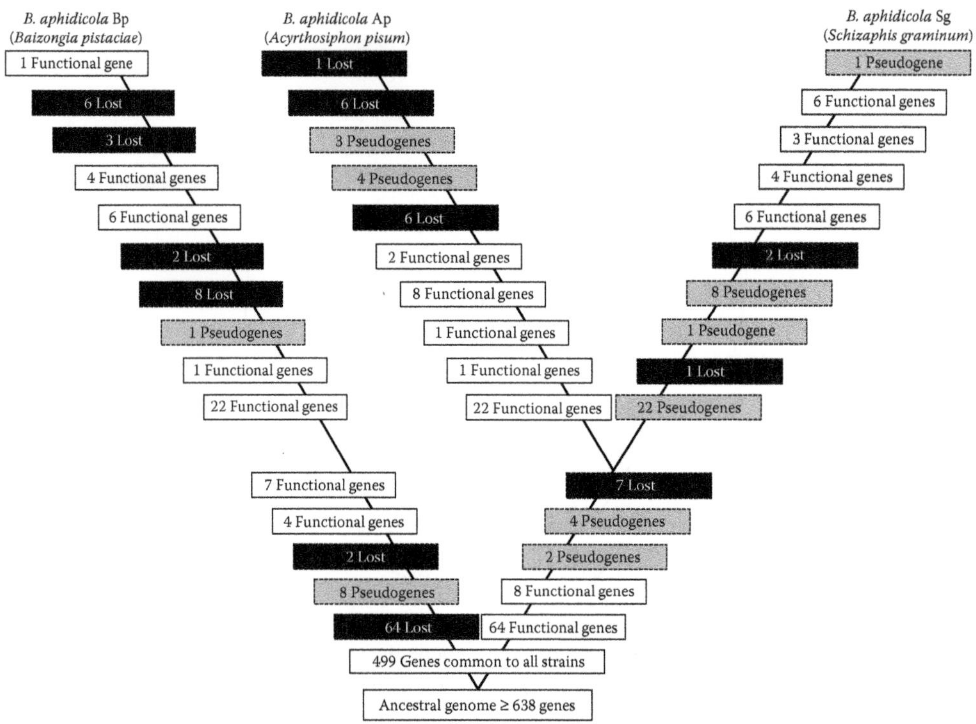

FIGURE 26.8 Changes in gene retention and loss in three strains of the aphid endosymbiont *Buchnera aphidicola*. The genomes of all three were analyzed to investigate the evolution of the three. The last common ancestral species was deduced to have at least 638 genes. Approximately 150–200 million years ago, the ancestral strain diverged into two separate strains. Both retained the same set of 499 genes (that remain common to all three strains). Following the divergence, the *B. aphidicola* BP strain lost 64 functional genes that have been retained by the *B. aphidicola* Ap and Sg strains. Several other genes were retained as functional versions (white boxes), became pseudogenes (gray boxes), or mutated out of existence (black boxes) in each of the two strains. Approximately 50–70 million years ago, a second divergence occurred, which generated the *B. aphidicola* Ap and Sg strains. Over the next 50–70 million years, additional genes were retained, became pseudogenes, or were lost in each of the three strains.

genes, and in others the genes have mutated so much that they have become unrecognizable or they have been expelled from the genomes.

Studies of closely related species have also helped to clarify both their taxonomic and phylogenetic relationships. For example, rather than being a single species, the taxonomic designation, *E. coli*, is a species complex and is closely related to species of *Shigella*. Comparative genomic studies of members of these taxa exhibit a great deal of similarity in their gene contents, but also show a large number of genes unique to each strain or species (Figure 26.9). The *E. coli* strain K-12 (which is nonpathogenic) is similar but very different from *E. coli* O157:H7 (a virulent pathogen), but they are more similar to each other than they are to *Shigella flexneri* 2a (a pathogen). Plant genomes have also been used in comparative genomics and phylogenomic studies. Plants have genomes that are generally from about one-eighth the size of the human genome to more than 1000 times larger than the human genome. They also have more genes, but less alternative splicing, and because of this, the total number of proteins that their genomes encode is less than for the human genome, but generally more than for other species. In a comparative genome study of three grass (monocot) species and a eudicot species (*Arabidopsis thaliana*), 55% of the gene families were in common to all four genomes (Figure 26.10). Another 13% of the gene

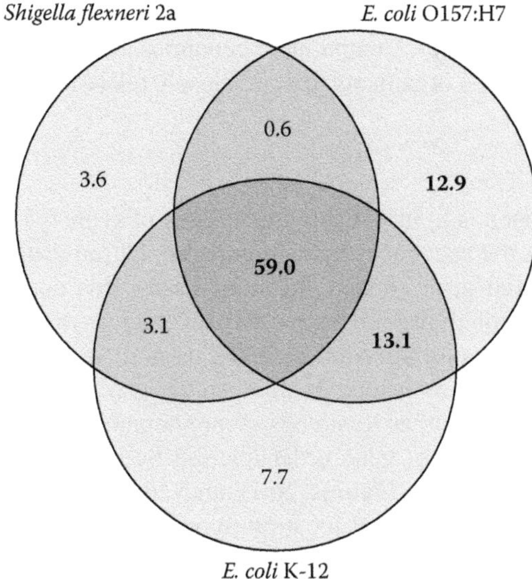

FIGURE 26.9 Venn diagram comparing the genomes of three related bacterial species. All three species have nearly 60% of their genes in common. Although the two *Escherichia coli* strains (K-12, a nonpathogenic strain and O157:H7, a pathogenic strain) differ from one another, they have more in common than with the *Shigella* strain. The total number of genes used in the analysis was 4878.

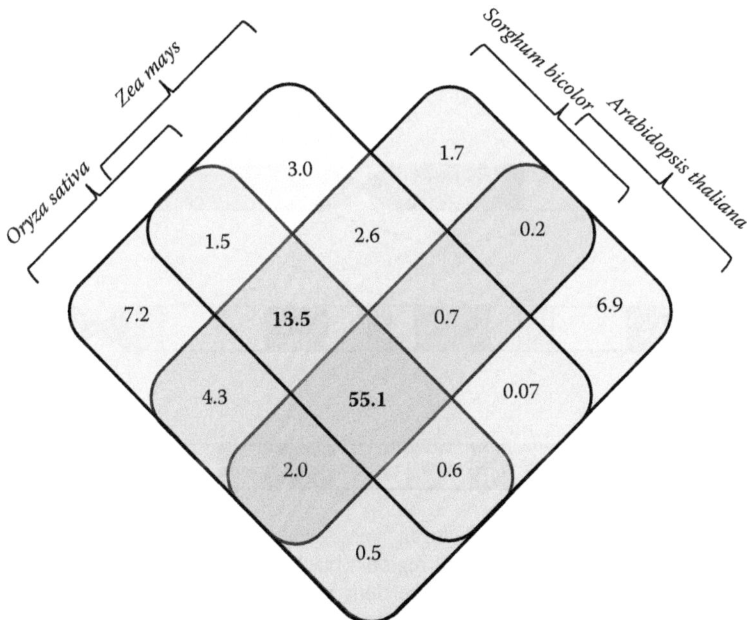

FIGURE 26.10 Venn diagram comparing gene families within four plant genomes. Three of the species are grasses, rice (*Oryza sativa*), maize (*Zea mays*), and sorghum (*Sorghum bicolor*), whereas the fourth is a eudicot, *Arabidopsis thaliana*. This is consistent with the number of gene families in common among all of the genomes, 55.1%, and another 13.5% of the genes are in common among the grass genomes, indicating their closer relationship to each other than to *A. thaliana*. The total number of gene families analyzed was 15,404.

families were in common only among the grass genomes, indicating their closer phylogenetic relationships relative to *A. thaliana*. Comparative genomics and phylogenomics can help to fine-tune our understanding of these organisms that are closely related.

SYNTENY

Another mode of comparison is to investigate the amount of synteny (identical gene order along the chromosomes) among the sequences being compared. Chromosomes change relatively rapidly in size, gene order, and gene content. In many cases, this does not appear to affect the expression of the genes, although in some extreme cases, such as in many cancers, chromosomal translocations are known to cause aberrant gene expression (see Chapter 18). The mechanism for chromosomal translocations is primarily unequal crossovers within a single chromosome, or between two chromosomes, either homologs or nonhomologs. The crossovers generate differences in gene order, reversals of gene order (caused by inversions), and transfers of genes from one chromosome to another (Figures 26.11 and 26.12). In addition to the transfers of genetic material from one chromosomal location to another, several other changes can occur. Duplications and deletions are frequently found, the positions of the centromeres can change, and chromosomes can split or fuse (Figure 26.12). These change the chromosome number, but not necessarily the number of genes or genome size. Although many of these changes do not change the expression of the genes in the genome, occasionally small to large changes occur. Often this is caused by movement of genes into regions where the chromosome architecture or the local promotors cause changes in the expression of genes in the immediate area of the chromosomal translocation. For example, some cancers are caused by translocations that move constitutive promoters close to normally tightly regulated genes.

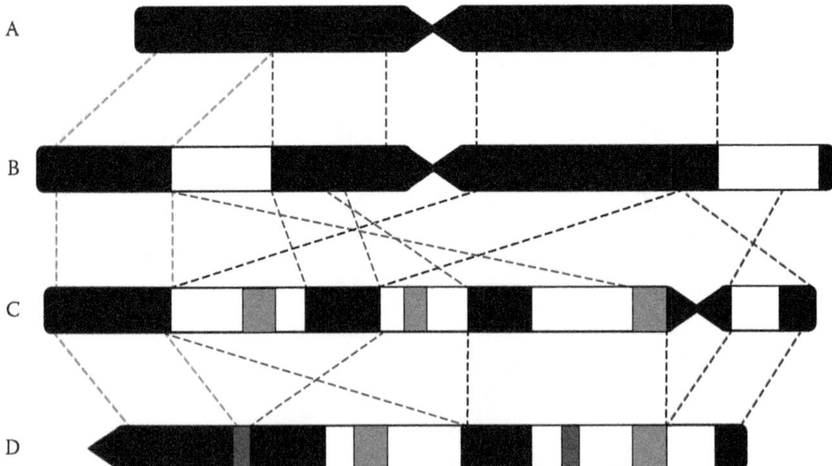

FIGURE 26.11 Hypothetical evolution of a chromosome, showing the degree of synteny within a chromosomal lineage. Chromosome B contains several regions (black) that are syntenous (identical in gene order) with chromosome A. However, several unique insertions have occurred in chromosome B (white regions). Chromosome C exhibits additional changes, including two inversions (red and brown crossed dashed lines), additional insertions (gray regions), and a shift of the centromere to create a chromosome with unequal arms (submetacentric). Chromosome D contains an inversion (blue dashed lines), a shift in the centromere creating an acrocentric or telocentric chromosome, a loss of some regions, and two insertions (dark gray regions).

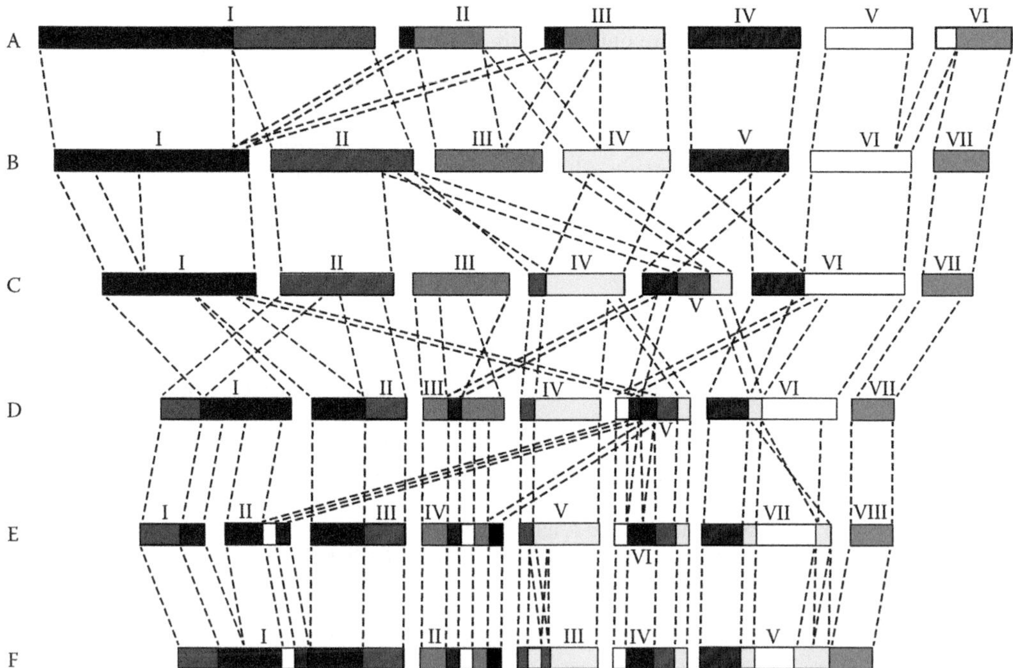

FIGURE 26.12 Hypothetical evolution of sets of chromosomes within a taxonomic group. From set A to B, there have been numerous chromosomal translocation events, as well as the splitting of chromosome I to generate a smaller chromosome I in set B. Although set A has fewer chromosomes than set B, the gene content and genome size are roughly equal for both sets. Although set C has the same number of chromosomes as set B, the sizes of the chromosomes, and therefore the genome, have decreased for set C. A large deletion has occurred in chromosome I in set C, and several chromosomal translocations have occurred such that chromosomes IV, V, and VI contain portions of chromosomes II, IV, V, and VI from set B. Additional translocations occur to generate chromosome sets D, E, and F. Finally, chromosome I in set D is split into two separate chromosomes (I and II) in set E, which has eight chromosomes (I–VIII); fusion of chromosomes I, II, and III, as well as VII and VIII from set E, generates set F that contains two larger chromosomes (I and V), as well as three smaller ones.

KEY POINTS

1. Advances in sequencing methods and phylogenetic analysis tools have allowed large portions of genomes from many organisms to be compared.
2. Phylogenomic analyses require first that the gene sequences from each of the genomes be evaluated and selected based on orthology, and that the alignments must be unambiguous.
3. In addition to using gene sequences for phylogenomic projects, SNPs, microsatellites, and minisatellites have also been used mainly to investigate relationships among closely related organisms.
4. Phylogenetic algorithms have been developed to handle the large datasets used in phylogenomics. NJ and Bayesian analyses are used most often to produce trees.
5. Phylogenomics can be used to test for selection, as well as to indicate instances of HGT.
6. Comparative genomics, and quantitative and qualitative analyses of two or more genomes, is often used in conjunction with phylogenomic studies.

7. Comparative genomic studies can be used to compare the genomes of organisms that are closely related or those that are only distantly related. However, the numbers of genes that can be used for comparisons are usually higher for comparisons of closely related taxa.
8. Chromosomes or parts of chromosomes can be used to investigate the amount of synteny (retention of gene order along the chromosomes) among sets of chromosomes.
9. During the evolution of organisms, chromosomes undergo frequent exchanges, insertions, deletions, segmentation events, and fusions. Some of these chromosomal translocations occur in many types of cancer cells.

ADDITIONAL READINGS

Higgs, P. H. and T. K. Attwood. 2005. *Bioinformatics and Molecular Evolution*. Carleton, Victoria, Australia: Blackwell Publishing.

Li, W.-H. 1997. *Molecular Evolution*. Sunderland, MA: Sinauer Associates.

Li, W.-H. and D. Graur. 1991. *Fundamentals of Molecular Evolution*. Sunderland, MA: Sinauer Associates.

Manning, G. 2005. Genomic overview of protein kinases. *WormBook*, ed. The *C. elegans* Research Community, *WormBook*, doi:10.1895/wormbook.1.60.1, http://www.wormbook.org.

Mount, D. W. 2004. *Bioinformatics: Sequence and Genome Analysis*, 2nd ed. Cold Spring Harbor, NY: Cold Spring Harbor Laboratory Press.

Prevsner, J. 2009. *Bioinformatics and Functional Genomics*, 2nd ed. Hoboken, NJ: Wiley-Blackwell.

Schnable, P. S., D. Ware, R. S. Fulton, J. C. Stein, F. Wei, S. Pasternak, C. Liang et al. 2009. The B73 maize genome: Complexity, diversity, and dynamics. *Science* 326:1112–1115.

Tupler, R., G. Perini, and M. R. Green. 2001. Human genome: Expression the human genome. *Nature* 409:832–833.

Van Hamm, R. C. H. J., J. Kamerbeek, C. Palacios, C. Rausell, F. Abascal, U. Bastolla, J. M. Fernández et al. 2003. Reductive genome evolution in *Buchnera aphidicola*. *Proc. Natl. Acad. Sci. USA* 100:581–586.

Wei, J., M. B. Goldberg, V. Burland, M. M. Venkatesan, W. Deng, G. Fournier, G. F. Mayhew et al. 2003. Complete genome sequence and comparative genomics of *Shigella flexneri* serotype 2a strain 2457T. *Infect. Immun.* 71:2775–2786.

Section VI

Genomes

27 RNA Viruses

INTRODUCTION

In Chapters 2, 15, 16, and 24, the ties between Bacteria, Archaea, and Eukarya have been made and are clear from the trees of life determined by ribosomal RNA genes, as well as other conserved genes. Even from earlier times, trees and taxonomic schemes recognized that all life on the Earth was related. This means that as we move backward on the tree of life, we find that humans have a common ancestor with chimpanzees. If we move further back, we share a common ancestor with dogs, and further back, we have a common ancestor with tomatoes. Eventually, we find that we have a common ancestor with *Escherichia coli* and *Salmonella*. And, of course, ultimately, we all have the original cell as our ultimate ancestor. It survived and produced progeny that produced all life on the Earth. We also find that eukaryotes are essentially endosymbiotic and mutualisitc cooperating consortia of bacteria, archaea, and sometimes other eukarya. The combination of microscopy, sequencing, and bioinformatics has led to these conclusions. However, in the past several decades, genomic studies have moved the investigations of evolution to another, and often deeper, level.

A genome is the total of the genetic material in the organism that is passed on to the next generation. For RNA viruses, this is normally a single RNA chromosome, although in some RNA viruses, such as influenza A virus, the genome consists of more than one RNA chromosome. For all other organisms, they contain one or more chromosomes composed of DNA that contain all of the genes necessary for life and propagation. For example, most bacteria have a single chromosome that is replicated, and each daughter cell receives one of the replicated chromosomes, in other words, one complete genome. For organisms having more than one chromosome, the genome consists of the sum of the DNA from all of the chromosomes (the haploid amount for diploid organisms). Many eukaryotes contain organelles that have their own genomes. Therefore, for eukaryotes there are nuclear genomes as well as organellar genomes for most species. For example, in humans, the nuclear genome consists of approximately 3.2×10^9 base pairs of DNA per haploid set of nuclear chromosomes, and 1.7×10^4 base pairs of DNA in the human mitochondrial genome.

Genomes range from a few thousand nucleotides in RNA viruses to hundreds of billions of nucleotide pairs in some eukaryotes (Figure 27.1 and Table 27.1). The genome size is also equivalent to the C-value. The C originally meant *constant* amount of DNA in a cell (for Bacteria and Archaea), or the haploid amount for Eukarya. However, as was mentioned in previous chapters, DNA values can change under certain circumstances. Many genomes have been determined by measuring the sizes of the nuclei because there is a direct relationship between the size of the nuclei and the amount of DNA they contain. The genomes of some extinct species (i.e., dinosaurs) have been determined in this way through microscopic examination of fossilized cells. For smaller genomes (including bacteria, archaea, organelles, fungi, and others), fluorescence microscopy, electron microscopy, restriction enzyme mapping, and other molecular biology methods have been used to determine genome sizes. A small subset of species has been subjected to genomic sequencing, where the sequence of the entire genome has been determined. In this chapter, and the following chapters, some of the characteristics of these genomes will be discussed.

The first bacterial genome, that of *Haemophilus influenzae*, was completed in 1995. This was followed by publication of other bacterial genomes (e.g., *Mycoplasma genitalium* and *E. coli*), archaeal genomes (e.g., *Methanocaldococcus jannaschii* and *Thermoplasma volcanium*), and eukaryal genomes (e.g., *Saccharomyces cerevisiae*, *Arabidopsis thaliana*, and *Homo sapiens*). Over the past decade, the number of genomes that have been sequenced has been expanding rapidly, especially

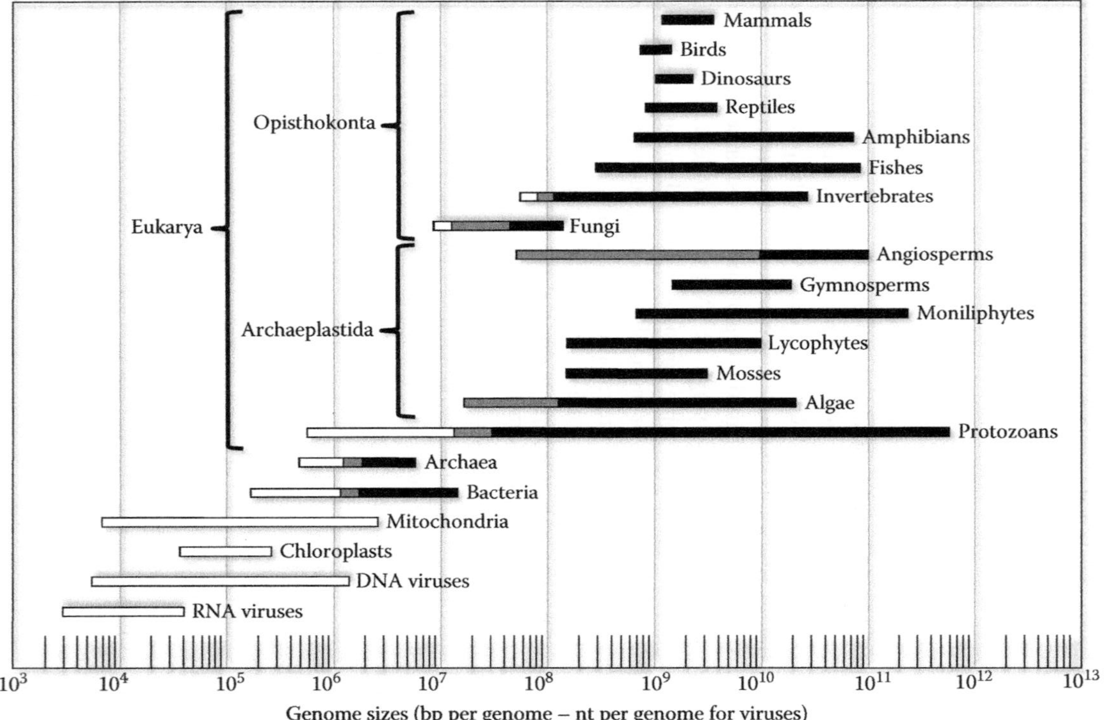

FIGURE 27.1 Genome sizes (log scale) of representative groups of organisms and endosymbiont organelles, arranged phylogenetically. All values are in base pairs, with the exception of viruses, which are in nucleotides (whether the viruses are single or double stranded). For diploid eukaryotes, haploid values are given. Protozoans include a broad range of eukaryotes (Excavata, Chromalveolata, Rhizaria, and Amoebozoa), which partly explains the wide range (over six logs) in genome sizes. The values are from a variety of sources. Dinosaur genomes were determined from fossil cell nuclei sizes, in comparison with analogous nuclei in birds and reptiles. White bars indicate that the organisms are obligate parasites. This includes all viruses and organelles, as well as a large proportion of bacteria, some archaea, and some fungi. Gray bars indicate a mixture of parasites and free-living organisms. Black bars indicate primarily free-living organisms.

TABLE 27.1
Genome Sizes and Number of Genes for Selected Species, Arranged in Order of Genome Size

Species	Taxonomic Group	Genome Size (bp)	Number of Genes	Gene Density (Genes per kb)
MS2	RNA virus	3.6×10^3	4	1.18
ϕX174	DNA virus	5.4×10^3	11	2.03
HIV	RNA virus	9.7×10^3	9	0.93
Influenza A	RNA virus	1.4×10^4	11	0.79
Homo sapiens	Mitochondrion	1.7×10^4	37	2.18
Sputnik virus	DNA satellite virus	1.8×10^4	21	1.17
Ebola	RNA virus	1.9×10^4	7	0.37
Toxoplasma gondii	Chloroplast	3.5×10^4	26	0.74
λ (lambda)	DNA virus	4.8×10^4	56	1.17
Carsonella ruddii	Bacterium	1.6×10^5	182	1.13
T4	DNA virus	1.7×10^5	288	1.69
Epstein–Barr virus	DNA virus	1.7×10^5	80	0.47
Nephroselmis olivacea	Chloroplast	2.5×10^5	156	0.62
Nanoarchaeum equitans	Archaea	4.9×10^5	537	1.10
Mycoplasma genitalium	Bacterium	5.8×10^5	485	0.84
Mimivirus	DNA virus	1.2×10^6	1,262	1.05
Methanococcus jannaschii	Archaea	1.7×10^6	1,783	1.05
Haemophilus infleunzae	Bacterium	1.8×10^6	1,738	0.97
Cucumis melo (muskmelon)	Mitochondrion	2.4×10^6	33	0.014
Mycobacterium leprae	Bacterium	3.2×10^6	1,604	0.50
Mycobacterium tuberculosis	Bacterium	4.4×10^6	3,959	0.90
Escherichia coli (K-12)	Bacterium	4.7×10^6	4,377	0.93
Methanosarcina acetivorans	Archaea	5.8×10^6	4,528	0.78
Saccharomyces cerevisiae	Fungus	1.25×10^7	5,770	0.46
Sorangium cellulosum	Bacterium	1.3×10^7	~4,500	0.35
Plasmodium falciparum	Protozoan	2.3×10^7	5,268	0.23
Neurospora crassa	Fungus	3.9×10^7	10,082	0.26
Caenorhabditis elegans	Nematode	9.7×10^7	19,000	0.20
Drosophila melanogaster (fruit fly)	Insect	1.2×10^8	13,601	0.11
Arabidopsis thaliana	Plant	1.2×10^8	~28,000	0.23
Tetraodon nigroviridis (pufferfish)	Fish	3.4×10^8	27,918	0.082
Oryza sativa (rice)	Plant	3.9×10^8	28,236	0.072
Canis lupus familiaris (dog)	Mammal	2.4×10^9	19,300	0.0080
Zea mays (corn, maize)	Plant	3.0×10^9	~50,000	0.017
Homo sapiens sapiens (human)	Mammal	3.2×10^9	~23,000	0.0072
Mus musculus (mouse)	Mammal	3.4×10^9	~23,000	0.0068
Triticum aestivum	Plant	1.6×10^{10}	~50,000	0.0031
Amoeba dubia	Protozoan	6.7×10^{11}	~16,000	0.000024

after the advent of the so-called next-generation sequencing technologies. However, the first genome that was ever sequenced was that of the RNA bacteriophage MS2 in 1976. The following year, the first DNA virus (ϕX174) was sequenced.

C-VALUE PARADOX

There is no strict correlation between the genome size and the number of different RNAs in an organism. On average, a gene is about a kilobase in length. For viruses, the genes are so condensed that there is more than one gene encoded for each kilobase of RNA or DNA. Some genes overlap other genes. In general, bacteria have larger genomes than viruses, although there is a great deal of overlap. Still, in bacteria, the genes are very close together, with very little space between genes. This is true for archaea as well. In all of these organisms, there is a correlation between genome size and organism complexity, as measured by the number of different RNAs and proteins that are produced (Table 27.1). However, the correlation is weak. For example, human immunodeficiency virus (HIV) has a genome of about 9.7 kilobases of RNA, which produces at least 15 proteins. Bacteriophage lambda has 48.5 kb of DNA that encodes 73 proteins (and an additional 20 genes that produce small controlling RNAs). Bacteriophage T4 has 169 kb of DNA that encodes about 300 proteins. Therefore, for these simple organisms, and even for bacteria and archaea, as genome size increases, RNA complexity also increases at a fairly steady rate, as do the numbers of proteins expressed. However, in eukaryotes, this correlation breaks down. For eukaryotes, the amount of DNA does not always correlate with the diversity of RNA molecules. The haploid genome for *H. sapiens* is about 3.2 billion base pairs in length (for females; for males it is closer to 3.1 billion base pairs, because of the small Y chromosome). Approximately 23,000–25,000 genes are included in this genome. Only about 1.5% of the genome contains genes. This is the case for most eukaryotes. There has been an increase in noncoding regions such that the majority of the genome contains no protein-coding genes. This is the so-called *C-value paradox*. Some of the DNA is structural, in that it organizes the DNA into individual chromosomes, including centromeres, telomeres, domains, replication origins, and other regions. Other parts contain no coding sequences. Some of these encode microRNAs, interfering RNAs, and other small RNAs that control gene expression of coding genes. However, many parts of eukaryotic genomes have no known function. Much of the DNA in eukaryotes, and especially those with very large genomes, is enigmatic. It may be a consequence of maintaining a particular cell size, because nuclear size and cell size are correlated. It may simply be that the nucleus has adapted to soaking up and integrating DNA in order to test out new gene possibilities. In plants, some of the additional DNA may shade other important parts of the genome from UV damage. However, beyond this speculation, not much is known about the function of the majority of DNA in most large genomes.

GENOMES AND GENOMICS

Genomics (the sequencing and study of whole genomes) has revolutionized investigations of all types of organisms during the past several decades. The intersecting fields of proteomics (study of the expression of proteins in cells), RNAomics (study of RNA functions in cells), and metabolomics (study of metabolic pathways in cells and in ecosystems) have led to a more detailed picture of how genes evolve and interact within and between organisms and environments. In the remainder of this chapter and in the following several chapters, genomes representing different groups of organisms and organelles will be presented, beginning with some of the smallest genomes in RNA viruses, and building up to some of the larger genomes. Details of gene expression and control are provided for some of the smaller genomes in order to illustrate similar processes in the larger genomes. As larger genomes are presented, interactions and evolutionary processes will be noted. By the time eukaryotic genomes are presented, a full appreciation of the enormous number of evolutionary events, including the speciation of organisms and the intricate interactions among sets of genes, will be gained.

RNA VIRUS GENOMES

As was stated in Chapter 13, viruses come in many forms, both RNA and DNA, including single-stranded, double-stranded, linear, and circular arrangements of their genomes. Also, the RNAs carried in virions may be plus (messenger RNA [mRNA]-like) or minus (reverse complement or antisense) strands. Those that infect bacteria and archaea need to recognize surface proteins on the cell walls, and then transfer their genomes, and some of the proteins, into the target cell. There, they may integrate into the chromosome, becoming a lysogen, or express other sets of genes to form new virions, becoming lytic. In eukaryotes, the virions must recognize the host cell wall or proteins on the cell membrane, traverse into the cell, uncoat their genome, and sometimes specific proteins (such as reverse transcriptase and RNA-dependent RNA polymerase) that are also packaged inside the virion, move into the nucleus for transcription and integration (for many, but not all, types of viruses), and move the transcribed RNA into the cytosol for translation and virion assembly, and then new virions are released by the cell, through either exocytosis or cell lysis. Although the viruses use many of the systems of the host cell, they require many additional functions that are unique to the viruses. Therefore, they must usually carry a substantial collection of genes in their genomes. Examples of RNA virus genomes are discussed in the sections that follow.

HUMAN IMMUNODEFICIENCY VIRUS

HIV (Figure 27.2) is the causative agent of acquired immunodeficiency syndrome (AIDS). It is within the Retroviridae (retrovirus) group of viruses, and its genome consists of 9.7×10^3 nucleotides of negative-sense (minus strand) single-stranded RNA. This RNA produces at least nine RNA transcripts that encode at least 15 proteins (Figure 27.3). HIV consists of at least two distinct genotypes, HIV-1 and HIV-2. Both infect humans, and both cause similar disease symptoms. The reason for the different numbers of transcripts and proteins is that some transcripts encode more than one protein. Some of the transcripts produce a polyprotein that is cleaved to produce the final proteins, whereas other transcripts are produced by alternative splicing. AIDS is a devastating disease, and related viruses also infect other primates, often with less devastating effects. However, in humans, the virus has some specific characteristics that make it deadly. It is amazing that a virus with only 9 genes encoding 15 proteins can be deadly to humans with more than 23,000 genes that produce more than 90,000 proteins. However, HIV infects a cell type in humans that is at the center of the immune response.

The HIV genome is very compact (Figure 27.3). It has long terminal repeats (LTRs) on the 5′ and 3′ ends, which are important for integration into the host chromosome. This is similar to LTR retrotransposons, and in Chapter 13, the similarities in the reverse transcriptase were also discussed. Therefore, this virus is related to a large group of genetic elements that include retrotransposons, group II introns, and some viruses. For HIV-1, the genes (in order from 5′ to 3′) are *gag*, *pol*, *vif*, *vpr*, *tat*, *vpu*, *env*, and *nef*. For HIV-2, the order is the same, except *vpu* is missing, and instead another version of the gene, *vpx* is present on the 5′ side of *vpr*. The genes overlap with one another, and *rev* and *tat* are both split and contain part of the *env* gene. Three of the genes (*gag*, *pol*, and *env*) are polygenes that encode larger proteins that are cut into several pieces to form the mature proteins. The *gag* gene is expressed initially as a large precursor protein, called P55. It contains the MA (matrix), CA (capsid), NC (nucleocapsid), and P6 proteins, but also acts in virus assembly. MA is the matrix protein that surrounds the capsid, which is composed of the CA and the NC proteins (Figure 27.2). The NC protein packages and coats the RNA, whereas the CA protein forms the primary conical structure of the mature virus particle. The MA protein stabilizes the membrane that surrounds the virus particle. The P6 protein mediates the assembly of the virus, along with the P55 and vpr proteins. Also, it aids in budding of the completed viruses as they exit the cell.

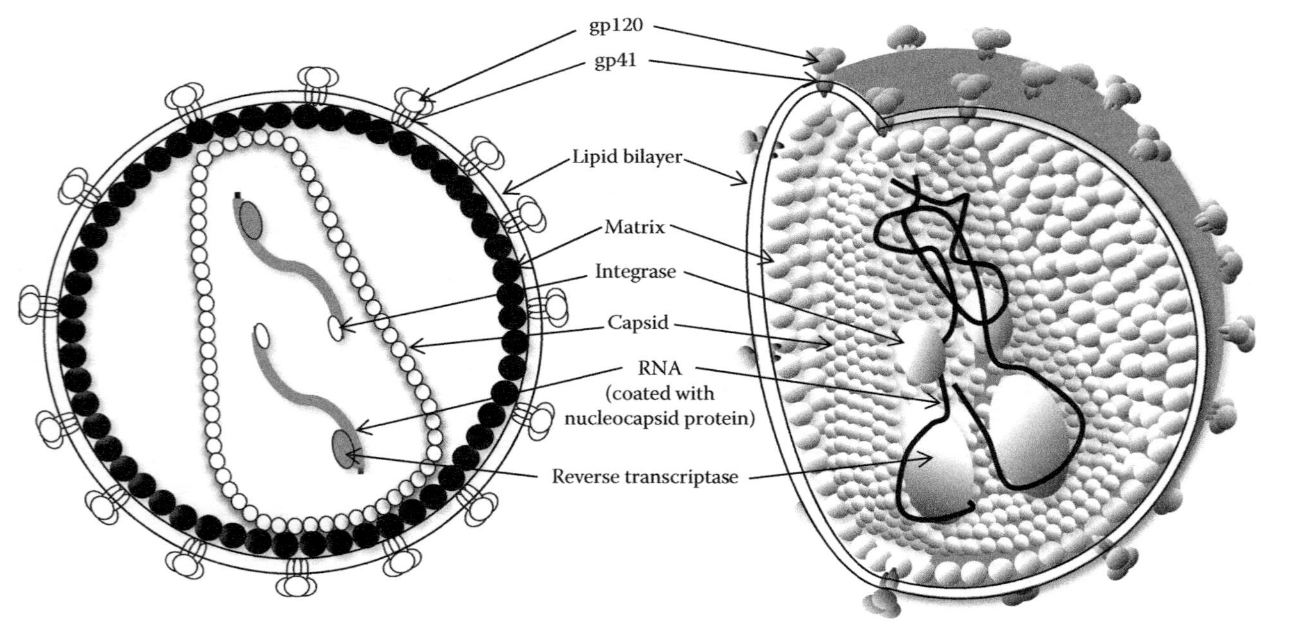

FIGURE 27.2 Cross section and three-dimensional diagrams of HIV. The center of HIV contains two molecules of the RNA genome, with attendant proteins, including reverse transcriptase and integrase. Surrounding these are capsid proteins, and surrounding these are matrix proteins. Immediately next to the matrix is a lipid bilayer that originated from the previous cell membrane. Embedded in the membrane are the gp120/gp41 trimer receptors that attach to the CD4 receptors on the cell surface of T lymphocytes.

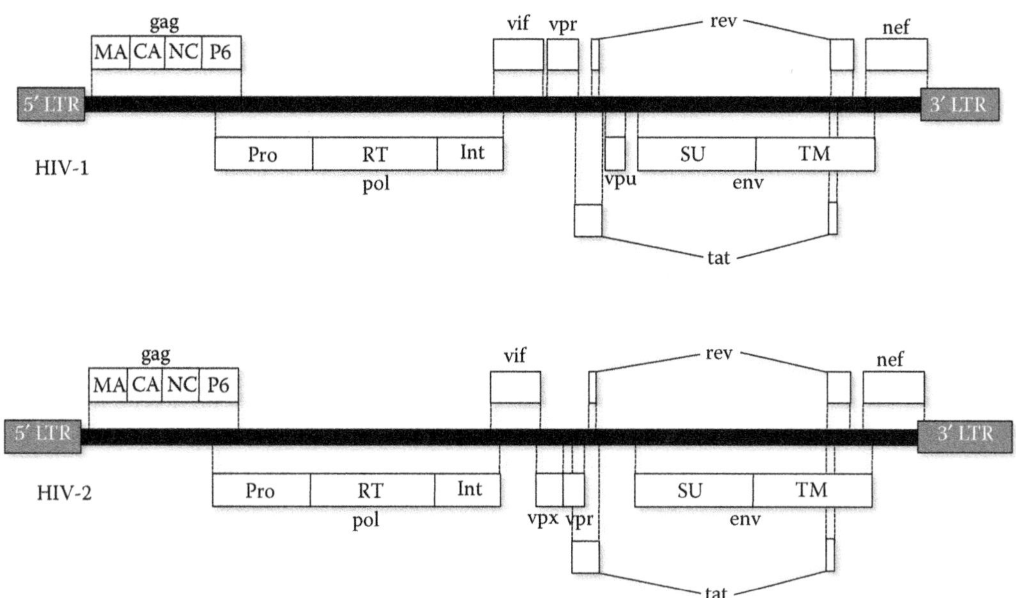

FIGURE 27.3 HIV genomes for HIV-1 and HIV-2. This virus is in the Retroviridae and carries its reverse transcriptase within the infectious virion. HIV-1 consists of 9749 nucleotides of single-stranded RNA. The genes are compact along the chromosome. The *gag* transcript encodes four primarily structural proteins: matrix (MA), capsid (CA), nucleocapsid (NC), and assembly protein (P6) that are separated after translation. The *pol* transcript encodes the three primary enzymes: protease (Pro), reverse transcriptase (RT), and integrase (Int). The *env* transcript encodes gp160 (glycoprotein 160). The protein is cleaved into the head subunit (SU) and the transmembrane (TM) portions. The *tat* and *rev* transcripts are produced by alternative splicing. They control expression and timing of expression of other genes during infection. The *vif*, *vpr*, *vpu*, and *nef* genes (as well as *vpx* in HIV-2) have other roles during infection, including movement of the viral RNA to the nucleus, and inhibition and degradation of CD4 receptors expressed by the T lymphocyte.

The *pol* gene encodes Pro (protease), RT (reverse transcriptase and RNase H), and Int (integrase). The protease is responsible for digesting the polyproteins into the final products. The RT produces the complementary DNA (cDNA) copy of the genome, whereas RNase H digests the original RNA molecule, followed by the second strand synthesis of the cDNA by the RT. The Int protein is responsible for integration of the viral cDNA into the host chromosome. Env (envelope) encodes gp160, a protein that is glycosylated in the Golgi before it is incorporated into the membrane that will surround the virus particle. Part of this protein is specifically recognized by the CD4 receptor on the host (human) helper T-lymphocyte cells. The gp160 protein is also digested by a cellular protease to form gp120 and gp41. The final env protein consists of the association of gp120 and gp41. The tat protein is necessary for transcription of the HIV genes. The rev protein helps to coordinate the timing of expression of the genes such that transcription and export of transcripts from the nucleus occur sequentially for efficient production of the proteins and assembly of virus particles. The nef protein is another regulator of transcription. It increases the rate of host CD4 receptor degradation, decreases expression of CD4 and class I major histocompatibility complex receptors, and may increase the expression of some of the HIV genes. The vpr protein is packaged into the virus particles and upon infection and helps to move the cDNA into the nucleus in preparation for integration into the host chromosome. The vpu protein is found in the membrane, and also aids in downregulation of the CD4 receptors of the host cell. This decrease in CD4 receptors in the infected cell is necessary prior to release of new virus particles so that the newly released viruses do not all end up attached to the CD4 receptors of the already infected cell. The vif protein counteracts a cellular antiviral factor so that replication of the virus can proceed during the infection process.

Although HIV has one of the smallest genomes of any organism, the function of the virus is nonetheless complex and coordinated, as indicated earlier. HIV begins by gaining access to the bloodstream of its host. It attaches to the helper T cells, which are central to signaling and control of infections by invading antigens (Figure 27.4). These cells communicate with other T cells, as well as B cells and other immune cells to attack, neutralize, and kill invading organisms. HIV attaches very specifically to T-cell CD4 receptors [via its two glycoproteins (GPs), gp120 and gp41] on the surface of the T cells. The membrane around the virus (that was part of the previous host cell membrane) fuses with the T-cell membrane, and the HIV capsid is released inside the cell. The matrix and the core proteins dissociate, and the HIV RNA, along with the associated reverse transcriptase, and int and vpr proteins are released into the cytosol. A DNA copy of the virus chromosome is made and transported into the nucleus, where the int protein helps to integrate the DNA version of the HIV genome into the host chromosome. Virus and cellular transcription factors then begin transcribing the HIV genes and exporting the RNAs out of the nucleus for translation on cellular ribosomes, followed by processing of the proteins and assembly and budding of the newly constructed virus particles. The cell has become a virus factory and its functions as a helper T cell begin to wane. As the cell ages, it also loses its ability to divide, but the DNA continues to replicate. This causes the cells to become multinucleate syncytia that

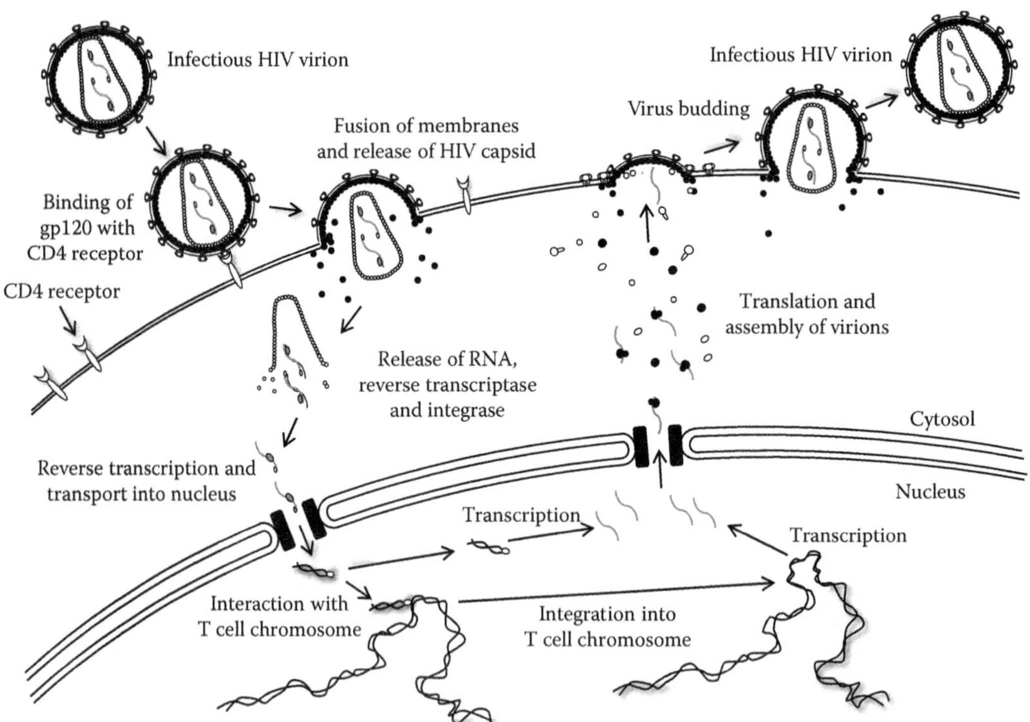

FIGURE 27.4 Progression of infection of a T lymphocyte by HIV. Initially, HIV attaches to the T lymphocyte through recognition of the gp120 protein by the CD4 receptor (and an attendant protein, not shown). The cell membrane then fuses with the HIV virion membrane releasing the capsid into the cytosol. The capsid begins to break down, and reverse transcription then begins, producing a cDNA of the virus genome. As reverse transcription continues, a double-stranded version of the genome is produced. Integrase integrates the HIV genome into the host chromosome, and transcription produces viral RNAs to form additional virions, which bud off from the host T lymphocyte. Eventually, the T cell becomes a multinucleated syncytium that produces large quantities of HIV virions but is incapable of any immune response.

continue to produce viruses, but are completely devoid of any immune response. As more of the T cells are similarly transformed by the virus, the infected host becomes more susceptible to infections by opportunistic microorganisms, and eventually may succumb to one of the infections. Therefore, with an extremely small genome and a handful of genes, this virus is capable of producing many more of its kind, while completely transforming a cell line within a huge genome with more than 23,000 genes, which eventually leads to disease and possibly to death of a very large and complex multicellular organism, a human being.

INFLUENZA A VIRUS

Influenza A is also a human pathogen but is primarily an avian virus that mutated long ago such that it now infects humans, swine, and a variety of other animals. There are at least 16 different subtypes based on one of the surface antigens, hemagglutinin, designated H1 through H16 (Figure 27.5); and at least 9 different subtypes of another surface antigen, neuraminidase, designated N1 through N9. Influenza A viruses have genomes of approximately 13,600 nucleotides that are separated on eight different pieces of negative-sense RNA, five of which encode single genes and three that encode 2 genes each, for a total of 11 genes (Figures 27.6 and 27.7, and Table 27.1). The largest RNA is approximately 2340 nt in length and encodes the PB2 protein, which is part of the RNA-dependent RNA polymerase responsible for initiating transcription of the viral RNA. The second RNA is also about 2340 nt and encodes the PB1 and PB1-F2 proteins.

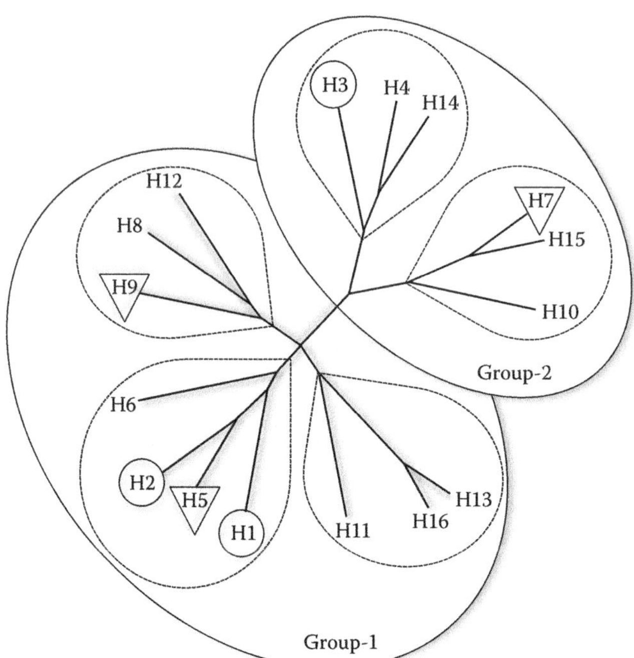

FIGURE 27.5 Unrooted phylogram of hemagglutinin (HA) genes. Sixteen subtypes of HA have been described and characterized. Two main groups are evident: Group 1 contains H1, H2, H5, H6, H8, H9, H11, H12, H13, and H16. Group 2 contains H3, H4, H7, H10, H14, and H15. In addition, there are five major branches (or clades): the first containing H3, H4, and H14; the second containing H7, H10, and H15; the third containing H11, H13, and H16; the fourth containing H1, H2, H5, and H6; and the fifth containing H8, H9, and H12. Circles indicate subtypes that can be transmitted among humans. Triangles indicate avian subtypes that occasionally infect humans to cause severe disease and death.

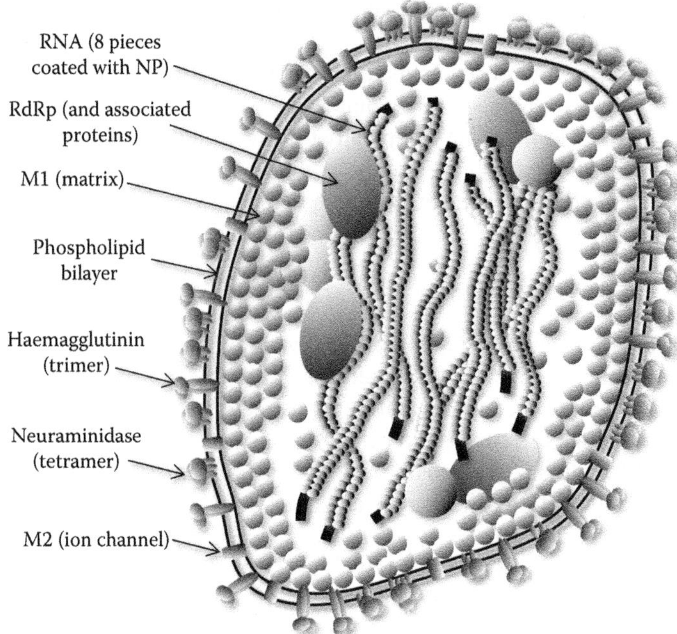

FIGURE 27.6 Diagram of influenza A virus. The center of the virus consists of eight pieces of RNA containing the genome of this virus. The RNA is stabilized and protected by a nucleocapsid protein. The virus RNA-dependent RNA polymerase (RdRp) molecules (and associated proteins) are also located next to the RNA. All of this is surrounded by matrix protein (M1). Immediately surrounding the matrix is a phospholipid bilayer of host origin. At least three proteins are embedded in the bilayer: hemagglutinin (that recognizes specific sialic acid groups on epitheilial cell surfaces—cell- and species-specific sites—and forms pores in the endosome for release of viral RNA into the cell), neuraminidase (a sialidase that digests the sialic acid groups during release of the virions), and ion channel M2 (responsible for acidification of the endosome, which releases the virus capsid into the cytosol of the host cell).

PB1 is responsible for elongation of transcription, whereas the PB1-F2 protein migrates to the mitochondria and shuts down their functions leading to cell death. The third RNA is approximately 2235 nt in length and carries the PA gene, which encodes yet another part of the RNA polymerase. The fourth RNA is about 1780 nt and codes for the hemagglutinin protein (HA), which is displayed on the membrane that surrounds the virus particle. The membrane is derived from the infected cell membrane, and the subtype of the hemagglutinin determines which cells will be susceptible to infection, because the hemagglutinins recognize and attach to sialic acid cell receptors that are specific to certain cells (Figure 27.8). This is what determines which species can be infected, as well as which cells of that species are infected. The fifth RNA is about 1565 nt in length and carries the NP gene, which is a nucleoprotein (NP) that attaches to the viral RNA in the virus particle, and is responsible for transport of the viral RNA in the infected cells. The sixth RNA is approximately 1415 nt in length and encodes the NA (neuraminidase) protein. This is another major membrane-associated protein that is responsible for release of the virus during infection by cutting the salicins from the host cell surface. The seventh RNA, which is approximately 1030 nt in length, encodes the two matrix proteins M1 and M2 which form the major virion capsid and an ion channel that aids in uncoating the virus upon infection, respectively. The last RNA is only 890 nt in length and encodes NS1 and NS2, called nonstructural proteins. NS1 is important in RNA transport into the nucleus, splicing, and translation, whereas NS2 is important for export of viral RNA from the nucleus.

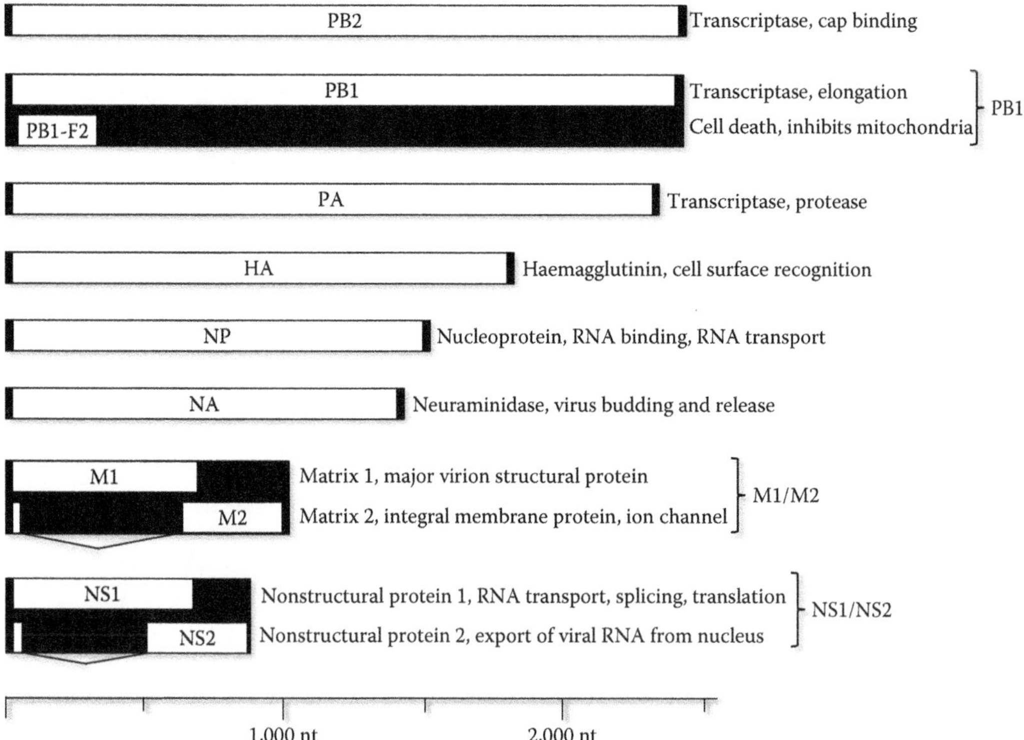

FIGURE 27.7 Influenza A genome. This genome consists of eight RNAs, including three that form more than one mRNA (and M1/M2 may form three mRNAs). In each case, the black boxes indicate the extent of the viral genomic RNA, whereas the white boxes indicate the mRNAs produced by each of the viral RNAs. The PB1 gene forms two mRNAs: one forming the PB1 full transcript forming part of the RNA-dependent RNA polymerase and PB1-F2, a short transcript that forms a protein that kills the host cell mitochondria. The PA gene encodes a transcriptase that is part of the viral polymerase. The HA gene encodes the hemagglutinin that is important in binding to the host cell and for releasing the viral RNA into the cytosol. NP is the nucleoprotein gene. This protein stabilizes and protects the viral RNA. The NA gene encodes the neuraminidase protein, which is a sialidase. The M1/M2 RNA produces two transcripts: one for the major matrix protein and the other for the ion channel responsible for acidification of the endosome. The NS1/NS2 RNA also encodes two transcripts: one that encodes a protein necessary for RNA import into the nucleus, splicing, and translation, and the other is primarily involved in export of viral RNA from the nucleus. The scale, in nucleotides, is at the bottom.

The course of the infection begins with attachment of the influenza A virus to the sialic acid receptor on epithelial cells by the hemagglutinin protein (Figure 27.8). Again, the recognition and attachment are species and tissue specific. For example, in birds, normally the influenza A virions attach to intestinal epithelial cells, and the disease is usually asymptomatic. However, some strains attach to lung epithelia, which can cause significant disease and mortality in birds. All 16 hemagglutinin subtypes have been isolated from birds. Humans can normally transmit only H1, H2, or H3 subtypes, and the viruses may be specific to intestinal or lung epithelia, the latter causing more serious disease. Swine are normally susceptible to H1 and H3 subtypes. Once the virus has attached to the cell, it is enveloped into an endosome by the cell. The M2 ion channels then act to acidify the endosome, which releases the viral RNA and associated viral proteins. Also, hemagglutinin associates with the endosome membrane to form pores that allow the release of viral RNA into the cytosol. The viral RNAs are transported into the nucleus, where sense-strand RNAs (mRNAs) are transcribed by RdRp and where negative-sense RNAs (viral genomic RNAs) are replicated, also by RdRp.

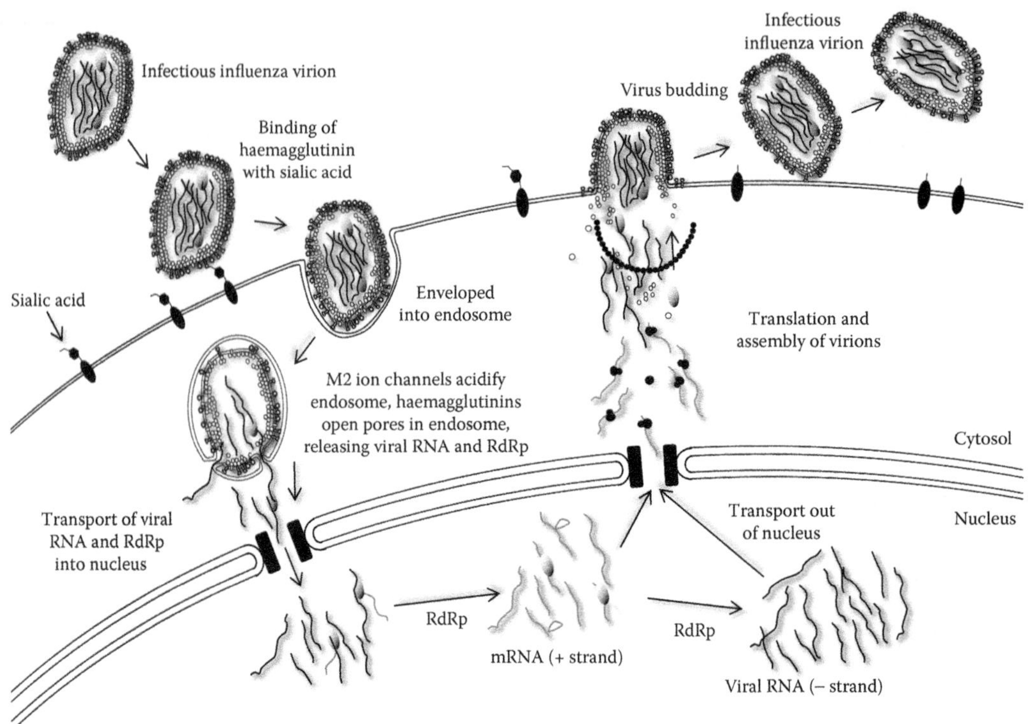

FIGURE 27.8 Progression of infection for influenza A virus. Infection begins with attachment of the hemagglutinin with the sialic acid moieties on cell surface proteins. This interaction is specific based on the molecule configuration, thus determining specificity of attachment. This determines which species and which cells are infected by this virus. Once there is recognition, the virion is enveloped by the cell membrane to form an endosome. The M2 ion channels cause a pH drop on the inside of the endosome that causes dissociation of the matrix proteins and release of the viral RNA. At the same time, the hemagglutinin interacts with the endosome membrane to form small pores in the membrane such that the viral RNA is released into the cytosol. From there, the viral RNA is transported into the nucleus where the RNA-dependent RNA polymerase transcribes mRNA from the viral RNA. Additional viral RNA is produced from the mRNA, also by the RdRp. The mRNA is transported out of the nucleus and is translated on cytosolic ribosomes. The viral RNA is transported out of the nucleus, and new virions begin to form. In the final steps, the neuraminidase cuts the sialic groups off of the membrane proteins so that the hemagglutinin does not attach to them, and the virion can be released from the host cell.

The mRNAs are transported from the nucleus, and the RNAs are translated on cytoplasmic ribosomes to produce the viral proteins. Some of the proteins (e.g., M1 and NS2) are transported into the nucleus where they associate with the virus genomic RNAs, and the complex is then transported out of the nucleus where it is joined by other viral proteins for final assembly and envelopment into hemagglutinin- and neuraminidase-coated host membrane, and then the newly constructed virus particles bud off from the cell to travel to uninfected cells or are excreted by the host organism, where they can infect other hosts.

A unique feature of the fragmented genome of influenza A is that the pieces of the genome can reassort. That is, if a pig becomes infected with both an avian influenza and a swine influenza, within infected cells, some of each type can be produced, and when the RNA pieces are packaged in the virions, they can include some avian pieces and some swine pieces. For example, the emerging virions might contain fragments 1–4 from birds and fragments 5–8 from swine. This would mean that they would carry an avian hemagglutinin, but a swine neuraminidase. The 2009 so-called swine flu, which was an H1N1 subtype, consisted of a mixture of avian, swine, and human influenza A genes. Apparently, several of the genes originated from birds, several were from swine, and at least one was from human sources. Additionally, there can be recombination events that occur during infection, such that even the individual genes can be hybrids. This means that influenza A viruses can mutate, reassort, and evolve rapidly. This is in addition to the fact that they are RNA viruses, which usually have high mutation rates. Occasionally, humans can contract some of the other subtypes of influenza A, including H5, H7, and H9. In some cases, disease symptoms are severe, and some of these infections have resulted in high rates of mortality. In recent years, the so-called bird flu or avian influenza, which is an H5N1, has been of concern, partly because it has killed large numbers of wild and domestic birds, but it has also been responsible for many instances of human disability and death. Although bird-to-human transmission was indicated in most cases, no confirmed human transmissions to other humans have been reported. Were that to occur, the virus would be much more dangerous.

Again, here is an organism with a very small genome of only about 13,600 nucleotides, encoding 11 genes, and it is able to infect a wide variety of species with very large genomes and body sizes to replicate itself. In doing so, it can cause a variety of disease symptoms from none at all to extremely severe, and in some cases death. The H1N1 influenza A that caused a pandemic in 1917–1919 caused the deaths of 20–100 million people worldwide. Rather than having a reassorted genotype, which was suspected, it appears to have been an avian strain that mutated so that it would attach to human lung epithelial cells. It could cause death in 2–4 days in otherwise healthy people, essentially by causing severe pneumonia and asphyxia.

EBOLA VIRUS

This virus emerges sporadically and has caused only a few epidemics. This is due to its nature of causing high rates of mortality in humans, such that once the disease symptoms are present, the patient is often bedridden, and thus cannot infect large numbers of other people. There have been outbreaks of Ebola in Africa, Europe, and Asia over the past several hundred years. Because of its debilitating symptoms of causing widespread bleeding in the infected person, normally only localized outbreaks occur. This negative single-stranded RNA virus has a relatively small genome of 19,000 nucleotides on a single piece of RNA encoding only seven genes (Figure 27.9). Unlike HIV where the genes overlap, the genes in the Ebola genome have intergenic sections that do not encode any proteins. The first gene on the 5′ end of the chromosome is the NP, which is the main protein of the ribonucleoprotein complex at the center of the virus particle. The second and fifth genes (VP35 and VP30, respectively) also are parts of the ribonucleoprotein complex, although VP30 binds weakly in the complex, so it might have other functions as well. The third gene on the chromosome (VP40) encodes the major matrix protein, whereas the sixth gene (VP24) encodes another matrix protein that forms the envelope surrounding the matrix. GP is the fourth gene along the chromosome. This protein produced called a peplomer, is glycosylated posttranslationally and embedded in the membrane that surrounds the virus. Finally, the L (long) gene encodes the RNA-dependent RNA polymerase. It produces sense-strand mRNAs, as well as replicates the genomic RNA.

Ebola attaches to cells via the GP peplomers embedded in its envelope. They appear to be able to attach to almost every type of cell in the body, except some types of bone cells. Once the virus enters the cell cytosol, transcription of sense-strand RNA and replication of negative-strand genomic RNA both begin using the RNA-dependent RNA polymerase of the virus. Cytosolic ribosomes are used for translation of the proteins, followed by posttranslational processes (primarily for the GP) and assembly of the virus. Ultimately, each infected cell becomes packed full of virus particles. The cell then dies and lyses. Blood cells clump together and form clots, cutting off circulation to various organs, which then die and leak their fluids. Eventually, fluids and blood may leak from many parts of the body prior to death. The blood of the infected patient

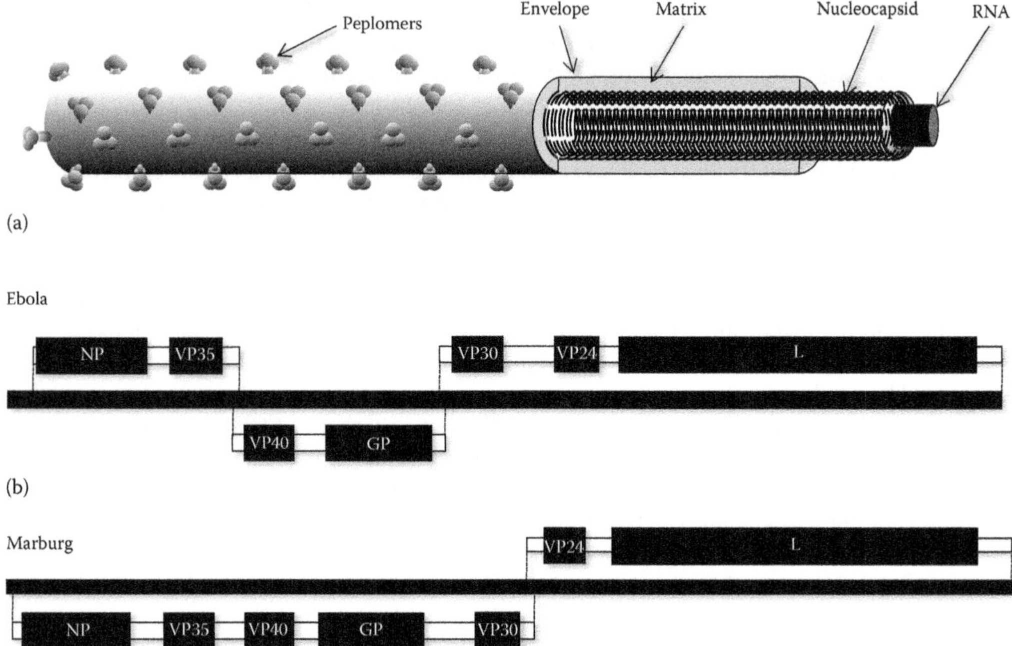

FIGURE 27.9 Diagram of an Ebola virion (a), as well as genomes of Ebola (b) and the related Marburg virus (c). The virion consists of an RNA coated with nucleocapsid proteins, which are coated by matrix proteins, and further encased in envelope proteins. They are studded with peplomers (glycoproteins encoded by the GP gene) that are necessary for attachment to host cell surfaces. The nucleoprotein (NP) as well as viral proteins, VP 30 and VP35, form the ribonuclear nucleocapsid, responsible for protection and transport of the viral RNA. The matrix protein is VP40, whereas VP24 encodes the envelope protein. The longest gene, designated L (for long) encodes the RNA-dependent RNA polymerase.

becomes concentrated with virus particles and dead or dying cells, leading to death in 40%–90% of those that are infected. Although the Ebola virus has not been reported in the nuclei of the host cells, the genomes of several mammals, including humans and guinea pigs, contain some of the genes from this virus. It is unknown how or when they integrated into these chromosomes.

Again, as with the other RNA viruses, the Ebola virus has a very small genome and only seven genes, but it can take over much larger and more complex cells to transform them into virus factories. In doing so, they rapidly kill the cells, which is ultimately not the best strategy for a pathogen. These RNA viruses illustrate that a collection of only a few genes can have a very large affect on larger genomes and organisms. They also indicate that some very complex functions can result from the combination of only a very small number of genes.

KEY POINTS

1. Genomics is the determination and study of the entire genomes of organisms.
2. Genomes of organisms range from just a few thousand nucleotides that encode four proteins to hundreds of billions of nucleotides that encode tens of thousands of proteins.
3. C-value (i.e., genome size) is not strictly correlated with RNA complexity, although there is some correlation for small genomes.
4. Genomics, proteomics, RNAomics, and metabolomics have revolutionized the study of evolution.

5. HIV is an RNA virus with a genome size of 9.7 kb, encoding at least 15 genes. Although it has a small genome, it can cause significant disease and death in humans.
6. Influenza A is an RNA virus with genome size of 13.6 kb, which is organized into eight pieces. Its 11 genes allow it to specifically infect a large diversity of animals (including humans).
7. Ebola is an RNA virus with a genome size of 19 kb. When it infects humans, it rapidly debilitates the patient by killing many types of cells. Blood cells are affected, which causes clotting and loss of blood supply to major organs.
8. Host organisms with very large genomes and tens of thousands of genes can be infected, disabled, and killed by organisms with very small genomes and only few genes.

ADDITIONAL READINGS

Alberts, B., D. Bray, K. Hopkin, A. Johnson, J. Lewis, M. Raff, K. Roberts, and P. Walter. 2013. *Essential Cell Biology*, 4th ed. New York: Garland Publishing.

Alberts, B., D. Bray, A. Johnson, J. Lewis, M. Raff, K. Roberts, and J. D. Watson. 1994. *Molecular Biology of the Cell*. New York: Garland Publishing.

Baron, S., ed. 1995. *Medical Microbiology*, 4th ed. Galveston, TX: University of Texas Medical Branch.

Becker, W. M., J. B. Reece, and M. F. Poenie. 1996. *The World of the Cell*, 3rd ed. New York: The Benjamin/Cummings Publishing Company.

Beer, B., R. Kurth, and A. Bukreyev. 1999. Characteristics of Filoviridae: Marburg and Ebola viruses. *Naturwissenschaften* 86:8–17.

Futuyma, D. J. 1998. *Evolutionary Biology*, 3rd ed. Sunderland, MA: Sinauer Associates.

Gregory, T. R., ed. 2005. *The Evolution of the Genome*. London: Elsevier Academic Press.

Hall, B. K. and B. Hallgrimsson. 2008. *Strickberger's Evolution*, 4th ed. Sudbury, MA: Jones & Bartlett Publishers.

Hedestam, G. B. K., R. A. M. Fouchier, S. Phogat, D. R. Burton, J. Sodroski, and R. T. Wyatt. 2008. The challenges of eliciting neutralizing antibodies to HIV-1 and influenza A virus. *Nature Rev. Microbiol.* 6:143–155.

Organ, C. L., A. M. Shedlock, A. Meade, M. Pagel, and S. V. Edwards. 2009. Origin of avian genome size and structure from non-avian dinosaurs. *Nature* 446:180–184.

Russell, R. J., P. S. Kerry, D. J. Stevens, D. A. Steinhauer, S. R. Martin, S. J. Gamblin, and J. J. Skehel. 2008. Structure of influenza haemagglutinin in complex with an inhibitor of membrane fusion. *Proc. Natl. Acad. Sci. USA* 105:17736–17741.

Takata, A. and Y. B. Kawaoka. 2001. The pathogenesis of Ebola hemorrhagic fever. *Trends in Microbiol.* 9:506–511.

Tropp, B. E. 2008. *Molecular Biology: Genes to Proteins*, 3rd ed. Sudbury, MA: Jones and Bartlett Publishers.

Weiss-Schneeweiss, H., J. Greilhuber, and G. M. Schneeweiss. 2006. Genome size evolution in holoparasitic *Orobanche* (Orobanchaceae) and related genera. *Am. J. Bot.* 93:148–156.

28 DNA Viruses

INTRODUCTION

In Chapter 27, some of the smallest genomes were discussed. The RNA viruses have only a few genes, but can infect more complex cells and utilize parts of those cells to produce more viruses. On average, DNA viruses are larger than RNA viruses (Figure 27.1), but they are also parasites that take over the host cell to produce more virus particles. However, viruses can be more than parasites. They are vital in some environments. In parts of the oceans, nutrient levels are extremely limiting, and therefore, few organisms can survive. However, viruses make it possible for a diverse set of organisms, especially microbes, to live. When the viruses infect cells, many of them lyse the host cell as a way to release the newly constructed virus particles. In doing so, they also release all of the organic contents of those cells. These organics become valuable sources of nutrients for the other organisms in the vicinity. Therefore, viruses are important components of nutrient cycling in the oceans and in other environments where nutrient levels are low.

DNA viruses have from about 10 to approximately 1000 genes with genomes ranging from a few thousand to over a million bases of DNA. Therefore, they range from viruses that are about as complex as many simple RNA viruses to those that are as complex as some Bacteria and Archaea. However, all are obligate parasites and need host cells to replicate and produce more virus particles. The Acanthamoeba Polyphaga Mimivirus (APMV) can actually be infected by another virus (a satellite virus, or virophage, called Sputnik), and thus, the line between living and nonliving biological entities is further blurred by the presence of these complex viruses that can be parasitized. It is a further indication that the progression from simple molecules through enzymatic/replicating molecules through viruses through cellular organisms is a smooth one with no absolute lines between each of the stages.

BACTERIOPHAGE ϕX174

Although bacteriophage ϕX174 is a small virus with a small genome, it has an interesting and complex life cycle (Figure 28.1). The virus first adsorbs to the host cell and then discharges its single-stranded DNA (ssDNA) into the cell through the function of protein H. Protein A synthesizes the other DNA strand to produce a double-stranded version of the virus chromosome. Replication, transcription, and translation begin almost immediately thereafter, and proteins A, A*, B, C, F, and G appear within the cell. The double-stranded version of the viral chromosome is replicated by protein A, and protein C then binds to these molecules to convert them to another replicative form that produces ssDNA by rolling circle replication. The coat proteins (encoded by gene *F*) and the major spike proteins (encoded by gene *G*) begin to form the initial virus capsid, aided by protein B, which forms a scaffold for assembly. Next, protein D helps to form the capsid into its final shape, followed by packaging of the ssDNAs into the developing capsid after being condensed by protein J and cut by protein A. Protein D then completes the head, and the minor spike proteins (H) are added to the capsid. Protein E increases in concentration until it causes lysis of the host cell, which releases the completed virus particles. These can then attach to another bacterial cell to start the process once again.

Bacteriophage ϕX174 was the first DNA virus to be sequenced. Its genome consists of a small (5368 nucleotides) circular ssDNA (Figure 28.2 and Table 27.1). Some of its 11 genes overlap with one another such that the gene density is very high. In fact, genes *A**, *B*, and *K* overlap with gene *A*, whereas gene *K* overlaps with gene *C*, and gene *E* overlaps with gene *D*. Thus, this virus has one of the most condensed genomes known, having an average of more than two genes per

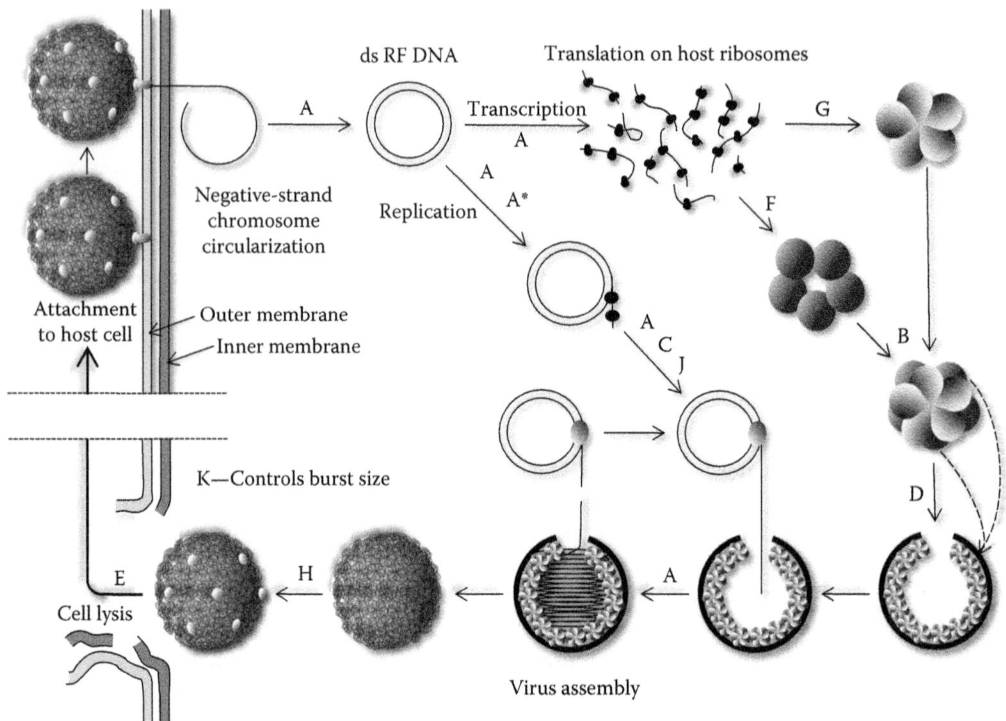

FIGURE 28.1 Life cycle of bacteriophage ϕX174. After attachment of the virus to the host cell (left), the ssDNA chromosome is injected into the cell through the outer and inner membranes of the host cell. The DNA circularizes and is converted to the double-stranded replicative form (RF) by protein A. This protein also begins to produce RNAs from the chromsome. Additionally, A, A*, and C produce additional copies of the replicative form, while beginning to produce linear concatemers that become packaged into the bacteriophage heads. The heads are produced from the assembly of the major head protein F and the major spike protein G, which is facilitated by proteins B and D. The linear ssDNA is threaded into the heads by protein J, which also compacts the DNA. Protein A participates in this process by cutting the DNA when a full genome has been loaded into the head. Finally, the minor spike protein, H, is attached to the surface of the virion, and the lysozyme (from gene *E*) builds up to a level that causes lysis of the host cell to release the newly constructed virus particles. The absolute function of protein K is unknown. However, when it is inactivated or missing, the burst size (number of virions released by each cell) decreases.

kilobase (Table 27.1). As mentioned before, gene *A* encodes the viral polymerase that synthesizes each strand of the DNA separately, and also synthesizes the single-stranded genomes via rolling circle replication. Gene *A** is a shortened version of gene *A* but has a very different function. It shuts off replication of the host chromosome. Gene *C* produces a protein that binds to the viral DNA such that it can be packaged into the virus heads by the protein encoded by gene *J* (the core protein). Also, it aids in the transition from the double-stranded replication to rolling circle replication. Genes *B* and *D* encode proteins that act as scaffolds to form the capsids of the virus particles, whereas gene *F* produces the main coat protein. The virion also has major (encoded by gene *G*) and minor (encoded by gene *H*) spike proteins that are important in adsorption of the virus to the host cell. Additionally, protein H is responsible for threading the viral DNA into the host cell to initiate infection. Gene *E* produces a protein that lyses the host cell. The function of gene *K* is unknown. However, when it is mutated so that no K protein is made, the viruses still can replicate and release viable phage from the host, but the number of virus particles is greatly reduced.

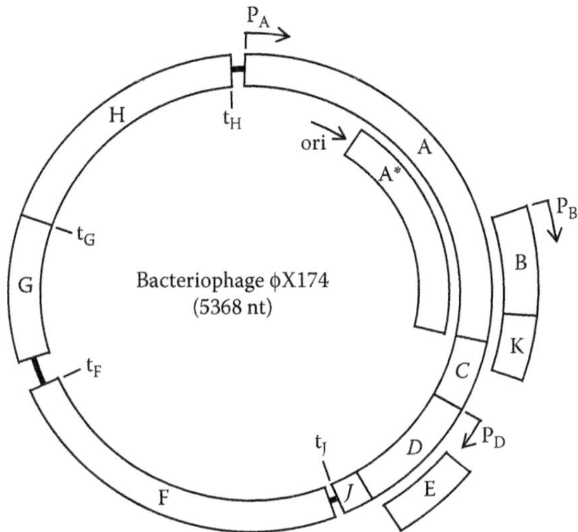

FIGURE 28.2 Genome of bacteriophage ϕX174. This genome consists of 5368 nt of DNA with 11 genes, some of which are overlapped with other genes. It has three promoters (P_A, P_B, and P_D) and four leaky translation terminators (t_J, t_F, t_G, and t_H), meaning that transcription does not always stop at those points. Protein A is the major viral DNA polymerase, which produces several different replicative forms of the DNA, including rolling circle replication to form the single-stranded chromosomes that are packaged into the newly constructed phage heads. The same region also produces the A* protein that stops replication of the host chromosome. Additionally, transcription from P_A continues through genes *C*, *D*, and *J* (that encode proteins for DNA binding, assembly of heads, and packaging of DNA, respectively), as well as read through past t_J to produce mRNAs for F and G, the major capsid and spike proteins, respectively. At the same time, transcription from P_B produces mRNA for protein B that is responsible for initial assembly of the F and G proteins in the prehead and protein K that affects the number of virus particle released from the cell (i.e., burst size). Additional read-through of the major transcript produces mRNA for the minor spike protein (H), which is added to the outside of the coat and which is responsible for recognition and binding to the host cell and for aiding in the injection of the bacteriophage DNA into the host cell. Transcription from the D promoter (P_D) produces mRNA for proteins D and E (responsible for prehead assembly and the lysozyme that lyses the host cell, respectively). Ori is the origin of replication.

BACTERIOPHAGE LAMBDA (λ)

This bacteriophage is probably one of the most studied and most utilized viruses on the Earth. It was isolated in the mid-twentieth century and was investigated using genetic experiments based on mutant genotypes and phenotypes. From this information, many of its functions and genes were identified even before it was sequenced. Once it was sequenced, more genes were discovered. Beginning in the 1970s, it was found that it could be reconstituted *in vitro* using DNA and the viral proteins. In other words, it was capable of self-assembly *in vitro*. This led to its use as a cloning vector. It was found that one large section in the middle of the genome could be completely deleted without affecting the life cycle of the virus. Scientists began putting other fragments of DNA into this area and found that they could produce millions of copies of the recombinant sequences in a short time by infecting *Escherichia coli* cells with the recombinant lambda virions. It became a heavily used cloning vector, and because of this, it was studied in hundreds of laboratories over several decades. Therefore, much is known about this virus.

The lambda genome consists of an average of 48,500 base pairs of DNA (9 times larger than the genome of ϕX174), arranged as a linear double-stranded molecule in the virus particle, which circularizes after entering the host cell (Figure 28.3). Circularization is facilitated by cohesive ends (cos sites), which are complementary 12-nucleotide single-stranded regions on the ends of the linear

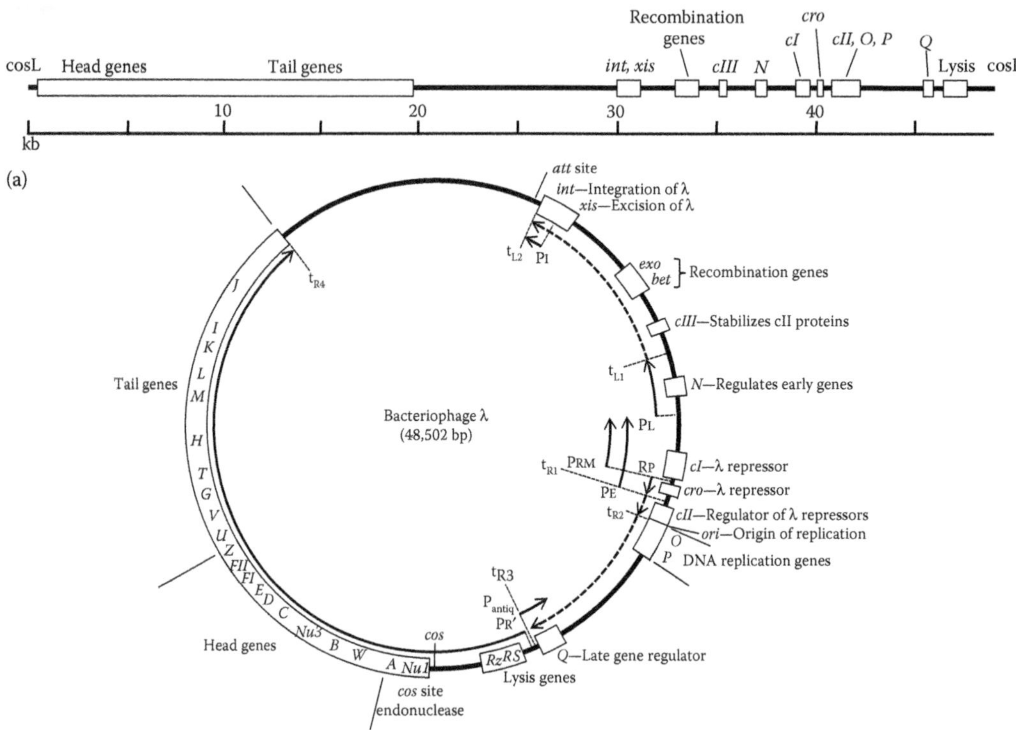

FIGURE 28.3 Bacteriophage lambda (λ) genome. The linear dsDNA chromosome (a). The single-stranded complementary *cos* sites (cosL and cosR) are located at the ends of the chromosome. They cause circularization of the chromosome once the DNA has been injected into the host cell (b). The 56 genes of the λ genome are arranged in groups along the chromosome according to their functions. All of the head and tail genes, as well as the lysis genes, are grouped in one region, whereas the genes for lysogeny (*bet*, *exo*, *xis*, and *int*) are grouped in a different location. The genes controlling expression (*cI*, *cro*, *cII*, *cIII*, and *N*) are also grouped (although some are somewhat distant from one another). There are seven promoter regions (PR, PRM, PL, PE, PR′, PI, and Pantiq), one of which has opposing promoters (PM and PRM), and regions (called operators) where two repressors, cI and cro, can attach to control gene expression (see Figure 28.5 for details). Also, there are six transcription terminators (t_{L1}, t_{L2}, t_{R1}, t_{R2}, t_{R3}, and t_{R4}) that are ignored when anti-terminators (e.g., N protein) are attached at those sites.

chromosomes. There are 56 genes (as well as at least eight additional open reading frames) arranged along the chromosome according to their functions. Lambda has two different pathways after entering the bacterial cell (Figure 28.4). It can go through a lytic cycle, in which it uses the cellular systems and its own gene products to produce more phage particles (Figure 28.5), or it can integrate its chromosome into the host chromosome, entering a lysogenic cycle.

A specific site on the lambda chromosome (the att site) recognizes a specific site on the *E. coli* chromosome (another att site) between the *gal* and *bio* genes, where integration occurs. Although a lysogen, each time the bacterial chromosome replicates, it replicates the section that is the virus DNA just as if it were a normal part of its own chromosome, because it is. The virus transcribes a few of its genes, although a lysogen, those that are responsible for inhibiting the genes involved in the lytic cycle, as well as those involved in continuing a lysogenic phase (Figure 28.5). Primarily repressor protein cI is produced at low levels. This occupies the PR, PL, and PRM promoters, which essentially blocks transcription of most of the λ genes.

Infection by λ begins with recognition and attachment to specific outer membrane proteins. Then, the DNA is injected through the outer and inner bacterial membranes and into the cell (Figure 28.4). The DNA circularizes, and transcription begins at the PR (right) and PL (left)

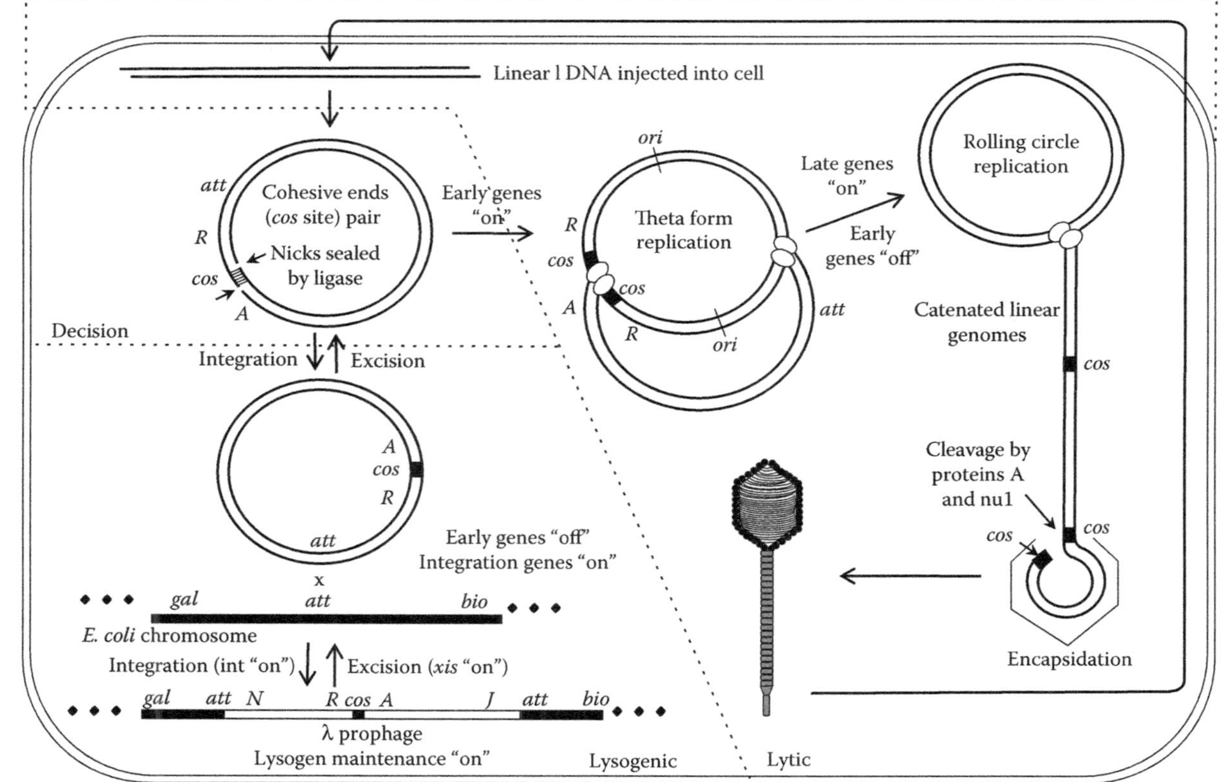

FIGURE 28.4 The two pathways for bacteriophage λ. Upon injection of the λ DNA into the host cell, intricate controls (decision phase) lead to the bacteriophage to proceed through a lytic phase or a lysogenic phase. During the initial stages of transcription and gene expression, if cro is produced in higher amounts than cI (Figure 28.5), then the early genes are expressed and lytic growth to produce new virus particles proceeds (Figure 28.6). + However, when cI concentrations are greater than cro concentrations (Figure 28.5), the leftward transcription from P_{RM}, P_E, P_L, and P_I predominate, and the λ chromosome integrates into the host chromosome, becoming a prophage. This is maintained by low levels of gene expression from the P_{RM}, P_E, P_I, and Pantiq promoters (see also Figure 28.3). This process is reversible. During times of cell stress, cI begins to break down and the chromosome can be excised from the host chromosome by the xis protein. The bacteriophage can then proceed through the lytic pathway to produce more virus particles and lyse the host cell.

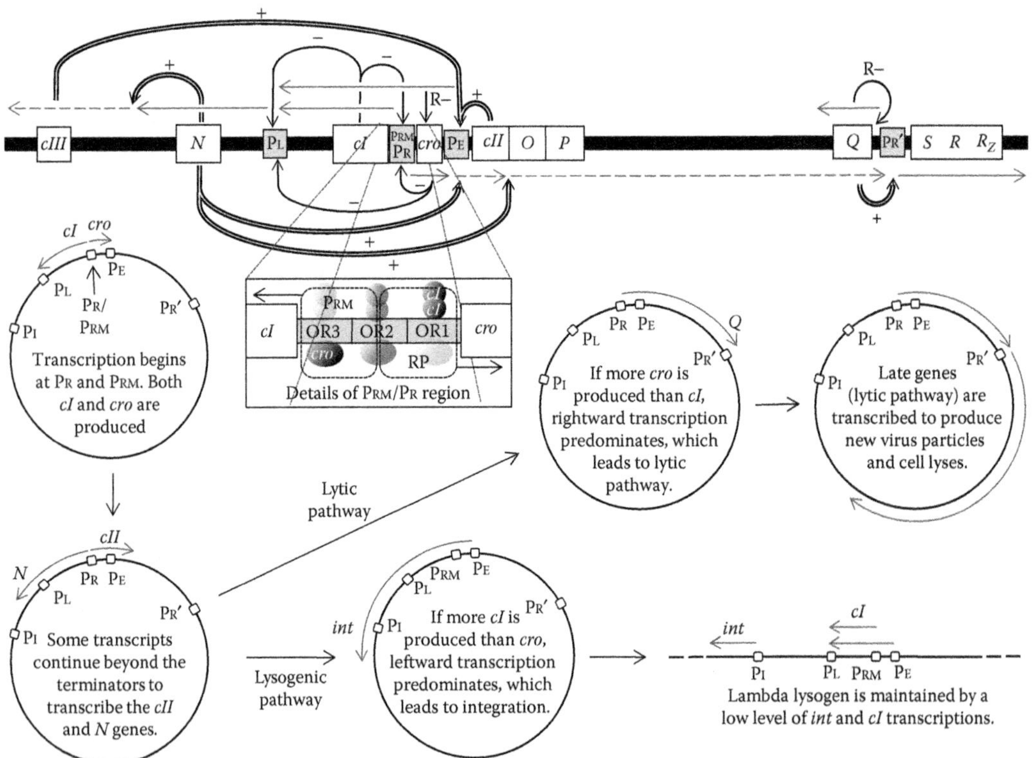

FIGURE 28.5 Details of gene expression control in λ. Transcription begins at the PR and PRM promoters. Minus (–) indicates downregulation of expression, whereas plus (+) indicates upregulation of a particular gene by proteins. The cI and cro proteins are repressors that block RNA polymerase binding. The cII protein aids in RNA polymerase binding to the PE promoter, whereas the cIII protein stabilizes the cII protein. The N protein is an anti-terminator of transcription. It acts by removing inhibitory secondary structures in the DNA at the terminator sites. At least two antisense RNAs, anti-cro and anti-Q (designated R-) are produced. These bind to the *cro* and *Q* transcripts, respectively, blocking translation of the transcripts and marking them for degradation by cellular enzymes. Once the cI and cro proteins begin to accumulate, they compete for the same sites in the promoter region. Three sites, called operators, OR1, OR2, and OR3, are present in this region. However, *cI* and *cro* bind in reverse order onto the operators. A dimer of *cI* binds first to OR1, and then to OR2 and OR3, whereas *cro* binds first to OR3, and then to OR2 and OR1. When the repressor proteins are on the promoters, RNA polymerase cannot bind to those regions, and therefore, transcription is blocked. At the beginning of infection, some transcription continues through gene *N* (which encodes an anti-terminator of transcription protein) and *cIII* (which helps to increase transcription from PE). When concentrations of cro are higher, additional leftward transcription (from PRM and PL) is blocked. Rightward transcription increases and is extended because of the anti-terminator N protein, thereby initiating the lytic pathway. Initially, this increases the number of λ genomes, because *O* and *P* are expressed, which are responsible for replication of λ DNA. Transcription proceeds through *Q*, which encodes a protein that promotes transcription from PR′, thus expressing the lysis, head, and tail genes. However, if *cI* concentrations are higher, then PR will be blocked, which will allow transcription from PRM and PE that will inhibit production of *cro* by blocking its transcription, as well as producing an antisense RNA (designated R-) that will mark any *cro* transcripts for degradation. Eventually, the *int* gene will be expressed, and the λ prophage (lysogen) will be formed by integration of the λ chromosome into the host chromosome. Lysogeny is maintained by low levels of transcription from PRM, PE, PI, and Panitiq (which produces an antisense RNA that hybridizes to the *Q* transcript to block its expression and mark it for degradation).

promoters (Figures 28.3 and 28.5). The immediate product from rightward transcription is messenger RNA (mRNA) that encodes cro, a repressor protein that inhibits leftward transcription from the PRM promoter. The PRM promoter is responsible for transcribing the gene that encodes the repressor cI. This protein inhibits rightward transcription from PRM, and thus represses lytic genes and the lytic phase. The promoter PL transcribes the gene for the *N* gene. The N protein is an antiterminator of transcription. There are two rightward terminators (t_{R1} and t_{R2}) and one leftward terminator (t_{L1}), which are recognized by the N protein. When N is attached to these regions, transcription is not terminated, but continues past each. When transcription from PR extends past cro, cII is transcribed. The cII protein turns on the PE (early) promoter and transcribes both an anti-cro RNA (which inhibits expression of the *cro* gene) and the *cI* gene, which encodes the cI protein, which represses rightward transcription from PR, thus shutting of cro expression (Figure 28.5). If sufficient N protein is produced, transcription from PL will continue through t_{L1}, which will proceed to produce mRNA for the recombination and integration genes that are needed to insert the lambda DNA into the bacterial chromosome to form the lysogen. However, if *cII* does not build up rapidly enough to make sufficient *cI* to shut off *cro*, transcription will proceed through to the *O*, *P*, and *Q* genes (Figure 28.5), signaling the initiation of the lytic phase. Genes *O* and *P* encode the enzymes responsible for λ DNA replication. The Q protein is a positive regulator of the late genes, which encode the enzymes that lyse the bacterial cells and those that form the mature bacteriophage (head, tail, and packaging genes).

When λ enters into the lytic phase, one of the first signals that this is occurring is that copies of the lambda genome begin to be produced (Figure 28.4). This begins as theta-form replication to form additional copies of the genome for transcription, and then changes to rolling circle replication to produce linear λ DNA concatemers that are packaged into the phage heads later in the lytic phase. Once transcription proceeds past the *O* and *P* genes, next the *Q* gene is transcribed, which encodes a protein that stimulates transcription of the late genes (Figures 28.3 and 28.5). This includes (in order of transcription) S, R, and Rz, which encode enzymes that lyse the host cell. Also, expression of the virus assembly genes is initiated (Figures 28.3 and 28.6).

These include (in order of transcription) the *Nu1* and *A* genes (encoding proteins that recognize and cut genome-length pieces of DNA from the newly synthesized long concatemers of λ DNA), the genes for constructing heads (*W*, *B*, *Nu3*, *C*, *D*, *E*, *FI*, *FII*, and *Z*), and the genes for constructing tails (*U*, *V*, *G*, *T*, *H*, *M*, *L*, *K*, *I*, and *J*).

When λ is in the lysogenic phase, lysogeny is maintained by a low level of leftward transcription from the PRM promoter, which leads to the production of a constant low level of cI (Figures 28.3 through 28.5). The cI protein represses the rightward transcription from the PRM promoter, which shuts off *cro* and all of the lytic genes. The cI protein also represses the expression of genes from the PL promoter. However, if the cell begins to produce symptoms of stress, often with increased protease activity, transcription from some of the other promoters can begin. If transcription from PL continues past t_{L1}, because of the action of the N protein, the recombination and excision (*xis*) genes can be produced, and the λ chromosome can exit from the chromosome. The lytic genes are then expressed, and the λ proceeds through the lytic phase. The expression of genes in the lytic phase of growth is highly regulated and coordinated (Figure 28.6). This allows the ordered assembly of virus heads, packaging of a single genome per head, construction of tails, attachment of tails, and finally lysis of the host cell to release the completed virions.

The genome of bacteriophage λ is 10 times larger than that of the smallest DNA virus genome, but it is about 200 times smaller than the largest virus genome. The intricate way in which it can switch its phases of growth and reproduction, as well as the complex assembly of new virus particles, can be used as a useful conceptual framework upon which to build up to other types of organisms, including microbes and multicellular organisms. As other genomes are discussed, remember the details of bacteriophage λ gene expression control and virus assembly. Most of the molecular mechanisms used by λ to control the expression of its genes in order to complete its life cycle are also utilized by other organisms.

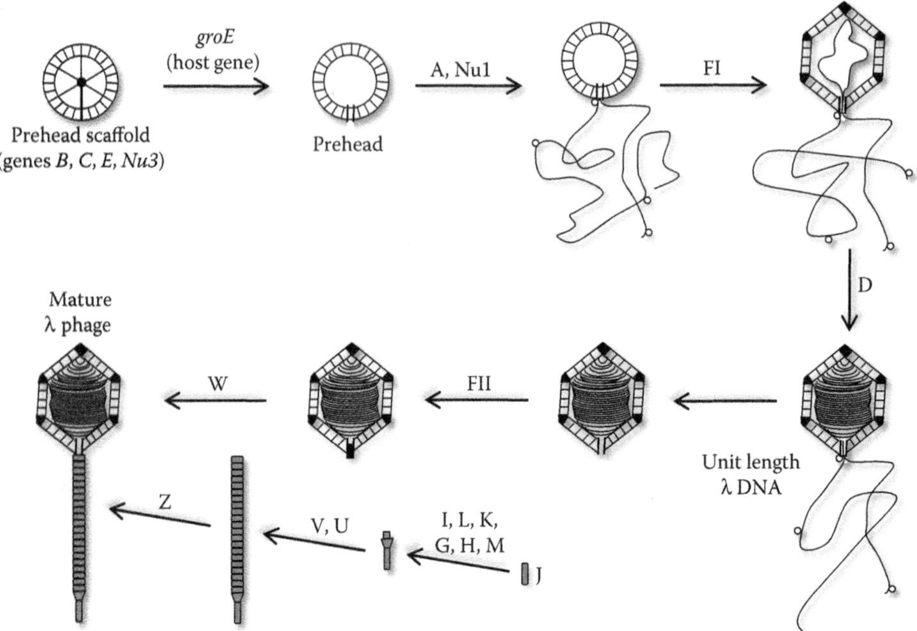

FIGURE 28.6 Production of λ virus particles in the lytic pathway. The virus prehead is constructed with proteins B, C, E, and Nu3, aided by *groE* produced by the *E. coli* cell. Then, proteins A and Nu1 recognize and attach to the cos sites on the concatemeric λ DNA. The FI and D proteins then thread the DNA into the phage head. Once the head is filled, A and Nu1 cut the DNA at the cos sites. The end of the DNA moves into the phage head, and the opening is sealed by protein FII. At the same time, the tails are being assembled. Initially, proteins G, H, I, J, K, L, and M form the base structure on the tail. This region is necessarily complex, because it interacts with the outer membrane of the *E. coli* cell, and then forms a channel through the outer membrane into the periplasm and delivers the DNA through the inner membrane and into the cytosol. The main tail proteins, U and V, are assembled, and then the tail is joined to the head through the actions of proteins W and Z. The bacteriophage is then complete and can infect *E. coli* cells.

BACTERIOPHAGE T4

The T-even phages (T2, T4, and T6) all are related phages, with T2 and T4 being the most studied. The genome of T4 is more than 3 times larger than the λ genome and has more than 5 times as many genes (Figure 28.7). The virion has a head and a tail as well as tail fibers (Figure 28.8). The tail fibers interact with proteins on the outer membrane of the bacterium. The head and the tail are both more complex than those of λ, but they have some similarities. Because of the added complexity of the head, the section of the genome that encodes genes for the heads occupies a length of about 20 kb of the chromosome, whereas the genes for the baseplate and tail occupy roughly 30 kb of the chromosome, and the section of the chromosome that contains all of the tail fiber genes is about 10 kb in length. Unlike bacteriophage λ, T4 only has a lytic phase and does not form a lysogen. Therefore, the genome lacks any genes for chromosome integration, and there is no region analogous to the PRM promoter region that is found in λ. However, T4 has a much more elaborate replication system, which makes replication much faster and much more accurate than the replication system in λ. This region consists of 10 genes that are included on a section of the chromosome that is about 20 kb in length. It also contains a gene for thymidylate synthase, a precursor for deoxythymidine triphosphate, which is used for DNA replication and repair. The gene also contains a group I intron, which is rare in virus genomes. The origin of this particular intron is unknown. If all of the genes for

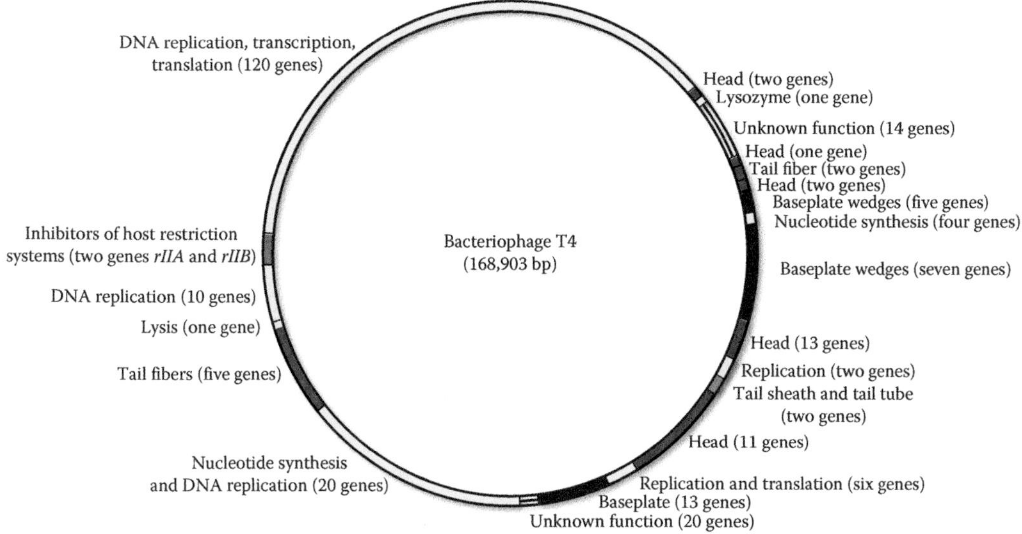

FIGURE 28.7 Genome of bacteriophage T4. Although T-even phages (T2, T4, and T6) lack a lysogenic phase. They are more than 3 times more complex than bacteriophage λ, based on the number of genes that they each possess. Roughly half of the genome (approximately 160 genes) encodes proteins responsible for DNA replication, nucleotide synthesis, transcription, and translation. By contrast, λ and ϕX174 have only two or three genes, respectively, that encode proteins involved in replication and/or transcription, and have no genes that have a role in translation. T4 has more head genes (29) than either ϕX174 or λ (3 and 9, respectively). It has also more tail genes (14) than does λ (10, ϕX175 has no tail), and has an additional seven genes that encode tail fibers. The arrangement of genes along the chromosome is somewhat mixed compared to the λ genome. For example, although many of the genes for head proteins are clustered, some are separated from other head genes by other interspersed genes. Tail baseplate and DNA replication genes are also split along the chromosome. This may indicate recombination events in the past that caused some disruption of an original arrangement in which genes for particular structures were clustered. The position of the tail fibers on a region of the chromosome distant from the head, tail, and baseplate genes might indicate evolutionarily more recent arrivals for these genes.

nucleotide synthesis, DNA replication, transcription, and translation are included, the region is about 100 kb in length with 160 genes.

Infection by T4 begins with recognition of cell surface proteins by the tail fibers. The tail fibers bind irreversibly to the cell surface proteins and they position the baseplate of the tail onto the surface of the bacterial outer membrane. The tail contracts, acting much like a hypodermic needle pushing through both the outer and inner membranes and delivering the T4 double-stranded DNA (dsDNA) chromosome into the cell. Transcription of the genes begins immediately setting up the progression of the infection. Some of the first proteins to appear are those for DNA replication, which is via a rolling circle mechanism. Long dsDNA concatemers of multiple genomes result from replication. Next, some of the initial proteins for assembly of the heads are synthesized, followed by collar proteins, then tail proteins, and finally tail fiber proteins and lysozyme. The heads, tails, and tail fibers are all assembled separately, and then joined together (Figure 28.8). The completed head is joined to the completed tails, and then the tail fibers are connected to the tails. One of the last steps in head assembly is to fill the head with the T4 DNA. There are no sites analogous to the cos sites in λ. The long concatemeric T4 DNA is threaded into the T4 head, and when the head is full, the DNA is cut. The cut sites appear to be random, such that each T4 head contains a different beginning and end to the chromosome. However, each head receives slightly more than a complete genome of T4, such that at least one copy of each gene is present in each T4 head.

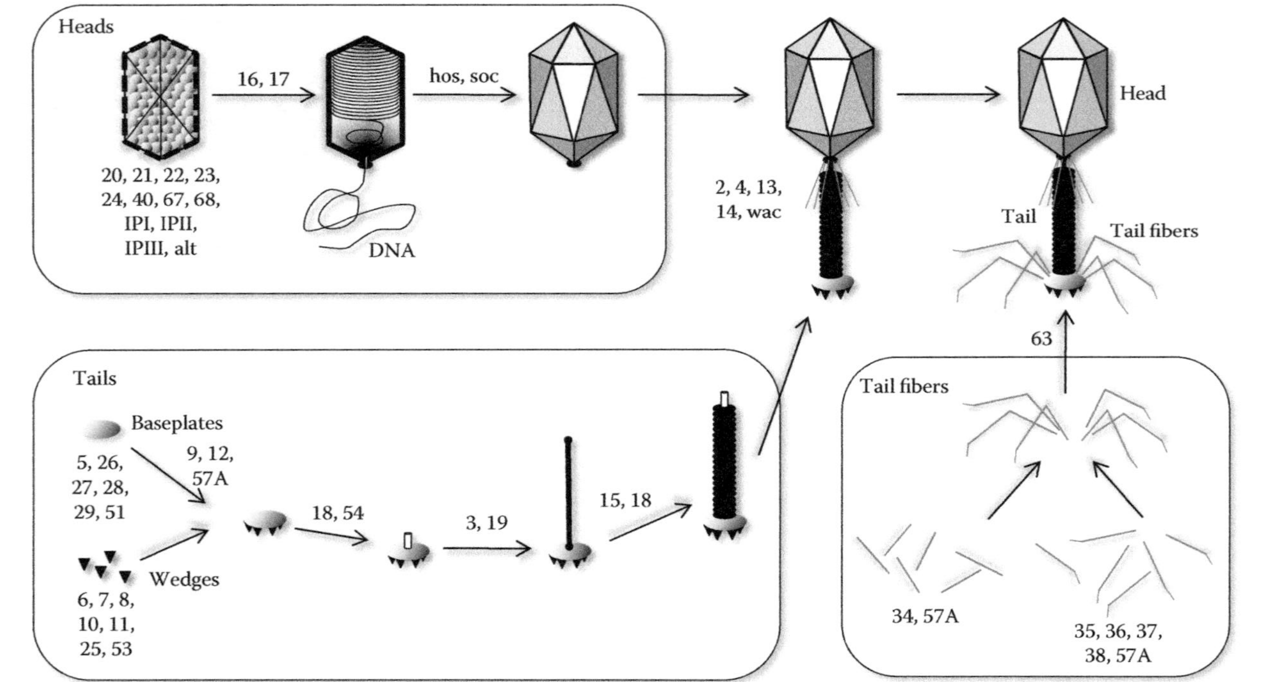

FIGURE 28.8 Pathway to production of T4 virus particles. As with other viruses, the prehead requires many genes. For this bacteriophage, 12 proteins (20, 21, 22, 23, 24, 40, 67, 68, IPI, IPII, IPIII, and alt) are required to construct a head that can accept DNA. Proteins 16 and 17 then thread the concatemeric DNA into the head and cut the DNA into genome-sized lengths. The hos and soc proteins then complete the head that is then ready to attach to a tail. As with other tailed bacteriophage, the baseplates are complex. The baseplate is composed of six different proteins (5, 26, 27, 28, 29, and 51), and the attached wedges are composed of seven different proteins (6, 7, 8, 10, 11, 25, and 53). This attests to their complex and vital role in attachment to the host cell membranes and injection of the virus DNA. The wedges are attached to the baseplates by proteins 9, 12, and 57A. Proteins 18 and 54 then form an attachment site for the tail tube (proteins 3 and 19) and the tail sheath proteins (15 and 18). The tail is then attached to the head through the actions of five proteins (2, 4, 13, 14, and wac). The tail fibers are initiated by proteins 34 and 57A to form the short tail fibers, and 35, 36, 37, 38, and 57A to form the long tail fibers. These associate with one another and are then attached to the baseplate of the tail by protein 63.

MIMIVIRUS

The APMV is a large enveloped DNA virus with one of the largest known genomes of any virus (Figure 28.9a). In fact, when it was first discovered in some ameba, it was initially thought to be a parasitic bacterium. The genome, nearly 1.2 Mb of linear dsDNA, is larger than some bacterial genomes. Most viruses carry genes for replication of their genome, a few genes for transcription, and genes that encode the components of their structures. Although APMV genome contains these types of genes, they are minor components of the genome. In addition, the genome contains many genes for protein translation, DNA repair, cell membrane synthesis, cell motility, posttranslational modification, polysaccharide synthesis, and other functions normally associated with cellular organisms (Figure 28.10). Specifically, it contains six transfer RNA (tRNA) genes, four aminoacyl tRNA transferases, translation initiation and elongation factors, and other components of translation. Genes that encode proteins for each of the major DNA repair pathways and both type I and type II topoisomerases are present. Again, this genome contains many of the genes that are normally restricted to cellular organisms. This blurs the distinction between living and nonliving biological entities. In fact, when several of the genes within APMV genome (e.g., aminoacyl tRNA sythetases, RNA polymerase II, 5′–3′ exonuclease) were used to perform phylogenetic studies compared to

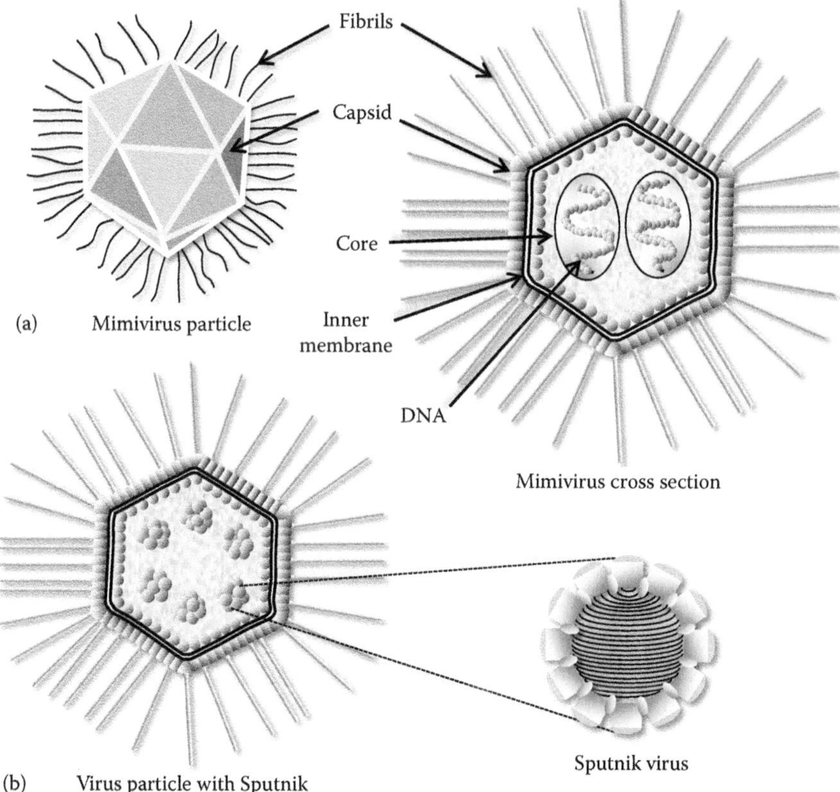

FIGURE 28.9 Diagrams of APMV or Mimivirus and Sputnik virophage. The APMV is a large (approximately 700 nm diameter) icosahedral virus that is enveloped with the membrane from the host (*Acanthamoeba*). (a) Capsid proteins surround the membrane and fibrils are attached to the capsid proteins. The large chromosomes, containing 1.2 Mb of DNA, are enclosed inside a central core. When Sputnik virophage coinfects with APMV (b), it uses the APMV proteins to produce more Sputnik virus particles. These are enclosed inside APMV virus particles to the exclusion of the APMV chromosomes. Sputnik virophage cannot infect and be replicated without coinfection with APMV.

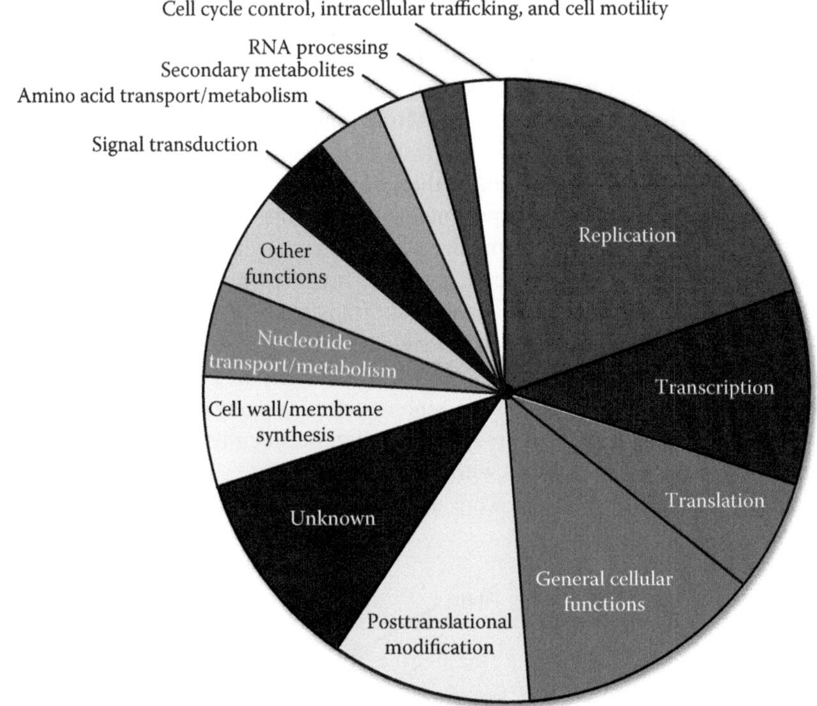

FIGURE 28.10 Summary of APVM genome. The total genome is 1,181,404 bp in length. A total of 1262 open reading frames have been identified, although only 911 appear to encode actual proteins. Of these, more than 200 have been categorized. Some of the unknown genes may be related to virus production. The pie chart is a summary of the categorized genes. Approximately 30% of the genes encode proteins responsible for replication, transcription, and translation. Approximately 13% of the genes are for general cellular functions, 11% encode genes specifically responsible for posttranslational modifications, whereas another 11% appear to encode authentic proteins, but their functions cannot be identified. The remaining genes (approximately 15%) encode proteins that are mainly considered to be for cellular function (e.g., cell wall/membrane synthesis, signal transduction, cell cycle control, and cell motility). Although the APMV genome lacks rRNA genes, it does possess several other genes central to mRNA translation, including tRNAs, aminoacyl tRNA sythetases, and tRNA modification genes. The genome of this virus resembles the genomes of bacteria that are obligate parasites, especially those that are endosymbionts. This virus appears to have incorporated many genes from its host(s) during its evolution.

several groups of cellular organisms, APMV branched approximately midway between Archaea and Eukarya, further blurring the distinction between cellular organisms and viruses.

Another characteristic of this virus is that it can be *infected*, or parasitized, by another virus. This is the virophage called Sputnik (Figure 28.9b). Its dsDNA genome is 18,343 bp in length and contains 21 genes (Figure 28.11), including three genes from AMPV, an integrase (necessary for movement of DNA segments), a primase–helicase (necessary for DNA replication and other functions), an adenosine triphosphatase (ATPase), an insertion sequence (used in transposition), a Zn ribbon, and the Sputnik capsid protein. This virus is within a class called satellite viruses, which depend on other viruses for their replication. One of the best studied viruses is satellite tobacco necrosis virus (STNV), which needs tobacco necrosis virus (TNV) for its replication. Both infect a broad range of plants, but were first identified and studied in *Nicotiana* (tobacco) species. STNV appears to be simply a smaller version of TNV, with fewer genes, and a smaller capsid protein encoded on the STNV chromosome that is closely related to the TNV capsid

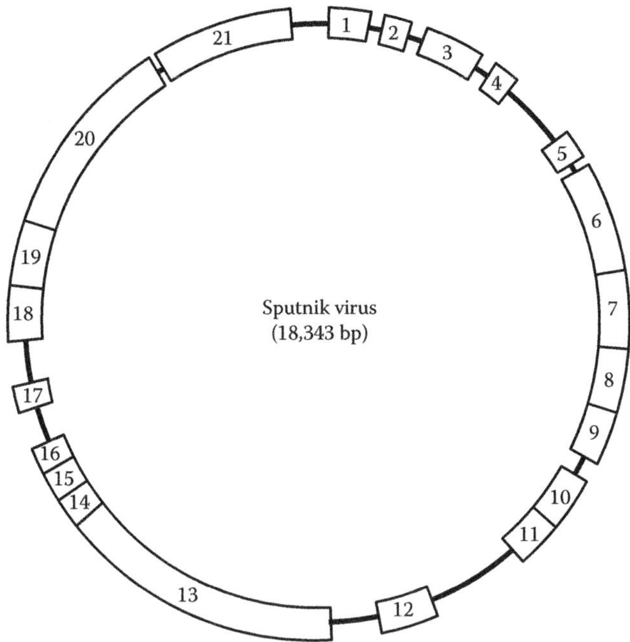

FIGURE 28.11 Genome of Sputnik virophage (satellite-like virus). The 18,343 bp dsDNA has 21 open reading frames, including three sections that appear to encode polyproteins. Gene 20 encodes the major capsid protein, whereas gene 10 encodes an integrase. In addition to these genes, six of the genes have homologies with other genes, including three from APMV, a primase–helicase, a virus-packaging ATPase, a Zn ribbon protein, as well as an insertion sequence normally associated with bacterial gene transposition.

protein. Although TNV can infect plant cells without STNV, STNV can only infect a plant cell that is also infected with TNV. The result of coinfection is that approximately 70% of the virus particles produced are STNV. Although the Sputnik virus acts somewhat like a satellite virus, it also has some variations that make it unique. Most satellite viruses move their chromosomes into the nucleus at sometime during infection, and their transcription is normally within the nucleus. There, they rely on genes from the cell and from the main virus for other functions (e.g., replication, translation, assembly). In the case of Sputnik, the virus chromosome moves into a specific compartment in the cytosol where APMV has begun to express its genes, replicate its chromosome, and assemble its virus particles. In essence, APMV sets up its own organelle. Sputnik takes over APMV systems and uses them to produce more Sputnik particles, at the expense of APMV production. Therefore, Sputnik is truly a parasite on the APMV parasite.

APMV and Sputnik comprise a system of viruses that crosses the lines of some definitions of life. In particular, APMV is as large as some bacteria, and its genome is also larger than the genomes of many bacteria. It can be infected by another virus, and therefore can be considered as a living organism. The genes contained in the APMV genome not only encodes proteins for the physical parts of the virus particle, but most are involved in the general processes to construct those proteins (e.g., transcription, translation). Previous to sequencing of the APMV genome, these functions had been thought to be restricted to cellular organisms. In general, collections of genes can be packaged in many ways to produce an organism capable of producing new copies of itself. Organisms package their genomes within a membrane (Bacteria, Archaea, and Eukarya), or within sets of membranes (i.e., in nuclei in eukaryotes, or within organelles), but they may also package them within protein coats with or without parts of the host membranes (i.e., in viruses). The types of genes within these genomes have evolved not based on a definition, but simply based on whether the particular combination of genes allows them to reproduce. APMV appears to be a virus that has many characteristics of a bacterium, and therefore may be considered to be living organism.

KEY POINTS

1. DNA virus genomes range from extremely small to the sizes of some small bacterial genomes.
2. Although viruses are all obligate parasites, the largest viruses resemble some bacteria.
3. The smallest virus genomes have few genes (e.g., the DNA bacteriophage ϕX174 has only 11 genes, the RNA bacteriophages MS2 and Qβ each have only 4 genes), whereas the largest APMV (i.e., Mimivirus) has at least 911 genes.
4. It has been suggested by some that viruses be classified as an additional domain of life.

ADDITIONAL READINGS

Alberts, B., D. Bray, K. Hopkin, A. Johnson, J. Lewis, M. Raff, K. Roberts, and P. Walter. 2013. *Essential Cell Biology,* 4th ed. New York: Garland Publishing.

Alberts, B., D. Bray, A. Johnson, J. Lewis, M. Raff, K. Roberts, and J. D. Watson. 1994. *Molecular Biology of the Cell.* New York: Garland Publishing.

Becker, W. M., J. B. Reece, and M. F. Poenie. 1996. *The World of the Cell,* 3rd ed. New York: The Benjamin/Cummings Publishing.

Calverie J.-M. and C. Abergel. 2009. Mimivirus and its virophage. *Annu. Rev. Genet.* 43:49–66.

Desjardins, C., J. A. Eisen, and V. Nene. 2005. New evolutionary frontiers from unusual virus genomes. *Genome Biol.* 6:212.1–212.3.

Futuyma, D. J. 1998. *Evolutionary Biology*, 3rd ed. Sunderland, MA: Sinauer Associates.

Green, M. R. and J. Sambrook. 2012. *Molecular Cloning: A Laboratory Manual*, 4th ed. Cold Spring Harbor, NY: Cold Spring Harbor Laboratory Press.

Gregory, T. R., ed. 2005. *The Evolution of the Genome*. London: Elsevier Academic Press.

Maniatis, T., E. F. Fritch, and J. Sambrook. 1982. *Molecular Cloning: A Laboratory Manual.* Cold Spring Harbor, NY: Cold Spring Harbor Laboratory Press.

Miller, E. S., E. Kutter, G. Mosig, F. Anisaka, T. Kunisawa, and W. Rüger W. 2003. Bacteriophage T4 genome. *Microbiol. Mol. Biol. Rev.* 67:86–156.

Raoult, D., S. Audic, C. Robert, C. Abergel, P. Renesto, H. Ogata, B. La Scola, M. Suzan, and J.-M. Claverie. 2004. The 1.2-megabase genome sequence of Mimivirus. *Science* 306:1344–1350.

Tropp, B. E. 2008. *Molecular Biology, Genes to Proteins*, 3rd ed. Sudbury, MA: Jones & Bartlett Publishers.

Wichman, H. A. and C. J. Brown. 2010. Experimental evolution of viruses: Microviridae as a model system. *Phil. Trans. R. Soc. B* 365:2495–2501.

29 Bacteria and Archaea

INTRODUCTION

Although they are mostly small single-celled organisms, members of domains Bacteria and Archaea represent extremely diverse groups of organisms. Some live miles underground, whereas others have been found miles into the atmosphere, and the majority live somewhere between those extremes. Some grow in hot pools of water, inside glaciers, under high pressures deep in the ocean, in saline soils, or on and inside plants and animals. In fact, just about anywhere on the Earth, one or more species of these microbes will be found. A few can survive on their own, but most require other species to supply some of their food and energy requirements. This diversity is reflected in the genomes of these organisms. As of 2014, approximately 4000 bacterial genomes had been completely sequenced, and nearly 200 archaeal genomes had been sequenced. Although this is a small proportion of the estimated hundreds of thousands of species included in both domains, the genomes do provide some insight into the evolution of these organisms. In this chapter, a few of these genomes will be discussed to highlight the similarities as well as the differences.

ESCHERICHIA COLI

Escherichia coli has been one of the most intensively studied of all bacteria. Additionally, it is not a single entity. There are many strains, and some of them are very different than the main group. Therefore, instead of being a single species, it may be a species complex or a group of related taxa. Nonetheless, it has been extremely useful in genetic studies, partly because of the ease of culturing this organism. Many of the genetic mechanisms that are known first were discovered and examined in this microbe. The genome of *E. coli* strain K12 is 4,639,221 bp in length, which contains 4289 protein-coding genes (Table 29.1 and Figure 29.1), 114 transfer RNA (tRNA) genes, and 7 ribosomal RNA (rRNA) operons (each containing a 16S, 23S, and 5S gene), plus an additional 5S gene. This bacterium grows in the intestines of animals, where conditions are anaerobic. However, it can survive for periods of time outside of the animals, and thus can live in aerobic conditions as well. The greatest proportion of the genome (by function) is devoted to energy metabolism. There are genes that produce proteins to allow *E. coli* to grow under both aerobic and anaerobic conditions. It has genes that encode proteins for energy metabolism, including glycolysis, adenosine triphosphate (ATP) synthase, electron transport, the tricarboxylic acid (TCA) cycle, sugar utilization, methanogenesis, fermentation, and others. It even has four genes that are normally used in photosynthesis, although they must be used in a different way by this nonphotosynthetic organism.

The portion of the genome that has the second highest number of genes is composed of transporters and binding proteins. In order to survive both inside and outside of the animals, *E. coli* has additional transporters, which allows a greater variety of molecules to be imported and exported, and allows greater flexibility as to where *E. coli* can grow. For example, it has a large number of anionic and cationic transporters, which allows it to grow under a wider range of pH conditions than species with fewer transporters. The third most numerous category of genes encode proteins responsible for transcription, including proteins involved in the regulation of transcription. Although the number of genes for RNA polymerase, helicase, and accessory proteins is about the same as that for bacteria with smaller genomes, the number of transcription regulators and factors is higher than in bacteria with smaller genomes. This allows *E. coli* to grow under many different conditions. It does this by sensing and responding to each of those conditions appropriately by turning on specific sets of genes while suppressing the expression of other genes. Transcriptional regulators and factors perform these functions.

TABLE 29.1
Categories, Numbers, and Proportions of Genes in the *Escherichia coli* Genome

Category/Function	Number of Genes	Proportion of Genome
Regulatory	178	4.15
Cell structure	182	4.24
Membrane proteins	13	0.30
Structural proteins	42	0.98
Phage, transposons, and plasmids	87	2.03
Transport/binding proteins	427	9.95
DNA replication, recombination, modification, and repair	115	2.68
Transcription and transcriptional control	55	1.28
Translation and protein modification	182	4.24
Nucleic acid biosynthesis	58	1.35
Amino acid biosynthesis	131	3.06
Fatty acid biosynthesis	48	1.12
Other processes	895	20.86
Unknown function	1,633	38.06
Total	**4,289**	**100.00**

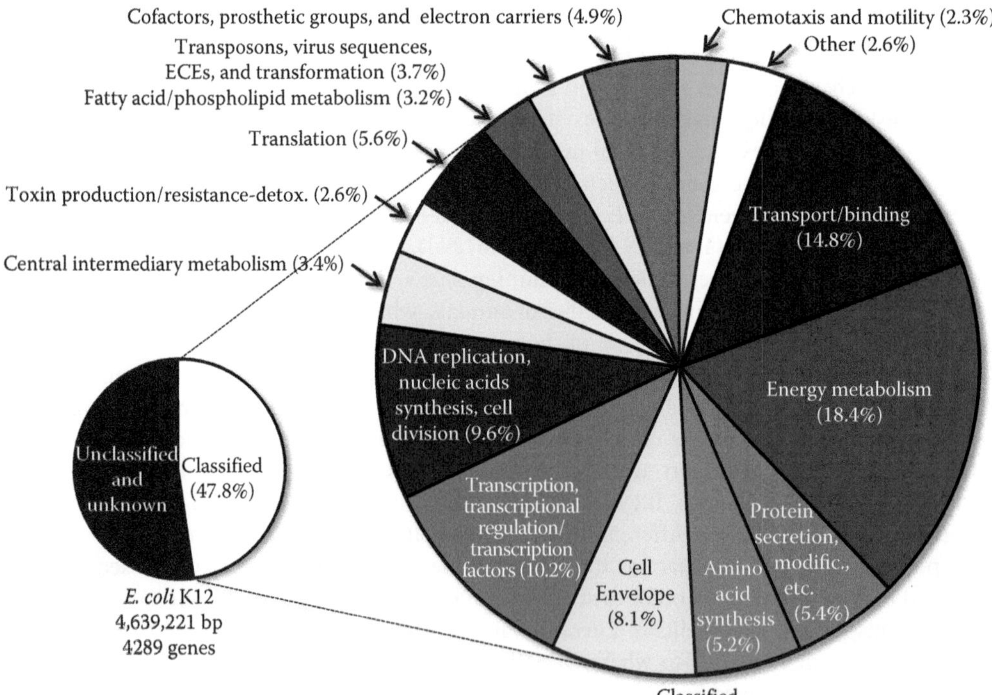

FIGURE 29.1 Summary of categories of genes in the *Escherichia coli* (strain K12) genome. The genome consists of 4,639,221 bp, encoding 4,289 genes, of which 2,059 (47.8%) have been classified. Over 30% of the classified genes encode proteins involved in DNA replication, transcription, and translation. More than 30% of the genome encodes proteins that build the structure of the cell, including the cell membranes, cell wall, membrane proteins, and transport proteins. Another 27% are involved in energy metabolism and central intermediary metabolism (both of which utilize cofactors, prosthetic groups, and electron carriers). The remaining genes (approximately 5%) include a large number of transposons and virus sequences, as well as genes that produce toxins to inhibit and kill other bacteria and genes that protect the *E. coli* cells from these toxins.

The fourth largest portion of the genome consists of genes involved in DNA replication and cell division. Compared to smaller bacterial genomes, there are more genes for nucleic acid synthesis, DNA replication, recombination, and repair. All of these increase the fidelity of DNA replication and help to maintain the larger genome of this organism. The gene envelope (cell wall, cell membrane, and membrane proteins) is encoded by approximately 8% of the genome. This gives structure to the cell, but also affects cell identity, including recognition by other *E. coli* cells, as well as by other organisms (including bacteriophage). Being a Gram-negative diderm bacterium, *E. coli* cells have two membranes separated by a periplast. In addition, they have a cell wall. This complex cell envelope requires a large number of genes and their products. Genes for synthesis of amino acids, translation, and protein secretion and trafficking each are encoded on about 5.5% of the genome. Each of these is higher than in smaller genomes, indicating the increase in complexity of *E. coli* cells and their ability to adapt to many different conditions. All other functions (e.g., genes for fatty acid metabolism, central intermediary metabolism, motility for motile bacteria, toxin production and resistance) comprise smaller portions of the genome and are roughly equivalent across the domain Bacteria.

One of the most surprising aspects of this genome is that it contains a large number of extrachromosomal elements (ECEs), transposons, and virus sequences. In fact, approximately 3.7% of the genome consists of these sequences. This equals approximately 170,000 bp of DNA, which is about the size of the entire T4 bacteriophage genome. This represents a major increase in these types of sequences in *E. coli* compared to genomes of lesser size. As was discussed in Chapter 13, transposons and some viruses have common origins, and therefore, this is an example of where these mobile genetic elements have influenced the evolution of a genome. One of the surprising things about the *E. coli* genome that was discovered during the first sequencing project is that a large proportion of the genes were completely new to science. That is, no one knew exactly what they were, and they still do not know what they are. More than 30% of the genes have yet to be identified. Scientists have determined that they are genes, but their functions are unknown. This is especially surprising because *E. coli* has been studied for decades in hundreds of laboratories worldwide. Thousands of mutants have been produced and characterized according to which genes have been mutated. Nevertheless, purely genetic methods appear to have missed a large proportion of the genes that exist in this species.

PHOTOSYNTHETIC BACTERIA

Cyanobacteria that are capable of carrying out oxygenic photosynthesis have existed on the Earth for at least 2.7 billion years. This is when clear signs of oxygenic photosynthesis appear on the Earth. The level of atmospheric O_2 began to rise, eventually transforming a reducing atmosphere into an oxidizing atmosphere. Most photosynthetic microbes have two photosystems that act in concert to fix carbon from CO_2 into usable organic molecules by splitting water and releasing oxygen as well as protons and nicotinamide adenine dinucleotide phosphate (NADPH). The protons are shuttled through ATP synthase, and the ATP and the NADPH are used to fuel carbon fixation (Calvin cycle). These cyanobacteria (as well as chloroplasts in plants) have both photosystem II (which produces protons and oxygen from water) and photosystem I. There are some cyanobacteria and other groups of photoautotrophs that use anoxigenic photosynthesis. In this case, photosystem II is missing or has mutant proteins. It is possible that cyanobacterial species existed long before 2.7 billion years ago, but that they (and other microbes) utilized anoxygenic photosynthesis. There are some cyanobacteria species currently living in some oceans that use this type of photosynthesis. Additionally, there are other bacteria that only have photosystem I, and there are bacteria that fix carbon using one of several other carbon fixation pathways or cycles. Photosystems I and II are similar in some of the core proteins that they contain, and therefore, it is likely that one photosystem was the result of gene duplication of the other photosystem.

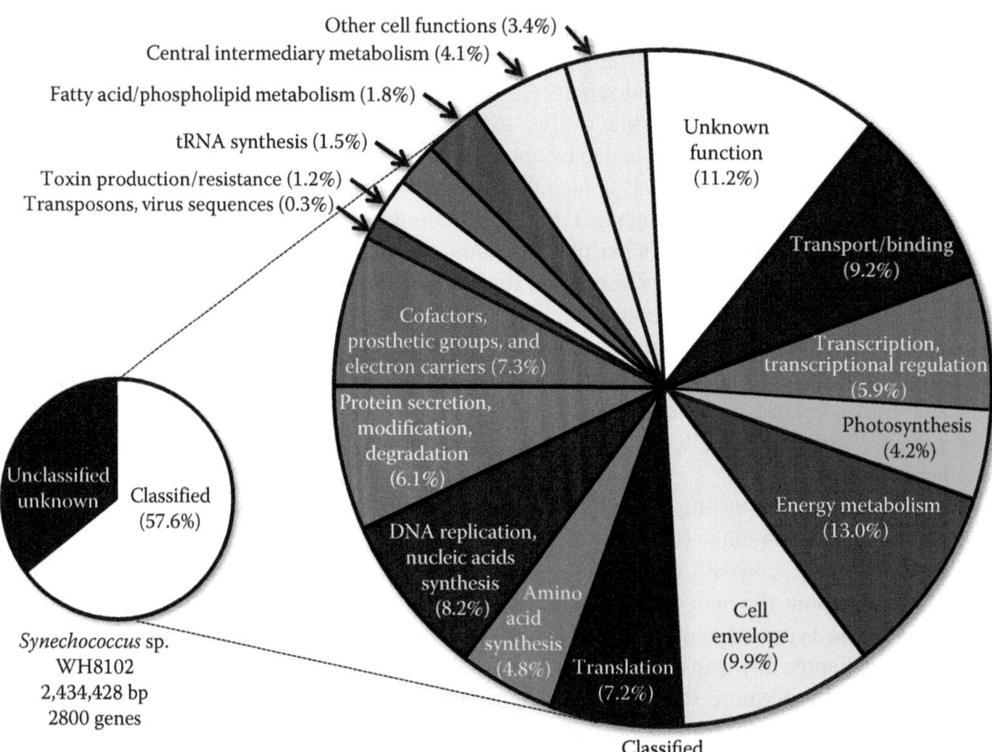

FIGURE 29.2 Summary of categories of genes in the *Synechococcus* sp. (strain WH8102) genome. The genome of 2,434,428 bp, consists of 2,526 genes, of which 1,455 (57.6%) have been classified. This genome is roughly half the size of the *Escherichia coli* genome. Nonetheless, it has about the same proportion of the genome that encodes proteins for DNA replication, transcription, and translation. However, because it is half the size, it has roughly half as many genes devoted to these processes. About 27% of the genome encodes proteins for the cell envelope, membranes, membrane proteins, and transporters. Almost 29% of the genome is responsible for metabolic processes. Part of this (4.2%) is specific to photosynthesis, and the number of genes encoding proteins responsible for the productions of cofactors, prosthetic groups, and electron carriers is 50% higher compared to *E. coli*, primarily because of an increase in genes that are needed to produce hemes and porphyrin, which are used in photosynthesis. Although the gene categories are in similar proportions to those in *E. coli*, the gene sequences themselves differ significantly (not shown), which is expected because *E. coli* and *Synechococcus* diverged from a common ancestor approximately three billion years ago.

Synechococcus sp. strain WH8102 is a marine cyanobacterium with a genome size of 2,434,428 bp (Figure 29.2), which includes 2526 protein-coding genes, 44 tRNA genes, and 2 copies each of the 5S, 23S, and 16S rRNA genes. Although the numbers of rRNA and tRNA genes are similar to those in *Haemophilus influenzae* and *Methanocaldococcus jannaschii*, the genome size and number of protein-coding genes are 30% and 60% higher, respectively. Most of these increases can be attributed to the process of photosynthesis. Of the genes that code for proteins, most support the central cellular functions, such as glycolysis, the TCA cycle, DNA replication, transcription, translation, electron transport, and other vital functions. As expected, an entire set of genes for photosynthesis, including the light-harvesting systems, as well as the carbon fixation systems are included in its genome. Additionally, proteins that sense high light intensities are present. Although these are photosynthetic organisms, if light intensities are too high, various molecules within the cell can be damaged. These cells can sense the high light intensities and produce shading pigments to decrease the amount of light that reaches the cytosol. They must also be able to sense low light conditions so that they can orient toward the highest intensities of light. In addition to all of the photosynthesis genes, additional mobile, extrachromosomal, and virus sequences are present in this genome. In general, this genome has the core set of bacterial genes as well as the set of photosynthesis genes, which is the primary reason that the genome is larger than those of many other bacteria.

AQUIFEX

The genus *Aquifex* has a very deep phylogenetic branch within the Bacteria (see Figure 2.4). This and other bacteria that are deeply branching (such as *Chloroflexi* and *Deinococcus*) are thought to contain genomes that resemble some of the earliest bacterial genomes. Members of the genus *Aquifex* are thermophiles, which grow in hot water near underwater volcanic vents. Some theories posit that life on the Earth began near these thermal vents. Therefore, the genome of *Aquifex* might represent a genome with a set of genes that is ancestral to most organisms. The genome of *A. aeolicus* is 1,551,335 bp in length (Figure 29.3) and is composed of 1522 protein-coding genes, 44 tRNA genes, and 6 rRNA genes (two each of the 5S, 23S, and 16S genes). One of the first notable

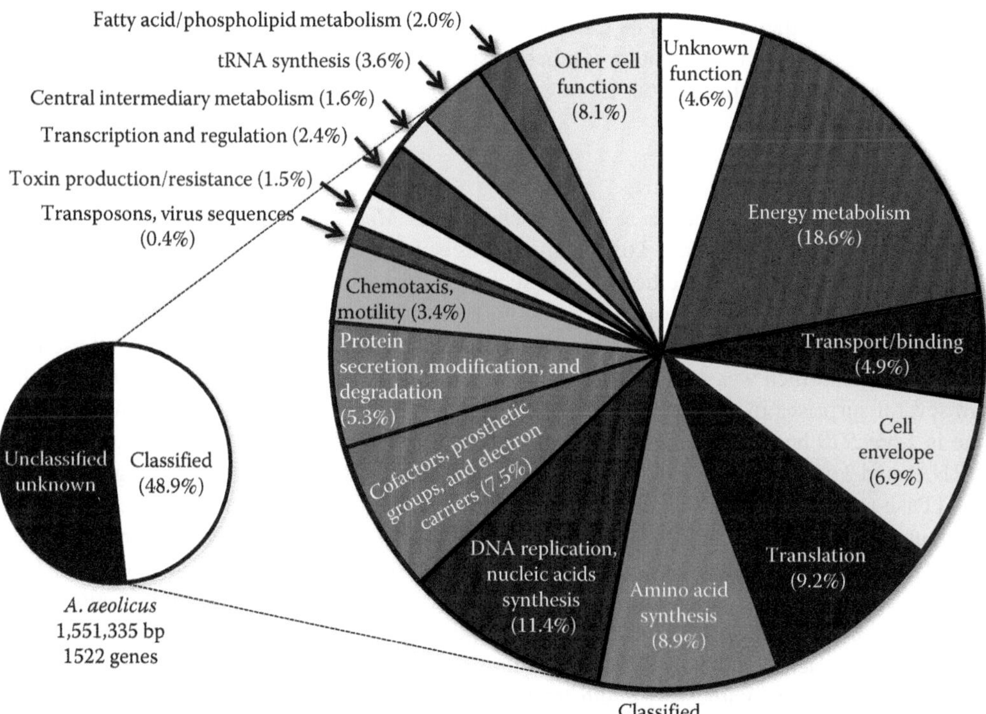

FIGURE 29.3 Summary of categories of genes in the *Aquifex aeolicus* genome. This species has one of the smallest genomes of a free-living bacterium. Its genome consists of 1,551,335 bp, containing 1,522 genes, of which 745 (48.9%) have been classified. It diverged from other bacteria very early in evolution (probably between 3.0 and 3.5 billion years ago), and may have experienced little change during that time. These bacteria grow near deep-sea thermal vents, which may be similar to those that have existed on the Earth for many billions of years. As with *Escherichia coli* and *Synechococcus*, approximately 30% of its genome encodes proteins involved in DNA replication, transcription, and translation. However, a smaller proportion of their genome consists of genes encoding proteins for construction of the cell membrane, cell wall, membrane proteins, and transporters. This may be a reflection of the simpler and more constant environment in which they live. If their environment has changed little over the past 3.5 billion years, then they would not have to be as adaptable as *E. coli* (for example), which is capable of growth in a wide range of conditions. Almost 30% of the *Aquifex* genome encodes proteins for metabolic processes and related pathways, which is a proportion that is similar to those of other bacteria.

findings from this genome was that approximately 16% of the genes in its genome were closest to those from Archaea. This confirms its spot nearer to the base of the Bacterial domain, which is closer to the branch point between Bacteria and Archaea. Alernatively, it may have acquired genes from Archaea by HGT. Compared to the *E. coli* genome, this genome contains most of the central functions, but has fewer genes for most of them. For example, it has only 70 genes for amino acid synthesis, whereas *E. coli* has 117. Interestingly, both have 29 aminoacyl tRNA synthetases. It also lacks the sophistication of *E. coli* with regard to nucleic acid synthesis and DNA replication and repair. It contains genes for many of the same energy metabolism pathways as *E. coli*, but there are fewer genes for most. However, it has additional genes for utilization of hydrogen, which it uses as an energy source from its thermal vent habitat. Interestingly, as with *E. coli*, it has a few genes normally associated with photosynthesis (five in this case). Their functions are unknown, but the fact that a few genes in the *A. aeolicus* and *E. coli* genomes are commonly associated with photosynthesis might indicate that they were originally used for other purposes in cells, and versions of them were utilized in photosynthesis by the phototrophic organisms. Alternatively, they may have been acquired via horizontal gene transfers from photosynthetic species. Although the genome of *A. aeolicus* contains fewer genes for RNA replication, repair, transcriptional regulation, transport/binding, and some other gene sets, its translation system is similar to that of *E. coli*. This indicates that the translation system (i.e., protein synthesis using ribosomes) has probably remained relatively stable for at least three billion years.

EURYARCHAEOTA

The first member of the Archaea to be sequenced was *Methanococcus jannaschii* (now calle, *Methanocaldococcus jannaschii*). This organism was isolated from mud on the ocean floor adjacent to a thermal vent on the East Pacific Rise. As such, this organism was living in complete darkness, under high pressure, in hot seawater. For energy production, it reduces CO_2 with H_2 to produce methane. Therefore, the genes that it possessed were expected to be very different than those found in *H. influenzae*, which was sequenced around the same time. Although they were different, there were several surprises in this genome. First, the genome includes three parts. The main chromosome is 1,664,970 bp in length (Figure 29.4), and includes 1785 protein-coding genes, as well as 37 tRNAs and 2 rRNA operons. It also has a large ECE that is 58,407 bp long with 46 genes (encoding ATPase-related proteins, two restriction enzymes, proteins involved in cell cycle control, and several unidentified genes) and a small ECE that is 16,550 bp long with 12 genes (identified only as hypothetical proteins, i.e., unknown genes). The core of this genome is similar in some ways to the genomes of *H. influenzae* and *E. coli*. It contains genes for import and export of nutrients (as well as ions), glycolysis/gluconeogenesis, electron transport, the TCA cycle, cell division, nucleic acid synthesis, amino acid synthesis, DNA replication, transcription, transcriptional control, translation, aminoacyl tRNA synthesis, protein trafficking, protein modification/repair, protein folding/stability, fermentation, and intermediary metabolism. However, the enzymes are thermostable versions, because *M. jannaschii* grows at temperatures between 48°C and 94°C (optimal temperature is 85°C). The genes are higher in GC pairs, and many of the proteins contain more cysteine residues, which can produce disulfide bridges to stabilize the secondary structures of proteins. Also, transport of molecules and water across the membranes must function at very high external pressures because this microbe grows on the seafloor. Therefore, although the proteins are similar in function, they are adapted to conditions in the extreme environments near the deep-sea thermal vents.

There are many differences between the *E. coli* and *M. jannaschii* genomes. First, *M. jannaschii* has genes for flagella (primarily flagellins), because it needs to be motile in its environment. Second, it has a large number of genes for proteins involved in methanogenesis and chemoautotrophy (primarily for reactions involving carbon monoxide and carbon dioxide). This allows it to harvest energy by reducing CO_2 with H_2. The genome contains several genes in the pathway for fixation

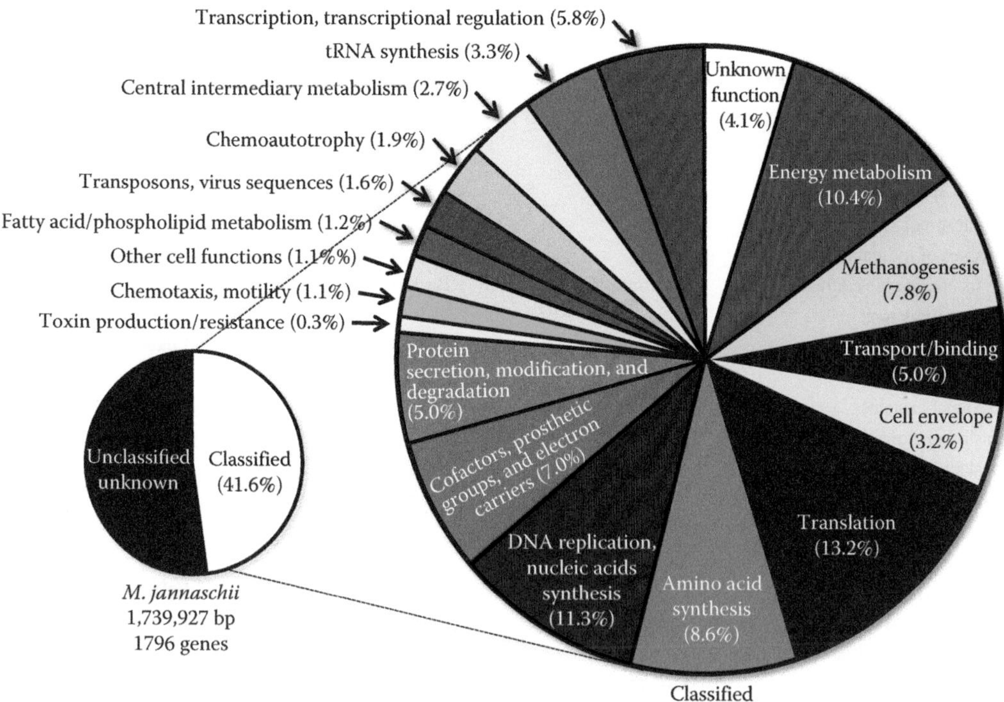

FIGURE 29.4 Summary of categories of genes in the *Methanocaldococcus jannaschii* (= *Methanococcus jannaschii*) genome. The genome of this member of the Euryarchaeota consists of 1,739,927 bp, encoding 1,785 genes, of which 743 (41.6%) have been classified. The genome is split between the main chromosome that is 1,664,970 bp in length and two ECEs, one of which is 58,407 bp in length and the other which is 16,550 bp in length. Although it has about the same proportion of its genome devoted to energy metabolism as *Escherichia coli*, almost 25% of that (7.8% of the total characterized sequences) is devoted to methanogenesis. The number and proportion of genes encoding genes involved in constructing the cell wall, membrane, membrane proteins, and transporters is about half that found in Bacteria with similar genome sizes. Conversely, the proportion of the genome involved in DNA replication, transcription, and translation is nearly 40%, indicating that these processes are more complex than those in Bacteria.

of nitrogen from N_2. However, it does not contain all of the genes, and when grown in laboratory conditions, it cannot fix nitrogen. These genes may be remnants from an ancestral lifestyle, but they may be functional in an undetermined process. The genome also has histones. These are the proteins that bind to and condense DNA, which controls gene expression in archaea and eukarya. Bacteria lack histones but use different basic DNA-binding proteins to accomplish the same purpose, whereas Archaea, such as *M. jannaschii,* have authentic histones. This was not a surprise, but when the sequences of many of the genes were examined phylogenetically, many were found to be closer to the members of the eukarya than to the members of the bacteria. This supported the theory of Carl Woese that Archaea were evolutionarily closer to Eukarya than Bacteria. However, there were also many similarities found that were closer to bacteria (Table 29.2), primarily attributable to the single-cell lifestyle. For example, transport of ions, polysaccharide biosynthesis, cell division proteins, and the number of rRNA genes all were similar for Bacteria and Archaea. However, *M. jannaschii* (and all archaea) are more similar to eukaryotes with regard to the genes and proteins involved in DNA replication, transcription, and translation. The genome of *M. jannaschii*, as well as sequences from other Archaea, have supported the theory that the archaea and eukarya diverged after their common ancestor had diverged from Bacteria.

TABLE 29.2
Comparison of Genes from a Member of Bacteria (*Escherichia coli*), Archaea (*Methanocaldococcus jannaschii*), and Eukarya (*Saccharomyces cerevisiae*)

Character	Bacteria	Archaea	Eukarya
Number of genes in genome	400–2,000	1,500–3,000	5,000–50,000
Chromosomes	Circular	Circular	Linear
Transport of inorganic ions	Bacterial	Bacteria-like	Eukaryotic
Polysaccharide biosynthesis	Bacterial	Bacterial	Eukaryotic
Ribosomal operons	1–15	1–5	1–23,000
Cell division proteins	Bacterial	Bacterial-like	Eukaryotic
Flagella	Bacterial	Bacterial	Eukaryotic
Ribosomal proteins	Common + bacterial	Common + eukaryotic	Common + eukaryotic
Translation elongation factors	Bacterial	Eukaryotic	Eukaryotic
Translation initiation factors	None	Eukaryotic	Eukaryotic
Aminoacyl tRNA synthetases	Bacterial	Archaeal + eukaryotic	Eukaryotic
RNA polymerases	Common	Common + eukaryotic	Common + eukaryotic
Transcription initiation	Bacterial	Archaeal + eukaryotic	Eukaryotic
Replication	Bacterial	Eukaryotic	Eukaryotic
Replication factor complex	None	Eukaryotic	Eukaryotic
DNA replication initiation	Bacterial	Eukaryotic	Eukaryotic
Histones	Basic proteins	Eukaryotic	Eukaryotic

CRENARCHAEOTA

Sulfolobus solfataricus is a member of the Crenarchaeota, and is therefore only distantly related to *M. jannaschii*, which is a member of the Euryarchaeota. It grows in hot acid pools, with temperatures of 75°C–80°C and pH in the range of 2.0–3.0. Therefore, it is both a thermophile and an acidophile. Its genome consists of 2,992,245 bp (Figure 29.5), containing 2960 protein-coding genes, 44 tRNA genes, and 4 rRNA genes (two 5S, one 23S, and one 16S). The size of the genome is approximately midway between the genomes of *M. jannaschii* and *E. coli*. As expected, the *S. solfataricus* genome has more similarities with the genome of *M. jannaschii* than with *E. coli*, but the number of genes for each of the function categories is higher than for *M. jannaschii*. For example, *S. solfataricus* has 88 genes for amino acid synthesis, whereas *M. jannaschii* has 65 and *E. coli* has 117. Similar relationships are found with many of the genes for core functions. However, *S. solfataricus* has an increased set of genes involved in central intermediary metabolism. Many act on derivatives of amino acids, and most appear to be involved with dealing with the acidic conditions where this organism lives. Thus, there are many dehydrogenases. Another obvious difference is found in the part of the genome that encodes the components of the electron transport chain. The genome contains the basic set of 36 genes for this function, whereas *E. coli* has 76. The primary function of electron transport is to couple the reaction of an electron donor and an electron acceptor with the transport of protons across the membrane. In this case, the protons are being pumped out of the cell. The additional gene products in *S. solfataricus* might be necessary because of the acidic conditions surrounding the cell. The concentration gradient from the outside to the inside is causing an influx of protons into the cell, and therefore, the cell must respond by pumping more protons out of the cell. Another notable difference is that this genome has a large number of transposons. There are approximately 122 open reading frames, hypothetical proteins, and transposase genes present in this genome, compared to the *E. coli* genome, with 42 such sequences, and *M. jannaschii*, with only 4 sequences. The significance of this difference is unknown.

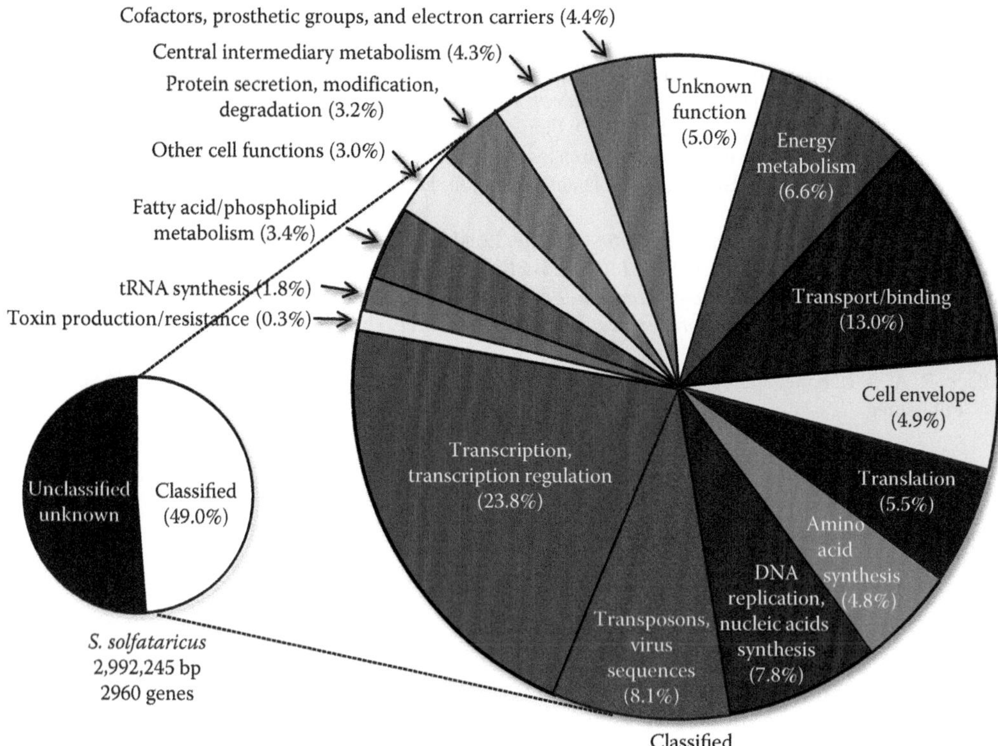

FIGURE 29.5 Summary of categories of genes in the *Sulfolobus solfataricus* genome. The genome of this member of the Crenarchaeota consists of 2,992,245 bp, encoding 2,960 genes, of which 1,449 (49.0%) have been classified. As with *Methanocaldococcus jannaschii*, the proportion of the genome-encoding genes involved in DNA replication, transcription, and translation is around 40%, and genes for transcription and control of transcription comprise more than 50% of this. Curiously, there are about the same number of genes that encode proteins involved in metabolism as there are in the genome of *M. jannaschii*, even though the genome is 75% larger. One area that is much expanded in the *S. solfoataricus* genome includes transposons and virus sequences. This comprises more than 8% of the genome, which converts to a few hundred thousand base pairs of DNA. One hypothesis is that eukarya are most directly related to the members of the Crenarchaeota than with other groups of Archaea. This might partly explain the great expansion of genes involved in transcription and the higher proportion of transposons and virus sequences, both of which are common characters in the members of the Eukarya.

KEY POINTS

1. Free-living bacteria have genomes that are usually greater than 1.5 Mb, which encode more than approximately 1500 genes.
2. The number of genes encoding proteins in Bacteria are approximately 30% DNA replication, transcription and translation; 30% cell envelope; and 30% metabolic functions.
3. Archaea generally have a higher proportion of their genes for transcription, transcriptional control, and translation than for metabolic processes.
4. Genomic studies support the rRNA phylogenetic conclusion that the domains eukarya and archaea diverged after the split of arcahea from bacteria.

ADDITIONAL READINGS

Alberts, B., D. Bray, K. Hopkin, A. Johnson, J. Lewis, M. Raff, K. Roberts, and P. Walter. 2013. *Essential Cell Biology*, 4th ed. New York: Garland Publishing.

Alberts, B., D. Bray, A. Johnson, J. Lewis, M. Raff, K. Roberts, and J. D. Watson. 1994. *Molecular Biology of the Cell*. New York: Garland Publishing.

Becker, W. M., J. B. Reece, and M. F. Poenie. 1996. *The World of the Cell*, 3rd ed. New York: The Benjamin/Cummings Publishing.

Blattner, F. R., G. Plunkett III, C. A. Bloch, N. T. Perna, V. Burland, M. Riley, J. Collado-Vides et al. 1997. The complete genome sequence of *Escherichia coli* K-12. *Science* 277:1453–1462.

Bult, C. J., O. White, G. J. Olsen, L. Zhou, R. D. Fleischmann, G. G. Sutton, J. A. Blake et al. 1996. Complete genome sequence of the methanogenic archaeaon *Methanococcus jannaschii. Science* 273:1058–1073.

Goffeau, A., B. G. Barrell, H. Bussey, R. W. Davis, B. Dujon, H. Feldmann, F. Galibert et al. 1996. Life with 6000 genes. *Science* 274:546.

Gregory, T. R., ed. 2005. *The Evolution of the Genome*. London: Elsevier Academic Press.

Lewin, B. 2008. *Genes IX*. Sudbury, MA: Jones & Bartlett Publishers.

Tropp, B. E. 2008. *Molecular Biology: Genes to Proteins*, 3rd ed. Sudbury, MA: Jones & Bartlett Publishers.

Venter, J. C. 2007. *JCVI comprehensive microbial resource.* http://cmr.jcvi.org/cgi-bin/CMR/shared/Genomes.cgi.

Wallace, R. A., J. L. King, and G. P. Sanders. 1981. *Biology, the Science of Life*. Santa Monica, CA: Goodyear Publishing.

30 Mutualists and Pathogens

INTRODUCTION

If organisms and genomes were completely separated from one another for all of their lives, then evolution would be relatively simple, as would be the discussion of evolution. However, almost all organisms live in close proximity to, attached to, or enveloped by other cells and organisms. Most could not survive without the presence of the other organisms. These interactions cause a variety of effects on evolution and genomes. The simplest interactions are those that involve an organism that becomes dependent on another organism for some biological process or particular molecules. For example, humans are dependent on gut microorganisms for much of their digestion of food, providing vital molecules, and for protection against pathogens. However, microbes also have analogous associations (Figure 30.1). Some microbes become dependent on other microbes for some of their nutrition. Some cells are in contact with cells of another species for much (or all) of their life cycles. These associations can range from coincidental to commensal to pathogenic. However, pathogens that do not kill their hosts ensure that a ready supply of host organisms will be available for infection by some of their progeny. Therefore, this evolutionary relationship is relatively stable over extended periods of time, although genetic changes are inevitable (Figure 30.1). Other symbiotic relationships can develop as well. Commensal relationships are common. In this form of symbiosis, one of the species gains from the other, but has no effect on the other organism. However, often, it becomes dependent on its host and begins to lose genes as they become unnecessary (Figure 30.1), through the processes of mutation, random drift, and/or selection.

Another form of symbiosis is when both organisms begin to depend on one another for their biological functions. This is called mutualism. In this case, often, both organisms begin to lose parts of their genomes as they become more dependent on one another, again through mutation, random drift, and/or selection. Some extreme examples of this have resulted in the formation of organelles in eukaryotes, which is discussed in Chapter 31. Each step from free-living organism to commensalism, mutualism, and/or parasitism leads to changes in the genomes. Occasionally, some of the host cell genes may be added to the invading cell genome, but more often there is a reduction in the number of genes in the genome as one or both organisms begin to rely more and more on each other for their biological functions. In addition to losing the genes, often many of the genes are transferred into the nucleus or other organelles where the genes are transcribed. The translated proteins are either shuttled back into the endosymbiont/organelle or utilized in another cellular compartment in a process called protein trafficking (discussed in Chapter 32). In many cases, multiple copies of the genes are transferred into the nucleus, although usually only one of them is expressed.

Many steps are involved in the pathway from parasite to obligate parasite to endosymbiont to organelle. The first step in the process is becoming an opportunistic parasite or commensal organism. That is, an organism seeks food and finds it wherever it can, including from multiple host organisms. Many bacteria and fungi are representative of this. For example, many species of *Aspergillus*, *Penicillium*, and *Phialophora* (all ascomycetes) can grow in the soil as saprobes, utilizing wood, dead insects, and other detritus for their nutritional needs. However, some can also survive on living organisms, including humans. Some fungi can be beneficial to young trees but may kill mature trees. Thus, they cross the border between mutualists and pathogens. One species of *Phialophora* (*P. americana*) can grow as a saprobe, can infect trees, and can infect on humans (and possibly other animals). Therefore, these fungi are generalists and have genes in their genomes to allow them to utilize such different food sources, but they also require nutrients from the organisms on which they associate, and therefore have probably lost some genes that are important in producing these nutrients.

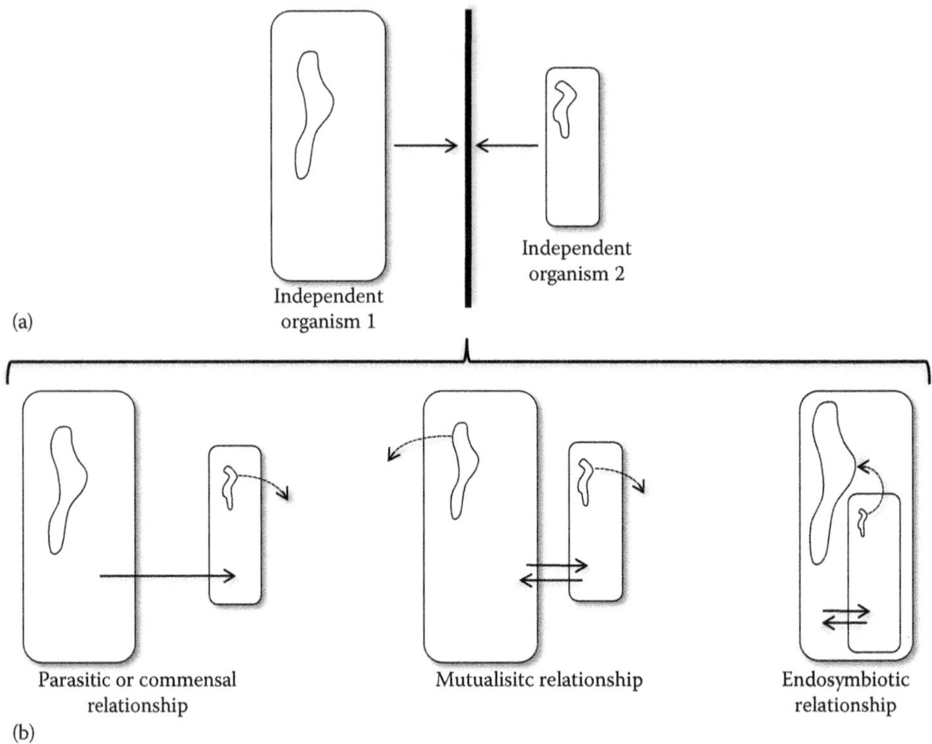

FIGURE 30.1 Interactions between organisms and genomes. Solid arrows indicate the flow of nutrients. Dashed lines indicate losses and movements of parts of the genomes. When organisms are independent of one another (a), there is little possibility that they will exchange nutrients or pieces of their genomes. However, most organisms interact with other organisms. In parasitic and commensal relationships (b, left), one organism obtains nutrients from the other. Because the organism receives some of its nutrients from the other organism, those compounds are in excess of the needs of that organism. Because of this, some of the genes involved in synthesis of these compounds can mutate, lose function, and eventually be deleted from the genome. In general, the genomes of these organisms decrease over time. In mutualistic relationships (b, middle), both organisms benefit by utilizing nutrients produced by the other organism. Because of this, there are redundant processes ongoing within the organisms, such that fewer synthetic processes are needed by each of the organisms. In this case, both genomes decrease in size. However, over time they become dependent on one another for their survival, such that independence becomes impossible. In endosymbiotic relationships (b, right), there are exchanges of nutrients (often in both directions). The endosymbiont often loses DNA, some or all of which is transferred into the host cell genome.

Some parasites have become specialists, only able to infect a narrow range of species, or a single species. For example, *Candida* species usually grow on certain mammalian species, including humans. In fact, in many cases, *Candida* is a symbiont that causes no disease symptoms. It is able to colonize the gut, and the immune system controls the population size. However, if the immune system is upset, then these organisms can multiply out of control and can begin to destroy parts of the body, becoming pathogenic. As these organisms specialize to grow on one host, or even a single cell type, some of their biosynthetic requirements lessen, and therefore, some of their genes become unnecessary because the host is providing those components. As evolutionary time continues, they lose more genes, such that they become totally dependent on the host cells and cannot survive without them. The minimum genome size for a free-living bacterium is about 1.3 Mb (Figure 30.2). Organisms with less than this amount all appear to be obligate parasites. Between genome sizes of approximately 1.2 and 2.0 Mb, there is a mixture of obligate parasites and free-living organisms. Above that amount, all appear to be free-living. For eukarya, the cutoff value for obligate parasite versus free-living organism is approximately 10–50 Mb (Figure 30.2).

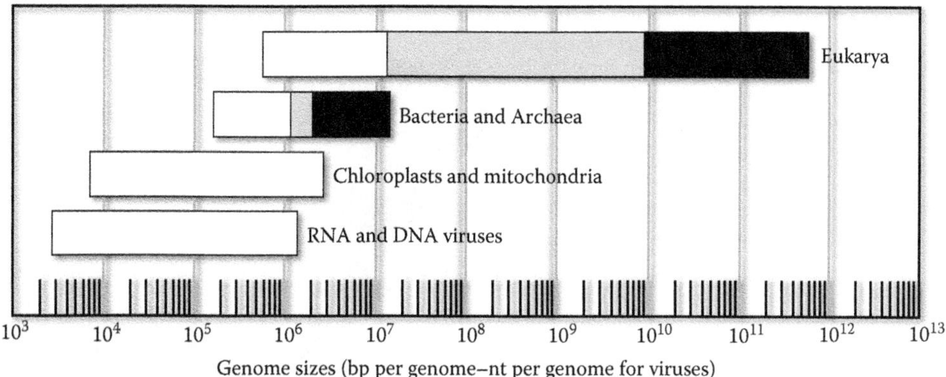

FIGURE 30.2 Comparisons of parasites/symbionts versus free-living organisms based on genome sizes. White bars indicate obligate parasites. Gray indicates a mix of parasites/symbionts and free-living organisms. Black bars indicate mainly free-living organisms. Viruses and organelles all are obligate parasites or symbionts. For Bacteria and Archaea, species with genomes smaller than about 1.2–1.3 Mb are parasites, commensals, or symbionts. Species with genomes between 1.2 Mb and approximately 2.0 Mb include a mixture of parasites, commensals, symbionts, and free-living organisms, whereas species larger than this are primarily free-living. The division point between parasites/symbionts and free-living species is broader for eukarya. Eukaryotic species with genomes smaller than about 10 Mb are parasites, commensals, or symbionts. However, between 10 Mb and about 10 Gb, there are both free-living and parasitic species. In particular, there are many species of nonphotosynthetic plants with large genomes that live as epiphytes and parasites on other plants.

TERMITE GUT MICROBES

Termite digestive systems contain consortia of approximately 200 species of microbes, including bacteria and eukarya. Without them, the termites would starve because they are unable to digest wood without the microbes in their gut. This is because termite genomes are lacking the genes for cellulose degradation. Each of the microorganisms in the termite gut is dependent on other microbes in the gut for their survival. For example, a protozoan is responsible for the enzymatic digestion of much of the cellulose that the termites eat. However, the protozoans are dependent on bacteria in the gut for the supply of some of their energy and biochemical requirements. If some of the microbes disappear, then the termite may starve. However, if the termite dies, then the microbes in the gut also die. They cannot survive outside of the termite gut because the termite cells are also providing some of their nutrition, as well as protecting from the outside environment. Although not all 200 species of organisms have been sequenced, a few have been sequenced, and they have smaller genomes than any of their free-living relatives. Each has lost certain genes or sets of genes. They rely on one or more of the other species to supply some of their nutrients. Some wood-eating fish may have similar systems. These fish defecate, and from time to time, they reinoculate themselves by eating some of their feces, which contain the microbes needed for digestion of the wood. When the feces are removed from their environment (e.g., from their fish tanks in laboratory experiments), often the fish die, even when there is plenty of wood available.

SMALLEST BACTERIAL GENOME

The smallest bacterial genome known is that of *Carsonella ruddii*. It has 182 genes on a chromosome that is only 159,662 bp in length (Figure 30.3). This makes it smaller than the genomes of many viruses, including bacteriophage T4, and similar in size to most chloroplast genomes. *C. ruddii* lives in the gut of an insect that lives on plant sap (called a psylid, which is related to aphids). The genes that it has in its genome are primarily restricted to the processes of DNA

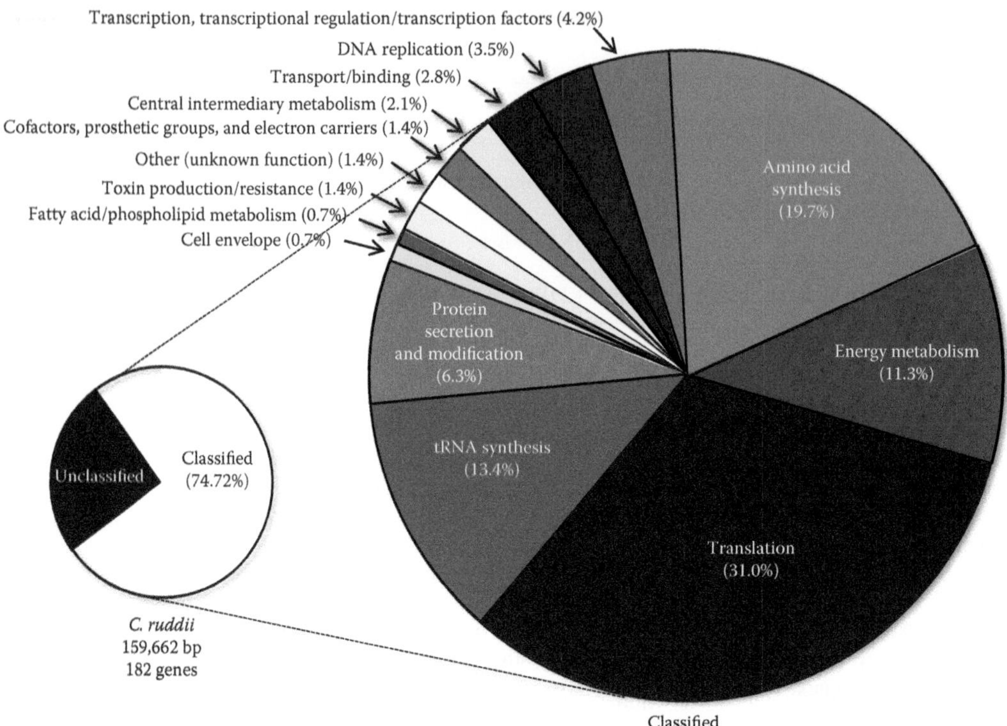

FIGURE 30.3 Summary of categories of genes in the *Carsonella ruddi* genome. The genome consists of 159,662 bp, encoding 182 genes, of which 136 (74.7%) have been classified. This genome has been greatly reduced from that of its free-living bacterial ancestor. Most functions necessary for a free-living lifestyle have been lost, leaving primarily the functions that allow it to express the few genes that remain. In particular, more than 70% of the genome encodes proteins that encode proteins necessary in DNA replication, transcription, and translation. Some important genes are missing, but this endosymbiont is still able to complete replication, transcription, and translation, possibly by importing many of the necessary proteins. Energy metabolism, probably needed to accomplish these functions, encompasses another 11.3% of the genome. The remaining 15% of its genome consists of a disparate set of genes for a variety of functions. However, none appear to be parts of complete pathways. Therefore, this endosymbiont may not be providing the insect (a psyllid that feeds on plant sap) with many nutrients, although its persistence in this insect indicates that it may be providing some vital functions to the host.

replication, transcription, translation, and cell division. Unlike most virus genomes, it does have ribosomal RNA genes (a single copy each of the small subunit, large subunit, and 5S genes) and ribosomal proteins (to produce its own ribosomes) and attendant proteins necessary for translation. The genome contains 28 transfer RNA (tRNA) genes (including at least one for each amino acid) but lacks the full complement of aminoacyl tRNA synthetases (which must be supplied by the host or by other microbes) for translation of its messenger RNAs (mRNAs). It has few complete pathways for amino acid and nucleotide synthesis, and therefore probably obtains these from the host, from the plant sap that is ingested by the host, or from coresident microbes. Other than genes for replication, transcription, and translation, the genome contains only a few genes for other functions. This bacterium, a member of the Bacteroides, probably has lived inside the guts of sap-feeding insects for hundreds of millions of years. During that time, the size and number of genes in its genome have decreased, such that only about 10% of the genome of the free-living ancestor remains in its genome. The bacterium has lost a free-living lifestyle to become a symbiont inside the guts of insects, which essentially deliver a food source (plant sap) to them. The insects benefit because they are ingesting a nutrient-poor food source that is being partly digested by the bacteria. Also, the bacteria are providing the insect with some needed biochemicals, including specific amino acids. In turn, the insects are producing a set of biological nutrients to the bacteria. Together, the parts of the necessary biochemical pathways from the organisms complement one another so that each organism can survive.

CORESIDENT SYMBIONTS

Another system, also in sap-eating insects (sharpshooters), the genomes of two species of bacteria, also with very small genomes, have been studied. *Sulcia muelleri*, a member of the Bacterioides, has the second smallest known genome for a bacterium. The 245,530 bp genome carries 228 protein-coding genes, 31 tRNA genes, and single copies of the 5S, 16S, and 23S ribosomal RNA (rRNA) genes, as well as genes for DNA replication, translation, and cell division. It coresides with the γ-proteobacterium *Baumannia cicadellinicola*. It has a genome of 686,194 bp with 650 genes, again including genes for replication, transcription, translation, and cell division. However, what *B. cicadellinicola* is missing (e.g., polyisoprenoids, methionine, histidine, several cofactors, and prosthetic groups) is supplied by either *S. muelleri* or the sap that is being ingested by the insect (Figure 30.4). Conversely, what *S. muelleri* is missing (e.g., several amino acids and a few cofactors) is supplied by *B. cicadellinicola* or

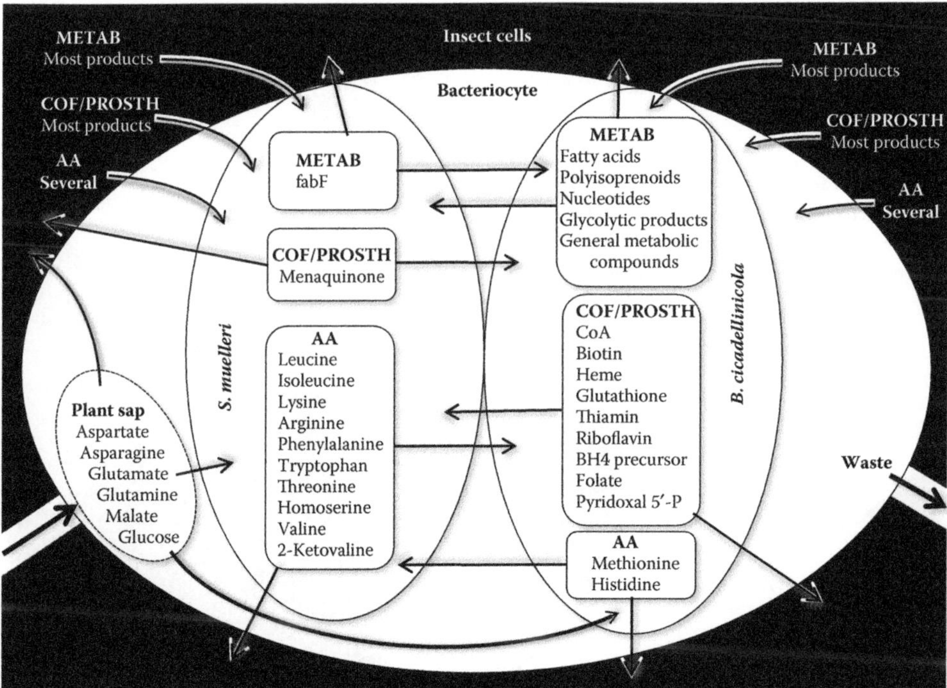

FIGURE 30.4 Bacterial symbiont interactions in the glassy-winged sharpshooter gut. This insect feeds on plant sap, which is poor in nutrition, containing primarily dilute glucose and a few amino acids. However, there are at least two bacterial symbionts (*Sulcia muelleri* and *Baumannia cicadellinicola*) that live in the insect gut, which provide additional nutrition for the insect. In return, the bacteria obtain nutrients from the plant sap as well as from the insect cells. In particular, *S. muelleri* produces a large number of amino acids that are used by the insect cells, as well as by *B. cicadellinicola*. The latter species produces a large number of prosthetic groups and cofactors, as well as other metabolic products, which it provides to *S. muelleri* and the insect cells. Each of the species is dependent on the other, such that if any one of the species are not present, the others cannot survive. AA, amino acids; COF/PROSTH, cofactors, prosthetic groups, and electron carriers; METAB, metabolic products.

the sap. In essence, they complement each other, as well as complement the host insect. For example, *B. cicadellinicola* has genes that encode proteins needed for nucleotide synthesis, which are lacking in *S. muelleri*, whereas the latter has genes that encode proteins for the synthesis of leucine, isoleucine, lysine, argenine, phenylalanine, tryptophan, threonine, homoserine, valine, and 2-ketovaline that are not synthesized by the former. Conversely, *B. cicadellinicola* can synthesize methionine and histidine, whereas *S. meulleri* cannot. The combination of the two microbes, the plant sap (which provides mainly aspartate, asparagine, glutamate, glutamine, malate, and glucose) and biochemicals provided by the host, constitutes all of the metabolic and catabolic functions needed by the organisms in this biological system. All of it emanates from the interplay between at least four genomes (including that of the plant), which produce gene products that carry out all of the necessary functions that allow the insect and its bacterial symbionts to survive. If just one of the organisms is missing, and that function is not provided by another organism, it is likely that the others will fail to survive.

ANIMAL PARASITE

The first bacterial genome ever to be sequenced was that of *Mycoplasma genitalium*, an obligate parasite that lives on the cells of primates. It cannot live without the host cells because it lacks many of the genes and gene products necessary for autonomous life. The genome of this microbe contains

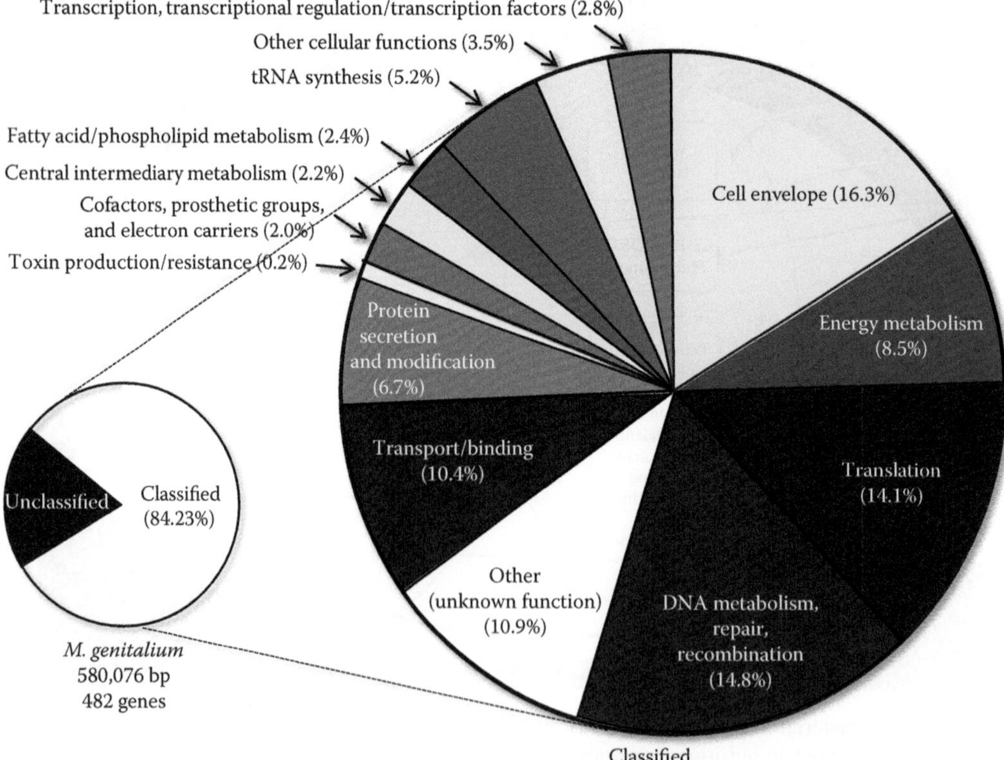

FIGURE 30.5 Summary of categories of genes in the *Mycoplasma genitalium* genome. The genome consists of 580,076 bp, encoding 482 genes, of which 406 (84.3%) have been classified. This bacterium is a parasite that lives on the cells of primates. Nearly 40% of the genome is necessary for DNA replication, transcription, and translation. More than 25% of the genome contains genes for proteins responsible for production of the cell envelope, including proteins that transport chemicals into and out of the cell. These are important because these cells import many vital components (including amino acids) that are produced by the host cells. Thus, most of its genome is necessary for its reproduction, gene expression, and acquisition of nutrients from the host cell.

TABLE 30.1
Comparison of the Number of Genes in the Free-Living *Haemophilus infuenzae* and the Obligate Parasite *Mycoplasma genitalium* Genomes

Cell Function	Number of Genes in *H. influenzae* Genome	Number of Genes in *M. genitalium* Genome	Difference (%)
Amino acid biosynthesis	68	1	−99
Cell envelope synthesis	84	17	−80
Energy metabolism	112	31	−72
Fatty acid metabolism	25	6	−76
Nucleic acid synthesis	53	19	−64
Replication	87	32	−63
Transcription	27	12	−56
Translation	141	101	−28
Transport and binding	123	34	−72
Total	**720**	**253**	**−65**

580,076 base pairs of DNA (Figure 30.5), which encodes 521 genes (482 protein-encoding genes). The genes are packed tightly on the chromosome. The genome of *M. genitalium* contains genes important in glycolysis, adenosine triphosphate synthesis, adenosine triphosphatase transporters, nucleotide synthesis, DNA replication, RNA transctiption, translation, antibiotic export, carbohydrate import, and phospholipid synthesis. However, *M. genitalium* must rely on the host organisms for its supply of amino acids and carbohydrates. In this case, the host appears to gain no benefits from the association with *M. genitalium* cells, and therefore, this is purely a parasitic association. When compared to a more complex parasite that is capable of growing under several different conditions, such as *Haemophilus influenzae*, many differences are observed (Table 30.1). The genome of *H. influenzae* is 1,830,140, which contains 1740 protein-coding genes, 58 tRNA genes, and 18 other RNA genes. *Escherichia coli* strain K-12 (which is nonpathogenic) has an even larger genome (4.639 Mb), which has 4288 genes, and it can grow under many different conditions. Therefore, during the process of becoming a parasite, *M. genitalium* has lost more than 65% of its ancestral genome. Specifically, the number of genes encoding proteins involved in amino acid synthesis has been reduced from 68 to only 1. Obviously, the parasite is using the host cell as a source of amino acids, possibly by using host cell proteins as amino acid sources. Many other functions have been reduced as well (Table 30.1). Pressure to reduce the genome was stronger than pressure to retain those particular parts of the original genome.

GENOME MIXING AND SORTING

The evolutionary steps that have occurred to produce parasites, mutualists, commensals, endosymbionts, and organelles all involve intricate interactions between gene products coming from at least two different genomes. In the first step, most of the time there may be conflicting reactions that cause damage to one or more of the species. If this relationship becomes one that is mutualistic, then the gene products, and therefore the genomes, are compatible or at least have few conflicting reactions occurring. At this point, one or more of the coexisting genomes begin to lose the functions of some genes. This is caused by mutation, random drift, and/or selection. Mutations occur constantly, but in this case random drift and/or selection act to retain only those functions that are necessary for survival and the production of progeny. For example, if one of the cohabitants is receiving an adequate supply of the amino acid leucine, then the cell no longer has to produce its own leucine, and therefore, a mutation that knocks out leucine production (which would be lethal for a free-living cell) is tolerated by that cell. However, this means that the cell now is dependent on the cells of the other species to continuously supply leucine. It is at this point that the cell has become either an obligate parasite or a mutualist. There can be mixtures of many different species that take part in these systems, as in the digestive systems of many insects and other animals. Each type of cell has survived because of the associations with the other cells, and therefore, the associations continue. Most often this means that independent survival is no longer possible. In another type of association, different species of cells can merge to form an intracellular endosymbiont surrounded by the host cell. Although these events probably have happened often over the past two billion years, there have been limited successes, due to the probability of incompatibilities of genomes, genes, and/or proteins. In each successful case that has been identified, the nucleus of these cells has incorporated much of the endosymbiont genome, whereas the endosymbiont genome had decreased in size, and in some cases may have disappeared completely. However, the mixture and interaction of the genomes probably is similar to the interaction when the endosymbiont first was established. When one organism interacts with another biologically, eats another organism, or infects another organism, there are often transfers of DNA, some of which become incorporated into the genome of that organism. There have been constant small and large transfers of DNA from one organism to another throughout the evolution of organisms on the Earth (see Chapter 15). They result in small to very large changes in genomes due to interactions of the gene products, random drift, and selective pressures both inside the cells/organisms and from the environment. In general, these changes tend to reduce the genome sizes of the symbionts but tend to increase the genomes of the ultimate host organisms. Remnants of these ancient transfers of genes and sometimes entire genomes are found in the nuclei of eukaryotes.

KEY POINTS

1. Parasitic and endosymbiotic events have led to the evolution of organelles, such as mitochondria and plastids.
2. As an endosymbiotic association develops, the invading cell loses parts of its genome. Some (or all) of these genes move into the nucleus.
3. DNAs have continually moved from the organelles into the nucleus, as well as from organelle to organelle.
4. Biochemical pathways in eukaryotic cells are chimeric, in that they may include genes from different origins, such as the nucleus, mitochondria, and plastids.
5. Many endosymbiotic events have led to a diversity of cell and organelle types.

ADDITIONAL READINGS

Alberts, B., D. Bray, K. Hopkin, A. Johnson, J. Lewis, M. Raff, K. Roberts, and P. Walter. 2013. *Essential Cell Biology*, 4th ed. New York: Garland Publishing.

Alberts, B., D. Bray, A. Johnson, J. Lewis, M. Raff, K. Roberts, and J. D. Watson. 1994. *Molecular Biology of the Cell*. New York: Garland Publishing.

Becker, W. M., J. B. Reece, and M. F. Poenie. 1996. *The World of the Cell*, 3rd ed. New York: The Benjamin/Cummings Publishing.

Breznak, J. A. 2002. Phylogenetic diversity and physiology of termite gut spirochaetes. *Integr. Compar. Biol.* 42:313–318.

Gregory, T. R., ed. 2005. *The Evolution of the Genome*. London: Elsevier Academic Press.

Hackett, J. D., D. M. Anderson, D. L. Erdner, and D. Bhattacharya. 2004. Dinoflagellates: A remarkable evolutionary experiment. *Am. J. Bot.* 91:1523–1534.

Lewin, B. 2008. *Genes IX*. Sudbury, MA: Jones & Bartlett Publishers.

McCutcheon, J. P. and N. A. Moran. 2007. Parallel genomic evolution and metabolic interdependence in an ancient symbiosis. *Proc. Natl. Acad. Sci. USA* 104:19392–19397.

Nakabachi, A., A. Yamashita, H. Toh, H. Ishikawa, H. Dunbar, N. Moran, and M. Hattori. 2006. The 160-kilobase genome of the bacterial endosymbiont *Carsonella*. *Science* 314:267.

Stingl, U., R. Radek, H. Yang, and A. Brune. 2005. "Endomicrobia": Cytoplasmic symbiosis of termite gut protozoa from a separate phylum of prokaryotes. *Appl. Environ. Microbiol.* 71:1473–1479.

Tamames, J., R. Gil, A. Latorre, J. Peretó, F. J. Silva, and A. Moya. 2007. The frontier between cell and organelle: Genome analysis of *Candidatus Carsonella ruddii. BMC Ecol. Biol.* 7:181.

Tropp, B. E. 2008. *Molecular Biology, Genes to Proteins*, 3rd ed. Sudbury, MA: Jones & Bartlett Publishers.

Venter, J. C. 2007. *JCVI comprehensive microbial resource.* http://cmr.jcvi.org/cgi-bin/CMR/shared/Genomes.cgi.

Wallace, R. A., J. L. King, and G. P. Sanders. 1981. *Biology, the Science of Life.* Santa Monica, CA: Goodyear Publishing.

31 Endosymbionts and Organelles

INTRODUCTION

Endosymbiosis differs from other types of symbioses because of the intimate and extensive mixing of genomes, genes and gene products, as well as other biochemicals. In the environment, as well as inside and on the surfaces of large and small organisms, like ourselves, microbes constantly interact with each other. The interactions take many forms, including establishment of commensal or parasitic relationships. Because the relationships require close proximity of the cells, some may develop into endosymbiotic relationships, where one cell is enveloped by another cell. In most cases, when a cell is phagocytized by another cell of a different type, it is dismantled by degradative enzymes, becoming food for the phagocytosing cell. In some cases, parts of the genome of the phagocytized cell may be incorporated into the genome of the phagocytizing cell. This generally amounts to incorporation of parts of a gene or a few genes at a time—a form of horizontal gene transfer. Occasionally, very large genomic sections may be transferred into the genome of the cell. However, throughout the evolution of life on the Earth, there are other interactions that generally begin by one cell being enveloped by another, but the enveloping cell does not degrade and consume the enveloped cell. The host cell maintains the internalized cell and is able to use it. When both cells benefit, usually by increased fitness of one or both, it can become a long-lived endosymbiotic relationship (Figures 31.1 and 31.2). At another level, many pathogens invade host cells but do not kill the cells that they invade. In fact, they may survive longer if they allow the invaded cells to survive. In this way, they may cause a reduction of function in the host cell, but the cell survives and continuously provides nutrients to the invading cell. After these relationships continue over evolutionary timescales, the invaders may become obligate parasites or endosymbionts, losing some of their pathogenic characters and becoming absolutely dependent on the host cells. During this process, these endosymbionts begin to lose parts of their genomes (Table 31.1), which is part of the reason for their absolute dependence on the host. This process has been occurring for billions of years (Figure 31.3) and likely continues to occur frequently. It has led to the gradual evolution of free-living species to become parasites and endosymbionts, and some of those have become organelles (Figure 31.4). During this evolution, both the genome sizes and the number of genes have also gradually decreased.

INTRACELLULAR ENDOSYMBIONTS

Wolbachia is an α-proteobacterium that is an obligate endosymbiont of *Brugia malayi*, a human filarial parasitic nematode (roundworms that infest human tissues). The genome of *Wolbachia* is 1,080,084 bp and contains 806 protein-coding genes, 34 transfer RNA (tRNA) genes, and single copies of the 5S, 16S, and 23S ribosomal RNA (rRNA) genes. Compared to free-living bacteria, it lacks about 300,000 base pairs of DNA in its genome, which is equivalent to about 300 genes. Presumably, over a period of many millions of years, it lost these genes by mutation, genetic drift, and selection because their functions are being supplied by the host cells. Additionally, many of the genes now exist in the host nuclear genome. It has all of the genes necessary for DNA replication, transcription, translation, and cell division, as do most of the smaller genomes. It contains all of the genes necessary to construct the glycolysis pathway as well as the tricarboxylic acid (TCA) cycle. It has the genes for a proton-motive adenosine triphosphate (ATP) synthase and phospholipid biosynthesis, as well as transporters for carbohydrates and several ions. In addition, it can produce nucleotides, riboflavin, flavin adenine dinucleotide (FAD), and heme that it provides to the

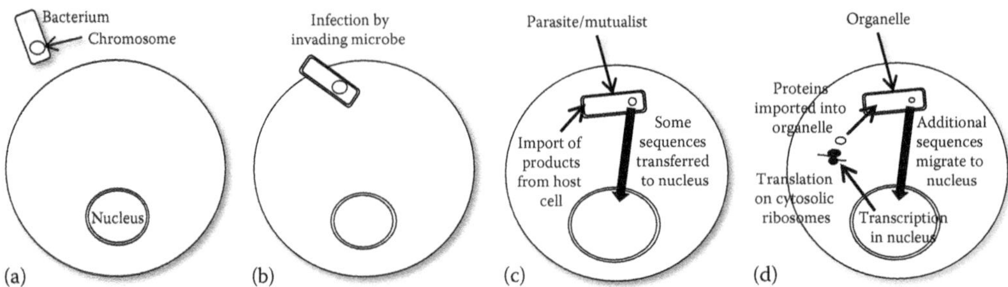

FIGURE 31.1 Pathway from free-living bacterium to organelle. (a) The process of evolution of a eukaryotic organelle begins with an interaction between a bacterial cell and a eukaryotic cell. The initial interaction often appears to be parasitic. (b) The bacterium moves into the host cell by endocytosis or by an analogous infection process. (c) Eventually, the bacterial cell becomes dependent on the host cell, becoming an obligate parasite or mutualist. In this process, it loses parts of its genome, because it uses host systems to fulfill those functions. (d) Some of the genome moves into the host cell nucleus, and a portion is expressed. As the bacterium becomes an organelle, more of the genome moves into the nucleus, and more of it is expressed from the nucleus. The genome of the organelle shrinks, whereas the nuclear genome grows. The proteins are translated in the cytosol on host ribosomes and then actively transported into the organelle.

nematode. However, the nematode provides *Wolbachia* with amino acids, some of which it clearly cannot synthesize (based on its genome). Therefore, just as with all of the other symbiotic associations, the *Wolbachia* genome has become smaller than its free-living ancestors and has benefitted by the association. Similarly, the nematode benefits from the association by gaining some additional biochemical products that it cannot synthesize. From a medical standpoint, the association presents some potential targets for curing a patient of the pathogenic nematode. Normally, it is safer to apply a drug that will kill a bacterium than one that attacks a eukaryote (i.e., the nematode), because bacteria are phylogenetically further from the host than is the nematode, and therefore, the drug targets would be more discriminatory than they were used to kill a eukaryote. Genes, proteins, and biochemical pathways that are closer to one another phylogenetically will react similarly to drugs. Therefore, it may be possible to use bacterial antibiotics for patients with this parasite that would kill the endosymbiont, which coincidentally would cause the death of the nematode.

MITOCHONDRIA

There are evolutionary steps beyond becoming obligate parasites. They have led to the establishment of organelles, such as mitochondria and plastids. It has long been hypothesized that mitochondria evolved from α-proteobacteria ancestors, and that plastids (e.g., chloroplasts) evolved from cyanobacterial ancestors. There are several reasons for this. First, sequence comparisons of these organelles definitely link them to these ancestors (see Figures 2.4 and 2.5). Second, each organelle has a double membrane (Figures 31.1 and 31.2), indicating a possible origin (the outer membrane from the phagoycytosing cell and the inner membrane originating from the phagocytosed cyanobacterial cell). Third, all have ATP synthases that are in the same orientation as those in bacterial membranes, consistent with a cyanobacterial origin (Figure 31.5). This process has taken billions of years, but also similar processes continue to occur today. Mitochondria are most closely related to α-proteobacteria. They are very ancient organelles and appear to have entered primitive eukaryotic cells as bacterial endosymbionts about 2.2–2.4 billion years ago. This was at a time when atmospheric oxygen levels were rising. Prior to this, the Earth had a reducing atmosphere, and oxygen was extremely toxic to the organisms living at that time. As oxygen levels rose, some genes mutated, which formed gene products that either sequestered oxygen or used it in chemical reactions. This became advantageous because energy could be harnessed to be used in other chemical reactions or the reactions could be coupled to other reactions. Although some eukaryotic cells also

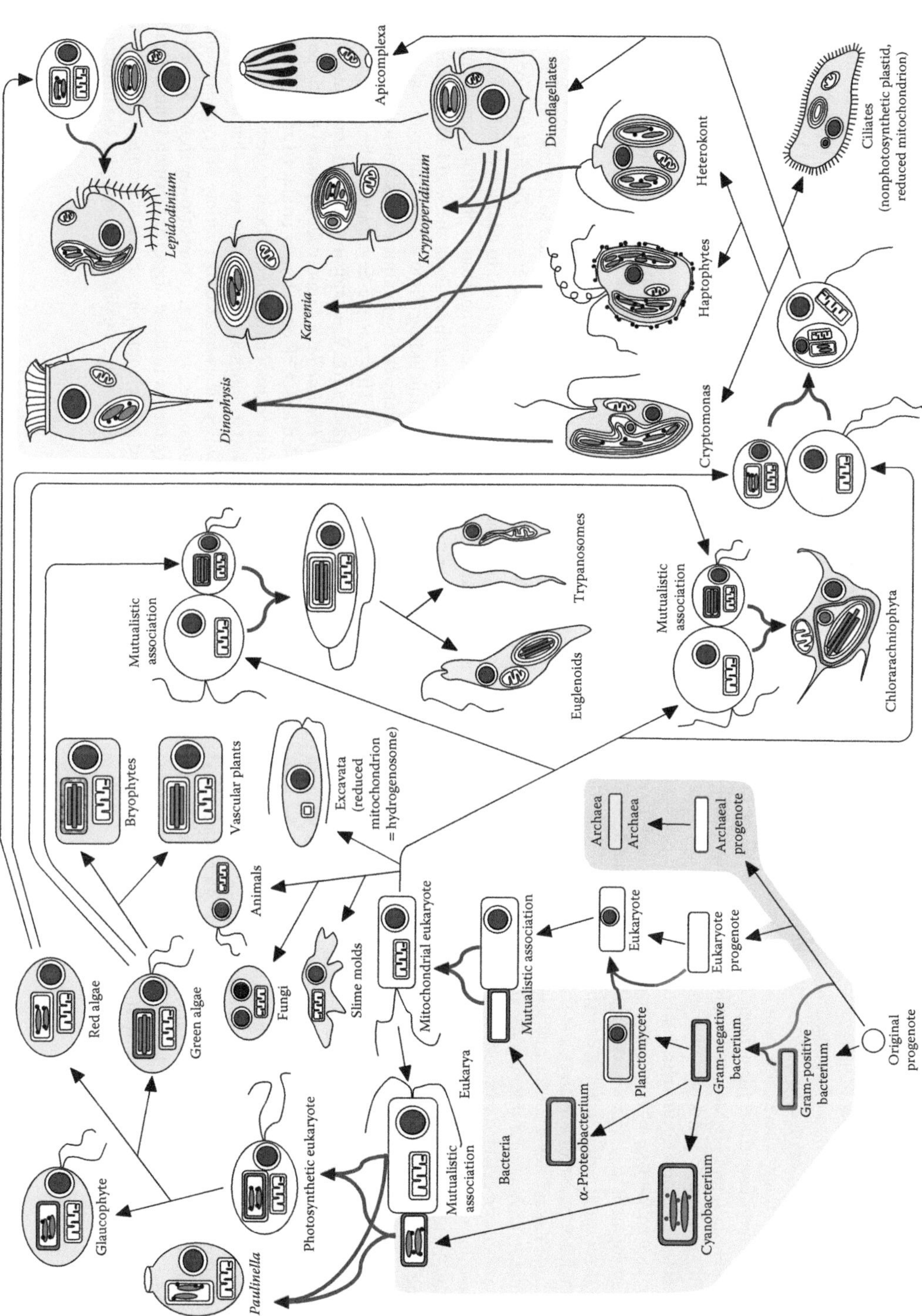

Apicomplexa
Dinoflagellates
Lepidodinium
Kryptoperidinium
Karenia
Dinophysis
Heterokont
Haptophytes
Cryptomonas
Ciliates
(nonphotosynthetic plastid,
reduced mitochondrion)
Mutualistic
association
Trypanosomes
Euglenoids
Mutualistic
association
Chlorarachniophyta
Bryophytes
Vascular plants
Excavata
(reduced
mitochondrion
= hydrogenosome)
Animals
Archaea
Archaea
Archaeal
progenote
Red algae
Green algae
Fungi
Slime molds
Mitochondrial eukaryote
Mutualistic association
Eukaryote
Eukaryote
progenote
Glaucophyte
Photosynthetic eukaryote
Paulinella
Eukarya
Mutualistic
association
Bacteria
α-Proteobacterium
Planctomycete
Gram-negative
bacterium
Gram-positive
bacterium
Original
progenote
Cyanobacterium

FIGURE 31.2 (Continued) Overall patterns of endosymbiosis and organelle evolution. Red lines indicate the path of the endosymbiont, whereas blue lines indicate the path of the host cell. White cells are hypothetical, whereas yellow cells are illustrations based on cells living today. Nuclei are indicated as red discs (and a blue disc in the case of dikaryotic fungi or orange discs for nucleomorphs). Chloroplasts are shown in green, whereas mitochondria are white with convoluted membranes. Membranes are indicated by black lines, whereas the pepidoglycan cell wall (e.g., surrounding Gram-positive bacteria) is indicated in blue. Endocytosis, gene transfers, and genome mixing have occurred many times throughout the history of life on the Earth. Evidence for this has come from microscopic examination of cells, including instances of multiple membranes, and phylogenetic studies of DNA sequences. Beginning in the lower left corner, the original progenote cell is assumed to have only one cell membrane. After millions of years of evolution, there would have been many different species of bacteria. There may have been an initial endosymbiotic event between two of these bacteria to form Gram-negative bacteria, which have two cell membranes. Modern Gram-negative bacteria only have the double membrane as evidence of this event. Some used the heat of the Earth or chemicals to build the compounds that they needed for growth and reproduction. Others harnessed sunlight energy for growth and development (e.g., cyanobacteria). One group of bacteria diverged from the main group to evolve into the Archaea. This group of single-celled organisms became specialized to growing in harsh environments. One of these diverged into the species that would lead to the evolution of eukaryotes. The origin of eukaryotes coincided with development of a nucleus and acquisition of a mitochondrion. There are three theories as to how this occurred, but each involved an endosymbiotic event. One proposes that eukaryotes were formed when an α-proteobacterium became an endosymbiont of an archaeal cell. This event led to the nucleus that was primarily archaeal and the mitochondrion that was the α-proteobacterium. Another theory proposes that there were two endosymbiotic events. The first (involving an archaeal and probably a bacterial cell) led to a cell with a nucleus, and the second (involving an eukaryote and an α-proteobacterium) led to addition of mitochondria. The third theory proposes that there was a single endosymbiotic event involving a planctomycete relative (because this group possess nucleomorphs that resemble nuclei, and they also have genes for microtubules that are similar to those in eukaryotes) and an α-proteobacterium. These cells spread throughout the world and became new species and groups due to selective pressures in the various habitats. Of course, one line eventually led to the Amoebozoa (including slime molds), fungi (including the Dikarya, dikaryotic fungi), and animals (including humans). Once the initial eukaryotic cell was established, there were many endosymbiotic events that followed many different evolutionary paths. One eukaryotic cell enveloped a member of the cyanobacteria (a photosynthetic bacterium). This was the beginning of the Archaeplastida, which diverged into the glaucophytes, red algae, green algae, and plants (including bryophytes and vascular plants; upper left). A second similar event involving a heterotrophic eukaryote and a cyanobacterium occurred more recently (probably within the past 500 million years) to form *Paulinella*, a single-celled photosynthetic eukaryote. One lineage of eukaryotes led to the Excavata, where the mitochondria became reduced, such that they either yielded less ATP or no longer produced any ATP. Other lineages began fusing with members of the red and green algae, probably to help in their metabolic processes. One of these associations led to both euglenoids and trypanosomes (middle). The former utilized the photosynthetic capabilities of the endosymbiont/organelle, whereas the latter retained some of the genes that it used in some of its biosynthetic pathways, but eliminated the organelle through mutation and selection processes. Other endosymbiotic events led to a variety of organisms and organelles in the Chromalveolata, including a wide variety of dinoflagellates (upper right), ciliates, haptophytes, cryptomonas, and heterokonts (lower right), as well as members of Rhizaria (e.g., Chlorarachniophyta; lower middle). Multiple membranes surround various parts of the organelles in these organisms, which is an indication of their evolutionary history. Some of the organelles still retain parts of the original nucleus (which have been reduced to nucleomorphs, shown as orange discs) and chloroplasts. However, many have lost the mitochondria within the organelle, which may indicate a conflict between the main mitochondria in the cell and the mitochondria in the endosymbiont/organelle.

TABLE 31.1
Progression from Autonomous Bacterium to Organelle

Organism/Organelle	Lifestyle	Genome Size (kbp)	Number of Genes in Genome
Bradyrhizobium japonicum	Free-living	9106	8317
Escherichia coli	Free-living	4639	4288
Haemophilus influenzae	Free-living pathogen	1830	1743
Chlamydia spp.	Obligate parasites	1000–1500	900–1100
Paulinella sp.	Chromatophore (photosynthetic—cyanobacterial origin)	1020	867
Mycoplasma genitalium	Obligate parasite	583	482
Plastids (cyanobacterial origin)			
Red algae	Organelle	120–191	140–209
Flagellates	Organelle (nonphotosynthetic)	128	46
Glaucophyte	Organelle (cyanelle)[a]	136	150
Green algae	Organelle	130–250	46–173
Plants	Organelle	151–174	81–87
Toxoplasma gondii (protist)	Organelle (nonphotosynthetic)	35	26
Mitochondria			
Animals	Organelle	15–18	35–40
Fungi	Organelle	18–156	30–40
Humans	Organelle	17	37
Plants	Organelle	167–11,300	62
Protists	Organelle	6–49	5–97
Yeast	Organelle	78	34

[a] The organelle in glaucophytes (e.g., *Cyanophora*) produces a bacterial cell wall component (peptidoglycan) that coats the organelle (called a cyanelle), which is very similar to the organization of a free-living bacterial cell wall.

developed mechanisms to deal with the toxic oxygen, and peroxisomes evolved for this purpose, the amount of energy that could be shuttled for other processes was much less. Eukaryotes that lack mitochondria or have reduced mitochondria exist, but they are inefficient in their utilization of oxygen compared with cells that contain mitochondria. Most live in environments with little or no oxygen.

The bacteria that tolerated oxygen 2.4 billion years ago could survive closer to the surface of the water where oxygen concentrations were higher, whereas the primitive eukaryotes were restricted to deeper anoxic areas (Figure 31.6). However, at some location during that time, an oxygen-utilizing bacterium was enveloped by a eukaryote, and a partnership began to develop, whereby the eukaryote could now survive higher in the water column, because it contained bacteria that could detoxify the oxygen in that region. This would open up entirely new niches for the eukaryote, allowing for its survival and propagation. Furthermore, it could acquire energy from oxidation reactions, which further improved its chances for survival in an atmosphere where oxygen levels continued to rise. Over the next few billion years, genes from the bacterial endosymbiont migrated into the nucleus, and today all but a few dozen genes from these mitochondria have migrated and integrated into the nuclear genomes.

There might have been a problem initially with expression of bacterial genes in the eukaryotic nucleus. Promotion of transcription differs between bacterial and eukaryotic genes. Therefore, as the pieces of DNA moved into the nucleus, initially most were probably silent, because the transcription machinery within the nucleus failed to recognize the genes. However, because this

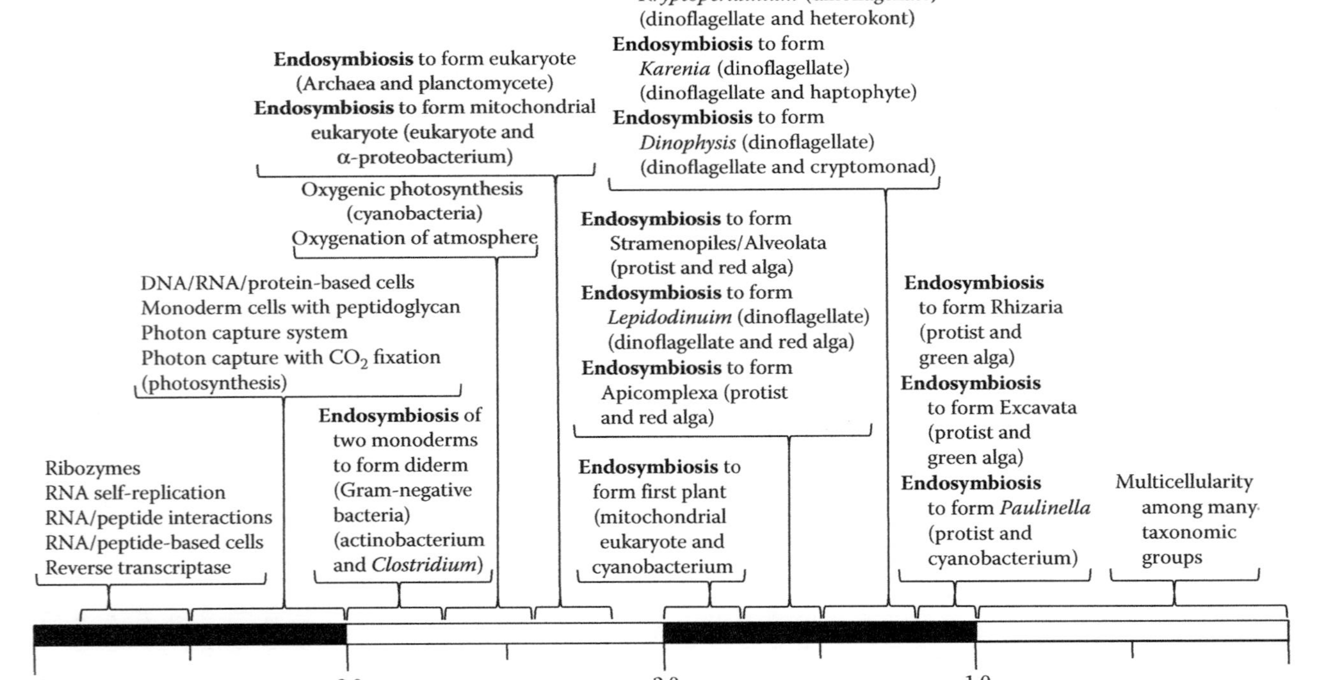

FIGURE 31.3 Significant events on the Earth leading to endosymbiotic events. During the initial 800 million years, life was becoming established on the Earth, beginning with RNA/protein-based life and developing into a DNA/RNA/protein life system. By 3.0 billion years ago, some organisms had begun to harness photon (solar) energy to fix carbon from CO_2. By about 2.7 billion years ago, monoderms (Gram-positive bacteria) and diderms (Gram-negative bacteria) were present. One hypothesis is that an endosymbiotic event occurred between two monoderms to form the first diderm. Genomic analysis supports an actinobacterium and a clostridium as the two monoderms that joined to form the diderms. Between 2.7 and 2.4 billion years ago, oxygen levels were increasing, likely caused by the activity of oxygenic photosynthetic cyanobacteria. By 2.4–2.2 billion years ago, the first eukaryotes and the first mitochondrial eukaryotes appear. This involved one or more endosymbiotic events, which included an archaea, an α-proteobacterium, and possibly a planctomycete. During the next billion years, there were many endosymbiotic events leading to the formation of novel groups of organisms, including plants, alveolates, stramenopiles, excavates, apicomplexans, rhizarians, and others. From about 1.0 billion years to the present, additional diversification of each of these groups occurred, including the evolution of many multicellular species. Endosymbioses are in bold type. Those events that have substantial evidence are underlined.

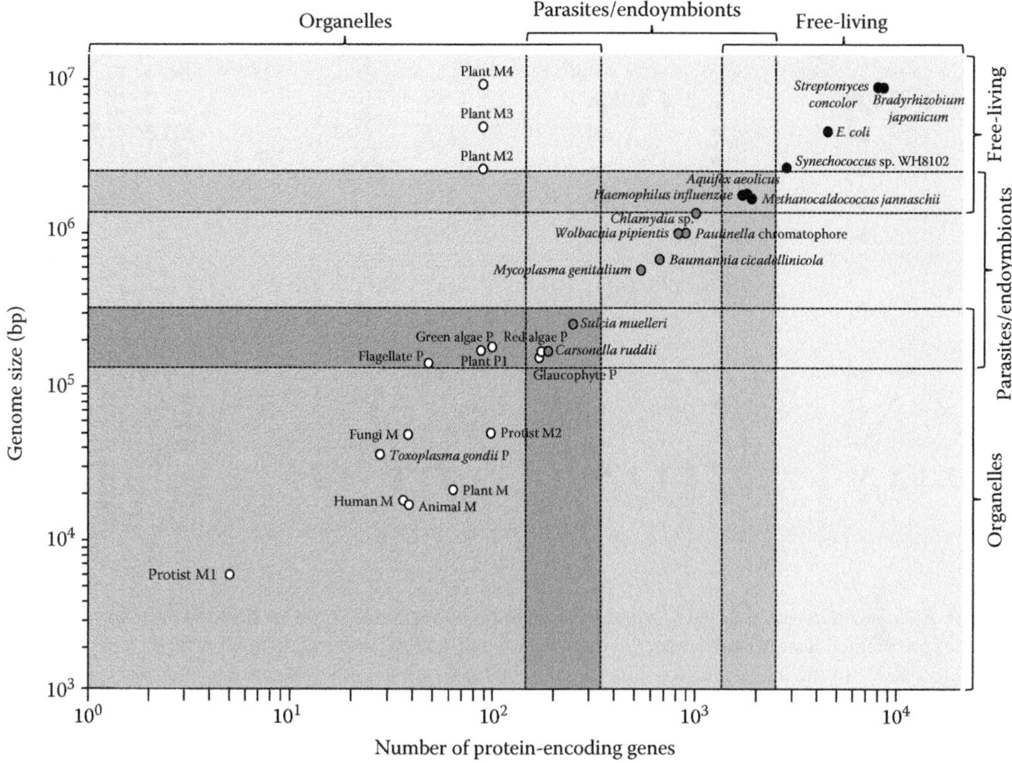

FIGURE 31.4 Genome sizes and number of genes for free-living, pathogens/endosymbionts, and organelles. Free-living organisms (black circles) all have genome sizes that are greater than approximately 1.2 Mb and have at least 1200 genes. Pathogens, commensals, and endosymbionts (gray circles) have genome sizes between about 120 kb and 2.4 Mb, with 120–2300 genes. Most organelles (white circles) all have genome sizes less than 310 kb (except some plant mitochondrial genomes), with fewer than 320 genes. Species names are provided for the first two categories, whereas general names for organelles are listed. M, mitochondrial; P, plastid. The range for protist mitochondria is indicated by the two extremes, Protist M1 and Protist M2.

occurred more than two billion years ago, the two types of promoters may have been more similar than they are at present. In either case, some gained functional promoters. This could happen if the genes initially were integrated next to a nuclear promoter, they gained promoter through a recombination process, the original bacterial promoters were functional in the nucleus, or the promoter region was generated through mutational processes. Whatever the mechanism, many of the immigrant genes and all of the functional genes have eukaryotic promoters in the eukaryotic cells in existence today. Free-living members of α-proteobacteria have genomes of approximately 4–9 Mb, containing about 3500–8500 genes. Parasitic members of this group have smaller genomes (1–3 Mb) and fewer genes (from about 500 to a few thousand). Mitochondrial genomes range from 6 kb to 2.4 Mb, but the genomes with higher amounts of DNA (found in higher plant mitochondria) have a great deal of noncoding DNA, and all of the mitochondrial genomes contain only 30–50 genes, regardless of genome size. This represents about 1% of the genes in their original genomes. Most of the remaining genes are involved in producing the translation machinery (i.e., tRNAs and rRNAs) for synthesizing a handful of proteins (Table 31.2). Therefore, over the two billion years of evolution, almost all of the genes migrated into the nucleus, and as they became expressed from the nucleus, the mitochondrial copies were lost due to mutation and elimination.

There is a core set of genes present in all mitochondrial genomes. These include genes for bacteria-like rRNAs, tRNAs, cytochrome b, cytochrome c oxidase I, and cytochrome c oxidase III,

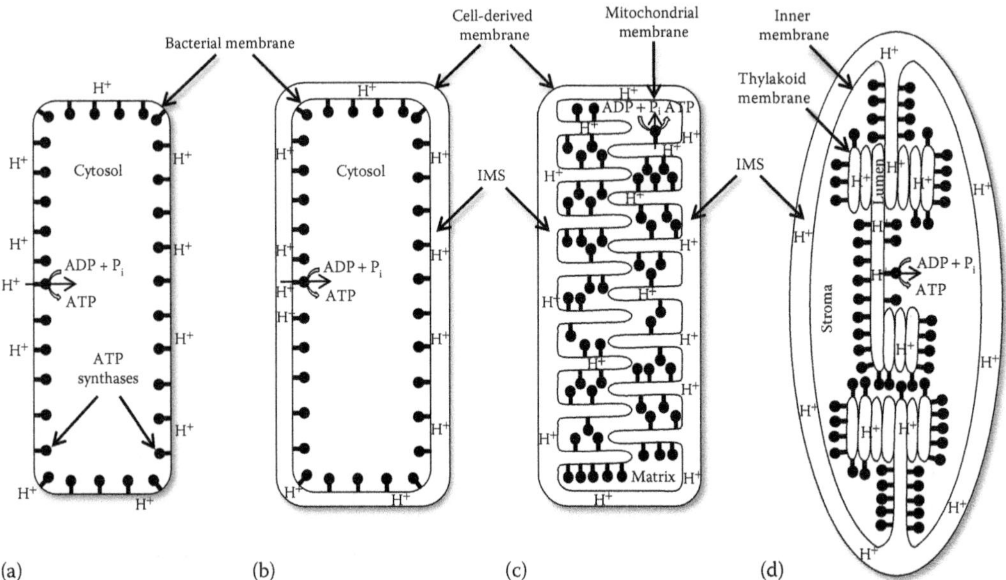

FIGURE 31.5 Conservation of ATP synthase orientation on membranes. (a) In Bacteria and Archaea, ATP synthases are pointed inward with respect to the formation of ATP. These organisms control the concentration of protons outside of the membrane (or inner membrane for Gram-negative bacteria) to increase the external concentration, and thus, a proton gradient forces the protons through the ATP synthase enzyme, which produces ATP from ADP inside the cell. (b) When a pathogen or obligate parasite is within a eukaryotic cell, two membranes are present, and protons are concentrated between the two. Again, the ATP synthase enzymes are pointed in the same direction. (c) In mitochondria, parts of the inner membrane (the original bacterial cell membrane) have invaginated, but the ATP synthases are on the same face of the membrane as with bacteria, and the ATP is formed within the matrix (analogous to the bacterial cytosol) of the organelle. (d) In chloroplasts, the ATP synthases are on the thylokoid membranes, which are formed from invaginations of the inner membranes. Again, the ATP synthases are on the same side of the membrane, and the ATP is formed within the stroma (analogous to the bacterial cytosol) of the organelle. IMS, intermembrane space.

and secondarily genes for cytochrome c oxidase IV, nicotinamide adenine dinucleotide (NADH) dehydrogenase, and adenosine triphosphatase. The smallest mitochondrial genome is in *Plasmodium falciparum*, the protozoan parasite that causes malaria in humans. The genome is only 5967 kb in length containing only three protein-coding genes and fragments of the rRNA genes. Apparently, it is completely dependent on the nuclear genome for all of its functions. Another organism that has had a different evolutionary path is *Cryptosporidium*, a genus of protozoans called Apicomplexans. They are parasites of humans and other animals. Although they have mitochondria that are apparently somewhat functional, they have no mitochondrial DNA. They are always found in close contact with the nuclear membrane, and therefore, transport of RNA as well as proteins from the nucleus and cytosol, respectively, might be occurring. One more variation in this process has occurred with protozoans within the Excavata, Amoebazoa, Opisthokonta, and Chromalveolata. In these organisms, mitochondrial-like organelles are found, which are the remnants of ancestral mitochondria. There are at least two different types, based on their characteristics: mitosomes (that function in the assembly of sulfur proteins but do not produce ATP) and hydrogenosomes (that produce small amounts of ATP by capturing the energy from conversion of protons into hydrogen gas under anaerobic conditions). They appear to have lost all of their mitochondrial DNA and are incapable of performing oxidative phosphorylation. Therefore, a number of different scenarios have occurred to produce mitochondria and related organelles. Their wide phylogenetic distribution indicates that the evolutionary processes that produced these reduced mitochondrial relatives have occurred several times in separate lineages of organisms.

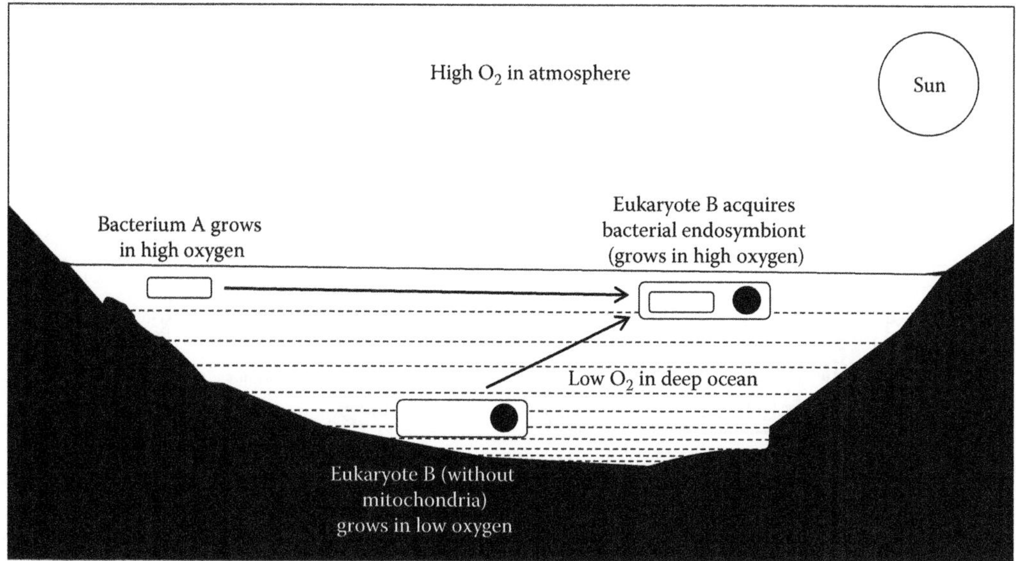

FIGURE 31.6 Proposed origin of the initial eukaryote with mitochondria. Bacterium A adapted to growing in higher oxygen levels. It is thought that the primary initial function of the enzymes was to sequester and detoxify the oxygen, but evolved to use the oxygen as an electron acceptor to form water. Eventually, the bacterium was enveloped by an anaerobic eukaryotic cell, and it became a major supplier of energy in the cell, as well as detoxifying oxygen. This opened up new niches for the eukaryotic cell, which became successful and diversified into all of the extant eukaryotes.

TABLE 31.2
Number and Types of Genes in Mitochondrial Genomes (Including Those for Plants, Fungi, and Animals)

Genes	Number of Genes
rRNA genes	2–5
tRNA genes	20–30
Oxidative phosphorylation	5–29
ATP synthase subunits	2–3

HOW MANY GENES MAKE A FUNCTIONAL MITOCHONDRION?

Mitochondria carry out extremely important functions in most eukaryotic cells. Pyruvate, fatty acids, and amino acids are imported from the host cell, which are fed into the TCA cycle that produces NADH and succinate, which are used in the electron transport chain of oxidative phosphorylation to produce large amounts of ATP. All of this requires a large number of proteins and cofactors, which in turn requires a large number of genes. Figure 31.7 shows an idealized mitochondrion, showing the processes that occur in standard mitochondria, as well as the minimum number of genes required to produce the proteins and cofactors needed for each part of the process. One of the functions of the mitochondria is to produce ATP from the movement of protons across the mitochondrial membrane in the process of oxidative phosphorylation, whereas another function is to produce NADH, protons, and succinate in the TCA cycle, which are needed for oxidative phosphorylation. Fatty acids, pyruvate, and amino acids are imported into the mitochondrion through the mitochondrial outer and inner membranes through specific transporters (Figure 31.7, upper right).

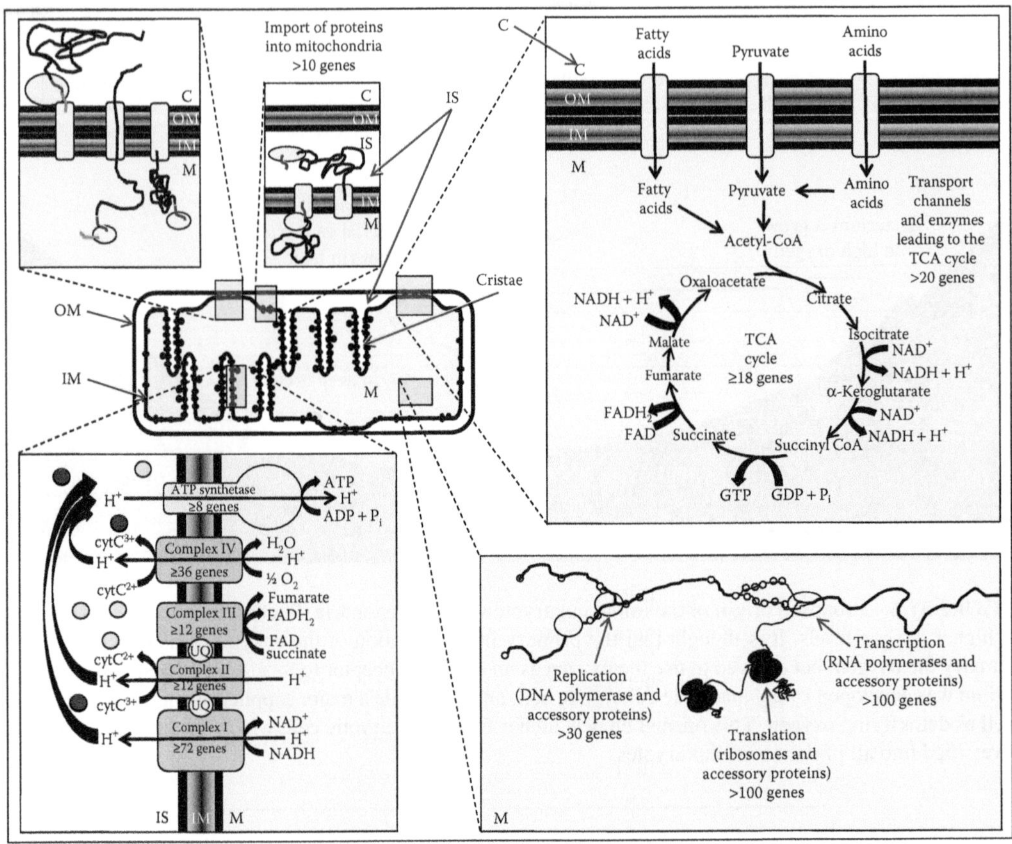

FIGURE 31.7 Processes and minimum numbers of genes needed for a functional mitochondrion. Although mitochondrial genomes contain approximately 30–50 genes, the number of gene products (proteins) needed for a fully functional organelle is minimally 400–500. Because some of the proteins are encoded within the mitochondrial genome, proteins and other factors are needed for DNA replication, RNA transcription, and translation of proteins (lower right) within the organelle. More than 200 genes are required for these functions. Relatively few of the mRNAs needed by mitochondria are encoded in the mitochondrial genome. Most of these genes exist in the nuclear genome. They are transcribed in the nucleus and translated on cytoplasmic (eukaryotic) ribosomes. The proteins must be imported into the mitochondria and end up in the proper compartment within the mitochondria for full function. There are proteins that recognize signal peptides on the ends of the proteins and mark them for transport through channels in the mitochondrial membranes (upper left). They recognize specific signal sequences on the polypeptide (i.e., protein) are their tickets into the mitochondria (indicated in red and green, upper left). The transport channels traverse both membranes at once, thus delivering the proteins into the matrix of the mitochondrion. If the proteins are destined to end up in the intermembrane space, some have yet another signal sequence that is exposed after the first signal is clipped off the end (upper left). This will allow specific movement through the inner membrane into the intermembrane space (upper middle). A minimum of 10 gene products are needed for these processes. The main function of mitochondria is to produce ATP that can be used in many cellular metabolic processes. The mitochondria employ the process of oxidative phosphorylation to produce ATP. The first products that are needed are NADH and oxygen. The first is produced in the TCA cycle (also known as the Krebs cycle, upper right). Acetyl-CoA is the starting product in this cycle, which can originate from fatty acid degradation, pyruvate from glycolysis, or from amino acid degradation (upper right). NADH is formed from NAD, and then the NADH is used by protein complex I on the mitochondrial cristae to shuttle energized electrons to ubiquinone (UQ), as well as move protons through the membrane from the matrix into the intermembrane space (lower left). *(Continued)*

Synthesis of these biomolecules occurs in the cytosol. Pyruvate is one of the products of glycolysis, which is a necessary process in free-living bacteria, but all of the genes for this process are absent from mitochondrial genomes. Therefore, mitochondria rely on the host cell for the products of glycolysis. Pyruvate is converted into acetyl coenzyme A (acetyl-CoA) utilizing at least 13 gene products. Acetyl-CoA is then fed into the TCA or Krebs cycle within the matrix. This consists of a set of enzymes encoded by at least 18 genes (some enzymes consist of more than one protein). The primary result of this is the formation of NADH, as well as some FAD, guanosine-5′-triphosphate, protons, and other compounds. Similar processes also occur in the cytosol of bacteria. Next, the NADH is oxidized by complex I (NADH/coenzyme Q reductase) in the mitochondrial inner membrane (Figure 31.7, lower left). This is the first complex in oxidative phosphorylation. It consists of many different proteins and subunits encoded by at least 72 genes. Energetic electrons are stripped from the NADH and shuttled to additional protein complexes (complex II, succinate/coenzyme Q reductase; complex III, coenzyme Q/cytochrome c oxidase; and complex IV, cytochrome c oxidase). Each has several proteins that are encoded by several genes. (Complex II consists of several proteins encoded by at least 12 genes, complex III by at least 12 genes, and complex IV by at least 36 genes.) In the final complex, oxygen is combined with protons and electrons to form water. In three of the complexes, as a part of the process, protons (H^+ ions) are moved from the matrix side of the inner membrane to the inter membrane space. This acidifies the inner membrane space. This creates a *proton-motive force* (i.e., a high concentration of protons on one side of the membrane), which then flows through ATP synthases that stud the inner membrane, which consist of at least eight proteins (encoded by at least eight genes). The catalytic parts of the ATP synthases are on the matrix side of the membrane, and thus convert adenosine diphosphate (ADP) to ATP, which is then used in the mitochondrion or transported through the membranes into the cytosol.

In addition to these genes, mitochondria require bacterial-like ribosomes and tRNAs for protein synthesis within the organelle (Figure 31.7, lower right). Bacterial ribosomes consist of at least 55 proteins and three different rRNAs (which are structural and enzymatic). Additionally, specific pores, made of proteins embedded in the membranes, are needed, each of which are encoded by one or more genes. Because DNA replication and RNA transcription also occur within mitochondria, genes are needed to make each of the proteins involved in each of these processes. Adding all of these together, the absolute minimum number of genes necessary is between 300 and 400. As stated before, the range of genome sizes for free-living α-proteobacteria is from 4 to 9 Mb. At least 1500 genes are necessary for a free-living organism, and approximately 450–500 genes appear to be necessary for most cellular obligate parasites. Mitochondrial genomes contain only 30–50 genes (Tables 31.1 and 31.2). Of these, 5–29 genes encode proteins involved in oxidative phosphorylation, 2–3 encode ATP synthase subunit genes, 2–5 are rRNA genes, and 20–30 are tRNA genes (Table 31.2). All of the other gene products needed (e.g., those for oxidative phosphorylation, the TCA cycle, RNA transcription, DNA replication, protein synthesis, transport through membranes, and other processes) are encoded by genes in the nuclear genome, most of which have migrated there from the mitochondrial genome over a period of a few billion years.

FIGURE 31.7 (Continued) The electrons are shuttled through complex II, where additional protons are moved through the membrane, and then into complexes III and IV, where more protons are moved through the membrane, cytochrome c is reduced, and the electrons are combined with oxygen and hydrogen to form water. The increase in the concentration of protons within the intermembrane space creates a *proton-motive force* that forces the protons through ATP synthase where ADP and inorganic phosphorus are joined to produce ATP. In addition to the 20 channels needed for movement of fatty acids, pyruvate, amino acids, and other compounds through the membranes, at least 18 gene products are needed for the TCA cycle, and at least 140 gene products are needed for complexes I–IV and ATP synthase. C, cytosol; IM, inner membrane, OM, outer membrane, IMS, intermembrane space; and M, matrix.

Although most of the tRNA genes are on the mitochondrial chromosome, the aminoacyl tRNA synthetases, which add the correct amino acids to the tRNAs, are not. All of these enzymes are encoded in the nuclear genome. Another problem with this process is that all of the proteins encoded by nuclear genes must be translated on eukaryotic cytoplasmic ribosomes, and then find their way through the two mitochondrial membranes (and those that end up in the inner membrane space actually cross the inner membrane a second time; Figure 31.7, upper left). This is accomplished by having signal sequences on the proteins that are specifically recognized by other proteins that take them to the protein-based pores (discussed in Chapter 32). The signal sequence is cut off of the protein as it is threaded through a protein pore in the membrane. The origin of these signal sequences is unknown, but they are specific for each type of membrane and assure that the proteins end up within the proper compartment of the cell. Thus, mitochondria probably have evolved through a process that included mutualism (or parasitism) endosymbiosis and organelle evolution. This included the movement of nearly all of the genes into the nucleus, which became the eventual controller of expression of the mitochondrial genes.

CHLOROPLASTS

The pathway from a parasite or a mutualist to an endosymbiont to an organelle is a continuous one, and many organisms have taken this pathway. Most chloroplasts originated from a single endosymbiotic event between a eukaryotic cell and a cyanobacterium that occurred approximately 1.9 billion years ago. This led to the evolution of glaucophytes, red algae, followed by green algae (about 1.2 billion years ago), and plants (starting about 700 million years ago). Cyanelles represent one of the evolutionary branches in the process (Table 31.1). These are organelles found in glaucophytes (e.g., *Cyanophora*). The cyanelles are very much like chloroplasts but appear to be yet an earlier version of the evolution of this group of organelles. They have more genes than chloroplasts (including at least one ribosomal protein gene), but also have peptidoglycans on the outside of their inner menbranes. Peptidoglycans are cell wall constituents of bacterial cells. Therefore, cyanelles appear to demonstrate one of the steps between an endosymbiont and a chloroplast. Certainly, they are very close to chloroplasts, but they indicate that this has been a step-wise series of events that led to the plastids present in most algae and plants.

Another endosymbiotic event has occurred more recently between a heterotrophic Rhizaria and a cyanobacterium to produce *Paulinella chromatophora*. Because this event has happened more recently, it provides an interesting comparison with respect to the number of genes that are transferred from the organelle to the nucleus over time (Table 31.1). Glaucophytes, red algae, and green algae have chloroplast genomes of 120–250 kb that contain 46–209 genes. Plant chloroplast genomes are all around 120–165 kb, with 80–114 unique genes. The lower diversity in genome size and number of genes for plants probably indicates the relatively recent evolution of higher plants (first appearing about 600–700 million years ago) relative to the algae. By contrast, the chromatophore (analogous to a chloroplast in plants) genome of *P. chromatophora* is 1.02 Mb of DNA and includes 867 genes. Approximately 0.5% of its nuclear genome is composed of cyanobacterial genes, whereas approximately 15% of the nuclear genome of some higher plants is of cyanobacterial origin. Nevertheless, the chromatophore in *P. chromatophora* is completely dependent on the host cell and is on its way to becoming an organelle, or may already be considered an organelle. The genomes of free-living cyanobacteria range from 1.7 to 15 Mb, and therefore, the chromatophore genome size has been reduced in comparison. Also, cyanobacterial genomes contain about 2500 genes. Therefore, the chromatophore genome has lost at least 70% of the genes it likely possessed at the initiation of the endosymbiotic relationship.

The core chloroplast genome for plants consists of genes for the small subunit of ribulose-1,5-bisphosphate carboxylase/oxygenase (RuBisCO, the major enzyme that fixes carbon in photosynthesis), many of the other proteins involved in photosynthesis, subunits for ATP synthase, proteins involved in metabolism, ribosomal proteins, approximately 30 tRNAs, and rRNAs.

For the latter, the rRNA genes are located on short inverted repeats along with several other genes in the chloroplasts of higher plants. For algae, both the large and small subunits of RuBisCO are in the chloroplast genome, as is the case for *P. chromatophora* as well. Additionally, the rRNA genes for both are single copy. No short inverted repeats containing the rRNA genes, or any other genes, exist for either group.

The nuclear genomes of members of the Archaeplastida have incorporated copies of the chloroplast genes such that some contain more than a single copy of each gene. In fact, in *Arabidopsis thaliana* (an angiosperm), there are on average two to three copies of each chloroplast gene in the nuclear genome, although many are not expressed (Figure 31.8). Thus, it appears that one of the initial stages may be that the nucleus incorporates multiple copies of the endosymbiont genome, many of which are not expressed until they integrate into a position that allows transcription by the eukaryotic transcription systems. Next, the messenger RNAs (mRNAs) would have to be processed and exported to the cytosol where they would be translated. If translated into protein, then the protein would have to be imported into the organelle, which requires a signal peptide on the protein. Only after all of this occurred would the corresponding chloroplast gene be redundant, and thus, it would be released from any selective pressure to remain functional in the chloroplast genome. The genes remaining in the chloroplast genome must be those that could not be replaced in this manner, and thus have been retained. It is clear that as long as some protein-coding genes remain in the chloroplasts, DNA replication, transcription, and translation must also be retained. DNA polymerase,

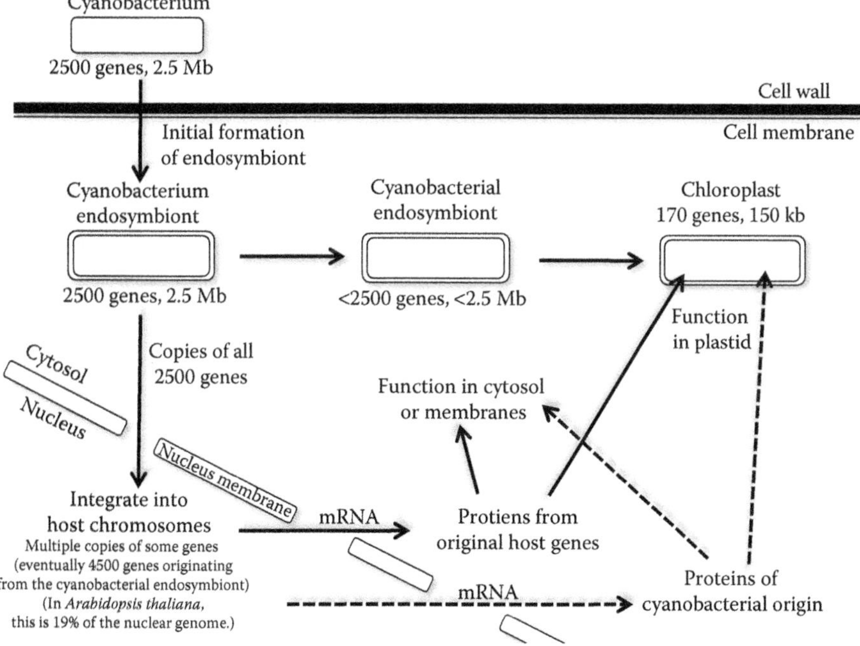

FIGURE 31.8 Movement of plastid genes into the nucleus from the original cyanobacterial cell. The original cyanobacterium probably was very similar to cyanobacteria living today. Genomes are between 2 and 3 Mb, and contain approximately 2500 genes. Once the cyanobacterium entered the host cell (possibly as a parasite or pathogen at first), it began to lose genes, as those functions were supplied by the host. In addition, parts of the genome moved into the nucleus and became integrated into the host chromosomes. Some became functional. Some sequences were duplicated, either by nuclear processes or by repeated movement of pieces of DNA into the nucleus. The cell became a testing ground for new gene combinations, and eventually through selection processes within the cell, successful gene combinations. In some cases, cytoplasmic versions of proteins functioned more successfully that the plastid versions, and vice versa, producing a cell whose genes were blended in an evolutionary sense. Eventually, the plastid genome was reduced to the present levels.

RNA polymerase, rRNA genes, tRNA genes, and ribosomal protein genes have remained in all chloroplast genomes. Some of the genes for metabolism, transport (in and out of the organelle) and photosynthesis also remain in these genomes, but it is unclear why these have never been successfully expressed from the nuclear copies, nor why they have been retained in the chloroplast genome.

Because chloroplasts originated about 1.9 billion years ago, one might expect that the number of genes that have been transferred from the plastids into the nucleus would be less than for mitochondria, which originated about 500 million years earlier. Generally, this is true, but further explanation is needed. Actually, all of the genes have been transferred to the nucleus. This was first reported in pea (*Pisum sativum*), and was confirmed and detailed in genomic studies of *A. thaliana*. In *A. thaliana,* approximately 4500 genes from the original cyanobaterial genome (equivalent to about two cyanobacterial genomes) have moved into the nucleus (Figure 31.8). This means that there are several copies of genes of cyanobacterial origin within the nuclear genome of *A. thaliana*. In fact, 19% of the nuclear genome of *A. thaliana* consists of genes that have a cyanobacterial origin. Currently, most of the cyanobacterial genes are transcribed in the nucleus, the resulting RNA is translated in the cytosol, and the proteins are transported into the chloroplasts. However, interestingly, some other genes are expressed, and the resultant proteins function in the cytosol or in other cellular compartments. Therefore, once the genes are transferred into the cell, mutation, random drift, and selection act to produce cellular pathways and processes that allow the organism (and genome) to survive, regardless of the sources of the genes.

The process of movement of functional genes to the nucleus is an ongoing process. This is indicated by the fact that both the small subunit and large subunit genes for RuBisCO are on the plastid genomes of red algae, but in plants only the large subunit gene is found in the plastid genome. The small subunit genes are in the nucleus. Because the first simple plants appeared about 700 million years ago, the transfer of the small subunit gene into the nuclei of plants probably occurred about the same time, because all plants have a similar arrangement of the two genes.

HOW MANY GENES MAKE A FUNCTIONAL CHLOROPLAST?

Chloroplasts have distinct compartments and membranes that separate processes (Figure 31.9). The inner portion of the chloroplast is called the stroma. This is analogous to the cytosol of a bacterial cell and is where all of the so-called dark reactions take place in this compartment. The primary reactions are within the Calvin (reductive pentose phosphate) cycle, where the carbon from carbon dioxide is covalently attached (i.e., fixed) to sugars (Figure 31.9, upper right). These reactions take place independent of light, which is the reason for their designation as the *dark reactions*. More than a dozen enzymes are required for this process, and several are encoded by more than one gene. The proteins responsible for the *light reactions* are embedded in the thylakoid membranes (Figure 31.9, lower left). These membranes begin as invaginations of the inner plastid membrane but become specialized because of high concentrations of photosystem proteins and ATP synthase. This also creates a third compartment in the chloroplasts. The inner portion of the thylokoid is called the lumen, which becomes acidified when the plastids are exposed to light.

When light is present, the light-harvesting complex energizes sets of electrons and shuttles them into photosystem II. The components of the light-harvesting complex are encoded by at least six genes. The components of photosystem II are encoded by at least 30 different genes. Photosystem II is responsible for splitting water to release oxygen and protons into the lumen. It also produces an electron that enters photosystem II and is used to energize a protein called plastoquinone. The electron is transferred next to heme molecules within a complex called cytochrome b/f, consisting of about 10 proteins. In addition to moving electrons from photosystem II to photosystem I, this complex moves protons from the stoma into the lumen, thus acidifying the lumen. The electrons are then transferred to photosystem I, where they are further energized by photosynthetic pigments. As with photosystem II, it is surrounded by light-harvesting complexes, and its components are encoded by

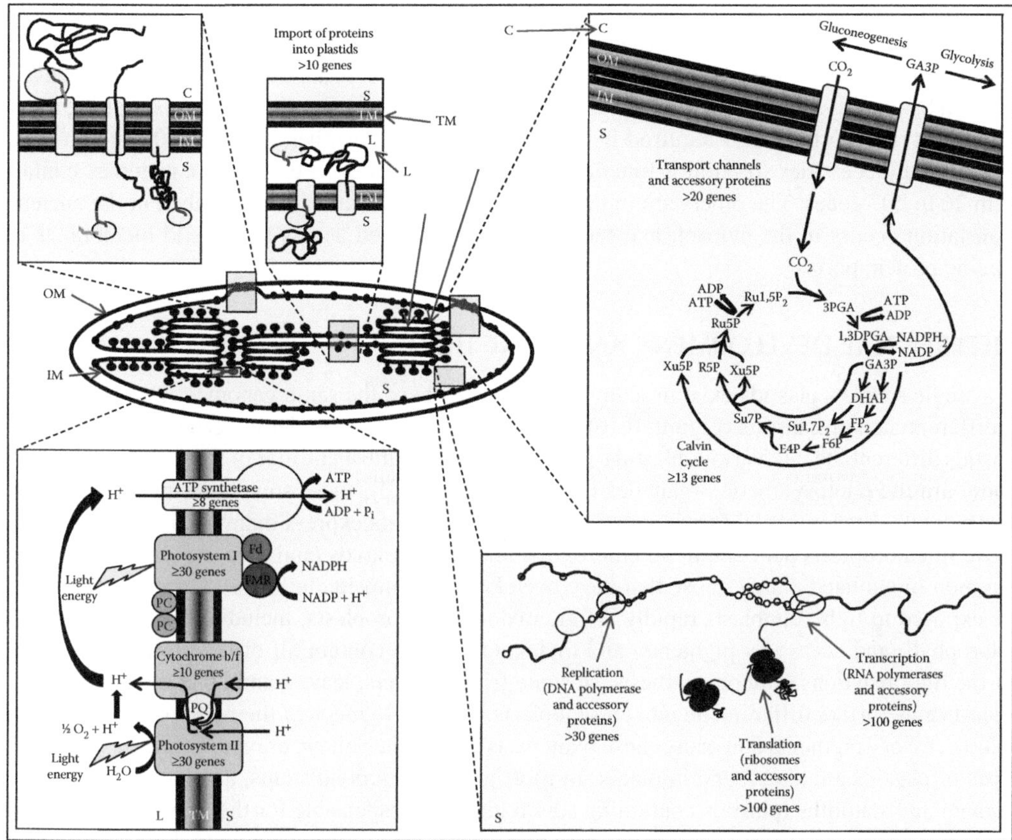

FIGURE 31.9 Processes and minimum numbers of genes needed for a functional chloroplast. Although plastid genomes contain approximately 50–200 genes, the number of gene products (proteins) needed for a fully functional organelle is probably 500–1000. As with mitochondria, some of the proteins are encoded within the plastid genome, whereas most genes exist within the nuclear genome. Genes are needed to encode proteins involved in the light (lower left) and dark (upper right) photosynthetic reactions. More than 100 genes are needed. Also, genes are needed for DNA replication, RNA transcription, and translation of proteins (lower right) within the plastid. More than 200 genes are required for these functions. Most are transcribed in the nucleus and translated on cytoplasmic (eukaryotic) ribosomes. The proteins must be imported into the chloroplast and end up in the proper compartment within the plastid for full function. There are proteins that recognize these proteins and usher them to the transport channels on the plastid membranes (upper left). They recognize specific signal sequences on the polypeptide (i.e., protein) that are their tickets into the organelle (upper left). The transport channels traverse both membranes at once, thus delivering the proteins into the stroma of the plastid. If the protein is destined to end up in the thylakoid (site of the light reactions of photosynthesis), they must have yet another signal sequence that is exposed after the first signal is clipped off the end (upper left). This will allow specific movement through the thylakoid membrane into the lumen (upper middle). As with mitochondria, a similar system is also used to deliver specific proteins into space between the inner and outer membranes. A minimum of 10 gene products are needed for these processes. The main function of chloroplasts is to fix carbon (from carbon dioxide), that is, to add carbon to form sugars. To do this, light energy is used to produce ATP and NADPH that are both needed as energy and phosphorus sources (lower left) for the Calvin cycle (upper right), where the carbon is fixed. One of the major products is GA3P, which is transported out of the chloroplast into the cytosol, where it can be used in glycolosis for ATP production, or into gluconeogenesis to produce glucose and starch for storage (upper right). In the light reactions, energy is used to split water into oxygen (which is liberated), protons, and electrons (lower left). Also, protons are shuttled across the thylakoid membrane into the lumen of the thylakoid. Ultimately, the electrons and some of the protons are added to NADP to produce NADPH, whereas the protons in the lumen are pushed through the ATP synthase to produce ATP. *(Continued)*

more than two dozen genes. The high concentration of protons in the lumen creates a proton-motive force that causes protons to move through the ATP synthases that synthesize ATP from ADP in the stroma. The ATP is then used in the dark reactions in the stroma. In addition, NADPH is formed and released by photosystem I, which is also used in the dark reactions. Replication, transcription, translation, and transport are also required in chloroplasts. Added together, at least 400 gene products are absolute necessities to build a functional chloroplast. However, chloroplast genomes contain from 26 to 207 genes. The others are in the nuclei of these cells and are transcribed in the nucleus. Translation occurs in the cytosol, and the proteins are shuttled across the plastid membranes by specific protein ports.

DIFFERENTIAL DEVELOPMENT AND FUNCTION

In a single species, plastids exist in many forms, but all have the same genome. They all begin as undifferentiated proplastids (Figure 31.10). In different tissues and under differing conditions, proplastids differentiate into various plastid types. Of course, the most familiar of these are chloroplasts. These are the photosynthetic organelles in all plants. Actually, these sometimes proceed through intermediate developmental forms, called etioplasts. Etioplasts express many of the genes that are active in chloroplasts but contain no chlorophyll and no thylakoids (and thus are white). These are common in etiolated plants (those that have been kept in complete darkness). Once the plant cells are exposed to light, etioplasts rapidly differentiate into chloroplasts, including the appearance of chlorophyll (and accessory pigments) and thylakoids, which contain all of the proteins necessary for the light reactions of photosynthesis. In some fruits, flowers, leaves, and other structures, chloroplasts may further differentiate into chromoplasts. These lose much of their chlorophyll but retain their accessory pigments that cause these plant parts to become yellow, orange, or red. Several other types of plastids are also found in plants. In roots and underground stems, amyloplasts (for starch storage) and statoliths (plastids containing starch granules responsible for the gravitropic response in plant roots, which cause the roots to grow downward) are common. Chromoplasts, leucoplasts, elaioplasts (oil storage), and proteinoplasts (protein storage) are found more often in aboveground portions of the plants. Elaioplasts and proteinoplasts store fats and proteins, respectively, in seeds, including nuts. Again, for a given plant, all of these organelles contain exactly the same plastid genome, even though their morphologies and functions are very different.

CHIMERIC PATHWAYS

Evolution has favored the movement and expression of genes to be primarily housed in the nucleus. This has been a complex process because the genes have had to move into the nucleus and gain sequences necessary for transcription in the nucleus. Also, the messages have had to be exported from the nucleus, recognized and translated on cytoplasmic ribosomes, and shuttled through membranes to reach the proper cell compartment in order to function correctly. When researchers began carefully studying the genes and proteins that were used in the Calvin cycle (the dark reactions of photosynthesis, in which carbon is fixed), they were a bit surprised by the fact that the cycle itself was chimeric (Figure 31.11), that is, the genes and proteins involved in the Calvin cycle in plants were from several different ancestral sources. Many of the genes were similar to those in the

FIGURE 31.9 (Continued) More than 100 gene products are needed for the light reactions, whereas at least 13 gene products are needed for carbon fixation in the Calvin cycle. $Ru1,5P_2$, ribulose-1,5-bisphosphate; 3PGA, 3-phosphoglycerate; 1,3DPGA, 1,3-diphosphoglycerate; GA3P, glyceraldehyde-3-phosphate; DHAP, dihydroacetone phosphate; $F1,6P_2$, fructose-1,6-diphosphate; F6P, fructose-6-phosphate; E4P, erythrose-4-phosphate; Xu5P, xylulose 5-phosphate; $Su1,7P_2$, sedoheptulose 1,7-bisphosphate; Su7P, sedoheptulose 7-phosphate; R5P, ribose-5-phosphate; Ru5P, ribulose-5-phosphate; C, cytosol; IM, inner membrane, OM, outer membrane, S, stroma; L, lumen; and TM, thylakoid membrane.

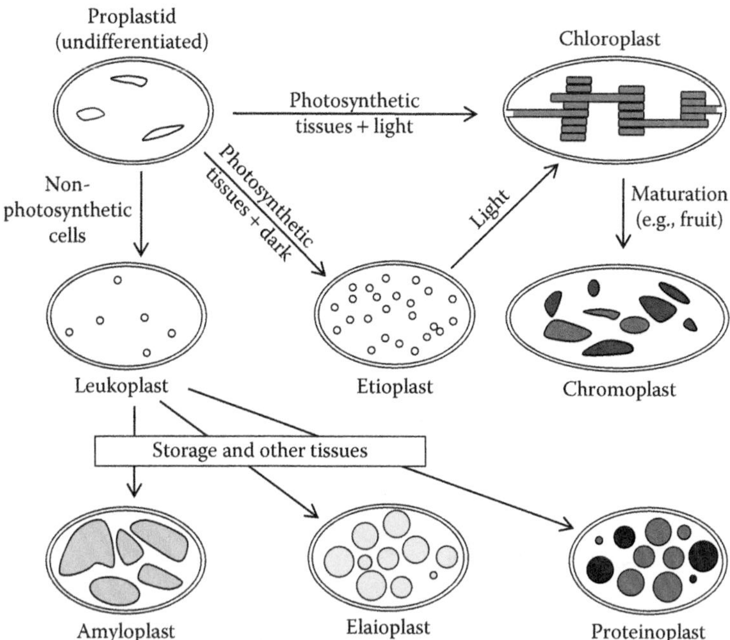

FIGURE 31.10 Differential development of plastids in plants. Plastids in plant embryos begin as undifferentiated proplastids. If germinated in the dark, the shoot will be completely white, because the plastids differentiate into etioplasts, which are partially mature chloroplasts. If the embryos are germinated in the light, or the etiolated plants are moved into the light, the plastids in all of the photosynthetic tissues will mature into chloroplasts. In some fruits (e.g., tomatoes), the chloroplasts undergo further development to become chromoplasts, in which the chlorophyll and associated proteins break down, and other pigments that are red and/or yellow remain. In nonphotosynthetic tissues (e.g., roots, tubers, bulbs, and some stems), the proplastids differentiate into leucoplasts, which are nonphotosynthetic, but perform other functions in the cells. During maturation of various tissues (e.g., roots, tubers, endosperm, cotyledons, bulbs, and other tissues), the leucoplasts differentiat further into amyloplasts (including those that form statoliths in root cells, which are important for gravitropism—roots growing downward), elaioplasts (which produce and store fats, such as in seeds), and proteinoplasts (which produce storage proteins, such as those found in seeds).

cyanobacterial ancestor. However, some appeared to have mitochondrial origins, and still others were of unknown origins (but probably were from the ancestral host cell, which had archaeal and bacterial origins). Furthermore, one of the main enzymes used in glycolysis in the cytosol, phosphoglucokinase (PGK), was of cyanobacterial origin. Essentially, it had replaced the original PGK of the host cell. The understanding that emerges from this information is that genes and proteins have been passed back and forth from organelle to organelle, and from organelle to nucleus, during the evolutionary processes. When several versions of genes (i.e, isoforms) are present in the same cell compartment, evolutionary processes act to favor some new combinations of enzymes within pathways over some of the original combinations, eventually leading to chimeric pathways. This complicates attempts to reconstruct phylogenies of organisms and genes, because more than one version of each gene resides within each cell during the initial phases of this process. When the genes are sequenced and compared by researchers, it is often difficult to discern simple sequence divergences from horizontal gene transfers, followed by selection of one version over another, coupled with additional sequence changes.

Other studies have shown that genes have moved from mitochondria into the nucleus, from the plastids to the nucleus, and from plastids to mitochondria, and vice versa (see Figure 13.10). It is less clear whether any sequences have moved from the nucleus into those organelles. It appears that growth of the nuclear genome has been tolerated in most organisms, whereas shrinkage of

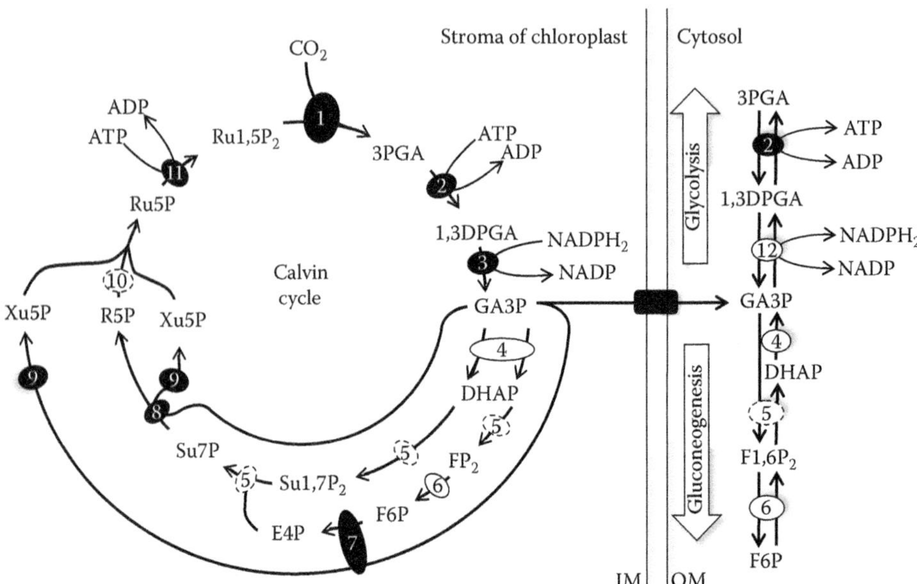

FIGURE 31.11 Origins of the enzymes used in the Calvin cycle and glycolysis in spinach (*Spinacia oleracea*). Enzymes shown as elipses: Black indicates a cyanobacterial origin, white with solid lines indicates a mitochondrial (α-proteobacterial) origin, and white with dashed line indicates an unknown origin (although most are probably from the genome of the original host cell). As expected, most of the genes for the Calvin cycle originated in cyanobacteria. However, a few of the enzymes (e.g., triosephosphate isomerase, enzyme number 4; fructose bisphosphatase, enzyme number 6) originated in the mitochondria, as well as a few (enzymes 5 and 10) with unknown origins (likely to be from the nuclear genome). Glycolysis in the cytosol also exhibits a complex evolutionary pattern, where some enzymes that originated in the mitochondrial lineage (enzymes 4, 6, and 12), some from the nuclear lineage, and at least one (enzyme number 2) from the cyanobacterial lineage have produced the glycolysis pathway in higher plants. This indicates that enzymes from different cell compartments and origins have undergone natural selection within the cell to yield chimeric pathways that include enzymes of mixed origins. Enzyme designations: 1, ribulose bisphosphate carboxylase/oxygenase; 2, phosphoglycerate kinase; 3, glyceraldehyde phosphate dehydrogenase (cyanobacterial origin); 4, triosphosphate isomerase; 5, fructose bisphosphate aldolase; 6, fructose bisphosphatase; 7, transketolase; 8, sedoheptulose bisphosphatase; 9, ribose phosphate epimerase; 10, ribose phosphate isomerase; 11, ribulose phosphokinase; 12, glyceraldehyde phosphate dehydrogenase (mitochondrial origin). Abbreviations for compounds are provided in the caption of Figure 31.9.

the organelle genomes has been the rule. Other organelles (peroxisomes, endoplasmic reticula, Golgi, etc.) may have been gained in eukaryotes in the same way, but normally they have only single membranes, indicating that their origins may be different from nuclei, plastids, and mitochondria, because they all have double membranes. Also, they lack DNA, indicating that they may not have evolved in ways analogous to mitochondria and plastids. Alternatively, their evolutionary pathways may have been different over longer periods of time. The cell, then, can be considered as an evolutionarily diverse environment, where cooperation, competition, genetic drift, and selection are constantly occurring.

ENDOSYMBIOSES LEADING TO OTHER ORGANELLES

Although the most well-studied DNA-containing organelles are mitochondria and chloroplasts, there are many others. Among the single-celled eukaryotes, there have been multiple endosymbiotic events that have led to the formation of organelles. Each has had a bearing on the genomes of each of the cells (and organelles) involved. For example, *Euglena* and trypanosomes are related and were produced when their common ancestor engulfed a green algal cell, which became an endosymbiont and then became an organelle. The combination of the genomes has resulted in a mixed genome for *Euglena*. The nuclear genome consists of genes from a heterotrophic ancestor as well as a photoautorophic ancestor (Figures 31.2 and 31.3). Originally, this led to some confusion about its position on phylogenetic trees, and the mixture still causes some problems, but it is now placed in a relatively basal position among eukaryotes on the universal tree of life (see Figures 2.4 and 2.5). Trypanosomes are also basal on the universal tree and were never thought to have contained chloroplasts. However, upon examination of their genomes, it was found that their nuclear genomes contained genes related to green algal chloroplast genomes. Other nuclear genes indicated a heterotrophic ancestor. This meant that their ancestors were an amalgamation of a heterotrophic eukaryote and an autotrophic eukaryote. However, in this case, the autotrophic lifestyle had been abandoned in favor of a parasitic lifestyle. After the parasitic lifestyle was adopted, the chloroplasts were not used. Mutation, genetic drift, and selection led to their elimination, but parts of the chloroplast genome are still present in the nuclear genomes of these parasites.

Other events have also led to the formation of additional organelles in the two major groups of single-celled eukaryotes. Members of Chlorarachniophyta (within Rhizaria) are good illustrations of this (Figure 31.2). Similar to trypanosomes and euglenoids, these organisms were formed from an ancient endosymbiotic event between a heterotrophic host organism and a green algal cell. Although a permanent resident of the host cell, the eukaryotic green algal descendant (organelle) is surrounded by two membranes, has a chloroplast surrounded by two membranes, and has remnants of a nucleus (called a nucleomorph). The mitochondrion is absent from the organelle. However, the main nucleus contains genes from the mitochondrion, chloroplast, and nucleus from the original algal genome, whereas the nucleomorph has lost the majority of its genome. In each of these associations, genes have moved preferentially from each of the cellular compartments to the nucleus. However, in some cases, they have taken indirect routes. For example, some genes from the chloroplast of the endosymbiont have moved first into the endosymbiont nucleus (nucleomorph), and then into the host cell nucleus. Others have moved directly to the host nucleus (see Figure 13.10). Members of the Chromalveolata have taken endosymbiosis to another level. There have been secondary, tertiary, and possibly tertiary endosymbiotic events to form unique organelles and genomes in this group (Figures 31.2 and 31.3). Some of the organisms may have more genes in their nucleus from the endosymbionts than they have of their original ancestral genomes. Ciliates are within the Chromalveolata. They are nonphotosynthetic, have reduced mitochondria, and have normally two nuclei (a micronucleus and a maxinucleus). The micronucleus contains the genome that is passed on to daughter cells, whereas the macronucleus is polyploid, containing multiple copies of several portions of the genome. These are the expressed genes. One ciliate, *Tetrahymena thermophila* has more than 27,000 genes in its nuclear genome (about 4000 more than humans have in their genome). Another, *Paramecium tetraurelia,* has approximately 30,000 genes in its nuclear genome. Ciliates were produced either by an endosymbiotic event between a heterotrophic eukaryote and a red algal cell, or possibly by two heterotrophs (at least one of which contained some chloroplast genes). They are primarily symbiotic with other eukaryotes, but a few are parasites. Cryptomonas, haptophytes, and heterokonts (including diatoms, brown algae, golden algae, and yellow-green algae, which were all once simply classified as algae) were produced by one or more endosymbiotic events between a heterotrophic host and a red algal cell (Figures 31.2 and 31.3). Haptophytes have two chloroplasts per cell, each surrounded by two membranes, whereas the two chloroplasts in heterokont cells are surrounded by three membranes. Cryptomonas exhibit a more complex cell structure that is more indicative of the endosymbiotic event. The organelle is surrounded by two membranes, inside of which is a nucleomorph (with two membranes) and a chloroplast surrounded by two membranes. Genes from each of these cellular compartments exist within the cell nucleus, again indicating the direction of transfer of genes in these associations. Finally, the dinoflagellates have undergone a number of endosymbiotic events that have led to the formation of organelles and the movement of genes from one genome to another. In some dinoflagellates, the host cell itself was a dinoflagellate that enveloped a haptophype. In other cases, the host cell was either a cryptomonad or a heterokont that enveloped a dinoflagellate. One final possible event appears to have occurred where a dinoflagellate that already contained red algal genes from a previous endosymbiotic event, engaged in another endosymbiotic event with a different red alga to form *Lepidodinium*, a dinoflagellate with red algal chloroplast genes from two sources. Again, the resulting genomes mix, especially in the main nucleus, the nuclear genomes become larger, and the genomes in the other compartments decrease in size.

KEY POINTS

1. Parasitic and endosymbiotic events have led to the evolution of organelles, such as mitochondria and plastids.
2. As an endosymbiotic association develops, the invading cell loses parts of its genome. Some (or all) of these genes move into the nucleus.
3. DNAs have continually moved from organelles into the nucleus, as well as from organelle to organelle.
4. Biochemical pathways in eukaryotic cells are chimeric, in that they may include genes from different origins, such as the nucleus, mitochondria, and plastids.
5. Many endosymbiotic events have led to a diversity of cell and organelle types.

ADDITIONAL READINGS

Alberts, B., D. Bray, K. Hopkin, A. Johnson, J. Lewis, M. Raff, K. Roberts, and P. Walter. 2013. *Essential Cell Biology,* 4th ed. New York: Garland Publishing.

Alberts, B., D. Bray, A. Johnson, J. Lewis, M. Raff, K. Roberts, and J. D. Watson. 1994. *Molecular Biology of the Cell.* New York: Garland Publishing.

Becker, W. M., J. B. Reece, and M. F. Poenie. 1996. *The World of the Cell*, 3rd ed. New York: The Benjamin/Cummings Publishing Company.

Bendich, A. J. 2010. Mitochondrial DNA, chloroplast DNA and the origins of development in eukaryotic system. *Biol. Direct* 5:42.

Bidlack, J. and S. Jansky. 2010. *Stern's Introductory Plant Biology*, 12th ed. Columbus, OH: McGraw-Hill Education.

Darnell, J., H. Lodish, and D. Baltimore. 2007. *Molecular Cell Biology*, 6th ed. New York: W. H. Freeman and Company.

Gregory, T. R., ed. 2005. *The Evolution of the Genome.* London: Elsevier Academic Press.

Hackett, J. D., D. M. Anderson, D. L. Erdner, and D. Bhattacharya. 2004. Dinoflagellates: A remarkable evolutionary experiment. *Am. J. Bot.* 91:1523–1534.

Lewin, B. 2008. *Genes IX.* Sudbury, MA: Jones & Bartlett Publishers.

Martin, W. and C. Schnarrenberger. 1997. The evolution of the Calvin cycle from prokaryotic to eukaryotic chromosomes: A case study of functional redundancy in ancient pathways through endosymbiosis. *Curr. Genet.* 32:1–18.

Reece, J. B., L. A. Urry, M. L. Cain, S. A. Wasserman, P. V. Minorsky, and R. B. Jackson. 2010. *Campbell Biology*, 9th ed. New York: The Benjamin/Cummings Publishing Company.

Tropp, B. E. 2008. *Molecular Biology, Genes to Proteins*, 3rd ed. Sudbury, MA: Jones & Bartlett Publishers.

Venter, J. C. 2007. JCVI comprehensive microbial resource. http://cmr.jcvi.org/cgi-bin/CMR/shared/Genomes.cgi.

Wallace, R. A., J. L. King, and G. P. Sanders. 1981. *Biology, the Science of Life.* Santa Monica, CA: Goodyear Publishing Company.

32 Protein Trafficking

INTRODUCTION

Cell membranes and internal membranes protect cells and organelles by excluding components that will damage the enclosed cell or compartment, while including the components that are vital to the function of the cell or compartment. Small molecules, such as water and ions, move through ports specific for each, with some being controlled by conformational changes and/or adenosine triphosphate (ATP). Proteins cannot normally traverse membranes. Additionally, most proteins that are embedded in membranes cannot embed themselves in the correct membrane, or in the correct conformation. If these processes were left simply up to diffusion, the process would be inefficient and ineffective, and the cells would cease to exist. This creates a dilemma for transfer of proteins from one cell compartment to another, or for embedment into a membrane. Cellular systems have evolved to shuttle each protein to the proper location in the cell and to aid them in their interactions with membranes. However, most proteins need signal sequences that tag them to be positioned within the cell and traverse a membrane, or to become embedded in one. For example, proteins that are to end up on the outside surface of the cell membrane, or are to be excreted from the cell, have specific signal sequences on the amino ends, called signal peptides (Table 32.1). These are recognized by signal recognition particles (SRPs), which shuttle them to specific cell membranes (Figure 32.1). Other proteins then deliver them to the correct cell compartment or membrane for the proper functioning of that portion of the cell, usually by transporting them through specific protein channels that are embedded in the membrane (Figure 32.2). This is protein trafficking.

The first indication of protein trafficking was in molecular biological research of the proteins that are encoded in nuclei, but that are transported into mitochondria and chloroplasts. It was discovered that when these proteins were isolated from the organelles, often they were shorter than the same proteins isolated from the cytosol, or identified from their nuclear genes. Short pieces of their amino ends were missing. Additionally, many of the mitochondrial proteins that were coming from the cytosol had identical amino acid sequences on their amino ends prior to being transported into the organelle. The situation was similar for chloroplast proteins that were encoded in the nuclei. Eventually, it was concluded that the peptides on the amino ends, called signal peptides (or target peptides), were necessary for transport of the proteins into the organelles. Subsequently, it was discovered that some organellar proteins had two different signal peptides on their amino ends. A protein with two such signal peptides was found to traverse at least two membranes, whereas proteins with only one signal peptide usually traversed only one membrane. The exception to this is in cases where two membranes come close to (or touch) one another, and the transport channel traverses both membranes. In these cases, only one signal peptide is required to traverse both membranes in a single process.

SIGNAL PEPTIDES IN BACTERIA

Cells are ordered and organized into compartments surrounded by membranes, and all cells have a cell membrane that encompasses the entire cell. In order for proteins to reach the proper compartment, many proteins must traverse a membrane. Signal peptides that are found on many proteins are recognized by SRPs that shuttle the proteins with signal peptides to the correct membrane. Prior to extensive sequencing studies, signal peptides were thought to have first evolved in

TABLE 32.1
Examples of Signal Peptides

Kingdom	Destination	Amino Acid Sequences[a]
Bacteria	Inner membrane	Met–Lys–Lys–Ser–Leu–Val–Leu–Lys–Ala–Ser–Val–Ala–Val–Ala–Thr–Leu–Val–Pro
		Met–Lys–Lys–Leu–Leu–Phe–Ala–Ile–Pro–Leu–Val–Val–Pro–Phe
	Periplasm	Met–Lys–Gln–Ser–Thr–Ile–Ala–Leu–Ala–Leu–Leu–Pro–Leu–Leu–Phe
		Met–Lys–Ala–Asn–Als–Lys–Thr–Ile–Ile–Ala–Gly–Met–Ile–Ala–Leu–Ala–Ile
		Met–Ser–Ile–Gln–His–Phe–Arg–Val–Ala–Leu–Ile–Pro–Phe–Phe–Ala–Ala–Phe
	Outer membrane	Met–Lys–Ala–Thr–Lys–Ley–Val–Leu–Gly–Ala–Val–Ile–Leu–Gly
		Leu–Arg–Lys–Leu–Pro–Leu–Ala–Val–Ala–Val–Ala–Ala–Gly–Val
		Met–Met–Ile–Thr–Met–Lys–Lys–Thr–Ala–Ile–Ala–Ile–Ala–Val–Ala–Leu–Ala–Gly–Phe–Ala
Archaea	Plasma membrane	Met–Ala–Met–Ser–Leu–Lys–Lys–Ile–Gly–Ala–Ile–Ala–Val–Gly–Gly–Ala–Met–Val–Ala–Thr–Ala–Leu–Ala–Ser–Gly–Val–Ala
		Met–Lys–Val–Phe–Glu–Phe–Leu–Gly–Lys–Arg–Gly–Ala–Met–Gly–Ile–Gly–Thr–Leu–Ile–Ile–Phe–Ile–Ala–Met–Val–Leu–Ala–Ala–Val–Ala
		Met–Asn–Thr–Tyr–Leu–Ser–Thr–Leu–Leu–Val–Leu–Thr–Thr–Ile–Phe–Ala–Leu–Ser–Ile–Ile–Ala–Tyr
Eukarya	Endoplasmic reticulum	Met–Met–Ser–Phe–Val–Ser–Leu–Leu–Val–Gly–Ile–Leu–Phe–Trp–Ala–Thr–Glu–Ala–Glu–Gln–Leu–Thr–Lys–Cys–Glu–Val–Phe–Gln
	Retain in endoplasmic reticulum	Lys–Asp–Glu–Leu–COOH
	Mitochondrial matrix	Met–Leu–Ser–Leu–Arg–Gln–Set–Ile–Arg–Phe–Phe–Lys–Pro–Ala–Thr–Leu–Cyc–Ser–Ser–Arg–Tyr–Leu–Leu
		Met–Leu–Ser–Ala–Arg–Ser–Ala–Ile–Lys–Arg–Pro–Ile–Val–Arg–Gly–Leu
		Val–Leu–Pro–Arg–Leu–Tyr–Thr–Ala–Thr–Ser–Arg–Ala–Ala
	Nucleus	Pro–Pro–Lys–Lys–Lys–Arg–Lys–Val
	Peroxisome	Ser–Lys–Leu–COOH
		Arg–Leu–X–X–X–X–X–His–Leu
	Cell surface	Met–Leu–Ser–Phe–Val–Asp–Thr–Arg–Thr–Leu–Leu–Leu–Leu–Ala–Vale–Thr–Ser–Cys–Leu–Ala–Thr–Cys–Gln–Ser
	Chloroplast	Met–Val–Ala–Met–Ala–Met–Ala–Ser–Leu–Gln–Ser–Ser–Met–Ser–Ser–Leu–Ser–Leu–Ser–Ser–Asn–Ser–Phe–Leu–Gly–Gln–Pro–Leu–Ser–Pro–Ile–Thr–Leu–Ser–Pro–Phe–Leu–Gln–Gly

[a] Sequences with COOH at the end are located on the carboxyl end of the protein. All others are located on the amino end of the proteins. "X" indicates variable amino acid.

eukaryotes, which have numerous cell compartments separated by cell membranes (Figure 32.3). However, it is now known that bacteria have extensive arrays of signal peptides, and therefore, they probably first evolved much earlier than previously thought. More than 175,000 signal peptides have been reported in bacteria (Table 32.1). There are at least 31 types of signal peptides in *Escherichia coli*, and approximately 10% of its 4,200 genes (i.e., more than 420 of them) encode signal peptides on the amino ends of their proteins. For monoderm (i.e., mostly Gram positive) Bacteria, there is only one membrane, the cell membrane. However, even for these bacteria, the proteins that are either traverse the cell membrane or embedded within it have signal peptides that assure the proteins will efficiently and accurately reach their final destination in the cell. Many of the proteins in diderm (i.e., mostly Gram negative) Bacteria must traverse at least one membrane (the inner cell membrane) in order to enter the periplasm or become embedded within the

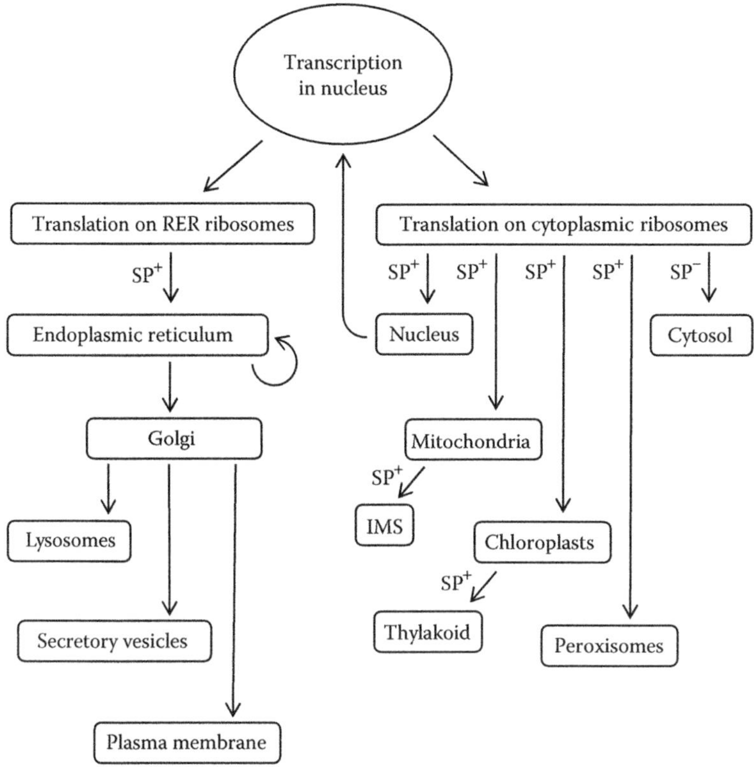

FIGURE 32.1 General pathways of protein trafficking in eukaryotes. As mRNAs are leaving the nucleus, they are immediately joined to ribosomes, and translation commences. As the polypeptide begins to emerge from the pore of the large ribosomal subunit, a signal peptide (SP^+) that is recognized by an SRP (also called chaperone) destined for the endoplasmic reticulum emerges. The SRP delivers the ribosome with partially completed protein to the surface of the RER. There, the polypeptide is threaded through a transport pore into the lumen of the RER as translation continues. Eventually, the completed protein is transferred to secretory vesicles, to the plasma membrane, or to lysosomes. If other signal peptides are on the protein, translation continues in the cytosol. Once the protein is completely synthesized, or nearly so, the SRPs (chaperone proteins) attach to the peptide and deliver the protein to the nucleus, mitochondria, chloroplasts (or plastids), or peroxisomes, depending on the signal peptide. Proteins lacking signal peptides (SP^-) remain in the cytosol.

inner membrane. Some must traverse both the inner and outer membranes. Each requires specific structures, called translocon protein complexes, which carry out the translocation (i.e., transport) of the proteins from one side of the membrane to the other, threading them through, while aiding in folding of the proteins on the other side (Figure 32.2). Enzymes cleave off the signal peptide once the protein has been translocated. However, for some transmembrane proteins, the signal peptide is never cleaved from the end. The proteins that comprise the translocon complex often have long hydrophobic regions that span the membrane, as well as a few charged amino acids on both ends, which assure proper entry and exit of the proteins being translocated. Although monoderm bacteria have only a single membrane, they also have a relatively large array of signal peptides, primarily for proteins that are exported out of the cell or that are embedded in the cell membrane. Archaea also have sets of signal peptides, some of which are similar to those found in bacteria, whereas others are more similar to those found in eukarya, indicating their divergence from bacterial ancestors, but also indicating their contribution to eukaryotic genomes. Although fewer archaeal genomes than bacterial genomes have been completed, more than 10,000 genes

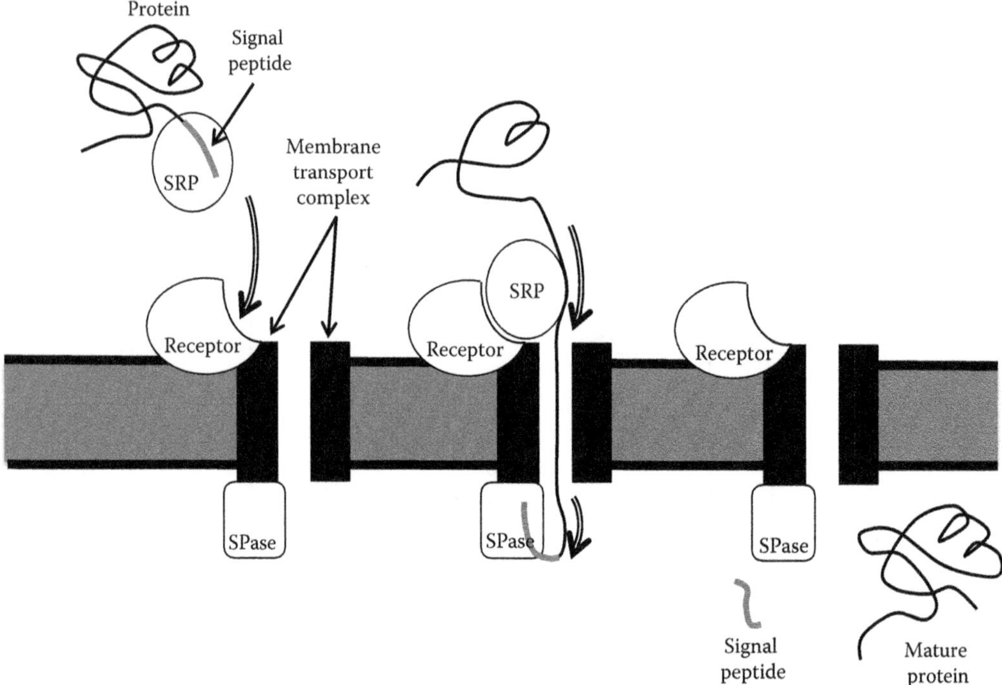

FIGURE 32.2 General mechanism of transport of a protein to the appropriate membrane and/or cell compartment. The signal peptides attach to the SRPs, which shuttle the protein to the appropriate membrane, where it docks with a receptor protein. The protein is then released to the membrane transport complex. The protein is threaded through the complex, starting with the amino end. Once on the other side of the membrane, an enzyme (signal peptidase [SPase]) cleaves the signal peptide from the end of the protein, and the protein is folded into its final conformation (often aided by other proteins).

have been reported to include sequences that encode signal peptides on the amino ends of the proteins, thus demonstrating their universal presence in all cellular organisms. In general, signal peptides are as small as 3–11 amino acids in length, whereas some are more than 60 amino acids in length. Planctomycetes have sets of unique signal peptides that are not found in other bacterial groups, but some are similar to those in eukaryotes. It has been hypothesized that these are required to deliver proteins to the numerous membrane-bound organelles that are present in planctomycetes (Figure 32.3).

SIGNAL PEPTIDE SYSTEMS IN EUKARYA

Eukaryotes have many compartments, all of which are surrounded by membranes (Figure 32.3). Mitochondria and plant plastids have two membranes, and some of the organelles in some protists have more than two membranes surrounding them, because of the multiple endosymbiotic events that occurred during their evolution (see Figure 31.2). Each of the compartments of a eukaryotic cell carries out specific reactions, processes, and functions. Many of these are incompatible with reactions, processes, and functions within other cell compartments. Mitochondria primarily carry out the process of oxidative phosphorylation, which exposes the interior of the mitochondria to high concentrations of reactive oxygen species. These would cause a great deal of damage to vital molecules in the cytosol, and in the nucleus they would increase mutation rates (they do cause higher rates of mutation in mitochondrial DNA [mtDNA]). Similarly, photosynthesis splits water to produce oxygen, protons, and electrons. Most of these products are toxic to the parts of the cells that are

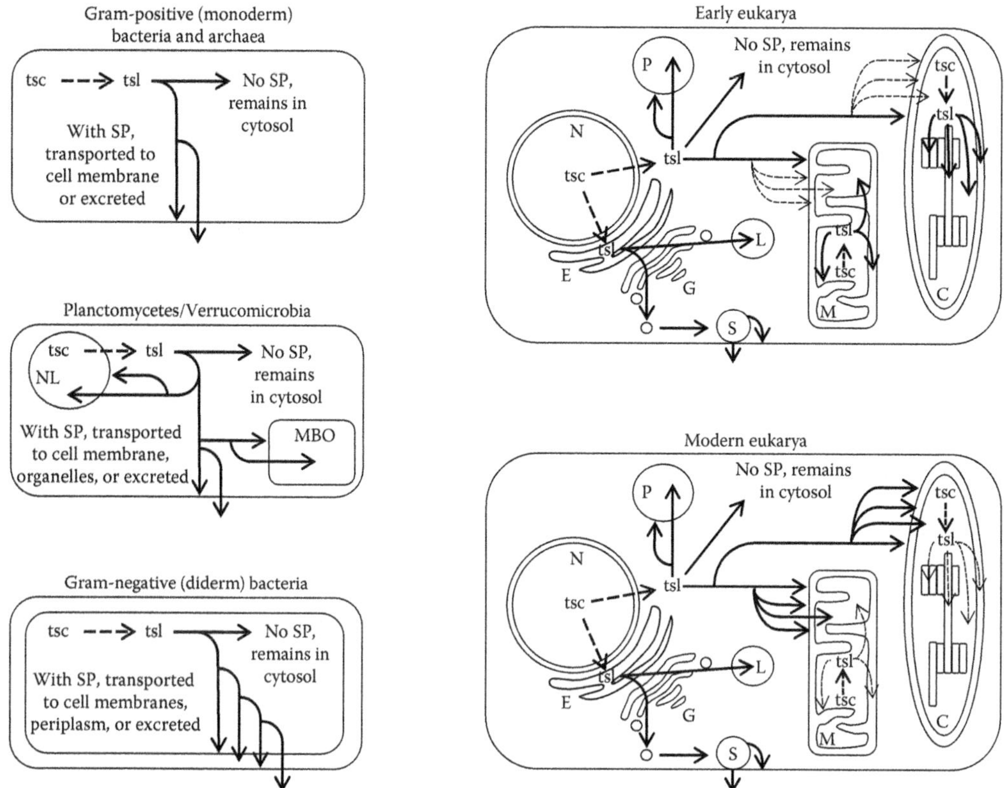

FIGURE 32.3 Summary of the transport of proteins in various types of cells. In monoderm (i.e., Gram positive) bacteria and archaea (upper left), there is a single membrane (the cell membrane), and therefore, the system for either transport of a protein across the membrane, or embedment in the membrane, is somewhat simple. For members of the Planctomycetes/Verrucomicrobia/Chlamydiae superphylum (middle left), the system for protein trafficking is more complex, because of the multiple membrane-bound compartments/organelles (MBO) present in the cells. In diderm (Gram negative) bacteria (lower left), the system is also more complex because of multiple compartments and membranes. As endosymbiotic relationships developed in early eukarya (upper right), the pathways for protein transport increased in number and complexity. Initially, more transport occurred in the organelles from gene expression within the organelles (thick lines), whereas smaller amounts were directed from the cytosol into the organelle (dashed lines). However, as genes moved into the nuclei during evolution, protein trafficking became primarily directed from the nucleus into the organelles (lower right; thick lines), although small amounts of protein trafficking were still directed by the organelles (dashed lines). Therefore, in modern Eukarya, there is a larger amount of protein trafficking into the organelles, than strictly within the organelles. tsc, transcription; tsl, translation; SP, signal peptide; NL, nucleus-like organelle; MBO, membrane-bound organelle; N, nucleus; P, peroxisome; E, endoplasmic reticulum; G, Golgi; L, lysosome; S, secretory vesicle; M, mitochondrion; C, chloroplast.

mainly anaerobic. Peroxisomes, organelles with single membranes, aid in detoxification of the reactive oxygen species. Thus, they soak up the damaging oxygen molecules and chemically alter them so that they are less toxic. However, DNA could not survive long if housed inside a peroxisome.

Nuclei have a single membrane that is folded back on itself, such that two membranes and an intermembrane space are present, and must be traversed by molecules being imported into, or exported out of, the nucleus. Nuclear pores control the export of RNAs and ribosomal subunits, whereas chaperones and transport proteins deliver proteins destined for the nuclear membrane or the nucleoplasm to the nuclear pores for transport through the membranes and into the nucleus. The endoplasmic reticulum, which has a single membrane, is where many proteins are synthesized, folded, and modified, which will end up either on or within, or be excreted from, a membrane. Many of these proteins are transferred into the Golgi (also with single membranes) where the proteins are further modified, often by attachment of sugar or fatty acid groups. Lysosomes also have single membranes and contain degradative enzymes that will digest any foreign particles that are brought into the cell, or simply to recycle cellular components. In addition to plastids, most plant cells have vacuoles, which have single membranes. These aid in the maintenance of cell turgor pressure, osmotic pressure, and pH, and also carry out some of the processes of photosynthesis and other metabolic processes. The cytosol is surrounded only by the cell membrane (i.e., plasma membrane). Inside the cytosol, the major metabolic processes occur, including most of the protein synthesis of the cell. The proteins that remain in the cytosol have no signal peptides on the amino ends of their proteins, but most other proteins do. Therefore, the major logistical problem is to assure that all of the proteins that need to traverse a membrane are delivered to the appropriate locations. Protein trafficking systems have evolved for this purpose.

PROTEIN TRAFFICKING IN MITOCHONDRIA

As described in Chapter 31, mitochondria have an outer membrane and an inner membrane, which becomes convoluted into cristae (Figure 32.3). The central portion of a mitochondrion is the matrix, and between the two membranes there is an intermembrane space. Although a few proteins are expressed from the mtDNA, most of the proteins are expressed from the nucleus, and therefore, the messenger RNAs (mRNAs) are translated on cytosolic ribosomes, and then they are imported into the mitochondria (Figure 32.4). Proteins that end up in the matrix, such as those in the TCA cycle, usually have a single signal peptide. A chaperone protein (i.e., an SRP) carries the proteins to the mitochondria and delivers them to the transport channel complexes that form controlled pores through the membranes. These transport channels span both the outer (the TOM, or mitochondrial outer membrane transport channel complex) and inner (the TIM, or mitochondrial inner membrane transport channel complex) membranes, such that the proteins are threaded through the channel from the cytosol directly into the matrix. Once there, an enzyme (a signal peptidase) cleaves off the signal peptide, and the protein folds into its final conformation. Some proteins that are destined for the inner membrane have a single signal peptide. These proteins are threaded through the TOM that forms a pore only through the outer membrane. The protein is delivered through the intermembrane space to assembly proteins on the inner membrane, and the protein is inserted into the membrane. Proteins that are destined for the intermembrane space often have two signal peptides on their amino end. The first is responsible for recognition and delivery to the outer mitochondrial membrane by a chaperone protein. The protein is threaded through the transport channel (TOM and TIM) into the matrix, and the first signal peptide is cleaved from the end. Then proteins recognize the second signal peptide and move the protein to the inner membrane transport channel, where it is threaded into the intermembrane space, and the second signal peptide is removed as the protein is folded into its final conformation. Proteins, such as cytochrome c, also have a heme added to the protein after arrival in the intermembrane space.

PROTEIN TRAFFICKING IN CHLOROPLASTS

Chloroplasts have an outer membrane, an inner membrane, and thylakoid membranes, which are formed from invaginations of the inner membrane. The membranes produce three compartments: an intermembrane space (between the outer and inner membranes), the stroma (the central portion of the chloroplast), and the lumen (enclosed within the thylakoid membranes). As with mitochondria, some chloroplast proteins have one signal peptide, and some have two (Figure 32.5). Proteins that are destined for the stroma or inner membrane have single signal peptides.

Proteins that are to be inserted into the inner membrane are transported through transport channel pores in the outer membrane, whereas proteins destined for the stroma (such as enzymes of the Calvin

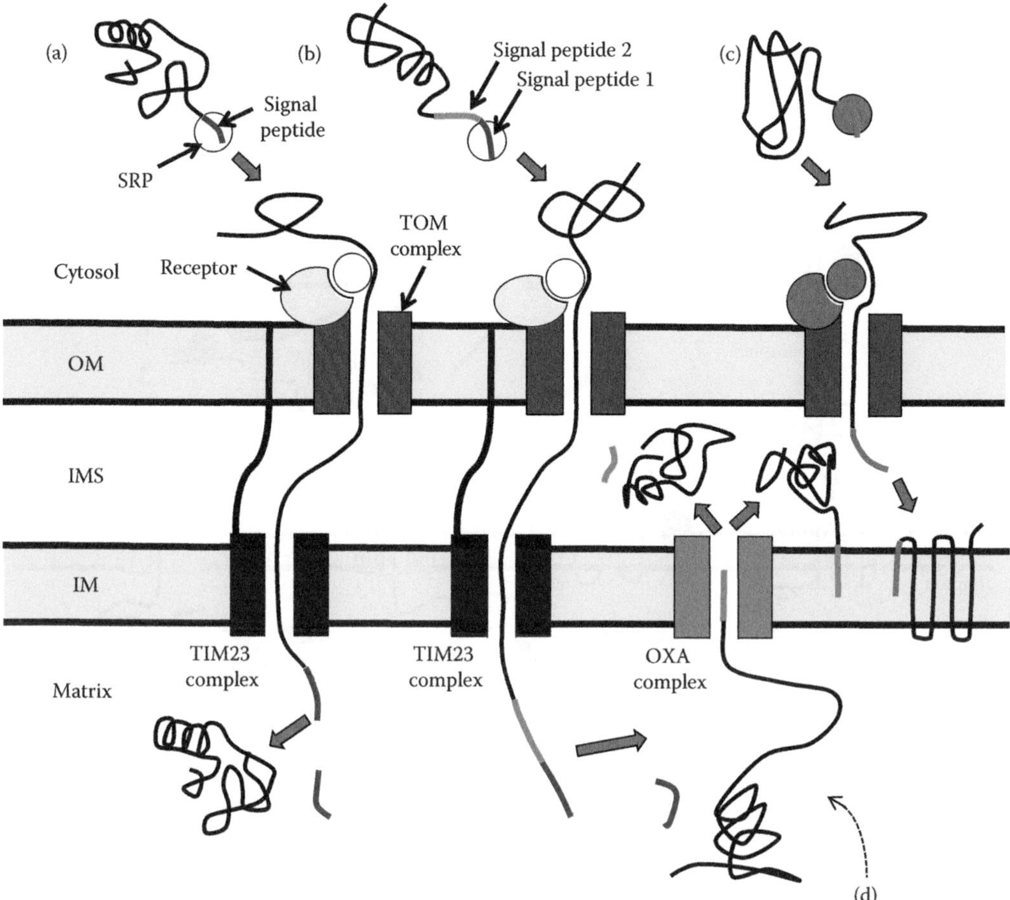

FIGURE 32.4 Modes of protein trafficking in mitochondria. Some proteins have single signal peptides on their amino ends (a) that are recognized by SRP/chaperone proteins that deliver the protein to the TOM. The protein is threaded through the TOM, and then through the mitochondrial inner membrane transporter 23 (TIM23), beginning from the amino end. Once it enters the matrix, the signal peptide is cleaved with a signal peptidase, and the protein assumes its final conformation. Some proteins have two sequential signal peptides (b). The first one causes the protein to be threaded through the TOM and TIM, as with single signal peptide proteins (as in a, above). The first signal peptide is then cleaved by a signal peptidase. The second signal peptide causes the protein to be threaded through the OXA transporter on the inner mitochondrial membrane (IM) from the matrix into the intermembrane space (IMS). If the protein is to remain in the IM, the signal peptide is cleaved off of the protein, and the protein is folded into its mature conformation. If the protein is to be embedded in the membrane, the signal protein acts to anchor the protein in the IM, and then the protein associates with the membrane, usually aided by other proteins, and is folded into its final form. Some proteins destined for the mitochondrial inner membrane have signal peptides that cause them to be shuttled through the TOM (c), which delivers them to the IMS. Once there, proteins on the IM insert them into the IM. The signal peptide remains attached to the protein. Some proteins are encoded by mtDNA (mitochondrial genome) and are translated on mitochondrial ribosomes. Although most remain in the matrix, a few are associated with membranes or are transported into the IM or IMS. For those, a signal peptide is present on the protein, which causes their transport through the OXA transporter. If the protein is destined for the IM, then the signal peptide is cleaved from the protein, and the protein is folded into its final conformation (for cytochrome c, a heme is added to the protein at this point). If the protein is to be inserted into the membrane, the signal peptide is recognized by membrane-bound proteins, which insert the protein into the membrane.

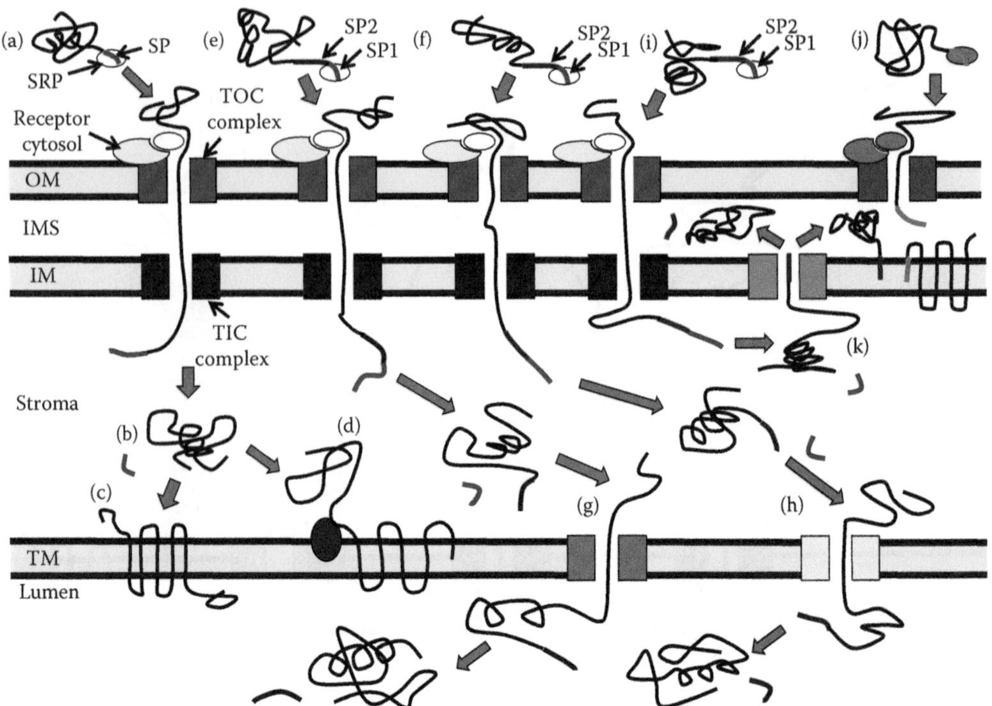

FIGURE 32.5 Modes of protein trafficking in chloroplasts. Although some of the transport systems in chloroplasts (and plastids, in general) are similar to those in mitochondria, they also have a few that differ from those in mitochondria, partly because of the presence of thylakoids. Some proteins with single signal peptides (a) are transported through the outer and inner membranes into the stroma via the TOC and the TIC. Although many of the proteins remain in the stroma (b), some can spontaneously insert themselves into the thylakoid membrane (c), whereas others require additional proteins that aid them in inserting into the thylakoid membrane (d). Many proteins that enter the chloroplasts have two signal peptides on their amino ends (e and f). These are destined for the lumen of the thylakoids. The first signal peptide causes their import from the cytosol, through the TOC and TIC, and into the stroma. There, the first signal peptide is cleaved by a signal peptidase, and the proteins are then shuttled to transporters, which move them through the thylakoid membranes. For some of the proteins, transport into the lumen is via an ATP electrochemical gradient transporter (g), whereas others proceed into the lumen through a proton gradient transporter (h), depending on their second signal peptides. Some proteins destined for the inner membrane (IM) also have two signal peptides (i): one that causes them to be transported through the TOC and TIC, and the other that causes them to be transported through the IM into the intermembrane space (IMS). Other proteins enter into the IMS through the outer membrane (OM), having a signal peptide specific for only the TOC (j). Many proteins are encoded by the cpDNA (chloroplast DNA; chloroplast genome) and are translated on chloroplast ribosomes within the stroma. Most of these proteins remain in the stroma, but a few have signal peptides and are transported through the IM by transport complexes (k), where they function in the IMS (after their signal peptides have been removed), or in the IM, where they retain their signal peptides.

cycle) are transported through channels that traverse both the inner and outer membranes in a single process (the TOC and TIC, chloroplast outer and inner, respectively, membrane transport channel complexes). Proteins destined for the lumen or thylakoid membranes (e.g., proteins involved in the light reactions or photosynthesis) have two signal peptides. The first is necessary for the proteins to traverse the inner and outer chloroplast membranes. Once in the stroma, the first signal peptide is cleaved from the protein. Then, recognition proteins attach to the second signal peptide and transport the protein to the thylakoid membrane, where they are moved through another transport channel into the lumen or into the membrane. There are at least four different modes of transport into the lumen or embedment into the thylakoid membrane (Figure 32.5, lower portion). One is an autonomous system, in which the protein inserts itself into the thylakoid membrane, based on hydrophobic regions of the protein. The second is similar but requires other proteins to assist the insertion of the protein into the thylakoid membrane. The other two are transport channel complexes that transport proteins destined to function in the thylakoid lumen. One requires an ATP electrochemical gradient, whereas the other uses the proton gradient set up by the light reactions of photosynthesis.

EVOLUTION OF PROTEIN TRAFFICKING SYSTEMS

For proteins that are destined to become incorporated into the plasma membrane, lysosomes, secretory vescicles, or Golgi, a signal peptide is recognized by an SRP, which delivers the newly forming peptide to the transport channel on the endoplasmic reticulum while it is still being synthesized by the ribosome (Figure 32.3). This may represent the ancestral eukaryotic protein trafficking system. The ribosomes on the outside surface of the endoplasmic reticulum appear as small dark bumps coating the endoplasmic reticulum, which is the reason it is called the rough endoplasmic reticulum (RER). All of the ribosomes are translating mRNAs on the cytoplasmic side into proteins that are being threaded into the interior of the RER as they are synthesized. These proteins are often modified in the Golgi, and are then split off of the main membrane into vesicles that become lysosomes, secretory vesicles, and other types of vesicles. All of the other proteins of eukaryotic cells are translated on cytoplasmic ribosomes that are dispersed throughout the cytosol. Many of them have signal peptides that are recognized by specific SRPs that deliver them specifically to the mitochondria, plastids, peroxisomes, or nucleus. Once delivered, translocation protein complexes thread them through the complexes to the other sides of the membranes. The exception to this occurs at the nuclear membrane. Here, the proteins are delivered to the nuclear pores, and the proteins are transported into the nucleus.

The two modes of transfer may point to separate evolutionary pathways for the two transport channel systems. Delivery of the protein (with attached signal peptide) to the endoplasmic reticulum is coupled with translation, such that the SRP moves the entire ribosome to the surface of the endoplasmic reticulum and delivers the entire complex to the transport channel where translation continues. The growing polypeptide chain is then threaded through the transport channel as it is being synthesized by the ribosome, producing the typical RER that can be seen in electron microscope preparations. Some ribosomes in bacteria also line up along the cell membrane, and the transport of the protein across the membrane occurs at the same time. This also occurs for some proteins that are translated within mitochondria and chloroplasts. Similar structures have been observed in some electron microscope preparations of planctomycete cells (e.g., *Gemmata obscuriglobus*); the ribosomes are found not only adjacent to the cell membrane but also adjacent to some of the internal membranes in the cells. These cells do have a number of unique signal peptides, and some of these may serve the same function in these bacteria as the signal peptides that target the RER in eukaryotic cells.

The other mode of transport of proteins across membranes is one in which translation and transport are uncoupled. That is, translation occurs in the cytosol, and once the protein is completed, a chaperone protein (e.g., an SRP) delivers the protein to the transport channel complex through the appropriate membrane. This mode of transport is used to deliver proteins to the mitochondrial, chloroplast, nuclear, and peroxisome membranes. Because of the presence of a double-membrane nucleus-like structure, as well as other membrane-bound internal structures, this type of system may have existed in the Planctomycetes/Verrucomicrobia superphylum long before the evolution of eukaryotes. Evolution of this mode of transport may have been advantageous for the evolution of eukaryotic cells, which may have made it possible for them to increase in size. An efficient means to deliver proteins across the larger cytosolic distances would have increased the survivability of these cells.

For endosymbiotic associations and the eventual evolution of organelles (e.g., mitochondria and chloroplasts), the transport systems would have to be compatible. Although this might not have been critical at first, for the eventual evolution of the host cell/organelle system to survive long term, the proteins in each of the transport systems would have to work together efficiently. For mitochondria and chloroplasts, the outer membranes were originally derived from the host that engulfed them. Therefore, the eukaryote-like transport channels would already have been present on those membranes, possibly being similar to the plasma membrane, a coated vesicle, or a lysosome. These would probably have been oriented to export proteins toward the endosymbiont/organelle, or to position proteins on the membrane. Similarly, for the bacterial (or eukaryotic) endosymbiont, transport channels would have also been present on those membranes. These would have probably been oriented to export proteins from the cell or to deposit proteins in the membrane. Therefore, the initial structure probably was not ideal for either organism, but with a bit of reorientation, changes to some components, and the loss of some of the transport mechanisms, the components evolved into the trafficking system present in modern cells.

Although the transport channel complexes may have been compatible, as the genes moved into the nucleus, the expressed proteins still had to reach the correct membrane, transport channels, and compartments. In recombinant DNA experiments, signal peptides have been experimentally changed so that a protein that is normally transported to one cell compartment is instead delivered to a different cell compartment. In these cases, usually the protein, and sometimes the organelle, fails to function normally. However, in other cases, the proteins appeared to function normally, and the organelles were not adversely affected. This implies that recombination within organisms could move sequences for one signal peptide onto other genes, or the sequence for a signal peptide could be removed from, or added to, a gene. Therefore, proteins could be shuttled into different compartments by changing their signal sequences. This appears to have occurred during the evolution of eukaryotes. For example, the chimeric pathways for the Calvin cycle and glycolysis/gluconeogenesis in plants are indicative of changes in signal sequences that have delivered proteins of diverse ancestry to different cell compartments, while halting delivery of analogous proteins to the same compartments (see Figure 31.11). Therefore, there is a mechanism for moving the signal peptides from one gene to another during evolution, and it has contributed to a complex evolutionary pattern for organisms with genomes that have experienced large influxes of foreign genes.

KEY POINTS

1. Because proteins cannot traverse membranes, protein trafficking systems have evolved to recognize signal peptides on the amino ends of the proteins and transport them across membranes.
2. Signal peptides first appeared in Bacteria, in order to embed specific proteins into membranes, as well as to secrete specific proteins, and in diderms to carry specific proteins into the periplasm.
3. In eukaryotes, more complex systems of protein trafficking have evolved in order to efficiently deliver proteins to their appropriate cell compartments.
4. One mode of protein trafficking carries the entire ribosome and emerging polypeptide to the surface of the RER, where translation continues, with the emerging polypeptide being transported directly into the lumen of the RER.
5. Another mode of protein trafficking involves a decoupling of translation from transport. The protein is delivered to the appropriate membrane (based on the signal peptide on the protein), and then threaded through a transport protein complex into the corresponding cell compartment (e.g., mitochondrial matrix or chloroplast stroma).
6. The two modes of protein trafficking may have evolved within Bacteria, and then become more complex once endosymbionts and organelles evolved in eukaryotes.
7. At least some of the protein trafficking mechanisms may have originated in the Planctomycetes/Verrucomicrobia superphylum. Many members of this group have internal membrane-bound compartments, as well as a double-membrane nucleus-like structure.

ADDITIONAL READINGS

Alberts, B., D. Bray, K. Hopkin, A. Johnson, J. Lewis, M. Raff, K. Roberts, and P. Walter. 2013. *Essential Cell Biology*, 4th ed. New York: Garland Publishing.

Alberts, B., D. Bray, A. Johnson, J. Lewis, M. Raff, K. Roberts, and J. D. Watson. 1994. *Molecular Biology of the Cell*. New York: Garland Publishing.

Becker, W. M., J. B. Reece, and M. F. Poenie. 1996. *The World of the Cell*, 3rd ed. New York: The Benjamin/Cummings Publishing Company.

Cox, M. M., J. A. Doudna, and M. O'Donnell. 2015. *Molecular Biology, Principles and Practice*, 2nd ed. New York: W. H. Freeman and Company.

Jarvis, P. and E. López-Juez. 2013. Biogeneisis and homeostasis of chloroplasts and other plastids. *Nat. Rev.* 14:787–802.

Lewin, B. 2008. *Genes IX*. Sudbury, MA: Jones & Bartlett Publishers.

Martin, W. and C. Schnarrenberger. 1997. The evolution of the Calvin cycle from prokaryotic to eukaryotic chromosomes: A case study of functional redundancy in ancient pathways through endosymbiosis. *Curr. Genet.* 32:1–18.

Schleiff, E. and T. Becker. 2011. Intra-organelles transport in mitochondria and chloroplasts. *Nat. Rev. Mol. Cell. Biol.* 12:48–59.

Tropp, B. E. 2008. *Molecular Biology, Genes to Proteins*, 3rd ed. Sudbury, MA: Jones & Bartlett Publishers.

33 Eukaryotic Genomes

INTRODUCTION

At their essence, eukaryotes are combinations of bacterial, archaeal, and viral genomes. Eukaryotic evolution has been one of parasitism, mutualism, endosymbiosis, and frequent horizontal gene transfers (HGTs). Although this began with endosymbiotic events between Bacteria and Archaea, it progressed into endosymbiotic events between Eukarya and Bacteria, as well as Eukarya and other Eukarya. In many ways, eukaryotes can be viewed as the products of the coordination (and sometimes competition) between two or more genomes. Eukaryotic evolution has been the result of endosymbiotic events that have mixed large numbers of genes into the progenote cells followed by mutation and selection that has led to combinations of genes and gene products that have resulted in successful cells and multicellular organisms. Additionally, HGT events are common in eukaryotes (see Chapter 15). They have occurred often in the past and continue to happen daily, although all of them do not lead to successful evolutionary events. These can be transfers of single genes or multiple genes by viruses, Bacteria, Archaea or other Eukarya. Alternatively, entire genomes can be transferred in the form of entire organisms, as was the case in the evolution of the organelles in eukaryotes.

Some organelles lack an intraorganellar genome, but some have suggested that these organelles in eukaryotes, such as peroxisomes, endosomes, and vacuoles, may have once been autonomous organisms that evolved into organelles, but that all of their genes in their genomes have been transferred into the nucleus. However, because they have single membranes, it is likely that they originated within the eukaryotic cells. Regardless of their evolution, it is clear that most of the ancestral mitochondrial (i.e., α-proteobacterial) genome has been transferred into the nucleus, and the evolution of the genes of the host genome and the mitochondrial genome has been affected greatly. Also, in the Archaeplastida (plants), almost all of the chloroplast (i.e., cyanobacterial) genome has been transferred into the nucleus, and a few pieces have moved into the mitochondrial genome as well. Here, the coordination between the three genomes (chloroplast, mitochondrial, and nuclear) is critical to the survival of those organisms. In many protists, additional symbiotic events have combined additional genomes into single cells, producing organelles that themselves contained multiple genomes. Because many of the gene products had analogous functions (e.g., genes for glycolysis and the tricarboxylic acid [TCA] cycle), evolution has selected for use of the best enzyme and gene for each compartment of the cell (see Chapter 31). Thus, evolutionary processes and outcomes in eukaryotes are complex and intricate.

Eukaryotic cells represent forms of life in which more than one genome has been brought together into the same cell. Additionally, most eukaryotic cells have more than a single chromosome. Although this adds more complexity to the process of cell division and gene regulation, it provides additional evolutionary and survival possibilities. One process that evolved in this group was mitosis, the accurate segregation of chromosomes into the two daughter cells. Once it was well established, another type of process developed that ensured mixing of the gene types among the progeny. This was the process of meiosis, which is a method for segregating chromosomes, but more importantly, it is a method for exchanging chromosomal segments such that there is a shuffling of gene alleles during the process (see Chapter 8). These shuffled chromosomes are joined with other shuffled chromosomes in the process known as sex. Early in the evolution of eukaryotes, there were simply different mating types that could join together to form a zygote.

However, later more elaborate processes led to alternate developmental patterns for the mating types to form male and female morphotypes, each of which became specialized in their contributions to maintenance of the species.

ORIGIN OF THE NUCLEUS AND MITOCHONDRION

Genomics has aided in attempts to determine the origin of eukaryotes. Explanation of the nucleus and intracellular membrane-bound organelles has been the topic of discussions and theories for decades. Although most agree that eukaryotes were the result of endosymbiotic processes and HGTs, disagreement on the set of events and even their order still exists. Based on paleontology, morphology, biochemistry, genomics, and bioinformatics, nuclei and mitochondria were established in a single event or the establishment of mitochondria occurred less than 200 million years after nuclei appeared in eukaryotes. The first eukaryotic fossils are from approximately 2.4 billion years ago. Mitochondria, or at least traces of mitochondria, have been found in all eukaryotes that have been examined for the presence of these organelles. However, estimates of their incorporation into eukaryotic cells range from 2.2 to 2.4 billion years, meaning that they could have been part of the original eukaryotic cell or they could have been incorporated at a later time. It is clear from phylogenetic analysis of the mitochondrial gene sequences that they originated from an α-proteobacterium (most likely, a Rickettsia - like species). However, this group of bacteria lacks the genes in the internal structure that resemble nuclear genes or a nucleus. They also lack genes that are similar to tubulins that are necessary to form microtubules, which form the spindles for the separation of the chromosomes. Therefore, this endosymbiotic event probably did not lead to the formation of a nucleus. The α-proteobacteria do have the genes for oxidative phosphorylation, which was why they were incorporated by eukaryotic cells. This allowed the eukaryotic cells to survive at a time when oxygen concentrations in the atmosphere and oceans were rising, due to the waste products from cyanobacterial oxygenic photosynthesis.

It is clear from genomic studies that the nuclear genes in eukaryotes have similarities to both Archaea and Bacteria (Table 29.2). Many of the genes for DNA replication, RNA transcription, translation of proteins on ribosomes, and condensation of chromosomes are more similar to those from Archaea, whereas a different set of genes is more similar to those from Bacteria (Figure 33.1). Therefore, it appears that there was an endosymbiotic event between an archaeal ancestor and a bacterial ancestor that predated the endosymbiosis that produced mitochondria. The nucleus, internal membranes, and the extensive use of microtubules characterize eukaryotic cells. Archaea have none of these (although some Archaea have genes that have some similarities to tubulins), and therefore, these characters likely originated from an endosymbiotic event with a bacterium (or they were acquired in stages by other events, which has also been proposed by other scientists). The only group of bacteria that have internal structures resembling nuclei and also contain tubulins that have similarities to eukaryotic tubulins are within the Planctomycetes–Verrucomicrobia–Chlamydiae (PVC) superphylum. Planctomycetes and Verrucomicrobia members have internal organelles separated by membranes, as well as tubulins. They have a double-membrane-bound region of their cells that enclose the chromosomes. Ribosomes appear to be attached to the membranes, which may be analogous to the rough endoplasmic reticulum system in eukaryotes. Some planctomycetes also contain a membrane-bound compartment called the anammoxosome where ammonia is converted to nitrogen gas. In doing so, the cell can capture energy from the reaction in order to produce adenosine triphosphate (ATP). Interestingly, in some electron micrograph sections of the anammoxosomes, there are membrane structures that resemble endoplasmic reticula and/or Golgi, which support a possible coincident origin for these eukaryotic organelles. The possible reason for separation of DNA, RNA, and proteins from the ammonia by membrane-bound compartmentalization in these cells is to protect the biological molecules from the damaging effects of the ammonia. Ammonia can cause rapid oxidation of RNA, and therefore, it is especially crucial to separate the RNA from the ammonia. Some archaea also utilize ammonia to produce nitrites (used by the planctomycete cells), and eventually nitrates (by other Archaea or Bacteria). The Archaea

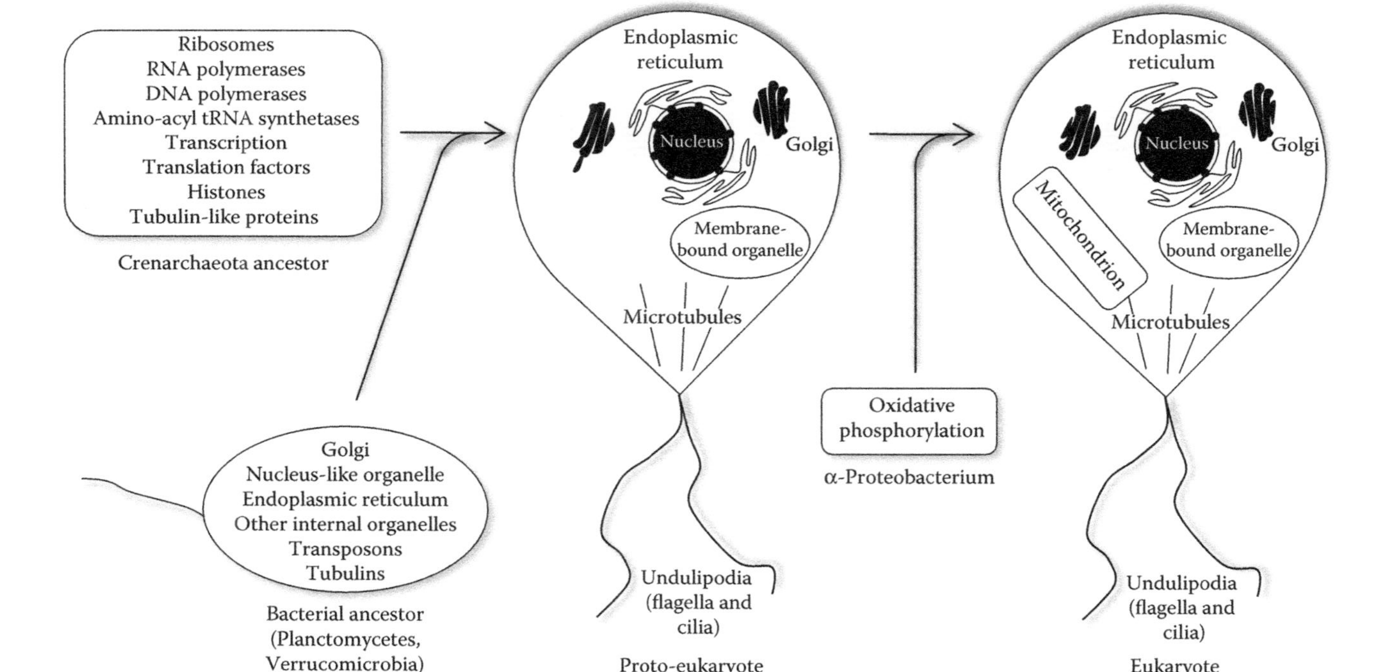

FIGURE 33.1 Proposed evolutionary pathway for the origin of the first eukaryotic cell. One of the most parsimonious pathways (based on the fewest number of events) is the development of an endosymbiotic relationship between a Crenarchaeota ancestor and a Planctomycete or Verrucomicrobia ancestor. This could explain the formation of a nucleus and other intracellular membrane-bound organelles (including the endoplasmic reticulum and Golgi) and tubulins, as well as transposons, that would have come from the Planctomycetes/Verrucomicrobia symbiont, and all of the other characters that would come from the Crenarchaeota cell. A second endosymbiotic event between the proto-eukaryote and an α-proteobacterium would yield the ATP-producing mitochondrion. All of this occurred by about 2.2 billion years ago.

that convert ammonia into nitrites do so aerobically, but the oxygen is damaging to the planctomycetes. However, if it associated with the archaea, it would be somewhat protected from the oxygen, which was on the increase at that time. If this relationship became an endosymbiotic one, then all of the eukaryotic characteristics, except for mitochondria, would be incorporated into a single cell. DNA packaging (i.e., histones), DNA replication, RNA transcription, and translation would be contributed by the archaeal genome; and the nucleus, intracellular membrane-bound organelles, and microtubules would be contributed by the bacterial cell. As oxygen levels continued to rise, association with an α-proteobacterium that could detoxify oxygen to produce much more ATP would have been vital to these cells.

Eukarya use microtubules for mitosis, meiosis, cytokinesis, construction of cilia, and other functions. Bacteria and Archaea do not have tubulin (with the exception of the PVC superphylum), but they have phylogenetically similar proteins, called Fts (and related proteins). The genomes of planctomycetes often do not have genes for these proteins, but tubulin genes are found in the genomes of members of the sister group, Verrucomicrobia. Some archaeal genomes have Fts-like genes, including those in the Crenarchaeota (proposed ancestors of eukaryotes). Therefore, there was more than one set of genes that could be used to transport components within cells. However, neither planctomycetes nor Archaea possess the genes for oxidative phosphorylation that are present in mitochondria today. It is well established that these were provided by an α-proteobacterium ancestor, and therefore, the event that led to the appearance of the eukaryotic nucleus probably was separate from the event that led to the evolution of the mitochondrion.

On average, the planctomycete genomes of members of the Planctomycetes and Verrucomicrobia are larger than those of the Archaea (e.g., the planctomycete genome of *Gemmata obscuriglobus* is 9.29 Mb, whereas the genome of the archaea *Sulfolobus solfataricus* is 2.99 Mb). Together, the total would be approximately 13 Mb, which is coincidentally in the size range of the smaller genomes of some free-living eukaryotes (e.g., Saccharomyces cerevisiae). In addition, the genome of *G. obscuriglobis* contains a large number of transposons, common among eukaryotic genomes. As mentioned in Chapter 13, these allow movement of sequences from one part of the genome to another, and some of the transposons are related to some types of viruses. These would facilitate the movements of genes from one cellular compartment to another. However, if the association between the archaeal cell and the planctomycete was mutualistic or parasitic instead of endosymbiotic, the genes could have been transferred over a longer period of time to lead to the initial eukaryotic cell. However, this was not the end of the evolutionary process for eukaryotes. Oxygen levels continued to rise, and detoxification or sequestration of the reactive oxygen molecules became an increasing problem for these organisms. However, at least one group of bacteria (the proteobacteria) had evolved to deal with this stress. At first, these bacteria may have used hemes and heme-containing proteins simply to attach to the oxygen so that it would not oxidize vital compounds in the cells. However, eventually, protein complexes evolved that could take the oxygen and combine it with protons and electrons to form water, which not only would detoxify the oxygen but would convert it into a useful compound. This reaction yields a great deal of energy, which was used to pump protons across the cell membrane and out of the cell. Because of the concentration gradient that this caused, the protons would then move back inside the cell through a different set of proteins, the ATP synthase, to produce ATP, which could be used in a wide variety of reactions that required energy. In a world where oxygen levels were increasing, this was a valuable system and a valuable set of genes. The acquisition of mitochondria by the eukaryotic cell involved the establishment of an endosymbiotic relationship between the eukaryotic cell and the oxygen-detoxifying/energy-producing bacterial cell (an α-proteobacterium). From that point onward, the eukaryotic lineages diverged. Some established additional endosymbiotic relationships, both with Bacteria (e.g., with cyanobacteria to form plastids) and with other Eukarya cells (e.g., with red and green algae in members of the Excavata, Chromalveolata, and Rhizaria). Other eukaryotes became specialized and many became parasites, eventually losing most of the genes from their genomes, and many losing some of their organelles, leading to species that lack mitochondria and/or chloroplasts.

MULTICELLULARITY

Another trait that has made eukaryotes so successful is the evolution of multicellularity and cell specialization. This requires that each of the cells can sense where it is in the larger organism and that its gene expression is consistent with its position in the organism. Multicellularity requires elaborate gene expression programming during development and growth of the organism so that the correct cell types grow where they are needed and that inappropriate cells (e.g., cancer cells) do not proliferate. By the time eukaryotes evolved, Bacteria and Archaea already possessed genes that accurately controlled gene expression under varied conditions. Gene expression patterns changed based on the needs of the organism in specific environments (e.g., switching from aerobic to anaerobic conditions). In multicellular organisms, more of these genes were needed, and they had to control sets of genes that would cause specific gene expression patterns in the cells, which would essentially lead to cooperation among the cells of the multicellular organism. One class of these regulatory genes are called homeobox genes, e.g., HOX genes (see Chapter 16). They can be found in all multicellular organisms, and surprisingly, parts of the genes are well conserved in organisms that have been separated by hundreds of millions of years of evolution. In insects, some of these cause segmentation, whereas others determine where legs or eyes arise on the animal. They perform analogous functions in humans as well. Some determine where your thumb and fingers will appear on your hand, whereas others determine where your eyes will be. If the expression of these genes changes, the organization of cells and organs in those animals can change drastically.

CHROMALVEOLATA

As mentioned earlier and in Chapter 31, dinoflagellates have a complex evolutionary history. Several endosymbiotic events have led to the formation of this diverse group of organisms, as deduced from genomic studies. All of these events occurred after the initial acquisition of the plastids to produce the glaucophytes and red algae about 1.9 billion years ago. Cytological examinations, as well as genomic sequences, indicate a red algal source for the plastids, and therefore, the endosymbiotic events may have occurred sometime between the appearances of red algae (approximately 1.9 billion years ago) and green algae (approximately 1.2 billion years ago). In the first endosymbiotic event, a red alga became the endosymbiont of a heterotrophic eukaryote (see Figures 31.2 and 15.17). A similar, but separate event, produced the Chlorarachniophyta. This event produced some of the dinoflagellates as well as the apicomplexans, both within the Chromalveolata (dinoflagellates being on the Chromista branch and apicomplexans being on the Alveolata branch). This also led to the cryptomonas (e.g., *Cryptomonas*), haptophytes (e.g., coccolithophores), and heterokonts (aka Stramenopiles; e.g., diatoms, brown algae, water molds), as well as ciliates (e.g., *Tetrahymena*). In each of these, the endosymbiont has been reduced, such that the mitochondria in the organelle are absent. However, they all have chloroplasts that are surrounded by two to four membranes, which is also indicative of the endosymbiotic origin of these organelles. Additionally, there is a remnant of the organelle nucleus in the cryptomonas. This much reduced organelle, called a nucleomorph, still has retained a small genome. If we consider this to be a secondary endosymbiotic event, the first being the one that led to the acquisition of chloroplasts 1.9 billion years ago, then there is a series of tertiary endosymbiotic events that occurred next. One was an endosymbiosis between a primitive dinoflagellate and a heterokont, where the latter became the organelle, which has a nucleomorph and chloroplasts. The second was between a primitive dinoflagellate and a haptophyte to form another type of dinoflagellate. However, in this case the dinoflagellate became the organelle. The third was between a primitive dinoflagellate and a cryptomonad, with the cryptomonad becoming the organelle. The fourth event was an endosymbiosis between a primitive dinoflagellate and a red alga. In essence, this organism has two sets of genes from red algae that were acquired in endosymbiotic events that probably were hundreds of millions of years apart. In each case, the endosymbiosis created mixing of genes and gene products to produce organisms that sometimes have very different morphologies and characteristics. However, all retained a basic set of core genes that performed all of the main necessary functions in all cells.

As soon as the ancestral cells became dependent on one another, their genomes began to change. Genes from the original organelle of the main cell continued to move into the main nucleus (see Figure 15.10). Also, genes continued to move from the organelles of the endosymbiont into its nucleus. However, genes from the nucleus of the endosymbiont also moved into the nucleus of the main cell, and some genes from the organelles of the endosymbiont may have moved directly into the nucleus of the main cell. Eventually, all of the mitochondrial DNA in the endosymbiont was lost from the mitochondria, and the mitochondria also disappeared. However, in most cases the chloroplasts remained within the endosymbiont. However, in ciliates and trypanosomes, the chloroplast genome was transferred into the nuclei and the chloroplasts either are small degenerate nonsynthetic compartments or have disappeared completely. As for the nuclei, in most cases, the endosymbiont nucleus is absent, but in several cases, its remnants are present as an organelle called a nucleomorph. Therefore, evolution of this complex set of organisms has been characterized by several endosymbiotic events that led to the formation of several different organelles. The mix of genomes has produced novel successful combinations of genes. Some dinoflagellate genomes contain more genes acquired from endosymbionts than genes from their ancestral genomes.

OPISTHOKONTA

Saccharomyces cerevisiae

The first eukaryotic genome to be determined was that of common brewers' or bakers' yeast, *Saccharomyces cerevisiae.* Yeasts grow as single-celled organisms and divide by budding, which is a form of mitosis and cytokinesis (Figure 10.5). The most recent genome annotations calculate that the nuclear genome is 12,156,677 bp (Figures 33.2 and 33.3) spread among 16 chromosomes, including 6575 open reading frames (open reading frames [ORFs]; regions that could conceivably encode a protein), of which at least 4862 appear to be authentic genes, 912 are uncharacterized or unknown, and 801 may not be genes. Also, *S. cerevisiae* contains a mitochondrial genome that is 85,779 bp in length containing 28 ORFs, of which at least 17 are authentic genes and plasmid (called the yeast 2μm plasmid) that is 6318 bp in length and contains four genes. The chromosomes range in size from 230,208 bp (the size of some viruses) for chromosome I to 1,531,919 bp (the size of some free-living bacteria) for chromosome IV. The number of ORFs per chromosome ranges from 117 on chromosome I to 836 on chromosome IV. There are 275 transfer RNA (tRNA) genes spread among all of the chromosomes and another 24 on the mitochondrial chromosome. All of the 140 tandem copies of the nuclear ribosomal RNA (rRNA) genes (including the 5S, small subunit [SSU], 5.8S, and large subunit [LSU] genes) are on chromosome XII, and there are two rRNA genes on the mitochondrial chromosome. The number of nuclear rRNAs in the genome varies. There are approximately 140 rRNA tandem gene repeats (each containing a 5S, SSU, 5.8S, and LSU gene) on chromosome XII. However, this number can vary from 40 to more than 200. This was first noticed when the chromosomes were separated on pulse-field agarose gels. Each of the chromosomes was well resolved on the gels, with the exception of chromosome XII, which was always somewhat blurry, indicating chromosomes of many different lengths. As was mentioned in Chapter 5, the reason for this is that frequent unequal crossovers in this region generate chromosomes with different numbers of rRNA repeats, and therefore, the overall lengths of each of the XII chromosomes vary from cell to cell.

The genome encodes all of the genes in the central metabolic pathways that would be expected of a free-living organism (Figure 33.3). That is, these organisms can replicate their genomes, transcribe RNA from the genes, and translate the messenger RNAs (mRNAs) into proteins. They produce proteins for aerobic, anaerobic, and fermentation systems, and they are able to construct their cell envelope with other sets of genes (e.g., cell membrane, membrane proteins, cell wall, cell surface proteins). The *S. cerevisiae* genome also has a large number of transposons, with

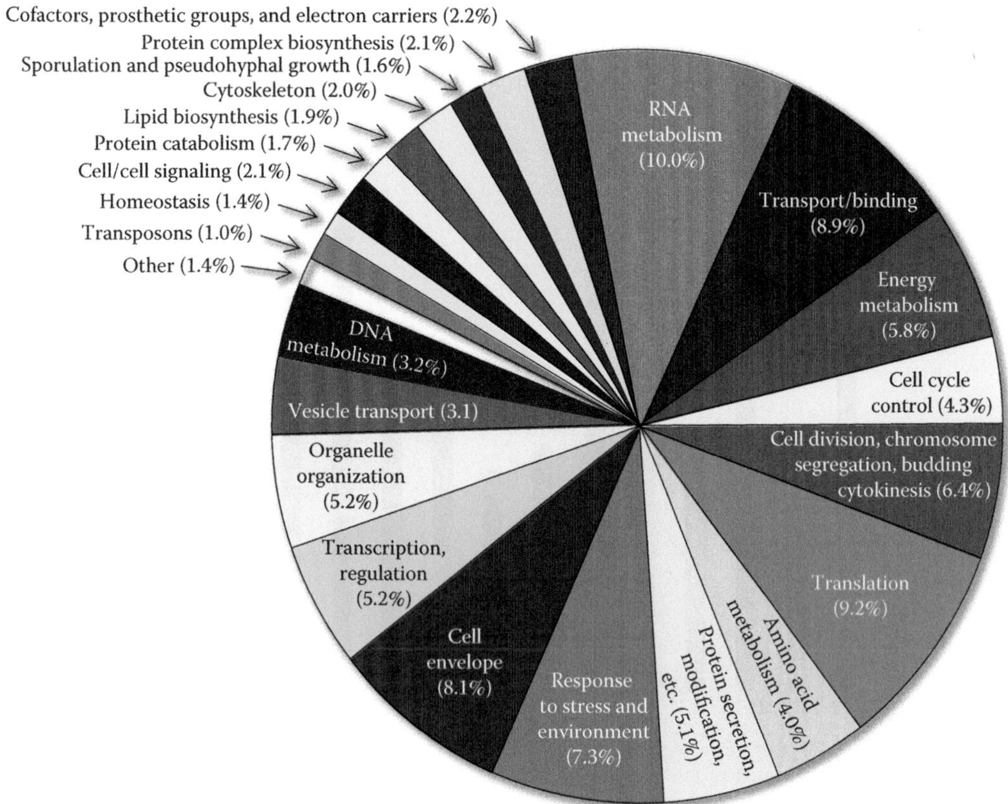

FIGURE 33.2 Summary of categories of genes in the *Saccharomyces cerevisiae* genome. The genome consists of 12,156,677 bp spread among 16 chromosomes, consisting of 4862 genes. Approximately one-third of the genome contains genes for nucleic acid metabolism, DNA replication, transcription, and translation. This would comprise approximately 1500 genes, indicating the increased complexity of these central processes in Eukarya, compared to Bacteria and Archaea. Cell cycle control and cell division are encoded by more than 10% of the genome (approximately 500 genes), again illustrating the increased complexity of both processes. Cell-to-cell signaling shows up as only 2.1% of the genome, but this is the far greater number of genes involved in this process than is seen in bacteria and archaea. Although energy metabolism occupies a smaller proportion (5.8%) of the genome compared to Bacteria and Archaea, the genome is larger, and thus, the total number of genes responsible for encoding proteins for energy metabolism is greater than in Bacteria and Archaea.

sequences comprising approximately 1% (approximately 1 Mb of DNA) of the genome. Compared to bacteria, they have a larger set of genes for response to stress and response to chemicals in the environment. Many of these have to do with signal transduction. This is the process in which a chemical or condition outside the cell is sensed by specific cell surface proteins, usually by binding of a ligand or a change in conformation. The changes in these proteins cause them to interact with other proteins inside the cell, often those that phosphorylate (i.e., kinases) other proteins, which causes a cascade of other changes in the cell to activate specific pathways that have evolved to react to certain environmental cues. This makes them very adaptable to many types of environments, and indeed in general fungi are adaptable organisms. Some species of fungi can live on substrates as different as soil, plants, and animals. There are some additional gene sets present in fungi that are either lacking or in low abundance in bacteria. These are genes for cell-to-cell interactions (i.e., communication between cells), cytoskeleton (which can be extensive in eukaryotes), organelle organization, and transport of molecules across internal membranes (e.g., vesicle membranes). Therefore, although the genome of this eukaryote is larger than the majority of bacterial

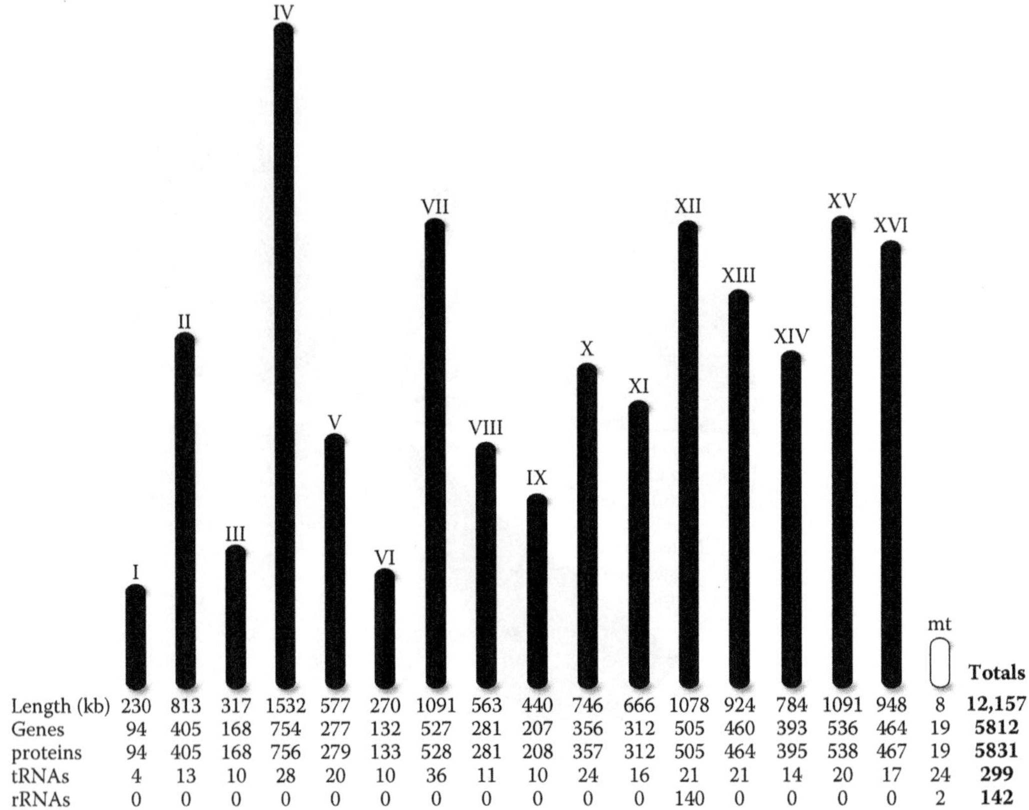

	I	II	III	IV	V	VI	VII	VIII	IX	X	XI	XII	XIII	XIV	XV	XVI	mt	Totals
Length (kb)	230	813	317	1532	577	270	1091	563	440	746	666	1078	924	784	1091	948	8	**12,157**
Genes	94	405	168	754	277	132	527	281	207	356	312	505	460	393	536	464	19	**5812**
proteins	94	405	168	756	279	133	528	281	208	357	312	505	464	395	538	467	19	**5831**
tRNAs	4	13	10	28	20	10	36	11	10	24	16	21	21	14	20	17	24	**299**
rRNAs	0	0	0	0	0	0	0	0	0	0	0	140	0	0	0	0	2	**142**

FIGURE 33.3 Representation of each of the lengths of each of the 16 *Saccharomyces cerevisiae* chromosomes, as well as the mitochondrial chromosome. The exact lengths and the numbers of genes and proteins encoded by those genes are shown at the bottom. Some of the genes encode more than one protein, because of alternative splicing, and therefore, the number of genes and proteins sometimes differs for some chromosomes. The numbers of tRNA and rRNA genes are also shown. Although tRNA genes appear on all of the chromosomes, the rRNA genes all are at one locus on chromosome XII. The number of rDNA repeats varies from approximately 40 to 200, with an average of about 140 tandem copies. There are two rRNA gene copies (mean value) in the mitochondrial genome.

genomes, most of the genome encodes the central metabolic pathways that also exist in Bacteria and Archaea, and because of this many of the genes are in common. Some genes and sets of genes appear to be different, although when their phylogenies are examined, in most cases a bacterial or archaeal origin can be traced (see Table 29.2). Some of these have been inherited vertically, but many appear to have been inserted into the genome more recently HGT. As with other eukaryotes, there are also many noncoding RNAs (ncRNAs) that are RNAs that control gene expression by binding to other RNAs or to DNAs.

Caenorhabditis elegans

The small nematode (1 mm in length) *C. elegans* lives for about 2–3 weeks and can go through a generation in about 4 days. It has two adult forms, a hermaphrodite that fertilizes and lays eggs, and a male that can also fertilize the eggs of the hermaphrodite (see Figures 16.6 and 16.7). The fates of all of the cells have been extensively mapped, such that it is known that there are 959 somatic cells (as well as sperm and eggs) in each adult hermaphrodite and 1031 somatic cells (as well as sperm cells) in an adult male. Multicellularity offers distinct advantages over single-celled organisms, but there are also significant challenges to a multicellular lifestyle. Multicellular organisms have complex developmental patterns in which cells differentiate for specific purposes and positions in these animals. In order to do this, the cells have to proceed through developmentally regulated patterns of gene expression. There must be tight control on when and in which cells the different sets of genes are expressed. Otherwise, the organization of cells and cell types in the organism can be negatively affected. In addition, the cells must communicate with one another in the organism. They must react to cells surrounding them so that gene expression matches their positions in the organism. They must react to cells that are in other parts of the organism. Also, they must react to signals that indicate the need for food, appropriate temperatures, and when it is time to mate.

The genome of *C. elegans* is just over 100 Mb in length and is divided among six chromosomes, as well as a small mitochondrial chromosome (Figures 33.4 and 33.5). From 20,100 to almost 25,000 protein-coding genes have been identified. The number of proteins that are produced is higher than the number of genes by more than 1000. This is caused by alternative splicing, which is common among multicellular organisms. In addition, it has been estimated that about 16,000 RNA genes are present in this genome. Of these, 55 are rRNA genes (SSU, 5.8S, and LSU genes), 110 are 5S rRNA genes, and 630 are tRNA genes. The others are small ncRNAs (small nuclear RNA, small nucleolar RNA, microRNA, and short interfering RNA). These act to control the expression of the protein-coding genes. Although alternative splicing and interfering

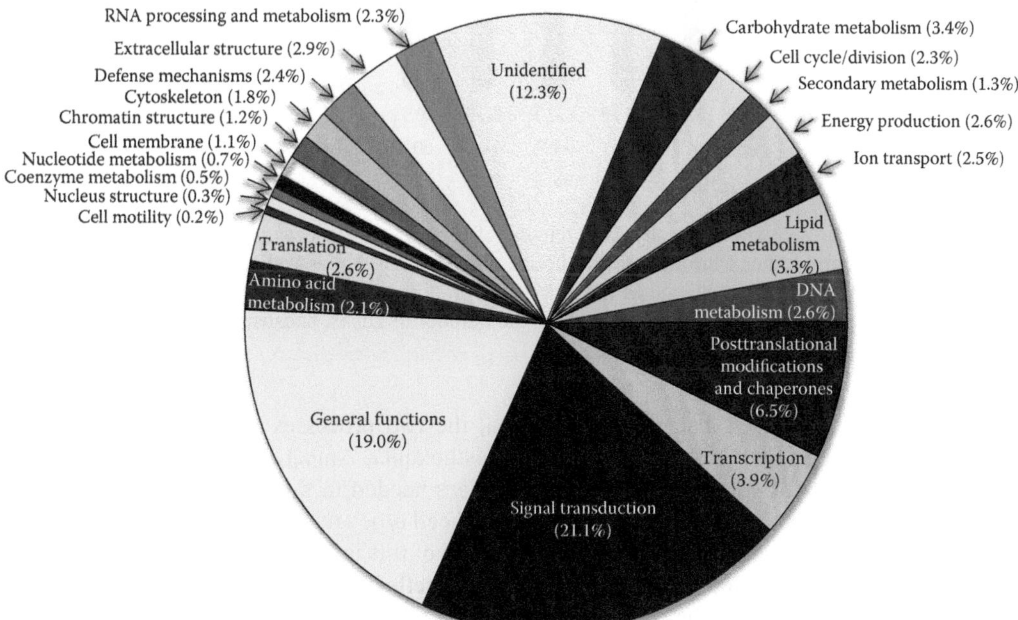

FIGURE 33.4 Summary of categories of genes in the *Caenorhabditis elegans* genome. The genome is 100 Mb in length, spread on six chromosomes (I through V, and an X sex chromosome). Genes for nucleic acid metabolism, DNA replication, transcription, transcription control, and translation occupy less than 15% of the genome. However, the genome is 8 times larger than that of *S. cerevisiae*, and therefore, there are actually more genes for each of these functions than in the yeast genome. Genes for energy metabolism also comprise a smaller percentage of the genome, but there are more genes for these functions than in *S. cerevisiae.* However, there is a category of genes that is almost nonexistent in yeast that encompasses more than 21% of the *C. elegans* genome. These are genes for signal transduction. These are extremely important for short- and long-range communications among cells in multicellular organisms. The cells must sense the cells that are next to them, as well as those that are elsewhere in the organism, and respond appropriately to maintain the form and functions of the organism.

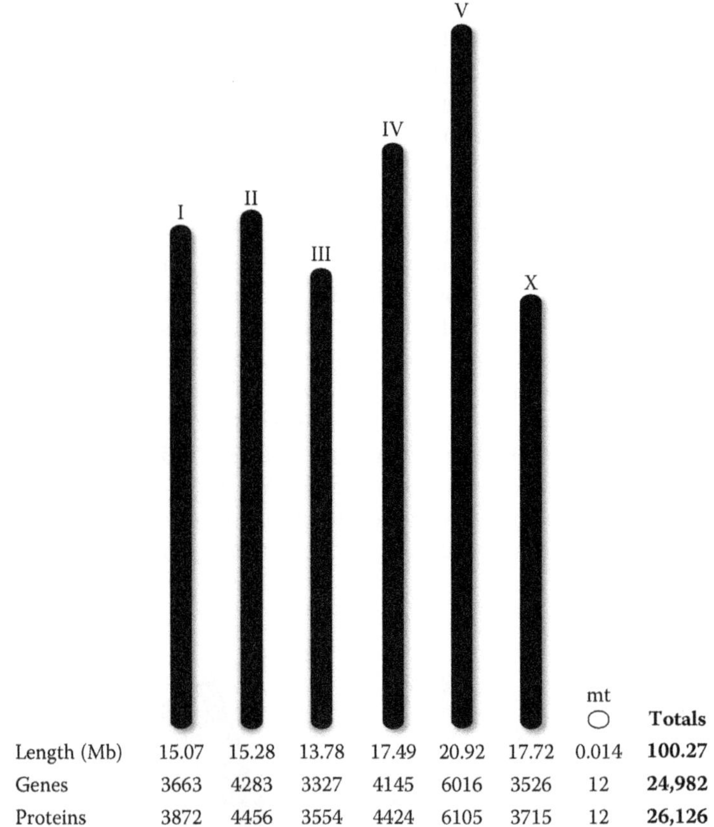

	I	II	III	IV	V	X	mt	Totals
Length (Mb)	15.07	15.28	13.78	17.49	20.92	17.72	0.014	**100.27**
Genes	3663	4283	3327	4145	6016	3526	12	**24,982**
Proteins	3872	4456	3554	4424	6105	3715	12	**26,126**

FIGURE 33.5 Representation of each of the lengths of each of the six *Caenorhabditis elegans* chromosomes (somatic chromosomes I through V, and the sex chromosome, X), as well as the mitochondrial chromosome. The number of genes, as well as the number of proteins encoded by those genes, is shown at the bottom. There are 1,144 more proteins formed than the number of genes, because of alternative splicing in some of the genes.

RNAs also exist in viruses, bacteria and archaea, the two processes are greatly expanded in eukarya, especially in multicellular organisms (Opisthokonta, Amoebozoa, and archaeplastida). The additional levels of control of gene expression are needed in multicellular organisms. The positions and gene expression patterns of each cell and cell type are controlled by combinations of proteins, RNAs, and interactions with other cells. Some of this is signaled by the specific patterns of cell division during development. For example, some cells in *C. elegans* are destined to die following cell division, whereas the other daughter cells will differentiate and become a necessary cell in the organism (see Figure 16.7).

Cell division and partitioning are not the only ways in which cells are programmed for specific patterns of gene expression. Communication with other cells and cell products is another way, and this is accomplished by signal transduction. This process is mediated by specific cell receptors. These consist of transmembrane proteins (or sets of proteins) that have receptors facing outward from the cell membrane and signaling domains on the cytosol side of the membrane. When a ligand attaches successfully to the receptor, the part of the molecule on the inside of the cell changes shape, which reacts with other proteins to bring the signal into the cell. In many cases, this signal is received by kinases or G-proteins. This causes the activation of various sets of proteins to change the biochemistry of the cell or the gene expression pattern of that cell. The number of genes that encode proteins for transduction and cell-to-cell signaling is much larger in *C. elegans* than in the single-celled yeast, *S. cerevisiae*. Although *S. cerevisiae* has just over 100 genes that encode proteins that are involved in cell-to-cell signaling and signal transduction (approximately 2.1% of its genome), *C. elegans* has more than 5000 such genes in its genome (approximately 21.1% of its genome). In addition, there are thousands of RNA genes that also contribute to control of gene expression to regulate cell development and characteristics.

Drosophila melanogaster

Insects are one of the most successful groups of organisms on the Earth, having diverged into many millions of different species. They live throughout the world in terrestrial and aquatic habitats. Some burrow through soils, crawl on top of the ground, and climb tall objects, whereas others fly. As with nematodes, they are multicellular and have complex developmental patterns in which cells differentiate for specific purposes and positions. Their cells proceed through developmentally regulated patterns of gene expression (see Chapter 16), and their cells must communicate with one another in the organism. Their central nervous system must be able to recognize external stimuli and send signals to appropriate cells in the organism to react to those stimuli. All of these complex interactions are mediated by gene expression patterns, some of which occur early in development, and others that are only fully functional in the adult animal.

The genome of *D. melanogaster* consists of 14,450 genes that encode 15,392 proteins, spread among five somatic chromosomes and one sex chromosome (Figure 33.6). It also has thousands of RNA genes, including those for rRNAs (approximately 140 copies), tRNAs, and ncRNAs. As with the *C. elegans* genome, a large number of gene products are involved in regulation of gene expression during development and in cell-to-cell communication and signal transduction. Another set of genes that controls development in insects, as well as in all multicellular organisms, is those containing HOX genes. Some of these had been studied prior to genomic sequencing of *D. melanogaster*, but more such genes and gene families were discovered after genomic sequencing of organisms as diverse as mouse, human, fish, sea urchins, nematodes, and plants. The similarities were especially striking among animals, where many of these genes have been characterized (including HOX, C2H2 zinc finger, helix-loop-helix binding domain, basic leucine zipper domain, forkhead transcription factor, EZ6 transformation specific [ETS] domain, retinoic acid receptor, and methyl-cytosine phosphatic guanosine [CpG]-binding domain). When the number of each of these homeotic (homeodomain) genes is compared among *S. cerevisiae*, *C. elegans*, *D. melanogaster*, and *Homo sapiens*, there is a steady progression from fewer to more of these genes.

In early genetic studies, it was recognized groups of homeotic proteins controlled distinct developmental pathways, and that they bound to DNA. Some contained several variable domains in addition to highly conserved domains. Furthermore, several patterns emerged from these studies. One set of these genes sets up different expression patterns very early to differentiate the head end from the tail end of the embryo. Another set of homeotic genes then turns on another set of genes that produced segments in the embryo. Another set of genes acts to bisect each of the segments, and still other sets of genes cause differential expression to set up dorsal and ventral parts of the animal. Still other sets determine where the eyes, legs, mouth parts, wings, and other organs will be placed. If these genes are expressed inaccurately, then body parts will be misplaced. For example, in one fruit fly mutant, legs appear adjacent to the eyes. These genes are arranged in tandem along a chromosome and are expressed temporally in the same order as they appear on the chromosome (Figure 16.10). Furthermore, their effects appear linearly along the organisms. In other words, in most cases the genes on one end of the HOX gene cluster are expressed in the anterior region of the animal, whereas those at the other end of the cluster are expressed in the posterior end. For invertebrates, there is a single gene cluster consisting of 9–13 genes. However, in vertebrates, there have been gene duplication and evolution events, such that some fish have six or seven gene clusters, whereas mammals (such as humans) have four clusters that have from 4 to 10 genes per cluster.

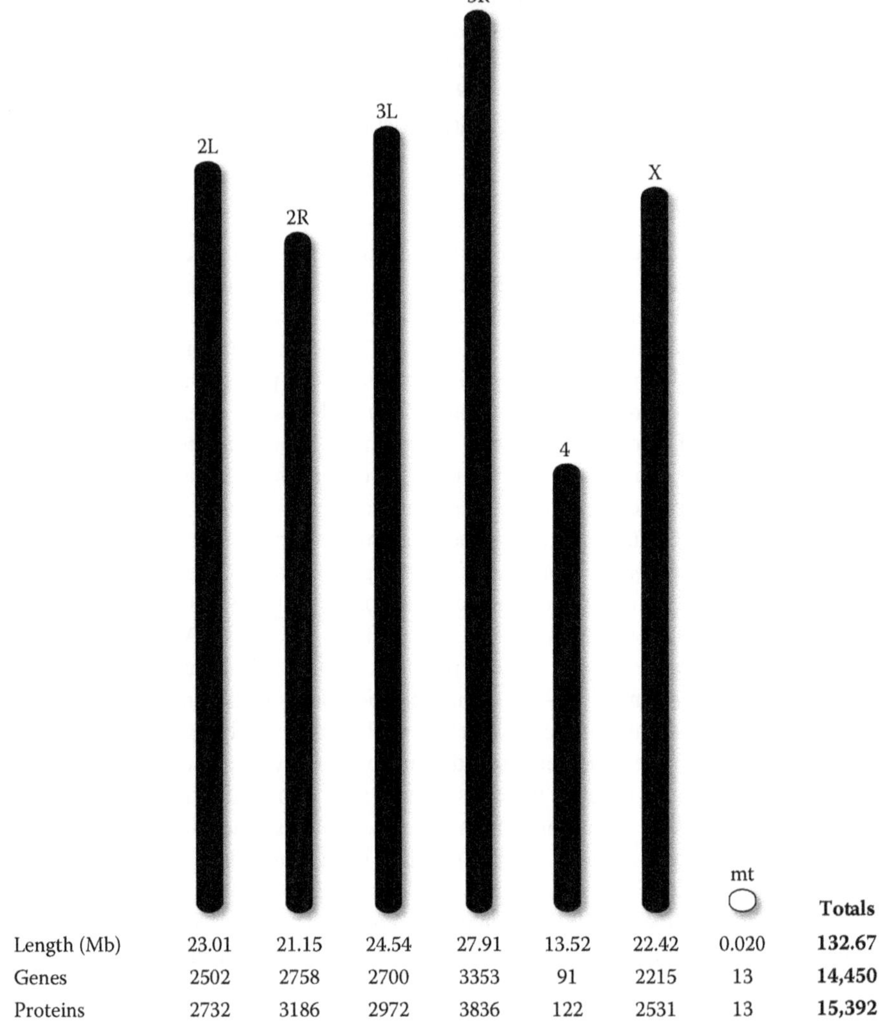

3R
3L
2L
X
2R
4
mt
Totals
Length (Mb) 23.01 21.15 24.54 27.91 13.52 22.42 0.020 132.67
Genes 2502 2758 2700 3353 91 2215 13 14,450
Proteins 2732 3186 2972 3836 122 2531 13 15,392

FIGURE 33.6 Representation of each of the lengths of each of the six *Drosophila melanogaster* chromosomes (somatic chromosomes 2L through 4, and the sex chromosome, X), as well as the mitochondrial chromosome. There are approximately 1000 more protein products than genes, due to alternative splicing of some genes.

ARCHAEPLASTIDA

Arabidopsis thaliana

This small annual plant has been a model organism for all plants for more than half a century. It was selected because of its ease of manipulation in the laboratory, primarily due to its small size and short life cycle (6 weeks). It also has a small genome (157 Mb, distributed on five chromosomes), which facilitated its sequencing (Figures 33.7 and 33.8). Although the genome is large compared to bacterial genomes, it is one of the smallest plant genomes, some of which are more than 1000 times larger. However, it contains almost all of the same gene components in common with all plants, and is therefore useful as a starting point to discuss plant genomes. As expected, it contains a large

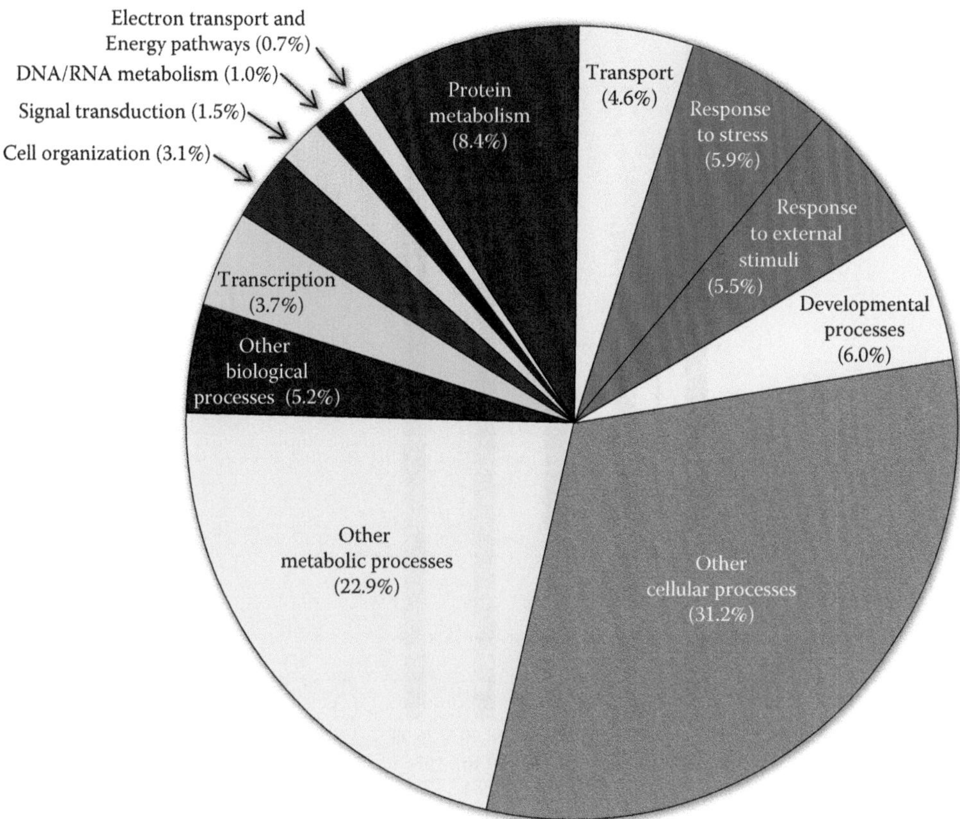

FIGURE 33.7 Summary of gene function categories for genes in the *Arabidopsis thaliana* genome. About 13% of the genome consists of genes that encode proteins for nucleic acid metabolism, DNA replication, transcription, and translation. Genes for energy metabolism, including photosynthesis, comprise almost 23% of the genome. More than 11% of the genome includes genes that produce proteins to help these organisms deal with stressful conditions and respond to external stimuli. More than 10% of the genome encodes proteins involved with cell organization, developmental processes (including embryogenesis), and signal transduction, some of which is for cell-to-cell communication.

set of genes devoted to the organization and function of the chloroplasts, and plastids in general. More than 12% of the genome (more than 3000 genes) encodes proteins that end up in the chloroplasts or plastids (Figure 33.9). This is understandable, given the fact that approximately 19% of the genome consists of sequences that originated from the cyanobacterial endosymbiont that became the chloroplast. This contrasts with 2.7% (approximately 750 genes) and 5.8% (approximately 1600 genes) of the genome whose protein products are sent to the mitochondria and nucleus, respectively (Figure 33.9).

Development in plants is complex, and therefore, homeotic genes and genes encoding proteins for signal transduction are present. However, because of the simpler development and the lack of nervous, endocrine, and immune systems, these genes comprise less of the genome than in animals. Signal transduction and developmental regulatory genes comprise about 7.5% of the genome (approximately 2000 genes). This is less than half the number of analogous genes in the *C. elegans* genome. As with all larger genomes, the percentages of genes encoding proteins for nucleic acid metabolism, transcription, and translation are smaller than for smaller genomes (e.g., *S. cerevisiae*

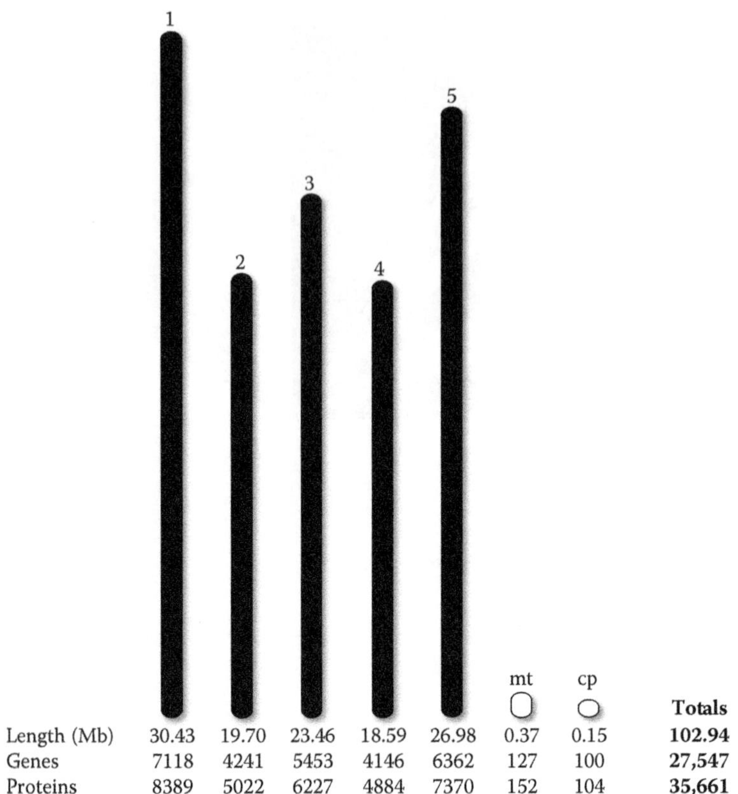

FIGURE 33.8 Representation of each of the lengths of each of the six *Arabidopsis thaliana* chromosomes (somatic chromosomes 1 through 5), as well as the mitochondrial and chloroplast chromosomes. There are more than 8000 more proteins produced than genes. As with the fungal and animal genomes, this is due to alternative splicing of some of the mRNAs. Also, there are thousands of known ncRNA genes that are not indicated on the diagram.

and *E. coli*). This is because the number of other types of genes is higher in the larger genomes. In reality, the number of genes encoding proteins for DNA replication, repair, recombination, transcription, and translation in the larger genome is greater than those in the small genomes, but other regions of the genome have expanded to a greater extent than those involved in some of the central pathways. Two parts of the genome that are expanded in plants are responses to stress (5.9% of the genome, or approximately 1600 genes in *A. thaliana*) and responses to external stimuli (5.5% of the genome, or approximately 1500 genes in *A. thaliana*). Most plants are photosynthetic, and therefore, they need their leaves to be in the sunlight. Also, most plants are stationary. They do not move once they have become established at a site. Therefore, they are exposed to the damaging wavelengths of light emitted by the sun (as well as the useful wavelengths). They are exposed to heat, cold, freezing, thawing, wind, rain, droughts, insects, microbes, and number of other stressors. They have survived, evolved, and diversified partly because they possess sets of genes that help them to overcome these stressors.

Oryza sativa

Rice is a member of the family Poaceae (formerly the Graminae), or the grass family. It is one of the most economically important plant families because it includes rice, wheat, maize, oats, sugarcane,

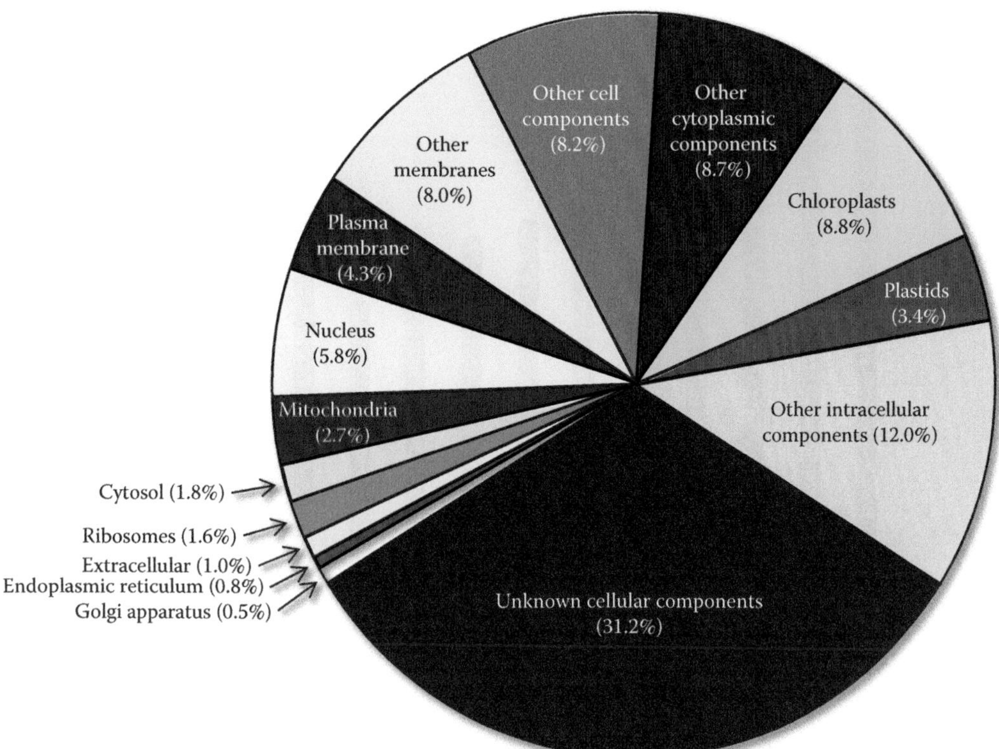

FIGURE 33.9 Summary of gene product destinations for the *Arabidopsis thaliana* genome. The proteins are targeted for a number of cell compartments, mainly through the actions of signal peptides (see Chapter 32). Nearly 9% (more than 3100) of the proteins end up in the chloroplasts, and another 3.4% end up in other plastids (e.g., chromoplasts and amyloplasts). Almost 6% are destined for the nucleus and 2.7% are sent to mitochondria. Only about 1.8% (approximately 450) end up in the cytosol. Most of the others are sent to a variety of cellular locations, including the cell membrane and other membranes that surround the organelles.

millet, and sorghum. Because of its economic importance, sequencing of a grass genome gained importance. However, many of the genomes of these grasses are tetrapoid (e.g., maize), hexaploid (e.g., wheat), or polypoloid (e.g., sugarcane), and some of the genomes are very large. It became important to approach sequencing in a careful and multilevel fashion. Rice has the smallest genome of these species. Even so, at 383 Mb, it is 3 times larger than the fruit fly genome and more than 80 times larger than the *E. coli* genome. Additionally, plant genomes usually have a great deal of repetitive DNA in their genomes. This complicates sequencing projects. Nonetheless, because it had the smallest genome of all the economically important grasses, its genome was sequenced. The 383 Mb of DNA is spread among 12 nuclear chromosomes, a mitochondrial chromosome, and a chloroplast chromosome (Figure 33.10). As expected, the contents of its genome are similar to those of the *A. thaliana* genome. This was expected because both are angiosperms and both utilize C3 photosynthesis. Additionally, central metabolic pathways and developmental pathways are similar between the two genomes.

Another way to examine these genomes is to look for homologies along the chromosomes (Figure 33.11). This is also known as synteny, meaning the degree of linear similarity along the chromosomes. The chromosomes of the major economic grasses have been compared, by both DNA hybridization studies and sequencing in some cases. A great deal of synteny is found among the chromosomes, although a great deal of chromosome rearrangement is also

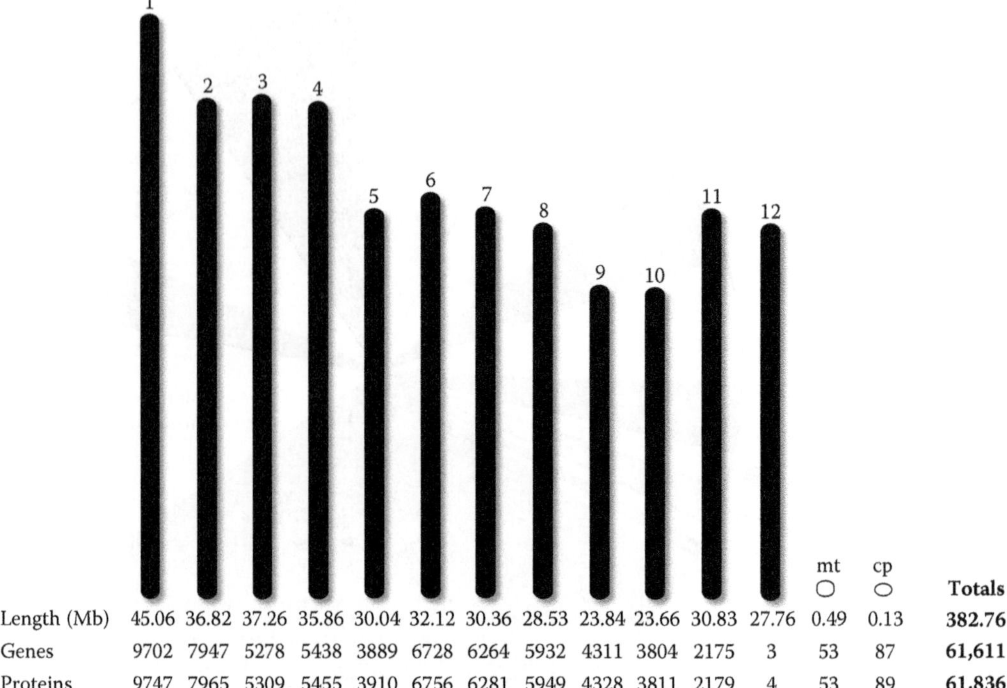

	1	2	3	4	5	6	7	8	9	10	11	12	mt	cp	Totals
Length (Mb)	45.06	36.82	37.26	35.86	30.04	32.12	30.36	28.53	23.84	23.66	30.83	27.76	0.49	0.13	**382.76**
Genes	9702	7947	5278	5438	3889	6728	6264	5932	4311	3804	2175	3	53	87	**61,611**
Proteins	9747	7965	5309	5455	3910	6756	6281	5949	4328	3811	2179	4	53	89	**61,836**

FIGURE 33.10 Representation of each of the 12 *Oryza sativa* chromosomes (chromosomes 1 through 12), as well as the mitochondrial and chloroplast chromosomes. Rice has a larger genome than *Arabidopsis thaliana*, and its genome encodes almost twice as many proteins as the *A. thaliana* genome, although alternative splicing is more limited, accounting for only about 220 additional proteins. Chromosome 12 has low density of genes.

evident. Furthermore, the comparisons illustrate the types of changes that have occurred during the evolution of grasses and beyond. In particular, the vast majority of genes are adjacent to the same neighboring genes among all of the grasses, that is, they are linked in a collinear fashion. However, several inversions have been documented. The order of some genes has been reversed. Many of the inversions involve more than one gene, and some involve extensive parts of the chromosome. In some cases, individual chromosomes have fused into single larger chromosomes, as is the case of chromosomes 7 and 9 in rice, which exist as a large fused chromosome in other members of the Poaceae (e.g., chromosome II in foxtail millet). Between *Oryza sativa* and *Avena sativa* (oat, genome = 11,300 Mb), many translocations, fusions, splitting, and recombination events appear to have occurred, such that even though the genome of *A. sativa* is 30 times larger than that of *O. sativa*, the haploid chromosome number for *A. sativa* is 7, whereas that of *O. sativa* is 12. This can occur within a genus as well. For example, broad bean, *Vicia faba*, has a haploid chromosome number of 6, with a genome size of approximately 13,500 Mb. All other species within the genus have smaller genome sizes and seven chromosomes. Additionally, their chromosomes all are approximately the same length. Chromosome 1 in *V. faba* is twice the size of any of the other five chromosomes. Hybridization studies have shown that in *V. faba*, two chromosomes present in the other species have fused to form a single large chromosome. Similar events have been illustrated when comparing the human genome with the mouse genome.

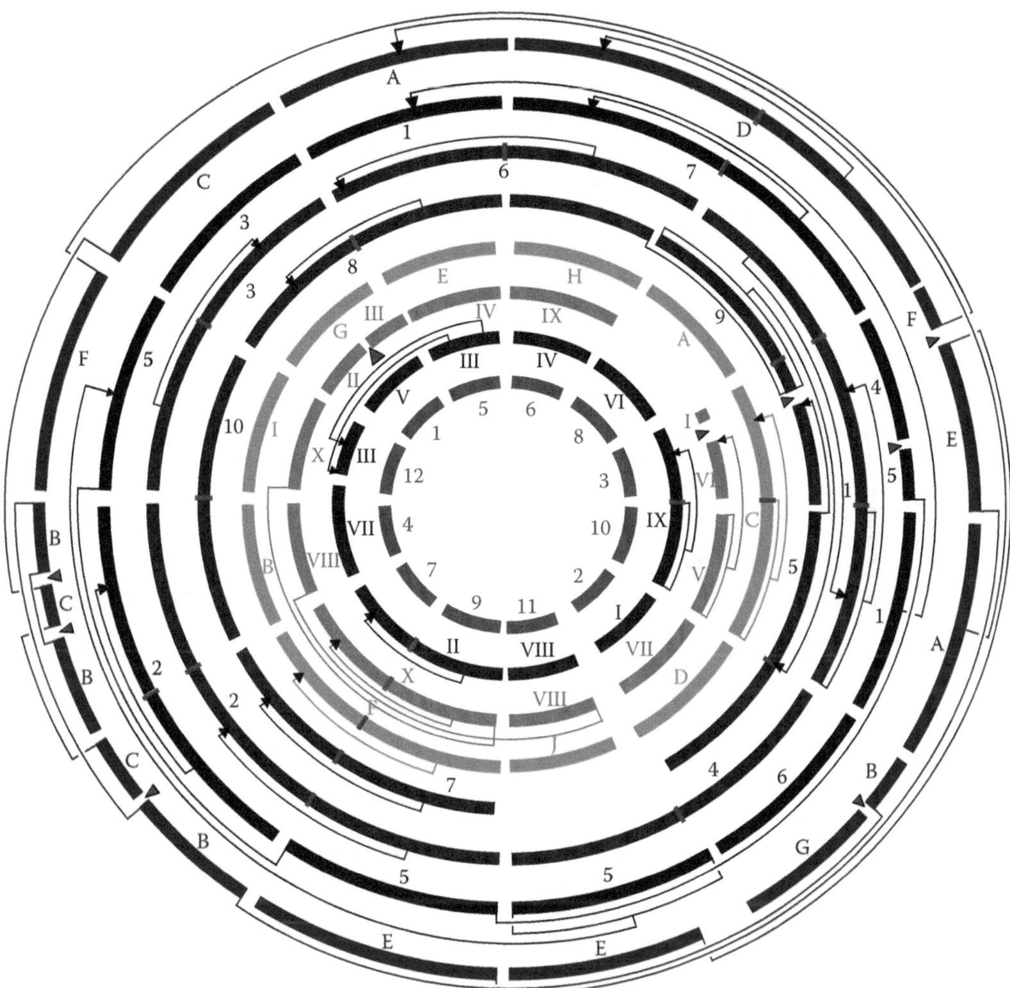

FIGURE 33.11 Circular alignment of the chromosomes for several grasses to demonstrate synteny and chromosomal rearrangements. The species with the smallest genome (*Oryza sativa*, rice, 483 Mb, with 12 chromosomes, in red) is in the center and progressing outward by increasing genome size. The genomes are (from inside to outside) as follows: foxtail millet (*Setaria italica*, 490 Mb, with nine chromosomes, in blue), sugarcane (*Saccharum officinarum*, 1000 Mb, with 10 chromosomes, in orange), sorghum (*Sorghum bicolor*, 1600 Mb, with 10 chromosomes, in green), maize (*Zea mays*, 2500, with 5 + 5 chromosomes [this species has experienced a doubling of the chromosomes and is tetraploid, with subsequent rearrangements and losses of some chromosomal fragments], in purple), wheat (*Triticum aestivum*, 6000 Mb [this is a hexaploid species, only the D chromosome set shown], with seven chromosomes, in black), and oat (*Avena sativa*, 11,300 Mb, with seven chromosomes, in dark red).

KEY POINTS

1. Eukaryotes contain mixtures of genes from bacterial and archaeal ancestors.
2. The evolution of eukaryotes has included multiple instances of endosymbiosis, organelle formation, and genome interactions.
3. The nucleus appeared approximately 2.4 billion years ago and may have originated from a similar structure found in planctomycetes.

4. Planctomycetes developed internal membranes probably to protect their nucleic acids from toxic compounds.
5. Mitochondria probably became established in an endosymbiotic process separate from the establishment of the nucleus in eukaryotes.
6. Multicellularity and genes for cell-to-cell communication, development, and ncRNA increased together during eukaryote evolution.
7. Evolution of eukaryotes has included chromosome homologies (i.e., synteny), chromosomal splitting, chromosomal fusing, chromosomal losses, inversions, and recombination, as well as mutations to gene sequences.

ADDITIONAL READINGS

Alberts, B., D. Bray, K. Hopkin, A. Johnson, J. Lewis, M. Raff, K. Roberts, and P. Walter. 2013. *Essential Cell Biology*, 4th ed. New York: Garland Publishing.

Becker, W. M., J. B. Reece, and M. F. Poenie. 1996. *The World of the Cell*, 3rd ed. New York: The Benjamin/Cummings Publishing Company.

Bermudes, D., G. Hinkle, and L. Margulis L. 1994. Do prokaryotes contain microtubules? *Microbiol. Rev.* 58:387–400.

European Bioinformatics Institute. EMBL-EBI. http://www.ebi.as.uk/integr8.

Fuerst, J. A. 2010. Beyond prokaryotes and eukaryotes: Planctomycetes and cell origin. *Nat. Educ.* 3:44.

Futuyma, D. J. 1998. *Evolutionary Biology*, 3rd ed. Sunderland, MA: Sinauer Associates.

Gale, M. D. and K. M. Devos. 1998. Comparative genetics in the grasses. *Proc. Natl. Acad. Sci. USA* 95:1971–1974.

Gregory, T. R. 2005. *The Evolution of the Genome.* San Diego, CA: Elsevier Academic Press.

Hall, B. K. and B. Hallgrimsson. 2008. *Strickberger's Evolution*, 4th ed. Sudbury, MA: Jones & Bartlett Publishers.

Hartl, D. L. and E. W. Jones. 2009. *Genetics, Analysis of Genes and Genomes*, 7th ed. Sudbury, MA: Jones & Bartlett Publishers.

Makarova, K. S. and E. V. Koonin. 2010. Two new families of the FtsZ-tubulin protein superfamily implicated in membrane remodeling in diverse bacteria and archaea. *Biol. Direct.* 5:33.

Monteiro, A.S. and D. E. K. Ferrier. 2006. Hox genes are not always collinear. *Int. J. Biol. Sci.* 2:95–103.

Moore. G., K. M. Devos, Z. Wang, and M. D. Gale. 1995. Grasses, line up and forma circle. *Curr. Genet.* 5:37–739.

Pilhofer, M. and K. Rappl. 2008. Characterization and evolution of cell division and cell wall synthesis genes in the bacterial phyla Verrucomicrobia, Lentisphaerae, Chlamydiae, and Planctomycetes and phylogenetic comparisons with rRNA genes. *J. Bacteriol.* 190:3192–3202.

Tropp, B. E. 2008. *Molecular Biology, Genes to Proteins*, 3rd ed. Sudbury, MA: Jones & Bartlett Publishers.

van Niftrik, L. A., J. A. Fuerst, J. S. Damsté, J. G. Kuenen, M. A. M. Jetter, and M. Strous M. 2004. The anammoxosome: An intracytoplasmic compartment in anammox bacteria. *FEMS Microbiol. Lett.* 233:7–13.

34 Human Genome

INTRODUCTION

The completion of the human (*Homo sapiens*) genome still remains a work in progress. By the time the first draft was reported in 2003, it was clear that there were major obstacles to be overcome to complete detailed analyses of the genomic sequence data. One obstacle was that only about 1.5%–2.0% of the DNA in the genome contained protein-encoding genes (Figure 34.1).

Therefore, out of the 3.0–3.3 billion base pairs comprising the total genome, only about 50 million of the base pairs encoded proteins. Most of the first draft of the genome was focused only on that small percentage of the genome. The second obstacle was that many of the genes contained introns, and some of the introns were very large, much larger than the exons. The initial searches for open reading frames (ORFs) found many partial genes. It was unknown at the time whether they represented pseudogenes, whether the sequence data contained errors, or if the genes were authentic. Eventually, many of these were pieced together, and it was discovered that some genes were much longer than anyone had anticipated. Some were over a megabase in length. For example, the dystrophin gene is 2.4 Mb in length (the longest gene known), which includes 79 exons and 78 introns. To transcribe the entire gene requires 16 h *in vivo*. After processing, the mature mRNA is only 14 kb in length. Therefore, while all of the exons together are 14 kb in length, the introns comprise more than 99% of the gene length (Figure 34.2). A set of shorter transcripts are also produced from internal promoters. All of the resulting proteins are involved in muscle function. Mutations in the dystrophin gene cause partially functional or nonfunctional proteins (Figure 34.3), causing either Becker muscular dystrophy (BMD; a mild form) or Duchene muscular dystrophy (DMD; severe form). The dystrophin genes in BMD patients are missing one or more exons, although the reading frame of the protein is unaffected, so a shorter, but somewhat functional, protein is produced. The dystrophin genes in DMD patients are also missing exons, but the deletions cause reading frame shifts, which causes premature truncation of the proteins because of the presence of stop codons within the normal exon. These proteins are nonfunctional, causing muscle cell dysfunction and severe disability in those patients.

In addition to detailing the human genome, sequencing studies of various indigenous peoples throughout the world have been undertaken. When combined with archaeological evidence, detailed investigations of the migration of humans over the past several hundred thousands of years have been made, with some surprising results. Mummified corpses and other human remains have been used to study past civilizations, migration, and culture. Sequence data has been used extensively in medical fields. Specific mutations, as well as combinations of single-nucleotide polymorphisms (SNPs; or snips) have been used in the diagnosis and treatment of various diseases, including certain cancers. Also, SNPs and sequence information have been used to determine the genealogy of families and individuals. Of course, genetic data has also been used in forensic studies of crime scenes. This valuable application has led to the conviction of many criminals, and has helped to exonerate the innocent.

THE HUMAN GENOME

The human genome was determined by many different groups around the world and was accomplished by using several different strategies. The first draft genomes were released in 2003 to coincide with the 50th anniversary of the 1953 Watson and Crick paper in Nature reporting the structure of DNA. The genome was determined using standard methods to sequence fragments of the genome, as well as by using copies of the expressed portions of the genome. Subsequently, various groups have gone back to sequence individual chromosomes in great detail. The first surprise of the first genome determinations is that there were fewer genes than were expected. Estimates had ranged

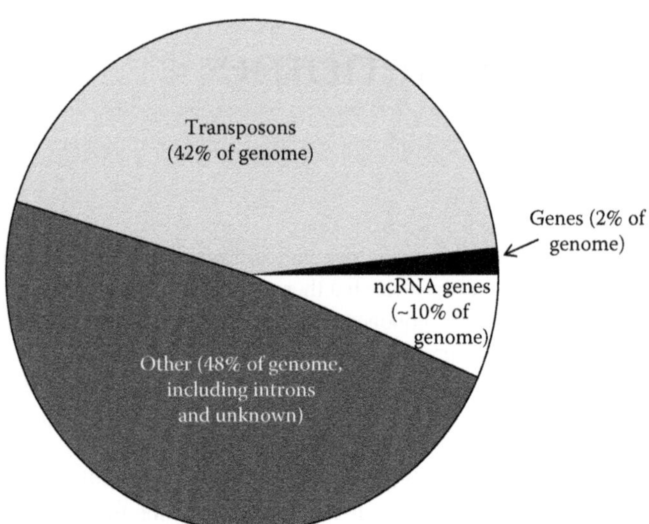

FIGURE 34.1 Pie chart of the general categories of sequences within the human genome. Approximately 1.5%–2.0% of the human genome consists of protein-encoding genes. Approximately 10% of the genome is composed of ncRNA genes. Transposons (primarily SINES and LINES) make up 42% of the genome, and the remainder consists of a mixture of introns and a large number of regions of unknown content and function.

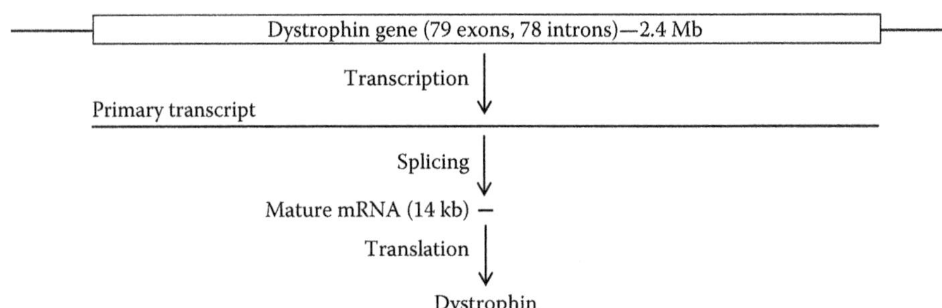

FIGURE 34.2 Characteristics of the dystrophin gene. The gene is 2.4 million base pairs in length, which is composed of 79 exons and 78 introns. The exons comprise 14 kb of the sequence, while the introns make up the remainder of the gene. The primary transcript is also approximately 2.4 million bp in length, which requires about 16 h of continuous transcription. Following processing (including splicing), the mature mRNA is 14 kb in length.

from 50,000 to 200,000 genes, but in the end the number was between 20,000 and 30,000. Current estimates range from 20,000 to nearly 23,000. The reason for the discrepancy in this genome and others is that it is difficult to know what is a gene and what is not. Open reading frames (ORFs) of triplets of nucleotides that include a stop codon at the 3′ end are considered genes. However, if they are short, they are excluded or questioned. If they contain introns, they may be missed, and if the introns are very long, some of the exons might be missed. This is true for the human genome. There are many introns, and some of them are extremely long. Also, when a gene is found, if it is novel or its function cannot be determined, it is unclear whether it is an authentic gene or not.

The genome of *H. sapiens* consists of over 3 billion bp that includes approximately 21,000 genes (estimates range from 20,000 to 23,000), which encodes approximately 90,000 to 100,000 proteins, spread among 22 somatic and two sex chromosomes, as well as a small mitochondrial genome (Figure 34.4). The nuclear genome also has thousands of RNA genes including those for rRNAs,

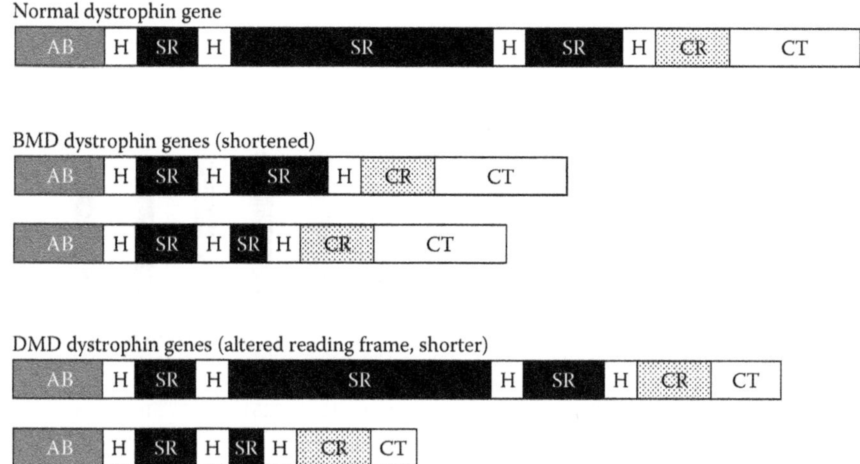

FIGURE 34.3 Normal and mutant dystrophin genes. Normal dystrophin genes encode several different regions of the protein, including an actin-binding domain (AB) on the amino end, four hinge regions (H), three regions with spectrin-like repeats (SR), a cysteine-rich region (CR), and a carboxy-terminus domain (CT). In the dystrophin protein, the last two domains bind several proteins involved in muscle cell structure and function. In BMD patients, one or more of the internal exons have been deleted, resulting in a shorter dystrophin protein, but one that still has partial function. In DMD patients, the deletions can be small or large, but they all change the reading frame, such that a stop codon results, which produces a dystrophin protein that is missing much of the carboxy-terminus domain. This protein is nonfunctional.

tRNAs, and other ncRNAs (noncoding RNAs). In the progression from single-celled organisms to complex multicellular organisms with complex tissues, organs, and developmental patterns, several characteristics emerge. First, the genomes of the more complex organisms have grown very large in most cases. Second, the number of genes has increased as organisms have become more complex. Third, much of the increase in the number of different proteins has been due to alternative splicing rather than increases in the number of genes. Much of the increase in complexity in eukaryotes is coincident with increases in the number of transcriptional regulatory genes, as well as genes involved in cell-to-cell communication.

An adult human has approximately 37 trillion (3.7×10^{10}) cells. These are organized into organs, which are composed of many tissue types, composed of many types of cells. Some cell types must interact with the outside or inside environment, as well as interacting with adjacent cells. Some cell types are primarily for locomotion, or other movements. Others produce large amounts of one or more compounds necessary for specific functions in the organ and organism. Still others move in the blood to carry oxygen, patch holes in blood vessels, and attack foreign organisms. Possessing a large genome is absolutely vital for survival. Some sets of genes will be expressed in one cell type, but not another. In other cells, only subsets of those genes will be expressed. And for most functions of these cells and cell types, sets of genes act to control the expression of other genes. The coordination within a large genome, such as in humans, must function properly, otherwise parts of the individual may fail causing the death of the entire individual. This can be observed when a disease causes tissue or organ failure, which either is debilitating or fatal.

The human genome is similar to the genomes of other mammals that have been examined. While the human genome is approximately 3.0–3.3 billion base pairs in length, the genome of chimpanzees is about the same length. Also, the gene set is almost identical, although mutations among some of the genes define the differences between the two species. The chromosomes contain the genes in the same order (they have high synteny) in both species. This is expected, because the two species diverged recently, approximately 2.0–3.0 million years ago. However, while humans have 22 pairs of chromosomes, plus the sex chromosomes, chimpanzees have an additional chromosome, which is

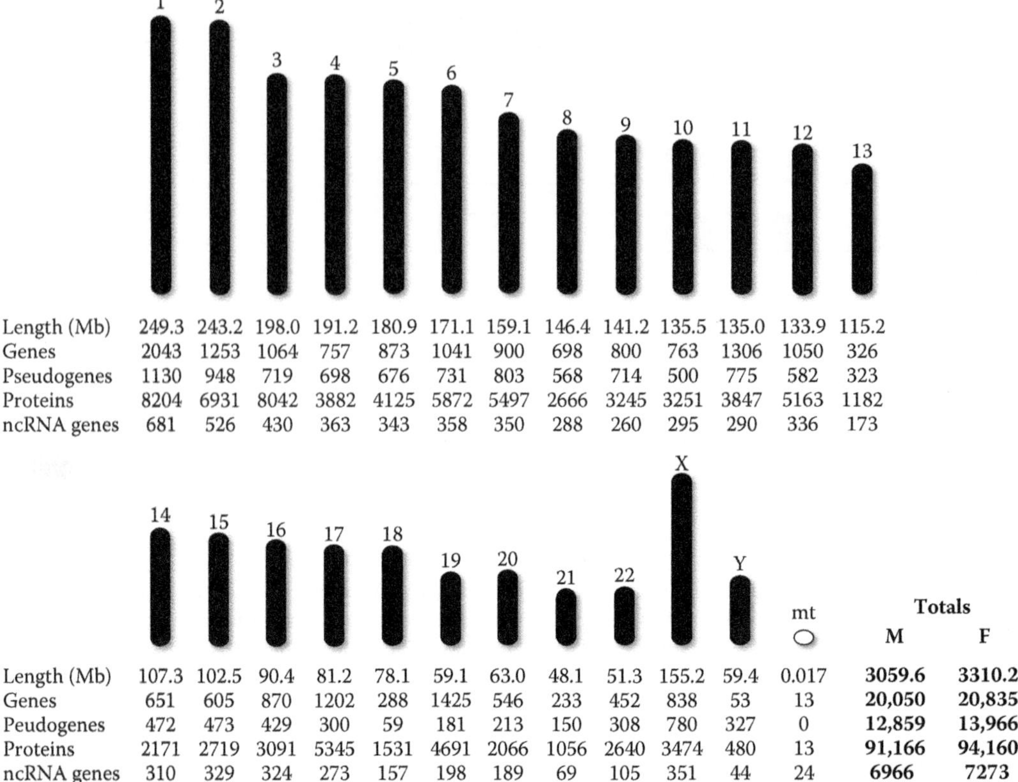

	1	2	3	4	5	6	7	8	9	10	11	12	13
Length (Mb)	249.3	243.2	198.0	191.2	180.9	171.1	159.1	146.4	141.2	135.5	135.0	133.9	115.2
Genes	2043	1253	1064	757	873	1041	900	698	800	763	1306	1050	326
Pseudogenes	1130	948	719	698	676	731	803	568	714	500	775	582	323
Proteins	8204	6931	8042	3882	4125	5872	5497	2666	3245	3251	3847	5163	1182
ncRNA genes	681	526	430	363	343	358	350	288	260	295	290	336	173

	14	15	16	17	18	19	20	21	22	X	Y	mt	Totals M	Totals F
Length (Mb)	107.3	102.5	90.4	81.2	78.1	59.1	63.0	48.1	51.3	155.2	59.4	0.017	**3059.6**	**3310.2**
Genes	651	605	870	1202	288	1425	546	233	452	838	53	13	**20,050**	**20,835**
Peudogenes	472	473	429	300	59	181	213	150	308	780	327	0	**12,859**	**13,966**
Proteins	2171	2719	3091	5345	1531	4691	2066	1056	2640	3474	480	13	**91,166**	**94,160**
ncRNA genes	310	329	324	273	157	198	189	69	105	351	44	24	**6966**	**7273**

FIGURE 34.4 The human genome, divided by chromosome. The 22 somatic chromosomes, the X and Y chromosomes, and the mitochondrial genome are shown. The lengths of the chromosomes, number of protein-encoding genes (either confirmed or predicted), the number of pseudogenes, the number of proteins produced from the authentic genes, and the number of ncRNA genes are presented. Totals (lower right) are provided for male (M) and female (F) genomes.

caused by a split of chromosome 2. That is, while humans have a long chromosome 2, chimpanzees have two chromosomes, that when aligned to the human chromosome 2, line up with the two arms of that chromosome. Therefore, during evolution, either the large chromosome 2 has split into two separate chromosomes in chimpanzee, or the two smaller chromosomes have fused together to produce the human genome. When organisms that diverged from humans earlier, the gene sets are similar, but gene synteny is lower. For example, mice (*Mus musculus*) diverged from humans between 30 and 90 million years ago. While mice have a gene set very similar to those of humans, the order of genes on the chromosomes have changed greatly (Figure 34.5). Also, gene expression and alternative splicing patterns differ between the two species, which causes the differences in morphology and behavior.

MEDICAL GENETICS

DNA sequencing analyses are being used with increased frequency in the field of medicine. There are now many tests for specific diseases (Table 34.1), such that the sequence information can be used in diagnosis and treatment. The treatments, procedures, and pharmaceuticals can be tailored specifically to treat the diseases, and in most cases this has led to increases in the effectiveness of the treatments. Molecular methods have also been used to screen healthy individuals to determine their risks for certain diseases. For example, individuals can be tested for the presence of specific genes that indicate a high risk for certain cancers, Huntington's disease, lupus erythematosus, colon

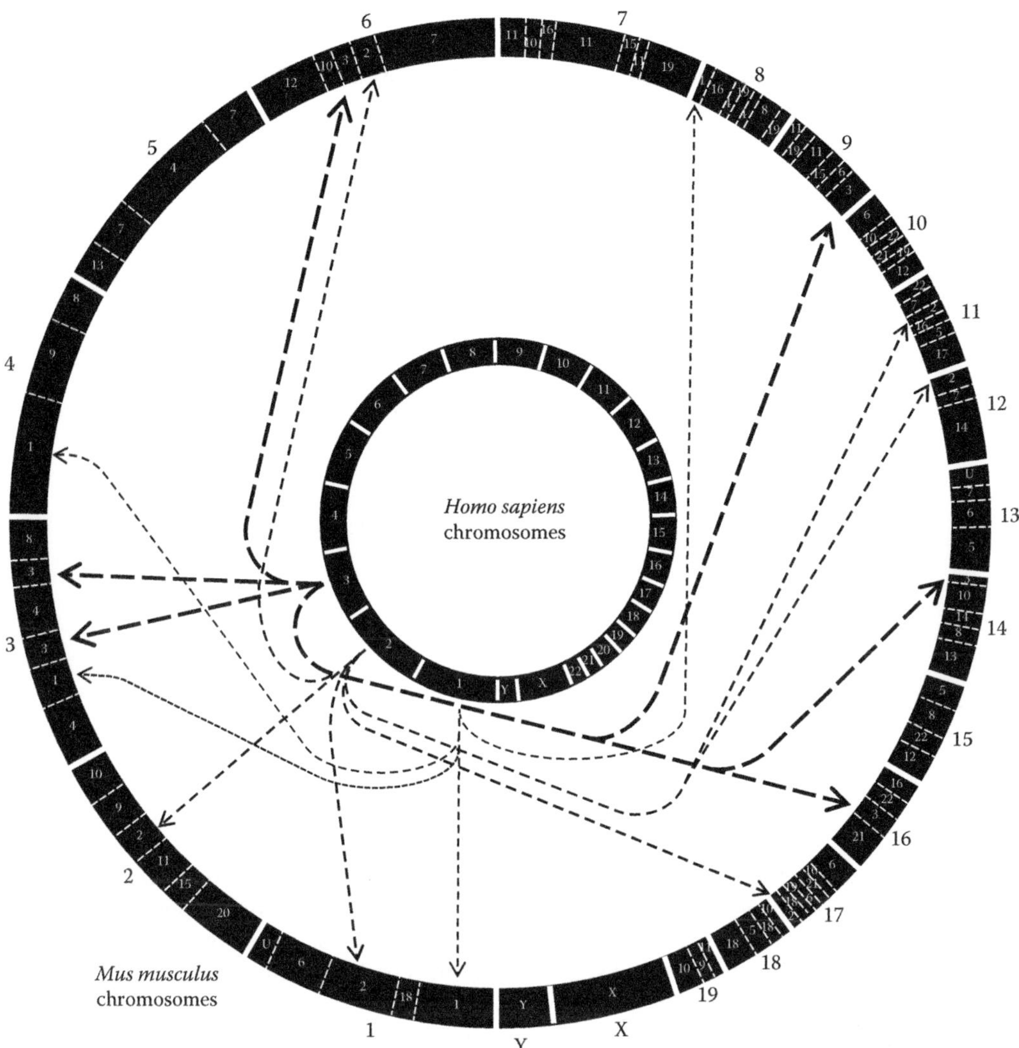

FIGURE 34.5 Diagram showing some of the rearrangements of genes along human versus mouse chromosomes. The human chromosome set (22 somatic chromosomes, plus X and Y) is represented in the inner ring. The mouse (*Mus musculus*) chromosome set (19 chromosomes, plus X and Y) is the outside ring, with numbers corresponding to the same gene regions on the human chromosome set. Arrows indicate the rearrangements of the regions of chromosomes 1 (thin dashed lines), 2 (medium dashed lines), and 3 (thick dashed lines). (Rearrangements of the regions of other chromosomes are not shown for simplicity.)

cancer, and others. By screening patients, doctors can perform more frequent exams to detect early signs of disease. In some cases, patients may want to take preventative actions by surgical removal of organs that are at high risk of becoming diseased (e.g., breast and ovaries).

Screening for other purposes has also increased in frequency. For example, then drugs are to be administered to patients, in some instances the genetic background of the patient is analyzed so that the proper dosages can be administered for the most effective treatment of the ailment. Patients with different genetic backgrounds can react differently to certain drugs, in terms of adverse reactions, as well as in the speed of metabolizing or catabolizing the pharmaceutical agents. Prenatal and newborn screening often is performed when any abnormalities are detected, or in cases where particular genetic diseases are present in a family. This can greatly affect the outcome of the birth and

TABLE 34.1
Examples of Diseases that Can Be Diagnosed through Genetic Identification of Specific Gene Mutations

Disease(s)	Gene Mutations
Bipolar	ANK3
Breast cancers	BRCA
Celiac	HLA-DQ
Colon cancers	APC, BRCA, MLH1, MSH1, MSH2, MSH6, PMS1, PMS2
Cystic fibrosis	CFTR
Fragile X chromosome	FMR1
Down syndrome	Amyloid, superoxide dismutase, ETS-2
Hemophilia	Clotting factors
Huntington's	Huntingtion
Lupus erythematosus	BANK1, BLK, CDKN1K, IRF5, ITGAM, PTPN22, STAT4, TNSFS4
Macular degeneration	ABCR
Muscular dystrophy	Distrophin
Obesity	FTO
Ovarian cancers	BRCA
Parkinson's	LRRK2
Psoriasis	HLA-C
Sickle cell anemia	HBB
Thalassemia	α-globins and β-globins

development of the child. Many individuals are tested prior to a pregnancy to determine whether or not they carry specific alleles of certain genes that might adversely affect a potential child. In some cases, embryos can be tested in vitro prior to implantation in the female in order to assure that specific mutant alleles are absent, and that a normal birth is probable. One other genetic test that is performed frequently is to determine paternity of a child.

Some diseases are directly caused by defects in the mitochondria. These may cause diabetes, heart disease, deafness, and or liver disease in children who inherit the faulty mitochondria. These mutations are easily detected using molecular biology methods, and there are ways to replace the mitochondria so that the child will be born without the disease-causing mutations. Because mitochondria are inherited exclusively from the egg cell of the mother, an egg cell from a donor with normal mitochondria can be obtained, and the nucleus can be replaced by one from the potential mother's egg cell. The engineered egg cell then carries all of the potential mother's nuclear genes, but carries the donor's mitochondrial genes. When this cell is fertilized with the father's sperm, the resulting zygote will have genes from three individuals, and the child that results from this will not carry any of the defective mitochondrial genes that might have caused serious disease in the child. In this way, medical genetics can predict an adverse outcome, and medical procedures can correct the problem such that a positive outcome is assured.

SINGLE-NUCLEOTIDE POLYMORPHISMS

Within the human population of over seven billion individuals, mutations are abundant. SNPs have been mapped in order to study human variation, heredity, diseases, and other phenomena. These are simply single-nucleotide changes that have occurred randomly throughout human evolution. Some are within genes, although most are found in regions that do not encode any protein.

The large majority of them cause no phenotypic changes in individuals with those mutations. They occur about once every 300 base pairs in the human genome, such that there are approximately 10 million SNPs in the human genome. They vary between populations, and among individuals, such that ancestry can be traced using some of these. Most mutations represented by SNPs are harmless. However, they can be used are identifiers of certain genotypes, and a few of them do indicate an increased risk of diseases (e.g., cancers, diabetes, heart disease) in specific genes. They are being used in medical studies to attempt to relate them to the risk of specific diseases. They also can be used in the diagnosis of some of the diseases presented in Table 34.1.

FORENSICS

Because the genome of each person is unique (and each cell is actually unique), their DNA is as unique (or more so) than a fingerprint. Many genetic loci can be used to determine identity from blood, semen, sweat, hair, bones, teeth, tissue fragments, fingernails, fingerprints, and other remnants from individuals. Of course, when the sample sizes are very small and/or they are old, degradation and contamination become confounding issues. DNA is the molecule most often used for identification. While the quantity of RNA is higher in each cell, most of it is rRNA, and it is rapidly degraded in the environment. Also, DNA is more constant from one cell to another, and it contains all of the genetic information for each individual.

Microsatellite and minisatellite DNAs have been used effectively for decades in forensic investigations. These occur in high frequency in the human genome (as well as in many other eukaryotic genomes). They consist of variable numbers of short repetitive elements (2–5 bp for microsatellites; and 6–100 bp for minisatellites) that are flanked by unique sequences. There are several ways to detect them. One is to digest the DNA, separate it on a gel (either agarose or polyacrylamide), blot the gel, and then probe the gel using a microsatellite or a minisatellite probe (labeled with radioactivity or a nonradioactive system). Bands appear at various locations on the gel, and they are unique to an individual, although related individuals have similar (or relatable) patterns. Another method of testing, especially when only small amounts of tissue and DNA are available is to use PCR. In this case, the unique DNA bordering the repetitive region is used, which again will produce bands on the gel of varying lengths, indicating the number of repeats for that band. Also, each individual will have a unique pattern of bands. These methods have been used to identify an individual from DNA collected from single fingerprints left at crime scenes, although this is not always successful.

HUMAN MIGRATION

Human migration and history have been studied in detail for centuries. Human remains in the form of fossils, desiccated tissues, and artifacts have led to many theories about where *H. sapiens* originated and spread throughout the world. During the past two decades, molecular methods, including sequencing studies using ancient DNA, have augmented these studies, often providing a more complete picture of what happened in the past. From the fossil and molecular data, it is clear that modern humans originated in Central Africa. Much of the data indicates that they diverged from *H. heidelbergensis*, approximately 200,000–300,000 years ago. Prior to that, approximately 500,000 years ago, two other species also diverged from *H. heidelbergensis*, namely *H. neanderthalensis*, and another species, currently known only as the Denisovans (the name was derived from the location where bones were found, called Denisova, in Siberia). These species were distributed in Europe and Asia before *H. sapiens* migrated out of Africa; *H. heidelbergensis* and *H. neanderthalensis* were found mainly in Europe and parts of Arabia, and the Denisovians were distributed in Asia (Figure 34.6). In the half-million years since the divergence of these species, mutations accumulated independently in each species. Therefore, while they each had essentially the same sets of genes, there were allelic differences, and these have been demonstrated in recent genomic studies

FIGURE 34.6 Human migrations as indicated by archaeological and molecular genetic studies. Prior to the appearance of *H. sapiens*, an ancestral archaic species, *H. heidelbergensis*, had spread into Europe, and diverged into at least three species, *H. neanderthalensis*, *H. sapiens*, and the Denisovians, and possibly a fourth species (indicated by a question mark in Africa). Approximately 200,000–150,000 years ago, the *H. sapiens* populations began to grow (A) and migrate throughout Africa (B). *(Continued)*

(using ancient DNA from bones) of members of each species. By 200,000 years ago, the *H. sapiens* population was concentrated in Central Africa. Approximately 150,000 years ago, the *H. sapiens* population grew to such a size that portions of the population began to migrate to other areas in Africa. This migration is supported not only by fossil evidence, but the molecular sequence data (primarily from mtDNA) indicates that Africa has the greatest genetic diversity of anywhere on Earth, due to the early migration, and then isolation of the various populations that moved away from the other populations.

The next notable migration occurred approximately 60,000–70,000 years ago, when one population moved into modern Arabia, India, Southern China, Malaysia, and Australia. However, as they migrated, they encountered some of the archaic humans, and interbred with them. Approximately 1%–2% of the alleles in the genomes of non-African individuals are from the *H. neanderthalensis* genome. Also, from 3% to 5% of the alleles found in the genomes of Melanesians and Aboriginal Australians are from the Denisovian genome. This indicates that individuals of *H. sapiens* were interbreeding with members of the other archaic human species. About 40,000–50,000 years ago, humans had begun to migrate throughout Europe. As with previous migrations, they encountered archaic humans on their way. They also gained alleles from those archaic humans, because from 1% to 2% of the alleles in individuals native to Europe are from the *H. neanderthalensis* genome. One other example of interbreeding in human history is known. In Cameroon, the Y chromosomes of certain populations contain dozens of alleles that are found in no other human population. It is thought that these alleles were introduced into this population by interbreeding with another species of human, possibly more than 100,000 years ago. However, the identity of the archaic human species is unknown.

Approximately 25,000–35,000 years ago, the Earth was in an ice age (one of many), which caused more water to be frozen in glaciers, which lowered sea levels by up to 100 m. This exposed large areas of the continental shelves, one of which created a land bridge between Eastern Asia and Western North America. Because of some morphological, cultural, and technological differences (primarily in tool and pot making) among the various populations of Native Americans, it was concluded that there had been two to four (or more) migrations from Asia into North America. However, recent DNA sequencing studies are consistent with either one or two migrations approximately 20,000–25,000 years ago, followed by spread throughout North and South America during the ensuing 10,000–15,000 years. One surprising outcome of the DNA studies was that the population of humans that migrated was already a genetic mixture of alleles from Melanesia/Australia and another set from Central Asia. The only exception to the timing is for the Inuits (i.e., Eskimos), who migrated from Northeastern Siberia into Northern North America more recently, possibly about 5000 years ago.

One other hypothesis regarding the migration of humans into North America was recently dealt a setback based on molecular data. In 1996, archaeologists unearthed the remains of a man on the banks

FIGURE 34.6 (Continued) One group in present day Cameroon, interbred with an archaic human species, indicated by a hybridization event (H1). From 60,000 to 70,000 years ago, at least one population of *H. sapiens* migrated out of Africa through the Arabian Peninsula. Some of the members appear to have interbred with members of *H. neanderthalensis* (H2), because all non-African populations have 1%–2% *H. neanderthalensis* alleles within their genomes. These populations then continued to migrate through Asia (C). While in Asia, they interbred with another group of archaic humans, known as the Denisovians (H3). Then, they continued to migrate through Malaysia and Micronesia, and eventually into Australia (C). From 40,000 to 50,000 years ago, the H. sapiens populations migrated into western and northern Europe (D), interbreeding with members of *H. neanderthalensis* (H4), whose numbers began to decrease, eventually going extinct, probably due to the increased competition with *H. sapiens*. About 20,000–25,000 years ago, *H. sapiens* populations from central Asia (E, F) and from Melanesia/Australia began to migrate across a land bridge formed by lowered sea levels during a major ice age. There may have been one or two such migrations, which extended to South America by 14,000 year ago (G), and further east into North America by about 12,000 years ago (H). An additional migration (I) occurred about 5000 years ago, when northern Asians moved into northern North America, to become the Inuits (i.e., Eskimos).

of the Columbia River, near the city of Kennewick, Washington. He was nicknamed *Kennewick Man*. The bones were carbon dated to be about 9000 years old. Also, some of the features of the skull and skeleton appeared to have more European than Native North American characters. This appeared to support a theory that there had been proposed previously that some very early migration had occurred from Europe into North America. The remains had been removed from Native American land, and the Umatilla tribe wanted the remains to be returned so that they could bury them again. On the other hand, the scientists wanted to study them to determine whether or not these remains were from a European. Legal maneuvering continued for years, and for part of the time the remains were unavailable to those on both sides of the issue. Finally, after nearly 20 years, a small amount of DNA from the bone marrow of the remains was extracted and analyzed using sequencing methods. The results matched the DNA sequences of Native Americans, and failed to match those of any known European genetic group. The conclusions based on the morphological characters of the bones were shown to be erroneous, and proof was provided that the bones were from a Native North American.

KEY POINTS

1. The human genome consists of approximately 3.0 (males) to 3.3 (females) billion base pairs of DNA per haploid cell.
2. While the first draft of the human genome was reported in 2003, estimates on the total number of genes are still being made.
3. Only about 1.2%–2.0% of the human genome contains protein-encoding genes, and most genes contain multiple introns, and many of the RNAs are subject to alternative splicing.
4. Organisms closely related to humans have nearly the same set of genes, and they are arranged on the chromosomes in the same order (i.e., high degree of synteny).
5. Mammals that are more distantly related to humans have nearly the same set of genes, but the order of the genes along the chromosomes has changed (i.e., moderate to low degrees of synteny).
6. DNA sequence information is useful in diagnosing human diseases, determining ancestry, forensic investigations, human migration, and for other purposes.

ADDITIONAL READINGS

Alberts, B., D. Bray, K. Hopkin, A. Johnson, J. Lewis, M. Raff, K. Roberts, and P. Walter. 2013. *Essential Cell Biology*, 4th ed. New York: Garland Publishing.

Alberts, B., D. Bray, A. Johnson, J. Lewis, M. Raff, K. Roberts, and J. D. Watson. 1994. *Molecular Biology of the Cell*. New York: Garland Publishing.

Becker, W. M., J. B. Reece, and M. F. Poenie. 1996. *The World of the Cell*, 3rd ed. New York: The Benjamin/Cummings Publishing Co.

Blake, D. J., A. Weir, S. E. Newey, and K. E. Davies. 2002. Function and genetics of dystrophin and dystrophin-related proteins in muscle. *Physiol. Rev.* 82:291–329.

Dillehay, T. D. 2009. Probing deeper into first American studies. *Proc. Natl. Acad. Sci. USA* 106:971–978.

European Bioinformatics Institute. EMBL-EBI. http://www.ebi.as.uk/integr8.

Futuyma D. J. 1998. *Evolutionary Biology*, 3rd ed. Sunderland, MA: Sinauer Associates.

Gibbons, A. 2014. Three-part ancestry for Europeans. *Science* 345:1106–1107.

Hall, B. K. and B. Hallgrimsson. 2008. *Strickberger's Evolution*, 4th ed. Sudbury, MA: Jones & Bartlett Publishers.

Hartl, D. L. and E. W. Jones. 2009. *Genetics, Analysis of Genes and Genomes*, 7th ed. Sudbury, MA: Jones & Bartlett Publishers.

Rassmussen, M., S. L. Anzick, M. R. Waters, P. Skoglund, M. DeGiorgio, T. W. Stafford Jr., S. Rassmussen et al. 2014. The genome of a Late Pleistocene human from a Clovis burial site in western Montana. *Nature* 506:225–229.

Reich, D., N. Patterson, M. Kircher, F. Delfin, M. R. Nandineni, I. Pugach, A. M.-S. Ko et al. 2011. Denisovan admixture and the first modern human dispersals into Southeast Asia and Oceania. *Am. J. Hum. Genet.* 89:516–528.

Tamaki, K. and A. J. Jeffreys. 2005. Human tandem repeat sequencesing forensic DNA typing. *Legal Med.* 7:244–250.

Templeton, A. 2002. Out of Africa again and again. *Nature* 416:45–51.

The International HapMap Consortium. 2007. A second generation haplotype map of over 3.1 million SNPs. *Nature* 449:851–861.

Tropp, B. E. 2008. *Molecular Biology, Genes to Proteins*, 3rd ed. Sudbury, MA: Jones & Bartlett.

Index

Note: Page numbers followed by f and t refer to figures and tables, respectively.

2′-deoxyribonucleic acid, 147
2′-deoxyribose adenosine monophosphate, 45f
5-methylcytosine (m5C), 48f
5S rRNAs, 57, 63
    transcription, 61
5′ untranslated region (5′ UTR) of mRNA, 163
6-methyladenine (m6A), 48f
8-hydroxy-guanosine (8-OH-G), 177
16S rRNA, 57
18S rRNA, 57
21st biological amino acid, 53
23S rRNA, 57
25S rRNA, 57
454 pyrosequencing, 362
    NGS using, 363f

A

Ableson murine leukemia virus (A-MuLV), 309
Abl gene, 305, 309
ABO blood group genotypes, 137, 137f
Acanthamoeba Polyphaga Mimivirus (APMV), 12, 461, 471, 472f, 473
Ac element. *See* Activator element
Acetyl coenzyme A (acetyl-CoA), 505
Acquired immunodeficiency syndrome (AIDS), 449
Actinobacteria, 28, 243
Activator element, 205
Adenines, 47, 49, 177
Adenocarcinomas, 301
Adenosine, 203
Adenosine 5′-phosphosulfate (APS), 362
Adenosine triphosphate (ATP), 26, 42, 115, 342, 362, 528, 530
    in nucleotides, 45f
    synthase, 496, 502f, 503, 505, 510
A-form DNA, 45, 47f
Agarose gel electrophoresis, 324–329. *See also* Polyacrylamide gel electrophoresis (PAGE)
    alternative methods of, 329f
    DNA separated by, 325f
    fundamentals of, 326f
    movement of DNAs during, 330f
Agouti, 139, 139f
Agrobacterium tumefaciens, 233, 233f
Alignment algorithms, 368
Alleles, 34, 132, 303
    Hardy–Weinberg equilibrium, 148–149, 149f
    inheritance analysis with, 136f
    in population, 148f, 150–152
    Punnett square analysis, 148, 149f
    through time, 159–170
        factors affecting allelic proportions, 170
        gene flow, 167, 169–170
        levels of selection, 162–165, 162f
        mating and dispersal, 166–167
        natural selection, 160–161
        proportion changes, 160f, 169f
        random genetic drift, 165–166, 165f
        trends in proportions, 161f
    time course of proportions, 132f
α (alpha) globin gene, 218
α-proteobacterial genome, 235, 527–528, 530
Alternative splicing, 187, 203–204, 205f
Amino acids, 52–53
    acidic and basic, 53
    characteristics compared with nucleotide, 85t
    frequencies in ancient proteins, 90–91, 90f
    on genetic code, 84
    hydrophobic, 53
    and peptide bonds, 52f
    sequence, DNA, 366, 370f
    structures of side chains, 53f
    universal code indicating categories, 88, 88f, 89f
Aminoacyl tRNA synthetases (aaRS), 65, 85–86, 86f
Amphibian oocytes, 79, 79f
    extrachromosomal amplification of, 80, 80f
    formation in animals, 79, 79f
Amplification, 121f, 122
    cancer, 305–306
        effects on survival, 308f
        gene amplification levels, 307f
    extrachromosomal
        of rRNA genes in amphibian oocytes, 80, 80f
        of rRNA genes in *Tetrahymena,* 78f
Amyloplast, 510
Anammoxosome, 528
Anastamosis, 120f
Anastral mitosis, 114
Ancestral mitochondrial genome, 527
Angiosperm, 275
    fertilization and development of, 275f
    inflorescences, 289f
    leaf arrangements exhibited by, 281f
Angstrom, 13
Animal(s)
    life cycles for, 154f
    mating choice, 166
    parasite, 490–491
    vs. plants, 287–291
Animal cell, cell cycle for, 100f
Animalia, 251, 251f
Anoxigenic photosynthesis, 477
Antennapedia, 254
Antibiotic resistance genes, 231, 233–234
Antipodals, 275
Anti-Shine–Dalgarno sequence, 67
APC (adenomatous polyposis coli) gene, 298
Aphids, life cycle for species of, 154, 155f
Apical initials, 276–277
Apicomplexans, 502, 531

APMV. *See* Acanthamoeba Polyphaga Mimivirus (APMV)
Apoptosis, 300
in *Caenorhabditis elegans,* 253f
triggered by p53, 253
Applied Biosystems, 359
Apurinic/apyrimidinic (AP) endonucleases, 181
Aquifex, 479–480
Aquifex aeolicus, 479–480
Arabidopsis thaliana, 237, 507–508, 538–540, 541f
Archaea, 34, 231, 247
cell cycle for, 97, 98f, 113f
central dogma, 69
kingdom, 396
polyadenylation, 163
replication pathways for, 112f
rRNA genes in, 58–59, 59f
variations on DNA segregation, 111–113
Archaeal genome, 436
Archaeal intron, 203
Archaeal splicing mechanism, 204f
Archaebacteria, 34
Archaeplastida, 271, 274
Arabidopsis thaliana, 538–540
Oryza sativa, 540–543
phylogram of major groups in, 272f
Archegonium, 275
A′ repeats sequence, 78
Aristotle (384–322 BC), 30, 32
Arthropod development, cells, 261–264
Ascomycetous fungi, 183
A (amino acid)-site, 75–76
Asterisk activity, 335
Astral mitosis, 114
AUG codon, 83
Autonomous bacterium to organelle progression, 499t
Auxin biosynthesis, 244
Avena sativa, 542
Avian leucosis virus (ALV), 308
Avian myeloblastosis virus (AMV), 309
Axin protein, 298

B

Back mutation, 174
Bacteria, 12, 34
cell cycle for, 97, 98f, 113f
central dogma, 69
chains of cells, 247
comparing genomes of, 439f
genomic studies of, 436
kingdom, 396
media for growing, 341
polyadenylation, 163
quorum sensing in, 249f, 250f
replication pathways for, 112f
rRNA genes in, 58–59, 59f
signal peptide in, 515–518
species, 393
variations on DNA segregation, 111–113
Bacterial artificial chromosomes (BACs), 373–374
Bacterial chromosomes, 234
Bacterial conjugation, 231, 232f
Bacterial diderms. *See* Diderms
Bacterial flagella gene, 222–223
Bacterial genomes, 429, 432, 445, 487–489
Bacterial phyla, 243
Bacterial ribosome, 73
Bacterial sex, 125
Bacterial virus genomes, 234
Bacteriophage
ϕX174, 461–462
genome of, 463f
life cycle, 462f
genome, 234
integration, 189, 191f
lambda (λ), 448, 463–468
gene expression control in, 466f
pathways of, 464, 465f
virus production in lytic pathway, 467, 468f
Bacteriophage M13, 210
Bacteriophage MSZ, genomic sequence of, 358f
Bacteriophage T4, 448, 468–470, 470f
Bacterium possessing, 233
Baltimore classification system, 192
Basic Local Alignment Search Tool (BLAST) algorithm, 367, 367f
Bateson, William, 131
Baumannia cicadellinicola, 489
Bayesian phylogenetic analysis, 415
B-cell lymphomas, 304
Bcl-2 protein, 304
Beadle, George, 34
Beak dimensions in birds, 395
Becker muscular dystrophy (BMD), 545, 547f
Belon, Pierre, 32
β-catenin, 298
β-galactosidase gene, 340
β (beta) globin gene, 218
Beta thalassemia, 187
Bicoid (*bcd*) gene, 261
Biological evolution, 19, 28
Biological molecules, 389–390
agarose gel electrophoresis. *See* Agarose gel electrophoresis
extraction
of nucleic acids using CTAB, 319–321
of proteins, 329–330
of RNA, 322–323
overview, 315–319
PAGE, 331–332, 331f
purification of organellar DNA, 321–322
quantification
of nucleic acids, 324, 324f, 325f
of proteins, 330
using centrifugation, 322f
Biological organisms, 41
Biosynthetic pathways for steroid hormone, 311f
Birds, beak dimensions in, 395
Bisbenzimide, 321
Biston betularia, 160
Bithorax gene, mutation in, 254
Blastomas, 301
Blunt-ended molecules, 342
BMD. *See* Becker muscular dystrophy (BMD)
Bony fish, 28
Bootstrapping method, 415–416, 416f
Bovine spongiform encephalopathy (BSE), 10, 196

Bradford method, 330
Branch-and-bound search strategy, 414f
Brca1 breast cancer gene, 302, 302f
Brca2 breast cancer gene, 302, 302f
Broad bean (*Vicia faba*), 77, 142, 142f, 542
Brown, Robert, 33
Brugia malayi, 495
Bulge–helix–bulge structure, 200f, 203
Burkitt's lymphoma, 304
Bypassing fission, 113f

C

Caenorhabditis elegans, 6, 155, 251–252, 534–537
  anatomy of, 252f
  apoptosis in, 253f
Calvin cycle, 238, 477, 508, 510
  enzymes origin, 512f
  in spinach, 238f
Cambrian explosion, 28, 37
Cancer
  annual US mortality (2008), 294f
  causes of mutations in carcinogenesis, 301–311
  cells, 293, 295
    characteristics, 293, 300
    documented chimeric genes in, 306f
    vs. normal cells, 296f
  classification, 301
  during diagnosis, 301
  gene categories, 295f
  genes involved in, 298–301
  hormones, 311–312
  overview, 293–295
  progression of, 295–298, 296f
  proteins involved in, 299f
  retroviruses cause, 310f
  stages of, 301
  translocations, 304f
  types of, 301
  world age-standardized rate, 294f
Cancer-inducing genes, 234
Candida, 486
Canis genus, 393, 394f
Canis lupus familiaris, 394
Cap, 163
Cap-binding proteins (CBP), 163
Capsid, 192, 461, 462
Carbohydrates, 42, 55, 55f
Carcinogenesis, 389
  causes of mutations in, 301–311
    amplification, 305–306, 307f
    DNA viruses, 309–311
    point mutations, 302
    recombination, 302–305
    viruses, 306–309
  changes in chromosomes and mutated genes, 297f
  genes involved in, 298
  mechanism of, 300f
Carcinomas, 301
Carsonella ruddii, 487–488
Caudal (*cad*) gene, 262
Caulobacter crescentus, 97, 247
  cell cycle of, 248f
C box, 64, 64f
CD4 receptor, 451
Cell(s), 247, 248f
  arthropod development, 261–264
  cancer. *See* Cancer, cells
  and cellular components, 15f
  cycle
    for animal cell, 100f
    for archaea, 97, 98f, 113f
    for bacteria, 97, 98f, 113f
    changes in DNA amount through, 120, 121f, 122
    for eukarya, 97, 99f
    for eukaryotes, 122f
    for single-celled bacteria and archaea, 98f
    variations in mitosis and, 116–117, 118f
  development in animals, 251
  development in plants, 274–277
    change shape, 274
    communication of plant cells, 273
    formation of leaves and floral organs, 279–287
    gene expression, 277–279
    morphology, 273–274
  development in vertebrates, 264–266
  diploid layers of, 276
  healthy, 300
  histogenic layers of, 276
  homeotic genes
    hierarchy and evolution of, 266–268
    and proteins, 253–261
    structure of, 255f
  membranes, 12
  nematode development, 252–253
  normal, 295
  nucleus, 14, 33
  quorum sensing, 249–251, 249f, 250f
  stalked, 247
  surface receptor, 278
Cellular dimensions, 13–14, 16t
Cellular life on Earth, 6
Cellular oncogenes, 234
Cellular organisms
  DNA for, 41
  genomes of, 196, 429
  ribosomes, 6, 73
Cenozoic Epochs, 27f
Central dogma of molecular biology, 6, 6f, 42
  beyond, 69
  defined, 57
  expanded version of, 58f
Cesium chloride (CsCl) isopycnic gradients, 321
Cetyltrimethylammonium (CTA^+) cation, 320
Chagas disease, 210, 239
Chain termination method, 358
Chaperones, 71, 520, 523
Charophyta, 271
Chelating agent, 315
Chemical sensory mechanisms, 247
Chemiluminescent methods, 349
Chemotaxis (CEM) cells, 252–253
Chimeric pathway, 510–512
Chimpanzees *vs.* human chromosome, 547–548
Chloramphenicol, 340
Chlorarachniophyta, 513, 531
Chlorophyta, 271

Chloroplasts, 399, 506–508
acquisition of, 28
functional, 508–510
protein trafficking in, 520, 522–523
rRNA transcription, 79
Choanoflagellates, 251
Chromalveolata, 238, 245, 513, 531–532
Chromatids, 115, 123–124
homologous, 115
during meiosis, 126f
Chromatin, 206
Chromatophore, 28
Chromosomal arrangements of *Hox* gene, 256f
Chromosomal recombination events, 303f
Chromosomal translocations, 302–304, 304f, 440
Chromosome(s), 14, 115–116, 440, 445
bacterial, 234
crossovers, 182–185, 183f, 184f
exchange, 34
in genetic loci, 182f
in human somatic cells, 108
hypothetical evolution of, 440f, 441f
integration, 191–192
metaphase, 14
movements during mitosis, 101f
number variations, 118–120, 121f
packaging for eukaryotes, 108f
polytene, 122
replication of, 107–108, 107f
Chronic myelogenous leukemia, 305
Chymes, 287
cI protein, 464, 467
cII protein, 467
Ciliates, 513
Circular phylogenetic tree, 25f
Circular phylogram, 427f
Circular plasmid, 209f
Circular trees, 402
CK1 (casein kinase 1), 298
Cladogram, tree, 401
Clamp, 120f
Clamp connection, 119–120
Cloning, 339
PCR products with TOPO TA vectors, 342
steps in molecular, 341
using ligation, 341f
using PCR amplicons, 342f
vector, 463
Clostridia, 28, 243
Cluster analysis of pairwise alignments (CLUSTAL), 368
Cnemidophorus, 394
Codominance, 137–138
ColE1, 232–233
Colicin, 233
Colorectal cancer, 298
Colorimetric probes, 349
Commensal organism, 485
Comparative genomics, 435–440
analysis of domains, 436
comparing bacteria, archaea, and eukarya, 436f
research, 436
Compatible end molecules, 342
Complementary DNA (cDNA), 451
Complex ribozymes, 6
c-oncogenes (cellular oncogenes), 308, 309f
Conidia, 151
Conjugation, 232
Conjugative plasmid, 210
Contigs, 363
Contour-clamped homogeneous electric field gel electrophoresis (CHEF), 327
Coomassie Brilliant Blue G-250 (blue dye), 330
Coresident symbionts, 489–490
Correns, Carl, 33, 131
Corymbs, 287
Cosmids, 339
Cotyledons (seed leaves), 276
CpG (cytosine–phosphate–guanosine) sites, 45, 47
Creighton, Harriet, 34
Crenarchaeota, 203, 482, 483f
Creutzfeldt–Jakob disease, 12
Crick, Francis, 42
Cro protein, 467
Cryptomonas, 513, 531
Cryptosporidium, 502
CTAB (cetyltrimethylammonium bromide), 315
extraction of nucleic acids using, 319–321, 320f
C-terminal amino acid, 366
Cuvier, Georges, 32
C-value
paradox, 445, 448
of species, 118
Cyanelles, 506
Cyanobacteria, 230, 477
Cyanobacterial endosymbiont, 237
Cyanobacterial genome, 527–528, 539
Cysteines, 53
Cytochrome b/f, 508
Cytochrome c, 520
Cytokinesis, 117
Cytosines, 47, 177, 357
Cytosol, 505, 507– 508, 510, 515, 518, 520, 523
Cytosolic ribosomes, 457

D

D-amino acids, 52
Dark reaction, 508, 510
Darwin, Charles, 32–33, 131, 160, 393–395, 400
Darwin, Erasmus, 32
Dataset, size of, 432
Daughter cell, 445
DCC (deleted in colorectal carcinoma) tumor suppressor protein, 298
Denisovan, 551
Densitometer, detection apparatus, 351
DEPC (diethylpyrocarbonate), 323
Descent of Man (Book), 33
Detergents, 315, 383
Detoxification, 230
Devonian period, 272
De Vries, Hugo, 33, 131
Diakinesis, 124
Diderms, 242, 243f, 516, 519f
Dikarya, life cycle for, 157, 157f
Dikaryon, 120f
Dinoflagellates, 117, 531
Dinosaurs, 28, 261

Dipeptides, 92–93, 92f
Diploid human genome, 180
Diploid layers of cells, 276
Disaccharides, 55
Disease diagnosis by genetic identification, 550t
Disk flowers, 287
Dissociation element, 205
Divergent morphology, 395
D-loop, 102f, 184
DMD. *See* Duchene muscular dystrophy (DMD)
DNA (deoxyribonucleic acid), 42, 147, 453–454
  A-form, 45, 47f
  amount through cell cycle, changes, 120, 121f, 122
  ancient DNA in phylogenetic studies, 407f
  binding capacities, 345
  for cellular organisms, 41
  chemical changes to, 174
  dimensions, 16t
  double-helical nature of, 34
  effects in, 177
  end-labeling of, 346f
  ethidium bromide on, 326–327, 327f
  in eukaryotes, 448
  to extract high-purity, 318f
  foreign, 339
  forms of, 47f
  fragments, 362, 364
  as genetic material, 41
  and histone modifications, 388f
  levels of selection, 162, 162f
  ligase, 100
  metabolism, 41
  in molecular biology and genetics, 315
  molecular mass of, 329
  molecules, migration of, 328f
  movement during agarose gel electrophoresis, 330f
  mutations, 176t
  nick translation for labeling, 347, 347f
  nucleic acids, 42–45, 47–51
  nucleotides, 43
    components of, 44f
    examples of, 45f
    general form of, 43f
    hydrogen bonded pairs, 46f
  oncoviruses, 306
  oxidation of, 177
  polymerase, 180–182
  polymerization, 99
  preservation under various conditions, 317f
  purification of organellar, 321–322
  radiation, 177
  random primer labeling of, 347f
  recombinant DNA methods, 338–343
  recombination events, 184
  repair mechanisms, 307–308
  repair system, 180–182, 181f
  segregation in bacteria and archaea, 111–113
  separated by agaorse gel electrophoresis, 325f
  sequence analysis, 7
  single-stranded, 362
  in slot-blotting, 351
  subject to degradative processes, 176f
  transfer of, 420
  types of mutation in, 174f
  for viewing in TEM, 354f
  viruses, 308f, 309–311, 461
  Watson–Crick interactions, 49
  Z-form, 45, 47, 47f
DNA replication, 34, 47
  of chromosomes, 107–108, 107f
  circular DNA molecule, 102f
  fidelity of, 97, 99–101
  lagging strand in, 100
  leading strand in, 100
  summary of, 102f
  topology during, 106–107
  variations of, 101, 103, 105
DNA–RNA hybrid region, 99
DNA sequencing methods, 339, 357
  aligning sequences, 368–370, 369f
  alignment scores, 367
  automated sequencing, 362f
  development of, 357–360
  high-throughput technologies, 360–361
  membrane-based method for, 366f
  NGS, 361–365
  protein sequencing, 365–366
  sequence homology searches, 367–368
Documented chimeric genes in cancer cells, 306f
Dogs, 394
Domain Eukaryota, 396
Dot blots method, 351
Double-stranded DNA (dsDNA), 469
  Sputnik, 472, 473f
  virus, 234f
Down's syndrome, 183
Drosophila melanogaster, 251, 253, 261, 436, 537–538
  cancer, 293
  eye color in, 138, 138f
  gene activation and inactivation in, 266f
  genome, 6
  life cycle of, 254f
  maternal gene products and gap genes in, 267f
DS (sodium dodecyl sulfate), 315
DsDNA–reverse transcriptase virus, 194
Ds element. *See* Dissociation element
Duchene muscular dystrophy (DMD), 545, 547f
Dystrobrevins, 385
Dystrophin gene, 385, 545
  characteristics, 546f
  normal and mutant, 547f

E

Earth
  cellular life on, 6
  chemical/biological evolution, 5f
  evolutionary steps to life on, 20f
  evolution of organisms on, 28
  first life on, 3
  history, 22–30
    acquisition of chloroplasts, 28
    amino acids, 23–24
    bony fish, 28
    dinosaurs, 28
    eukaryotes, 25–26, 25f, 26f
    extinction, 29
    flowering plants and mammals, 28

Earth (*Continued*)
Homo sapiens sapiens, 29–30
length of time for organisms, 29, 29f
marine invertebrates, 28
Moon, 22
natural, 21f
oldest rocks, 23
as one year, 35, 36f, 37
oxygenic photosynthesis, 24
periods of, 33
solar system, 22
stromalites, 23
volcanoes, 22–23
presence of membrane, 12
viruses on, 9
Ebola virus, 457–458, 458f
EDTA (ethylenediaminetetraacetic acid), 315
Electroblotting methods, 343–344
Electron transport, 482
Electrophoresis, 332, 383–384
one-dimensional gel, 383
two-dimension gel, 384
Embryogenesis, 220
Emulsion polymerase chain reaction (emPCR), 362
End-labeling of DNA, 346f
Endomitosis, 122
Endonuclease, 187, 192, 198
Endoplasmic reticulum/reticula, 519–520, 523, 528
Endoreduplication, 112f, 113f, 121f, 122
Endoribonuclease, 203
Endosperm, 276
Endosymbionts, 35
to organelles, 236f
Endosymbiosis, 495
to organelles, 512–514
patterns of, 497f, 498
Endosymbiotic event, 500f, 527–528, 530–531
Energy metabolism, 475
Enhancer elements (EEs), 266–267
Entities, species, 393
Enveloped virus, 192
Env gene, 308–309
Enzymatic RNA, 198
Enzymes, 181
Epigenomics, 388–389
Epistasis, 138–140
pigment pathway, 140, 141f
Punnett square analysis
of alleles coat color in rodents, 139, 139f
in sweet pea flower color, 139, 140f
Epstein–Barr virus (EBV), 310
Error-prone process, 97
Escherichia coli (*E. coli*), 152, 191, 463, 475–477, 480, 482, 491
F plasmid in, 232
high frequency of recombination (Hfr) strains of, 232
K-12, 61–62, 435, 438
ribosomes, 71
E (exit)-site, 75–76
Estrogen receptors, 311
Ethidium bromide on DNA, 326–327, 327f
Etioplast, 510
Euchromatic regions, 113, 206
Euglena, 239, 513
Eukarya, 396
cell cycle for, 97, 99f
central dogma, 69
introns, 163
ribosomal assembly in, 61f
rRNA genes in, 58–59, 59f
signal peptide system in, 518–520
Eukaryotes, 6, 12, 25–26, 25f, 26f, 35, 231, 271, 445, 448
5S genes, 59
cell cycle for, 122f
chromosome packaging for, 108f
DNA in, 448
genetic loci in diploid, 141
methylation, 47
with mitochondria, 503f
mitosis in, 113
relative numbers of transcription factors among, 437f
RNA polymerase I, 57
rRNA genes in
organization of, 62, 62f
processing of, 63f
transcription, 76
virions, 449
Eukaryotic cell
origin of, 529f
RNA, 323
Eukaryotic genomes, 373, 435–436
Archaeplastida
Arabidopsis thaliana, 538–540, 541f
Oryza sativa, 540–543
Chromalveolata, 531–532
evolution, 527
multicellularity, 531
Opisthokonta
Caenorhabditis elegans, 534–537
Drosophila melanogaster, 537–538
Saccharomyces cerevisiae, 532–534
organelle in, 527
Euryarchaeota, 480–481
Evolution, 19
biological, 19, 28
of genetic code, 83–86, 91f
of globin gene, 221f
and hierarchy of homeotic genes, 266–268
of histone gene, 224f
history of study of, 30, 32–35
Darwin's ideas, 32–33
human civilization and historical events, 31f
inheritance, 30
inventions and information, 30, 32–34
marine fossils, 30
periods of Earth, 33
integral parts of translation, 95f
Lamarckian, 111–112
of mobile genetic element, 190f
multigene family, 224–227
organelle, 497f, 498
organisms on Earth, 28
peptide-producing reactions, 92, 92f, 93f
of polyploid genome, 144, 145f, 146
of ribosomes, 73f
steps to life on Earth, 20f
of triplet codon, 89f
of tRNA, 94f

Evolutionary trees, 399, 401f
Excavata, 242f, 245
Exon, 192, 198, 200, 203
Extinction, 29
Extrachromosomal element (ECE), 477, 480

F

Female gametophyte in gymnosperm, 275
Femtogram, 13
Fermentation, 33
Fibonacci patterns, 282
Fibrillar portion, 113f
Firmicutes, 243
Flagellated cell, 248
Flax (*Linum usitatissimum*), 142, 145f
Floral organs, formation of, 279–287, 285f, 286f
Flowering plants and mammals, 28
Fluorescence-tagged primers, 338
Fluorescently labeled 2′,3′-dideoxynucleotide triphosphates (ddNTPs), 361f
Fluorimeters, 324
Fluorimetry method, 324
Foreign DNA, 339
Founder effect, 170
Fox, George, 34, 71
F plasmid in *E. coli*, 232, 232f
Fragment assembly, process of, 376f
Fragments, DNA, 362, 364
Franklin, Rosalind, 34
Free-living bacterial genes, 237
Fruit flies. *See Drosophila melanogaster*
Fts protein, 530
Functional chloroplast, 508–510
Functional genomics, 389–390, 390f
Functional mitochondrion, genes, 503–506
Fungi
 life cycles for, 154f
 species, 393

G

Gag gene, 308–309, 449
Gametophyte-dominant lifestyle (haploid organisms), 272
Gap genes, 262
Garden peas (*Pisum sativum*), 131
 dominant and recessive characters, 136, 136f
 flower color
  inheritance of, 135f
  phenotypes and genotypes of, 133f
 phenotypic analysis, 134f
 Punnett squares, 133, 133f, 134f
Gels
 IEF, 383
 one-dimensional, 383
 SDS, 383
 Southern hybridization, 343, 344f
 two-dimensional, 383
Gemmata obscuriglobus, 530
Gene(s). *See also specific genes*
 breast cancer, 302
 changes in gene retention, 438f
 common to genomes, 430f
 comparison, 482t
 conversion, 217
 copy number, 349–351, 352f
 envelope, 477
 expression, 531
 flow, 167, 169–170
  introgression, 169
  between populations, 168f
  predictions, 168f
 hierarchies, 35
 involved in cancer, 295f, 298–301
 local duplication of, 267
 loss, 196
 movements among organisms, 237
 overlap for genomes, 431f
 for selected species, 447t
 transfer, 232
Genetic bottleneck, 166
Genetic code, 83
 amino acid categories, 89f
 with canonical and alternative codes, 85f
 complexity of, 93
 evolution of, 83–86, 91f
 first, 90–91
 for organisms and organelles, 84f
 rearranged, 87f
 symmetry of, 91
 universal, 83–84
Genetic drift, 147
Genetic engineering project, steps, 338
Genetic loci, 551
 chromosomes in, 182f
 in diploid eukaryotes, 141
Genetic recombination, 182–185, 182f
Genetics, 131
 medical, 548–550
 population. *See* Population genetics
 principles of, 34
Genome(s), 429
 analysis, problem of, 430
 archaeal, 436
 bacterial, 429, 432, 445
 comparing, 434
 composed of mRNA genes, 433
 dimensions, 16t
 diploid human, 180
 eukaryotic, 435–436
 genes in human, 375
 and genomics, 448
 haploid human, 180
 HIV, 448–449, 451f, 452
 human, 429
 mixing and sorting, 491–492
 nuclear, 445
 organellar, 445
 overlap of genes for, 431f
 plant, 438, 439f
 reduction, 28
 RNA chromosome, 445
 RNA virus, 449
 size, 445, 446f, 447t
 structure of species, 34

Genomics, 6, 251, 373–376, 528
comparative, 435–440
analysis of domains, 436
comparing bacteria, archaea, and eukarya, 436f
research, 436
defined, 373
genomes and, 448
locations of, 374f
sequence/sequencing, 430
projects, 375f
testing for selection, 434
Genotrophs, 142, 145f
Germ cell tumors, 301
Giardia lamblia, 25
Giardiasis, 25
Glassy-winged sharpshooter gut, 489f
Glaucophytes, 37, 531
Global control regions (GCRs), 266–267
Globin gene
evolution of, 221f
family, 218–222
Glycolysis, 505
in animals, 238
in cytosol, 238, 238f
pathway, 495
in spinach cells, 239f
Glycoproteins (GPs), 452, 457
Glycosylated proteins, 390
GNRA (guanosine–nucleotide–purine–adenine) loops, 49
Golgi, 520, 523, 528
Gould, John, 33
gp160 protein, 451
G-protein, 536
Grade 2 cells, 301
Grade 3 cells, 301
Grade 4 cells, 301
Gram-negative bacteria, 28, 242–243, 426
Gram-positive bacteria, 28, 426
Gray, Asa, 33
Green algae, 37, 239
Group I introns, 63–64
Group II introns, 67
GSK3-β (glycogen synthase kinase 3-β), 298
Guanosine, 198, 200, 202
methylation, 47
mRNAs, 43, 46f
Guide tree, 368
Gut microorganism, 485
Gymnosperms, 28, 275

H

H1N1 influenza A, 457
Haemophilus influenzae, 435, 445, 478, 480, 491
Hanahan, Doug, 343
Haploid genome for *H. sapiens,* 97, 180, 448
Haptophytes, 513–514
Hardy–Weinberg equation, 148, 161
Hardy–Weinberg equilibrium, 148–150, 149f
Hedera helix (English ivy)
chloroplast chimera of, 276, 277f
photomicrograph and cell patterns of, 278f
Hemagglutinin (HA) genes, 454–455
unrooted phylogram of, 453f
Hemi-methylated state, 47
Hemizygous, 207
Hemoglobin, 218, 220, 222
abnormal, 187
oxygen saturation curves, 219f
Hermaphrodites, 252–253, 534
Hermaphrodite-specific egg-laying neuron (HSN) cells, 253
Heterochromatin, 113
Heteroduplex region, 182, 185
Heteronuclear RNA (hnRNA), 58, 196
Hexadecyltrimethylammonium bromide, 315
High frequency of recombination (Hfr) strains of *E. coli,* 232
High-throughput technologies, 360–361
Hippocrates (460–370 BC), 30
Histogenic layers of cells, 276
Histone gene, 419, 481
evolution of, 224f
family, 223–224
Hoechst 33258, 321
Holocentric (holokinetic) chromosomes, 117, 117f
Homeobox genes. *See Hox* genes
Homeobox/homeodomain, 254
Homeotic genes, 537
hierarchy and evolution of, 266–268
and proteins, 253–261
structure of, 255f
Homing intron, 198
Hominids, 37
Homologous chromosomes, 101f, 302
crossovers, 183
Homo heidelbergensis, 551–553
Homo neanderthalensis, 551–553
Homo sapiens, 545, 551–553
genome, 547–248
proteome of, 385
Homo sapiens sapiens, 29–30, 37
Hoogsteen edge, 49, 49f
Horizontal gene transfers (HGTs), 187, 419, 422–423, 422f, 429, 527
defined, 229–231
Gram-negative bacteria, 242–243
introns, 244–245
parasites and pathogens, 239–242
phylogram of, 231f
plasmids, 231–234
signs of, 243–244
symbionts and organelles, 235–239
types of, 230f
viruses, 234–235
Hormones, 311–312
Housekeeping genes, 376
Hox genes, 226–227, 531
Drosophila melanogaster, 537
expression, 255, 258f
alleles, 261
changes in mammals and birds, 261
chromosomal arrangements of, 256f
evolution of limbs in relation to, 260f
model for evolution of, 268f
produce nerve neuronal, 262f
produce somatic tissues, 263f–264f
regulation of *HoxD* locus, 259f

of tunicate *Oikopleura dioica,* 265f
zinc-finger DNA-binding protein, 259f
Human genome, 545–548
forensics, 551
project, 373
with protein-encoding genes, 545, 546f
Human immunodeficiency virus (HIV), 187, 190, 449–453
cross section and three-dimensional diagrams of, 450f
genomes, 448–449, 451f, 452
genotypes, 449
infection of T lymphocyte by, 452f
Human migration, 551–554
Human papilloma virus, 309
Human somatic cells, chromosome in, 108
Humans, taxonomic classifications for, 396
Hunchback (*hb*) gene, 261
Hybridization
events, 394
methods, 349
Southern, 343–349
basic steps in, 344f
labeling and detection for, 345t
labeling of nucleotides, 346f
membranes used, 345
species, 423–424, 423f
Hydrogenosomes, 502
Hydrolysis of purines, 177
Hydrolytic reaction, 177
Hydrophobic amino acids, 53
Hydroxyl group, 323

I

Illumina method, NGS using, 364f
Incomplete dominance, 137–138
Incongruent trees, 434–435, 435f
Indel mutation, 173
Inflorescences, 287
angiosperm, 289f
sunflower, 291f
types, 290f
Influenza A virus, 192, 453–457
diagram, 454f
fragmented genome of, 456
genome, 455f
human, 455, 457
progression of infection for, 456f
Inheritance mechanism, 30
Interbreeding, 394
Internal guide sequence (IGS), 198
International Space Station (ISS), 378
Int (integrase) protein, 451
Intracellular endosymbiont, 495–496
Introgression, 169
Introns, 67, 196–204, 205f, 206f, 244–245, 468, 545–546
Inversion mutation, 173
Ionic detergent, 320
Ion-sensitive field-effect transistors (ISFETs), 364
Ion Torrent system, 364
Isoelectric focusing (IEF) gels, 332, 383
Isoelectric point, 383

J

Jukes–Cantor model, 409, 434
Jumping genes, 204
Jurassic period, 28

K

Kaposi sarcoma-associated herpesviruses (KSHV), 310–311
Kelvin, Lord, 33
Kennewick Man, 554
Kinases, 203, 298, 533, 536
Kinesins, 115
Kinetochore microtubules, 116, 117
Kinetoplast DNA (kDNA), 240–241
Kingdoms of life on Earth, 396, 398f
Kissing bugs, 241
KNAT1 (knotted-1-like 1) gene, 279
Kölreuter, Joseph Gottlieb, 32
Krebs cycle, 505

L

Labyrinthodonts, 261
Laccase gene, 223
LacZ gene, 340
Lagging strand in DNA replication, 100
Lamarckian evolution, 111–112
Lamarck, Jean-Baptiste, 32
L-amino acids, 52
Lamins, 116
Large subunit (LSU), rRNA, 57, 59, 71, 200, 203
Late bombardment, 13
LB (lysogeny broth) medium, 341
Leading strand in DNA replication, 100
Leaves, formation of, 279–287
Lepidodinium, 28, 514
Leucine, 491–492
Life
classification of, 396–399, 397f
continuum of, 11f
defining, 9–13
on Earth
evolutionary steps, 20f
first life, 3
kingdoms, 398f
examples of, 11
imagining cellular and molecular dimensions, 13–14, 16t
reconstruction of evolutionary history, 399
RNA and, 6–9
signs of, 7
before translation, 92–93
Light microscopy, 351, 353f
Linear plasmid, 209f
Linnaeus, Carolus, 32
Lipidomics, 389
Lipids, 42, 54–55, 389
Lipoproteins, 390
Little striped whiptail, 394–395
Living fossils, 19
Long terminal repeats (LTRs), 449
LRP5/6 (low-density lipoprotein receptor) protein, 298

Lumen, 508, 520, 522–523
Luria–Bertani medium, 341
Lyell, Charles, 32–33
Lymphomas, 301
Lysis, 461–462
Lysogen, 464, 467
Lysogenic cycle, 189, 191, 464, 467
Lysosome, 520, 524
Lytic cycle, 191, 464, 467

M

M2 ion channels, 455
MA (matrix protein), 449
Macronucleus, 78, 513
Mad cow disease, 10, 12
Major and minor grooves, 45
Mammoth, 32
Marine invertebrates, 28
Maternal proteins, 262
Mating and dispersal, 166–167
Maury, Matthew Fontaine, 33
Maxam and Gilbert method, 358
  sequence determination using, 359f
Maxicircles, 240
Maximum likelihood (ML) method, 414–415, 415f, 431
Maximum parsimony (MP), 411–414, 431
McClintock, Barbara, 34, 204–205, 209
Measles virus, 190
Medical genetics, 548–550
Meiosis, 120, 123–124, 253, 527
  advantage to organisms, 123
  chromatids during, 126f
  diagram of, 124f
  stages of, 123f
  synaptonemal complex during prophase in, 125f
Melanomas, 301
Membrane-based method for sequencing, 366f
Membrane bound, 12–13
Mendel, Gregor, 33–34, 131
Mendelian genetics, 131
Mendelian inheritance, 131, 204
  with alleles, 136f
  basics of, 132–133, 135–137
  of flower color, 135f
Mesorhabditis, 155
Messelson, Mathew, 34
Messenger RNA (mRNA), 6, 9, 65, 67, 189, 377–378
  5′ UTR of, 163
  guanosine, 43, 46f
  organization of eukaryotic, 68f
  transcription, 65, 66f, 67
  translation, 51, 51f
Metabolomics, 389
Metagenomic analysis, 7, 9, 9f
Metagenomics/metatranscriptomics, 378, 379f
Metaphase chromosome, 14
Metaphase plate, 100f, 115
Metastasis, 301
Methanococcus jannaschii, 478, 480–482
Methylation, 47, 48f
Mice chromosome, 548, 549f
Micelles, 54
Microbiomics, 378–381
  human microbiome results, 381f
  vaginal microbiomes, 380
Microgram, 13
Micrometer, 13
Micronucleus/micronuclei, 78, 297, 513
Microorganisms, 11, 32
MicroRNAs, 68
Microsatellite DNA, 551
Microsatellites, 433–434
Microscopy, 351–353, 353f
Microtubule, 528, 529f, 530
Microtubule organizing center (MTOC), 114
Miescher, Johann, 33
Miller, Stanley, 3, 23
Miller–Urey apparatus, 4f
Mimiviruses, 192, 196, 471–474
Minicircle, 210
Minisatellite DNA, 551
Minisatellites, 433–434
Mismatch repair system, DNA, 181
Misshapen tissue, 207
Mitochondria, 271, 399, 496–503
  functional, 503–506
  protein trafficking in, 520, 521f
  replacement, 550
  rRNA transcription, 79
Mitochondrial genomes, 321–322
Mitochondrion origin, 528–530
Mitosis, 113–116, 527
  anaphase, 115, 116f
  in dikaryotic filamentous fungi, 119, 120f
  in fungi, 119f
  phases in, 113, 119
  time course of cell components, 116f
  variations and cell cycle, 116–117, 118f
Mitosomes, 502
Mobile genetic element, 231, 477
  evolution of, 190f
  plasmids, 231
  transposons, 10, 204
Mobile introns, 229
Molecular biological research, procedures, 316f
Molecular biology
  central dogma of, 6, 6f, 42
    beyond, 69
    defined, 57
    expanded version of, 58f
  dimensions, 13–14, 16t
  method, 335
Molecular cloning, steps in, 341
Molecular mechanisms, cells, 247
Monoderms, 242, 516, 519f
Monosaccharides, 55
Monovalent cations, 342
Moreau, Pierre-Louis, 32
Morphology, 396
Mortality
  by cancer (2008), annual US, 294f
  due to Ebola virus, 457
Motifs, 386–387
Mouse (*Mus musculus*) chromosome, 548, 549f
M repeats sequence, 78
Müller, Otto, 32

Mullis, Kary, 335
Multicellularity, 247
 in eukaryotes, 531, 534–535
 as haploids, diploids, and polyploids, 152, 154f
Multicellular organisms, 244, 271
 cancer, 293
 first appearance of, 26, 27f
 non-Mendelian segregation, 142
 nonrandom mating, 166
Multigene family
 evolution, 224–227
 generation of, 215, 216f
Multiple alignments using fast Fourier transformation (MAFFT), 368
Mutant dystrophin gene, 547f
Mutation(s), 19, 147, 150, 162, 164, 491
 causes in carcinogenesis, 301–311
  amplification, 305–306, 307f
  DNA viruses, 309–311
  point mutations, 302
  recombination, 302–305
  viruses, 306–309
 causes of, 174–179, 175t
 classes of, 173–174
 in DNA, 174f
 DNA repair, 180–182, 181f
 factors influencing, 175f
 in gene, 254
 genetic recombination, 182–185, 182f
 models of, 400, 409–410
 during replication, 179–180, 179t
Mutualism, 485
Myb gene, 309
Myc gene, 304
Mycoplasma genitalium, 435, 490–491
Myelomas, 301
Myoglobin, 218, 222
 oxygen saturation curves, 219f

N

NADH. *See* Nicotinamide adenine dinucleotide (NADH)
Nanogram, 13
Nanometer, 13
Nanos (*nos*) gene, 262
National Institutes of Health (NIH) sequencing project, 374
Natural selection
 alleles, 159–161
 processes, 19, 22f
NC (nucleocapsid) protein, 449
Neanderthals, 315
Negative single-stranded RNA virus, 457
Neighbor joining (NJ), 410–411, 412f, 431
Nematode, 495–496
Nematode development, cell, 252–253
Neutral theory, 165
New Mexico whiptails, 395
Next-generation sequencing (NGS) method, 9, 361–365, 429
 additional, 365f
 using 454 pyrosequencing, 363f
 using Illumina, 364f
Nick translation for labeling DNA, 347, 347f
Nicotinamide adenine dinucleotide (NADH), 502–503, 505
Nitrocellulose membranes, 345
Noncoding RNAs (ncRNAs), 57, 68
Non-Mendelian mechanisms, 420
Non-Mendelian traits, 141–142, 144, 146
Nonphotosynthetic protist, 271
Nonradioactive methods, 349, 350f
 detection of probe using, 351f
 labeling, 359
Nonstructural proteins, 454
Nontranscribed spacer (NTS), 62
Normal cells, 295, 300
 vs. cancer cells, 296f
Normal dystrophin gene, 547f
Northern blotting/hybridization, 343
Novel alleles, 159
N protein, 467
N-terminal amino acid, 366
Nuclear genomes, 321, 445
Nucleic acids, 42–51
 extracting, 317t
  from plants and fungi, 320f
  quantification of, 324, 324f, 325f
  steps in, 318f
  using CTAB, 319–321
 sequencing of, 357
Nuclein, 33
Nucleolar dominance, 62
Nucleolus, 14, 59, 61
Nucleomorph, 513, 531–532
Nucleoprotein (NP) gene, 454
Nucleotide(s), 229
 amino acid characteristics compared with, 85t
 DNA and RNA, 43
  components of, 44f
  edges on, 49f
  examples of, 45f
  general form of, 43f
  hydrogen bonded pairs, 46f
 excision repair system, 182
 labeling of, 349f
 mutation, 173
 in RNA viruses, 445
 substitution
  observable and nonobservable, 409f
  rates of, 405f, 408f
 in triplet codon, 84
Nucleus origin, 528–530
Nylon-based membranes, 345

O

Octamer, 223
Okazaki fragments, 100
Omega (ω) introns, 192, 198
Omics
 categorization of, 384f
 epigenomics, 388–389
 functional genomics, 389–390, 390f
 genomics, 373–376
 metabolomics, 389
 metagenomics/metatranscriptomics, 378, 379f
 microbiomics, 378–381

Omics (*Continued*)
overview, 383
proteomics. *See* Proteomics
RNAomics, 387–388, 387f
structural genomics, 386–387
transcriptomics, 376–378
Oncoviruses
DNA, 306
infection and integration, 235f
retroviruses, 235
RNA, 306
One-dimensional gels for electrophoresis, 383
Oomycetes, 239, 244
Open reading frame (ORF), 545–546
Operational taxonomic units (OTUs), 401–402, 410–411
joining, 412f, 414
unrooted trees for, 413f
Opisthokonta
Caenorhabditis elegans, 534–537
Drosophila melanogaster, 537–538
Saccharomyces cerevisiae, 532–534
Organellar DNA, purification of, 321–322
Organellar genomes, 445
Organellar sequence analyses, 235
Organelle(s), 13
endosymbiosis to, 512–514
in eukaryotes, 527
evolution, 497f, 498
genetic code for, 84f
replication in, 105
Organisms, 23, 393
chromosomes, 445
circular phylogenetic tree, 25f
fossil evidence of, 3
genetic code for, 84f
and genomes interactions, 486f
meiosis advantage to, 123
pathway to, 3
rRNA in, 58
The Origin of Species (Book), 33, 394–395
Orthogonal-field-alternation gel electrophoresis (OFAGE), 327
Orthologous genes, 224, 243
Oryza sativa, 540–543
Osmotrophs, 244
Overdominance, 137–138
Oxidation of DNA, 177
Oxidative phosphorylation, 235, 503, 505, 528, 529f, 530

P

P6 protein, 449
P16 tumor suppressor protein, 298
P53 proteins, 253, 300
loss of, 298
P55 protein, 449
Pair-rule genes, primary, 262
Pappus, 287
Paralogs, 224
Paramecium tetraurelia, 513
Parasites and pathogens, 239–242
Parthenogenesis, 394
Passenger pigeon, 159
Pasteur, Louis, 33
Paulinella, 28
Paulinella chromatophora, 506
Peplomer, 457
Peptides, 92, 92f, 93f
Peptidoglycans, 506
Percoll (colloidal silica), 321
Periplast, 477
Peroxisomes, 519
Phagocytized cell, 495
Phanerozoic Eon, 27f
Phenotypic analysis, 134f
Phialophora, 485
Philosophie Zoologique (Book), 32
Phosphate buffer, 319
Phosphoglucokinase (*PGK*) gene, 240f, 511
Phospholipid bilayers, 54, 54f
Phosphorylation, 298, 390, 533
Photolyases, 180
Photosynthetic bacteria, 477–479
Photosynthetic organelles, 28
Photosystem II, 508
Phyllotaxy by gene, control of, 280f
Phylogenetic lines, 225f
Phylogenetic models for enzymes, 240
Phylogenetic networks
advantages of, 420–422
examples of reticulate evolution events, 426–427
HGTs, 419, 422–423
overview, 419–420
reassortment, 425–426
recombination, 424–425, 424f
species hybridization, 423–424
transposition, 425
Phylogenetics, 400–401
analyses
considerations when performing, 408–409
of reticulate events, 420
standard, 431
analyzing aligned sequences, 410
ancient DNA in, 407f
Bayesian phylogenetic analysis, 415
bootstrapping method, 415–416, 416f
choosing genomic region for, 402–408
ML, 414–415, 415f
models of mutation, 409–410
MP, 411–414
NJ, 410–411
UPGMA, 410
vertical *vs.* horizontal evolutionary events, 416–417
Phylogenetic trees, 229, 230f, 231f, 368, 419, 421f, 422f
of life based on SSU rRNA sequence, 71, 72f
of organisms, 245f
terminology, 401–402, 401f, 402f
construction, 411f, 412f
evaluation using parsimony analysis, 413f
evolutionary tree, 401f
MP, 411
OTU, 401–402
parts of, 402f
rooting, 402, 404f
rotation of branches, 404f
types of trees, 403f

Phylogenomics
  alignment of sequences for, 430f
  to compare, 432–434
  gene loci, 431
  improvements in sequencing and, 429–432
Phylogram, 427f
  with gene duplications, 424–425, 425f
  sequences and, 406f
  tree, 401
Phylum Cercozoa, 28
Phylum Euglenozoa, 239
Phytoalexin, 223
Phytophthora species, 244
Picogram, 13, 324
Pilus (pili), 209–210, 232
Pisum sativum, 508
Planctomycetes, 25, 113, 528, 530
Planctomycetes–Verrucomicrobia–Chlamydiae (PVC) superphylum, 528, 530
Plant(s), 395–396
  cell development, 14, 274–277
    change shape, 274
    formation of leaves and floral organs, 279–287
    gene expression, 277–279
    surrounded by adjacent cells, 274f
  exhibit phyllotaxy arrangements, 282f
  genomes, 438, 439f
  life cycles for, 154f
  mating choice, 166
  morphology, 273–274
  phylogeny of, 288f
  sap, 487, 490
  sexual reproduction in vascular, 156f
  vs. animals, 287–291
Plasmids, 112, 125, 209–210, 231–234
  encode proteins, 233
  ligated, 343
  variety of, 340f
Plasmodium falciparum, 502
Plastic genotroph, 142
Plastid
  differential development in plant, 511
  genomes, 321
Plastochron, 277
Plastoquinone, 508
Poaceae family, 540, 542
Point mutations, 302
Polar bodies, 79f
Polar microtubules, 115
Polar nuclei, 275
Pol gene, 308–309, 451
Pollen grains, 166
Polyacrylamide gel electrophoresis (PAGE), 331–332, 331f. *See also* Agarose gel electrophoresis
Polyadenylation, 163
Polyethylene glycol (PEG), 342–343
Polymerase chain reaction (PCR), 335–338, 551
  advantages of, 336
  amplification, 7, 8f
  labeling of probes using, 348f
  qPCR, 338
  reporter primers, 338
  theory and practice of, 337f
Polymerization, 362, 364
Polymers, 319, 360
Polymorphisms, single-nucleotide, 433
Polypeptides, 6, 52–53, 230
Polyploid genome, 144, 145f, 146
Polyploidization, 224–227
Polysaccharides, 55
Polysomes, 71
Polytene chromosomes, 122
Polyteny, 122
Poly-U RNAs, 83
Population genetics
  Hardy–Weinberg equilibrium, 148–150
  life histories, 150f, 151–154
    for animals, plants, and fungi, 154f
    for Dikarya, 157, 157f
    of *Saccharomyces cerevisiae,* 152, 153f
    for species of aphids, 154, 155f
  modes of reproduction, 154–158, 156f
  population size, 150–151
  time course of proportions of alleles in, 148f
Post-replication repair systems, 180
Posttranslational modifications, 390
Prasinophyta, 271
Pre-mRNA, 58f
Prenatal screening, 549–550
Preserved materials, 316f
Primordial processes, 92
Primordial soups, 41
Principles of Geology (Book), 32
Prions, 10, 12, 196
PRM promoter, 467
Programmed cell death, 300
Proline, 53
Protein(s), 6, 383, 385
  analysis, 353
  electrophoresed in native gel, 331
  extraction of, 329–330
  involved in cancer, 299f
  maternal, 262
  nonstructural, 454
  quantification of, 330
  sequencing, 365–366
  structures found in, 386–387, 386f
  synthesis, 73
  trafficking, 485, 515
    in chloroplasts, 520, 522–523
    in mitochondria, 520, 521f
    system evolution, 523–524
Protein–protein interactions, 390
Proteobacterium, 230
Proteomics, 383–386
  analysis, 385f
  categorization, 384f
  categorization of proteins in, 385–386
  of *Homo sapiens,* 385
Protista kingdom, 396
Protists, 32
Proton-motive force, 505, 510
Proto-oncogenes, 308
Protoribosomes, 73
Protozoan, 487
Prunus species, 284
Pseudogene, 215, 218, 545
  types of, 217f

P (polypeptide)-site, 75–76
Psylid, 487
Puccinia monoica, 167
Pulsed-field gel electrophoresis (PFGE), 327
Punctuated evolutionary events, 433
Punnett square analysis
  of allele combinations, 148, 149f
  of epistasis
    alleles coat color in rodents, 139, 139f
    in sweet pea flower color, 139, 140f
  eye color in *Drosophila melanogaster,* 138, 138f
  in garden peas, 133, 133f, 134f, 135
Pyrimidines, 173, 177, 357
Pyruvate, 505

Q

Q protein, 467
Quantification
  of nucleic acids, 324
    fluorimeter for, 325f
    UV spectrophotometer for, 324f
  of proteins, 330
Quantitative PCR (qPCR), 338
Quantitative trait loci (QTL), 140
Quorum sensing, 249–251, 249f, 250f

R

Radiation, DNA, 177
Radioactivity on blots, detection of, 348f
Random events, 19
Random genetic drift, 165–166, 165f
Random primer labeling of DNA, 347f
Ray flowers, 287
Ray, John, 32
rDNA, need of copies, 76–77
RdRp. *See* RNA-dependent RNA polymerase (RdRp)
Reads, DNA sequence, 363
Recombinant DNA methods, 338–343
Recombination and linkage, 141
Recombination-dependent replication (RDR), 103, 105, 105f, 106f
Recombination events, 184, 424–425, 424f
  cancer, 302–305
    chromosomal, 303f
    generalized, 305f
  frequent, 432
Reconstruction of evolutionary history, 399
Rectangular phylograms, 402
Red algae, 37
Repetitive regions, 432
Replication
  mutations during, 179–180, 179t
  topoisomerases in, 107
  topology during, 106–107
Repressor protein cI, 464–467
Reproduction, modes of, 154–158, 156f
Reptiles, 28
RER. *See* Rough endoplasmic reticulum (RER)
Restriction enzymes, 336f
Retrotransposon, 203–204, 206
Retroviruses, 194, 235
  cause of cancer, 310f
  Retroviridae, 449
Reverse transcriptases, 194, 202, 205
  phylogenetic tree of, 188f
Reverse transcriptase PCR (RT-PCR), 335
Rhizaria, 28
Ribosomal RNA (rRNA), 6–7, 9, 50, 57–64, 113, 192, 242
  arrangement of, 59f
  eukaryotes
    organization in, 62, 62f
    processing of, 63f
  extrachromosomal amplification of genes
    in amphibian oocytes, 80, 80f
    in *Tetrahymena,* 78, 78f
  gene(s), 396, 419, 434
    family, 217–218
    per haploid genome, 77, 77t
    sequence variation within, 405f
  gene copies
    among progeny, 143f
    in broad bean (*Vicia faba*), 142, 142f
    changes by recombination, 144f
    number, mechanisms for increasing, 77–80
  in organisms, 58
  transcription, 57, 60f, 61, 79
  translation, 51, 51f
Ribosomes, 7, 14, 23, 50, 57, 71, 352
  bacterial/archaeal and eukaryotic, 72f
  complexity of, 80–81
  Escherichia coli, 71
  evolution of, 73f
  interfaces between ribosomal subunits, 74, 74f
  origin of, 72–74
  positions for tRNAs on, 75
  as ribozymes, 71–72
Ribozymes, 4, 57, 64, 198
Ribulose-1,5-bisphosphate carboxylase/oxygenase (RuBisCO), 506, 507
RNA (ribonucleic acid), 24, 315
  base triples, 49, 50f
  classes of, 51
  to extract high-purity, 318f
  extraction of, 322–323
  and life, 6–9
  molecular mass of, 329
  molecules, 71
  mutation, 147
  nucleic acids, 42–45, 47–51
  nucleotides, 43
    components of, 44f
    edges on, 49f
    examples of, 45f
    general form of, 43f
  oncoviruses, 306
  polymerases, 9
  protocells population, 147
  types of, 57
  viruses
    genomes, 449
    overview, 445
RNA-dependent RNA polymerase (RdRp), 192, 194
RNAomics, 387–388, 387f
Robinet, Jean-Baptiste, 32

Rolling circle replication, 101, 104f, 461–462, 467, 469
Rooting, tree, 402, 404f
Rough endoplasmic reticulum (RER), 523
Rous sarcoma virus, 309

S

Saccharomyces cerevisiae, 6, 152, 153f, 198, 532–534, 537
S-adenosyl methionine, 177
Sanger method, 357, 359, 362
    sequence determination using, 360f, 429
SAPR (somatostatin–angiotensin-like receptor protein) gene, 305
Saprobe, 485
Sarcomas, 301
Satellite tobacco necrosis virus (STNV), 472–473
Satellite virus, 196, 472–473, 473f
Saturated lipids, 389
Scaffold, 108f
Schizophrenia, 183
Schwann, Theodor, 33
Scrapie disease, 12
Search tree, 414
Secondary constrictions, 62f, 114
Seed ferns, 28
Selenocysteine, 53
Semiautomated sequencing methods, 359
Semi-conservative replication, 34, 47
Semi-exhaustive method, 411
Seminomas (testicular cancer), 301
Septa, 120f
Sequence homology searches, 367–368, 367f
Sexual reproduction, 125–127
Shigella, 438
Shine–Dalgarno sequence, 67
Signal peptide
    in bacteria, 515–518
    system in eukarya, 518–520
Signal recognition particle (SRP), 515, 517, 520, 523
Signal transduction, 299
Silica polymers, 319
Single-celled bacteria and archaea, cell cycle for, 98f
Single-celled eukaryote, 6
Single-nucleotide polymorphisms (SNPs), 433, 550–551
Single-stranded DNA (ssDNA), 184, 335, 362, 461
Single-stranded RNA virus, negative, 457
Sister chromatids, 114
    crossovers, 183
    misalign, 302
    recombination, 302, 303f
Sister group, 402
Slot-blotting process, 351
    to determine gene copy numbers, 352f
Small interfering RNAs (siRNA), 68
Small nuclear RNAs (snRNA), 68
Small nucleolar RNAs (snoRNA), 68
Small subunit (SSU), rRNA, 57, 59, 71, 200, 201f, 202–203, 403, 420–421
Sodium dodecylsulfate (SDS) gels, 383
Solar system, 22
Soluble proteins, 383
Somatic cell, 180
Southern, Edwin M., 343
Southern hybridization/blotting, 343–349
    basic steps in, 344f
    labeling and detection for, 345t
    labeling of nucleotides, 346f
    membranes used, 345
Species, 393–396
    for bacteria, 393
    comparison of number of, 400f
    C-value of, 118
    defined, 396
    entities, 393
    for fungi, 393
    hybridization, 423–424, 423f
    plants and animals, 393
    problem with defining, 395
Spectrophotometry, 330
Spermatopsida, 271
Spermidine, 342
Sperm nuclei, 275
Spinach (*Spinacia oleracea*)
    enzymes used in, 238f
    glycolysis in, 239f
Spirochaetes, 113
Spliceosomal introns, 67, 203, 245
Splicing mechanism, 67, 199f, 200f
Sporophyte-dominant lifestyle (diploid organisms), 272
Sputnik, 196, 461
    virophage, 471f, 472–473, 473f
Src gene, 309
SRP. *See* Signal recognition particle (SRP)
SSU ribosomal RNA (SSU rRNA), 71
    phylogenetic tree of life based on, 71, 72f
Stahl, Frank, 34
Stalk/stalked cell, 97, 247
Star activity, 335
Statolith, 510
Stern, Curt, 34
Steroid hormones, 311–312, 311f
STM (SHOOT MERISTEMLESS) gene, 279
Stroma, 508, 520, 522
Stromalites, 23
Structural genomics, 386–387
Sucrose, extraction, 321
Sugar edges, 49, 49f
Sulcia muelleri, 489
Sulfolobus solfataricus, 482, 483f
Sunflower inflorescence, structure of, 291f
Supercoiling, 106
Swarmer cells, 97, 247–248
Syconium, 287
Symbionts
    and organelles, 235–239
    vs. free-living organisms, 487f
Synechococcus, 478
Synteny, 440–441, 541, 543f

T

Tail fiber, 468–469
Tatum, Edward Lawrie, 34
TA vectors, plasmids, 342
Taxonomic classifications for humans, 396
Taxonomic scheme, 192
Telomerase, 101, 103f

Telophase, 116, 119, 124
Temperate bacteriophage, 234
Tepals, 284, 287
Terminal branches, 401
Termite gut microbes, 487
Testicular cancer, 301
Tetrahymena, 78, 78f
Tetrahymena thermophila, 88, 513
Tetrapods, 265
T-even phages, 468
Theophrastus, 30, 32
Thermophile, 479
Thermotoga, 229
Theta-form replication, 467
Thylakoid membrane, 508
Thymidine (pyrimidine) dimer, 179f
Tiger whiptail (*C. tigris*), 395
Tobacco mosaic disease, 33
Tobacco necrosis virus (TNV), 196, 472–473
Tonoplast, 274
Topoisomerase (TOPO cloning), 342f
Topoisomerases in replication, 107
Topology during replication, 106–107
TOPO TA vectors, plasmids, 342
Transcriptomics, 376–378, 377f
Transduction, 234
Transfer RNA (tRNA), 6–7, 9, 64–65
  aaRSs, 86
  characteristics of, 64
  evolution of, 94f
  general structure of, 65
  model of parts of, 93, 94f
  positions on ribosomes, 75
  transcription of, 64–65, 64f
  translation, 51, 51f
Transformation process, 343
Transitions, mutation, 173, 174f
Translation, 51–52, 51f, 74–76, 83
  evolution of integral parts, 95f
  life before, 92–93
  schematic of process of, 75f
Translocon protein complex, 517
Transmission electron microscopy (TEM), 352
  DNA for viewing in, 354f
Transposable elements, 34, 204–209
Transposase, 205, 208f
Transposons, 10, 204, 209, 229
  types of, 207f
Transversions, mutation, 173, 174f
Tree terminology, phylogenetic, 401–402
  construction
    using NJ, 412f
    using UPGMA, 411f
  evaluation using parsimony analysis, 413f
  evolutionary tree, 401f
  MP, 411
  OTU, 401–402
  parts of, 402f
  rooting, 402, 404f
  rotation of branches, 404f
  types of trees, 403f
T-region, (T-DNA), 233
Tripartite attachment complex, 240
Triplet codon, 50, 87–90
  evolution of, 89f
  nucleotide in, 84
Tris(hydroxymethyl)aminomethane buffer, 319, 330
TRIZOL reagent, 323
Trypanosoma cruzi, 239, 241f
Trypanosomes, 240, 241, 513
Tubulin, 528, 530
Tumor-inducing (Ti) plasmids, 233
Tumors, ranking, 301
Tunica-corpus model, 276
Twintron, 198
Two-allele systems, 161
Two-dimensional gels, 332, 383
Tyrosine kinase, 309

U

UCUCUCUCUC synthesized RNA, 83
UGA codon, 88, 90
Ultraviolet (UV) irradiation, DNA, 177
Umbels, 287
Underdominance, 137–138
Underreplication, 121f, 122
Unequal crossover, 173
Unicellular organisms, 247, 293
Universal genetic code, 83–84
Unmanned Mariner probes, 7
Unweighted pair group method with arithmetic mean (UPGMA), 410–411, 411f
Uracil glycosylase, 181
Urey, Harold, 23
User-supplied trees, 415
Utrophin, 385
UV spectrophotometer, nucleic acids, 324, 324f

V

Vacuole, 520
Vacuum blotting systems, 343
Vaginal microbiomes, 380
Vegetative shoot apices, 284
Venter, Craig, 374
Vernalization, 283
Verrucomicrobia, 530
Vertical *vs.* horizontal evolutionary events, 416–417
Vesicle, 523
Vicia faba. See Broad bean (*Vicia faba*)
Vif gene, 449, 451f
Viral oncogenes, 234
Virions, 462, 468
  in eukaryotes, 449
  influenza A, 455
Viroids, 12, 196
Virophage, 196
Virus(es), 9–10, 12, 189, 192–196, 234–235
  cancer, 306–309
  cycle, 194f, 195f
  Ebola, 457–458, 458f
  genomes, 373
  HIV, 449–453
  influenza A. *See* Influenza A virus

integration, 189f
RNA virus genomes, 449
Volcanoes, 22–23
vapor blown out, 3
v-oncogenes (viral oncogenes), 308
c-oncogene into, 309f
Von Tschermak-Seysenegg, Erich, 33, 131
Vpr gene, 449, 451f
Vpu gene, 449, 451f

W

Wallace, Alfred Russel, 33, 160
Wasps, 287
Watson–Crick base pairs, 44, 46f, 64
Watson–Crick interactions, 49, 49f
Watson–Crick orientation, 99
Weismann, August, 33
Western blotting, gel, 384
Whorl pattern, 282
Woese, Carl, 34, 71
theory of, 481
Wolbachia, 495
Word, sequence homology, 368
WOX (WUSCHEL-related homeobox) gene, 278, 279f
WUS (WUSCHEL) gene, 277, 279f

X

Xenophanes of Colophon (570–480 BC), 30
Xenopus laevis, 80
Xeroderma pigmentosum, 177
X-gal, 340–341
X-ray crystallography, 387

Z

Zea mays (maize), 204
Z-form DNA, 45, 47, 47f
Zonation model of shoot, 276
Zygote, 276, 527

Printed and bound by CPI Group (UK) Ltd, Croydon, CR0 4YY

01/11/2024

01782601-0012